McGRAW-HILL YEARBOOK OF
Science &
Technology

1994

McGRAW-HILL YEARBOOK OF
Science & Technology

1994

Comprehensive coverage of recent events and research as compiled by
the staff of the McGraw-Hill Encyclopedia of Science & Technology

McGraw-Hill, Inc.

New York   San Francisco   Washington, D.C.   Auckland   Bogotá   Caracas   Lisbon   London   Madrid
Mexico City   Milan   Montreal   New Delhi   San Juan   Singapore   Sydney   Tokyo   Toronto

McGRAW-HILL YEARBOOK OF SCIENCE & TECHNOLOGY
Copyright © 1993 by McGraw-Hill, Inc.
All rights reserved. Printed in the United States of America.
Except as permitted under the United States Copyright Act of 1976,
no part of this publication may be reproduced or distributed in any
form or by any means, or stored in a database or retrieval system,
without prior written permission of the publisher.

1 2 3 4 5 6 7 8 9 0   DOW/DOW   9 9 8 7 6 5 4 3

Library of Congress Cataloging in Publication data

McGraw-Hill yearbook of science and technology.
1962–   . New York, McGraw-Hill Book Co.

    v. illus. 26 cm.
    Vols. for 1962–   compiled by the staff of the
McGraw-Hill encyclopedia of science and technology.
    1. Science—Yearbooks. 2. Technology—
Yearbooks. 1. McGraw-Hill encylopedia of
science and technology.
Q1.M13   505.8   62-12028

ISBN 0-07-051517-4
ISSN 0076-2016

International Editorial Advisory Board

Editorial Staff

Sybil P. Parker, Editor in Chief

Arthur Biderman, Senior Editor
Jonathan Weil, Editor
Betty Richman, Editor
Ginger Berman, Editor
Patricia W. Albers, Editorial Administrator
Frances P. Licata, Editorial Assistant

Ron Lane, Art Director
Vincent Piazza, Assistant Art Director

Joe Faulk, Editing Manager
Ruth W. Mannino, Editing Supervisor

Thomas G. Kowalczyk, Production Manager
Suzanne W. Babeuf, Senior Production Supervisor

Suppliers: Electronic Technical Publishing Services Company, Portland, Oregon, generated the line art, and composed the pages in Times Roman, Helvetica Black, and Helvetica Bold.

The book was printed and bound by R. R. Donnelley & Sons Company, The Lakeside Press at Willard, Ohio.

Consulting Editors

Consulting Editors (continued)

Prof. Randy Moore. *Chairman, Department of Biological Sciences, Wright State University, Dayton, Ohio.* PLANT ANATOMY.

Prof. Conrad F. Newberry. *Department of Aerospace and Astronautics, Naval Postgraduate School, Monterey, California.* AERONAUTICAL ENGINEERING AND PROPULSION.

Dr. Everett C. Olson. *Professor of Zoology, Emeritus, Department of Biology, University of California, Los Angeles.* PALEONTOLOGY (VERTEBRATE).

Dr. Gerald Palevsky. *Consulting Professional Engineer, Hastings-on-Hudson, New York.* CIVIL ENGINEERING.

Prof. Jay M. Pasachoff. *Director, Hopkins Observatory, Williams College, Williamstown, Massachusetts.* ASTRONOMY.

Dr. William C. Peters. *Professor Emeritus, Mining and Geological Engineering, University of Arizona, Tucson.* MINING ENGINEERING.

Prof. Don S. Rice. *Director, Center for Archaeological Investigations, Southern Illinois University, Carbondale.* ANTHROPOLOGY AND ARCHEOLOGY.

Prof. D. A. Roberts. *Plant Pathology Department, Institute of Food and Agricultural Sciences, University of Florida, Gainesville.* PLANT PATHOLOGY.

Prof. W. D. Russell-Hunter. *Professor of Zoology, Department of Biology, Syracuse University, New York.* INVERTEBRATE ZOOLOGY.

Dr. Andrew P. Sage. *First American Bank Professor and Dean, School of Information Technology and Engineering, George Mason University, Fairfax, Virginia.* CONTROL AND INFORMATION SYSTEMS.

Prof. Susan R. Singer. *Department of Biology, Carleton College, Northfield, Minnesota.* DEVELOPMENTAL BIOLOGY.

Prof. William A. Steele. *Department of Chemistry, Pennsylvania State University, University Park.* THERMODYNAMICS.

Prof. Thomas N. Taylor. *Department of Plant Biology, Ohio State University, Columbus.* PALEOBOTANY.

Prof. Marlin U. Thomas. *Chairman, Department of Industrial Engineering, Lehigh University, Bethlehem, Pennsylvania.* INDUSTRIAL AND PRODUCTION ENGINEERING.

Prof. Joan S. Valentine. *Department of Chemistry and Biochemistry, University of California, Los Angeles.* INORGANIC CHEMISTRY.

Prof. Frank M. White. *Department of Mechanical Engineering, University of Rhode Island, Kingston.* FLUID MECHANICS.

Prof. Richard G. Wiegert. *Department of Zoology, University of Georgia, Athens.* ECOLOGY AND CONSERVATION.

Prof. Frank Wilczek. *Institute for Advanced Study, Princeton, New Jersey.* THEORETICAL PHYSICS.

Prof. W. A. Williams. *Department of Agronomy and Range Science, University of California, Davis.* AGRICULTURE.

Prof. George S. Wilson. *Department of Chemistry, University of Kansas, Lawrence.* ANALYTICAL CHEMISTRY.

Dr. Richard E. Wyman. *Retired; formerly, Vice President of Research, Canadian Hunter Exploration, Ltd., Calgary, Alberta, Canada.* PETROLEUM ENGINEERING.

Contributors

A list of contributors, their affiliations, and the titles of the articles they wrote appears in the back of this volume.

Preface

The *1994 McGraw-Hill Yearbook of Science & Technology* continues a long tradition of presenting outstanding recent achievements in science and engineering. Thus it serves both as an annual review of what has occurred and as a supplement to the *McGraw-Hill Encyclopedia of Science & Technology*, updating the basic information in the seventh edition (1992) of the Encyclopedia. It also provides a preview of advances that are in the process of unfolding.

The Yearbook reports on topics that were judged by the consulting editors and the editorial staff as being among the most significant recent developments. Each article is written by one or more authors who are specialists on the subject being discussed.

The *McGraw-Hill Yearbook of Science & Technology* continues to provide librarians, students, teachers, the scientific community, and the general public with information needed to keep pace with scientific and technological progress throughout our rapidly changing world.

Sybil P. Parker
Editor in Chief

McGRAW-HILL YEARBOOK OF
Science & Technology

1994

Acoustooptics

The field of acoustooptics originated in 1932, when experimental results on the interaction of light and ultrasound were reported. The term acoustooptics, however, appeared only about 30 years later, when the invention of lasers, which are coherent optical sources, rekindled interest in the area. Research has continued and has led to advances in the theory as well as to the discovery of many novel applications, such as light modulators and deflectors, signal convolvers and correlators, spectrum analyzers, acoustic imaging devices, algebraic matrix processors, and hybrid bistable devices.

To some extent, the phenomenon of the scattering of light by ultrasound resembles that of x-ray diffraction in crystals. In the latter case, the atomic planes scatter an incident electromagnetic wave, and the scattered waves in turn interfere constructively for certain angles of incidence of the original wave. In acoustooptics, light may be viewed as being scattered (or diffracted) from the planes of compression and rarefaction caused by the sound waves. In reality, the sound comprises propagating sinusoidal waves, in contrast to the stationary planes in a crystal, causing the diffracted light to be Doppler-shifted in frequency, an effect commonly exploited in signal spectrum analysis and other modern heterodyned signal-processing applications.

The interaction between light and sound occurs in an acoustooptic modulator (also called an acoustooptic cell or sound column), which may consist of an acoustic medium, such as glass, water, quartz, or tellurium oxide (TeO_2), to which a piezoelectric transducer is bonded. The transducer converts an electric signal (usually at frequencies between 1 MHz and 1 GHz) to propagating sound waves whose frequency spectrum matches, within the bandwidth limitations of the transducer, that of the electrical excitation. The sound velocity varies from 0.6 mi/s (1 km/s) in water to 2.6 mi/s (4.2 km/s) in tellurium oxide.

Basic principles. A heuristic derivation of the direction of the diffracted light and the amount of Doppler shift is obtained by treating the light and sound as streams of photons and phonons, respectively, and by invoking the principles of conservation of energy and momentum. By using the energy conservation principle, it follows that Eq. (1) is satisfied, while conservation of momentum (**Fig. 1**) demands that Eq. (2)

$$\omega_1 = \omega_0 + \Omega \tag{1}$$

$$\mathbf{k_1} = \mathbf{k_0} + \mathbf{K} \tag{2}$$

be satisfied; here ω_0, ω_1, and Ω are the angular frequencies of the incident (or zeroth-order) light, first-

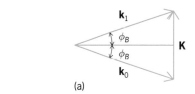

(a)

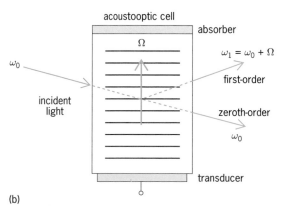

(b)

Fig. 1. Upshifted acoustooptic diffraction. (*a*) Wave-vector diagram. (*b*) Typical experimental configuration showing sound wavefronts.

order light, and sound, respectively, and $\mathbf{k_0}$, $\mathbf{k_1}$, and $\mathbf{K}$ represent the corresponding wave vectors. A part of the light incident at the Bragg angle ϕ_B, given by Eq. (3),

$$\phi_B = \frac{K}{2k_0} = \frac{\lambda_0}{2\Lambda} \qquad (3)$$

where λ_0 ($= 2\pi/k_0$) and Λ ($= 2\pi/K$) are the wavelengths of light and sound, respectively, is diffracted into the first order, with the angle between the two orders being equal to $2\phi_B$. The assumption here is that both the light and sound are plane waves. This type of interaction is termed Bragg diffraction.

In reality, interaction occurs between light and sound beams, which diffract during propagation, and may be tracked through their angular plane wave spectra, or a collection of wave vectors around the nominal directions of propagation. The Bragg regime is thus characterized by the condition $Q \gg 1$, where Q, given by Eq. (4), is the Klein-Cook parameter, with L

$$Q = \frac{\lambda_0 L}{\Lambda^2} \qquad (4)$$

representing the width of the sound column. Correspondingly, the condition $Q \ll 1$ implies a narrow sound column, which supports strong diffraction of the sound during propagation. This diffraction, in turn, gives rise to multiple diffracted orders of light, as a result of the acoustooptic interaction, with frequencies ω_m given by Eq. (5), where m is the order of diffrac-

$$\omega_m = \omega_0 + m\Omega \qquad (5)$$

tion. These orders are also separated from each other by an angle $2\phi_B$. This type of interaction is called Raman-Nath diffraction and is similar to the diffraction of light from a thin phase grating.

A more mathematical analysis of the acoustooptic interaction starts from the wave equation describing optical field in a slowly varying inhomogeneous medium, where the inhomogeneity is introduced by the propagating sound wave. In the simplest case of ideal Bragg diffraction ($Q \to \infty$), the wave equation is decomposed into a pair of coupled equations between the undiffracted (zeroth) and diffracted (first) orders. This analysis, which is similar to that of the diffraction of light from a thick grating, shows that the diffraction efficiency η, defined as the ratio of the diffracted and incident light intensities, I_1 and I_{inc}, is given by Eq. (6),

$$\eta = \frac{I_1}{I_{inc}} = \sin^2\left(\frac{\alpha_0}{2}\right) \qquad (6)$$

where α_0 is the peak phase delay of the light through the sound cell, proportional to the sound amplitude and the width L of the sound column. From Eq. (6) it follows that the maximum theoretical diffraction efficiency is 100% for $\alpha = \pi$.

Many applications of the acoustooptic effect assume operation in the Bragg regime. However, the finite width of the incident light beam (usually gaussian) and the boundedness of the sound column (implying a finite value of Q) translate to a departure from ideality, which in turn reduces the diffraction efficiency and is responsible for a change in the transverse profile of the diffracted (and undiffracted) light. Rigorous analyses restricted to two dimensions have been performed

since the late 1970s. Interaction between focused light beams and focused sound beams (continuous-wave or pulsed) is under investigation by researchers, and the extension of the theory of three dimensions is still an open problem.

Applications. Applications include modulators, deflectors, spectrum analyzers, Bragg diffraction imaging, signal convolvers and correlators, and hybrid bistable devices.

Modulators. The obvious application of accoustooptics is light modulation, since it is clear that the first-order light is frequency-shifted by the frequency of the sound. In practice, the sound (or radio-frequency) carrier may itself be modulated by a baseband (or video) signal.

Deflectors. Another application is in the realization of optical beam deflectors. Introduced in 1965, beam steering is achieved by realizing that the deflection angle of the first-order light (with respect to the zeroth-order) depends on the Bragg angle, which can be changed by varying the electronically addressable sound frequency. Beam deflectors are used in optical scanners, and the number of achievable resolvable angles (equal to the product of the frequency range of the sound and the transit time of the sound across the light beam) may be enhanced by using a traveling-wave lens.

Spectrum analyzers. Another application based on the same principle is signal spectrum analysis, where the electrical signal to be analyzed modulates the radio-frequency signal introduced into the acoustooptic cell. Different spectral components simultaneously steer the first-order light into different angles about the nominal direction corresponding to the sound-carrier frequency. A novel demonstration is the implementation of a frequency-modulation (FM) radio receiver where a split diode photodetector is placed over each diffracted spot corresponding to an FM channel, and its electrical output is amplified and fed to a speaker.

Bragg diffraction imaging. A shown by the momentum conservation relation, there exists a 1:1 correspondence between the interacting plane waves of sound and light. Furthermore, the amplitude of the diffracted light is proportional to the sound amplitude in the acoustooptic cell. Thus, if the sound comprises a collection, or spectrum, of plane waves traveling in different directions or angles about the nominal direction of propagation, the diffracted light must also have an angular plane wave spectrum that bears the signature of the sound field. This method of visualization of sound fields, called Bragg diffraction imaging, can be effectively used to make a sonogram optically visible.

Convolvers and correlators. Acoustooptic devices have also found applications in the area of signal convolution and correlation. The process of convolution involves physically sliding one function, $f(t)$, over another, $g(t)$, and determining the common overlap area as a function of the relative shift between the functions. Since sound waves propagate in an acoustooptic cell, radio-frequency signals modulated by the signals f and g may be propagated through

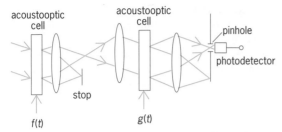

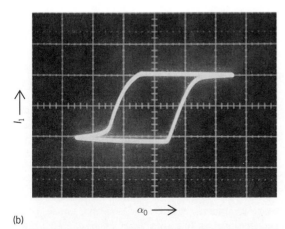

Fig. 2. Space-integrating two-cell acoustooptic correlator.

two acoustooptic cells, as shown in **Fig. 2**. The spatial correlation of f and g (assumed to be real for simplicity) is obtained at the pinhole on the focal plane of the last lens by exploiting the spatial-Fourier-transforming property of lenses. The photodetector output, which contains this correlation as a function of time, is displayed on an oscilloscope. This approach is the basis of a space-integrating correlator. If the phase of the correlation is desired, optical heterodyning techniques using a reference beam at the pinhole may be used. Correlators based on time integration rather than space integration have also been realized by using acoustooptic devices and a photodetector array.

The time- and space-integrating correlators described above have found recent applications in the adaptive processing of received radar signals from a target in the presence of noise (such as interfering signals, scatter, multipath returns, and jammers). The number of independent noise sources that can be nulled depends on the number of weighted antenna elements in the radar system. The nulling takes place by adap-

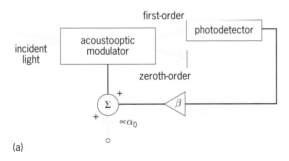

(a)

(b)

Fig. 3. Hybrid acoustooptic bistable device. (a) Typical experimental setup. (b) Oscilloscope display of the input-output (α_0-I_1) characteristics.

tively imposing complex weights to the antenna elements to suitably modify their collective radiation pattern. Multichannel acoustooptic cells are used as delay lines, and the required operation for adaptive nulling is achieved by using acoustooptic time- and space-integrating correlators and a spatial light modulator.

Hybrid bistable devices. Bistability refers to the existence of two stable states of a system for a set of input conditions. Bistable optical devices have gained much attention in recent years because of their potential application in digital optical computing. In general, nonlinearity and feedback are needed to achieve bistability. **Figure 3a** shows an acoustooptic hybrid bistable device, where the diffracted light is detected, amplified (gain factor β), and fed back in phase to the transducer driving the acoustooptic cell. In this arrangement, the effective peak phase delay α that determines the diffraction efficiency depends on the diffracted intensity I_1 through Eq. (7). If α_0 is

$$\alpha = \alpha_0 + \beta I_1 \qquad (7)$$

slowly increased from zero, the output (diffracted intensity) switches to a higher stable state from the lower stable state at the upper threshold, as shown in the oscilloscope plot in Fig. 3b. If α_0 is now slowly reduced, the switching from the upper to the lower stable state occurs at the lower threshold. The device is thus bistable, and exhibits hysteresis similar to that in a magnetic material or in an electronic Schmitt trigger. A similar effect is observed if the undiffracted light is fed back as part of the incident light. The acoustooptic feedback device also exhibits period-doubling bifurcations and chaos for high values of the feedback gain. The feedback effect has also been exploited to improve on the diffraction efficiency of acoustooptic devices, and there is recent experimental evidence that the diffracted light may have a better signal-to-noise ratio than the incident light, implying that squeezed states of light may be easily generated in this way. SEE SQUEEZED QUANTUM STATES.

For background information SEE ACOUSTOOPTICS; CHAOS; OPTICAL BISTABILITY; OPTICAL INFORMATION SYSTEMS; OPTICAL MODULATORS; PERIOD DOUBLING; SQUEEZED QUANTUM STATES; X-RAY DIFFRACTION in the McGraw-Hill Encyclopedia of Science & Technology.

Partha P. Banerjee

Bibliography. P. P. Banerjee and T. C. Poon, *Principles of Applied Optics*, 1991; P. K. Das and C. M. DeCusatis, *Acousto-optic Signal Processing*, 1991; A. Korpel, *Acousto-optics*, 1988; A. Korpel (ed.), *Selected Papers on Acousto-optics*, SPIE Milestone Series, 1990.

Acquired immune deficiency syndrome (AIDS)

Type 1 and type 2 human immunodeficiency viruses (HIV-1 and HIV-2), the causative agents of acquired immune deficiency syndrome (AIDS), must convert their genetic material, composed of single-stranded ribonucleic acid (RNA), into double-stranded deoxyribonucleic acid (DNA) after infection. This conversion is accomplished by the enzyme reverse transcriptase. Because of the multiple functions displayed by reverse transcriptases, they remain attractive targets for antiviral drugs aimed at controlling spread of AIDS.

A complex between reverse transcriptase and a host-encoded transfer RNA (tRNA) primer is formed prior to initiation of the replication events. Subsequent DNA synthesis from both RNA and DNA templates yields the minus (−) and plus (+) strands, respectively, of the viral DNA. During these polymerization events, at least one and possibly two ribonuclease activities (RNaseH and RNaseD) are utilized to degrade the viral RNA genome. The virion-associated reverse transcriptases catalyzing these functions are heterodimers of 66- and 51-kilodalton polypeptides (p66 and p51). Although a role for p51 remains to be assigned, the association of p51 into heterodimer with p66 appears to be a prerequisite for all enzymatic functions.

The availability of substantial quantities of enzymatically active HIV-1 and HIV-2 reverse transcriptase from recombinant bacteria has proven invaluable for detailed genetic and biochemical characterization of these processes. In combination with crystallographic data now available for both intact heterodimer HIV-1 reverse transcriptase and a recombinant polypeptide representing the RNaseH domain of p66, a foundation has been laid for development of alternative therapeutic strategies against this highly versatile enzyme.

Primer binding. Retroviral replication initiates from the $3'$ terminus of a host tRNA hybridized to the primer binding site, a specific sequence near the $5'$ end of the single-stranded viral RNA genome. DNA sequence analysis of the HIV-1 and HIV-2 genomes has implicated a lysine-coding transfer RNA $(\text{tRNA}^{\text{Lys},3})$ as the replication primer. The interaction between HIV-1 reverse transcriptase and its replication primer, mediated in part through bases derived from the tRNA anticodon domain, has been demonstrated by several investigators. Heavily modified bases within the anticodon domain of $\text{tRNA}^{\text{Lys},3}$ may contribute to the specificity or affinity in this complex, possibly functioning to correctly position the primer.

A potential therapeutic strategy has been to develop small oligonucleotides that mimic the anticodon domain of $\text{tRNA}^{\text{Lys},3}$ so that they are recognized and stably bound by the HIV reverse transcriptase. Competition of these synthetic oligonucleotides with the natural tRNA replication primer would thus prevent HIV reverse transcriptase from proceeding into minus (−) strand DNA synthesis. The feasibility of this approach has been demonstrated: a purified oligoribonucleotide derived from the anticodon domain of $\text{tRNA}^{\text{Lys},3}$, when complexed with the reverse transcriptase, was capable of inhibiting RNA-dependent DNA synthesis. Furthermore, cordycepin analogs of $2'-5'$ oligoadenylic acid have been shown to impair binding of $\text{tRNA}^{\text{Lys},3}$ to the HIV-1 reverse transcriptase. Analogs of $2'-5'$ oligoadenylic acid, which may mimic the tRNA primer, effectively compete for binding to reverse transcriptase.

Polymerization. Reverse transcription of the viral RNA genome results in synthesis of (−) strand DNA. Plus-strand DNA synthesis initiates from a second RNA primer located at the $3'$ end of the viral RNA. This second-strand synthesis is also accomplished by reverse transcriptase, which exhibits DNA-dependent as well as RNA-dependent DNA polymerase activity. Although each subunit of the heterodimer contains amino acids constituting the active site, analysis of reconstituted, selectively mutated HIV-1 heterodimers indicates that both polymerase functions are restricted to the 66-kDa subunit. The absence of a binding cleft and inaccessibility of proposed catalytic residues of p51 have been proposed to account for these observations.

Inhibition of the polymerase function of reverse transcriptase via incorporation of chain-terminating nucleotide analogs has been the most successful anti-HIV therapy to date. Examples of these analogs include azidothymidine (AZT), dideoxycytidine (ddC), and dideoxyinosine (ddI). Unfortunately, an increasing problem with administration of these drugs is the rapid emergence of resistance, mediated through spontaneous amino acid alterations in the polymerase function of reverse transcriptase. Several extremely potent nonnucleoside analogs (such as R82150, nevirapine, and the pyridinone derivative L697,639) have also been reported. However, in addition to the problem of drug resistance, these compounds suffer from a further disadvantage in that their activity is restricted to the HIV-1 reverse transcriptase. Despite these problems, the continued development of potent and selective inhibitors of polymerase function remains a critical antiviral strategy. Since the chain-terminating and nonnucleoside inhibitors have different target sites on reverse transcriptase, combination therapy with these or related compounds may prove useful in combating drug resistance.

Ribonuclease H. Reverse transcription events are accompanied by coordinated degradation of the viral RNA genome. In addition, the primers of (−) and (+) strand DNA synthesis must be excised to generate duplex viral DNA uninterrupted by ribonucleotides. An RNaseH activity, catalyzing degradation of RNA within RNA/DNA duplexes, has been assigned to the C terminus of p66. Although RNaseH is a ubiquitous enzyme, loss of the virus-coded function severely impairs propagation of recombinant viruses; at the molecular level, this loss most likely reflects disruption of events whereby nascent (−) strand DNA is translocated from the $5'$ to the $3'$ end of the RNA genome.

While several lines of evidence suggest an interdependence between the polymerase and RNaseH domains of HIV reverse transcriptase, preliminary laboratory experiments have suggested that selective inhibition of the RNaseH activity might be achieved. Illimaquinone, a secondary metabolite from the Red Sea sponge *Smensopongia* sp., is an effective inhibitor at concentrations that do not severely impair polymerase function of either the HIV reverse transcriptase or DNA polymerase α of the host cell. Similarly, HP 0.35, a degradation product prepared from a cephalosporin drug, displays preferential activity against HIV RNaseH. Since both illimaquinone and HP 0.35 are derived from classes of pharmacological agents presently in clinical use, it should be possible to screen related compounds rapidly for their anti-HIV activity. Furthermore, recent crystallization of *Escherichia coli* RNaseH, as well as a recombinant polypeptide representing the RNaseH domain of HIV-1 reverse transcriptase, should prove valuable in guiding rational design of drugs specific for this retroviral function. Recent evidence has suggested that the RNaseH domain of p66 might also contain an activity for hydrolysis of duplex RNA. Mutagenesis experiments have suggested that the RNaseH catalytic center also controls this activity, designated RNaseD. Although the significance of RNaseD activity remains to be established, RNaseD may provide the opportunity for alternative therapeutic strategies.

Dimerization. The biologically significant form of HIV reverse transcriptase is a heterodimer of p66 and p51. These subunits are the product of the same gene, but differ in that the smaller arises through partial C-terminal maturation of p66 by the virus-coded proteinase. Alternative explanations for this partial maturation have been proposed, the most likely being asymmetric organization of 66-kDa subunits in a homodimer, thereby making a single C-terminal domain available for proteolysis. The catalytic component of heterodimer HIV-1 and HIV-2 reverse transcriptase is the p66 subunit, while a role for the p51 subunit remains to be established. Possibilities have included enhancing processivity of the catalytic subunit during DNA synthesis, maintaining p66 in a conformation permitting coordination of polymerase and RNaseH functions, and docking of the tRNA primer for extension into (−) strand DNA synthesis from the primer binding site. Thus, the possibility of developing chemotherapeutic agents targeted to the dimer interface of the enzyme might be considered.

At present, there are insufficient structural data on HIV reverse transcriptase to assign specific amino acids that participate in dimerization. A periodic array of leucine residues, analogous to the so-called leucine zipper motif mediating subunit interactions of several transcriptional factors, has been noted. However, its importance for association of the p66 and p51 subunits remains to be established. The RNaseH domain of p66 most likely contributes to organization of the heterodimer, since data from both x-ray crystallography and neutron diffraction indicate that this domain is involved in both intra- and intersubunit contacts. Research in this area should be facilitated by recent elucidation of the three-dimensional structure of HIV-1 reverse transcriptase; as this structure is refined, it will be possible to more precisely locate domains contributing subunit interactions.

For background information SEE ACQUIRED IMMUNE DEFICIENCY SYNDROME (AIDS); RETROVIRUS; REVERSE TRANSCRIPTASE; RIBONUCLEASE in the McGraw-Hill Encyclopedia of Science & Technology.

Stuart F. J. Le Grice

Bibliography. J. F. Davies et al., Crystal structure of the ribonuclease H domain of HIV-1 reverse transcriptase, *Science*, 252:88–95, 1991; Jacobo-Molina and E. Arnold, HIV reverse transcriptase structure-function relationships, *Biochemistry*, 30:6351–6361, 1991; L. A. Kohlstaedt et al., Crystal structure at 3.5Å resolution of HIV-1 reverse transcriptase complexed with an inhibitor, *Science*, 256:1783–1790, 1992; S. F. J. Le Grice, Human immunodeficiency virus reverse transcriptase, in A. M. Skalka and S. P. Goff (eds.), *Reverse Transcription*, Cold Spring Harbor Monograph Series, 1993.

Adaptive optics

Light propagating through the atmosphere can be severely aberrated by atmospheric turbulence. Turbulence-induced aberration is what causes stars to twinkle. This distortion of starlight impedes astronomical investigations. Consequently, the new generation of 8–10-m (26–33-ft) telescopes, although having impressive light-gathering capabilities, will have visible-light resolution no better than that of an amateur astronomer's 10–15-cm (4–6-in.) telescope.

Formerly, the only way to overcome the limitation of atmospheric turbulence was to place, at great expense, space telescopes such as the Hubble in orbit above the atmosphere. However, information from previously classified United States military research has demonstrated that adaptive optics systems using synthetic beacons (also called laser guide stars) can successfully compensate atmospheric distortions on ground-based astronomical telescopes. A new generation of large astronomical telescopes equipped with adaptive optics systems should provide unprecedented views of the cosmos.

Stars as optical beacons. The relatively simple situation in which the star being imaged is also used as the beacon for the adaptive optics system is illustrated in **Fig. 1**. Light from a star passes through the atmosphere, where it is distorted by atmospheric turbulence, and is collected by a telescope. The light then reflects from a deformable mirror, which typically consists of a thin glass faceplate behind which are piezoelectrically driven actuators that can deform its surface by distances corresponding to a few wavelengths of light. From the deformable mirror, part of the light goes to an imaging camera and part to a wavefront sensor. The wavefront sensor measures an array of local phase gradients, or wavefront tilts, which are then processed in a wavefront reconstructor (usually a special-purpose computer) to derive a phase map of the incoming wavefront. The phase values are used in a multichannel servo system to drive the deformable mirror so as to flatten the incoming wavefront, resulting in a star image that is nearly diffraction-limited.

The required size and speed of the adaptive optics system are set by fundamental atmospheric-turbulence parameters. For good correction, the measurement-correction channels should be no more than r_0 apart, where r_0 is the turbulence coherence length (corresponding to the maximum diameter over which the star image would be diffraction-limited without correction). For typical conditions, r_0 ranges from 5 to 15 cm (2 to 6 in.) for visible light; thus, an adaptive optics system for a 1-m (40-in.) telescope requires about 100 channels, whereas a system for an 8-m (26-ft) telescope requires many thousands of channels. These values are for visible-light correction; since r_0 is proportional to $\lambda^{6/5}$, where λ is the wavelengthof light, correction in the infrared requires far fewer channels. The required correction bandwidth, which is a measure of how rapidly the correction must be updated, depends on both the turbulence strength and the wind speed. For typical conditions, the required bandwidth is of the order of 100 Hz for correction in the visible wavelengths. The bandwidth scales as $\lambda^{-6/5}$, so again the system requirements are much less demanding in the infrared. For typical turbulence conditions, the deformable mirror must correct for an optical path difference of approximately 3 micrometers (0.0001 in.) for a 1-m (40-in.) telescope and $16\mu m$ (0.0006 in.) for an 8-m (26-ft) telescope, independent of the wavelength being corrected. Since more than half of the required correction is tilt, adaptive optics systems usually have a separate tilt-correction mirror (Fig. 1) to relieve the stroke requirements on the deformable mirror.

Although the adaptive optics scheme shown in Fig. 1 can be effective, it requires a very bright star. At visible wavelengths, for instance, a star of approximately 5th magnitude is needed to run the wavefront sensor at the required update rate and spatial resolution. This requirement restricts the technique to observation of only about a thousand of the brightest objects in the heavens. The situation is somewhat better in the infrared, but the number of objects that can be successfully compensated is still small.

The technique shown in Fig. 1 may be extended by using a bright star to compensate for a dim astronomical object nearby. The bright star is known as a guide star. The guide star must be separated from the object to be imaged by no more than the isoplanatic angle θ_0, within which the turbulence-induced phase aberration is approximately constant. For visible wavelengths the isoplanatic angle is small, only about 10–25 micro-radians (2–5 arc-seconds). The small angle restricts the use of the natural guide-star technique to a small portion of the sky.

Synthetic beacons. To overcome the limitations of natural stars as beacons and to obtain potentially full-sky coverage for adaptive optics, synthetic beacons (also called laser guide stars) can be used (**Fig. 2**). A ground-based laser beam is sent in the direction of the astronomical object to be imaged, and a synthetic beacon is generated, either by Rayleigh backscatter or by resonant backscatter from the mesospheric sodium layer. By using Rayleigh backscatter, which is scatter from atmospheric nitrogen and oxygen, a synthetic beacon can, in principle, be generated

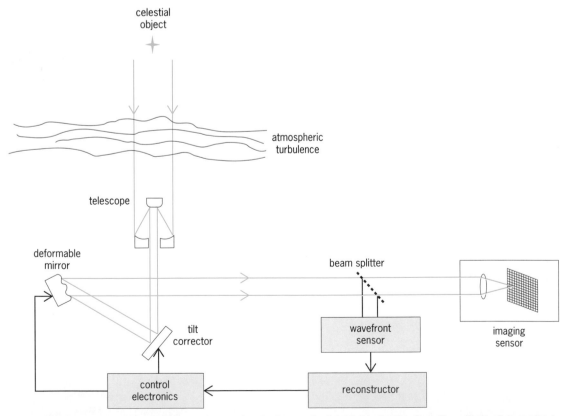

Fig. 1. Adaptive optics arrangement for compensating the image of a bright star for the degrading effects of atmospheric turbulence. (After D. P. Greenwood and C. A. Primmerman, Adaptive optics research at Lincoln Laboratory, Lincoln Lab. J., 5(1):4–24, 1992)

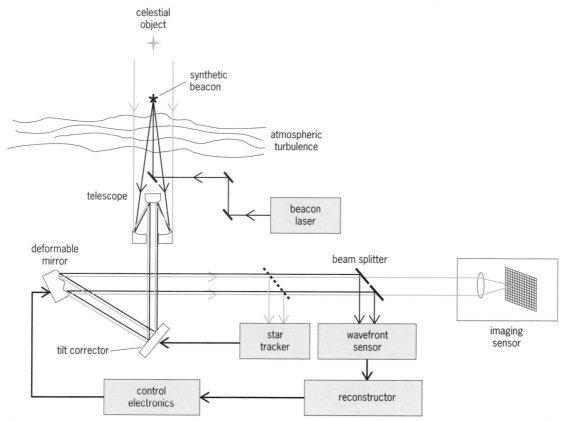

Fig. 2. Adaptive optics arrangement for compensating the image of an astronomical object by using a synthetic beacon.

at any altitude. However, the decreasing atmospheric density with altitude and practical considerations of illuminator laser power effectively limit Rayleigh beacons to altitudes of less than about 25 km (15 mi). The mesospheric sodium layer comprises atomic sodium about 15 km (9 mi) thick centered at an altitude of about 90 km (55 mi); by tuning the illuminator to the 589-nanometer sodium D_2 line, it is possible to receive a strong resonant backscatter signal.

Light from the synthetic beacon along with light from the astronomical object of interest passes through the atmosphere and reflects from the deformable mirror. The light from the synthetic beacon is sent to the wavefront sensor, where its wavefront is measured and used to control the deformable mirror. The light from the astronomical object is corrected by the deformable mirror and sent to an imaging camera or other astronomical instrument. Use of a synthetic beacon decouples the brightness of the astronomical object from the performance of the system, thus allowing adaptive optics correction on very dim objects. Since the synthetic beacon can be aimed anywhere in the sky, complete coverage is possible for the adaptive optics.

Use of a synthetic beacon solves the problem of obtaining atmospheric correction at any angle in space but introduces several new problems. First, a measurement error, termed focal anisoplanatism, results from the beacon's being at a different range from that of the astronomical object. Focal anisoplanatism has two components: (1) There can be unsensed turbulence above the synthetic beacon. (2) Below the synthetic

beacon the optical rays from the beacon do not follow exactly the same path through the atmosphere as do the essentially collimated rays from the astronomical object (Fig. 2). Because the rays increasingly diverge as the telescope size increases, focal anisoplanatism limits the size of the telescope that can be corrected with a synthetic beacon at a given altitude. Focal anisoplanatism can be reduced by increasing the altitude of the synthetic beacon or by using multiple synthetic beacons, where each beacon is used to sense the error over a portion of the telescope aperture, and the multiple-beacon signals are then stitched together to correct over the full telescope. Since the multiple-beacon stitching process introduces its own limitations, it is desirable to have the beacon altitude as high as possible; thus, generation of the beacon in the sodium layer appears most attractive.

In addition, because the synthetic beacon is projected up through the turbulent atmosphere its angular position relative to the astronomical object is not precisely known. Thus, the synthetic beacon cannot be used for tracking and tilt correction. Instead, as shown in Fig. 2, a separate tilt-correction system using light from the astronomical object or a nearby natural guide star must be employed. This requirement brings back the problem with regard to natural guide stars, but there are several advantages in using these stars for overall tracking information as compared to using them for full adaptive optics correction. First, the tracker collects light over the full telescope aperture, not just over a 10–15-cm (4–6-in.) subaperture, al-

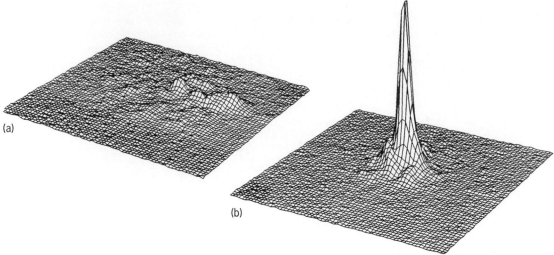

Fig. 3. Images of the star Procyon at a zenith angle of 20° taken with a 64 × 64 charge-coupled device camera. (*a*) Uncompensated image. (*b*) Image compensated by using a 241-channel adaptive optics system and a synthetic beacon formed by Rayleigh backscatter. (*After C. A. Primmerman et al., Compensation of atmospheric optical distortion using a synthetic beacon, Nature, 353:141–146, 1991*)

lowing much dimmer guide stars to be used. Second, the tilt anisoplanatic angle (the angle over which the overall tilt is sensibly constant) is much larger than the isoplanatic angle for finer-scale atmospheric distortions, allowing guide stars farther away from the astronomical object of interest to be used. As a consequence, it is still possible to achieve full-sky coverage by using a synthetic-beacon adaptive optics system.

Synthetic-beacon results. As yet, no synthetic-beacon adaptive optics system is operational on a working astronomical telescope. However, several experiments have conclusively demonstrated the effectiveness of the technique. **Figure 3** shows a result from one of those experiments obtained with a 241-channel adaptive optics system used with a 60-cm (24-in.) telescope. Images of the star Procyon both uncompensated and compensated are shown. The uncompensated image is ill-formed and smeared; in fact, it slightly exceeds the field of view of the imaging camera. The compensated image, in contrast, is a well-formed spot with considerably increased peak intensity. The width of the corrected beam at half maximum, which may be taken as the system resolution, is essentially diffraction-limited for the telescope aperture. The peak intensity is 0.46 times the diffraction limit. The adaptive optics system has formed a nearly perfect central lobe, but some energy has been thrown into the wings. This behavior is typical for adaptive optics systems.

The results shown in Fig. 3 were obtained by using a single low-altitude Rayleigh backscatter beacon. Atmospheric correction has also been performed by stitching together multiple synthetic beacons, and atmospheric phase measurements have been made by using synthetic beacons in the sodium layer.

Prospects for astronomy. The experimental results support the feasibility of deploying adaptive optics systems using synthetic beacons on astronomical telescopes. To achieve near-diffraction-limited imaging at visible wavelengths for a current-generation 3.5-m (12-ft) telescope, an adaptive optics system with 500 channels and a single sodium illuminator operating at 200 W might be required. Such a system is well within current capabilities. For the next generation of 8–10-m (26–33-ft) telescopes, one to four sodium illuminators each operating at 200 W and an adaptive optics system with 2500 channels might be required. This requirement is only slightly beyond current capabilities.

As astronomers discover its utility, an adaptive optics system will become an essential feature on almost every ground-based telescope. It is likely that every large telescope built in the future will incorporate such a system as an integral part.

For background information SEE ADAPTIVE OPTICS; OPTICAL TELESCOPE; SCATTERING OF ELECTROMAGNETIC RADIATION in the McGraw-Hill Encyclopedia of Science & Technology.

Charles A. Primmerman

Bibliography. R. Q. Fugate et al., Measurements of atmospheric wavefront distortion using scattered light from a laser guide star, *Nature*, 353:144–146, 1991; R. A. Humphreys et al., Atmospheric-turbulence measurements using a synthetic beacon in the mesospheric sodium layer, *Opt. Lett.*, 16:1367–1369, 1991; D. V. Murphy et al., Experimental demonstration of atmospheric compensation using multiple synthetic beacons, *Opt. Lett.*, 16:1797–1799, 1991; C. A. Primmerman et al., Compensation of atmospheric optical distortion using a synthetic beacon, *Nature*, 353:141–143, 1991.

Adaptive sound control

Sound fields radiated by vibrating elastic structures often create noise problems. Active structural acoustic control is a technique in which vibrating active forces or strains are applied to the vibrating structure while the radiated sound pressure field is sensed and minimized. In essence, the active structural inputs create anti-vibrations in the structure that cancel those vibrations that are radiating sound. The phenomenon is related to the interference pattern obtained from two coherent light sources. The active structural inputs can be in the form of point-force shakers, piezoelectric elements, or more advanced surface-strain actuators attached to the structure. The radiated pressure can be sensed with conventional transducers such as microphones or estimated from the output of distributed structural sensors such as piezoelectric films, or of arrays of point sensors such as accelerometers. The heart of the active structural acoustic control technique is an adaptive controller that adjusts the magnitude and phase of the control actuators until the radiated pressure field is driven to a low value. The advantage of the active structural acoustic control technique is that effective control can be achieved in the low- to mid-frequency range, where passive control techniques are ineffective, with relatively few control actuators, and the transducers can be arranged to create a compact, lightweight system.

Control of aircraft interior noise. Early work in active structural acoustic control demonstrated that sound transmission into closed elastic cylinders could be controlled by active point forces applied to the cylinder wall. The aim of the work was to develop advanced techniques for controlling interior noise within propeller-driven aircraft. In general, sound is transmitted to the aircraft interior in the following manner. The propellers generate noise (due to their motion through the air), which radiates toward the fuselage and excites it into motion. The vibration of the aircraft fuselage subsequently radiates sound to the interior space. The fuselage vibration occurs in set patterns or structural modes at typical propeller sound frequencies. It was shown that only certain structural modes of the fuselage vibration couple or radiate sound to the interior space, and thus it is necessary to control only these modes out of the complete set of response patterns. This control effect, called modal suppression, requires a low number of active inputs. Since the structural motion that gives rise to the interior sound pressure response is being directly controlled, the interior sound field is reduced globally (that is, through an extended volume), independent of the fuselage modal response shape.

An example of such control is shown in **Fig. 1**, which presents results of a test on noise reduction in a full-scale DC-9 fuselage. Figure 1*a* shows a plan view of the rear of the aircraft. Microphone locations are numbered, and the position of noise and control inputs are indicated. Attenuations of the order of 10 dB were obtained for structure-borne interior noise transmitted through the engine pylons and simulating engine vibrations. Two point-force active inputs were applied directly to the fuselage near the pylon attachment points. Four microphones, marked with asterisks in the figure, were used to provide information to the controller, which adjusted the outputs of the active forces. The overall global reduction in sound pressure level was of the order of 10 dB. The performance obtained was far better than that using conventional passive treatments such as damping materials.

Weight, mounting considerations, and control spillover effects have led to recent cooperative work to study the use of piezoceramic active actuators bonded to fuselage structural members. On a full-scale composite fuselage section, a four-actuator, six-sensor sys-

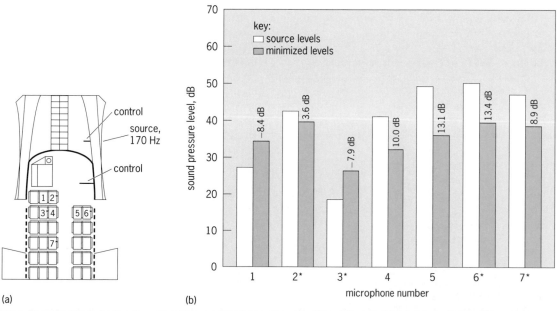

(a)   (b)

Fig. 1. Test of noise reduction through active structural acoustic control in a full-scale DC-9 fuselage. (*a*) Plan view of rear of aircraft, showing locations of noise source (frequency = 170 Hz), forward and aft active control shakers, and microphones indicated by numbers 1–7. The microphones marked with asterisks are error sensors. (*b*) Measured noise levels at the microphones before and after noise reduction. The numbers above the bar give the noise reduction at each microphone. Averaged noise reduction was 10.3 dB. (*After M. A. Simpson et al., Full-scale demonstration tests of cabin noise reduction using active vibration control, J. Aircraft, 28(3):208–215, 1991*)

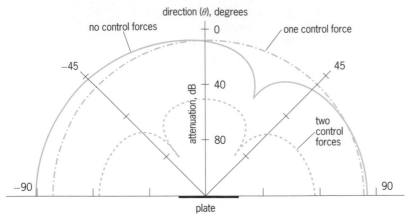

Fig. 2. Directivity patterns of radiated sound pressure fields from a vibrating clamped circular plate. Patterns of noise in absence of control forces and in presence of one and two control forces are shown, with $k_0 a = 0.45$, where k_0 is the wave number and a is the radius of the plate. (*After C. R. Fuller, Active control of sound transmission/radiation from elastic plates by vibration inputs: I. Analysis, J. Sound Vibr., 136(1):1–15, 1990*)

tem has achieved interior global attenuations of 8–15 dB over a range of test conditions. Optimization of the size, distribution, and location of the control transducers is expected to produce further improvements. The active structural acoustic control technique is currently being studied by aircraft companies for potential use in production airplanes.

Control of hull-radiated sound. Controlling the sound radiated from the hulls of marine vessels is another important application of active structural acoustic control. Previous work in this area studied the control of free-field radiation from panels immersed in a light fluid medium such as air. **Figure 2** shows directivity patterns for sound radiated from a vibrating clamped circular panel with one and two control inputs. **Figure 3** presents the corresponding amplitude of vibration of the same plate.

The plate is being excited near the frequency of the second mode of vibration. Figure 2 shows that one

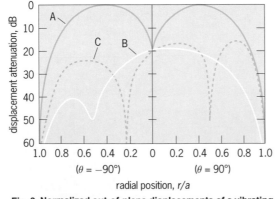

Fig. 3. Normalized out-of-plane displacements of a vibrating clamped circular plate. Displacements are shown for noise in absence of control forces (A) and with one (B) and two (C) control forces, with $k_0 a = 0.45$, where k_0 is the wave number and a is the radius of the plate. Displacements are shown along a diameter of the plate, and radial position is given in units of r/a, where r is the distance from the center of the plate. (*After C. R. Fuller, Active control of sound transmission/radiation from elastic plates by vibration inputs: I. Analysis, J. Sound Vibr., 136(1):1–15, 1990*)

control force produces very little attenuation of sound, while two control forces give global attenuations of the order of 45 dB. Correspondingly, Fig. 3 reveals that increasing the number of active control inputs from one to two does not cause a significant reduction in the panel vibration amplitude; however, the response pattern becomes more complex.

In other work, a reduction in sound radiation was observed to occur in most cases where the system was being driven off-resonance, with little or no change in the spatially averaged structural response, similar to the above effect. It was concluded that the residual or controlled structural response has a lower radiation efficiency (loosely defined as sound level per unit of structural response) for the same overall level of response, thus leading to lower sound radiation. This effect was termed modal restructuring, since the modes of the system are restructured so that the controlled system has an overall lower radiation efficiency. More recently, it has been shown that modal suppression corresponds to a decrease in all structural wave-number components. In wave-number analysis, a spatial Fourier transform is applied to the structural response, decomposing it into a set of infinite, differing wavelengths λ or wave numbers $k(= 2\pi/\lambda)$. Only those wavelengths that are longer than the free wavelength of sound in the surrounding medium can radiate sound. Modal restructuring has been demonstrated to correspond to a reduction in the radiating, supersonic wave-number components, while the subsonic components are left largely unaffected or even increased.

The significance of modal restructuring is that large attenuations of radiated sound can be achieved by an appropriate change in the shapes of the controlled modes without affecting overall amplitude. This approach has been shown to require significantly less control energy.

This work has now been extended to include the influence of heavy fluid loading, such as water, on the structural motion as well as to consider radiation from nonplanar surfaces such as closed cylinders. Experiments performed in a water tank have confirmed that the active structural acoustic control technique still provides high global attenuations of radiated sound when the radiation loading induces significant modal coupling.

Actuator configuration. Other work on active structural acoustic control of sound radiation into a free field has centered on using optimally configured piezoelectric sensors and actuators bonded or embedded in the structure (resulting in adaptive, intelligent, or smart structure). Emphasis has been placed on shaping the distributed sensors so that only the radiating components of the structural motion are observed, as previous work has demonstrated that driving the structural response to zero at discrete points can lead to an increase in sound radiation. In essence, a distributed, shaped piezoelectric sensor acts as a structural wave-number filter and observes only the radiating components of vibration. Such approaches have led to good reductions in radiated sound levels for both on- and off-resonance excitation frequencies. Analyses and experiments have demonstrated that optimizing the configuration (that is, the shape, size, location, and so forth) of the actuators is of the same importance in terms of control performance as increasing the number of control channels.

Control strategy. In order to implement the active structural acoustic control technique, a control strategy is required. Initial work has concentrated on using single-frequency, time-domain, adaptive least-mean-square algorithms implemented on fast-signal-processing chips. The advent of these fast digital-signal-processing chips has enabled active control approaches to be implemented in real time; because of the high computational speed of the chips, the controller can converge and adapt faster than the characteristic noise time scales of the system to be controlled. More recent work has extended the least-mean-square control approach to multifrequency and broad-band disturbances with multichannel input-output configurations. Control of transient disturbances has also been demonstrated by utilizing state-space feedback approaches in conjunction with radiation filters to model the structural acoustic coupling (the sound radiation).

Prospects. In summary, the active structural acoustic control technique has demonstrated much potential in aerospace and marine applications. However, there remain many areas to be investigated and technical problems to be resolved for the approach to be generally applicable. Future work will focus on extending these techniques to more complex disturbances and structures as well as on improving modeling capabilities to study and predict control performance. Potential applications apart from those discussed above include many common distributed elastic systems excited in the low- to mid-frequency range such as transformers, factory machines, and household appliances. A multidisciplinary approach is required to synthesize a design procedure that integrates the elements of structural acoustics and transducer and control technology. The payoff will be reduced noise levels as well as a significant cost and weight savings, often accompanied by performance improvements.

For background information SEE ADAPTIVE SOUND CONTROL; CONTROL SYSTEMS; MECHANICAL VIBRATION; SOUND; TRANSDUCER in the McGraw-Hill Encyclopedia of Science & Technology.

Chris R. Fuller

Bibliography. C. R. Fuller, Active control of sound transmission/radiation from elastic plates by vibration inputs: I. Analysis, *J. Sound Vibr.*, 136(1):1–15, 1990; M. A. Simpson et al., Full-scale demonstration tests of cabin noise reduction using active vibration control, *J. Aircraft*, 28(3):208–215, 1991; B. Widrow and S. D. Stearns, *Adaptive Signal Processing*, 1985.

Adipose tissue

Living organisms break down nutrients and use the released energy for the multiplicity of functions inherent to life. A major fraction of the energy contained in foodstuffs is lost as heat in a process called obligatory thermogenesis. There is a precise ambient temperature, called thermoneutrality, at which obligatory thermogenesis is sufficient to maintain body temperature in homeothermal animals. When the temperature is reduced below thermoneutrality, mechanisms are activated to reduce the dissipation of heat from the body and to increase heat generation. Although shivering is an effective mechanism for increasing heat production acutely, sustained production of heat involves nonshivering mechanisms in a process called nonshivering facultative thermogenesis or, simply, facultative thermogenesis. Facultative thermogenesis is a regulated form of energy dissipation.

Brown adipose tissue. In mammals, brown adipose tissue (brown fat) is a specialized tissue that is a site of facultative thermogenesis. There are abundant mitochondria in brown fat, and the brownish color derives from the mitochondrial heme-containing cytochromes. This fat is a form of readily available energy in multilocular storage, which morphologically differentiates it from white adipose tissue. The capacity of brown fat to produce heat resides in uncoupling protein. Uncoupling protein, located in the inner membrane of the brown fat mitochondria, is an ionic transporter that dissipates the proton gradient generated during mitochondrial respiration. In the mitochondria of any cell, the energy released from the oxidation of substrates is transiently accumulated in a proton gradient across the inner mitochondrial membrane. The buildup of this gradient slows mitochondrial respiration, but respiration proceeds as the energy of the gradient is transferred to adenosinetriphosphate (ATP) by the enzyme ATP synthetase, also present in the inner mitochondrial membrane. Thus, the proton gradient couples respiration to adenosinediphosphate (ADP) phosphorylation. In brown fat mitochondria, uncoupled protein dissipates this gradient, bypassing the ATP synthetase. As a consequence, when uncoupling protein is activated to dissipate the gradient, respiration is markedly accelerated and the energy directly dissipated as heat.

The sympathetic nervous system activates brown fat thermogenesis via the transmitter norepinephrine. By interacting with brown fat surface receptors, norepinephrine triggers the hydrolysis of the stored triglycerides, with release of fatty acids, which then activate uncoupling protein and serve as fuel for the production of heat. Substantial experimental evidence indicates that the thermogenic potential of brown fat is determined by the mitochondrial concentration of uncoupling protein. Increased uncoupling protein synthesis is an integral part of brown fat thermogenic response. Within 15 min of sympathetic stimulation of brown fat, it is possible to demonstrate an increase in uncoupling protein gene transcription, followed shortly by the accumulation of uncoupling protein messenger ribonucleic acid (mRNA). Another integral part of the brown fat response, in order to cope with the increased fuel demands that rapidly deplete the fat stores, is the activation of glucose and free-fatty-acid uptake from circulating triglycerides and the stimulation of local lipogenesis.

Energy balance. Although facultative thermogenesis is most potently stimulated by cold, it is also linked to food intake. Following food intake, in addition to the energy expended to process the food, there is a distinct increase in heat production called diet-induced thermogenesis. Fasting reduces the sympathetic activation of brown fat, whereas food intake results in increased brown fat thermogenesis. The role of brown fat in diet-induced thermogenesis is proportional to the amount of food ingested but is also influenced by the type of nutrient. In studies on genetic obesity using mice and rats whose sympathetic activation of brown fat is impaired, the animals are not able to maintain their body temperature in the cold, and also develop a striking obesity as they grow older.

Hormones and thermogenesis. Several hormones stimulate thermogenesis. In the case of sex steroids, adrenal steroids, and growth hormone, the increase in thermogenesis is basically the result of the stimulation of specific metabolic pathways and growth; that is, it represents obligatory energy dissipation. Insulin, glucagon, and epinephrine also stimulate heat production in an obligatory fashion, since they stimulate the mobilization and transformation of substrates. Thyroid hormones have widespread effects in accelerating metabolic pathways, thereby increasing obligatory energy dissipation.

Animals with severe thyroid hormone deficiency cannot survive in cold environments, and cold intolerance is a symptom in human hypothyroidism. This inability to withstand cold has been considered the consequence of reduced obligatory thermogenesis. Brown fat from hypothyroid rats shows evidence of increased stimulation by the sympathetic nervous system, but measures of uncoupling protein activity and concentration indicate that neither increases in response to cold exposure. This lack of response can be corrected by the administration of thyroid hormones, but the induction of hyperthyroidism was observed to cause a paradoxical reduction in uncoupling protein, leading to the conclusion that thyroid hormones are permissive for the function of brown fat.

Synergistic interactions. The knowledge of the intimate mechanisms involved in brown fat activation suggests a more important role for thyroid hormone in brown fat function. Of the two major circulating thyroid hormones, thryoxine (T_4) and 3,5,3′-triiodothyronine (T_3), the latter is responsible for most of the biological activity. Under physiological circumstances, T_4 is largely a prohormone that is converted in tissues to the much more active T_3. This process is catalyzed by two enzymes that are present in several tissues, including brown fat. One of the enzymes provides active thyroid hormone for local consumption, while the other (5′-deiodinase II) produces T_3 for circulation. In brown fat, 5′-deoiodinase II is uniquely activated by the sympathetic nervous system. Within 1–2 h of cold exposure or injection of norepinephrine into rats, a tenfold or greater activation is observed with a severalfold increase in the concentration of T_3 in brown fat.

The local increase in T_3 following the adrenergic stimulation of 5′-deiodinase II is crucial for the response of uncoupling protein to cold. It has been possible to manipulate the content of T_3 locally in brown fat and dissociate the tolerance to cold from the effects of thyroid hormone in other tissues, that is, from obligatory thermogenesis. Experiments have demonstrated that contrary to the old concept that thyroid hormone plays a marginal role in facultative thermogenesis, the inability of animals to withstand cold in hypothyroidism is due to a major deficiency in the function of brown fat, that is, to facultative thermogenesis. These experiments have also provided support for the relevance of brown fat in adaptation to cold. It has been demonstrated, for example, that the ability of hypothyroid rats to maintain body temperature in the cold correlates with the restoration of the uncoupling protein response to cold, and that both the uncoupling protein and thermogenic responses to cold correlate with the amount of T_3 bound to its receptor in brown fat. Actually, this T_3 may induce marked differences in brown fat responses to sympathetic stimulation. It also has been shown that 5′-deiodinase II is stimulated by insulin and by food intake, both of which increase brown fat thermogenesis as well.

The response of uncoupling protein to cold or exogenous norepinephrine is due to an increase in the transcription of the uncoupling protein gene. In the absence of thyroid hormone, this response is insufficient to protect the animals from hypothermia in a cold environment. The administration of thyroid hormone dramatically amplifies the transcriptional effect of norepinephrine, leading to levels of uncoupling protein compatible with a normal response to cold. Although T_3 may be important for the strength of the norepinephrine signal in brown fat, most of the synergistic interaction between norepinephrine and T_3 takes place at the level of the uncoupling protein gene.

Obesity and brown fat in humans. A growing accumulation of evidence demonstrates that reduced energy expenditure is a risk factor for obesity. The thermogenic response to overfeeding varies markedly among individuals, and this variation is genetically determined.

Brown fat is present in all mammals, but its importance and level of activation depend on the size of the animals and the degree of thermal stress under which they live. In rats, the thermoneutrality temperature is 82°F (28°C). Rats are small animals with a relatively large ratio of surface area to volume; therefore, living at temperatures below thermoneutrality represents a substantial thermal stress. Brown fat level of activity in rats at room temperature, 72°F (22°C), is significant, as evidenced by the reduction in brown fat activity when they are placed at thermoneutrality. Brown fat is also physiologically important in human newborns, but when uncoupling protein gene expression has been measured in humans of all ages, it is evident that brown fat function decreases with age. However, brown fat is responsive to adrenergic stimulation. Patients with tumors of the adrenal medulla (pheochromocytoma) have high plasma concentrations of norepinephrine, highly activated brown fat, and increased heat production. Likewise, individuals dying of conditions involving thermal stress, such as hypothermia, show clear signs of brown fat activation.

One of the arguments for attributing a merely permissive role in brown fat to thyroid hormone function is the observation that, being necessary for a normal response to cold, thyroid hormone in excess causes a reduction in uncoupling protein. A logical explanation is that, in the hyperthyroid state, obligatory thermogenesis is exaggerated, creating a downward shift in thermoneutrality and a consequent reduced need for brown fat thermogenesis to maintain body temperature. A similar situation can be created simply by acclimating animals at high ambient temperatures. These observations may be relevant to the condition in the adult human. Thus, a larger body size, that is, reduced ratio of body surface area to volume, and the ability of humans to control the temperature of their environment lead to a reduction in brown fat.

For background information SEE ADIPOSE TISSUE; HYPOTHERMIA; METABOLISM; MITOCHONDRIA; THERMO-REGULATION in the McGraw-Hill Encyclopedia of Science & Technology.

J. Enrique Silva; Antonio C. Bianco

Bibliography. A. C. Bianco and J. E. Silva, Intracellular conversion of thyroxine to triiodothyronine is required for the optimal thermogenic function of brown adipose tissue, *J. Clin. Invest.*, 79:295–300, 1987; B. Cannon and J. Nedergaard, The biochemistry of an inefficient tissue: Brown adipose tissue, *Essays Biochem.*, 2:110–164, 1985; D. G. Nicholls and R. M. Locke, Thermogenic mechanisms in brown fat, *Physiol. Rev.*, 64:1–64, 1984; S. Rehnmark et al., Transcriptional and post-transcriptional mechanisms in the uncoupling protein mRNA response to cold, *Amer. J. Physiol.*, 262:E58–E67, 1992.

Agricultural health

There are an estimated 5 to 6 million farmers and farm workers in the United States, and more than twice this number of individuals may be at risk from exposures in the agricultural setting. According to traumatic occupational fatality rates, agriculture is the most hazardous industry in the United States. The National Safety Council estimates that there are approximately 50 deaths per 100,000 agricultural workers, or about five times the rates for all industries. Children in agricultural settings are also at risk, with approximately 300 deaths and more than 23,000 injuries of individuals under the age of 16 per year.

In addition to traumatic injuries and fatalities, numerous acute and chronic injuries and illnesses result from the diverse exposures and work practices in agriculture. Hazardous exposures involve physical agents such as noise, vibration, sunlight, heat, and cold; chemical agents such as pesticides, fertilizers, petroleum products, and gases from animal wastes; and biological agents. Many illnesses and chronic diseases are associated with specific agricultural exposures or with the farm environment. Most prominent are respiratory diseases, dermatoses, infectious illnesses, cancer, and mental health problems.

Traumatic injuries. A major cause of both fatal and nonfatal injuries is accidents involving farm machinery. Tractors cause the majority of these injuries with tractor roll-overs accounting for more than half of the total deaths and injuries in this category. Because a farm worker frequently works alone and away from immediate aid, survival after an accident is further impacted by remote location and lengthy transport to treatment sites. As much as 19% of seriously injured farm workers die during transport to local hospitals.

Respiratory disease. A wide variety of potentially toxic respiratory exposures exist in agricultural settings, including chemicals as well as organic and inorganic dusts, endotoxins, fungal proteins, diesel exhaust, and welding fumes. In addition, animal-confinement practices result in a buildup of toxic gases including hydrogen sulfide, carbon dioxide, carbon monoxide, ammonia, and methane. Exposure to nitrogen oxides from silage may result in acute pulmonary edema, bronchiolitis obliterans, and death. Obstructive airway disease and bronchitis may result from exposure to respiratory irritants. Allergic symptoms, including asthma, may result from exposure to many organic antigens, such as pollens, grain particles, bacterial and fungal cell-wall components, and animal danders.

Farmer's lung is a pneumonitis that results from hypersensitivity to fungal spore antigens in moldy hay. Other sources of fungal antigens, such as moldy sugarcane, may also result in pneumonitis. While the acute symptoms of coughing, shortness of breath, and fever in this syndrome usually resolve spontaneously, chronic exposure may result in fibrotic (restrictive) lung disease. Organic dust toxic syndrome is clinically similar to farmer's lung but results from a nonimmunologic inflammatory response to inhaled dusts. Other lung diseases are caused by pesticides. For example, interstitial lung disease is caused by the herbicide paraquat, and a silicosislike disease is caused by the copper sulfate found in pesticides used by vineyard sprayers in Portugal.

Dermatoses. Skin diseases are common disorders, the vast majority of which are due to plant or chemical exposures. Poison oak and poison ivy are common sources of dermatitis in agricultural populations, but other plants and agricultural crops may also result in contact dermatitis. Pesticides and other agrichemicals may cause contact dermatitis by irritation or allergic sensitization. For example, the pyrethrum-containing pesticides are allergic sensitizers, while the synthetic pyrethroids cause irritant dermatitis.

Infectious agents. Respiratory diseases as well as a wide range of other illnesses are caused by infectious agents present in the agricultural environment.

Anthrax is caused by the spore of *Bacillus anthracis*. Humans may contract this disease by skin contact with infected tissue, including hides and wool. Humans can also contract anthrax by ingestion or by inhalation of spore-containing air.

Brucellosis is caused by organisms of the genus *Brucella*. Different species of *Brucella* occur in different animals. The disease is contracted by ingestion or by dermal contact with secretions or excretions of infected animals, and is most common in livestock producers and meat-packers.

The animal reservoir for Western or Eastern equine encephalitis is the horse. The disease is spread by mosquitoes, and agricultural workers are at increased risk of exposure.

Leptospirosis is caused by bacteria of the genus *Leptospira*. Many animals, including dogs, cats, wild rodents, and domestic farm animals, can be infected. Human infection is caused when drinking water is contaminated by the urine of infected animals.

Q fever is caused by the rickettsia *Coxiella burnetii*. Humans can contract the disease by exposure via the respiratory tract, digestive tract, or skin. The most frequent exposure occurs during the birthing of lambs and kids.

Rabies, a virus carried almost exclusively by carnivores, is usually spread by a bite from a wild animal. Infection can also be caused by excretions or secretions from infected animals coming in contact with the mucus membranes, the cornea, broken skin, or the respiratory tract.

Rocky Mountain spotted fever is caused by *Rickettsia rickettsii*, transmitted to humans via tick bites. The animal reservoir for the organism includes rabbits, mice, and dogs.

Staphylococcus aureus is ubiquitous and is found in particularly high concentrations on dairy farms. It can cause food poisoning or skin diseases such as boils or impetigo.

Other potential infectious disorders among workers in agricultural settings include tularemia (caused by *Francisella tularensis*), tetanus (*Clostridium tetani*), and Lyme disease (*Borrelia burgdorferi*).

Cancer. Multiple studies in numerous countries indicate an increased incidence of certain cancers in farmers. Hodgkin's disease, non-Hodgkin's lymphomas, lip cancer, leukemia, stomach cancer, skin cancer, and connective tissue malignancies are consistently linked to agricultural exposures. In most cases, the exact cause of these cancers is unknown. One exception is the consistent association of phenoxyacetic acid herbicides with non-Hodgkin's lymphomas. Sunlight is also a known occupational hazard for agricultural workers, increasing the risk of skin cancer.

Stress-related disorders. Multiple factors in the agricultural environment may cause or exacerbate stress-related disorders. These factors include the dangerous nature of farm work, lack of control over the weather and other environmental factors, long and inflexible work schedules, and lack of help during illness or injury. Surveys show that stress is a major concern of farmers. Manifestations of stress among farmers have not been well documented, but range from anxiety and somatic complaints to suicide.

Other disorders. Cumulative trauma disorders such as carpal tunnel syndrome are caused by repet-itive motions. Many agricultural processes are likely to result in the occurrence of cumulative trauma disorders, but little is known about these disorders in the agricultural setting. Vibrations from machines, particularly from power hand tools such as chain saws, cause a Raynaud's-phenomenon–like vibration white finger, and vibration of the entire body from operating machinery may damage the intervertebral discs. Increased exposure to noise from multiple agricultural sources causes hearing loss. Audiographic studies of farmers consistently demonstrate hearing loss greater than in the general public.

Remedies. There is a great need for active educational programs on agricultural health and safety hazards through county extension offices, universities, and farmer and farm worker organizations. Issues that need to be addressed include closed space entry, prevention and recognition of zoonoses, machinery operation safety, and prevention of illness from chemical hazards.

Another area of primary prevention that can reduce injuries and deaths involves engineering controls on farm equipment. Tractors, the most common cause of traumatic injuries, can be engineered for safer operation. Possible tractor modifications include mandatory roll bars, provision for passenger safety, kill switches at each possible work station, and possibly a deadman switch on the seat. The science of ergonomics should be applied to farm equipment and basic farm tools to reduce the cumulative trauma disorders that are common in farming.

For background information SEE INDUSTRIAL HEALTH AND SAFETY; ZOONOSES in the McGraw-Hill Encyclopedia of Science & Technology.

Marc B. Schenker; Patrick H. Thorpe

Bibliography. D. H. Cordes and D. F. Rea (eds.), *Occupational Medicine: State of the Art Reviews*, vol. 6, no. 3: *Health Hazards of Farming*, 1991; J. A. Dosman and D. W. Cockcroft (eds.), *Principles of Health and Safety in Agriculture*, 1989; I. J. Selikoff (ed.), Special issue on agricultural occupational and environmental health, *Amer. J. Indus. Med.*, 18(2–4):100–526, 1990.

Agricultural soil and crop practices

The high average annual rainfall, high solar radiation, and mild temperatures of the Gulf and southeastern-Atlantic Coastal Plains of the United States are ideal for growing crops. Yet, some other major geographical areas, such as the Midwest, consistently outproduce these areas. For crop management, and related soil management, the fundamental problem in the Coastal Plains is alternating periods of high rainfall and drought during the growing season. The soils of the area further complicate agricultural production with their low water-holding capacities and, in some places, shallow rooting zones. Thus, the Coastal Plains are characterized by variable agricultural production from year to year, and consequent rural economic instability. Actively controlling the depth of the soil water table has been shown to alter this situation so that these areas can produce consistently high yields.

The idea of controlling a water table is not new. Some archeologists believe that the Mayans in the Yucatan peninsula used a primitive form of water-table control combined with intercropping to feed 250–350 people per square mile (96–135 people per square kilometer). More recently, water-table control has been used in the coastal areas of the Netherlands. The water-table management being developed in the Coastal Plains of the United States is on a large scale and is more actively managed than earlier versions. It is being developed at a time of enhanced ecological awareness in terms of protection of groundwater from fertilizers and pesticides and of protection of wetlands.

Climate. Ironically, the droughty conditions and poor plant growth in the Coastal Plains was a result of trying to cope with spring floods. These coastal areas experience occasional heavy rains on the order of 3 in. (7.6 cm) per hour. Because these areas are flat, water does not drain rapidly. The consequent coastal flooding is a nuisance to agriculture and to the general public as well. As a result, Congress enacted Public Law 566 to provide drainage for these areas. Ditches were dug deep to allow for proper slopes to drain the flat land; however, the ditches permitted too much of the water to drain off.

It is paradoxical that water is a limiting factor in an area that needs to be drained, an area that averages 43 in. (109 cm) of rain per year. However, rainfall distribution is not uniform. It comes in intense storms that alternate with long dry periods. This pattern is limiting because, as the growing season becomes hot, plants require a steady source of water.

Soils. In a region of soils with a large water-holding capacity, a few large rainfalls might be sufficient to provide adequate water for plants. However, in the southeastern Coastal Plains, many of the soils are sandy in texture, with a characteristically low water-holding capacity, often only enough for a few days' growth.

Maintaining the soil water table at a desired level is accomplished by controlling the depth of the water in the ditches. Water from the soil water table migrates upward by capillary action. This water can be taken up by the roots, and then more water can move into the root zone. If the water table drops too far below the root zone, water is unable to move upward. Controlling the water table prevents it not only from dropping below the root zone but also from rising to such a level that it drowns roots.

Several researchers have shown that controlling the water table can provide adequate water for crops, and have recommended optimum depths. These depths range from 0.5 to 4 ft (0.15 to 1.2 m), depending on the crop and the texture of the soils. Changes in rainfall from year to year make specific depth recommendations difficult.

Drain tiles. Many farmers drain their fields not with ditches, which would interfere with traffic, but

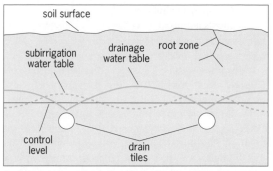

Drainage and subirrigation. The control level is the targeted soil water level. The actual soil water levels are the subirrigation water table when water is flowing into the soil, and the drainage water table when water is flowing out of the soil.

with drain tiles. Ceramic tiles were used formerly, but perforated plastic tube tiles are used in newer installations. These tiles are buried in parallel lines at a depth of about 3 ft (0.9 m), which is far enough below the soil surface to be out of the way of tillage equipment. Drainage within the fields during the wet spring months dries them out to permit tillage and planting. Subsequently, these tiles can be used to carry water back to the fields to supply the plant roots from below; this method is known as subirrigation.

For subirrigation, drain spacings should be one-half to three-fourths of that needed for drainage alone. One reason for the closer spacing is that during subirrigation the water table is lower between the tiles (see **illus.**); the water table assumes a shape like a string hanging between two rods. If the tiles are too far apart, the depth of the water table between them becomes too great, and water moves upward too slowly or not at all. If the drain spacings become too close (30–45 ft or 9–13.5 m), economic concerns outweigh the physical considerations. Installation of many closely spaced drain tiles may become too expensive.

Subirrigation sites. In the southeastern United States, 7×10^6 acres (2.8×10^6 ha) of farmland are suitable for subirrigation. However, subirrigation is being used for only 1.5×10^6 acres (600,000 ha). Suitable sites are flat with normally high water tables or impermeable layers. These conditions minimize water loss due to deep percolation. Although deep percolation may be an important factor, other limitations provide an overriding concern. Leaching of nutrients and pesticides into the groundwater can be more harmful than losing the water.

Another requirement of a controlled-water-table site is a highly permeable surface soil, which is characteristic of the surface sandy soils. Sands permit water to flow rapidly through the soil for subirrigation.

It is recommended that a preliminary investigation of the site and soil characteristics be completed before installing a subirrigation system. Computer models are available to facilitate the choice of appropriate sites and to map out a plan for the installation.

Implementation. Water tables can be controlled on a watershed basis or a field-by-field basis. The problem with the watershed basis is that homeowners may not want to share in the expense of an implementation that mainly benefits agriculture. Also, homeowners may not want the higher water tables, which can lead to decreased rainwater drainage.

Water tables can be controlled on a field-by-field basis by placing flashboard risers (small wooden dams of adjustable height) within the ditches. Water levels in the ditches are then maintained at a specific depth. Water tables in the soils are fed by the water from the ditch by flow through tile lines and through the soil to the plant root zone. If water from the streamflow alone is not enough to maintain a sufficient depth of water in the ditches, water from wells can be added.

More sophisticated systems involve a network of computers, solenoids, valves, pumps, and tile lines to maintain the water levels in individual fields. Higher management pays off in higher yields. Ongoing research by state and federal agencies involves the development of controls and management systems for controlled drainage subsurface irrigation.

Water from the drainage ditches affects water tables within 300–700 ft (90–210 m) of the ditch (or tiles). However, a large pool of surface water (in the deep ditches) and subsurface water (in the soil) exists as a result of the controlled water level. This water can be used by surface irrigation systems such as center pivots and sprinklers. The water is easily accessible, since the irrigation inlet can be placed in the stream or ditch where the water level is controlled.

Water quality. Controlled drainage and subirrigation tend to increase the residence time of water by reducing drainage outflow. The result is that fertilizers and pesticides have more time to become depleted in the root zone, where they were intended to reside.

Several design and management techniques can further reduce the outflow of fertilizer and pesticide. One technique to increase residence time of fertilizers and pesticides is to balance the need for narrow drain spacing with wide spacing, which also has the advantage of being cheaper to install. The drainage outlet depth can be adjusted during or immediately after, but not before, a rainfall event to prevent flushing of the system. A buffer zone, such as a riparian area, can be maintained to filter out chemicals and sediments before they get into streams and lakes. The water table can be maintained at the maximum optimum depth for growth, thus providing more storage for rain and less potential for runoff of water, fertilizers, and pesticides. Finally, management of fertilizer, such as by split applications of nitrogen, can ensure that not all of it is available for leaching.

Some excellent computer models are already developed to help researchers trace fertilizers and pesticides through controlled drainage and subirrigation systems. These models, along with the ongoing research projects, will provide a clearer picture of the fate of the fertilizers and pesticides and a better understanding of how to manage these systems for maximum agricultural production and environmental protection.

Prospects. Controlled drainage and subirrigation have the potential to benefit millions of acres of already cropped land in areas where production is erratic. However, many problems need to be solved. Most problems deal with system management and range over pesticide application techniques, suitable cropping systems, riparian water rights, and the best combination of computers, solenoids, and pumps for water-table control. Agricultural engineers, physicists, chemists, and others are actively addressing these issues.

For background information *SEE AGRICULTURAL SOIL AND CROP PRACTICES; CANAL: DROUGHT; IRRIGATION (AGRICULTURE)* in the McGraw-Hill Encyclopedia of Science & Technology.

Warren J. Busscher

Bibliography. C. W. Doty, Crop water supplied by controlled and reversible drainage, *Trans. Amer. Soc. Agr. Eng.*, 23:1122–1126, 1980; R. W. Skaggs, Water table movement during subirrigation, *Trans. Amer. Soc. Agr. Eng.*, 16:993–998, 1973; series of articles on water table control in *J. Soil Water Conserv.*, 47:47–74, 1992.

Aircraft

Aircraft survivability is the ability of an aircraft to avoid and/or withstand hostile environments. The normal environment for aircraft consists of the forces and events involved in flying, such as the loads associated with takeoff, cruising, and landing. Hostile environments are the result of such phenomena as severe turbulence, lightning strikes, on-board fire or explosion, midair collisions, and crashes. In wartime, a human-made hostile environment is created by the enemy's air defense, whose goal is to destroy aircraft.

Development of survivability. The survivability of military aircraft in combat has been a consideration in both aircraft design and mission planning throughout their history. In World War I, many aircraft had armor plating around the areas housing the crew and other critical components. Tactically, the pilots kept outside the range of antiaircraft guns and gunsights of enemy fighters as they flew their missions.

In World War II, many survivability features were added to aircraft. Armor was common on nearly all aircraft; some aircraft had fire and explosion protection for their fuel tanks and engines, and electronic countermeasures in the form of chaff and noise jammers were introduced. Bombers with .50-caliber machine guns pointing in all directions flew in large formations at high altitude to deter attacking enemy fighters. Fighter escorts further increased the bombers' survivability. In spite of these efforts, approximately 4700 B-17 bombers were destroyed in combat.

Interest in survivability waned shortly after the war. As a result of the limited consideration given to survivability from the late 1940s through the 1950s, most of the United States aircraft that were used in the Vietnam conflict in the 1960s were not designed to survive the guns and missiles used by the North Vietnamese. As a consequence of the loss of more than 5000 United States aircraft in southeast Asia between 1963 and

1973, combat survivability has emerged as a formal design discipline for military aircraft.

Definition of combat survivability. Aircraft combat survivability has been defined as the capability of an aircraft to avoid and/or withstand a human-made hostile environment. Survivability depends upon the design of the aircraft, the weapons it carries, its speed and flight path, and its environment, that is, the combination of friendly supporting forces, threats to the aircraft, and other factors such as aircraft speed and flight path, time of day or night, and weather. Survivability is relative; some aircraft are more survivable than others in a given scenario. The survivability of an aircraft in a given situation is not deterministic; that is, no prediction can be made at the beginning of a mission as to whether an aircraft will survive. Rather, survivability is a stochastic attribute that can be measured by the probability that the aircraft will survive the mission, P_S.

Survivability is dependent upon two negative attributes of aircraft, susceptibility and vulnerability. The inability of the aircraft to avoid the damage-causing mechanisms that make up the human-made hostile environment, such as bullets, high-explosive blasts, and warhead fragments, is referred to as the aircraft's susceptibility. Susceptibility can be measured by the probability that an aircraft will be hit during its mission by one or more damage-causing mechanisms, P_H. Aircraft with large signatures or observables that fly slowly at medium altitude through the lethal envelopes of enemy air defense weapons without electronic countermeasures or support from friendly forces are very likely to be hit, and hence are very susceptible. The inability of an aircraft to withstand hits is referred to as vulnerability. Vulnerability can be measured by the probability that the aircraft will be killed if it is hit, $P_{K/H}$. Aircraft with no fire or explosion protection for the fuel system, only one engine, no ballistic protection for the pilot, and no separation of redundant hydraulic power components for flight control surfaces are easily killed when hit, and hence are very vulnerable. Aircraft susceptibility and vulnerability are related to aircraft survivability by the equation below. Thus, survivability is enhanced when

$$P_S = 1 - P_H P_{K/H}$$

susceptibility and vulnerability are reduced.

Enhancing survivability. The susceptibility and vulnerability of an aircraft to the threats of combat, such as guns, guided missiles, directed-energy weapons, chemical-biological weapons, and nuclear weapons, are reduced by using one or more of 12 survivability enhancement concepts. Six are susceptibility reduction concepts, namely, threat warning, noise jamming and deceiving, signature reduction, expendables, threat suppression, and tactics, and six are vulnerability reduction concepts, namely, component redundancy (with separation), component location, passive damage suppression, active damage suppression, component shielding, and component elimination.

Susceptibility reduction. Threat warning consists of the use of electronic or other sensors to warn the crew of potential danger, such as an approaching missile. Noise jamming and deceiving uses onboard or off-board electronic equipment to prevent the enemy's detection and tracking sensors from seeing the aircraft or to deceive them as to the aircraft's position. Signature reduction is the intentional design (and possibly painting) of the aircraft to reduce the signatures or observables, such as the radar signature and the infrared signature, used by the enemy's detection and tracking sensors. Expendables are relatively inexpensive devices that are ejected from an aircraft to screen it from detection and tracking sensors or to act as decoys. Threat suppression consists of the physical degradation or destruction of the enemy's air defense equipment. The tactics employed, such as ingress and egress flight paths, aircraft speeds and altitudes, and the supporting forces, have a major impact on aircraft survivability.

Vulnerability reduction. Vulnerability reduction through component redundancy (with separation) consists of the availability of more than one component to perform an essential function, such as lift, thrust, and control for flight. An example of this concept is the placement of two widely spaced engines and redundant sources of hydraulic power for the flight control surface actuators such that a single hit anywhere on the aircraft cannot destroy the aircraft hydraulic power. Proper location of critical components within the aircraft can reduce vulnerability. For example, placing possible ignition sources outside the spaces where flammable vapors can accumulate after damage due to a hit reduces the possibility of a fire or explosion in those spaces. Passive and active damage suppression reduces vulnerability by constraining or minimizing the damage caused by a hit. An example of passive damage suppression is a self-sealing fuel tank. The self-sealing material prevents fuel from leaking through a hole caused by a bullet or fragment into adjacent dry bays and reduces the risk of fire. An example of active damage suppression is a self-repairing flight control system that has the ability to reconfigure the flight control laws to maintain stable flight after damage due to a hit. Component shielding is the use of materials to prevent damage to critical components, such as armored crew seats. Component elimination is the removal of components that contribute vulnerability, such as elimination of the pilot on a crewless air vehicle, or the replacement of a particular component with one that is less vulnerable, for example, the substitution of nonflammable for flammable hydraulic fluid.

Requirements and organizations. The United States armed forces have established a formal program for ensuring that survivability is considered in aircraft design. The survivability program for each major system acquisition is reviewed. Survivability has been defined as a critical system characteristic, and requirements have been established for survivability throughout the life cycle of the aircraft. Each service has survivability organizations to provide guidance to program managers on improving the survivability of aircraft.

In 1987, Congress passed the Live Fire Test Law, which mandates realistic survivability testing of a covered system before it can proceed beyond low-rate initial production. Realistic survivability testing is defined as testing for vulnerability of the system in combat by firing munitions likely to be encountered in combat at the system configured for combat. A waiver from the full-scale full-up tests can be obtained if such tests are unreasonably expensive and impractical.

The Joint Technical Coordinating Group on Aircraft Survivability (JTCG/AS) was established at the end of the Vietnam conflict to promote the survivability as a design discipline. It co-sponsors the Survivability and Vulnerability Information and Analysis Center (SURVIAC), a centralized information and analysis center for all aspects of nonnuclear survivability and weapon lethality of United States systems.

Survivability and Desert Storm. In contrast to the situation in Vietnam, many of the aircraft used in the 1991 Desert Storm operation were specifically designed to survive in combat. In particular, the Air Force's F-117 stealth fighter was a major contributor to the success of the operation because of its survivability. As a result of the improved design of the United States military aircraft and the tactics used in mission planning, the loss rates of United States aircraft in Desert Storm were significantly lower than those suffered by the United States in the Vietnam conflict.

Commercial aircraft survivability. Commercial aircraft are designed to survive the hostile environments within which they operate by incorporating adequate strength, electromagnetic interference shielding, collision avoidance systems, and fire-fighting equipment. However, the worldwide availability of guns and human-portable antiaircraft missiles poses a new hostile threat. In addition, commercial aircraft can be threatened by explosives smuggled on board by terrorists. The combat survivability design discipline that has evolved for military aircraft is applicable to commercial aircraft. Efforts are under way to adapt this technology to the commercial industry. After the terrorist bombing of PanAm flight 103 in 1988, the President created the Commission on Aviation Security and Terrorism to examine the effectiveness of current airport security and aircraft vulnerability to terrorist bombs. One of the Commission's recommendations was that the Federal Aviation Administration (FAA) initiate a program to determine the ability of current commercial airliners to survive internal explosions and to develop hardening techniques to minimize the damage from these explosions. The FAA asked the Air Force to establish the Commercial Aircraft Hardening Program. Many of the combat survivability organizations from the other two services are contributing to this program.

For background information SEE ELECTRONIC WARFARE; MILITARY AIRCRAFT in the McGraw-Hill Encyclopedia of Science & Technology.

Robert E. Ball

Bibliography. D. B. Atkinson, Survivability requirements, *Aerosp. Amer.*, p. 37, August 1992; R. E. Ball, *The Fundamentals of Aircraft Combat Survivability Analysis and Design*, 1985; J. Barnes and R. L. Peters, The challenge of commercial aircraft survivability, *Aerosp. Amer.*, pp. 55–56, August 1992; J. F. O'Bryon, Live fire testing requirements: Assessing the impact, *Aerosp. Amer.*, pp. 35–36, August 1992.

Antenna (electromagnetism)

The corrugated horn is one of the most important of a class of microwave or millimeter-wave devices used in the transmission or reception of signals by a reflector antenna. It is used in satellite ground terminals; in spacecraft for communications, broadcast, and remote sensing; in some terrestrial microwave communications systems; and very widely in radio telescopes. Its use leads to improved antenna efficiency, low cross polarization (which allows for a doubling in communication capacity), and low levels of radiation in directions away from that of maximum antenna gain, the so-called side-lobe region of the antenna pattern.

The corrugated horn belongs to a class of passive microwave devices that support hybrid modes. Modes of this kind, and in particular the lowest-order or dominant mode, exhibit the ideal transverse form to ensure the radiation characteristics described above. They also propagate within the structure with very low attenuation. It is possible to synthesize a hybrid field from more than one pure mode in a conventional smooth-wall waveguide, but the operating bandwidth is then restricted. If bandwidth is not an important consideration, such as in a domestic-satellite receiving antenna, these alternative feeds are commonly used because they are compact and relatively inexpensive.

Development of corrugated feeds. The corrugated horn was devised independently and almost

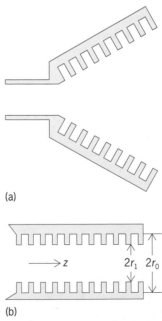

Fig. 1. Corrugated feeds. (*a*) Horn. (*b*) Waveguide. r_0 = outer radius, r_1 = inner radius, z = direction of propagation.

simultaneously in the 1960s by A. F. Kay in the United States and by H. C. Minnett and B. Thomas in Australia. Both investigations were concerned with feeds for large reflector antennas. **Figure 1a** shows the type of feed investigated by Kay, while Fig. 1b shows the type developed by Minnet and Thomas for use with the Parkes radio telescope. For the latter, both high efficiency and low cross-polarization properties were important. Following these discoveries, a number of antenna researchers became involved in the analysis and design of corrugated horns. These devices can now be reliably designed and manufactured to give highly predictable performance at wavelengths from almost the longest used in microwave systems through to the shortest of millimeter wavelengths. The corresponding frequencies lie between approximately 1 and 300 GHz.

Operation of corrugated feeds. A simple way to understand the advantages of a hybrid feed for a reflector antenna is to consider the reflector as part of a receiving system.

Focal field of a paraboloidal reflector. Figure 2a shows a plane electromagnetic wave incident upon a paraboloidal reflector. The x and y axes are chosen parallel to the directions of the electric-field and magnetic-field components, E_i and H_i, of the incident plane wave, and the z axis is the axis of the reflector. From the laws of geometric optics, reflection of a ray such as QR, which lies parallel to the direction of propagation of the incident plane wave, reaches the focal point of the paraboloidal reflector at P. The focal length of the reflector is f and its diameter D. In practice, the diameters of antennas used in the microwave and millimeter-wave spectrum are between about 30 and 300 wavelengths, although larger and smaller antennas are used in some specialized applications. However, because the reflector size is not very large in wavelength terms the reflected fields and associated energy are not concentrated entirely at P but are distributed across the focal plane, which contains P and lies normal to the reflector axis.

The main contribution to the focal field behaves as a Bessel function $J_0(x)$, which, for small values of x, is similar to the gaussian function. The sum of all the contributions to E_x (the x component of the electric field) coming from the reflector is given by Eq. (1).

$$E_x = A \left(\frac{\pi D}{4 f \lambda} \right)^2 \frac{J_1(U)}{U} \qquad (1)$$

Here, A is a constant, λ is the wavelength, $J_1(U)$ is a Bessel function of the first kind and order 1, and U is given by Eq. (2). The quantity r_f is shown in Fig. 2a to

$$U = \frac{2 \pi r_f}{\lambda} \sin \theta_0 \qquad (2)$$

be the radial distance of the point of observation from P, the geometric optics focal point; and θ_0 is the angle subtended at P by $D/2$ (half the diameter of the reflector). The appearance of the electric field in a quadrant of the focal plane is shown in Fig. 2b. Approximately 84% of the total power of the incident wave reflected by the paraboloid passes through the first circular region and is essentially linearly polarized.

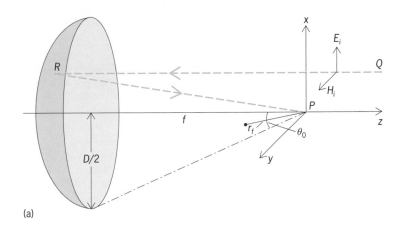

(a)

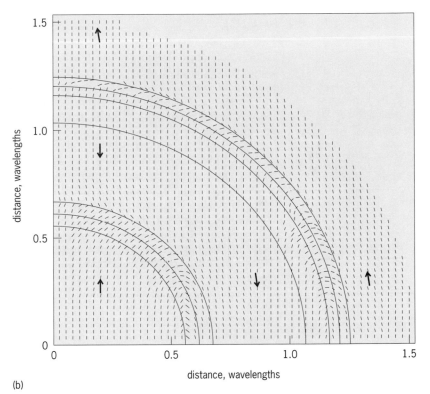

(b)

Fig. 2. Paraboloidal reflector. (*a*) Plane electromagnetic wave incident on the reflector. The geometric optics ray *QR* is reflected to the focal point *P*. Other symbols are explained in the text. (*b*) Electric field distribution in the focal plane of the reflector. The arrows indicate the direction of the electric field. The axes represent the distance from the focal point in wavelengths.

Feeds used with reflectors. G. B. Airy identified the properties of the focal field, and the first circular region is often called the first Airy disk. This disk corresponds to the region where the aperture of a feed should be placed in order to capture most of the available energy incident on the reflector. To do so requires a waveguide or horn in which the electric field is nearly linearly polarized across its entire transverse plane and whose orthogonal magnetic field is related to the electric field through Eq. (3), where H_y

$$\frac{E_x}{H_y} = Z_0 \qquad (3)$$

is the y component of the magnetic field and Z_0 is the free-space wave impedance.

Conventional smooth-wall metal waveguides or

horns do not support modes with these properties. For example, the transverse electric field of the dominant mode of a circular waveguide or conical horn not only is curved around part of the periphery but also does not satisfy the impedance relationship referred to above. By contrast, the transverse electric and magnetic fields in a corrugated waveguide, such as the waveguide of Fig. 1b, have a $J_0(x)$ dependence and satisfy the above conditions precisely, making them desirable as feeds. The similarity to the field of the first Airy disk is evident.

If the pitch of the corrugations is at most one-third of a wavelength and the depth about one-quarter of a wavelength, then both the electric and magnetic fields in the waveguide form loops, and their transverse fields are nearly linear and are at right angles to one another. At the inner boundary of the waveguide ($r = r_1$ in Fig. 1b), the longitudinal magnetic field is finite, as is the longitudinal electric field, at least across the gaps between the corrugations. This form of so-called hybrid field couples perfectly to the focal-region field of the reflector, ensuring that a hybrid-mode feed is the best for receiving or transmitting signals from a reflector.

Feeds used as radiators. When, alternatively, the corrugated waveguide or horn is considered as a radiator, the radiation pattern is found to have very weak side lobes and to be very nearly free from cross polarization. This latter property is especially important in modern communication and broadcast systems. It permits two different signals to be transmitted or received at the same carrier frequency but in opposite senses of linear or circular polarization and without appreciable cross coupling. A typical radiation pattern for a corrugated feed has a gaussianlike shape, and the peak level of cross polarization (which occurs in planes at an angle of $\pm 45°$ to the principal planes) is usually only about −45 dB relative to the peak level of the copolar radiation on boresight. This very low level is sufficient for all communication and broadcast applications.

Another favorable property of the corrugated feed is its weak side lobes. The gaussianlike transverse electric field in the aperture of the feed falls to a very low value at the inner boundary of the horn. This drop guarantees a very low level of the peak of radiation away from the horn main lobe, and a level of less than −30 dB relative to the main lobe is typical for the first side lobe of a horn. Thus, a very low level of energy spillover beyond the reflector is assured when the horn is used as a feed at the secondary focus of a dual reflector. This property is desirable when corrugated-horn antennas are used in high-performance Earth stations. There, the energy that enters the feed side lobes contributes to the thermal noise of the antenna. This noise, together with the antenna gain, forms the critical parameter in assessing Earth-station antennas and radio-telescope performance.

Construction of corrugated feeds. As mentioned, corrugated feeds are used over a wide range of microwave and millimeter-wave frequencies. In feeds located at the prime focus of a reflector, a waveguide form is normally used with a flared corrugated flange at the aperture. In dual-reflector antennas, corrugated horns are normally used, and the semiflare angle may vary from about 10 to 45°. In these applications the horn is usually machined under computer control, although a preformed casting may be used with wider-angle horns to speed the manufacturing process and reduce wastage. To reduce the reflection at the horn entrance or throat, it is usual to vary the depth of the corrugations between throat and aperture. Computer programs can predict the corrugation dimensions precisely; these programs can be used to prepare milling instructions. Corrugated waveguides are also made by using electroforming, and in the case of very high frequency horns just a few millimeters in length, spark-erosion techniques have been used to form the tiny corrugations. At 180 GHz, for example, the corrugation depth and pitch are of the order of only 0.4 mm. Nonetheless, horns and waveguides have been constructed successfully for operation at and above these frequencies.

For background information SEE ANTENNA (ELECTRO-MAGNETISM); BESSEL FUNCTIONS; DIFFRACTION; ELECTROMAGNETIC WAVE TRANSMISSION; ELECTROPLATING OF METALS; MICROWAVE; WAVEGUIDE in the McGraw-Hill Encyclopedia of Science & Technology.

Peter Clarricoats

Bibliography. P. J. B. Clarricoats and A. D. Olver, *Corrugated Horns for Microwave Antennas*, 1984; A. F. Kay, *A Wide Flare Angle Horn: A Novel Feed for Low Noise Broadband and High Aperture Efficiency Antennas*, U.S. Air Force Cambridge Res. Lab. Rep. 62-757, 1962; H. C. Minnett and B. Thomas, A method of synthesising radiation patterns with axial symmetry, *Trans. IEEE*, AP-14; 654–656, 1966.

Anthracnose disease

Dogwood anthracnose, a disease caused by the fungus *Discula destructiva* Redlin, affects dogwood trees (*Cornus* spp.) throughout North America. In the Pacific Northwest, dogwood anthracnose was first observed in 1976 in western Washington. In the eastern United States, it was first observed in 1977 in southeastern New York. Since then, dogwood anthracnose has been identified in Oregon, Idaho, and British Columbia in the Pacific Northwest, throughout the northeastern United States, and down the Appalachian mountain range into northern Georgia and Alabama (see **illus.**). In the southeastern United States alone, an estimated 5.7×10^6 acres (2.4×10^6 hectares) were affected by December 1990. In areas where the disease has become established, it has demonstrated the capability to destroy many trees over a period of several years. Because of the destructive nature of dogwood anthracnose and its wide distribution, the disease presents a serious threat to native flowering dogwood trees in North America.

Symptoms. Symptoms first appear as leaf spots consisting of necrotic brown tissue, often surrounded by a reddish brown–purple zone. Symptoms usually appear following prolonged periods of cool, wet weather. A shot-hole appearance may develop if the initial infection is followed by hot, dry weather. When environmental conditions remain conducive for disease development, leaf spots coalesce to form olive-brown necrotic blotches, irregular in outline and often bounded by a reddish brown–purple zone. As the infection progresses, the entire leaf becomes necrotic (blighted). Some blighted leaves remain attached to branches through the winter and into the following spring.

The infection can proceed from the leaves into shoots through the petioles, causing a shoot dieback. Discoloration and shriveling of bark tissue and darkening of terminal buds are characteristic of shoot dieback. Dark cankers are often formed where infected branches join the main trunk. A proliferation of new shoots usually occurs along the scaffold branches and trunk of infected dogwoods. This combination of blighted leaves, proliferation of new shoots, and trunk cankers is diagnostic of dogwood anthracnose in the field.

Early attempts to implicate a causal organism or link the symptoms observed on both coasts of the United States were inconclusive. In 1988 it was clearly demonstrated that dogwood anthracnose was caused by a fungus belonging to the imperfect genus *Discula* that was responsible for causing the disease on both coasts. Further attempts to identify the species were unsuccessful, and it was not until 1991 that the fungus was officially described as a new and unique species of *Discula* and given the name *D. destructiva* Redlin.

Details concerning the pathogenic life cycle of the fungus are still unclear. Initial inoculum in the spring originates from blighted leaves or overwintering shoot infections. Seed can become infected and serve as an additional dispersal mechanism. All reproduction and dissemination of fungus is presumed to take place via the asexual stage (anamorph). Optimum germination of the asexual spores (conidia) occurs at 68–75°F (20–24°C) in vitro. To date, the sexual stage (teleomorph) has not been identified.

It is often very difficult to reproduce disease symptoms under experimental conditions. Mechanical wounding and simulated acid rain increased the susceptibility of inoculated leaves in controlled studies. Once penetration of host tissue has taken place, extensive necrosis of palisade and spongy parenchyma cells is followed by a further proliferation of fungal tissue into the epidermis, mesophyll, and leaf vascular tissue. The time lag between infection and appearance of symptoms (latent period) may be as short as 6 days. Entire trees may be killed in as few as 3 years after the first leaf symptoms appear.

Etiology and epidemiology. Environmental conditions at the microsite and macrosite levels strongly influence disease development. For example, at the microsite level canopy microclimate within individual trees has a significant effect on disease. The incidence and severity of disease are lower in canopies with an evaporative potential greater than 0.25 g H$_2$O/h. At the macrosite level, disease in naturally forested areas of the southern Appalachians is influenced by the surface orientation (azimuth), elevation, and density of the native dogwood stands. Multiple infection periods that may occur during the growing season are correlated with the number of precipitation events.

Control. The eastern flowering dogwood (*C. florida*), western dogwood (*C. nuttallii*), and pagoda dogwood (*C. alternifolia*) are all very susceptible to dogwood anthracnose. No resistance in germ plasm native to North America has yet been identified. Imported Asian dogwood (*C. kousa*) can become infected, but the disease does not progress beyond the initial leaf spots. The potential exists for a genetic cross between native and imported Asian *Cornus* species that will have the flowering and growth characteristics of native species and the disease resistance of Asian species.

Preventive application of fungicides can significantly reduce the incidence of disease. Propiconazole, tebuconazole, benomyl, and chlorothalonil have demonstrated significant levels of control when applied on a regular basis. However, the logistics and

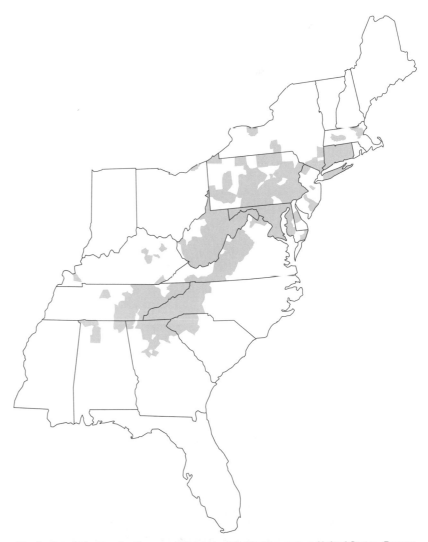

Distribution of dogwood anthracnose (areas in color) in the eastern United States, December 1990. (*USDA, Forest Service, Forest Pest Management*)

labor involved in repeated applications of fungicides to mature trees render this approach impractical for naturally forested areas and very difficult to carry out in urban landscape settings. In addition, environmental concerns preclude the widespread use of fungicides in urban areas.

Within the urban landscape setting, trees that are well maintained will have lower incidence of disease than trees that are stressed. Cultural practices recommended to minimize the impact of disease include mulching, pruning, irrigation during dry periods, and a timely application of insecticide for control of the dogwood borer.

Prospects. The initial appearance of dogwood anthracnose near two major ports of entry, its rapid and massive spread in native dogwood species, and the apparent resistance of imported Asian dogwoods suggest that *D. destructiva* is an introduced pathogen. There is mounting evidence that the ecological range of dogwood anthracnose may preclude establishment of the disease in areas where it is not already present. In areas where the disease has become established, the potential exists for managing the disease in urban or residential landscape settings through an integrated approach incorporating the development of interspecific crosses with enhanced disease resistance, manipulation of canopies through pruning to increase the evaporative potential, cultural practices to minimize plant stress, and one or two timely applications of a fungicide. In naturally forested areas where the disease has become established, the prognosis is not as good. Perhaps over time an ecological balance may be established between the pathogen and the host. This balance can occur if organisms with biological activity toward *D. destructiva*, through microsite competition, mycoparasitism, or antibiosis, become established. Possibly, resistant germ plasm may yet exist in the wild and, through natural selection, become more prevalent in naturally forested areas.

For background information SEE FUNGI; PLANT PATHOLOGY in the McGraw-Hill Encyclopedia of Science & Technology.

Dan O. Chellemi

Bibliography. D. O. Chellemi and K. O. Britton, Influence of canopy microclimate on incidence and severity of dogwood anthracnose, *Can. J. Bot.*, 70: 1093–1096, 1992; D. O. Chellemi, K. O. Britton, and W. T. Swank, Influence of site factors on dogwood anthracnose in the Nantahala mountain range of western North Carolina, *Plant Dis.*, 76:915–918, 1992; C. R. Hibben and M. L. Daughtrey, Dogwood anthracnose in the northeastern United States, *Plant Dis.*, 72:199–203, 1988; S. C. Redlin, *Discula destructiva* sp. nov., cause of dogwood anthracnose, *Mycologia*, 83(5):633–642, 1991.

Aral Sea

The Aral Sea, a huge lake in the Commonwealth of Independent States (C.I.S., former Soviet Union) is rapidly shrinking. Severe environmental and human problems have accompanied the sea's desiccation. Efforts are under way to mitigate the most serious negative consequences and to partially restore the sea and the surrounding region to their previous condition.

The Aral Sea is located among the deserts of Central Asia (**Fig. 1**). Its level is fundamentally determined by the balance between surface inflow from two large rivers, the Amu Dar'ya and Syr Dar'ya, and evaporation from its surface. Over 80% of the sea's drainage basin lies in the C.I.S., with the remainder in Afghanistan and Iran. Five independent states of the C.I.S. lie in the basin of the Aral: Uzbekistan, Kazakhstan, Kyrgyzstan, Turkmenistan, and Tadzhikistan.

Changes since 1960. The Aral Sea was the world's fourth largest lake in the area in 1960 (**Fig. 2**). It had a productive fishery and served as a major regional transportation route. The deltas of the Syr Dar'ya and Amu Dar'ya had major ecologic and economic significance. The sea moderated the climate of an adjacent strip of land 50–300 km (30–180 mi) wide. Since 1960, the sea has steadily shrunk and salinized (Fig. 2) because of dwindling discharge from the Amu Dar'ya and Syr Dar'ya, caused primarily by the expansion of irrigation in their basins. A secondary cause has been the drier climatic condition in the 1970s and 1980s compared to preceding decades. Average annual river inflow to the sea for the period 1981–1990 was around 4.5 km^3 (1.08 mi^3), only 9% of the average 55 km^3 (13.2 mi^3) estimated for 1911–1960.

In 1987, the Aral divided into two water bodies, which are developing separately with their own water balances. The Syr Dar'ya flows into the small northern sea, whereas the Amu Dar'ya enters the large southern sea. A connecting channel between the two was blocked by a dike in 1992 to prevent further water loss from the small to the large sea. The **table** shows the estimated parameters for the two seas in the year 2000, assuming a continuation of recent average annual inflows.

The environmental, economic, and human degradation from the Aral's desiccation has been severe. The fishery was ruined by the early 1980s as native species unable to adapt to rapidly changing habitat conditions disappeared and the shoreline receded tens of kilometers from fishing towns and villages. The Aral also lost all its transportation significance as efforts to keep the navigation channels open to the major ports of Aral'sk in the north and Muynak in the south were abandoned. Floral and faunal communities in the Syr Dar'ya and Amu Dar'ya deltas are undergoing species extinction and ecological simplification. Irrigated agriculture has been damaged by constrained water supplies and the rising salinity of river flow. Animal husbandry has suffered from the diminishing productivity of pastures affected by spreading desertification.

People living in the near-Aral region suffer from a variety of health problems. Some of these problems are directly linked to the sea's recession, such as respiratory afflictions and possibly cancer from inhalation of blowing salt and dust, whereas others are results of environmental pollution associated with irrigation and third-world medical, health, and hygienic condi-

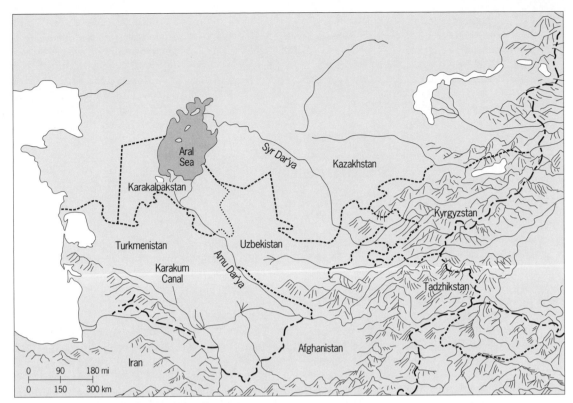

Fig. 1. Aral Sea and its basin. Karakalpakstan, a republic of the former Soviet Union, is now part of Uzbekistan.

tions. Poor-quality, frequently polluted drinking water is implicated in high rates of typhoid, paratyphoid, viral hepatitis, and dysentery. This water problem and other factors such as high fertility, poor medical care, poor diet, and lack of sewage systems contribute to the highest rates of general mortality and morbidity and of infant mortality and morbidity in the C.I.S.

Improvement attempts. The Aral problem became a major national issue in the Soviet Union by the late 1980s. Reacting to public pressure, the Central Government promulgated several decrees on the Aral between 1988 and the breakup of the Soviet Union in 1991. The decrees proposed wide-ranging programs of environmental, economic, and health and medical improvement. Total cost of the most comprehensive of these programs was estimated at nearly 60 billion (1990) rubles. Various water-saving measures were also to be pursued, primarily in irrigation, that were to free nearly 32 km^3 (7.7 mi^3) of additional water

per year for partially refilling the Aral and restoring the deltas of the Syr Dar'ya and Amu Dar'ya. Implementation of some parts of these programs began, but even if the Soviet Union had remained intact, full realization would have been daunting, considering the cost, the amount of water to be freed for the Aral, and the planned short period of realization.

Salt and dust blowing from the former sea bottom is carried as far as 500 km (300 mi) and settles over a considerable area adjacent to the Aral Sea, damaging both native vegetation and crops. Summers in the zone adjacent to the sea have become warmer and drier, winters cooler, spring frosts later, and fall frosts earlier.

The breakup of the Soviet Union has affected the handling of the Aral Sea problem. The states are adamant that they control their own affairs independently of Moscow. They are still interested in technical and other financial aid from Russia to cope with

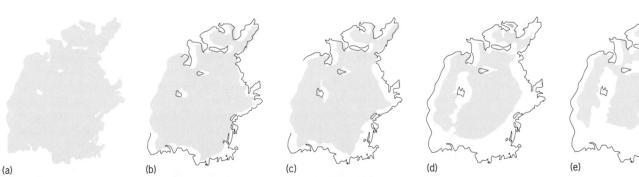

Fig. 2. Changing profile of the Aral Sea. (*a*) 1960. (*b*) 1971. (*c*) 1976. (*d*) 1993. (*e*) 2000 (projected).

Changing characteristics of the Aral Sea*

Year	Average level, m	Average area, km^2	Average volume, km^3	Average salinity, g/liter
1960	53.41	66,900	1090	10
1971	51.05	60,200	925	11
1976	48.28	55,700	763	14
1993 (Jan. 1)		33,642	300	
Large sea	36.89	30,953	279	~37
Small sea	39.91	2,689	21	~30
2000 (Jan. 1)		24,154	175	
Large sea	32.38	21,003	159	65–70
Small sea	40.97	3,152	24	~25

* 1 m = 3.28 ft. 1 km^2 = 0.386 mi^2. 1 km^3 = 0.240 mi^3. 1 g/liter = 0.0334 oz/qt.

the Aral Sea situation, but they reject any attempt to continue centralized management. The C.I.S. has promised help, and an agreement on the Aral was signed that includes provisions for financing research and improvement efforts. However, the C.I.S. is fragile, powerless, and financially impoverished.

A high degree of cooperation among the new states of the Aral Sea basin is necessary to cope with the situation. An important first step was taken in February 1992 with the signing by all five states of an agreement on the joint management and protection of interstate water resources. The agreement obligates them to work jointly to resolve ecological problems connected with the desiccation of the Aral Sea. It also creates an intergovernmental water management coordinating commission for overseeing regulation, efficient use, and protection of interstate water courses and bodies.

The Aral basin states are seeking international help. Projects are under way or under discussion with private and governmental organizations in the United States, Japan, Germany, Turkey, and other countries. In November 1991, Uzbekistan's Foreign Minister sent a letter to the United Nations General Secretary formally requesting assistance. The World Bank sent an evaluation team to the Aral Sea region in September 1992 as a first step in devising an aid strategy.

Restoration programs. With a concerted and cooperative effort among the Aral basin states and real help from Russia and the international community, there is hope that the problems in the Aral Sea region can be alleviated. Immediate efforts should focus on improving health and medical conditions, providing clean drinking water, and slowing rapid population growth. Reconstruction of irrigation systems is also a priority but will take decades to fully implement. This type of project could easily cost several tens of billions of rubles (1990), and it would probably save only 10 km^3 (2.4 mi^3) per year, not the much higher figures that some have suggested.

Preservation of the deltas of the Amu Dar'ya and Syr Dar'ya is also important. An intensive effort to stabilize the dried bottom of the Aral is critical to reduce the deflation of salt, dust, and sand. Efforts to partially restore the small Aral Sea on the north seem feasible. Its level could be raised with a relatively small increase of inflow, compared to the situation for the large Aral. With proper management, the transportation and fishery significance of the small Aral Sea could be revitalized.

Restoration of the Aral Sea to pre-1960s conditions would be very difficult. Average annual discharge to the sea would need to be brought back to 55 km^3 (13.2 mi^3). This level would require a 12-fold increase in inflow compared to the amount for 1981–1990, requiring not only sizable improvements in irrigation efficiency but also around a 40% reduction in the irrigated area. Such a strategy not only would take decades but would also require huge expenditures and a fundamental reorientation of the economy, social structure, and culture of the region.

For background information SEE BLACK SEA; LAKE in the McGraw-Hill Encyclopedia of Science & Technology.

Philip P. Micklin

Bibliography. W. S. Ellis, A Soviet sea lies dying, *Nat. Geog.*, 177(2):73–93, February 1990; P. P. Micklin (guest ed.), Special issue on the Aral Sea crisis, *Post-Soviet Geog.*, 33(5):269–332, 1992; P. P. Micklin, *The Water Management Crisis in Soviet Central Asia*, Carl Beck Papers in Russian and East European Studies, no. 905, Center for Russian and East European Studies, Pittsburgh, 1991.

Archeology

Recent research has utilized geophysical techniques in prospecting archeological sites, and gas chromatography and mass spectrometry techniques in analyzing trace organic residues of archeological materials.

Geophysical Techniques

The archeological heritages of countries throughout the world are under increasing threats because of the expansion of urban and agricultural areas. In addition, excavation of an archeological site destroys it for future study. Hence, there is a demand for noninvasive methods of obtaining information on subsurface content. The archeological community has begun to adapt noninvasive tools and methods from exploration geophysics to obtain subsurface information. Of many such methods, those most suitable to archeology are resistivity prospection, magnetic prospection, and the use of ground-penetrating radar.

Resistivity prospection. One of the earliest geophysical methods used on archeological sites, resistivity prospection, depends on the differences in the strength of an electric current passing through a feature of interest and through the surrounding soils. Resistance is a measure of the inability of a conductor to pass current. It depends on the geometry of the conductor as well as the material of the conductor. For a given applied voltage, high resistance results in low current, and low resistance results in high current. Resistivity is a related property; it is independent of the geometry and depends only on the material of the conductor.

Resistivity of soils depends primarily on the water content. Other factors influencing resistivity are the ionic content of the water and the structure and porosity of the soils. Anthropogenic changes in soil resistivity involve intrusive features such as foundations and wells, and altered soil structures such as middens, storage pits, and hearths.

To find the resistance of a volume of soil, the usual way is to insert two probes into the ground, pass a current between them, and measure it. Then two other probes are inserted, and the voltage difference between them is measured. The effective or average resistance is the ratio of voltage to current for a volume of soil determined by the geometry of the probe positions. This ratio can be converted to an effective resistivity by using an appropriate factor determined by the geometry.

In the most commonly used configuration of probes, the Wenner array, four probes are equally spaced along a straight line. The soil volume over which the resistance is averaged is to a depth somewhat greater than the inter-probe interval. In operating this array on an archeological site, the four probes are inserted in a line centered about each grid point, and a measurement is made. This procedure is very time-consuming for a large site. In another configuration, the twin-probe array, one pair of probes is fixed in the ground at a considerable distance from another pair. The second pair is moved from grid point to grid point and inserted for a measurement, resulting in a more rapid operation. Since the effective depth of the volume of measurement is determined by the probe separation, increasing the separation results in inclusion of more of the vertical volume in the averaging of the resistance. Thus, information can be obtained on the depth of a feature.

Figure 1 is an isoresistance contour map obtained by using a twin-probe array resistance meter with a probe separation of 2.5 ft (75 cm). The site is the backyard of a historic house built in 1869; the yard is currently empty but until 1923 was occupied by the rear wing of the house. **Figure 2** shows the position of the building foundation and other features as revealed by excavation after the resistance survey.

Another way to measure soil resistance is with an electromagnetic conductivity meter. (Conductivity is the inverse of resistivity.) This device has two coils: one emits a radio signal, the other receives it. The phase and amplitude of the received signal are altered

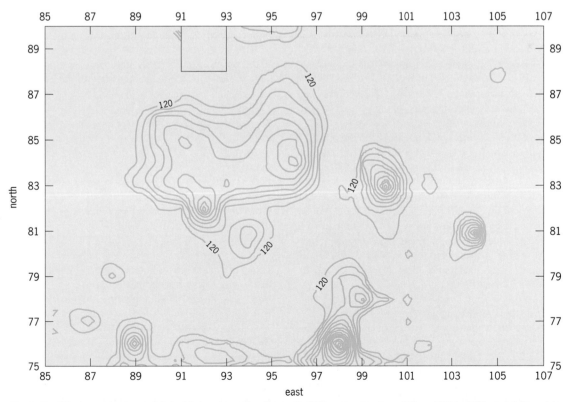

Fig. 1. Resistance contour map (plot of lines of equal resistance) of the rear-wing foundation of historic Kennard House in Lincoln, Nebraska. Contour interval is 8 ohms; contours below 120 ohms are eliminated. Coordinates are given in meters (1 m = 3 ft).

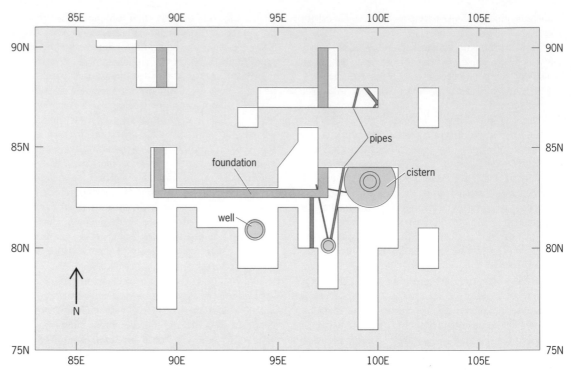

Fig. 2. Foundation and other features of the rear wing of Kennard House revealed after excavation. Coordinates are given in meters (1 m = 3 ft).

by the intervening soils. From the nature of the received signal, the conductivity (and therefore the resistivity) of the soil is determined. The effective volume is determined by the separation of the coils.

Magnetic prospection. For many years, exploration geophysicists have used sensitive magnetometers to measure the strength of the Earth's magnetic field in order to locate magnetic minerals. With the development of portable proton magnetometers in the mid-1950s, magnetic methods were successfully applied to locate buried kilns in England. Since then, the instruments and methods applied to archeology have been continually improved.

Measurements of the total magnetic field of the Earth can give information of subsurface soils because soils contain various iron minerals, some of which are slightly magnetic. Human activities as well as natural processes can alter these minerals and therefore their magnetic state. The magnetic field of the Earth induces magnetization in these minerals. Thus, if there is a localized concentration of a magnetic mineral, a weak local magnetic field adds to the external field of the Earth. A very sensitive instrument can measure the small local contribution to the total field.

Two processes that contribute to the enhanced magnetization of anthropogenic features are the firing of soils, which tend to lock in the induced magnetization as they cool down, and the chemical reduction of weaker magnetic minerals in topsoils to stronger magnetic minerals. Typical anthropogenic features that can give rise to magnetic anomalies are fired kilns and hearths, ditches filled with topsoil, and storage pits filled with organically rich topsoil. On historic sites, foundation features having a magnetic contrast

with the surrounding soils can be observed, and ferrous metal objects can give rise to strong anomalous signals.

Magnetometers can measure the field of the Earth to 1 part in roughly 500,000. The usual method is to measure the total field on a grid of points laid out over the region of interest, with the sensor of the magnetometer a short distance above the site. Because the field of the Earth varies throughout the day, with typical changes several times larger than anomalies of interest, it is necessary to compensate with a second magnetometer that monitors the changes. The corrected field values are then mapped or profiled to help identify meaningful anomalies.

Another way to measure the magnetic field is to place two sensors one above the other and to operate both simultaneously. The upper reading is subtracted from the lower reading, and the difference is plotted. With this configuration, known as a gradiometer, the diurnal changes are canceled, as in the previous method, and long-range trends, usually caused by geological features, are suppressed. The reason for this is that the lower sensor is more strongly influenced by near-surface features than the upper sensor, whereas both sensors are about equally influenced by deep features.

Ground-penetrating radar. This method has been used by engineering geophysicists for some years and is gaining interest in the archeological community. If a high-frequency radio signal is directed into the Earth, some of the signal will return as an echo bounced off interfaces in the soil where electrical properties of the soil change abruptly. By timing the echo, information on the depth of the interface causing it can be obtained. Anthropogenic features revealed by radar include foundation walls and floors, compacted or stone roads, and earthen features, such as mounds, that have a compacted floor. This method works well in dry sandy soils but poorly in moist clay soils.

John W. Weymouth

Trace Organic Residue Analysis

Archeometry, the application of chemical and physical analyses to archeological questions, is gaining acceptance. One application is trace organic residue analysis, which uses gas chromatography and mass spectrometry to identify the microscopic organic residues that have adhered to or been absorbed into the structural matrix of archeological materials. Analysis of these residues can provide information about the function of the artifact in, and the subsistence strategies of, a past culture.

Although differential preservation of organic materials can result in the misinterpretation of their relative importance in analyses of past cultures, it has been shown that some components of organic materials can survive in archeological contexts for at least 2 millennia. There is usually enough preserved organic material absorbed into the matrix of porous materials (such as ceramics, sandstone, and limestone) or adhered to protected surfaces (nooks and crannies of nonporous materials such as flint or chert) to allow for in-depth analysis of the organic residues.

Lipid analysis. Most archeological approaches to trace organic residue analysis focus on lipids extracted from the artifacts. Lipids are substances that are present in or derived from living organisms, are insoluble in water but soluble in organic solvents such as chloroform and ether, and contain long-chain hydrocarbon groups. There are several different types of lipids, including alcohols, aldehydes, fatty acids, hydrocarbons, and sterols. The presence, absence, and concentrations of different lipids can be used to identify the parent organic materials from which the residue originated.

A gas chromatograph coupled to a mass spectrometer can provide a very powerful tool for analyzing the lipids extracted from ancient residues. Gas chromatography is an instrumental technique to separate and quantify organic compounds. Different organic compounds are separated on the basis of differential interaction between volatized components and a stationary phase within the gas chromatograph column. As the sample components flow out of the column in the carrier gas, they are detected and their signal strength is plotted on a strip chart recorder (or stored in a computer data file), producing a chromatogram. Each peak in the chromatogram represents a different compound if all of the compounds are completely separated.

Although gas chromatography is an excellent separation technique, it does not identify the compounds. Possible identifications can be suggested, but for true identification a mass spectrometer needs to be coupled to the gas chromatograph. As the compounds enter the mass spectrometer, they are bombarded by electrons, causing them to ionize and fragment. The ionization and molecular fragmentation produce a so-called fingerprint of the sample component, from which the identity and structure of the compound can then be derived.

Problems. Several problems, including degradation of the lipids, contamination from postdepositional factors, and analysis of the large amount of data acquired from the techniques, inhibit data interpretation. However, depending on the nature of the material from which the residue is being extracted, these problems can be greatly reduced. If the material of the artifact is porous, such as in ceramics or in limestone or sandstone cooking slabs, and a small section of the artifact can be destroyed during analysis, then about 0.04 in. (1 mm) of the surface (interior surface of ceramic vessels) can be ground off and the lipids extracted from the area below. This procedure removes contaminants such as fingerprints, ink, and other materials added after excavation. It also removes the area with the most highly oxidized lipids, allowing analysis of a sample with a smaller percentage of degraded lipids. Both interior and exterior surfaces can be sampled to show a gradient in the lipid concentration, which generally indicates that the residue is the result of use rather than postdepositional factors. For nonporous artifacts or those that cannot be damaged, such as chipped-stone tools, it is much more difficult to be assured of a pure ancient residue. It is best to sample unwashed artifacts for which the handling history is known. Extraction of residues from small specific points on the tools (such as the working edges) that may have had less handling also reduces contamination and allows for comparison of results from multiple samples from the same artifact.

One major problem with utilization of fatty acids and other lipids in archeological studies is structural deterioration (diagenesis) of the lipid form through processes such as oxidation. Nevertheless, although susceptible to deterioration, archeological samples analyzed in several studies contained unsaturated fatty acids, which are the most likely to undergo diagenesis. These results indicate that the residues have some stability. In addition, as some lipids are less stable than others, they can be isolated or less heavily weighted in the interpretation of the data.

Interpretation. Data interpretation is the most difficult aspect of trace organic residue analysis. Even if all of the lipids in the residue can be identified correctly and there are no problems resulting from degradation, interpretation of the enormous amounts of data is not trivial. Different techniques help in data interpretation, including use of characteristic ratios and, more recently, pattern recognition programs. Characteristic ratios are calculated between lipids that are generally associated with a specific class of parent material. For example, polyunsaturated fatty acids are generally found in plants, whereas saturated fatty acids and some of those with branched chains are associated with faunal parent materials. By calculating the ratios between combinations of these two types of lipids (sometimes expressed in terms of percent of saturated fatty acids), a general characterization of the residue can be determined. However, this type of analysis does not provide identification of the specific parent material from which the residue derived.

Pattern recognition programs offer a much greater potential for identification of specific parent materials. These programs have been used successfully to identify different harvest years and provenances of modern wines and olive oils and to quantify the adulteration of orange juice with less expensive components such as grapefruit juice. Preliminary experiments indicate that this technique has good potential for the interpretation of trace organic residues from archeological samples.

The data provided by trace organic residue analysis, especially those deriving from major studies of several hundred analyzed samples, can be used to study aspects of economic relationships. By thus examining the distribution of food resources, and the contexts in which they were used, it should be possible to reconstruct subsistence strategies.

Even small sample sizes can provide information useful for interpretation of the archeological record. For example, the residues from a ceremonial smash of keros (drinking cups) at the site of Tiwanaku in the Bolivian altiplano indicate that the cups contained faunal elements in addition to those of corn beer. The faunal residue could have been llama blood, given that llama blood is caught in cups as part of the ritual activities in several modern ceremonies. In another example, the residues from a sandstone slab excavated in northern Texas indicate that fish were processed on it. The top side of this artifact (from which the fish residues were extracted) was burned, and the bottom was unburned (and had a much smaller concentration of residues). Gas chromatography analysis leads to the interpretation that a fire (or coals) was placed on the top side of the slab and fish was cooked on it (the coals may have been swept off so the slab could serve as a form of frying pan).

Although still experimental, and with limits not yet fully known or explored, trace organic residue analysis has great potential in the reconstruction of past diets and artifact usage. Through a complete study of the intrasite variation of raw materials, finished products, and trace organic components, and the resultant interpretation of diet and exchange systems, it should be possible to reconstruct the economic elements of the society from which social aspects, one of the major goals of the archeologist, may be derived.

For background information SEE ARCHEOLOGY; GAS CHROMATOGRAPHY; GEOMAGNETISM; GEOPHYSICAL EXPLORATION; MASS SPECTROMETRY in the McGraw-Hill Encyclopedia of Science & Technology.

Michael L. Marchbanks

Bibliography. M. Aitken, *Physics and Archaeology*, 1974; L. W. Aurand, A. E. Woods, and M. R. Wells, *Food Composition and Analysis*, 1987; M. Kates, *Techniques of Lipidology: Isolation, Analysis, and Identification of Lipids*, 1986; T. F. M. Oudemans, Molecular archaeology: Analysis of charred (food) remains from prehistoric pottery by pyrolysis gas chromatography/mass spectrometry, *J. Anal. Appl. Pyrol.*, 20:197–227, July 1991; A. C. Roosevelt, *Moundbuilders of the Amazon: Geophysical Archaeology on Marajo Island, Brazil*, 1991; M. Schiffer (ed.), *Advances in Archaeological Method and Theory*, vol. 9, 1986; I. Scollar (ed.), *Archaeological Prospecting and Remote Sensing*, 1990.

Ascidiacea

Ascidian tadpole larvae are the simplest animals that have a chordatelike structure; they have sometimes been regarded as chordate prototypes. They swim by oscillating a finned tail; and while recent work has gone some way to link their structure to the manner in which they swim, it does not support the hypothesis that they show primitive chordate characters.

Most sessile ascidians produce eggs that hatch into tailed tadpole larvae that swim actively before settling at a site where they metamorphose. The larvae are designed for short-range site selection, and they swim for only a few minutes to several hours. There are considerable differences between ascidian species in the size of the larva and in the proportions and morphology of the trunk, but the tail is essentially of the same chordatelike structure in all species. A central notochord and dorsal nerve cord are surrounded by rows of muscle cells, and there are dorsal and ventral fins composed of test material. This chordatelike structure of the tadpole larva tail, first recognized in 1867, aroused much interest among zoologists and led to the alliance of the Tunicata with the Chordata.

Anatomy. **Figure 1** shows the anatomy of the tadpole larvae of the pleurogonid *Dendrodoa* and the enterogonid *Ciona*. Although placed in two different orders, they are very similar in organization. Apart from size, the main differences between them are the absence of a photoreceptor in *Dendrodoa* and the absence of caudal ventral sensory cells in *Ciona*. It is reasonable to suppose that their locomotor design is typical of all tadpole ascidian larvae. In the tail, the central notochord is surrounded by three tiers of rounded muscle cells (four tiers in some other species); these cells are electrically coupled by gap junctions, so that changes in membrane potential in any one cell are transmitted to neighboring cells, and each side of the tail contracts as a single unit. The muscle cells have a thin myofibrillar cortex surrounding an inner mitochondria-rich cytoplasm with much glycogen.

Motor axons, partially overlain by ciliated sheath cells, run along the dorsal surface of the notochord, and form the dorsal nerve cord; so far as is known, the cell bodies of these motor axons lie in the ganglia of the trunk and none are found along the nerve cord. Primary sensory cells are scattered in the epithelium of the tail, and their cilia pass into the test material of the tail fins. The sensory axons run anteriorly in small bundles under the epithelial cells without contacting the motor axons of the dorsal nerve cord.

All the muscle cells of the dorsal tier are innervated by the motor axons of the dorsal nerve cord; but only the most anterior of the ventral tier, and none of the middle tier, are innervated. There are, then, two ways in which the caudal musculature can be driven by the tadpole nervous system: by the anterior ventral axons, and by the axons supplying the entire range of dorsal muscle.

Swimming movements. The larvae show two types of movements: long or short bursts of symmetrical oscillations at frequencies between 16 and 36 Hz, which drive them forward in a spiraling motion, and asymmetric tail flicks at lower frequencies, which cause them to start from rest at various angles to the resting position or to circle on the bottom. The asymmetric flicks may be interpolated within swimming bursts, thus producing a change of direction. Both kinds of locomotor activity are usually rhythmic, occurring at intervals of 5–10 s.

Figure 2 shows successive positions of the two types of larvae swimming forward. The movements of the tail are eellike, so that the tail tip makes large-amplitude excursions from the axis of progression, as does the tip of the trunk, which yaws 55–80° from the axis of progression. These movements result from the interaction of caudal muscle contractions with the resistance of the surrounding water; the transverse propulsive waves that pass down the tail at a velocity of around 0.8 mm/s in *Ciona* and 42 mm/s in the larger *Dendrodoa* are not caused by sequential activation of the caudal muscle cells. The tip of the trunk and the center of mass (approximately at the tail-trunk junction) move alternately backward and forward along the axis of progression, so that there are large variations in forward velocity during the tail-beat cycle. In both larvae, the mean forward velocity is only about one-tenth that of the propulsive wave down the tail, with the maximum forward velocity in both being about 10 body lengths per second, that is, just over 10 mm/s in *Ciona* and 25 mm/s in *Dendrodoa*. In line with the high mitochondrial content of the muscle cells, indicating an aerobic metabolism, swimming bursts may be prolonged for several minutes, though they are usually shorter. The asymmetrical tail flicks may occur singly, as when the larvae start from rest or change the direction of swimming, or they may occur in short trains when the larvae circle on the bottom. In *Dendrodoa*, circling behavior is more frequent the longer the larvae have been released, and presumably is related to site selection on the substrate after an initial swimming period for dispersal.

Muscle cell activity. Records of muscle activity in restrained larvae show that the two types of swimming movement are accompanied by different types of potential; those in the symmetrical swimming bursts are smaller than those corresponding to the flicks to one side. Larval tails amputated from the trunk continue to show rhythmic movements for long periods. The neuromuscular transmitter is acetylcholine, and movements in both intact larvae and isolated tails are reversibly abolished by the cholinergic blocker D-turbocurarine. Thus, in isolated tails, the continued activity is under the nervous control of motor axons divorced from their cell bodies. If the tail is cut from

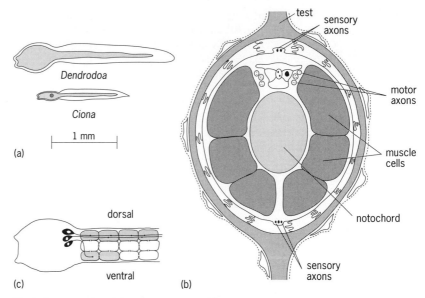

(a)

dorsal

ventral

(c)   (b)

Fig. 1. Anatomy of ascidian tadpole larvae. (*a*) *Dendrodoa* and *Ciona*. (*b*) Cross section of midregion of tail (fins cut short) of *Ciona*. (*c*) Innervation pattern of tail of both species; innervated cells are in color. (*After Q. Bone, On the locomotion of ascidian tadpole larvae, J. Mar. Biol. Ass. U.K., 72:161–186, 1992*)

the trunk close to its base, both swimming movements and asymmetric tail flicks are seen, whereas isolated tails cut farther away from their junction with the trunk show only single flicks. It thus seems that symmetrical swimming results from ventral innervation,

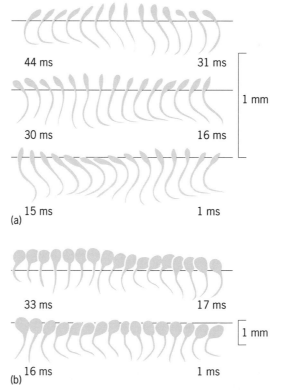

Fig. 2. Successive positions (1 millisecond apart) of a larva swimming forward. (*a*) *Ciona*, (*b*) *Dendrodoa*. (*After Q. Bone, On the locomotion of ascidian tadpole larvae, J. Mar. Biol. Ass. U.K., 72:161–186, 1992*)

while single tail flicks are driven by dorsal innervation.

Significance of swimming patterns. The ascidian tadpole locomotor system differs from that of chordates not only in scale but, notably, in the close coupling of the muscle cells along each side of the tail, the absence of motor neuron cell bodies along the tail, and the absence of proprioceptive input during swimming. Two kinds of nearly simultaneous contractions of the muscle cells along one side of the tail are apparently driven by two different types of innervation. This is an entirely different kind of locomotor design from that of such chordates as fish and urodele amphibia, where the tail is oscillated by muscle fibers in serial myotomal muscle blocks that are not electrically coupled. The myotomes are sequentially activated along the body by a much more complex set of neurons whose cell bodies lie in the spinal cord. These differences are so fundamental that the ascidian larva seems an unlikely candidate for a prototype chordate.

More promising candidates are perhaps the superficially similar larvacean or appendicularian tunicates, which filter-feed in the complex "houses" that they secrete. Appendicularian status has been long debated. One view is that forms ancestral to modern appendicularians gave rise both to the other tunicates and to the chordates. The other view is that appendicularians arose by neoteny from ascidian tadpolelike ancestors long after the divergence of the tunicates and chordates. Recent evidence from sperm structure favors the former view.

Unfortunately, the design of the appendicularian locomotor system, like that of the ascidian tadpole, is not much like that of present-day chordates. A single row of 10 coupled muscle cells along the tail is innervated by two types of motor neurons whose cell bodies lie along the dorsal nerve cord in a more or less segmental arrangement. These and other differences between appendicularians and ascidian tadpoles are sufficiently striking to make it improbable that appendicularians can have arisen from ascidians by neoteny. Tail movements are more complex than in ascidian tadpoles, since appendicularians use tail oscillations not only to swim but also to expand their houses and to drive water through their filters. These movements have not been examined in any detail, and their control is not well understood.

Although appendicularian tunicates may possibly be the closest living forms to the ancestral chordate, their locomotor systems, and the way in which they are controlled, are very different from those of living chordates. Either a radical change must have taken place in the chordate line after the two separated, or living appendicularians have become more simplified from the ancestral forms.

For background information SEE APPENDICULARIA; ASCIDIACEA; CHORDATA; TUNICATA in the McGraw-Hill Encyclopedia of Science & Technology.

Q. Bone

Bibliography. Q. Bone, On the locomotion of ascidian tadpole larvae, *J. Mar. Biol. Ass. U.K.*, 72:161–186, 1992; G. O. Mackie and Q. Bone, Skin impulses and locomotion in an ascidian tadpole, *J. Mar. Biol. Ass. U.K.*, 56:751–768, 1976.

Atmosphere

Interest in the middle atmosphere, lying roughly between 10 and 100 km (6 and 60 mi) altitude, has been fueled by increasing concerns over the Earth's environment. The discovery of the ozone hole in the Antarctic lower stratosphere, together with predictions of greenhouse warming, has greatly accelerated interest in both scientific and public sectors. The term ozone hole refers to the severe depletion of the ozone layer over Antarctica each year during spring (late August to early October; **Fig. 1**). Among the most difficult and important tasks in atmospheric science today are unequivocal determinations of the magnitude and direction of long-term changes in both the amount of ozone and the greenhouse-induced global temperature. Both factors are intimately related to the composition, structure, and dynamical motions in the middle atmosphere.

The Earth's climate is determined by a balance between incoming solar and outgoing Earth thermal radiative energy, both of which must pass through the middle atmosphere. The lower portion of the middle atmosphere, the stratosphere, contains many greenhouse gases (ozone, water vapor, carbon dioxide, methane, nitrous oxide, chlorofluorocarbons (CFCs), and others); it is predicted that the stratosphere is cooled at the same time as the lower atmosphere is warmed by the greenhouse effect. The middle atmosphere is also a focus for effects of emissions from proposed commercial fleets of stratospheric aircraft. In addition to the chemistry involved, dynamical transport modeling and measurements are needed to predict the widespread transport of these important trace gases and emissions over the globe.

Middle-atmosphere structure. The stratosphere extends from about 10 to 50 km (6 to 30 mi) altitude; from about 50 to 80 km (30 to 48 mi) or so lies the mesosphere. The location of the base of the stratosphere (the tropopause) depends on meteorological conditions, varying on average from about 10 km (6 mi) in altitude at the poles to about 15 km (9 mi) at the Equator. Actually, the middle atmosphere is not so far removed from everyday experience. The flat-topped anvils of thunderstorms are one example. They are formed when rapidly ascending air flattens out like a pancake against the bottom of the stratosphere; further vertical penetration is inhibited by negative buoyancy effects due to the rising temperature with height in the stratosphere. A second example is experienced by jet aircraft travelers on middle-latitude or polar routes that often are flown near or just below the tropopause; a traveler looking up into the darker sky from the aircraft window is peering into the stratosphere. A third example involves eruptions of volcanoes. Often after a major volcanic explosion, a beautiful twilight purple glow can be seen a few minutes after sunset, caused by dust ejected into the stratosphere perhaps half a world away and transported by strong winds in the stratosphere. Because rain does not occur in the stratosphere, the tiny dust particles may remain there for a year or more.

In equatorial regions the tropopause is characterized by, and usually defined by, a sharp change in the rate of temperature decrease with altitude. This definition is not as useful at high latitudes, because there the temperature structure is different. A more general, dynamical definition of the interface between the lower and the middle atmosphere is in terms of a quantity called potential vorticity, which has large values in the stratosphere but small values in the lower atmosphere. In many cases of interest, the potential vorticity is also an approximately conserved quantity, making it useful for tracing motions of air parcels.

The majority (about 90%) of the Earth's ozone shield lies in the stratosphere. Ozone is produced mainly at low latitudes (near the Equator) by solar ultraviolet radiation; however, larger concentrations of ozone are found at middle and polar latitudes (Fig. 1). The resolution of this apparent anomaly lies in transport from the equatorial source to the polar regions by dynamical processes in the stratosphere. For example wave-motion-induced transport moves significant amounts of stratospheric ozone to the Antarctic polar region; furthermore, the stratosphere transports the ozone-destroying (mainly chlorine) compounds to the polar regions. Without such transport there would not be a significant Antarctic ozone hole. SEE ATMOSPHERIC OZONE.

Understanding the dynamical and radiative physics underlying middle-atmosphere transport is thus critical for several aspects of predicting global climate change.

Middle-atmosphere dynamical features.

Internal gravity waves (not to be confused with the gravity waves of general relativity) result from the combined forces of gravity and pressure gradients. Typical characteristics are transverse polarization, vertical wavelengths of 0.1–10 km (0.06–6 mi) horizontal wavelengths of 1–100 km (0.6–60 mi) or more, and periods in the range of 5 min to several hours. These waves may be excited by airflow over orography (mountains) as standing lee waves, by growing clouds, and by large-scale storm complexes in the lower atmosphere; and then they propagate up into the middle atmosphere.

Planetary-scale Rossby waves are large and slowly moving waves affected by the Coriolis effect due to the Earth's rotation. They arise from the variation with latitude of the component of the Earth's rotation vector normal to the planetary surface, and they propagate against the planetary rotation (that is, westward) with periods of several days. They have vertical wavelengths of the order of 10 km (6 mi) and horizontal wavelengths from hundreds of kilometers up to as large as wave 1, where one wavelength fits around the Earth's circumference (wave 2 refers to two wavelengths fitting around the Earth's circum-

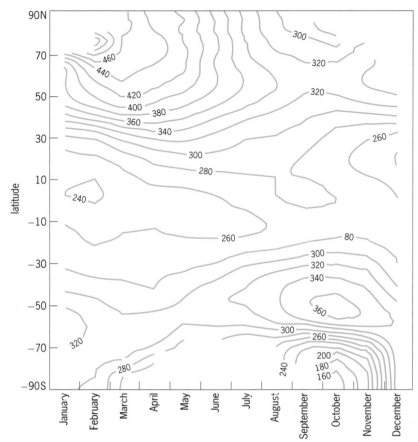

Fig. 1. Time-latitude section showing seasonal variation of total ozone in Dobson units (contours), based on the Total Ozone Mapping Spectrometer (TOMS) satellite instrument data for 1987. The maxima in middle to high latitudes of each hemisphere's respective spring are due to dynamical transport from the equatorial source region. The very low values of ozone in October near 75–90°S represent the Antarctic ozone hole. (*After World Meteorological Organization Global Ozone Research and Monitoring Project, Report of the International Ozone Trends Panel, Rep. 18, vol. 1, 1988*)

ference, and so forth). Rossby waves are common at middle latitudes in winter, where they can propagate up into the middle atmosphere from excitation regions below.

Near the Equator, hybrid mixed-Rossby gravity waves and also Kelvin waves (a special class of eastward-propagating internal gravity waves having no north-south velocity component) have been observed in the middle atmosphere.

A variety of global-scale normal-mode oscillations (somewhat like the vibrations of a kettledrum head) are also found in the middle atmosphere, prominent examples being wave 1, westward-moving waves with periods of about 5 and 16 days, and wave 3, a westward moving feature with a period of about 2 days. Another observed oscillation in the middle atmosphere has been found with periods in the range of 1–2 months (propagating up from the lower atmosphere).

Waves resulting from fluid-dynamical instabilities are also observed: medium-scale (waves 4–7) eastward-moving waves with periods of 10–20 days (actually the tops of lower-atmosphere phenomena) can dominate the circulation of the summer Southern Hemisphere lower stratosphere. There are other possible instabilities as well.

Phenomena under study. There are a number of outstanding questions involving the middle atmosphere.

Perhaps the most subtle and complex middle-atmosphere phenomena are related to nonlinear effects of large-amplitude waves. An example, still poorly understood, is the contrast between theory and observations of winds in the stratospheric-mesopheric jet stream. This striking feature of the middle atmosphere resembles a donut-shaped symmetric vortex with strong winds that maximize near the top of the stratosphere and can reach high wind speeds (**Fig. 2**). The winds blow primarily toward the east in the winter hemisphere and toward the west in the summer hemisphere. The primary cause of the jet is a combination of the Equator-to-pole difference in solar heating and the Coriolis acceleration due to the Earth's daily rotation. Maximum observed winds (Fig. 2) are usually less than about 100 m/s (around 200 mi/h), several times smaller than predicted by radiative-transfer models without dynamics. The mechanism that closes the jet (preventing extremely large winds) is thought to result from vertical propagation of internal gravity waves, their subsequent growth of amplitude, and their eventual dissipation in the mesosphere through small-scale instabilities. This mechanism is important in transporting momentum into the mesosphere and providing a brake on the formation of extremely strong jet-stream winds that would form in the absence of dynamical processes. Relatively few observations are available, and many more are needed to guide numerical modeling work on this topic.

Sudden stratospheric warming. One of the most dramatic events in the stratosphere occurs about once every other year when the Northern Hemisphere winter polar stratosphere suddenly experiences dramatic warming in a few days. Such a sudden stratospheric warming leads to a rapid reversal in direction of the stratospheric winter jet, changing it from its usual winter eastward motion (Fig. 2) to westward. Only once has a sudden stratospheric warming been observed in the Southern Hemisphere, the difference between hemispheres apparently being related to stronger planetary-scale waves (waves 1 and 2) forced by the larger orography and land-sea contrasts in the Northern Hemisphere. Theoretical understanding of the phenomenon of sudden stratospheric warming is founded on vertical propagation and absorption of planetary-scale waves (Rossby waves). Although modeling studies have been carried out for over two decades, the understanding of all the observed details and necessary conditions leading to formation of sudden stratospheric warming is still incomplete.

Periodic oscillations. An unusual, almost periodic zonally symmetric quasi-biennial oscillation is found in the equatorial stratosphere, manifested as slowly descending winds at low latitude, switching between westward and eastward directions with an irregular period averaging about 27 months (**Fig. 3**).

The basic mechanism of the quasi-biennial oscillation involves a subtle internal oscillation resulting from wave interactions with the mean stratospheric jet flow. Vertically propagating Kelvin and Rossby-gravity waves are absorbed or radiatively damped

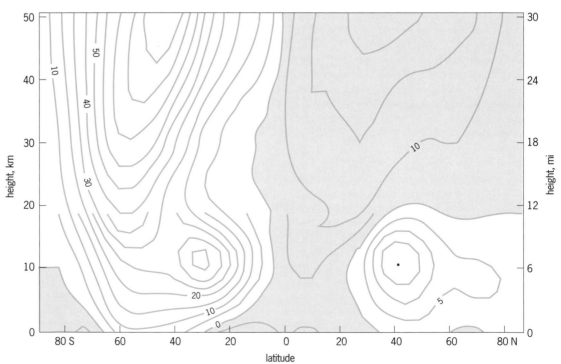

Fig. 2. Mean east-west wind for July as a function of altitude and latitude. Wind values (contours) are in meters per second. Note the strong stratospheric jet stream in winter (Southern Hemisphere), exceeding 90 m/s from the west in the upper stratosphere [45–50 km (27–30 mi) altitude]. In the summer stratosphere, the winds blow from the east (shaded). 1 m = 3.3 ft. (*After W. J. Randel, Global atmospheric circulation statistics, 1000–1 mb, Tech. Note NCAR/TN-366+STR, Nat. Cen. Atm. Res., 1992*)

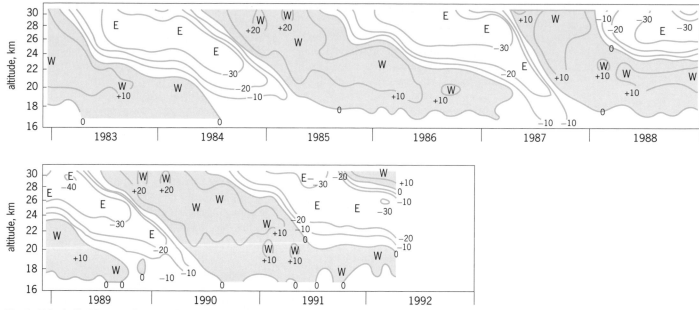

Fig. 3. Altitude (in kilometers) versus time section showing monthly mean winds (contours in meters per second) along the Equator in the lower stratosphere. The quasi-biennial oscillation is seen as descending wind regions alternating from the west (W, shaded) and east (E) with an irregular period of about 27 months. 1 km = 0.6 mi. (*After B. Naujokat, Free University, Berlin*)

in the lower stratosphere, leading to the alternating wind regimes of the quasi-biennial oscillation. Although the main idea and beauty of the fundamental principle is clear, aspects of the observations, especially why the two regimes of the quasi-biennial oscillation are not symmetric, have yet to be elucidated.

Another middle-atmosphere phenomenon is the semiannual oscillation, with a 6-month period. The equatorial semiannual oscillation, like the quasi-biennial oscillation, is thought to involve vertical propagation and damping of Kelvin waves. However, it appears that additional mechanisms may be required to explain the observations. More research is needed.

Waves due to instabilities. The dominant wave features in the Southern Hemisphere winter stratosphere are large-scale (wave 2) and move eastward with periods of 2–3 weeks. These features are postulated to originate from instabilities that draw their energy from spatial variations in the background wind field. Instabilities have also been predicted to account for curious long-lived perturbations (which have been called quasi-nondispersive features) observed in the winter polar stratosphere. Several aspects of both these phenomena remain to be explained.

Transport of constituents. Finally, a primary focus of middle atmosphere dynamics is to predict and explain the transport of constituents such as ozone, chlorofluorocarbons, and various other greenhouse gases. While progress has been made with two-dimensional models, much more work with three-dimensional models is required. Performing this modeling as well as analyzing and interpreting the vast quantities of data from the *Upper Atmosphere Research Satellite* (*UARS,* launched in 1991) and planned *Earth Observing System* (*EOS*) satellites will require input

from many scientists in the coming decade.

For background information SEE ATMOSPHERE; AT-MOSPHERIC OZONE; GREENHOUSE EFFECT; JET STREAM; STRATOSPHERE in the McGraw-Hill Encyclopedia of Science & Technology.

J. L. Stanford

Bibliography. D. G. Andrews, A stratospheric transport system, *Phys. World*, 4(11):41–46, November 1991; D. G. Andrews, J. R. Holton, and C. B. Leovy, *Middle Atmosphere Dynamics*, 1987; S. Solomon et al., Tracer transport by the diabatic circulation deduced from satellite observations, *J. Atm. Sci.*, 43:1603–1617, 1986.

Atmospheric ozone

Recent research involving atmospheric ozone includes studies of ozone depletion and enhancement through observation and modeling of the processes involved, and studies of the stratospheric aerosol ejecta from Mount Pinatubo (Philippine Islands) and their possible effect on global ozone depletion.

Ozone Depletion and Enhancement

In the stratosphere at altitudes above about 15 mi (25 km) molecules of oxygen (O_2) are weakly photodissociated by short-wave solar ultraviolet radiation. The resulting oxygen atoms quickly combine with O_2 to form a layer of ozone (O_3). Ozone itself strongly absorbs ultraviolet radiation of wavelengths longer than those that are absorbed by O_2. Through this absorption the ozone layer heats the stratosphere and reduces the flux of ultraviolet radiation at the surface of the Earth. The ozone layer is thus important both for establishing the temperature distribution of the atmosphere and for protecting the biosphere from biologically harmful ultraviolet radiation.

Despite its fundamental role in the global climate system, ozone is a very minor atmospheric constituent. Even at the peak of the ozone layer, the mixing ratio of ozone (that is, the ratio of the number of ozone molecules to the total number of atmospheric molecules in a given volume) seldom exceeds 10 parts per million. If the entire content of atmospheric ozone were brought to sea level, it would form a layer of gas with a thickness of about 0.1 in. (3 mm).

Ozone destruction. The mechanism for photochemical production of ozone has been known since about 1930, but only since about 1970 has it been appreciated that the major loss occurs through a number of catalytic reaction cycles. The simplest of these cycles can be expressed symbolically as reactions (1)–(3), where X is a catalytic molecule that remains un-

$$X + O_3 \rightarrow XO + O_2 \qquad (1)$$

$$XO + O \rightarrow X + O_2 \qquad (2)$$

$$\text{Net}: \quad O + O_3 \rightarrow 2O_2 \qquad (3)$$

changed at the end of the cycle. The rate at which such a catalytic cycle can destroy ozone is limited not only by the reaction rate constants but also by the concentrations of the catalyst and of oxygen atoms. The most important catalyst in the natural stratosphere is nitric oxide (NO), which accounts for the majority of ozone loss at altitudes below 22 mi (35 km). The main source for stratospheric NO is nitrous oxide (N_2O), which is produced by biological processes at the surface and transported to the stratosphere by atmospheric motions. There N_2O slowly reacts with oxygen atoms to produce NO.

Between 22 and 25 mi (35 and 40 km) nitrogen and chlorine catalysis are both important. Above 25 mi (40 km) the major catalyst for ozone destruction is the chlorine atom (Cl). The source for chlorine in the natural stratosphere is the upward transport by atmospheric mixing of methyl chloride (CH_3Cl), which is produced mainly by the world oceans. Methyl chloride reacts with the hydroxyl radical (OH) to release chlorine atoms for participation in catalytic ozone destruction. Because OH attacks methyl chloride in the troposphere as well as the stratosphere, most methyl chloride is destroyed before it can reach the stratospheric ozone layer. This cleansing effect of OH regulates the amount of chlorine available for ozone destruction in the natural stratosphere.

Tropospheric OH does not, however, attack the chlorofluorocarbons (CFCs), such as CF_2Cl_2 and $CFCl_3$. These synthetic gases are almost completely inert in the troposphere; thus they have sufficiently long atmospheric lifetimes that they may easily reach the stratosphere. There they are photolyzed by ultraviolet radiation to release chlorine atoms, which can then destroy ozone catalytically. Owing to the wide use of CFCs in refrigeration, air conditioning, foam blowing, and other industrial processes, the atmospheric burden of chlorine has been increasing by about 5% per year since around 1967. Chlorine re-

leased by human activities now constitutes the bulk of stratospheric chlorine, as indicated in **Fig. 1**, where the horizontal line shows the amount of chlorine in the stratosphere generated by natural processes. Human sources account for around 80% of the total chlorine in the stratosphere.

Ozone holes. By the mid-1970s it was recognized that the large anthropogenic perturbation of the natural chlorine content of the stratosphere might have significant effects on the ozone budget. Detailed chemical models based on reactions (1)–(3) and other similar catalytic cycles suggested that the first signs of ozone loss should be at altitudes of 21–24 mi (35–40 km), where the gas-phase chlorine catalysis is most effective. Because most ozone molecules lie below this region, the models predicted that with current rates of CFC production a total column change in ozone of about 5% could be expected by the middle of the twenty-first century.

Antarctic. It was thus a complete surprise when, in 1985, it was discovered that column ozone was depleted by about 50% over the Antarctic continent during the Austral spring (Fig. 1). Later work showed that there was nearly complete removal of ozone in the altitude range of 10–14 mi (16–22 km). Although the occurrence of this so-called ozone hole was clearly correlated with the increase in chlorine in the stratosphere, gas-phase catalytic cycles involving chlorine were not expected to be effective in the lower polar stratosphere, because theory predicted that most of the chlorine in that region would be bound up in the rather inactive compounds hydrochloric acid (HCl) and chlorine nitrate ($ClONO_2$).

Some scientists recognized, however, that heterogeneous reactions on ice particles might convert chlorine into reactive forms. Although the concentration of water vapor is too low for water-ice clouds to form in most of the stratosphere, polar stratospheric clouds composed of solid nitric acid trihydrates ($HNO_3 \cdot$

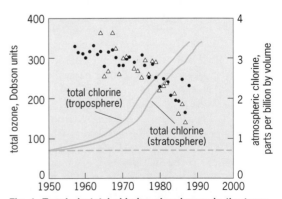

Fig. 1. Trends in total chlorine abundances in the troposphere and the stratosphere. The horizontal broken line indicates the abundance in the natural stratosphere, not inclusive of human-caused increase in chlorine content; this amount has not changed since 1950. Data points (circles for Halley Bay and triangles for the South Pole) indicate the trends in October mean total ozone measured at two Antarctic stations [100 Dobson units = 1-mm-deep column at standard temperature and pressure]. (*After S. Solomon, Progress towards a quantitative understanding of Antarctic ozone depletion, Nature, 347:347–354, 1990*)

3H$_2$O) can form when temperatures drop below 195 K (−108°F). Such cold conditions occur frequently in the Antarctic stratosphere during the winter and early spring, and less commonly in the Arctic winter.

Observations confirm the widespread occurrence of polar stratospheric clouds in the Antarctic winter stratosphere, and laboratory studies have confirmed that reaction (4) proceeds rapidly on solid surfaces,

$$HCl + ClONO_2 \rightarrow Cl_2 + HNO_3 \qquad (4)$$

but not in the gas phase. The result of reaction (4) is to release chlorine from the reservoir species HCl and ClONO$_2$. At the same time, nitrogen is removed from the gas phase and stored as solid HNO$_3$, reducing the availability of reactive nitrogen for converting chlorine monoxide (ClO) back into ClONO$_2$ and thus increasing the potential for chlorine catalysis to destroy ozone.

The molecule Cl$_2$, which is produced in reaction (4), is in turn dissociated in the presence of sunlight to form Cl atoms. The catalytic cycle in reactions (1)–(3) cannot, however, operate effectively in the lower polar stratosphere owing to the very low concentration of unattached oxygen atoms in that region. Rather, the catalytic cycle in reactions (5)–(9), where M is a

$$2(Cl + O_3 \rightarrow ClO + O_2) \qquad (5)$$

$$ClO + ClO + M \rightarrow Cl_2O_2 + M \qquad (6)$$

$$Cl_2O_2 + sunlight \rightarrow Cl + ClO_2 \qquad (7)$$

$$ClO_2 + M \rightarrow Cl + O_2 + M \qquad (8)$$

$$Net: \quad 2O_3 + sunlight \rightarrow 3O_2 \qquad (9)$$

catalytic molecule, appears to be the primary source of ozone depletion in September in the Antarctic. The rate at which this cycle can destroy ozone is limited by the rate at which ClO reacts with itself in reaction (6). This reaction in turn clearly depends on the local concentration of chlorine monoxide.

Ground-based measurements at McMurdo Sound (Antarctica) in spring 1986, which detected the presence of ClO and a closely related molecule (OClO), provided some indirect support for this chlorine catalysis cycle. The definitive proof came, however, in September 1987 during the Airborne Antarctic Ozone Expedition. A high-altitude research aircraft, the NASA ER-2, simultaneously measured ozone and chlorine monoxide during several flights from South America to the Antarctic continent. The data from these flights (**Fig. 2**) showed a dramatic inverse correlation between the concentrations of ozone and chlorine monoxide. At lower latitudes, where temperatures were in excess of those required for formation of polar stratospheric clouds, the concentration of ClO was low and ozone concentrations were normal. Poleward of about 68°, however, the ClO abundance rose dramatically and the ozone abundance decreased by more than a factor of 2. Observations of temperature and cloud particles during the Airborne Antarctic Ozone Expedition confirmed that conditions of high abundance of ClO occurred only when temperatures were sufficiently low for formation of polar stratospheric clouds.

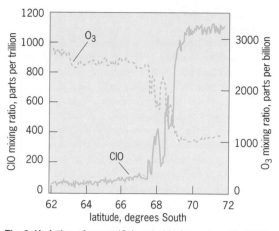

Fig. 2. Variation of ozone (O$_3$) and chlorine monoxide (ClO) across the edge of the Antarctic ozone hole on September 16, 1987, showing the strong anticorrelation between the concentrations of ClO and O$_3$. (*After J. G. Anderson, W. H. Brune, and M. H. Profitt, Ozone destruction by chlorine radicals within the Antarctic vortex: The spatial and temporal evolution of ClO-O$_3$ anticorrelation based on in situ ER-2 data, J. Geophys. Res., 94:11465–11479, 1989*)

Satellite observations of ClO and O$_3$ by the Microwave Limb Sounder on the *Upper Atmosphere Research Satellite (UARS)* during September 1991 confirmed that the correlation of high concentration of ClO and low concentration of O$_3$ is a continent-wide phenomenon in the Antarctic. These observations also confirmed that the ozone hole is confined to the polar vortex region, where winds blow primarily from west to east and there is little mixing with air from lower latitudes until the vortex breaks up in the late spring. Other satellite measurements have shown that after the vortex breakup, air with low concentration of ozone is mixed into the midlatitudes of the Southern Hemisphere. Reductions of the total ozone column by 10–15% were observed over Australia in December 1987 as a result of this process.

Arctic. Airborne observations were also carried out in the Arctic during January and February 1989 and for the entire winter of 1991–1992. These expeditions, together with observations from the *UARS*, confirmed that elevated levels of ClO do occur in the Arctic winter. However, the polar winter vortex in the Arctic is far less isolated than that of the Antarctic. Heat transfer from lower latitudes by meteorological disturbances prevents the temperatures in the Arctic stratosphere from remaining below 195 K (−108°F) for the long periods characteristic of the Antarctic. Observations in 1992 confirmed that the stratosphere warmed to about formation temperatures for polar stratopheric clouds before sunlight returned to the Arctic. Chlorine could then return to the reservoir species (especially ClONO$_2$), so that ozone losses of only a few percent occurred.

The winter 1991–1992 was not, however, an especially cold one in the Arctic stratosphere. Temperatures cold enough for the formation of polar stratospheric clouds have been more widespread and longer-lasting in some previous winters. As the total chlorine loading of the stratosphere increases, the chances of significant ozone depletion during the Arctic winter will also increase.

Middle latitudes. In addition to the dramatic polar depletion discussed above, there is evidence of ozone depletion of a few percent at middle (but not tropical) latitudes since about 1980. This depletion has occurred primarily at altitudes below 15 mi (25 km) and is present at all seasons. It cannot be accounted for by gas-phase chemistry, nor is the seasonal distribution of this observed depletion consistent with mixing of ozone-depleted polar air into midlatitudes. Current evidence suggests that sulfate aerosols can facilitate heterogeneous chemical reactions that lead to conversion of chlorine into reactive forms in a manner analogous to that of the polar stratospheric clouds. Models that include the effects of sulfate aerosols appear to be consistent with observed midlatitude ozone losses. More research is necessary to confirm that such aerosols are in fact responsible for the observed midlatitude ozone depletion.　　　　　　　　　　　　*James R. Holton*

Stratospheric Aerosols from Volcanic Eruption

The eruption of Mount Pinatubo, a volcano at about 15°N latitude in the Philippine Islands, during June 12–16, 1991, has been classed as one of the largest of the twentieth century in terms of the expected effects on the global environment. These effects include cooling caused by reflection of sunlight into space and the depletion of ozone associated with the surface chemistry of the particles that form in the stratosphere.

Effects of volcanic eruptions. A major explosive volcanic eruption injects both dust and gases into the atmosphere, sometimes to heights of 25 mi (40 km). The most important eruptions, in terms of global atmospheric effects, occur in the tropics because the general circulation in the stratosphere at altitudes of approximately 6–25 mi (10–40 km) is from the Equator to the poles. Thus, although major volcanic eruptions have occurred in other latitudes, for example in Alaska, their effects are not felt around the world as are those of tropical eruptions. Steady, zonal winds in the stratosphere result in the circumnavigation of the globe at the Equator by the volcanic cloud in about 2 weeks. Much weaker poleward winds require months to complete global distribution.

Almost everything known about the atmospheric effects of volcanic eruptions began with observations following the relatively small eruption of Mount St. Helens, in the state of Washington, in May 1980. Prior to 1980 it was believed that the volcanic dust layer in the stratosphere caused the brilliant twilight displays through reflection of sunlight and lasted for years after a major eruption. In fact, it is now known that the so-called dust from a volcanic eruption is composed of relatively large, mainly silicate, particles that fall out of the atmosphere within months of the eruption. However, most eruptions inject large quantities of sulfurous gases, generally in the form of sulfur dioxide (SO_2), into the stratosphere, where they undergo chemical conversion to sulfuric acid (H_2SO_4) vapor, which rapidly condenses into droplets, 0.000005 in. or 0.1 micrometer in diameter, composed of H_2SO_4 and water (H_2O). It takes years for these droplets, which are the real cause of the atmospheric effects, to fall out of the stratosphere.

Ozone layer perturbation. In April 1982, the volcano El Chichón erupted in Mexico at almost the same latitude as the Pinatubo eruption and caused a stratospheric effect about 50 times larger than that of Mount St. Helens. Stratospheric ozone was observed to be nearly 10% lower than normal at northern midlatitudes following the eruption, and for several years debate centered on whether the reduction was caused by normal variations in atmospheric transport or by chemical depletion associated with the eruption. The homogeneous chemical (reacting of a molecule with another molecule in the gas phase) catalytic destruction of ozone by chlorine was known at this time, but the role of heterogeneous chemistry (reaction of a molecule with another molecule in the condensed phase) was not.

Other possible volcanic-related perturbations to the ozone layer can arise from the increased scattering of solar radiation, which perturbs the photochemistry of ozone formation, and from stratospheric heating caused by the presence of H_2SO_4/H_2O droplets. Although the droplets do not absorb solar radiation, they do absorb upwelling infrared radiation from the Earth, acting as a greenhouse gas. The heating causes the air to rise, increasing vertical circulation and resulting in the increased transport of ozone away from the region of the volcanic aerosol, so that there is an apparent reduction in ozone.

Following revelation of the Antarctic ozone hole in 1985 and the theoretical explanation, which involved heterogeneous chemistry on the surface of particles of polar clouds, it was hypothesized that a major volcanic eruption might cause global ozone depletion because volcanic eruptions also create clouds in the stratosphere and not only in the polar regions, where polar clouds form because of the low winter temperatures in the stratosphere. Laboratory measurements of relevant heterogeneous chemical reactions on the surfaces of the H_2SO_4/H_2O droplets at low temperatures revealed that although the reactions did not proceed as rapidly as in the case of polar stratospheric clouds they nevertheless could perturb ozone chemistry.

Crucial in heterogeneous chemistry in the normal stratosphere are reaction (10)–(11), where the asterisks

$$N_2O_5 + H_2O^* \rightarrow 2HNO_3 \qquad (10)$$

$$ClONO_2 + H_2O^* \rightarrow ClOH + HNO_3 \qquad (11)$$

indicate molecules that are in the condensed phase. Although reaction (10) does not involve chlorine, it effectively places normally reactive nitrogen into a nonreactive reservoir. This step reduces NO_2, formed from dinitrogen pentoxide (N_2O_5), and thus repartitions chlorine, increasing the active ClO component by reducing the effect of reaction (12), which forms

$$ClO + NO_2 \rightarrow ClONO_2 \qquad (12)$$

chlorine nitrate ($ClONO_2$). These reactions are favored

in winter, because in the presence of sunlight HNO_3 can be easily photolyzed, returning nitrogen to the reactive phase.

Laboratory studies indicate that the probability of reaction (10) occurring is about 1 in every 10 collisions of an N_2O_5 molecule with an H_2SO_4/H_2O droplet, essentially independent of temperature. Modeling indicates that reaction (10) is important even for the relatively small amount of surface area presented by the background stratospheric aerosol droplets. Reaction (11) is potentially more important for ozone depletion than reaction (10), because it is capable of producing free (reactive) chlorine in sunlight through the photolysis of ClOH while sequestering reactive nitrogen. Laboratory measurements of reaction (11) on surfaces of H_2SO_4/H_2O droplets indicate that the probability of reaction is only about 1 in 5000 at typical midlatitude stratospheric temperatures of about $-70°F$ ($-57°C$) but increases to nearly 1 in 10 as temperatures drop below $-115°F$ ($-82°C$), as is common in the polar winter stratospheres. Thus, the enhanced surface area available for chemical reactions in the stratosphere following major volcanic eruptions (up to 100 times background for Mount Pinatubo) is capable of perturbing reactive nitrogen levels globally and of reducing ozone, at least in the polar regions. It is fortunate for the stratospheric ozone layer that reaction (13),

$$ClONO_2 + HCl^* \rightarrow Cl_2 + HNO_3{}^* \qquad (13)$$

which is very effective in producing free chlorine on solid nitric acid trihydrate particles in the Antarctic winter stratosphere, is not effective for the H_2SO_4/H_2O aerosol particles. The apparent reason is that the latter are liquid droplets and HCl does not reside solely on the surface, where it could react.

Mount Pinatubo eruption. From satellite measurements of the eruption of Mount Pinatubo, it has been estimated that as much as 2×10^7 tons of SO_2 was injected into the stratosphere—about three times as much as was injected in the El Chichón eruption. However, the aerosol-droplet growth process is nonlinear; and it appears from preliminary data obtained by balloons, aircraft, and satellites that the effect of Pinatubo, in terms of the surface area of stratospheric particles in the Northern Hemisphere, will be about only 1.5 times as large as for El Chichón. Nevertheless, the droplet layer will have profound effects on stratospheric chemistry.

In contrast to the El Chichón aftermath, numerous relevant measurements are being made in the case of Pinatubo. Preliminary results relating to stratospheric ozone depletion indicate that NO_2 was reduced globally following the eruption. Record low concentration of ozone was observed over Europe during the winter of 1991–1992, but the relative magnitude of dynamical and chemical effects has not yet been sorted out. Satellite ozone detectors found that tropical ozone was significantly reduced below normal shortly after the Pinatubo eruption (**Fig. 3**). Although the lower tropical stratosphere is colder than the midlatitude stratosphere, which would increase the reaction rate

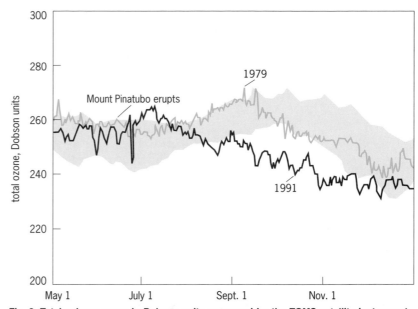

Fig. 3. Total column ozone in Dobson units measured by the TOMS satellite instrument, which detects ozone by measuring its absorption of solar ultraviolet radiation reflected from the Earth and from cloud tops, averaged globally between latitudes of 12°S and 12°N. The shaded region gives the range of minimum values observed for all years from 1979 to 1990. The black curve defines the 1991 minimum values that begin to deviate from previous years during August, following the eruption of Pinatubo. (*M. Schoeberl, NASA Goddard Spaceflight Center, Greenbelt, Maryland*)

of (11), the large amounts of sunlight present in the tropics result in the photolysis of HNO_3, freeing gas-phase reactive nitrogen so that it can return reactive chlorine to the inactive $ClONO_2$ reservoir and prevent ozone depletion. Thus, the apparent reduction of ozone in the tropics following the Pinatubo eruption was probably not the result of heterogeneous chemistry but was due to perturbations in normal ozone photochemistry and to enhanced circulation related to aerosol heating of the stratosphere.

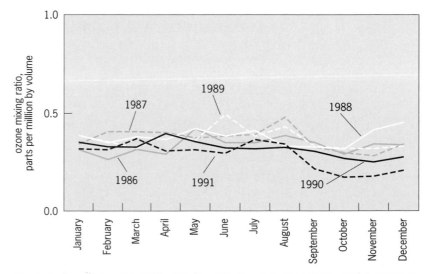

Fig. 4. Average ozone mixing ratio between altitudes of about 7 and 8 mi (11 and 13 km) measured by ozone sensors on balloons flown from the U.S. Amundsen-Scott Station at the South Pole since 1986. Note the lower values in 1991 beginning in late September. (*After D. J. Hofmann et al., Observation and possible causes of new ozone depletion in Antarctica in 1991, Nature, 359:283–287, 1991*)

The record low amount of ozone during the formation of the 1991 Antarctic ozone hole is believed to be at least partially due to not only the Pinatubo eruption but also the smaller eruption of Mount Hudson in August 1991 in Chile. The eruption produced a dense aerosol cloud in the lower stratosphere that rapidly encircled Antarctica and was observed to enter the polar vortex in September; this condition apparently caused an unusual ozone reduction during formation of the normal ozone hole, but at a lower altitude. Temperatures in the lower Antarctic stratosphere were low enough at this time to make reaction (11) very effective. As indicated in **Fig. 4**, by late October 1991 the ozone mixing ratio between altitudes of about 7 and 8 mi (11 and 13 km) at the U.S. Amundsen-Scott research station at the South Pole was reduced by 50% below values observed during the same period each year since 1986. Similar observations of unusually low amounts of ozone were made at the South Pole in September 1992. These effects are likely associated with Pinatubo aerosol that has descended into the lower stratosphere in the winter polar vortex.

The eruption of Pinatubo has presented the scientific community with a rare opportunity for a global experiment in stratospheric ozone depletion. The analysis of a great wealth of data on Pinatubo's effect on ozone during the next several years will reveal the nature of volcanic ozone depletion and allow prediction of future environmental effects of major eruptions as stratospheric chlorine levels continue to rise. The analysis will also provide information on the validity of present models that attempt to explain the Antarctic ozone hole, and the accuracy of laboratory measurements of heterogeneous chemistry on H_2SO_4/H_2O aerosol droplets.

For background information SEE ATMOSPHERIC OZONE; GREENHOUSE EFFECT; STRATOSPHERE in the McGraw-Hill Encyclopedia of Science & Technology.

David J. Hofmann

Bibliography. G. Brasseur and C. Granier, Pinatubo aerosols, chlorofluorocarbons, and ozone depletion, *Science*, 257:1239–1242, 1992; D. J. Hofmann et al., Observation and possible causes of new ozone depletion in Antarctica in 1991, *Nature*, 359:283–287, 1992; S. Solomon, Progress towards a quantitative understanding of Antarctic ozone depletion, *Nature*, 347:347–354, 1990.

Atom

The manipulation of neutral atoms with laser light has progressed dramatically in recent years. Advances in the field have included the development of new techniques to cool atoms to a few microkelvins above absolute zero; the invention of a wide variety of optical, magnetic, and magnetooptic traps; and the demonstration of atom "optical" components such as mirrors, diffraction gratings, and lenses. More recently, several of these technological advances have been combined to create new instruments such as prototype atomic clocks and atom interferometers that can serve as extremely sensitive inertial sensors.

Particle interferometers. In 1802, Thomas Young performed an experiment that showed that light passing through a double slit (**Fig. 1**) would create a diffraction pattern of alternately light and dark regions. The interference pattern is calculated by adding the electric-field amplitudes due to light from each of the slits and then squaring the result to get the light intensity. The interference of electric-field amplitudes is central to the wave properties of light.

Much later, it was shown that if the light intensity is decreased to the point where the average number of photons in the apparatus is less than one, an interference pattern can still be observed after a sufficient amount of time. Every time a photon is detected, it appears as a well-localized particle; but if all the detected particles are summed over a long period of time, a histogram of the counts will display the same interference fringe pattern. According to quantum mechanics, the photon is described as the sum of two amplitudes that account for the passage of the particle through both slits at the same time. Each photon interferes with itself.

Quantum-mechanical interference has been observed with electrons, neutrons, and most recently with atoms. By using mechanical slits made with electron-beam lithography techniques, Young's double-slit experiment was performed with helium atoms in 1991. A similar experiment using diffraction gratings instead of slits was reported at the same time.

Interferometry with light pulses. An alternate approach to atom interferometry combines well-known ideas from nuclear magnetic resonance and the fact that well-defined changes in momentum can be given to atoms by light. When a photon is used to induce a transition from one atomic state to another, its momentum is transferred to the atom. By adjusting the intensity, tuning, and duration of the applied light pulse, the excitation probability can be controlled. In the most successful version of the light-pulse interferometer, a two-photon Raman transition is used to induce the transition from one atomic state to the other state. There are a number of advantages in using this type of transition, particularly when the interferometer is based on a so-called atomic fountain. (The atomic fountain is made by laser-trapping and cooling a cloud of atoms and then launching the atoms gently upward so that gravity will cause the atoms to turn around. During the long free-fall time, the atoms are in a perturbation-free environment, and very precise measurements can be made.) First, the Raman transition is between two states of the atom that are stable against radiative decay, so that long drift times of the atoms through the interferometer are possible. Second, the frequency difference between the two laser beams, rather than the absolute frequency of the light, has to be stable during the flight time of the atoms. The resolution of the interferometer depends on the measurement of the frequency shift of a very narrow resonance line, and while there are straightforward techniques to phase-lock two frequencies relative to each other, very precise control of absolute frequency of the light is much more difficult. Finally, the two counterpropagating beams transfer twice the momentum kick of a single photon, so that the separation of the two parts of the atom is increased. Indeed, the momentum transfer Δp is given by the equation below, where h is Planck's constant and λ_1 and λ_2

$$\Delta p = \frac{h}{\lambda_1} + \frac{h}{\lambda_2}$$

are the wavelengths of the two light beams. **Figure 2**a shows how atoms are exposed to a series of light pulses. The first light pulse when the atom is in position 1 is adjusted so that the excitation probability is $1/2$. Quantum-mechanically, the atom is actually in both atomic states at once. The part of the atom in the second state has acquired an additional momentum component equal to the momentum of the absorbed photon. Thus, the light pulse is the matter analog to a beam splitter used in optics.

If a second light pulse of twice the duration of the first is applied to the atom, the probability of making the transition is unity. If such a pulse is applied after the atom has time to separate, the part of the atom in the first state (position 2) is put into the other state with a momentum kick along the direction of the propagating light beam, while the part in the second state (position 3) is stimulated to emit a photon and return to the first state. Thus, the second pulse is the matter analog to a mirror in that it reverses the direction of the atomic trajectories. A final pulse identical to the first (position 4) can then be used to put the atom back together again after the trajectories of the separated atom overlap again. By adjusting the phase of the third pulse relative to the first two, the atom can be left in either of the two atomic states. Interferometers of this type combine an interference of the internal state of the atom with an interference due to the physical separation of the atom. Figure 2b shows how the atoms separate when the pulses are parallel to the initial velocity of the atoms.

A related interferometer can be made if the atoms are illuminated by a sequence of four pulses whose duration has been adjusted so that the probability of excitation of all the pulses is $1/2$. In this configuration, there are actually two interferometer paths that take the form of trapezoids.

Both the atom interferometer of the types shown in Fig. 2a and b and the interferometer based on a four-pulse sequence have been demonstrated. In the four-pulse work, the laser induced optical transitions between a ground state and a metastable excited state, which takes a relatively long time to radiatively decay to another state.

Phase shifts and inertial sensing. It can be shown that the quantum-mechanical phase accumulated by the atom as it moves along each path of the interferometer shown in Fig. 2 is identical as long as the environments of the atom along each path are identical; that is, there are no gradients in any external field that is applied to the atom.

Once the atoms are launched and are in free fall, the laser apparatus will, in general, accelerate and ro-

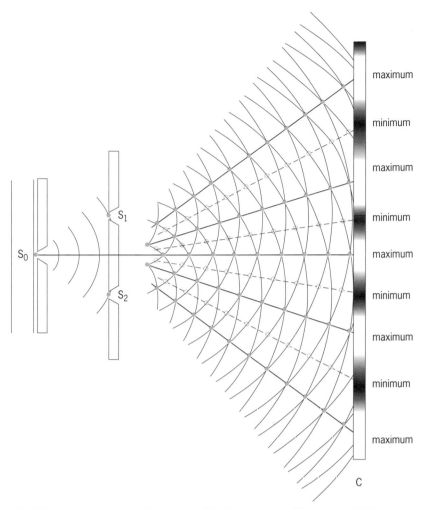

Fig. 1. Schematic diagrams of Young's double-slit experiment. The narrow slit S_0 acts as a line source of light that illuminates slits S_1 and S_2. The interference of the two sources produces a fringe pattern on screen C. (*After R. A. Serway, Physics for Scientists and Engineers, 3d ed., Saunders College Publishing, 1990*)

tate with respect to the atoms. Thus, the frequency of the atomic transition is Doppler-shifted relative to the laser, and this Doppler shift appears as a phase shift in the interferometer signal. Because the observation time in an atomic fountain can approach 1 s, the effective linewidth of the Raman transition may be as narrow as 1 Hz. However, since the Doppler shift of a Raman transition between ground states of a sodium atom will be 3.3×10^7 Hz, the frequency shift is over 10^7 times greater than the linewidth of the transition.

A version of this apparatus is shown schematically in **Fig. 3**a. Atoms from an atomic beam are first slowed by a laser beam opposing their motion and then are confined in a magnetooptic trap. Next, the magnetic field is turned off, and the atoms are further cooled to a temperature of $\sim 3 \times 10^{-5}$ K before being launched upward. The three Raman pulses are then applied to the upward-moving atoms, and the number of atoms that make the transition from the original ground state to another selected ground state is monitored by resonantly ionizing the atoms from the second ground state. The interference fringe shift shown in Fig. 3b made it possible to measure the acceleration of gravity on an atom to a precision of 3 parts in 10^8. The resolution of the experiment was limited by the degree of vibration isolation of the mirror at the bottom of the vacuum that was used to retroreflect the Raman laser beam. With a more sophisticated vibration isolation system, the resolution should be improved by another factor of 10^3.

To give a scale for the sensitivity of this measuring device, the acceleration of gravity, g, changes by parts in 10^7 because of the same gravitational forces that give rise to the ocean tides. If the apparatus were raised by 3 mm (0.1 in.), the change in g would be of the order of 1 part in 10^9. Potential applications of this ultrasensitive accelerometer include tests of the equivalence principle (for example, whether sodium atoms and cesium atoms fall at the same rate), measurements of changes in land or sea height, and oil and mineral exploration. Oil deposits change the local density of the Earth, and less sensitive mechanical g meters are currently used for exploration. The same principles used to make a g meter can also be used to design an exquisitely sensitive gyroscope, and work is under way to construct such a device.

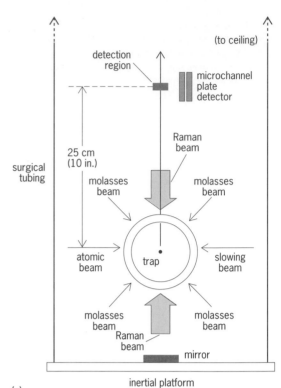

(a)

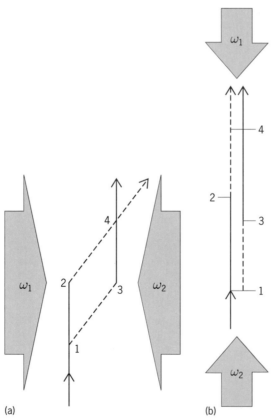

(a)   (b)

Fig. 2. Three-pulse interferometer. (*a*) Configuration with laser beams perpendicular to the initial motion of the atom. Light at frequencies ω_1 and ω_2 is turned on for brief periods of time when the atom is at position 1, when part of the atom is at position 2 and part at position 3, and when the atom is at position 4. This geometry is useful if the interferometer is to be used as a gyroscope. (*b*) Configuration in which the Raman beams of frequency ω_1 and ω_2 are aligned vertically and parallel to the initial motion of the atom. The acceleration due to gravity can be measured accurately in this geometry. (*After M. Kasevich and S. Chu, Measurement of the gravitational acceleration of an atom with a light-pulse atom interferometer, Appl. Phys. B, 54:321–332, 1992*)

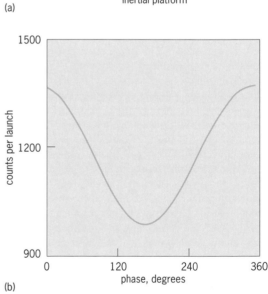

(b)

Fig. 3. Measurement of the acceleration due to gravity. (*a*) Measurement apparatus. (*b*) Example of an interference fringe observed with an interferometer. The great precision of this instrument comes from the fact that over 300,000 fringes preceded this one and that the position of the last fringe was determined to about 1/300 of the width of the last fringe. (*After M. Kasevich and S. Chu, Measurement of the gravitational acceleration of an atom with a light-pulse atom interferometer, Appl. Phys. B, 54:321–332, 1992*)

For background information SEE INTERFERENCE OF WAVES; LASER COOLING; PARTICLE TRAP; QUANTUM MECHANICS; RAMAN EFFECT in the McGraw-Hill Encyclopedia of Science & Technology.

Steven Chu

Bibliography. S. Chu, Laser manipulation of atoms and particles, *Science*, 253:861–866, 1991; S. Chu, Laser trapping of neutral particles, *Sci. Amer.*, 266(2):

70–76, February 1992; M. Kasevich and S. Chu, Measurement of the gravitational acceleration of an atom with a light-pulse atom interferometer, *Appl. Phys. B*, 54:321–332, 1992.

Audio compression

Recent advances in low-bit-rate digital audio coding technology are based on the application of advanced human auditory system models and the underlying perceptual limitations of the ear. The best low-bit-rate audio coders deliver sound quality comparable to digital audio compact disks at a data rate of 128 kilobits per second per channel, about one-fifth the data rate of compact disks. Digital signal processing and custom large-scale integrated circuit devices provide the required computational power and economical means of coder implementation for professional and consumer applications in telecommunications, broadcasting, high-definition television, optical and magnetic recording, and computer multimedia products.

A primary goal in the development of low-bit-rate audio coders is to provide for cost-effective transmission and storage of high-quality digital audio. In contrast to speech-coding techniques where intelligibility is the key goal and assumptions can be made regarding the limited variety of sounds emanating from the vocal tract, low-bit-rate audio coders must be designed to work with an unlimited variety of natural and synthesized sounds with no loss of fidelity. The common element in perception of coded sounds is the human ear, and low-bit-rate audio coders are designed to reduce the data rate by exploiting perceptual limitations in the human auditory response. Despite the complexities involved and incomplete knowledge as to exactly how the ear responds to complex audio signals, audio coder developers have recently made substantial gains in the development of mathematical models that characterize auditory limitations and provide the foundation for advanced low-bit-rate audio coder design.

Auditory masking. The critical-band concept, introduced by H. Fletcher in 1940, and the psychoacoustic principles of auditory masking are fundamental to the design of effective low-bit-rate audio coders. Auditory masking describes the phenomenon whereby a loud signal tends to hide the presence of other quiet signals nearby in frequency. Masking is a consequence of an increase in the ear's threshold of perception in the frequency range of the loud signal that leaves the ear deaf to quieter signals at or near the same frequency. Results of experiments at low signal levels reveal that a quiet signal masked by a louder tone nearby in frequency remains inaudible until the frequency spacing between them exceeds a certain threshold bandwidth, the so-called critical-band spacing.

Critical-band model. The critical-band model of the ear as a parallel bank of narrow-band filters was developed as a means of conceptualizing measured variations in threshold bandwidth or masking selectivity as a function of frequency. In the model, theoretical critical-band filters are approximately 100 Hz in width below 500 Hz, and are of constant fractional bandwidth, that is, one-fifth of center frequency, above 500 Hz. This model serves as a measure of the minimum frequency selectivity required to take maximum advantage of the ear's masking characteristics. The important principle to audio coder developers is that the ear functions much like a 25-band real-time spectrum analyzer with bandwidths and sensitivity thresholds that vary somewhat over the frequency range from 20 Hz to 20 kHz. Single-tone masking experiments indicate that masking effects are minimal within the first 30 decibels above the threshold of hearing, that is, near the quietest sound levels the ear is capable of perceiving. At progressively louder levels, however, masking occurs over a broader frequency range encompassing an increasing number of critical bands, particularly in the frequency range above the masking signal.

Masking trends. Although the body of published data on masking is derived largely from experiments involving sine waves and narrow-band noise, these data represent applicable upper limits on the thresholds of audibility with more complex audio signals, and are therefore relevant to audio coder design. Loud low-frequency signals effectively mask the presence of quieter low-frequency signals and provide a masking effect that broadens into the midfrequency range as signal loudness increases. Loud midfrequency signals best mask quieter mid- and upper-frequency signals; however, this masking effect falls off rapidly just below the frequency range of the masking signal. Loud high-frequency signals effectively mask quieter high-frequency signals but provide very little masking at middle frequencies and no masking at low frequencies. The exact degree of masking is a complex function of the amplitude and distribution of the frequency components of the audio signal, and much remains to be learned in quantifying these complex masking effects.

Transient signals. An additional form of masking that must be considered in optimized low-bit-rate audio coder design is provided by transient signals. Under steady-state signal conditions the frequency resolution of the ear is excellent, but it takes the ear a finite time to tune in to signal changes, thus implying inherent limitations in time resolution. However, actual measurements in the key time interval just prior to the onset of high-level transient test signals confirm a time resolution of under 5 milliseconds. Interestingly, masking of quieter signals can occur before, during, and after the occurrence of a transient signal. Pretemporal masking, which occurs just prior to the transient, is strongest up to 10 ms before the transient. The masking effect, understandably, is strongest during the transient and falls off over a period of 50–200 ms thereafter.

Low-bit-rate audio coding. The fundamental process of low-bit-rate coding includes generation of a frequency-domain representation of the audio signal, variable quantization of the signal's frequency components to a reduced accuracy based on an auditory masking model, allocation of bits to meet the varying demands of the quantizer, and resynthesis of an approximation of the original time-domain waveform following transmission or storage of the coded data. Generation of the frequency-domain representation of the audio signal is accomplished through use of a multifrequency-band filter bank. Two different frequency-division techniques, one based on the discrete Fourier transform and one based on polyphase digital filters have emerged as popular methods. Digital signal processing and large-scale integration technology are employed to implement the filter bank, frequency analysis, masking threshold calculation, quantizer, and bit-allocation functions.

Audio signals consist of nearly stationary signals and of transients that change rapidly with time. Signals that change slowly in time are best coded by using a filter bank with a high degree of frequency selectivity such that the spectrum of coding errors may be confined to the spectral region of the signal and masking may be exploited to best advantage. Transient signals, however, are best coded by using a filter bank that has a time resolution equal to that of the ear, thus avoiding coding errors that spread in time beyond the audibility limits set by the ear's temporal masking characteristics. As excellent frequency selectivity and short-time resolution are mutually exclusive requirements, filter-bank design can involve a compromise between time and frequency resolution with sufficient bit rate

allocated to meet temporal and spectral masking constraints, or a filter bank with time-varying optimization for either time or frequency resolution depending on the characteristics of the signal to be coded. Both techniques are employed in low-bit-rate audio coders.

A conceptual block diagram of a multifrequency-band low-data-rate audio coder is shown in **illus**. *a*. A sampled and quantized time-domain input signal consisting of low-, middle-, and high-frequency components, *A, B,* and *C,* is converted to a frequency-domain representation by using an appropriate filter-bank technique with critical-band-frequency resolution. The frequency-domain representation of the signal is shown in illus. *b*. Once the frequency components of the audio signal are identified, the masking thresholds are estimated on a band-by-band basis by direct calculation or by comparison with a preprogrammed model of the ear. The masking thresholds are established based on the loudest signal components present in each of the frequency bands, and other nearby signal components are then analyzed to determine whether the loudest signals provide sufficient masking to render quieter signals within that same critical band inaudible. Once this determination is made, the bit-rate reduction portion of the process can be completed. Bit-rate reduction involves quantizing the

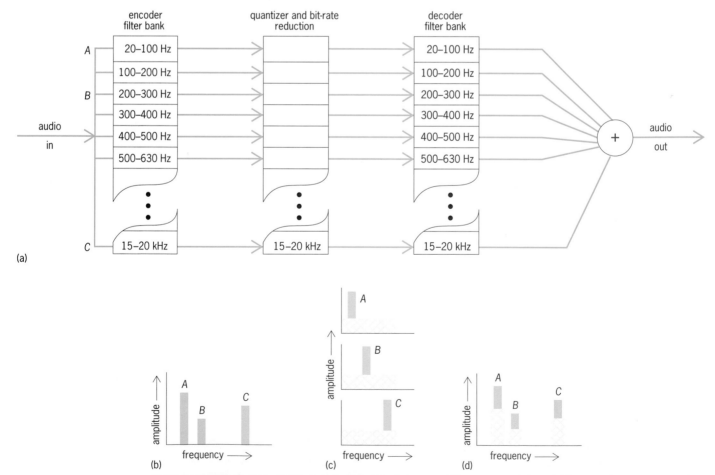

Multifrequency band coder. (*a*) Block diagram and (*b–d*) outputs of (*b*) encoder filter bank (frequency-domain representation of the audio signal), (*c*) quantizer and bit-rate reduction process, and (*d*) decoder filter bank. Cross-hatched areas in *c* and *d* represent quantization noise.

frequency components in each of the individual filter bands with sufficient accuracy to keep the quantization noise just below the calculated in-band masking thresholds, and amplitude scaling of the signals to normalize their peak levels to make optimum use of the dynamic range of the digital signal processor.

Although a low-bit-rate representation of the signal has been created, a wide-band noise component is introduced as a result of quantizing the frequency-domain signals. The quantization noise added to the signal components, just below the in-band masking thresholds, is indicated in illus. *c*. Based on masking criteria discussed above, this wide-band noise component would not be effectively masked by the signal components if no further action was taken, and the desired high fidelity would not be achieved. However, an additional process takes place in the decoder after transmission or storage of the coded representation of the signal, where a filter bank identical to that used in the encoder is employed. Quantized frequency-domain data are received by the decoder and passed through the filter bank. This refiltering process leaves the frequency components of the signal intact, while it tightly constrains the unwanted quantizer-introduced noise to a narrow frequency range below the masking threshold (illus. *d*). As long as the filter-bank frequency selectivity and out-of-band signal rejection are sufficient, masking thresholds may be conservatively applied with sufficient bit rate allocated to keep quantizer noise below audible limits, and the reconstructed time waveform at the decoder output will sound subjectively equivalent to that of the input signal.

The design of high-quality low-bit-rate audio coders involves a trade-off between the degree of bit-rate reduction and subjective audio quality. Systems currently available achieve near-perceptual transparency for 20 Hz to 20 kHz bandwidth audio signals at a data rate of 128 kilobits per second per channel. Coder development work continues toward total transparency at current data rates and equivalent sound quality at lower data rates.

For background information SEE COMPACT DISK; INFORMATION THEORY; MASKING OF SOUND in the McGraw-Hill Encyclopedia of Science & Technology.

Steven E. Forshay

Bibliography. M. Bosi and G. Davidson, High quality, low-rate audio transform coding for transmission and multimedia applications, *93d Convention of the Audio Engineering Society*, San Francisco, October 1–4, 1992; G. Stoll and Y. F. Dehery, High quality audio bit-rate reduction system family for different applications, *Proceedings of the IEEE International Conference on Communications*, Atlanta, April 1990.

Barnacle

Barnacles are the only truly sessile crustaceans. With the exception of a few parasitic forms, barnacles are suspension feeders, securing food by filtering small particles from the ambient water by using their six pairs of thoracic legs (cirri) with attached setae as a net or sieve (cirral basket). Common acorn barnacles feed on particles that range in diameter from a few micrometers (phytoplankton) to several hundred micrometers (invertebrate larvae, copepods, and other planktonic crustaceans).

Active versus passive suspension feeding. Acorn barnacles are highly responsive to local flow environment. Unless ambient currents are very slow, they orient the cirral basket by rotating it at its base so that the concave side faces upstream. A few acorn barnacles such as *Balanus nubilis* simply hold the cirri out in the ambient currents. Such passive suspension feeders presumably invest little metabolic energy in moving water through the cirral basket, and are entirely dependent on ambient currents for feeding. Most acorn barnacles, however, can actively sweep the cirral basket through the water. Although active suspension feeding is probably more energetically expensive than passive suspension feeding, the animal need not depend on the magnitude of ambient currents for feeding.

Most intertidal and shallow-water acorn barnacles exhibit plasticity in feeding behavior. When ambient currents are slow, the barnacles actively suspension-feed by sweeping the cirral baskets through the water. When current speed exceeds some threshold value (typically a few centimeters per second), the animals switch to passive suspension feeding, everting the cirral basket, orienting it to the flow, and then holding it stationary as the water passes through. This behavioral plasticity presumably allows these barnacles to feed with a minimum expenditure of energy in higher flows but permits active feeding (at greater metabolic expense) in low flows.

Barnacle clusters. Acorn barnacles may occur in densities as high as 100,000 animals per square meter, and even when densities are lower they often occur in distinct clusters. These aggregations typically exhibit hill-shaped or hummocky forms, with highly elongated individuals in the center of the hummock and individuals of decreasing height moving out symmetrically toward the periphery. It has been suggested that the regular occurrence of these multianimal forms results either from the physical constraint to lateral growth of the central animals by the pressure of immediate neighbors or from differential feeding success of barnacles in different locations in the cluster. Differences in feeding success are likely to be due to modifications of the flow environment around individuals, modulated by behavioral plasticity of the animals.

The latter hypothesis can be tested by using physical models of barnacle aggregations to determine the effects of aggregate position on feeding success and by directly observing particle-capture success of living barnacles in clusters. Both lines of evidence are necessary to isolate the hydrodynamic effects of position in an aggregation from behavioral changes by the animals.

Physical models and direct observation. Barnacles engaged in passive suspension feeding can be modeled by using small plastic tubes with a triangular piece of fine-mesh plankton netting (representing the cirral basket) on one end; the feeding success of these mock barnacles can be measured by the capture of particles suspended in the water flowing past the model. Clusters of mock barnacles of various profiles can be produced simply by assembling groups of individuals. The **illustration** shows the relative capture success of mock barnacles in various positions in clusters of four different shapes: a flat array (the cirral baskets of all mock barnacles are at the same height), a symmetrical array (the peak in height lies in the center of the cluster), and two asymmetrical arrays (the peak in height is shifted toward either the upstream or downstream end of the cluster). For maximum feeding success it is advantageous for a barnacle to be located either at the leading edge of, or at the highest point in, a cluster.

The same pattern of relative capture success occurs in clusters of living barnacles feeding on live brine shrimp larvae, known as nauplii, from a unidirec-

tional current. Animals on the upstream edge of, or at the highest point in, the cluster capture significantly more brine shrimp than animals located downstream of the peak. Animals at the upstream cluster edge or peak feed exclusively by passive suspension; animals downstream (and therefore in a turbulent, low-velocity eddy) feed exclusively by active suspension feeding. Because downsteam animals are actively suspension-feeding and hence expending energy, net energy intake is further reduced over that of upstream and peak animals.

Evolution of cluster shape. If growth is correlated with feeding success, barnacle clusters should change shape over time. In a unidirectional flow, a flat array of barnacles will develop a peak at its upstream end. Both a symmetrical cluster and a cluster with its peak at the downstream end will transform into clusters that have a peak at the upstream end, as increased feeding success of animals upstream of

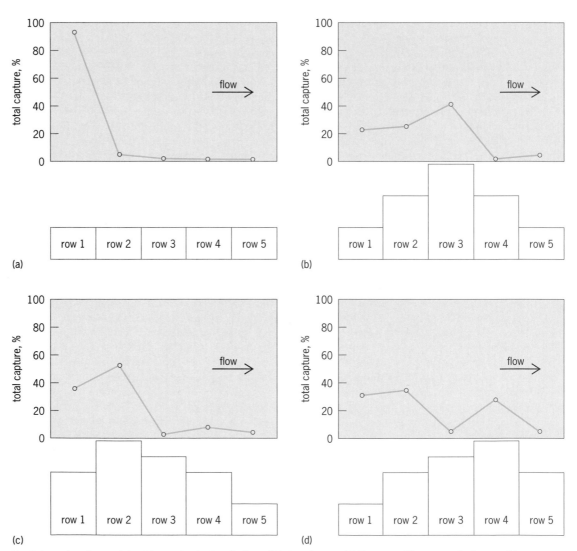

Particle capture by models of barnacle clusters in four different shapes: (*a*) flat array, (*b*) symmetrical array, (*c*) upstream asymmetrical array, and (*d*) downstream asymmetrical array. Values given for each row are the percentage of the total number of particles captured by the entire model. In general, the most upstream row and the row at the peak of the cluster (if there is a peak) capture the greatest percentage, and the rows downstream of the peak capture the smallest percentages, of particles. Since flow is from left to right, row 1 is the most upstream row. There is no peak in *a*, but for *b*, *c*, and *d* the peak is, respectively, row 3, row 2, and row 4.

the peak causes greater growth. However, strictly uni-directional flows are rare in the marine environment, particularly in the shallow waters where barnacles are most common. Rather, flows oscillating in direction with time scales from seconds (waves) to hours (tidal currents) are the norm. Even for high-frequency oscillations (as great as 0.65 cycle/s), barnacles are able to reorient the cirral basket rapidly enough to passively suspension-feed in the main flow. In an oscillating flow, an upstream animal becomes a downstream animal and a downstream animal becomes an upstream animal when the flow direction reverses; an animal on the peak of a cluster, however, remains on the peak. Peripheral animals will thus alternate between high and low capture success as flow direction reverses, while animals at the peak will always have high particle-capture success. Since growth occurs on a time scale much longer than the natural frequencies of changes in flow direction, the average feeding success (and thus growth) of animals at the cluster periphery will always be depressed relative to that of animals at the peak. Thus, in an oscillating flow any aggregation of barnacles that is not precisely flat should evolve (by differential growth of individuals) into a symmetrical cluster.

Symmetrical clusters are the form observed commonly in intertidal environments where oscillating flows predominate. Clusters of barnacles on ship hulls, where flow is predominantly unidirectional, presumably are asymmetrical with the peak of the cluster located near the upstream (bow) end. Unfortunately, there are no published descriptions of the shapes of barnacle clusters on ship hulls, nor is it clear that there is time for distinct clusters to develop between times when the hulls are scraped to reduce drag.

Although cluster form in acorn barnacles can be explained as the natural consequence of the interaction of hydrodynamics and feeding success, an additional advantage may accrue to the symmetrical cluster form. Unlike most sessile invertebrates, egg fertilization in barnacles occurs by direct insemination rather than by release of the sperm freely into the water. Barnacles thus must be in proximity for successful reproduction. Given that some form of aggregation must occur, it is interesting that model symmetrical clusters have 50–300% higher particle-capture success than other potential cluster forms. An individual barnacle in a symmetrical cluster increases its chances of securing more food and thus increasing its reproductive output.

For background information SEE BALANOMORPHA; BARNACLE; CIRRIPEDIA in the McGraw-Hill Encyclopedia of Science & Technology.

Michael LaBarbera

Bibliography. H. Barnes and H. T. Powell, The development, general morphology and subsequent elimination of barnacle populations, *Balanus cretanus* and *B. balanoides*, after heavy initial settlement, *J. Anim. Ecol.*, 19:175–179, 1950; J. Pullen and M. LaBarbera, Modes of feeding in aggregations of barnacles and the shape of aggregations, *Biol. Bull.*, 181:442–452, 1991; G. C. Trager, J.-S. Hwang, and J. R. Strickler, Barnacle suspension-feeding in variable flow, *Mar. Biol.*, 105:117–127, 1990.

Bioinorganic chemistry

Uptake and storage genes control the critical flow and use of iron (Fe) and copper (Cu) that living organisms exploit for respiration, cell division, photosynthesis, and nitrogen fixation. To maintain the balance necessary as the living organism is exposed to fluctuations of iron and copper, the genes encoding the uptake or storage proteins have evolved to sense the environmental concentrations of metal. The result is production of proteins to overcome iron or copper deficiency, on the one hand, and to protect against toxic effects of iron or copper excess, on the other. Protein production can be controlled at several sites, including transcription of the genes to messenger ribonucleic acid (mRNA), translation of the mRNA into protein, and protein turnover (breakdown/stabilization). Transcription can be regulated by both copper and iron. Examples include the genes for metallothionein (Cu) in yeast and animals; for ferritin (Fe); for plastocyanin (Cu), a protein that is important in electron transfer and photosynthesis in higher plants; and for iron uptake (the *Fur* gene) in bacteria. Repressor proteins, which can bind both to metals and to specific sequences in the iron- or copper-sensitive genes, transmit the metal signal as an altered interaction with deoxyribonucleic acid (DNA) in the presence of metal. Copper and iron also stabilize the uptake or storage proteins after translation.

Regulation of genes for metal metabolism is very sensitive to environmental fluctuations because of the diseases that result from either excess or deficiency of metals. Unraveling the different molecular types of gene regulation by metals is facilitated by the gene sensitivity and provides a model for other genetic signals.

Biological behavior of iron and copper. Iron and copper are needed in small amounts by animals, plants, and bacteria. Proteins with iron and copper are used in respiration and photosynthesis; the metals are in an ionic state bound to the protein by amino acid side chains, or in organic cofactors. Many biological reactions can be catalyzed by either iron or copper proteins, but iron proteins are much more abundant in animals and plants. Iron proteins are keys to the synthesis of DNA and to nitrogen fixation, the conversion of atmospheric nitrogen to ammonia that is characteristic of legumes. An important copper protein is human dopamine beta-hydroxylase, important in the production of neurotransmitters. Deficiencies of either metal cause diseases such as anemia in humans and chlorosis in plants. Excesses of iron or copper are toxic, largely because of reactions with oxygen species that produce free radicals such as superoxide [$O_2^-\cdot$; reaction (1)]

$$O_2 + Fe^{2+} \text{ or } Cu^+ \rightarrow O_2^-\cdot + Fe^{3+} \text{ (or } Cu^{2+}) \quad (1)$$

and hydroxyl [HO; reaction (2)]. The free radicals can

$$H_2O_2 + Fe^{2+} \text{ or } Cu^+ \rightarrow HO\cdot + OH^- + Fe^3 \text{ (or } Cu^{2+}) \quad (2)$$

damage DNA and membranes.

Most contemporary organisms have evolved genes for proteins that manage the incorporation and storage of iron and copper. The genes allow the cells to maintain relatively constant, safe amounts of the metals during environmental fluctuations. Changes in the concentrations of iron and copper alter the action of the genes through gene regulation.

Gene regulation. This process turns genes on or off to select the genetic information to be used. Stretches of DNA-encoding functional proteins or molecules of ribonucleic acid are known as genes. At any given point in time, only a subset of the genes are expressed at levels required for so-called housekeeping (specialized functions). When the environment changes or as development proceeds, groups of genes are activated or repressed in a precisely controlled fashion. Gene regulation can occur at the DNA blueprint or in the RNA working plan.

The major route for transfer of genetic information in the cell has two parts. The first is DNA to mRNA (transcription and processing); the second is mRNA to protein (translation, processing). Transcription is the synthesis of an RNA copy of the gene (DNA) sequence. RNA processing includes modification of bases, excision of unneeded segments and splicing, and transport to the site of translation. Translation uses the triplet nucleotide codons to align amino acids, resulting in the sequential polymerization of the aligned amino acids to produce the protein. Protein processing also includes modification of amino acids, binding of cofactors (often vitamins or metals), transport to the functional site in the cell, or secretion to the outside. Currently, copper and iron are known to regulate both transcription of mRNA and posttranslational stability of proteins for iron uptake or storage. Iron is also known to regulate translation and stability of mRNA.

Iron as a modulator. Iron acts as a modulator in both storage and uptake of iron.

Iron storage genes. These are regulated by environmental iron. The genes probably date to the time when the oxygen content of the terrestrial atmosphere began to increase because of photosynthesis. Oxidized iron (Fe^{3+}) is extremely insoluble in the neutral, salty water of cells and organisms. To maintain the iron in solution from one day's worth of old red cells, for example, a human would need to drink 10^{13} gallons of water or 5 gallons of orange juice. Instead, iron is concentrated and stored as ferritin, a large, complicated protein found in most cells. Sequences conserved (DNA, RNA, and protein) in animals, plants, and bacteria indicate that ferritin is an ancient protein. Iron increases the amount of ferritin in both plants and animals.

Recent investigations show that the same end result of ferritin gene regulation occurs by very different paths in plants and animals. In plants, iron signals target ferritin genes and increases the amount of ferritin gene transcription and RNA processing. The extra ferritin mRNA produces extra ferritin protein for iron storage. Detailed mechanisms of how the iron acts on ferritin genes have not yet been characterized, but the

control of copper uptake genes described below is a good model.

In animals, iron mainly controls ferritin mRNA translation; this process has been studied extensively not only to understand how metals regulate gene expression but also to understand how any mRNA functions. In the absence of iron, a regulator protein binds to a specific sequence [the iron regulatory element (IRE)] in ferritin mRNA, and translation is blocked, a process known as negative control. In the presence of iron, the regulatory region without the regulator protein stimulates translation and protein synthesis; this step is known as positive control. Iron can also stabilize both the iron storage and transport proteins.

Iron uptake genes. In animals these are regulated by iron in concert with the iron storage genes. Excess iron leads to decreased transferrin receptor protein (TfR), decreased iron uptake, and increased ferritin to increase stored iron. Iron deficiency leads to increased TfR protein and decreased ferritin. The main site of iron regulation of TfR is mRNA, as it is for ferritin; but the presence of iron results in degradation of TfR mRNA in contrast to activation of ferritin mRNA (see **illus.**). While it is good physiology for excess iron to decrease iron uptake and increase iron storage, it is not obvious how the iron works. Although the mechanism has not been completely elucidated, a great deal has been learned recently. When the mRNAs (and DNA) for ferritin and TfR are compared, the almost identical sequences are observed in the regulatory regions of the two mRNAs. Since the proteins and the coding region are different, the similarity of the regulatory sequences is emphasized; the conserved sequence, the iron regulatory element, is 28 nucleotides long and folds into a distinctive three-dimensional structure. Also, a specific regulator protein (IRE-BP) binds to the iron regulatory element (in analogy to regulation of DNA), and ferritin mRNA is bent into a shape that is not translated; presumably the TfR mRNA is stabilized.

Iron must change the IRE-BP binding to mRNA to allow translation (ferritin) or degradation (TfR), but the iron signal itself is a mystery. One clue may be the similarity of the IRE-BP to another protein, aconitase, that folds around a cluster of iron and sulfur. Forming the iron cluster on the IRE-BP may free ferritin mRNA for translation. Another clue may be the binding of heme (an iron porphyrin) to the IRE-BP, which also frees ferritin mRNA for translation and possibly TfR mRNA for degradation.

How the same RNA/protein complex produces two different effects is another puzzle (see illus.). Sequences that neighbor the iron regulatory elementss known to be different in ferritin and Tfr mRNAs, as well as possibly in other proteins, may contain the solution to the puzzles. In bacteria, in contrast to animals, iron uptake genes have been studied most extensively under conditions of iron deficiency. Iron regulates transcription of a group of genes that synthesize iron chelators (strong iron-binding compounds) and receptors for taking up the iron chelates. The regulation is indirect, since transcription of the chelator or siderophore genes are all controlled by a master

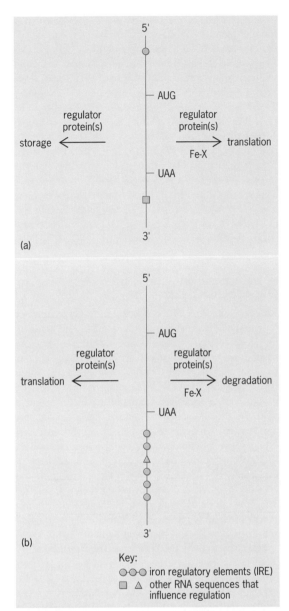

Effects of iron on the fate of (a) ferritin mRNA and (b) transferrin receptor mRNA. Fe-X = iron signal, A = adenine, G = guanine, U = uracil. (After E. C. Theil, Regulation of ferritin and transferrin receptor mRNAs, J. Biol. Chem., 265:4771–4774, 1990)

gene, *Fur*. (A siderophore is a molecular receptor that binds and transports iron.) Iron deactivates *Fur* gene transcription by binding to a specific DNA binding protein or repressor that blocks transcription. In the absence of iron, the *Fur* gene is transcribed, the mRNA is translated, and the *Fur* gene product activates transcription of siderophore and iron uptake genes. The medical importance of siderophores, which make infecting bacteria more virulent, accounts for the intense interest and elegant investigations of iron deficiency in bacteria.

Almost nothing is known about iron uptake genes in plants.

Copper as a modulator. It makes sense that when cells respond to changes in environmental copper levels they will alter the expression of genes that encode products involved in either the metabolism of copper or protection of the organism from copper toxicity.

Metallothionein genes. A particularly striking example of copper-regulated gene transcription is found in the metallothionein (MT) genes; they encode low-molecular-weight, cysteine-rich metal-binding proteins found in almost all eukaryotic and some prokaryotic organisms. Metallothionein proteins bind metals, such as Cu, cadmium (Cd), and zinc (Zn), efficiently and tenaciously through thiolate clusters, a critical aspect of the role of metallothioneins in metal detoxification. The stoichiometry of bound metals is dependent on the preferred coordination geometry for each metal; therefore, metals such as Cd and Zn, which are often coordinated tetrahedrally, are bound with a lower stoichiometry per metallothionein molecule than Cu, which is primarily trigonally coordinated.

The methallothionein genes of yeast have provided an excellent system with which to investigate Cu-activated gene transcription. In bakers' yeast, *Saccharomyces cerevisiae*, the single metallothionein gene (designated *CUP1*) is transcriptionally regulated by Cu through the action of two major elements: a cysteine-rich nuclear protein called ACE1, which is a DNA binding transcription factor, and a stretch of *CUP1* promoter DNA known as upstream activation sequence (UAS$_{CUP1}$), which harbors four specific binding sites for ACE1 protein. When yeast cells are exposed to high environmental levels of Cu, the Cu is bound by ACE1 protein, which is activated through a ligand-induced conformational change to bind the four target sites in UAS$_{CUP1}$. This conformational switch, which is inferred on the basis of the observation that protease-sensitive ACE1 apoprotein becomes highly resistant to protease as a metalloprotein, is driven by the process of Cu coordination. Recent extended x-ray absorption fine-structure (EXAFS) investigations have demonstrated that Cu is bound as Cu(I) to cysteine thiols through trigonal coordination at a stoichiometry of six or seven Cu atoms per protein molecule, with the resultant structure forming a Cu polynuclear cluster. It has been noted that Cu is bound to the yeast metallothionein protein with an almost identical stoichiometry and as a Cu(I) polynuclear cluster, suggesting that perhaps ACE1 was evolutionarily derived from metallothionein protein as the requirement arose to tightly regulate the levels of this protein in response to copper. In this copper-dependent transcription activator from *S. cervisiae*, the 225-amino-acid polypeptide contains 12 cysteine residues in the amino-terminal copper-activated DNA binding domain that trigonally coordinates Cu(I). The carboxyl-terminal domain contains a high proportion of acidic amino acids and is required for the transcription activation process.

In the absence of added Cu, *CUP1* mRNA levels are low but clearly detectable. These basal levels are due to many factors, including low-level activation by Cu-ACE1 formed from existing intracellular Cu and other transcription factors that are unaffected by Cu but that also activate *CUP1* transcription. Once Cu-ACE1 is bound to UAS$_{CUP1}$, initiation of the transcription of the *CUP1* gene is activated 10- to 50-fold over basal levels due to the interactions of Cu-ACE1, transcription factor IID (TFIID), RNA polymerase II enzyme, and other (as yet unidentified) ancillary transcription factors. Since metallothionein is an abundant cellular Cu-binding protein, as metallothionein accumulates to high levels, it is an effective competitor for available copper with ACE1; therefore, Cu-dependent ACE1-mediated transcription is reduced, but it remains at significant levels over the generation time of a yeast cell culture. In fact, the presence of metallothionein protein in yeast cells has also been shown to govern the availability of Cu for ACE1-mediated expression at normal levels of environmental copper. Therefore, in yeast, Cu-regulated gene transcription is a dynamic process that involves the interplay of a Cu-binding gene product (metallothionein) with the Cu-binding transcriptional activator ACE1. It is also interesting to note that Cu-ACE1 activates transcription of the yeast *SOD1* gene, encoding a (Cu,Zn)-superoxide dismutase known to destroy superoxide radicals (O_2^-). Because Cu interacts with superoxide anion to generate more toxic radical species, perhaps Cu-ACE1 induction of *SOD1* transcription is another line of defense against copper toxicity.

Eukaryotic genes. Due to their experimental tractability, lower eukaryotic organisms have provided a great deal of insight into the levels and mechanisms by which Cu regulates gene expression. In the eukaryotic algae *Chlamydomonas reinhardtii*, a copper-containing plastocyanin and a heme protein cytochrome c6 are interchangeably used as photosynthetic electron carriers. In this organism, Cu reciprocally regulates the levels of these gene products through two mechanisms. In the absence of Cu, the *cyt c6* gene is transcriptionally derepressed several hundredfold, which allows accumulation of cytochrome c6 for use in electron transfer. Under these same conditions, plastocyanin is rapidly turned over because of the instability of the apoprotein. Cells grown in the presence of Cu rapidly and completely repress *cyt c6* gene transcription with a concomitant stabilization of plastocyanin. Therefore, as a consequence of available copper, *C. reinhardtii* employs transcriptional and posttranslational mechanisms to regulate the availability of critical cellular components needed for energy generation.

Control of copper. Due to the essential yet toxic nature of copper, its levels in living organisms must be precisely controlled. At present, changes in gene regulation at the levels of transcription and protein stability are known to play crucial roles in this delicate homeostatic process. Future work in this area is likely to reveal the intricate molecular details for these forms of regulation and may well reveal that copper regulates gene expression at many biochemical levels.

For background information SEE DEOXYRIBONUCLEIC ACID (DNA); GENE; IRON METABOLISM; PROTEIN; RIBONUCLEIC ACID (RNA); SUPEROXIDE CHEMISTRY in the McGraw-Hill Encyclopedia of Science & Technology.

Dennis Thiele; Elizabeth C. Theil

Bibliography. S. Kaptain et al., A regulated RNA binding protein also possesses aconitase activity, *Proc. Nat. Acad. Sci. USA*, 88:10109–10113, 1991; A. M. Lescure et al., Ferritin gene transcription is regulated by iron in soybean cultures, *Proc. Nat. Acad. USA*, 88:8222-8226, 1991; M. C. Linder and C. A. Goode, *The Biochemistry of Copper*, 1991; S. Merchant, K. Hill, and G. Howe, Dynamic interplay between two copper-titrating components in the transcriptional regulation of cyt c6, *EMBO J.*, 10:1383–1389, 1991; E. C. Theil, Regulation of ferritin and transferrin receptor mRNAs, *J. Biol. Chem.*, 265:4771–4774, 1990; D. J. Thiele, Metal-regulated transcription in eukaryotes, *Nucl. Acids Res.*, 20:1183–1191, 1992; G. Winkelmann et al. (eds.), *Iron Transport in Microbes, Plants and Animals*, 1986.

Biomagnetism

Recent technological advances in the recording of magnetic fields provide a unique means of monitoring the electrical activity of biological processes, ranging from those of the single cell to large groups of cells in heart, muscle, nerves, and the human brain. These biological signals associated with the movement of ions can be 10^9 times weaker than the Earth's magnetic field. New devices, which include amplifiers operating at room temperature as well as sensors incorporating superconducting quantum interference devices (SQUIDs) operating in liquid helium at a temperature of only 4 K ($-452°$F), enable the imaging of the currents that are the sources of the magnetic fields. The term magnetic source image (MSI) has come to denote the kind of representation that relates biological activity, as manifested by magnetic fields, to anatomy.

Multichannel SQUID measurements. A relationship between electric currents and activity in nerve and muscle tissue has been appreciated since the experiments of L. Galvani and A. Volta in the eighteenth century. In the 1940s and 1950s, A. Hodgkin and A. Huxley produced the first mathematical model

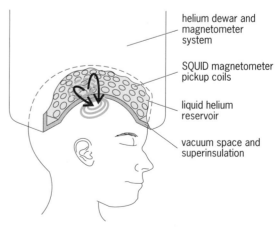

helium dewar and magnetometer system

SQUID magnetometer pickup coils

liquid helium reservoir

vacuum space and superinsulation

Fig. 1. Multichannel SQUID magnetometer positioned over the head of a subject to record the magnetoencephalogram (MEG). The arrows represent the magnetic field associated with cortical activity.

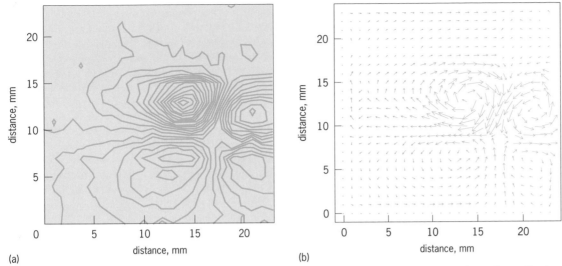

(a)   (b)

Fig. 2. Magnetic images of action currents propagating in a slice of cardiac tissue. (*a*) Isofield contours 6 ms after a stimulus, with 25-picotesla contour spacing. (*b*) Arrows showing the strength and direction of the currents in the slice, as determined from the magnetic field above the slice. (*After D. J. Staton, R. N. Friedman, and J. P. Wikswo, Jr., High resolution SQUID imaging of octupolar currents in anisotropic cardiac tissue, IEEE Trans. Appl. Superconduct., 3:1934–1936, 1993*)

of how a nerve propagates electric impulses. Although it has been known that electric currents produce magnetic fields since H. C. Oersted's discovery in 1820, the magnetic fields produced by the pulses of electrical current within the heart and brain were not detected until the 1960s. Only in the 1970s, with the application of SQUID sensors, was sufficient sensitivity available for quantitative measurement of the magnetocardiogram (MCG) from the heart and the magnetoencephalogram (MEG) from the brain. The early SQUID instruments had only a single sensor, thereby limiting magnetocardiogram and magnetoencephalogram studies primarily to repetitive signals recorded sequentially at different locations. Since the early 1980s, successively larger arrays have been developed. In 1992, one with 122 sensors was successfully introduced, and systems with 200 or more independent SQUIDs are under development (**Fig. 1**). Since the scalp and skull impose a barrier greater than 1 cm (0.4 in.) thick, it is sufficient to space the sensors for magnetoencephalogram recordings at intervals of several centimeters across the scalp.

Magnetoencephalogram recordings of the evolving field patterns provide insight into how the human brain responds to a wide variety of sensory stimuli, such as touch, sound, and light. Since neuronal sources can be readily located in three-dimensional space from the information provided by their magnetic field patterns, it is possible to determine where specific brain functions are located. For example, measurements of the acoustically evoked neuromagnetic field were used to construct a tonotopic map on the human cortex, wherein a different area of the auditory cortex responds to each stimulus frequency. It was found that the spacing between tones follows the logarithm of the sound frequency, just like the keys on the piano.

Studies of nerve and muscle bundles. It became clear in the 1980s that an understanding of the magnetocardiogram and the magnetoencephalogram requires knowledge of how the magnetic field

is produced by individual cells. However, the pickup coils of SQUID instruments devised for studies on humans are ill-suited for studies on isolated nerve and muscle tissue. Pickup coils of 2–3-cm (approximately 1-in.) diameter used in typical SQUID magnetometers were too large and could not be placed sufficiently close to resolve the fine spatial details of the cellular magnetic field. For the first successful measurement of the magnetic field from an isolated nerve bundle, the nerve was threaded through a small, ring-shaped sensing coil connected to a SQUID. Subsequently, it was found that the signal was sufficiently strong that the SQUID could be replaced by a specially designed amplifier operating at room temperature. Recent improvements have led to a device that can be used in an operating-room environment to assess the condition of human nerves. This technique for magnetic recording has several advantages over traditional cellular electric measurements: impalement by an electrode to detect a voltage or direct physical contact with the biological tissue is unnecessary; movement and stimulation artifacts are reduced; scanning is easily accomplished; and, most importantly, the currents inside the cell can be measured accurately.

MicroSQUID tissue studies. Recently, a new class of high-resolution SQUID magnetometers with millimeter-diameter pickup coils has been introduced. With an innovative cryogenic design, an array of such coils can be placed within a millimeter or two of the warm biological tissue. With such a microSQUID, it is possible to record and localize magnetic signals from a single motor unit (muscle fibers that fire as a group) in the thumb. This magnetic source image will make it possible to measure the number of fibers that are in a group without having to insert an electrode needle in the muscle while asking the patient to bend the thumb, which is an uncomfortable procedure.

Most importantly, such a system can obtain detailed

magnetic source images of current distributions in tissue, as is shown for an isolated slice of cardiac tissue in **Fig. 2**. The swirling current patterns are consistent with theoretical predictions. The spatial and temporal variations in the patterns are large, yet highly reproducible, and show that the configuration of the currents is determined by the anisotropy of the electrical conductivity of the cardiac tissue. Such images may allow quantitative determination of how changes in this anisotropy might affect the generation of cardiac arrhythmias, and may provide valuable, previously unobtainable biophysical information about cardiac tissue.

Combined MSI-MRI brain studies. On the larger scale of the intact human brain, magnetic source imaging has taken a major step forward with the marriage of the new, multisensor SQUID magnetometer systems and the well-established technique of magnetic resonance imaging (MRI). It is possible to make an accurate, noninvasive identification of the anatomical regions that give rise to electrical responses lasting only a few milliseconds. The combination of magnetic source imaging and magnetic resonance imaging led to the recent discovery that the brain's strongest response to a sound, once thought to originate in the primary auditory cortex (on the upper surface of the temporal lobe, above the ear), also has a contribution from a second source, in the association area of the temporal lobe (**Fig. 3**). Measurements can be made over the scalp to monitor each source individually. It was found that the primary cortex and assocation cortex respond to different physical aspects of a sound. The association cortex responds preferentially to the onset of the tone, whereas the primary cortex responds to both the onset and the cessation of the tone. Moreover, by monitoring how the response strength to a tone is reduced by an identical tone played at various earlier times, it was possible to determine the duration of the sound memory in each of the two auditory areas. The cortical response to a tone causes a sustained change in the pattern of synaptic connections between cells, and this activation trace was found to decay exponentially in time. The corresponding lifetime of the trace was several seconds longer in the association cortex than in the primary cortex. For the first time a characteristic lifetime was identified for activation traces in the human brain. Behavioral studies show that a subject's memory for the loudness of a sound also decays exponentially and has the same duration as the subject's neuronal activation trace. In this way it is possible to identify and characterize memory functions of the human brain.

The biomagnetic methods developed in basic research have also been translated into applications in clinical research. An example is the presurgical determination of both the location of an epileptic focus and a test of adjacent, unaffected cortex, thereby minimizing the risk of surgical damage to otherwise healthy brain tissue. The magnetoencephalogram has also been used to study the steady electric currents in the brain that are associated with migraine headaches.

Combination with EEG. A combination of biomagnetic with other recording methods may yield additional benefits. The electroencephalogram (EEG), a recording is on the surface of the scalp of the electrical field produced by brain activity, has been used since 1929 for numerous psychological and clinical studies. However, the low electrical conductivity of the skull in combination with the high conductivity of the scalp leads to a blurring of the scalp electroencephalogram as compared with the electrical signals that can be recorded intraoperatively on the surface of the exposed cortex. With the development of improved, multielectrode recording systems and mathematical data analysis techniques, a modern electroencephalogram may be able to provide, possibly at lowercost, much of the information that can now be obtained from the magnetoencephalogram. However, comparisons of the differing spatial sensitivities of the electroencephalogram and magnetoencephalogram, and of the differing responses to noise of the mathematical models used to interpret them, suggest that the two techniques are complementary and might best be used together with magnetic resonance imaging to obtain even more detailed images of the sources of the brain's electrical activity. In comparison to other techniques for recording brain function, such as positron emission tomography (PET) and functional magnetic resonance imaging, only the magnetoencephalogram and electroencephalogram have the requisite bandwidth to analyze the temporal variation of an individual evoked response.

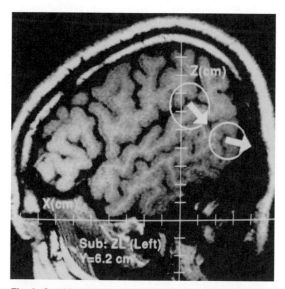

Fig. 3. Combination of a magnetic source image, showing the detected locations of the human brain's response about 0.1 s after the onset of a musical tone, with a left-side view of the cortical anatomy provided by a magnetic resonance image recorded just below the brain's surface, with the temporal lobe at the center. The base of the upper arrow denotes a location of activity in the primary auditory cortex, on the upper surface of the temporal lobe. The arrow indicates the direction of the ionic current flowing within active cells at the moment of peak activity. The base of the lower arrow denotes the location of activity in the association cortex, on the surface of the temporal lobe. Ellipses at the arrows indicate the uncertainty in locating the positions of the sources (99% confidence limit). (*Neuromagnetism Laboratory, New York University; Department of Radiology, New York University Medical Center; Silicon Graphics, Inc.*)

The potential benefits of magnetic source imaging are just now being realized, both at the scale of microSQUID systems optimized for recording from isolated nerve and muscle tissue and at the much larger scale of multichannel clinical SQUID systems. Whether signals are recorded from a slice of cardiac tissue or the intact human brain, the ability to relate patterns of biological activity deduced from magnetic field patterns to images of the underlying anatomy provides a new biophysical tool for an array of important applications.

For background information SEE *BIOMAGNETISM; ELECTROENCEPHALOGRAPHY; HEARING (HUMAN); MEDICAL IMAGING; SQUID; SUPERCONDUCTING DEVICES* in the McGraw-Hill Encyclopedia of Science & Technology.

John P. Wikswo, Jr.; Samuel J. Williamson

Bibliography. Z.-L. Lu, S. J. Williamson, and L. Kaufman, Behavioral lifetime of human auditory sensory memory predicted by physiological measures, *Science*, 258:1668–1670, 1992; D. J. Staton, R. N. Friedman, and J. P. Wikswo, Jr., High resolution SQUID imaging of octupolar currents in anisotropic cardiac tissue, *IEEE Trans. Appl. Superconduct.*, 3:1934–1936, 1993; J. P. Wikswo, Jr., and J. M. van Egeraat, Cellular magnetic fields: Fundamental and applied measurements on nerve axons, peripheral nerve bundles, and skeletal muscle, *J. Clin. Neurophysiol.*, 8(2):170–188, 1991; S. J. Williamson et al., *Advances in Biomagnetism*, 1989.

Biomimetic material

Biomimetics can be defined as a discipline in which synthetic systems are developed on the basis of information derived from biological systems. While the principle is straightforward, reduction to practice has been the stumbling block.

Knowledge of the biological processes responsible for producing tissues or organs or structures with specific properties and functions could provide the basis for the development of novel advanced materials, including synthetic biomaterials for replacing or restoring human body parts. These synthetic materials would be made by duplicating, or mimicking, the biological process (thus, the origin of the words biomimetics and biomimesis). This concept goes back several decades, especially among the researchers seeking biomaterials as replacements for bones and teeth. Advances in materials technology and the development of new instrumentation capable of probing properties on the micrometer and nanometer levels of structures have been responsible for the present elevated interest in and intensification of research on biomimetic concepts.

The possibility that new materials developed through biomimetics research would have significant impacts on technological processes in the next several decades has generated governmental interest. In the United States in October 1990, the National Science Foundation held a University/Industry Workshop on Biomolecular Materials in Washington, D.C., that in large part addressed biomimetics research, although this term was not used in the report of the workshop. Instead, the term biomolecular materials was used to express the same idea. Similarly, the agencies responsible for advancing technology in Japan—the Ministry of International Trade and Industry and the Agency of Industrial Science and Technology—cosponsored a Bionic Design International Workshop in Tsukuba in January 1992. The reason for this level of national interest in biomimetics and related research is the diversity of the potential products. New drug delivery systems; so-called smart (adaptive) materials; biosensors; and functional elements for chemical, electronic, and optical devices have been proposed as some of the potential results of biomimetic research. Increased research activity in the field is reflected by the appearance in 1992 of a new journal, *Biomimetics*.

Biological calcification. Most often cited as a target for biomimetics research are calcified tissues (sometimes known as mineralized tissues), such as bones, teeth, and shells. These tissues are complex hierarchical materials—structural composites. Thus, they provide an ideal paradigm for the design of new composite materials. Hydroxyapatite, the inorganic component of bones and teeth, has been used as the basis of several biomimetic approaches for bone augmentation. Composites of hydroxyapatite particulates dispersed in various polymers such as polyethylene have been developed. One such dispersed composite uses microbically produced polyhydroxybuterate as the matrix in order to obtain a biodegradable bone substitute. An alternate approach has been to use coral (porites) either in its natural form as aragonite (calcium carbonate) or chemically converted into hydroxyapatite. In either case, the interconnected porous structure is maintained, presumably facilitating bone ingrowth and attachment. However, in neither of these approaches is anything like the structure of bone achieved. Nor are the materials suitable for any present use other than in their biomedical applications.

An entirely different task has been the attempted development of a synthetic analog of the biomineralization process by controlling the crystallization of an inorganic component within a polymer matrix. Nanometer-scale cadmium sulfide (CdS) aggregates of amorphous particles have been used as the inorganic component. When dispersed in a poly(ethyleneoxide) matrix, cubic crystals of CdS, also on a nanometer scale, are formed throughout the film of poly(ethyleneoxide). Apparently, the CdS morphology can be altered by the choice of crystalline polymer matrix. Extension of this technique to higher concentrations of the inorganic component would be necessary before technological applications of such composites become practical. Full realization of the potential for the mineralization process will require increased understanding of the process by which bony structures are formed in nature.

Biomolecular electronics and photonics. Switching and transduction mechanisms are important both in nature and in technological applications. Thus, the biological materials serving such purposes provide another potential biomimetic approach for the development of advanced materials. Many biological systems exhibit a close association between deformation and electricity. Shear-induced piezoelectricity has been observed both in mammalian tissues such as hair, horn, bone (dry), tendon, teeth, shell, and in nonmammalian materials such as silk, wood, and hemp. A shear stress applied with respect to the direction of the material's oriented molecules produces electrical polarization perpendicular to the plane of the shear stress. For dry tendon, the appropriate piezoelectric constant ($d_{14} = -2.0$ pC/N) is comparable to that for one of the quartz constants ($d_{11} = -2.2$ pC/N). At various times, it has been suggested that the appearance of such behavior in biological systems has an important role in remodeling processes caused by applications of force. The biopolymers collagen and keratin both exhibit shear piezoelectricity. A synthetic polymer found to exhibit similar properties is poly-L-lactic acid. Because of the suggestion that some form of electrical stimulation may be the transduction mechanism for bone resorption and remodeling, piezoelectric polymer-based composites are being tested as bone plates and screws in order to accelerate healing fractures.

Lipid microtubules also provide a biomimetic approach in the electronics field. Tubules plated with gold can serve as miniature microwave cavities. Tubules made from lipids with side chains exhibiting conformational order have interesting electric and magnetic properties that could make them suitable as optical devices.

Another biomolecular material that might provide switching capabilities is bacteriorhodopsin, which shows some promise as the basis for optical memory systems in computers, a technology known as photonics or guided-wave optics. This application would involve green and red lights, which cause bacteriorhodopsin to be switched between two states. Other molecules being studied for their biomimetic potential include those involved in photosynthesis and visual response.

Molecular machines. Most biological systems are composed of materials made of complex polymers such as proteins, lipids, and polysaccharides. The so-called flagellar motor of bacterial species is the classic example of a molecular machine. These motors, the active agents for the swimming of bacterial species, are constructed from several different kinds of protein molecules. The structure of an entire flagellar motor is of such complexity that its reconstruction has not been successful to date.

Reconstruction of some parts of the flagellar motor from component proteins has been attempted. The screw of a flagellar motor is a long helical filament, with known pitch and diameter, composed of identical protein subunits known as flagellins. Filaments have been reconstructed from flagellins under physiological conditions. The joint (hook), which transmits the torque from the motor to the filament, has also been reconstructed out of hook proteins. However, the length of the reconstructed hook is variable rather than fixed as in native hooks. The rotor, which is composed of only one protein type and is involved in the conversion of electrochemical energy into the mechanical energy of rotation, has been reconstructed by genetic engineering. The shaft of the flagellar motor involves five different proteins, and reconstruction has been unsuccessful to date. Neither the bushing nor the stator of the flagellar motor has been reconstructed successfully as yet. When all the individual components are reconstructed in the laboratory, it should be possible to reconstruct the entire flagellar motor. This step will be significant in characterizing both the self-assembly mechanism for molecular machines and the mechanism of energy conversion in the motor.

Other molelcular machines being studied are actin and myosin. These motor proteins are the basis for muscle contraction via the relative sliding of filaments of actin and myosin. A complete understanding of the mechanism of the chemicomechanical energy transduction involved would be necessary in a biomimetic adaptation of the process.

Stimulus-responsive systems. Several hydrogels based on certain acrylamide polymers and copolymers are responsive to various stimuli such as pH, temperature, and electric or electromagnetic fields, so that the hydrogels either swell or collapse. Drug molecules can be incorporated into such hydrogels, for subsequent cyclical release within the body by cycling changes in temperature or electric fields. Similarly, enzymes incorporated in these stimulus-responsive hydrogels can be released under appropriate environmental stimuli to provide significant enhancement of bioreactor productivity.

These hydrogels also have the ability to act as permeation switches, controlling the permeation rate of a solution through a porous membrane whose pores are coated with these polymers. Cycling the temperature for such a switch could provide a system for reducing fouling by adsorbed proteins on the surfaces.

Another area where interactions of natural surfaces and synthetic materials are critical is in the design of cardiovascular and blood substitute materials. In the former case, biomimetic approaches are being used to develop materials with surfaces that will have negligible platelet depositon and complement activation. Such surfaces are critical for the development of membranes for long-term oxygenators for chronic respiratory distress as well as of the elastomers used in artificial heart pumps. In the case of blood substitutes, several approaches attempt to mimic the behavior of erythrocytes (red blood cells) in their life-sustaining function.

Genetically engineered polymers. Methods for recombined deoxyribonucleic acid (DNA) are now being used to develop polymeric materials with precisely determined uniformity of chain length, sequence, and stereochemistry. The process involves encoding the primary amino acid sequence of the desired polymer into a complementary sequence of DNA constructed by solid-phase organic synthesis and enzymatic ligation, followed by a series of genetic engineering steps, and culminating in fermentation and production isolation. A number of proteinlike polymers produced by this technique are being studied. In another approach, direct synthesis of protein-based polymers uses the methods of organic chemistry. Analogs of elastin, whose dimensions can be changed by variations in temperature or pH, have been developed by this technique. Cloning and expression of natural or artificial genes has been used to develop various synthetic structural and adhesive proteins. Modification of the genes can also be used to obtain desired changes in physical, chemical, and biological properties of the synthetic proteins.

Prospects. The examples of research in biomimetrics discussed above are by no means exhaustive. Biomimetics research is now on the initial portion of what predictably will be an exponential growth curve. The potential for the development of new advanced or smart materials based on biomimetics has brought this nontraditional approach to materials development into sharp focus. International competition for leadership in many technological areas depends upon advances in materials science. Thus, biomimetics is likely to be an active and rich area for future development. *See Smart Materials.*

For background information *See Biotechnology; Bone; Cilia and flagella; Genetic engineering; Medical chemical engineering; Molecular biology; Prosthesis; Transducer* in the McGraw-Hill Encyclopedia of Science & Technology.

J. Lawrence Katz

Bibliography. *Biomimetics,* quarterly; *Biomolecular Materials Report of the University/Industry Workshop,* National Science Foundation, October 10–12, 1990; *Proceedings of the Bionic Design International Workshop: Molecular Machine System and Biocrystallization and Adaptive Structure,* Tsukuba, Japan, January 29–30, 1992.

Biotechnology

Transgenic biotechnology has become a relatively routine process, introducing foreign genes into the germ line of an animal or plant species. Genetic engineers are currently creating breeds of sheep, goats, and cows that secrete human proteins in their milk and that thus serve as bioreactors. A variety of specific protein pharmaceuticals, new types of food proteins, and industrial enzymes can be made economically by this bioreactor approach.

Organization of genes. A gene in a chromosome of higher organisms comprises the structural gene, that is, the nucleic acid that encodes for the final gene product, plus regulatory regions that flank the gene both proximally ($5'$ end) and distally ($3'$ end). The structural gene either may be contiguous or, more commonly, may be made up of a number of individual segments. These segments of deoxyribonucleic acid (DNA) are initially transcribed in their entirety into messenger ribonucleic acid (mRNA). Subsequently, mRNA pieces (exons) are excised as the mRNA is processed into the final form that will be translated into a protein molecule. Usually, the regions that control

the expression of genes are located proximally from the structural gene. Occasionally, parts of both the structural gene and the distal regions have regulatory properties.

Ordered sequences of individual nucleic acids form a kind of language that scientists are beginning to decipher. Regulatory sequences typically extend over a few nucleotides and are the place where specific, regulatory proteins bind to the DNA molecule.

Gene constructs. The key issue in engineering genes for export into milk is inclusion of the appropriate regulatory sequences. The nature of the regulatory sequences that cause milk protein genes to be expressed only in female mammals during lactation is not yet fully understood. Nevertheless, appropriate temporal and developmental patterns of expression have been achieved for other proteins by using the control elements from milk protein genes. Since the regulatory features of milk proteins are shared among mammals, gene constructs for a particular protein can be tested in laboratory mammals such as mice, and the production of that protein can then be scaled up by putting the same gene constructs into mammals that have larger milk production capacities. The choice of species that might ultimately be targeted for commercial-scale production is determined by the amount of product needed. For pharmaceuticals whose total required amount of product may be in the hundreds or thousands of kilograms, sheep or pigs might be appropriate, but for quantities of thousands of metric tons (for example, infant formula proteins) high-producing dairy cattle are the choice.

Production of transgenic animals. The methodology for production of transgenic animals relies on technology that is adopted from human medicine. Oocytes can be collected through ultrasound scanning without the necessity of hormone treatment. When oocytes thus obtained are placed in tissue-culture medium, they can be matured and fertilized. Once oocytes have been fertilized, transgenic animals are produced by the direct injection of linear pieces of engineered DNA into the male pronucleus. For reasons that are not well understood, only a small percentage (approximately 1%) of the injected embryos will have incorporated one or more copies of the DNA into the genome. Although this insertion is random and causes a variation in the level of expression of the gene, it is not harmful to the animal because much of the genome is quiescent and nonfunctional. The genes incorporated into the DNA are passed on to offspring and inherited as normal dominant genetic characteristics. The techniques being developed for human gene therapy may also be applied in the production of transgenic animals and may lead to improvements in reliability of the gene incorporation procedure.

Expression of pharmaceutical and other proteins. The potential for expressing proteins in mammary glands of transgenic animals is virtually limitless. The expression of many different human proteins in the milk of transgenic animals has already been demonstrated. These proteins include alpha-1 antitrypsin (a protein that helps keep cells elastic); clotting factor IX; tissue plasminogen activator (converts plasminogen to plasmin); and interleukin 2 (a protein that promotes the growth of T lymphocytes). Many of the proteins expressed in transgenic animals have a defined medicinal value and cannot presently be produced in conventional ways.

In addition to use as pharmaceuticals, proteins are widely used as ingredients in foods. Sometimes these proteins possess undesirable features, but it is now becoming possible for food scientists to design food proteins so as to either introduce desirable features or remove undesirable ones. For example, one of the milk proteins can prevent the production of undesirable flavors in certain cheeses. It is anticipated that milk proteins with improved manufacturing and consumer characteristics will be produced in the future. An example might be the production of low-fat cheeses with the same textural and flavor characteristics of cheeses higher in fat content.

Catalysts of various types are used in industrial processes. Enzymes are protein catalysts, and many enzymes might also be produced in milk in quantities that would make them applicable to industrial processes. This methodology not only would provide new manufacturing opportunities but also would open the way to use of environmentally safe biological manufacturing systems.

Safety issues. Safety issues associated with the production and use of products from transgenic animals are not inherently greater than those for many products currently on the market. Until recombinant human insulin came on the market, insulin used by diabetics was isolated from pig and cattle pancreatic tissue; this insulin could generate immunological responses. The products to be produced through the use of transgenic livestock should have a lower risk than products historically produced by using animal tissue, since they should not generate immunological responses. The main concern with transgenic products is the possibility of disease transmission. Certain animal diseases, such as tuberculosis, brucellosis, and certain parasitic diseases, can be transmitted from animals to humans. However, normal animal veterinary care along with appropriate diagnostic testing should ensure the safety of future transgenic products.

For background information SEE BIOCHEMICAL ENGINEERING; BIOTECHNOLOGY; GENE; GENE ACTION; LACTATION in the McGraw-Hill Encyclopedia of Science & Technology.

Robert Bremel

Bibliography. R. D. Bremel, H. C. Yom, and G. T. Bleck, Alteration of milk composition using molecular genetics, *J. Dairy Sci.*, 72:2826–2833, 1989; K. Gordon et al., Production of human tissue plasminogen activator in transgenic mouse milk, *BioTechnology*, 5:1183–1187, 1987; R. Jimenez-Flores and T. Richardson, Genetic engineering of the caseins to modify the behavior of milk during processing: A review, *J. Dairy Sci.*, 71:2640–2654, 1988; J. P. Simons, M. McClenaghan, and A. J. Clark, Alteration of the quality of milk by expression of sheep β-lactoglobulin in transgenic mice, *Nature*, 328:530–532, 1987.

Bivalvia

Recent research on bivalves has focused on genetics of natural populations and on pedal feeding and the probable evolution of early bivalve mollusks.

Population Genetics of Marine Bivalves

During the 1960s, techniques were developed to separate the protein products of different alleles at gene loci by gel electrophoresis. This approach enabled the unambiguous determination, from the banding patterns on the gel, of the genotypes at structural gene loci, particularly those coding for enzymes because of the ease of staining with enzyme-specific substrates.

The population genetics of a wide range of plant and animal species have been investigated by electrophoresis. Statistical testing of observed genotype frequencies against those predicted by the Hardy-Weinberg model has generally confirmed the model. Nevertheless, electrophoretic data from studies of natural populations of marine mollusks, particularly bivalves such as mussels, oysters, clams, and scallops, tend to be characterized by a general deficiency of heterozygotes in comparison to numbers predicted by the model. Such deficiencies at individual loci are not always sufficient to cause a statistically significant deviation from the model, but across many loci the direction of the deviation is almost always toward too few heterozygotes.

Heterozygote deficiency. Possible explanations for heterozygote deficiency at a locus include inbreeding, the Wahlund effect, the existence of null alleles, and preferential mortality of heterozygotes. Various modifications or combinations of these explanations have also been proposed to account for heterozygote deficiencies but, as yet, no consistent explanation has emerged from studies on natural populations.

Inbreeding. If organisms that mate are closely related, that is, siblings or cousins, they are more likely to share alleles in common at a locus than organisms that mate at random with unrelated individuals. Thus, more of the offspring from inbred matings will be homozygous, and fewer will be heterozygous, than would be predicted by the Hardy-Weinberg model. An extreme example of inbreeding is self-fertilization by hermaphrodites (selfing). Self-fertilizing homozygotes will breed true: only half of the offspring of selfing heterozygotes will be heterozygous ($1/2$ *AB*); the other half will be homozygous ($1/4$ *AA*, $1/4$ *BB*). Therefore, in each generation there will be a reduction in the proportion of heterozygotes predicted by the model. After one or more generations of selfing or inbreeding, there are usually disadvantageous phenotypic consequences of this decreased heterozygosity (inbreeding depression), for example, reduced growth or viability.

The reproductive biology of most bivalve mollusks involves high fecundity, random external fertilization, and extensive larval dispersal. It is difficult to see how any inbreeding could possibly occur in such organisms, since the chances of two related individuals settling close to one another and their gametes then coming into contact is remote. Some bivalves, however, are hermaphroditic; in these cases, some degree of self-fertilization might be possible. Nevertheless, if self-fertilization were the cause of the phenomenon, higher levels of heterozygote deficiency in hermaphroditic species compared to species with separate sexes would be expected. This is not the case.

Wahlund effect. If the sampled population consists principally of immigrants (via larval dispersal) derived from two other populations with very different allele frequencies at a locus, the sampled population will have fewer heterozygotes at that locus than are predicted by the Hardy-Weinberg model. The degree of heterozygote deficiency depends primarily on the extent of the allele frequency differences between the source populations: the greater the differentiation, the more the heterozygote deficiency.

Generally, for most bivalves either with or without larval dispersal, subpopulation differentiation is insufficient to account for more than a fraction of the reported heterozygote deficiencies. Also, the Wahlund effect would cause heterozygote deficiencies at only those few loci where subpopulation differentiation was evident rather than across most loci, as is observed.

Null alleles. Mutations can occur at a locus that will result in an allele coding for a nonactive or noneffective protein, that is, a null allele. Clearly, an organism that is homozygous for a null allele at an enzyme locus will be unable to produce that enzyme in the cells of its body and will probably perish early in life. Nevertheless, some heterozygotes between a null allele and a normal allele do survive. When such individuals are run on electrophoretic gels, their banding patterns are not always distinguishable from those of a homozygote for the functioning allele; therefore they would be scored as homozygotes rather than as the heterozygotes that they actually are. Thus, the genotype frequencies observed would deviate from the Hardy-Weinberg model toward a deficiency of heterozygotes. If, as would be predicted, null-allele homozygotes do not survive, there is a maximum frequency that such alleles can attain within a population, and therefore a maximum deviation from the Hardy-Weinberg model. Calculations suggest that only a small fraction of the heterozygote deficiencies reported in studies of bivalve mollusks can be accounted for by the presence of undetected null alleles.

Preferential mortality. Possibly, heterozygote deficiencies are generated by the preferential mortality of heterozygotes. The difficulty with this explanation is that traditional selectionist theory argues that polymorphisms at most loci are maintained by either the general higher fitness of heterozygotes or the equality of the heterozygote fitness to the fittest homozygote. Selection against heterozygotes should not normally occur. The alternative neutralist approach proposes that, at most loci, alternative alleles are neutral in relation to fitness. Alleles arise by mutation and either spread through the population or are eliminated by the chance effect of random genetic drift rather than by the possession of advantageous or disadvantageous qualities. Accordingly, general selection for, or against, specific genotypes at many of the loci investigated would not be expected.

Heterosis. It has been extensively documented that overall high heterozygosity across the genome of a diploid organism is usually linked to high fitness (heterosis). The most common example involves crosses between closely related species that produce hybrid vigor. Studies on natural populations of marine bivalves have indicated that this effect can be demonstrated within a population even if only a small number of gene loci (6–20) are considered. On average, individuals heterozygous at many loci grow faster, live longer, expend less basal metabolic energy, and have a higher scope for growth than individuals homozygous at many loci.

These data may be explained on the basis of greater or equal fitness of heterozygotes to the fittest homozygote at the scored loci or at closely neighboring loci on the chromosomes. Alternatively, it has been argued that the phenomenon could be caused by decreased fitness of multiple-locus homozygotes because of the concomitant increased incidence of homozygosity for deleterious recessive alleles at loci (as is believed to happen in inbreeding depression). Both mechanisms probably operate, but the balance between the two is in dispute.

Paradox. Electrophoretic data from studies on bivalves often show both a deficiency of heterozygotes and a positive correlation between multiple-locus heterozygosity and, for example, growth. How can there be fewer heterozygotes than expected under the Hardy-Weinberg model at many loci while, on average, individuals that are heterozygous at many of those loci appear fitter than those that are homozygous?

So far it has proven unproductive to address this paradox by studying natural populations of adult bivalves because, for example, it is not always possible to entirely eliminate inbreeding, the Wahlund effect, or null alleles as contributors to heterozygote deficiency. Recent studies on hatchery- or laboratory-reared populations of bivalves are providing clues to the paradox. Laboratory-reared cohorts of bivalves are derived from just a few parents rather than from the random mixing of many parents expected in the wild. In such studies, there is never a detectable correlation between heterozygosity and growth, but a deficiency of heterozygotes against expectations is still evident in the data. Thus, the two phenomena, although occurring in the same data sets from wild populations, do not simultaneously occur in data from laboratory studies and may in fact be causally unconnected.

Laboratory trials. Inbreeding, the Wahlund effect, and null alleles can be excluded as causes of heterozygote deficiency in laboratory trials because individuals taken at random for mating are unlikely to be closely related, the Wahlund effect is a population effect, and the genotypes and their proportions in offspring can be checked for the existence of null alleles in the parents. Evidence from a recent study on laboratory-reared mussels, where offspring were genotyped at several loci soon after the larvae had settled out of the planktonic stage and again as juveniles, suggested that heterozygote deficiencies were actually generated between those two stages by preferential mortality of heterozygotes. Selection against heterozygotes has therefore been demonstrated as the cause of heterozygote deficiencies in laboratory trials, but it is still not confirmed as the explanation for the phenomenon in the wild. Nevertheless, evidence indicates that heterozygote deficiencies in natural populations tend to decrease with age; thus, presumably, heterozygotes survive better as they get older, allowing the Hardy-Weinberg model to be approximately correct among older individuals in the population.

The relative importance of the heterozygote-advantage or homozygote-disadvantage explanations for the positive correlation between multiple-locus heterozygosity and growth can not be easily tested by using natural populations. However, the argument can be made that the lack of such correlations in laboratory-reared offspring from restricted numbers of parents tends to point toward homozygote disadvantage rather than heterozygote advantage as the more important mechanism. It is not easy to see why heterozygote combinations that confer a fitness advantage on their carriers in wild populations should not also do so in offspring from restricted matings. The genetic background is derived from a greater range of genomes in wild individuals compared to offspring from restricted matings, possibly reducing or negating the fitness advantage of certain heterozygote combinations in laboratory-reared individuals. Although possibly a significant factor in single families (offspring from two parents), this reduction in fitness advantage is less likely to be important in laboratory matings involving several tens of progenitors.

In restricted matings, offspring are derived from few chromosome sets (genomes); therefore, the chances of inclusion of rare deleterious recessive alleles at loci within those participating genomes are very small. If homozygote disadvantage is the principal cause of the positive correlation between heterozygosity and fitness, the consistent lack of the correlation in offspring from restricted matings can be explained by the reduced frequency of deleterious recessive alleles linked to the scored homozygotes.

Although many questions remain concerning the full explanation of these contradictory phenomena of heterozygote deficiency and positive correlation of multiple-locus heterozygosity with fitness in marine bivalves, the use of laboratory trials is a new approach helping to resolve them.   *Andy Beaumont*

Pedal Feeding and the Evolution of Bivalve Mollusks

Most bivalves use their gills and labial palps for filter feeding. Although these organs are almost universal throughout the Bivalvia, there is enough variation to confuse the issue of ancestral bivalve anatomy and habit. One thing is certain: the most ancient bivalves, which begin to appear in the fossil record more than 500 million years ago, were invariably minute. Therefore, evolutionary clues might be found in extant small species and the juvenile stages of the larger bivalves. A striking universal feature of extant small species and juvenile stages is that almost all engage in pedal feeding.

Since most bivalves live buried in sediment or attached to the substrate, the siphons and feet are the only organs that confer behavioral plasticity. For example, in exceptional cases the foot can be used for swimming as well as burrowing; in the deep-sea thermal-vent clams the foot absorbs hydrogen sulfide for symbiotic metabolism; and in carnivorous bivalves the foot maneuvers captured prey into the mouth.

Pedal feeding in mature bivalves. Pedal feeding in mature bivalves is rare. The pedal cilia of the coral sand clam *Fimbria* draw up detrital matter from the sediment. Pedal feeding in small freshwater pea mussels and in the small marine bivalve *Mysella* has been misinterpreted as a cleansing process. It would seem to be a waste of energy to collect food and then get rid of it. *Mysella* can feed pedally both while moving and while stationary. Lying on its side at the substrate surface, it sweeps the foot from a posterior to anterior position, all the while drawing food particles with the pedal cilia to the mouth. Periodically, it retracts the foot and ingests the mucus-bound particles.

Corbicula is an introduced fresh-water clam that massively invaded the watersheds of the southeastern United States and drew attention by clogging up the cooling systems of nuclear power stations. It can suspension-feed on phytoplankton and also make complete use of all available organic matter by pedal feeding, thus continuing to grow and reproduce in the absence of phytoplankton. *Corbicula* can feed pedally while plowing through the sediment surface. While the clam is buried in the sediment, the foot is widely expanded, and the pedal cilia convey potential food particles to the food grooves of the gills. When the clam lies on its side on a hard surface with detrital deposits, the cilia of the underside of the foot pick up food particles. The tip of the foot levers the clam across the surface, and a cleared trail is left behind. It seems likely that a number of small venerid and galeommatacean bivalves are pedal feeders. This feeding habit has also been found in *Miodontiscus* (Carditidae) and *Modiolaria* (Mytilidae).

Modes of pedal feeding. It has long been suspected that in some bivalves a filter-feeding hiatus occurs between metamorphosis, when the veliger larva loses its suspension-feeding velum, and the time when the juvenile develops effective filtration gills. For almost 3 months after metamorphosis, juveniles of the geoduck, *Panope abrupta*, depend upon various pedal-feeding modes for growth and survival. These modes comprise locomotory feeding, probe feeding, sweep feeding, and interstitial pedal feeding (**Fig. 1**).

Locomotory feeding. The animal moves over the surface of the substrate, and pedal cilia collect potential food particles and pass them toward the large labial palps, which consolidate a food bolus while rejecting denser, nondigestible particles.

Probe feeding. The foot is protracted and retracted, working its way through a circular area. Whenever its tip comes into contact with potential food particles, mucus is secreted to form an adhering mass and the tip is drawn back to the mantle cavity. The food particles may be brushed onto the labial palps or thrust directly into the mouth.

Sweep feeding. The animal lies on its side at the surface of the sediment, protracts the foot posteriorly, and then sweeps it anteriorly. Meanwhile, particles of detritus are collected by mucus secretions and carried by the pedal cilia toward the labial palps. Early juveniles of Manila clams can be stimulated to vigorous sweep feeding by the presence of the microscopic alga *Chaetoceras gracilis.*

Interstitial pedal feeding. The foot is extended into the sedimentary substrate, and free particles are drawn into the mantle cavity. Interstitial pedal feeding has also been observed in several other genera, including *Corbicula.*

Panope demonstrates the importance of the labial palps as complementary structures for pedal feeding. In juvenile bivalves, the palps are proportionately larger than in the adults. Moreover, in small types of bivalves the mature animals retain large labial palps.

Evidence for pedal feeding. Some or all of these pedal-feeding modes have been found in juveniles of such different types as scallops (*Patinopecten yessoensis, Crassadoma gigantea*), giant clams (*Tridacna gigas*), and Manila clams (*Tapes philippinarum*), and their occurrence can be inferred in juveniles of Mytilidae, Semelidae, and Tellinidae. Such inferences were neglected by the original observers, who had no general hypothesis to underpin their observations.

Important evidence still lies in unpublished theses, such as a study of Nuculidae, belonging to the ancient bivalve group Protobranchia, which were traditionally taken as models for the primitive condition. Nuculidae have small, simple gills with only minor food-filtration functions. The ciliated labial palp proboscides are extended to convey large quantities of unsorted, detrital matter from the substrate surface to the palp lamellae and mouth. The anterior position of the inhalant currents is also taken to be a primitive condition. Adult nuculids cannot feed pedally, but in the early juvenile the foot is the only alimentary organ. In the later juvenile stages, before the palp-feeding apparatus has developed, the anterior gill bars pass food particles to the acceptance tract on the foot and around the mouth. This method of feeding is an interesting contrast to that of most other bivalves, where the labial palps help to collect food prior to the development of the mature filtration mechanism of the gills. On this evidence, mature Nuculidae are poor candidates for the archetypal bivalve condition, but juvenile nuculids may be more appropriate.

There must be many caveats about inferring the nature of primitive invertebrates from small types or from conditions in the larvae and juveniles. Many small bivalves are pedomorphic simplifications of their larger and more elaborate ancestors. Juvenile traits may be specializations for minuteness. For example, it has been argued that for very small bivalves the anterior location of the inhalant current and the posterior position of the exhalant current is a physical necessity to prevent mixing. However, in *Nucula* juveniles, both currents are posterior, and there is no fluid-dynamic difficulty. Only in the adults is the exhalant current posterior and the inhalant current anterior. The position of these currents is an evolutionary red herring: the direction of the feeding current is a reflection of the position and action of the major feeding mechanism, whether pedal, labial, or ctenidial.

Archetype. Given the caveats, the argument that the archetypal bivalve was a pedal feeder can still be made. Not only is pedal feeding universal throughout all bivalves studied (except for oysters and boring pholads), but also there is almost universal persistence of a range of distinctive and complex pedal-feeding modes in bivalves with a diversity of mature habits. The most parsimonious model of a small archetypal bivalve requires for respiration only a pair of simple, posterior sets of gill filaments similar to those of the early juveniles of modern bivalves, and engages in pedal locomotory, probe, and sweep feeding, supplemented by two labial palps that are homologs of the juvenile upper palps (**Fig. 2**). Such an organism would be able to function either on a hard substrate or at the surface of a marine sediment.

The bivalve form of this archetype potentiated the elaboration of the deposit feeding labial apparatus of the protobranchs and the divergent ctenidial suspension-feeding apparatus of lamellibranchiate bivalves. Both of these evolutionary divergences required an increase in size to accommodate organs large enough to

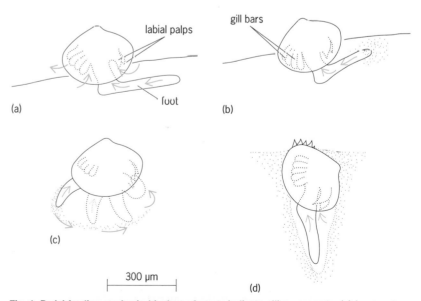

Fig. 1. Pedal-feeding modes in bivalves. Arrows indicate ciliary currents. (*a***) Locomotory feeding. (***b***) Probe feeding. (***c***) Sweep feeding. (***d***) Interstitial pedal feeding. (***After R. G. B. Reid, Feeding behavior of early juvenile shellfish, with emphasis on the Manila clam, in T. Y. Nosho and K. K. Chew, eds., Remote Setting and Nursery Culture for Shellfish Growers, Washington Sea Grant Program, Seattle, 1991***)**

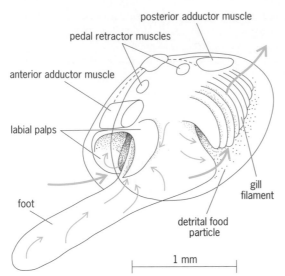

posterior adductor muscle

pedal retractor muscles

anterior adductor muscle

labial palps

gill filament

foot

detrital food particle

1 mm

Fig. 2. Hypothetical archetypal bivalve. Heavy arrows represent water currents; light arrows represent ciliary movement of food particles. The mouth is situated between the labial palps.

be effective. However, at the same time, the structure of the mantle chamber provided a system that was energy-efficient and fluid-dynamically sound.

Aquaculture. Primitive pedal-feeding behavior has persisted in most bivalve juveniles to bridge the gap between larval and late-juvenile filter feeding. Despite high mortalities in aquacultured juvenile clams at and just after metamorphosis, aquaculturists have been unaware of the pedal-feeding stage. Ongoing research on Manila clam juveniles shows that if culture chambers are designed with irrigation regimes that keep the food accessible to pedal feeding, growth and survival are improved remarkably.

For background information SEE BIVALVIA; ELECTROPHORESIS; HARDY-WEINBERG FORMULA; POPULATION GENETICS in the McGraw-Hill Encyclopedia of Science & Technology.

R. G. B. Reid

Bibliography. A. R. Beaumont, Genetic studies of laboratory reared mussels, *Mytilus edulis*: Heterozygote deficiencies, heterozygosity and growth, *Biol. J. Linn. Soc.*, 44:273–285, 1991; R. K. Koehn, W. J. Diehl, and T. M. Scott, The differential contribution by individual enzymes of glycolysis and protein catabolism to the relationship between heterozygosity and growth rate in the coot clam, *Mulinia lateralis, Genetics*, 118:121–130, 1988; B. Morton, The biology and functional morphology of the coral-sand bivalve *Fimbria fimbriata* (Linnaeus 1758), *Rec. Austral. Mus.*, 32:390–420, 1979; R. G. B. Reid et al., Anterior inhalant currents and pedal feeding in bivalves, *Veliger*, 35:93–104, 1992; E. R. Trueman and M. R. Clarke (eds.), *The Mollusca*, vol. 10, 1983; E. Zouros and D. W. Foltz, Possible explanations of heterozygote deficiency in bivalve molluscs, *Malacologia*, 25:583–591, 1984; E. Zouros, M. Romero-Dorey, and A. L. Mallet, Heterozygosity and growth in marine bivalves: Further data and possible explanations, *Evolution*, 42:1332–1341, 1988.

Black hole

General relativity predicts that when a massive body is compressed to sufficiently high density, it becomes a black hole, an object whose gravitational pull is so powerful that nothing can escape from it. Such objects are of great astrophysical interest as, for example, the presumed engines that power quasars. In addition, black holes are of great intrinsic theoretical interest. Although in classical physics a black hole is unable to shed any of its mass, quantum effects allow it to radiate and lose energy. This black hole evaporation poses a great potential paradox. Nothing seems to prevent the black hole from radiating away all of its mass and disappearing completely. Thus, all information about the object from which the black hole formed seems to become forever inaccessible.

It is not yet firmly established that black holes really destroy information in this way. But if they do, it will be necessary to find a new conceptual basis for all of physics. Conceivably, the discovery of black hole radiance portends a scientific revolution as profound as those that led to the formulation of relativity and quantum theory in the early twentieth century.

Black hole thermodynamics. In classical general relativity, a black hole can be characterized as a region of no escape bounded by a surface known as an event horizon. Whoever falls through the horizon is unable to return to the region outside the black hole, or even to send any signal that can propagate to an observer outside. In fact, general relativity predicts that one who enters the black hole is inexorably driven to a so-called singularity, a region of infinite space-time curvature and hence infinitely strong gravitational forces. The singularity signifies the breakdown of classical physics deep inside the black hole. Quantum effects become important there, and classical general relativity cannot predict what will happen.

The time-independent black hole solutions to the classical field equations of general relativity can be characterized by just a few parameters; a solution is uniquely determined once its mass, angular momentum, electric charge, and magnetic charge are specified. Thus, although the collapse of a star to form a black hole may be a very complex process, the end result is a remarkably simple object with very little structure. Once the collapsing body settles down to a stationary state, all detailed information about the body has become completely inaccessible to an observer who stays outside the black hole. In J. Wheeler's maxim, "A black hole has no hair."

In the case of a nonrotating, uncharged black hole, the event horizon is a sphere; its radius R is related to its mass M according to Eq. (1), where G is Newton's

$$R = \frac{2GM}{c^2} \qquad (1)$$

gravitational constant and c is the speed of light. (The radius R is defined by $A = 4\pi R^2$, where A is the area of the horizon.) Thus, a black hole of one solar mass has a radius of about 2 mi (3 km). According to the area theorem of general relativity, the total area of all event horizons can never decrease in any process involving any number of black holes. This result (along with conservation of energy) implies that a black hole cannot split up into smaller black holes. In classical general relativity, a (nonrotating) black hole is an absolutely stable object; it can accrete matter to become larger and heavier, but nothing can make it smaller and lighter. (Rotating black holes can lose energy by spinning down, and this property accounts for some of their interesting astrophysical effects.)

When quantum effects are included, this statement must be modified: a black hole can emit radiation and lose mass. In fact, semiclassical calculations show that the black hole emits like a thermal body with a characteristic temperature. For a nonrotating black hole of radius R, this temperature T is given by Eq. (2), where

$$k_B T = \frac{\hbar c}{4\pi R} = \frac{\hbar c^3}{8\pi GM} \qquad (2)$$

k_B is Boltzmann's constant and $\hbar$ is Planck's constant divided by 2π. Thus, the typical radiation quanta emitted by the black hole have a wavelength comparable to R. For a solar-mass black hole, the temperature is $T = 6 \times 10^{-8}$ K.

Since the emission is thermal, a black hole surrounded by a radiation bath at temperature T remains in equilibrium: it emits and absorbs at equal rates. Although this equilibrium is actually unstable, emission or accretion of radiation at temperature T can nonetheless be regarded as a reversible thermodynamic process, and the entropy of a black hole can therefore be computed (up to an unknown additive constant). The result is given by Eq. (3), where A is the area of the horizon and L_{Planck}, defined by Eq. (4), is the

$$\frac{S}{k_B} = \frac{1}{4} \frac{A}{L_{\text{Planck}}^2} \qquad (3)$$

$$L_{\text{Planck}} \equiv \left(\frac{\hbar G}{c^3}\right)^{1/2} \sim 10^{-35} \text{ m} \qquad (4)$$

Planck length that can be constructed from the fundamental constants. Thus, the area theorem of classical general relativity can be regarded as a special case of the second law of thermodynamics, which says that entropy is always nondecreasing. Indeed, the analogy between the area theorem and the second law inspired J. Bekenstein to anticipate Eq. (3) (up to a then unknown multiplicative constant of order one) even before S. Hawking discovered that black holes radiate. The entropy of a black hole is truly enormous—about 10^{78} for a solar-mass hole, some 20 orders of magnitude larger than the entropy of the Sun.

Difficulties. The discovery of black hole radiance established a deep and satisfying connection between gravitation, quantum theory, and thermodynamics. But it also raised some disturbing puzzles.

Interpretation of entropy. One puzzle concerns the interpretation of the black hole entropy. In other contexts, the statistical-mechanical entropy counts the number of accessible microstates that a system can occupy, where all states are presumed to occur with equal probability. If a black hole really has no (or very

little) hair, the nature of these microstates is obscure. Equation (3) invites the interpretation of the horizon as a quantum membrane with about one degree of freedom per Planck unit of area. But a more concrete conception of these degrees of freedom remains elusive.

Information-loss paradox. Even more distressing is a serious paradox raised by S. Hawking. In his semiclassical calculation of black hole radiance, he found that the emitted radiation is exactly thermal. In particular, the detailed form of the radiation does not depend on the detailed structure of the body that collapsed. The radiation carries little information because it is induced by the gravitational field of the black hole outside the horizon, and the black hole has no hair that records detailed information about the collapsing body. While the semiclassical approximation used by Hawking is not exact, it is highly plausible that a more accurate treatment would still find that the emitted radiation is only very weakly correlated with the state of the collapsing body. The key constraint comes from causality: once the collapsing body is behind the horizon, it is incapable of influencing the radiation.

Thermal radiation is in a so-called mixed state, not a pure (precisely known) state. Therefore, the state of the radiation cannot be precisely predicted; it is only possible to assign probabilities to different alternatives. But according to the laws of quantum mechanics, if the initial state of a system is pure, the state remains pure at later times. Information is never destroyed in principle (although it is often lost in practice). An observer outside a black hole who detects the emitted radiation would recover little information about the body that collapsed to form the black hole. Since information is conserved in principle, this observer would conclude that most of the information about the initial body must be retained inside the black hole.

But suppose that the black hole continues to evaporate until it disappears completely. Then it seems that an initially pure quantum state, by collapsing to form a black hole and evaporating completely, has evolved to a mixed state. In other words, even if the initial state is precisely known, it is not possible to predict with certainty what the final state will be. It is only possible to assign probabilities to different alternatives. This is the information-loss paradox. What is paradoxical is that an attempt to analyze the evolution of a black hole by using the usual principles of relativity and quantum theory leads to a contradiction, for these principles forbid the evolution of a pure state to a mixed state. Hawking concludes that the usual rules of quantum mechanics cannot apply in all situations, which means that the fundamental laws of physics must be reformulated.

Attempts to evade the paradox. Is there no way to escape this radical conclusion? It is conceivable that the semiclassical calculations are misleading and that detailed information about the collapsing body really is encoded in the emitted radiation. But as noted above, it is difficult to reconcile this possibility with causality, since the collapsing body is out of causal contact with the radiation once the body crosses the horizon.

Other suggestions for evading the paradox seem to have serious flaws. As a black hole evaporates and shrinks, it eventually becomes so small that the semiclassical picture of the evaporation process no longer applies. The usual notions of causality, and indeed the very concepts of space and time, break down at distances comparable to L_{Planck}; at this distance scale, space-time is subject to violent quantum fluctuations. Because these quantum gravity effects are not well understood, there is freedom to speculate about how a black hole with radius $R \sim L_{Planck}$ would behave.

Perhaps the black hole stops evaporating, leaving behind an absolutely stable black hole remnant. Such a remnant could serve as repository for most of the information about the collapsing body, thus resolving the paradox. This scenario, however, raises new problems. Since the initial black hole could have been arbitrarily massive, the remnant must be capable of carrying an arbitrarily large amount of information. Hence, there must be an infinite number of species of stable remnant, all with size comparable to L_{Planck}, and mass comparable to the Planck mass M_{Planck}, defined by Eq. (5). Most likely, a theory that admits this kind

$$M_{Planck} = \left(\frac{\hbar c}{G}\right)^{1/2} \sim 10^{-5} \text{ g} \qquad (5)$$

of infinite degeneracy does not make sense, because the calculated rates for many otherwise plausible processes are found to be infinite.

Perhaps when the black hole shrinks down to the Planck size, the information that has been stored in it can finally leak out. But if the amount of information is large, it is impossible for it to come out quickly. There are fundamental limits to the amount of information that can be encoded in a Planck energy's worth of radiation in a finite time: the more information, the longer the time. So, there would need to be Planck-size black holes that are arbitrarily long-lived, even if no given species is absolutely stable. This scenario then runs into the same difficulties as the stable remnants.

Outlook. So far, all attempts to reconcile the evaporation of black holes with the conventional laws of physics have failed. It seems increasingly likely that the information-loss paradox presages a scientific revolution. The puzzle of black hole evaporation remains one of the most important and challenging at the frontier of fundamental physics.

For background information SEE BLACK HOLE; NON-RELATIVISTIC QUANTUM THEORY; QUANTUM GRAVITATION; QUANTUM MECHANICS; RELATIVITY in the McGraw-Hill Encyclopedia of Science & Technology.

John Preskill

Bibliography. J. D. Bekenstein, Black-hole thermodynamics, *Phys. Today*, 33(1):24–31, 1980; N. D. Birrell and P. C. W. Davies, *Quantum Fields in Curved Space*, 1982; S. W. Hawking, *A Brief History of Time*, 1988; S. W. Hawking, The quantum mechanics of black holes, *Sci. Amer.*, 236(1):34–40, 1977.

Borehole logging

The ability to look through rock would allow mineral deposits to be described more accurately, hazards to miners to be detected more easily, and potential environmental problems related to mining to be identified more reliably. Since light cannot pass through rock, geophysical systems for examining underground features such as mineral deposits must use other forms of energy, including sound waves in seismic (acoustic) systems, electromagnetic waves with frequencies lower than light, and radiation.

Some geophysical methods are used on the surface; others require a borehole drilled into the ground. Surface geophysics is more suitable for exploration over a large area. Borehole systems allow the sensing probes to be closer to deep mineral deposits and thus can provide more detail. The most common type of borehole geophysics is geophysical well logging (often called wireline logging), in which a probe in a borehole measures conditions in or near it. The resulting well log is a graph or recorded digital output that describes conditions along all or part of the borehole depth. Idealized well logs are shown in the **illustration**. The term well logging derives from the petroleum industry; logging a borehole that is not a well is still called well logging. Borehole geophysics includes surface-to-borehole systems where either a transmitter or a receiver is on the surface and the other is in a borehole, and borehole-to-borehole (crosshole) systems where the transmitter probe is in one borehole and the receiver probe is in another. The **table** lists some common borehole geophysical methods and the information they yield, but there are many other types of logs.

Borehole geophysics. Much of the borehole geophysics developed in the United States has been for exploration, especially for coal. These geophysical methods can help to determine the size and shape of a deposit after its general location has been inferred from surface geophysics. This application is successful because an ore deposit often has geophysical properties different from those of the barren rock. For example, copper or zinc sulfide ores often have lower resistivity than the surrounding rock.

A deposit is usually examined with a combination of borehole geophysics and geologic core analysis. Core analysis is valuable, providing details about conditions at the borehole. However, it is expensive because it requires special drilling equipment and labor-intensive core examination; therefore borehole geophysics may be used to choose the depths for core analysis. The cost of drilling boreholes solely for geophysical logging is lower than for coring, and obtaining results requires less labor. Borehole geophysics examines more rock than core analysis, because a larger volume around the borehole affects the geophysical results. Cores are very small relative to the mineral deposit; thus core analysis can give unreliable estimates of ore grade in highly variable deposits.

Well logging is performed for many other reasons. One common use is gathering information for groundwater studies. Some logs show the condition of the borehole. Deviation logs are often performed to measure the straightness of the borehole, so that the position of the borehole at depth can be calculated. A caliper log measures the borehole diameter to check the condition of the borehole wall. More specialized methods include acoustic logs to measure the velocity of sound near the borehole and neutron logs to estimate porosity and moisture content.

Multicomponent probes. Well-logging companies usually use multicomponent probes containing several types of sensors. For example, a set may include sensors for resistivity, spontaneous potential, and natural gamma. The cable from the surface to the probe contains wires for each set of sensors. Many combinations are possible.

Electrical methods. A popular electrical method for identifying rock types of contrasting resistivity is the combination of short normal and long normal resistivity logs. The short and long logs measure the resistivity between electrodes 16 and 64 in. (0.41 and 1.63 m) apart, respectively. Current lines between the short-space electrodes flow in and very close to the borehole, while the current lines between the long-space electrodes flow more in the deposit. Thus, the short spacing allows borehole factors such as drilling mud (clay or similar material used to aid drilling) remaining on the borehole wall to be considered when interpreting the deposit resistivity from the long spacing. An alternative method is the guard or focused resistivity system, in which additional electrodes focus the current flow outward to obtain greater penetration. The single-point resistance system measures the total resistance from a point on the surface near the borehole to an electrode on a borehole probe. It is generally less effective than the resistivity methods for identifying rock types, because borehole factors strongly affect resistance measurements, but abrupt changes in resistance can indicate changes in rock type.

The spontaneous potential log is another common electrical method for distinguishing rock types. It indicates how much the deposit acts like a battery to generate voltages in different parts of the deposit. The voltage is usually measured between an electrode on

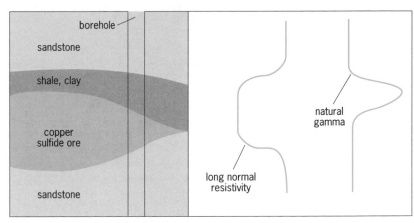

Geologic cross section and idealized resistivity and gamma logs.

the downhole probe and a reference electrode on the surface near the borehole. As an example, the top of a sulfide ore deposit is often at a different voltage than that of the surrounding rock. Another electrochemical property is measured by the induced polarization log. It indicates how strongly molecules are polarized (lined up electrically) when a voltage is applied across two electrodes spanning part of the deposit. Logging several boreholes can indicate the sizes of metallic sulfide mineral deposits.

Gamma log. The natural gamma log can distinguish certain rock types by their radioactivity. All rocks are radioactive to some degree. The probe contains a sensitive gamma-ray detector, typically a scintillator. Some gamma probes measure only the total counts. Other probes, known as spectral probes, measure the gamma rays from different elements. The most common spectral system measures uranium, thorium, radioactive potassium, and total counts. Clays and shales can be distinguished from clean sandstone, because the former generally give a higher gamma-count rate. In uranium deposits, the total gamma activity can be correlated with grade if the uranium has remained in equilibrium with its decay products. This method is not reliable if much of the uranium has been dissolved by groundwater and moved away from its less soluble radioactive decay products, which gen-

Borehole geophysical methods	
Method	Property detected or determined
Seismic (acoustic)	Seismic velocity near borehole, rock layers
Spontaneous potential	Electrochemical contrasts, metallic sulfides
Single point resistance	Resistance by borehole, leaks in plastic casing
Short normal resistivity	Resistivity in and very near borehole
Long normal resistivity	Resitivity in deposit near borehole, rock types
Induced polarization	Electrochemical contrasts, metallic sulfides
Fluid resistivity	Groundwater quality
Induction	Conductive rocks or fluids
Natural gamma	Uranium, thorium, potassium, shales, clays
Gamma-gamma	Density near borehole, coal seams
Neutron	Porosity, moisture content
Temperature	Flow of groundwater
Caliper	Borehole diameter and condition of walls
Borehole deviaiton	Straightness, position of borehole at depth
Televiewer	Fracture density and orientation

erate most of the gamma radiation. A related logging system, gamma-gamma density, contains both a source of gamma radiation and a detector. It determines rock density by measuring the gamma rays returning from the deposit near the borehole. The gamma-gamma log is often used for locating coal seams.

Hazard detection. Borehole geophysics has the potential to assist miners by indicating changing conditions in front of mining machines, such as a fault (a dislocation) of a coal seam where a coal-mining machine could hit hard rock. Borehole geophysics could also warn of hazardous conditions such as clay intruding into a coal seam where rock above the mining machine might fall, or fractured zones where the mine might collapse unless special precautions are taken to strengthen the sides and top of the mine opening. Methods for hazard detection include techniques for detecting rock types, since they can indicate clay layers and fractured rock. However, the boreholes would need to be closer together for effective hazard detection. In deep mines, the boreholes could be drilled from inside the mine to reduce drilling costs.

Patterns of fluid flow. For environmental protection, an important application of borehole geophysics is monitoring effluents such as acid mine drainage. With the proper equipment, downhole measurements of fluid conductivity, pH, oxidation-reduction potential, and temperature are more reliable and faster than collection and analysis of samples from boreholes. A downhole flowmeter and accurate measurements of the temperature can show the depth of fluid flow. Measurements in a network of boreholes can reveal the flow pattern.

When predicting fluid-flow patterns, the distribution of fractures is very important, because the fracture pattern often controls the flow pattern. A television camera in a borehole probe can provide a picture of the borehole wall. If the borehole is not cased and the fluid in the borehole is clear, the camera can show the frequency and direction of fractures. However, obtaining fracture orientations is often difficult: borehole walls may be coated with drilling mud, or the fractures may be obscured by murky borehole fluid. Unlike the television camera, a televiewer (an acoustic viewer) will work in drilling mud. The televiewer images borehole walls with high-frequency sound waves; the acoustic response can indicate fractures and their directions.

Research. Much of the research in borehole geophysics is conducted by petroleum companies, but the results have applications to mining. The mining community in the United States, including private companies, federal and state agencies, and universities, is also conducting research. An important research goal is to improve the ability to obtain two- and three-dimensional images. Testing of the experimental methods involves comparing results with known geologic structures or features of interest. The tests define the conditions under which the methods will work reliably and be cost-effective.

Tomography. One active research area is geophysical tomography. Crosshole tomography gives a two-dimensional image of the distribution of some parameter of interest between boreholes. The principles are similar to those of medical computer assisted tomography (CAT) scans, but seismic or nonionizing electromagnetic energy is used instead of x-rays. Seismic tomography usually gives the distribution of seismic velocity. For seismic crosshole tomography, a source in one borehole sends sound waves through the rock to receivers in the other borehole. The times for the sound waves to travel from many source locations to many receiver locations are measured and used to calculate the distribution of seismic velocity between the boreholes. The presence of rocks of contrasting seismic velocity can then be inferred, because seismic velocity is higher in hard rocks such as granite than in softer rocks such as sandstone. It can also show fractured regions, because the seismic velocity is lower in fractured rock than in intact rock of the same kind.

In principle, attenuation (how rapidly sound waves die out) can also be measured, but it is much more difficult to achieve. Electromagnetic tomography can yield an image of electromagnetic attenuation or velocity; it can locate rocks of contrasting resistivity and fractured regions filled with more conductive material. Clay-filled fractures have lower resistivity than hard crystalline rocks such as granite.

Electromagnetic induction. Additional research is being conducted in surface-to-borehole methods such as electromagnetic induction. Electromagnetic induction methods determine conductivity by measuring the magnetic fields produced by small currents that are induced in conductive material in the deposit. The small currents are induced by a large alternating current or by a quick shutoff of a large current in a transmitter. One advantage is that measurements can be made through plastic casing in dry boreholes, whereas direct resistivity measurements require electrical contact with the borehole wall through fluid or special electrodes. Conventional analysis assumes simple geology with uniform, one-dimensional layering. Current research is extending the analysis to two and three dimensions.

In situ mining. Research on in situ mining provides additional applications for borehole geophysics. In situ mining consists of injecting and recovering chemicals (leach solution) through boreholes to dissolve (leach) the desired material, which is then pumped to the surface. Single-point resistance logging finds leaks in plastic well casing by showing where the resistance is unusually low. For in situ uranium mining, gamma logging of injection and recovery wells helps to locate the zones of uranium ore. This information indicates the depths at which the leach solution should be injected and recovered. Borehole geophysics has potential for helping to monitor flow after injection. Monitoring methods would be similar to those used for monitoring effluents from other mines.

Combined methods. A combination of geophysical methods gives more reliable results than use of a single method. For most geophysical methods, changes in the readings can be caused by several different factors. The geologic cross section and corresponding idealized logs shown in the illustration convey the benefit of combining results from several methods. These simplified long-normal and gamma logs assume that the resistivity is similar for the shale and clay layer and for the copper sulfide ore, and that the gamma radiation is similar for the sandstone and the copper sulfide ore. The combination of logs distinguishes the different layers, but neither log by itself distinguishes all the layers. The increasing power of computers is allowing significant advances in combining results from several methods and determining what hypothesis best fits the combined data. The need for experience and recognition of the limitations of each system remain, however, and there is no substitute for the personal judgment of a geophysicist familiar with the systems.

For background information *SEE BOREHOLE LOGGING; COMPUTERIZED TOMOGRAPHY; GEOPHYSICAL EXPLORATION* in the McGraw-Hill Encyclopedia of Science & Technology.

Daryl R. Tweeton; Richard E. Thill

Bibliography. A. V. Dyck and R. P. Young, Physical characterization of rock masses using borehole methods, *Geophysics*, 50(12):2530–2541, December 1985; J. C. Hanson, *Applicability of Electrical Methods in Deep Detection and Monitoring of Conductive Lixiviants*, Bur. Mines Inform. Circ. 9308, 1992; M. N. Nabighian (ed.), *Electromagnetic Methods in Applied Geophysics*, vol. 2: *Application*, Society of Exploration Geophysicists, 1991; D. D. Snyder and D. B. Fleming, Well logging: A 25-year perspective, *Geophysics*, 50(12):2504–2529, December 1985.

Cell cycle

The cell division cycle is the process by which a cell duplicates its contents and then precisely divides these contents into two daughter cells. Although many aspects of this remarkable process remain unknown, the central controlling molecules regulating its initiation and completion have recently been identified, and much has been learned about how they work.

Regulatory molecule cdc2. The molecule called *cdc2* is a protein kinase, an enzyme that transfers a phosphate molecule from adenosinetriphosphate to another protein molecule. This covalent modification of the target protein will commonly affect enzymatic activity of the protein in diverse ways, depending on the specific target protein and on the location in the protein at which the phosphate is added. The *cdc2* kinase (or a close relative) is essential for multiple steps in the cell cycle: cycle initiation, replication of deoxyribonucleic acid (DNA), and actual cell division (mitosis).

Remarkably, the *cdc2* protein kinase is found in all eukaryotic cells examined. Where it has been possible to make detailed comparisons between different eukaryotic organisms (as different as humans and fungi), only minor differences have been found in how *cdc2* acts in controlling the cell cycle. Thus, conclusions drawn in one organism are likely to be broadly applicable to many others.

Since *cdc2* protein kinase is needed for many cell cycle steps, it appears likely that various target proteins need to be phosphorylated for these steps to occur; indeed, some such targets have been identified. However, the *cdc2* protein is inactive until it is bound by one of a growing class of proteins called cyclins.

Activation by cyclins. Members of a growing family of proteins appear to be able to bind to the *cdc2* protein and activate its protein kinase. All of these proteins are related, as determined by deducing the amino acid sequence from the DNA sequence of the genes; all contain a homologous region called the cyclin box. It is assumed that this region is critical for *cdc2* binding and activation.

Mitotic control. Unlike *cdc2*, cyclins appear restricted to controlling a single cell cycle step. One class of cyclins largely controls mitosis, the process in which the cell splits, dividing its duplicated chromosomes between two daughter cells. The activation of *cdc2* protein kinase activity by this class of cyclins has been well studied. Also, a few target proteins at this cell cycle stage have been identified. For example, one step in mitosis is the dissolution of the nuclear envelope. Nuclear lamin A, a nuclear envelope protein that is critical in making the protein coat on the inner face of the nuclear envelope, is phosphorylated by *cdc2*/cyclin protein kinase, resulting in the dissolution of the protein coat. This dissolution is believed to be required for overall dissolution of the nuclear envelope.

Cell cycle initiation. Another class of cyclins, the *CLN* cyclins (identified in yeast), appears likely to control cell cycle initiation. These cyclins most likely work by binding to and activating *cdc2*. Cell cycle initiation is an important regulatory step in yeasts; numerous external stimuli that control the cell cycle affect this step. In mammalian cells, there may be a similar regulatory step at cell cycle initiation; it is unclear if *cdc2* (or a related kinase) works at this stage in mammalian cells, and if any cyclin is involved. Still, the evolutionary conservation observed in cell cycle control makes it plausible that the lessons from yeast may be broadly applicable to mammalian cells as well, in which case *CLN*-equivalent activities might be expected. A protein called cyclin E may well perform this role in human cells, but further work is required to establish this point.

Cyclin/*cdc2* complexes may also be required for either initiation or continuation of DNA replication. It is not clear how DNA replication is controlled by cyclin/*cdc2* since no target proteins for *cdc2* phosphorylation have been identified in this process. However, cyclin A has been implicated in the control of DNA replication in a variety of experiments.

Thus, cyclins appear more specific in their action in the cell cycle than *cdc2*, acting at only one or a few steps. It is speculated that cyclins may provide substrate specificity to *cdc2*; that is, the *cdc2* protein kinase presumably phosphorylates many more different target proteins to trigger mitosis than to trigger DNA replication. These targets are the substrates of the *cdc2* enzymatic activity. Possibly the specific cyclin subunit promotes phosphorylation by *cdc2* of some substrates over others.

Regulation of cyclins in the cell. Most cyclins are regulated in amount or level through the cell cycle so that the level of the respective cyclin in the cell is at a maximum at the time of action of that cyclin. This regulation is most clearly seen for mitotic cyclins, whose level increases throughout the cell cycle, peaking in mitosis before dropping dramatically in the middle of cell division. The *CLN* cyclins show a different specificity: at least some of these proteins are regulated to peak in amount only at the beginning of the cell cycle, the time of action of these cyclins. Cyclin A, which may act in the middle of the cell cycle, shows an intermediate pattern of accumulation. The mechanisms by which cyclin level is regulated include control over cyclin gene transcription and cyclin protein degradation; different classes of cyclin are regulated in distinct ways.

Order in the cell cycle. It is important that activated *cdc2* not simply phosphorylate all of its targets at some time. One central aspect of the cell division cycle is that it is an ordered process. It is essential that mitosis not be undertaken before DNA replication is complete, or else the two daughter cells would not receive equal complements of chromosomes. If *cdc2* controls both DNA replication and mitosis, the potential for failure of ordering appears to exist. However, if cyclins provide substrate specificity, and if the levels of different kinds of cyclin are regulated differently by cell cycle progression, cell cycle events would be ordered.

The idea that cyclins confer substrate specificity is only a speculation at present, and experimentally engineered misexpression of cyclins does not have the drastic consequences to be expected if cyclin regulation were the only means of controlling order of cell cycle events. It is clear that other regulatory mechanisms affecting *cdc2* protein kinase activity do not involve regulating cyclin levels: *cdc2* is itself phosphorylated by other protein kinases, and these phosphorylations can block the action of *cdc2* toward its targets. Some evidence suggests that regulation of these phosphorylations may be critical for maintaining cell cycle order. In general, the cell cycle appears to maintain order by regulatory checkpoints, in which events that should occur late in the cycle are blocked from initiating until earlier events have been completed. Checkpoint control has been demonstrated experimentally; it remains to be seen how these controls interact with the cyclin/*cdc2* regulatory protein kinases.

For background information SEE CELL CYCLE; CELL DIVISION; CYTOKINESIS; MITOSIS in the McGraw-Hill Encyclopedia of Science & Technology.

Frederick R. Cross

Bibliography. F. Cross, J. Roberts, and H. Weintraub, Simple and complex cell cycles, *Annu. Rev. Cell Biol.*, 5:341–395, 1989; S. L. Forsburg and P. Nurse, Cell cycle regulation in the yeasts *Saccharomyces cerevisiae* and *Schizosaccharomyces pombe*, *Annu. Rev. Cell Biol.*, 7:227–256, 1990; L. H. Hartwell and T. A. Weinert, Checkpoints: Controls that ensure the order of cell cycle events, *Science*, 246:629–634, 1989; A. W. Murray and M. W. Kirschner, Dominoes and clocks: The union of two views of the cell cycle, *Science*, 245:614–621, 1989.

Cell membranes

Membrane fusion is of vital importance to all cells. It unites sperm and egg to initiate the development of a new organism. In the developing organism, individual muscle cells fuse into large multinucleated myotubes, thereby providing the functional basis for motility. Fusion of bone-degrading cells is necessary for bone growth and remodeling. Even yeast cells of opposite mating types fuse to form a diploid yeast cell. Furthermore, enveloped viruses (viruses surrounded by a membrane), such as the influenza virus or the human immunodeficiency virus (HIV), which causes acquired immune deficiency syndrome (AIDS), must fuse with their target cells to infect them and ensure viral propagation. These examples are extracellular events that involve fusion of plasma membranes of the fusion partners. Cells also support numerous but equally important intracellular fusion processes, such as the reassembly of intracellular organelles after dispersal during cell division, or vesicle fusion during intracellular transport or along the secretory pathway.

Under physiological conditions, protein-free lipid membranes or liposomes, with the same lipid composition as cells or viruses, do not fuse with one another. It follows that proteins are required to promote membrane recognition and fusion.

Despite the abundance of cellular membrane fusion events, more is known about viral membrane fusion systems because they are relatively simple and therefore easier to study. Unlike cells, enveloped viruses usually carry only very few different proteins on their surface, including special membrane fusion proteins. The general principles of protein-mediated viral fusion are used below as a paradigm to explain the conceptually similar requirements for cellular fusion, with an emphasis on sperm–egg fusion. The common denominator of all naturally occurring fusion events is that they are highly regulated and include a specific recognition and binding step followed by membrane fusion and the union of the fusing partners.

Viral fusion. Viral membrane fusion proteins are glycoproteins that are anchored in the viral membrane through a hydrophobic transmembrane domain. Usually, they are made as larger inactive precursors that must be proteolytically cleaved to become functional. The most important feature of viral fusion proteins is the presence of a second, short (20–30 amino acids) hydrophobic domain known as the fusion peptide. In addition to the fusion peptide, viral fusion proteins contain a domain involved in target-cell recognition and binding. It is thought that once a virus has bound to its target cell the fusion protein unfolds, allowing the fusion peptide to insert into the target cell membrane and induce membrane fusion.

Influenza hemagglutinin. The best-studied example of a viral fusion protein is influenza hemagglutinin, a trimeric membrane-anchored glycoprotein with a fusion peptide. Hemagglutinin derives its name from its ability to agglutinate, or cluster, red blood cells by binding to sialic acid residues present on their surface. Binding to sialic acid, which is also present on other cells, allows the influenza virus to recognize and bind its target cell, usually an airway epithelial cell. Once bound, the influenza virus enters the cell via the endocytic pathway, along which are compartments with increasingly lower pH. At a certain threshold (around pH 5), the hemagglutinin trimer unfolds, uncovering the previously buried fusion peptide. The fusion peptide then inserts into the endosomal membrane and induces membrane fusion, thus releasing the viral genome into the cytoplasm.

HIV gp120/40. The HIV fusion protein, gp120/40 (a glycoprotein with two subunits of molecular weight 120,000 and 40,000, respectively), is also an oligomeric membrane-anchored glycoprotein containing a hydrophobic fusion peptide. HIV gp120/40 binds to CD4, a receptor protein on the surface of T lymphocytes. It is not yet clear what triggers gp120/40 to induce fusion. The HIV can fuse with the cell surface without being endocytosed into a low-pH compartment. It is thought that binding to CD4 may somehow trigger a conformational change in gp120/40, resulting in exposure of the fusion peptide and membrane fusion. Regardless of the different infection pathways and the target-cell surface receptor used, the principal functions of viral fusion proteins are very similar. Recognition and binding is followed by a conformational change in the fusion protein, leading to exposure of the fusion peptide and ultimately membrane fusion.

Cellular fusion events. The simple principles outlined for viral fusion (recognition, binding, and fusion) are equally important for cellular membrane fusion, although the responsibility for the different steps may be shared between participating proteins. The task of recognition and binding in both intracellular and extracellular membrane fusion is performed by proteins referred to as receptors and their corresponding ligands (or counterreceptors). Recognition of and binding to the appropriate fusion partners precede fusion. In analogy to viral fusion, cellular fusion is thought to be mediated by specific membrane-anchored fusion proteins that contain a fusion peptide.

Intracellular membrane fusion reactions, which are not discussed in detail here, can be studied in yeast by using genetics, or in mammalian cells by using purified subcellular organelles. In both cases, intricate pathways with many genetically or biochemically distinguishable steps have been described. Experiments to date have identified steps leading up to the fusion event. These steps include vesicle budding, vesicle targeting, and the assembly of a so-called fusion machine, consisting of a number of different proteins that function in membrane binding and fusion.

Extracellular membrane fusion. An important extracellular fusion event is sperm-egg fusion. For sperm, recognition of and binding to the egg is a multistep process. Sperm are first attracted to the egg by chemoattractant factors. Once they have moved

through the layer of follicle cells surrounding the egg, they encounter the zona pellucida, a glycoprotein coat consisting predominantly of the glycoproteins ZP1, ZP2, and ZP3. Sperm carry receptors for ZP3 on their surface. Binding to ZP3 is very specific and is possible only between sperm and eggs of the same species. Upon binding to the zona pellucida, sperm receptors for ZP3 trigger the acrosome reaction, releasing proteolytic enzymes from a compartment of the sperm head called the acrosome. The released enzymes enable the sperm to burrow its way through the zona pellucida. Next, the sperm must recognize, bind to, and fuse with the plasma membrane of the egg. On guinea pig sperm, a potential fusion protein called PH-30 has been identified that may be responsible for those two final steps. It is very likely that a protein similar to PH-30 is present on the sperm of other animals and on human sperm.

PH-30 is considered to be a sperm-egg membrane fusion protein for several reasons. An antibody against PH-30 blocks sperm-egg membrane fusion. PH-30 is localized on the posterior aspect of the sperm head, which in mammals is the region that fuses with the egg. PH-30 is similar to known viral fusion proteins in that it is an integral membrane glycoprotein made up of two subunits, both made as larger precursors. Cleavage of one of the precursors correlates with the acquisition of fertilization competence in sperm. Most importantly, PH-30 has the two hallmark features of viral fusion proteins: a potential hydrophobic fusion peptide on one of its subunits, and a specific recognition and binding domain on the other.

When PH-30 was sequenced, the role of its binding domain was revealed through a remarkable similarity with short, soluble peptides called disintegrins, which are found in the venom of pit vipers. In the victim of a snake bite, disintegrins bind tightly to specific blood platelet receptors called integrins, thereby preventing a critical step in blood clotting. Integrins, in turn, are a large family of cell-surface receptors that are found on virtually every cell type. Integrins play important roles in cell adhesion and signaling. Since soluble snake venom disintegrins bind to integrins on platelets, it is thought that the membrane-anchored disintegrin domain present on PH-30 binds to an integrin on the egg. In analogy to viral fusion, binding to an egg integrin may trigger a conformational change in PH-30, expose the fusion peptide, and induce fusion. As receptor for a fusion protein, the egg integrin would serve a similar function in fertilization to the one served by the CD4 protein for HIV infection.

Prospects. Specialized proteins for recognition and binding, as well as for fusion itself, appear to be required for all naturally occurring membrane fusion events. Their study not only provides insight into the underlying principles of membrane fusion but may also lead to new ways to aid or block the process. Promoting membrane fusion may someday allow the delivery of healthy muscle cells to individuals with muscular dystrophy or may help cure certain types of infertility. Conversely, inhibiting membrane fusion may prevent viral infection or specifically block fertilization without the side effects of hormone-based contraceptives.

For background information *SEE CELL MEMBRANES; CELL-SURFACE DIFFERENTIATIONS; FERTILIZATION; VIRUS* in the McGraw-Hill Encyclopedia of Science & Technology.

Carl P. Blobel

Bibliography. B. Alberts et al., *Molecular Biology of the Cell*, 1989; C. P. Blobel et al., A potential fusion peptide and an integrin ligand domain in a protein active in sperm-egg fusion, *Nature*, 356:248–252, 1992; P. Primakoff et al., Identification and purification of a sperm surface protein with a potential role in sperm-egg membrane fusion, *J. Cell Biol.*, 104:141–149, 1987; P. M. Wassarman, Mouse gamete adhesion molecules, *Biol. Reprod.*, 46:186–191, 1992; J. M. White, Viral and cellular membrane fusion proteins, *Annu. Rev. Physiol.*, 52:675–697, 1990.

Cell walls (plant)

It is estimated that worldwide, 10% of grains and 20% of perishable crops are lost to microorganisms during postharvest handling and storage, largely as a direct consequence of plant wounding. A wound is any external or internal injury that disrupts the integrity or function of a plant and results in the destruction of cells in a specific area of tissue. Like most other organisms, plants face two major problems when wounded. First, tissue fluid and nutrient loss occur. Second, a wound site provides a pathway for possible infection by many disease-causing microorganisms such as bacteria and fungi. Healing is the process by which a plant partitions or sequesters the area of injured cells to maintain the integrity of adjacent healthy cells.

Sources of wounds. There are three general sources of plant wounds: external biotic sources, abiotic sources, and endogenous sources.

Biotic wounds result from the natural interaction of a plant with some other organism. Some of the common biotic wounds result from challenges by infectious pathogens such as fungi and bacteria, which derive their nutrition from the digestion of plant cells. Fungi and bacteria synthesize degrading enzymes such as pectinases, cellulases, hemicellulases, lipases, and proteases, resulting in dissociation of cells, degradation of walls, rupture of the plasma membrane, and death of the cells.

Abiotic wounds result from a plant's interaction with nonliving components in its environment. Many abiotic wounds are a consequence of physical damage from meteorological phenomena, such as wind, lightning, ice, drought, and dust. In agriculture, during planting, growing, harvesting, and processing, plants may be cut, bruised, or punctured. Environmental pollutants, such as acid rain, heavy metals, industrial emissions, and herbicides, can cause chemical injury of plant tissues.

Wounding endogenous to the plant occurs during normal ontogeny, that is, during dormancy, growth, development, maturation, and senescence. For example, the emergence of shoots and roots from a seed, development of lateral roots, and growth of a pollen tube are life processes that wound cells. Ontogenetic wounding may result from indirect or distal external environmental factors such as seasonal changes in photoperiod, temperature, or rainfall. These factors can cause the development or abscission of plant organs. More proximately, such wounding may be a consequence of a change in levels of hormones. Usually concurrent with endogenous wounding, plants initiate formation of protective cell layers at the developing wound site. For example, as days shorten and cool during fall in temperate regions, leaves of woody dicotyledons separate from the stem at the base of the petiole. During this process, an abscission zone develops consisting of two distinct layers—a top separation (abscission) layer of structurally weak cells where detachment of the petiole occurs, and a lower protective layer of suberized cells that remains as a leaf scar on the stem.

Structures and substances. Through processes of evolutionary selection, plant cells have developed molecular pathways that generate specialized structures and substances for protection against injury. Three widespread wound-barrier substances are suberin, lignin, and callose. Other substances appear to be specific to a particular plant species; for example, phytoalexins provide resistance to particular pathogens. Even between individual plants in a species there are genetically inherited, but environmentally influenced, qualitative and quantitative differences in the synthesis of the various wound-barrier substances.

General description of wound healing. Typically, as a consequence of cell wounding, the middle lamella, cell wall, and plasma membrane are disrupted and the protoplasmic organelles, structures, and fluids are released into the environment of the surrounding cells. Wound healing appears to be initiated within seconds after injury has been incurred. Completion may be within hours or extend into years depending upon many factors, including the existing age and condition of the plant, the extent of the wound, and other stress-related factors such as adverse weather conditions. Although the sequential processes associated with a wound response occur as a continuum, there are distinct phenomena that culminate in healing. For example, three distinct sequential events are associated with cortical and perimedullary cells in the healing of cut wounds in a potato tuber (*Solanum tuberosum*): (1) The middle lamellae and primary walls of wound-reaction cells become progressively lignified, and then suberin lamellae are deposited along the inner surfaces of their walls. (2) A wound (cork) cambium develops from a single layer of cells immediately beneath the wound-reaction cells. (3) Cork cells are produced by the wound cambium. The cork cells also become lignified and then suberized.

Wound signals. An injury response requires a signal or elicitor for activation of a transduction pathway that leads to the synthesis of a wound-reaction barrier. Signals must both affect wound-reaction cells immediately adjacent to the wound and initiate wound cambium formation several cell layers distant from the wound site. There appear to be two general pathways for signal reception in plants. One pathway includes either those factors associated with the plasma membrane that activate ion channels, or membrane receptor proteins that activate second messengers, thus initiating cytosolic enzyme synthesis of substances to be utilized in healing. Another pathway includes signals that traverse the plasma membrane and initiate intracellular signal systems that regulate gene transcription by binding to specific deoxyribonucleic acid (DNA) sites in chromosomes of the nucleus. The genetic messages are subsequently translated and processed into regulatory enzymes involved in the metabolic pathways concerned with wound healing. Both pathways may initiate multiple effects, which are further modulated by other regulators, that is, environmental factors, and may be specific to cell type and maturational status of the plant. Thus, metabolic pathways are initiated in response to wounding; they degrade damaged cell components, digest energy-rich nutrient stores of starch and oils, and synthesize molecules associated with wound repair and construction of protective barriers.

Identification and demonstration of a particular phenomenon as a wound signal directly linked to initiation of healing has been challenging. Historically, changes in physical, chemical, and electrochemical states have been investigated as possible sources by which plant cells perceive injury. Early investigators suggested that a change in turgor pressure or water loss initiates the wound response or that the release or oxidation of particular substances from within a wounded cell stimulates healing. Traumatic acid (*trans*-2-dodecenedoioic acid) was called a wound hormone after it was isolated and determined to initiate cell division and growth in bean endocarp cells. It was later found to be present in both healthy and wounded bean cells and thought to be a normal product of the fatty-acid breakdown of the plasma membrane. One hormone definitively associated with wound healing is ethylene, a growth regulator known to promote senescence and abscission of leaves, flowers, and fruits. The hormone auxin is known to initiate abscission but also appears to regulate ethylene synthesis. Auxin acts in conjunction with another hormone, cytokinin, to regulate cell division during healing. The growth regulator abscisic acid inhibits the effects of all three hormones and may act as a signal terminator.

Phytoalexins vary between tissue types, and each tissue can produce a variety of molecules. However, all known phytoalexins are polar lipids and may kill both pathogen and host cells by becoming integrated into the plasma membrane and disrupting permeability and transport properties. In the case of infection, some of the potential elicitors associated with the production of phytoalexins are nonspecific fungal polysaccharide wall fragments including chitosans, glycoproteins, and arachidonic acid. Some plant cells synthesize the enzymatic hydrolases chitinase and β-1,3-glucanase in response to specific fungal infections, and these cell hydrolases are known to degrade fungal cell walls.

Second messengers. The plasma membrane is a dynamic and variable structure. Its local asymmetry is determined in part by the distribution of particular membrane fatty acids, protein types, and hormone receptors and by its electrochemical potential, which can vary from -80 to -120 mV. During wounding, the membrane electrochemical potential is temporarily disrupted when there is a net cellular efflux of potassium (K^+) and chloride (Cl^-) ions and a net influx of hydrogen (H^+) and calcium (Ca^{2+}) ions. The Ca^{2+} ion has gained recognition as an important regulator or second messenger that complexes with a cytosol protein, calmodulin. The complex binds to specific sites of Ca^{2+}:calmodulin–sensitive enzymes linked to promotional regulators of metabolic and peroxidase activity, phytoalexin synthesis, membrane-associated enzymes such as callose synthetase, and several polysaccharidases. Calcium ions may also be important plasma membrane stabilizers assisting in maintenance of membrane integrity.

Ultrastructural changes. Ultrastructural changes associated with healing can be observed within minutes of wounding and include the presence of callose and the degradation of starch grains several cells distant from the wound, reflecting the high energy demands for increased metabolic activity. Cells immediately adjacent to the wound surface have increased nuclear activity, with increased numbers of dense nucleoli and ribosomes reflecting that signal pathways are being transformed into protein synthesis. Often the nucleus takes on a distinct ameboid shape associated with increased surface area for import-export exchange of substances. Within the cytoplasm, there is increased protein synthesis for export, as suggested by increased activity of the rough endoplasmic reticulum, Golgi apparatus, and vesicle formation. There is also increased activity at the plasma membrane as substances are imported and exported. Important information is lacking as to how the observed ultrastructural changes are directly related to the synthesis of specific substances associated with wound healing. For example, although lignin is one of the most important plant biopolymers known, specific organelle activity related to its synthesis has yet to be fully demonstrated.

Genetic regulation. Biotechnologists have begun to explore gene expression associated with wound healing. For example, in poplar trees wound-inducible production of particular messenger ribonucleic acid (mRNA) sequences coding for enzymes such as chitinases (fungal growth inhibitors) have been identified. Signal molecules that appear to assist in systemic plant resistance to pathogens and pests are also being investigated. One candidate is salicylic acid, which is widely distributed in dicotyledons and monocotyledons. Salicylic acid is produced via the shikimic pathway in response to pathogen proteins and travels through the phloem to other leaves. Another interesting hormone is systemin, a small polypeptide found in tomato leaves that can also travel through phloem. Systemin promotes the production of proteinase inhibitors that may protect against herbivory by decreasing the digestibility and nutritional quality of leaf proteins. Recently, it has been discovered that human touching of a plant can stimulate the activity of calmodulin, which then induces the expression of four specific touch genes that inhibit further plant growth. In response to touch, *Arabidopsis* plants develop with reduced height and shorter petioles and begin to flower sooner.

For background information SEE CELL WALLS (PLANT); PLANT GROWTH; PLANT HORMONES in the McGraw-Hill Encyclopedia of Science & Technology.

Norman Thomson

Bibliography. R. M. Bostock and B. A. Stermer, Perspectives on wound healing in resistance to pathogens, *Annu. Rev. Phytopathol.*, 27:343–371, 1989; C. J. Lamb et al., Signals and transduction mechanisms for activation of plant defenses against microbial attack, *Cell*, 26:215–224, 1989; N. G. Lewis and M. G. Paice (eds.), *Plant Cell Wall Polymers*, 1989; P. Raven, R. F. Evert, and S. Eichhorn, *Biology of Plants*, 1992.

Cellular adhesion

The majority of microbes probably exist in nature in close association with particular surfaces. The adhesive properties of microorganisms were first recognized at the beginning of the twentieth century. Since then it has been shown that bacterial adhesion is important in plant and animal hosts, pathogenesis, medical devices, aquatic and soil ecosystems, biodegradation, and industrial processes.

In human and animal pathogenesis, bacterial adhesion plays an important role in diseases such as urinary and gastrointestinal infections, respiratory ailments, wound infections, septicemia, infections at the site of artificial joints and other implants, dental plaque, and dental cavities.

Properties. The disease-provoking ability of pathogenic microorganisms can be assigned to four major properties of the organism, involving the ability to enter the host, resist or not stimulate host defenses, acquire essential nutrients for growth and multiplication in or on the host tissues, and damage the host. These properties can be ascribed to specific virulence factors, often involving surface components of the organism.

An important surface property of the bacteria is their ability to adhere to and subsequently colonize cell surfaces of the host. This property is an essential attribute of pathogens not normally found in or on the host animal. Nevertheless, it also seems to be an important aspect for commensals of the normal flora that are harmless, or even helpful, in their usual habitat but that may become significant agents of disease if they establish growth in other habitats. Disease is not caused by adhesion as such but by other virulence factors produced by the microorganism.

Entry of the microorganism into the host is more easily accomplished through the alimentary, respiratory, and urogenital tracts and the conjunctiva than through the intact skin. A number of host-defense mechanisms have developed to deal with the danger of bacterial colonization. Defenses against bacterial colonization include cleansing mechanisms such as coughing, sneezing, kidney flushing (cleansing action of urine), and bowel movements. In order to colonize, the microorganism has to bind to the host's epithelial cells, thus avoiding being swept away. Mechanisms by which bacteria maintain proximity to the host's cell surface can be roughly categorized as association, adhesion, and invasion, according to the degree of intimacy between the bacterial and epithelial surfaces.

Association. The localization of bacteria on a surface that does not involve any specific mechanisms is termed association. The term refers to the loose reversible attachment of bacteria to a surface. Association may precede specific adhesion. It is thought that bacteria maintain their position along a mucosal surface by establishing a number of noncovalent bonds to the mucosa by chemotaxis or motility. Chemotaxis enables the bacteria to exploit mucosal regions for optimal substrate and nutrient availability. Bacteria most commonly move by means of flagella, which are unbranched helical filaments of uniform thickness on the bacterial surface. Dispatching of motile cells from established colonies allows colonization of mucosal surfaces to progress in a direction opposite to physical cleansing mechanisms.

This initial reversible phase of attachment, or association, may be the result of the combined effects of van der Waals forces, hydrogen bond formation, and ionic and hydrophobic interactions.

Adhesion. The relatively stable attachment of bacteria to surfaces is known as adhesion. In contrast to association, adhesion requires numerous noncovalent bonds between complementary molecules of adjacent surfaces. A possible physicochemical explanation for the net adhesive interaction between two like charged surfaces is provided by the Derjaguin-Landau and Verwey-Overbeek (DLVO) theory. This theory of the physical interaction between colloid particles states that as two bodies of like charge approach each other they are affected by both attractive and repulsive forces, which vary independently with the distance between the bodies. In addition, cations such as Ca^{2+} (calcium), Mn^{2+} (manganese), and Fe^{3+} (iron) may act as ionic bridges between negatively charged surface molecules.

Bacterial adhesion is a highly specific phenomenon in which microorganisms display preference for certain tissue sites. Specific adhesion requires the interaction of specialized complementary molecules between surfaces of substratum and bacteria. Any bacterial structure or molecule responsible for adhesion is called an adhesin. Bacteria display an assortment of adhesins that are composed of proteins, polysaccharides, lipoteichoic acid, or conjugates of these. The majority of adhesins that have been characterized at the molecular and genetic level are surface proteins. These proteinaceous adhesins can be categorized as filamentous or amorphous (lacking definite shape and size).

All nonflagellar filamentous appendages are termed fimbriae or pili (from Latin for thread and hair, respectively). To avoid confusion, it has been suggested to limit the term fimbriae to the adhesive organelles (adhesins) and the term pili to the sexual appendages involved in transfer of genetic material between bacteria. The amorphous adhesins are referred to as nonfimbrial or afimbrial adhesins. The term receptor is used for both known and putative components on surfaces that bind to the corresponding adhesin during specific bacterial adhesion.

Fimbriae and nonfimbrial adhesins. Virtually every gram-negative bacterial species that has been examined, but only two species of gram-positive bacteria, produce fimbriae. Fimbriae differ in morphology, composition, and receptor specificity.

The filamentous structure of most fimbriae is readily apparent when negatively stained bacterial suspensions are viewed by electron microscopy (see **illus.**). Fimbriae are arranged in polar (at one end) or peritrichous (all around) fashion on the bacterial cell and vary in number. They are 2–7 nanometers in diameter and may extend up to 4 micrometers from the bacterial surface. Unlike flagella, which are thicker and curved, fimbriae are flexible and either straight or kinked. Biochemically, fimbriae are protein polymers, composed mainly of identical subunits (about 1000 subunits per fimbria). The molecular weight of a single subunit ranges from 15,000 to 30,000. The subunits are held together in stable threadlike structure via hydrophobic and electrostatic interactions.

In several fimbrial types, it has been shown that the actual adhesive molecule is a protein distinct from the fimbrial subunit. This molecule is integrated in the fimbrial stalk and is situated either on the tip or throughout the length of the filament. Genes encoding for the subunits, for the transport of the subunits through the cell membrane, for the anchoring of the subunits to the membrane, and for the regulatory genes have been first identified and described in detail in _Escherichia coli_ strains. The adhesins of nonfimbriated bacterial species tend to be outer membrane proteins or secreted proteins that remain loosely associated with the bacterial surface. For some bacterial species, the adhesin is an exopolysaccharide or a lipid component rather than a protein.

The production of bacterial adhesins is regulated in various ways. Regulation results in economizing bacterial resources and escaping host immune defenses by changing antigenic characteristics of the bacterial surface. Phase variation is a regulation system that involves a reversible on-off process between the states of fimbriation and nonfimbriation. Fimbriae production is also regulated by temperature. The optimal expression of adhesins is at physiological temperatures (89–99°F or 30–37°C). At lower temperatures (62–69°F or 18–25°C), few fimbriae are produced, and below 62°F (18°C) production ceases completely. The composition of the growth medium with regard to different nutrients also affects the expression of a given fimbrial type.

Receptors. The epithelial cell membrane and the overlying mucus layer contain a diverse group of glycoproteins and glycolipids. The glycoconjugates are anchored in the phospholipid layer of the cell membrane, and oligosaccharide side chains project from the cell surface to function as potential receptor molecules. A variety of pathogenic microorganisms have evolved mechanisms to utilize mucosal surface receptors as nutrients and thereby enhance colonization. Because of the molecular diversity provided by glycoconjugates, these substances are ideally suited for specific recognition of proteinaceous adhesins. The repertoire of receptor molecules of a mucosal surface is a major determinant of the number and kinds of bacteria capable of infecting the tissue and consequently the host. To date, very few actual receptors have been identified.

Invasion. The ability of some pathogenic bacteria to enter and multiply in the epithelial cells of the host is known as invasion. This intracellular colonization may provide a new source of nutrients for the bacteria as well as protection from many host defense mechanisms. Intracellular pathogens have mechanisms that enable them to enter the host, avoid being killed, multiply, and provide escape for their offspring to new cells.

To invade the epithelium the following sequence of events must take place: (1) attachment of the microorganism to the epithelial surface, (2) degeneration of the microvilli, (3) local dissolving of the cell membrane, (4) degeneration of cell junctions and cellular organelles, (5) engulfment of the bacterium in a membrane-bound vacuole, (6) bacterial multiplication, (7) escape from the vacuole into the cytoplasm, and (8) spread of the microorganism to other host cells.

Prospects. The correlation between adhesion and pathogenesis has led to a search for means by which adhesion can be prevented. Attempts are being made to produce antiadherence vaccines that will provide long-lasting protection from pathogens. The veterinary sector has obtained promising results, and some vaccines are available commercially that provide short-term protection.

For background information SEE BACTERIA; CELL MEMBRANES; CELLULAR ADHESION in the McGraw-Hill Encyclopedia of Science & Technology.

Karen A. Krogfelt

Bibliography. R. J. Doyle and M. Rosenberg (eds.), _Microbial Cell Surface Hydrophobicity_, American Society for Microbiology, 1990; K. A. Krogfelt, Bacterial adhesion: Genetics, biogenesis, and role in pathogenesis of fimbrial adhesins of _Escherichia coli, Rev. Infect._

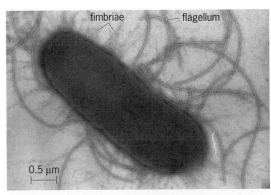

Electron micrograph of an _Escherichia coli_ bacterium possessing fimbriae and flagella. (_J. Blom, Department of Molecular Cell Biology, Statens Seruminstitut, Copenhagen, Denmark_)

Dis., 13:721–735, 1991; J. A. Roth (ed.), *Virulence Mechanisms of Bacterial Pathogens*, American Society for Microbiology, 1988.

Cellular immunology

The ability to combat infections of bacteria, viruses, and parasites critically depends upon the type of white blood cells called lymphocytes. These cells are uniquely able to recognize and distinguish among a large number of different molecules. Genetic diseases in which lymphocytes cannot be made or cannot function properly are invariably fatal within a year or two of birth unless the disease can be treated. Similarly, severe, life-threatening infections are common in patients with acquired immunodeficiency syndrome (AIDS). In this disease, a subset of lymphocytes, the helper T lymphocytes, are destroyed by the human immunodeficiency virus (HIV). Helper T cells play a central role in the body's immune defenses, as they are required for both cell-mediated immune responses and the production of antibody against many antigens. In the latter process, helper T cells play a regulatory role, providing a necessary stimulus to B lymphocytes, which then proliferate and secrete large quantities of antibodies. Recent investigation has provided considerable insight into the molecules responsible for the B-cell–T-cell interactions.

Antigen binding. Lymphocytes recognize antigens via antigen receptors, which have the ability to distinguish between different molecular structures. All of the antigen receptors expressed by an individual lymphocyte have the same specificity for antigen, whereas different lymphocytes recognize different antigens. In the case of the B cell, the antigen receptor is a membrane-bound form of the same antibody that is secreted by a B cell after it has become activated to make an immune response. The B-cell antigen receptor recognizes antigen in its unaltered form. T cells utilize an analogous set of molecules as the antigen receptor but recognize only antigen that has first been broken down by another cell and has then become bound to that cell's surface major histocompatibility complex (MHC) proteins.

B-cell activation. Antigen binding to B-cell antigen receptors plays a primary role in activating B cells, but generally this event is not sufficient to induce expansion of the B cells or to induce high-level antibody production. Rather, helper T cells must provide further activation signals in the form of secreted protein mediators (called cytokines or interleukins) and in the form of cell-contact-dependent signals. Helper T cells produce these additional signal molecules only after they contact the antigen that is recognized by their antigen receptors.

Classical experiments have revealed that a vigorous antibody response requires an antigen molecule in which the parts recognized by the B cells and the parts recognized by the helper T cells are present on the same molecule. If the B cells and the T cells recognized parts on different molecules, a much weaker response was seen. Recent studies have provided strong support for the following explanation of this phenomenon. First, the antigen binds to the antigen-specific B cell via the B cell's membrane antibody molecules. This antigen is then taken into the cell by endocytosis, a process in which small regions of membrane are budded off into the cytoplasm of the cell to create an endocytic vesicle or endosome. The center of the vesicle contains fluid and molecules derived from the exterior of the cell, including the antigen bound to the antigen receptor. The endocytosed antigen is degraded and, in the case of protein antigens, gives rise to short antigenic peptide fragments. These peptide fragments can bind to the class II MHC proteins present in the endosomes of the B cell, and the peptide/MHC class II complex can then go to the cell surface. This complex provides the structure that is recognized by the antigen receptors of T cells, and can bring the antigen-specific helper T cell together with the antigen-specific B cell. Recognition of antigen also promotes the activation of the helper T cell so it, in turn, produces the cytokines that enhance B-cell proliferation and antibody production.

Cell-contact-dependent signal. It has become clear that antigen stimulation of helper T cells also leads to the synthesis and appearance at the T-cell surface of a molecule that provides a strong cell-contact-dependent signal to the B cell. This molecule has recently been identified, and the gene that encodes it has been isolated. It is called CD40 ligand, as the molecule on the B cell that receives this signal was previously identified and named CD40. Initial studies of CD40 ligand indicate that it acts at an early stage to promote expansion of the antigen-specific B cell and that the cytokines act later to increase cell proliferation and promote production and release of antibodies.

Although the basic features of the interaction between the B cell and the helper T cell are clear, key questions remain. One of these relates to the timing of the interaction. As B cells and T cells specific for any single antigen are likely to be present in quite low numbers, probably no more than 1 in 10,000 lymphocytes under most circumstances, it is not clear how these rare antigen-specific B cells and T cells find each other. Does the interaction occur early in an immune response, or do the cells first need to proliferate to expand the number of antigen-specific lymphocytes? A clue to the answer may be the observation that helper T cells can recognize antigen taken up and processed not only by B cells but also by macrophages or dendritic cells. Presentation of antigen by these other types of cells can lead to proliferation of the antigen-specific T cells and increase in their numbers. Physiologically, this step may occur prior to the interaction of the helper T cell with the antigen-specific B cell.

Also related to the issue of the timing of the interaction of B cells and T cells is the observation that the signal-conveying molecules delivered by the helper T cell to the B cell are not stored within the T cell for rapid release but are made only after antigenic stimulation of the T cell. Appearance of CD40 ligand on the cell surface and production of cytokines appear to take at least 4 h. Experiments have been done to examine whether the helper T cell will still be interacting with the B cell at this time. Once the antigen-specific helper T cell recognizes the antigenic peptide/MHC class II complex on the surface of the B cell, it strengthens its attachment to the B cell. This coupling appears to last long enough to provide the cell-contact and cytokine signals to the B cell.

Second signals. Not only does the antigen-based interaction between the helper T cell and the B cell have important consequences for the B cell, but it can also play a critical role in regulating the behavior of the helper T cell. Just as the B cell requires more than an antigenic signal to become activated to make an antibody response, so the helper T cell requires more than just the antigenic peptide/MHC class II complex to be activated.

Second signals control the nature of the response of lymphocytes to antigen. In order to activate the helper T cell properly, the B cell must express a cell surface molecule called B7. This molecule interacts with a molecule on the T-cell surface called CD28, and this interaction provides the necessary second signal to greatly increase cytokine production by the antigen-stimulated helper T cell. In addition, the activated T cell synthesizes and displays on its cell surface the CD40 ligand molecule, which can interact with the CD40 molecule on the B cell to promote early activation events in the B cell. This second signal, along with some of the cytokines, induces the B cell to multiply and its progeny to become antibody-producing factories. The existence of several additional molecules that play roles similar to B7 and CD28 has also been suggested. In the absence of second-signal molecules, the antigen-recognizing helper T cell becomes inactivated; the process is thought to prevent immune responses to components of the host organism.

Since the immune system has powerful mechanisms to kill invaders, and the helper T cell is a critical regulator of the immune system, it is important to prevent T-cell responses to the body's own components. Curiously, the regulated cell provides the key instruction here. Clearly, the regulation of the production of these second signals is an area where further understanding is needed. It appears that various infections in some way induce the production of these second signals, thus providing vital direction to the helper T cell. Further study of the molecules that mediate regulatory signals exchanged between T cells and B cells will sharpen the picture of how immune responses are regulated. Ultimately, such understanding should prove useful in the development of methods for treating AIDS, autoimmune diseases, and other diseases involving the immune system.

For background information *SEE ACQUIRED IMMUNE DEFICIENCY SYNDROME (AIDS); CELLULAR IMMUNOLOGY; HISTOCOMPATIBILITY; IMMUNOGLOBULIN* in the McGraw-Hill Encyclopedia of Science & Technology.

A. L. DeFranco

Bibliography. A. L. DeFranco, Lymphocytes offer a helping hand, *Curr. Biol.*, 2:477–479, 1992; A. L. DeFranco, Tolerance: A second mechanism, *Nature*, 342:340–341, 1989; R. J. Noelle and E. C. Snow, T helper cells, *Curr. Opin. Immunol.*, 4:333–337, 1992; Y. Liu and P. S. Linsley, Costimulation of T-cell growth, *Curr. Opin. Immunol.*, 4:265–270, 1992.

Chemical synthesis

Recent advances in chemical synthesis include the development of a process known as template-mediated synthesis.

Template-mediated synthesis. Also known as molecular imprinting, template polymerization, or host-guest polymerization, template-mediated synthesis is designed to produce substrate-selective materials. The basic steps in this technique are (1) mixing the template molecule with monomers, allowing any intended interaction to occur before polymerization (preorganizing functionalized monomers with the template); (2) polymerizing the mixture; and (3) removing the template from the resulting material by extraction (see **illus.**). Polymerization thus freezes the geometry of the template into the polymer, yielding a material with the ability to selectively absorb the template. This selectivity can be attributed to a superior fit for the template compared to structural analogs, similar to the manner in which a substrate is recognized by the geometry of an enzyme's binding site. Template-mediated synthesis avoids the potentially difficult and complex step-by-step synthesis of an enzymelike substrate binding domain by using the substrate molecule as a template around which the polymer assembles. This process is analogous to producing a three-dimensional mold of an object.

Polymer systems. Several types of polymer have been investigated as the base material for template-mediated synthesis. Some examples are polystyrene, polyacrylates, and silica. In each case, the key ingredient is a cross-linking agent (such as divinyl benzene for polystyrene), which produces a rigid network into which the geometry of the template is cast. The rigidity of the polymer ensures that the geometry is preserved. In a loosely cross-linked network, the polymer chains are free to move relative to each other and would lose the imprint of the template. However, if the network is too inflexible, the binding domains may permanently trap the target molecule, preventing extraction after polymerization. Flexibility is controlled by the nature and proportion of the cross-linking agent.

The polymer network is believed to form cavities around the template. These cavities can provide some degree of selectivity based on steric interactions alone. Early studies using certain dyes in a styrene/divinyl benzene polymer exhibited this type of binding behavior. In these cases, the fit of the molecule into the cavity appears to be responsible for the selectivity observed.

In order to enhance the ability of these cavities to select for the template molecule, a monomer that interacts with (binds to) the template molecule can be added to the reaction mixture prior to polymer-

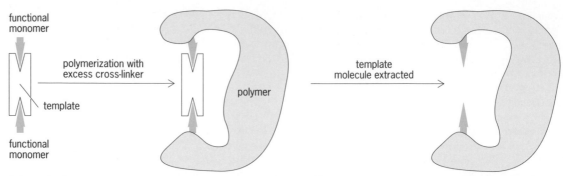

Schematic of template polymerization process. The template is preorganized with a functional monomer. Polymerization with excess cross-linker produces a polymer with selective binding cavities, and the template is extracted to leave the binding site available.

ization. In this way, information about both the geometry and chemistry of the template can be transcribed onto the polymer network. One of two different approaches is used: (1) the template is covalently, but reversibly, bound to a monomer, or (2) a monomer is selected that can interact noncovalently with the template.

Covalently mediated synthesis. Covalent, reversible binding requires a monomer that is capable of forming a labile covalent bond with a given template. There are several examples of resolution of optically active carbohydrates (sugars) on templated polymers through the use of boronate linkages, which are readily formed or broken under acidic conditions. The position of the hydroxyl groups (the point of attachment on the template) is imprinted on the polymer matrix through the inclusion of phenylboronic acid groups into binding cavities. After the sugar is eluted, the phenylboronic acid moieties retain their relative spacing on the matrix, providing a memory of the hydroxyl spacing. Other pairs of reactive groups have also been used with some success, including ketones with hydroxyls, and aldehydes with amines.

Although the above examples have exhibited some degree of selectivity, some factors limit the general applicability of covalent interactions. The number of readily reversible reactions is relatively small; even within this group, the rate of bond formation or cleavage can be quite slow if access to the site is hindered, as is often the case in a polymeric matrix. In addition, the conditions required to reverse the interaction may adversely affect either the polymer or the template. For example, some optically active compounds such as amino acids racemize under extreme basic or acidic conditions or at elevated temperatures.

Noncovalently mediated synthesis. The use of noncovalent interactions in template-mediated synthesis greatly expands the range of potential template systems. Useful interactions include hydrogen-bonding, electrostatic-bonding, charge-transfer, hydrophobic-bonding, and metal-ligand coordination. Although these interactions are, in general, weaker than covalent bonding, they have proven more useful for several reasons. The first important advantage follows directly from the comparative weakness of the interactions. The rapid kinetics, or lability, of bonding give

rise to superior performance in chromatographic separations.

Lability also plays an important role in choosing elution conditions. Noncovalent interactions can generally be broken under mild conditions such as a small change in pH, addition of salt to the buffer, or even a change of solvent. Weaker interactions also mean that elution can often be performed at or near ambient temperature, with little fear of chemically altering either the polymer or the template molecule.

The wide array of noncovalent interactions that have been successfully exploited to produce substrate-selective polymers illustrates another advantage. The number of useful, readily reversible covalent interactions is very limited. In contrast, useful noncovalent interactions are plentiful. For example, the only requirement for participation in electrostatic bonding is a charged site on the template molecule. If all the possible noncovalent interactions are considered, it is clear that the number of potential applications is almost limitless.

Some recent examples of template-mediated synthesis with noncovalent interactions as recognition elements have produced materials capable of recognizing subtle differences in molecular structure. Polymers formed in the presence of one enantiomer of a chiral amino acid, by using hydrogen bonding or ion pairing for formation of the recognition sites, have been utilized to achieve complete resolution of enantiomers from a racemic mixture when the polymers are used as chromatographic supports. Similar results have been seen for mixtures of chiral sugar derivatives. Copper-containing polymers synthesized in the presence of a positional isomer of a bisimidazole compound have been shown to display selectivity for the original template molecule in rebinding studies. The selectivity was shown to depend on the spacing between the imidazole nitrogen atoms in the template compared to other imidazole-containing substrates.

Applications. Templated polymers may eventually prove useful in a number of areas. The demonstrated ability of these materials to discriminate molecules that are close structural analogs or enantiomers offers a new approach to difficult separation problems. When these polymers are used in conjunction with high-performance liquid chromatography, analytical or preparative scale separations of previously unresolvable mixtures may be performed. In addition, eventual extension into use for separation of delicate and complex biomolecules such as proteins is possible. Current methods for protein purification generally use a series of separations of low-specificity to effect the high level of purity required for pharmaceutical, clinical, and analytical applications. Efforts to increase the specificity through the use of affinity chromatography and biospecific interactions (such as monoclonal antibodies) have proven difficult; factors such as cost, durability, and conditions of elution have limited these applications. Templated polymers, with relatively low cost, high durability, and mild elution conditions, have the potential to overcome these limitations while retaining significant selectivity.

Polymers capable of recognizing a specific compound may eventually find use in chemical-specific sensors. Used in conjunction with one of several thin-layer techniques (such as lipid bilayers or polymer coatings), templated materials could allow for detection of toxins in environmental settings, secondary metabolites or viruses in blood analysis, or nutrient levels in fermentations. For environmental or fermentation applications, these materials could allow for on-line, continuous monitoring of substances that were previously assayed in a laboratory setting. For medical applications, templated polymers could offer an alternative to expensive antibody tests currently used for viruses through recognition of virus coat proteins. In addition, it may prove possible to replace many steps in identification of secondary metabolites with a single separation step involving a templated polymer stationary phase.

Exploration of the template binding site might also provide some insight into biological recognition events. The ease with which a binding pocket can be created for virtually any substrate by using template-mediated synthesis offers a way of probing the importance of fit, flexibility, and noncovalent bonding for substrate recognition. Current research has shown that all these factors strongly influence selectivity and binding kinetics.

For background information SEE ENZYME; MOLECULAR RECOGNITION; POLYMERIZATION; RACEMIZATION; STEREOCHEMISTRY in the McGraw-Hill Encyclopedia of Science & Technology.

Sean D. Plunkett; Frances H. Arnold

Bibliography. P. K. Dhal and F. H. Arnold, Template-mediated synthesis of metal-complexing polymers for molecular recognition, *J. Amer. Chem. Soc.*, 113:7417–7418, 1991; B. Ekberg and K. Mosbach, Molecular imprinting: A technique for producing specific separation materials, *Trends Biotechnol.*, 7:92–96, 1989; K. J. Shea and T. K. Dougherty, Molecular recognition on synthetic amorphous surfaces: The influence of functional group positioning on the effectiveness of molecular recognition, *J. Amer. Chem. Soc.*, 108:1091–1093, 1986; G. Wulff, Molecular recognition in polymers prepared by imprinting with templates, *Polym. Reagents Catalysts*, ACS Symp. Ser., 308:186–230, 1986.

Chemical thermodynamics

A wide variety of theoretical and applied branches of science rely upon compilations of thermodynamic data for quantitative computation of stability relationships and modeling of reversible and irreversible processes. With the recent development of affordable and powerful computers in a desktop environment, the users of thermodynamic data have multiplied and the demands made of these data have become increasingly stringent. This article reviews the essential features of internally consistent thermodynamic data sets along with their geological applications.

Internally consistent data sets. At any pressure (P) and temperature (T), phase stability can be computed conveniently with the apparent Gibbs free-energy function, $\Delta_a G^{P,T}$, as shown in Eq. (1),

$$\Delta_a G^{P,T} = \Delta_f H^0 + \int_{298.15}^{T} Cp\ dT$$

$$-T\left[S^0 + \int_{298.15}^{T} (Cp/T)\ dT\right] + \int_{1}^{P} V\ dP \qquad (1)$$

where $\Delta_f H^0$ and S^0 are the enthalpy of formation from the elements and third-law entropy, respectively, at the reference state of 1 bar (10^5 Pa) and 298.15 K; V is the volume; and Cp is the heat capacity. The thermodynamic data needed to evaluate this equation are acquired by two distinct techniques: calorimetric/thermophysical measurement of individual phase properties and determination of equilibria among chemically equivalent phase assemblages. The latter does not determine any property of an individual phase directly; it only determines changes in the thermodynamic properties of a reaction, such as the Gibbs free energy of reaction, $\Delta_r G^{P,T}$, as given by Eq. (2),

$$\Delta_r G^{P,T} = \sum_j v_j \Delta_a G_j^{P,T} \qquad (2)$$

with v_j being the stoichiometric reaction coefficient.

The pioneering work of H. C. Helgeson and coworkers demonstrated that it is necessary to consider phase equilibrium data in order to achieve the greatest accuracy in derived thermodynamic parameters. The reason is that uncertainties of calorimetrically determined enthalpies of formation are typically 1.5–6 kJ mol^{-1}, whereas the uncertainty of a phase equilibrium determination may translate to as little as 1–2 kJ mol^{-1} distributed among all phases. A major disadvantage of phase equilibrium data, however, is that inaccurate properties for any one phase will be transferred to other phases that participate in the same phase equilibria.

Regardless of the source of thermodynamic data, the term internal consistency implies at a minimum that all data be derived by using one and the same set of reference values, such as the gas constant, temperature scale, and element or oxide properties. In recent geological literature, however, internal consistency has acquired a much more stringent connotation, so that thermodynamic data not only use the same set of reference values but also are compatible with all valid experimental (both calorimetric and phase equilibrium) data. Unfortunately, this connotation provides a wide range of latitude, and the nontrivial onus of evaluating the extent of internal consistency of a thermodynamic data set rests on the user.

Databases. Two different mathematical techniques have been used for the combined analysis of phase equilibrium and thermophysical data: mathematical programming and multiple linear regression. The merits and pitfalls of each have been argued extensively in recent geological literature. An encouraging sign is that the two most recent databases derived by each of these techniques—the UBC database and the HP data set—show a considerable degree of concordance. It appears now that the most significant part of the retrieval analysis is not the technique but the accuracy of the equations of state that are used to represent thermodynamic properties, because of the high degree of correlation among thermodynamic properties derived from phase equilibrium data. Of particular importance are the representations of the properties of supercritical fluids and polymorphic transformations, as well as the heat capacity, expansivity, and compressibility of minerals.

Both of the thermodynamic data sets for minerals discussed above are largely independent of the thermodynamic properties of aqueous species; that is, mineral properties have been retrieved from phase equilibrium data in which aqueous species do not participate. The UBC database is also almost entirely independent of the mixing properties of minerals, while derivation of the HP included analysis of some more complex chemical systems with assumed mixing properties of minerals. To date, there has been very little effort directed toward simultaneous evaluation of the thermodynamic properties of end-member minerals together with properties of solid solutions and aqueous species. However, several recent studies are addressing this problem, which lies at the core of most realistic applications in multicomponent chemical systems relevant to geologic processes. Although sophisticated software has now been developed for performing a variety of calculations that give insight into geologic processes, this insight and the nature of the hypotheses that can be tested is a direct function of the accuracy of the thermodynamic properties of all phases involved in the calculations. For this reason, a great deal of effort continues to be placed on improved determination of thermodynamic properties of earth materials.

Applications. An important application of internally consistent thermodynamic data is the calculation of phase diagrams illustrating the stability relationships of earth materials as a function of pressure, temperature, and compositional variations. Phase diagrams have classically been determined directly by experiment, but their calculation underscores a major advantage of the thermodynamic approach in providing the means not only to interpolate between existing experiments but also to extrapolate beyond these data. For example, the UBC database consists of 67 minerals whose properties were refined through analysis of phase equilibrium data for 180 different reactions.

The number of possible equilibria among 67 minerals in an 11-component system is given by Eq. (3), and the only limitation on the utility of calculating

$$\frac{67!}{(11+1)!55!} = 5.99 \times 10^{12} \qquad (3)$$

the position of any of these equilibria is the accuracy of the thermodynamic data involved. At present, a computer program known as Ge0-Calc appears to be the most sophisticated generalized software for calculation of phase diagrams involving minerals and aqueous species. The latest version of this software permits calculations with either of the mineral databases noted above, together with a database for aqueous species. The latter data can be used up to 5 kbar (5×10^5 Pa) and 600°C (1110°F), but recent work has extended both the pressure-temperature range of the equations of state as well as the number and nature of the aqueous species.

Petrogenetic grids. These phase diagrams for given rock compositions (such as pelites or basalts) commonly illustrate the effect on important phase boundaries of major compositional variations of minerals observed in nature. They are used extensively by petrologists in order to place pressure-temperature limits on the occurrence of observed mineral assemblages and thereby constrain parts of the pressure-temperature path that a rock has followed in its metamorphic and tectonic evolution. Recent geological literature contains examples of petrogenetic grids for both pelitic and basaltic rocks, constructed with internally consistent thermodynamic data. Their main limitations are the general lack of knowledge of the solid-solution characteristics in these complex chemical systems where many of the important minerals mix nonideally.

Inverse geological problem. In recent years, internally consistent thermodynamic data have been utilized increasingly to solve the so-called inverse geological problem: given a set of measured mineral compositions in a rock, at what pressure, temperature, and fluid composition did that rock equilibrate? Whereas the traditional technique of geothermometry and geobarometry involved application of a minimum of two experimentally calibrated equilibria in order to define pressure and temperature from their intersection, the use of thermodynamic data sets allows many more equilibria to be applied to a given mineral assemblage. Several different averaging techniques have been proposed as solutions to the overdetermined problem. The results provide both more robust pressure-temperature determinations and a means to question the underlying assumption of equilibrium among all minerals used in the calculations.

Petrological modeling. A critical advance in the use of thermodynamic data sets has been the development of various programs that perform forward petrological modeling, that is, determination of the stable phase assemblage and associated phase compositions at a specified pressure, temperature, and bulk composition. This type of modeling has been used to construct petrogenetic grids and to give geologists more confidence in tectonic reconstructions by comparing predicted assemblages and compositions for pressure-temperature paths with those inferred from mineral zoning profiles and other geologic evidence. Equilibrium assemblage calculations have also been used for a variety of other fundamental geologic problems, such as the petrologically important basalt to eclogite transition, upper mantle evolution, volcanic rock petrogenesis, and planetary formation from condensation of solar gas. An additional application is the planning of experimental studies that can be performed more efficiently by considering predicted stable assemblages and compositions over the range of desired laboratory conditions.

Reaction-path modeling. The most sophisticated thermodynamic modeling simulates geological processes as a series of equilibrium steps, with the stable phase assemblage computed at each step (reaction-path modeling). The incorporation of kinetics and material transport into such modeling is an exciting enhancement that is just in the initial stages of development. Some of the more important geologic problems currently being addressed with reaction-path modeling include nuclear waste disposal, contaminant transport and water treatment, physical and chemical controls on ore deposit formation and regional water geochemistry, evolution of magmatic systems, as well as a number of industrial processes such as recovery of hydrocarbons from oil sands. Resolution of these problems requires internally consistent thermodynamic data for minerals, gases, aqueous species, and melts. The ability to solve some of the most pressing geological problems will depend not only on modeling innovations but also on the continued measurement, integration, and refinement of the data that form the backbone of these calculations.

For background information *SEE CHEMICAL THERMODYNAMICS; MODEL THEORY; PHASE EQUILIBRIUM; STATISTICS; THERMODYNAMICS* in the McGraw-Hill Encyclopedia of Science & Technology.

Robert G. Berman

Bibliography. R. G. Berman, Internally consistent thermodynamic data for stoichiometric minerals in the system $Na_2O–K_2O–CaO–MgO–FeO–Fe_2O_3–SiO_2–TiO_2–H_2O–CO_2$, *J. Petrol.*, 29:445–522, 1988; T. H. Brown, R. G. Berman, and E. H. Perkins, PTA-system: A Ge0-Calc software package for the calculation and display of activity-temperature-pressure phase diagrams, *Amer. Mineralog.*, 74:485, 1989; H. C. Helgeson, D. H. Kirkham, and G. C. Flowers, Theoretical prediction of the thermodynamic behavior of aqueous electrolytes at high pressures and temperatures, IV. Calculation of activity coefficients, osmotic coefficients, and apparent molal and standard and relative partial molal properties to 600°C and 5 kb, *Amer. J. Sci.*, 281:1249–1516, 1981; T. J. B. Holland and R. Powell, An enlarged and updated internally consistent thermodynamic dataset with uncertainties and correlations: The system $K_2O–Na_2O–CaO–MgO–MnO–FeO–Fe_2O_3–Al_2O_3–TiO_2–SiO_2–C–H_2–O_2$, *J. Metamorph. Geol.*, 8:89–124, 1990.

Chemiluminescence

Electrogenerated chemiluminescence (electrochemiluminescence or ECL) refers to the light emission produced by an electrochemical reaction. Electrochemical reactions are electron transfer reactions. The light generation occurs at an electrode surface in an electrochemical cell when a voltage is applied to the electrode. The luminescent product is formed from a highly exothermic and energetic electron transfer reaction. The emission of light (fluorescence or phosphorescence) occurs after the luminescent product is formed. A typical electrochemiluminescence reaction scheme is shown in reactions (1)–(4), where R is the

$$R \rightarrow R^+ + e^- \text{ (oxidation)} \qquad (1)$$

$$R + e^- \rightarrow R^- \text{ (reduction)} \qquad (2)$$

$$R^+ + R^- \rightarrow R^* + R \text{ (ion annihilation)} \qquad (3)$$

$$R^* \rightarrow R + \text{photon (emission)} \qquad (4)$$

generalized starting material and luminophor; R^+ and R^- are electrochemically formed products; e^- is an electron; and R^* is the luminescent excited state, formed in an ion-annihilation reaction between the oxidized product R^+ and the reduced product R^-. When the exothermic energy of reaction (3) is greater than the luminescent excitation energy, the excited state is produced. The light emission step [reaction (4)] is the same process as in fluorescence or phosphorescence. Reactions (1), (2), and (3) are electron transfer reactions. Reactions (1) and (2) occur at the electrode/solution interface. Reaction (3) is a solution phase reaction occurring away from the electrode surface.

Electrochemiluminescence has attracted a great deal of research interest because it can facilitate study of the fundamental mechanisms and kinetics of electron transfer reactions. Traditionally, mechanistic information on electron transfer reactions was derived by studying the relationship between applied voltage and current. By using electrochemiluminescence, additional mechanistic information is available from the light intensity behavior. As a result of the fundamental studies, applications of electrochemiluminescence have evolved. Detection of the light emitted by a very few molecules is possible from such reactions. Measurement of light to extremely low levels is routinely done with simple instruments called luminometers. Combining the detection of light at low levels with electrochemiluminescence allows for an extremely sensitive analytical technique.

Fundamental studies. The first electrochemiluminescence reactions studied in the early 1960s generally involved organic fluorescors, such as the aromatic hydrocarbons 9,10-diphenyl anthracene and rubrene. In 1972 the ruthenium complex $Ru(bpy)_3^{2+}$ (bpy = 2, 2′-bypyridine) was introduced as an electrochemiluminescent luminophor and has been widely investigated. A number of alternative luminescent systems have also been evaluated: $Ru(bpz)_3^{2+}$ (bpz = 2, 2′-bipyrazine) and $S_2O_8^{2-}$; $Ru(bpy)_3^{2+}$ and oxalate, $Ru(bpz)_3^{2+}$ in acetonitrile; hexanuclear molybdenum

(Mo) and tungsten (W) clusters, $M_6X_8Y_6^{2-}$ [M = Mo, W; X, Y = chlorine (Cl), bromine (Br), iodine (I)] in dichloromethane; the micelle solubilized osmium complex [$Os(bpy)_3^{2+}$] in aqueous solutions; and ruthenium (II) 4,4′-biphenyl-2,2′-bypyridine and ruthenium (II)-4,7-diphenyl-1,10-phenanthroline in aqueous and acetonitrile solutions.

Electrochemiluminescence has been applied to new research. It has been used to image the topography of an electrode surface. The electrochemiluminescence of $Ru(phen)_3^{2+}$ (phen = 1,10-phenanthroline) and $Os(bpy)_3^{2+}$ was applied to investigate the interaction of the transition metal chelates with deoxyribonucleic acid (DNA). When $Ru(bpy)_3^{2+}$ is immobilized in a Nafion film coated to an electrode, a sensor for oxalate, amines, and nicotinamide adenine dinucleotide reduced form (NADH) is produced. Electrochemiluminescence from $Ru(bpy)_3^{2+}$ contained within Nafion films is also used to study charge and mass transport within polymer films. A regenerative electrochemiluminescence system, using a chemiluminescent polymer of tris(4-vinyl-4′-methyl-2,2′-bipyridine) ruthenium (III), is prepared by electropolymerization.

The properties of partially quaternized poly (4-vinyl pyridine) films coordinately attached to a luminescent probe, rhenium carbonate(phen), and a redox probe, $Ru(bpy)_2Cl^{2+/1+}$, was studied. A monolayer of the ruthenium complex, $Ru(bpy)_2(bpy-C_{19})^{2+}$, with a long hydrocarbon chain attached to the electrode surface, was characterized by electrochemiluminescence. The ruthenium complexes were compacted to assemble a Langmuir-Blodgett film on a surface. *SEE ELECTRODE.*

Application to diagnostic testing. An important and innovative commercial application of electrochemiluminescence has been developed. It forms the basis of a highly sensitive technique for the detection of biological materials, such as DNA, proteins, haptens, and therapeutic drugs, in the clinical laboratory. The technique uses a binding assay method, combined with electrochemiluminescence as a detection method. Such an approach integrates the electrochemical cell, light measurement system, and means for handling and preparation of the sample into one simple compact instrument.

In this system the electrochemiluminescence reaction uses $Ru(bpy)_3^{2+}$ as the luminophor and tri-*n*-propyl amine (TPrA) as the coreactant. The reaction is novel with respect to the use of amines, not with respect to the luminophor. The new reaction is more versatile than other electroluminescence systems. It can be carried out in aqueous or nonaqueous solvents, in the presence of biological media, over a range of solution pH including neutral pHs, and in the presence of dissolved oxygen, and is highly reproducible.

An alternative means is utilized in the technique to produce suitable precursors to participate in the energetic electron transfer of reaction step (3). Electrochemical oxidation of tripropyl amine at the electrode produces a short-lived tripropyl amine radical cation. This material loses a proton from the carbon atom adjacent to the nitrogen center of the amine. The deprotonated substance is a strong reducing agent. The result of these steps is equivalent to that of reaction step (2). Concurrently, the luminophor $Ru(bpy)_3^{2+}$ is oxidized to produce $Ru(bpy)_3^{3+}$ as in reaction (1). The reducing agent reacts with $Ru(bpy)_3^{3+}$ to generate $Ru(bpy)_3^{2+}*$; the product is the luminescent species as in reaction (3). The cation $Ru(bpy)_3^{2+}$ is regenerated many times in one measurement cycle, whereas the tripropyl amine is irreversibly consumed. The electrochemiluminescence reaction is very sensitive, because many photons are emitted from each luminophor during one measurement cycle. The limit for detection of $Ru(bpy)_3^{2+}$ is near 200 femtomolar. The intensity shows a wide dynamic range in response to luminophor concentrations.

The cation $Ru(bpy)_3^{2+}$ can be chemically modified with reactive groups on one of the bipyridine ligands. With the use of reactive groups, $Ru(bpy)_3^{2+}$ can be chemically linked with biological materials such as proteins, haptens, and DNA. The labeling of different biological materials with a luminophor possessing electrochemiluminescence activity allows measurement of concentration (assay). The light intensity is directly proportional to the concentration of the luminophor and the material to which it is labeled. Unlike other labeling agents such as enzyme labels, linking $Ru(bpy)_3^{2+}$ to biological materials is easy and rapid. The luminophor can be used as a label without affecting the biological materials' immunoreactivity, solubility, or ability to hybridize.

Advantages. Electrochemiluminescence has many distinct advantages over other detection methods. The traditional method for conducting binding assays uses radioisotopes for detection. The health and environmental hazards of radioisotopes make their use problematic. Electrochemiluminescence uses simple metal chelates that are not radioactive. The use of tripropyl amine is beneficial because of its stability compared to other detection methods, such as chemiluminescence. Stability of materials is very important for cost and shelf life. Detection limits are extremely low. The assay measurements are homogeneous and rapid. The detection range for label quantification extends over five orders of magnitude. Luminophors of differing wavelengths of emission can be employed. Multiple assays from one sample can be achieved through separate measurements of light intensity from two or more simultaneously emitting labels.

Other applications. Lasers based on electrochemiluminescence have been proposed. A chromatographic detector using this phenomenon has had some success. Its use as a display may prove practical in the future. Electrochemiluminescence with high efficiency can be obtained from microband array electrodes (a new type of microelectrodes) operated in the collector-generator mode. This device can also be used to explore electrochemiluminescence in solutions of low ionic strength, with minimal distortion from ohmic drip.

For background information *SEE CHEMILUMINESCENCE; LUMINESCENCE* in the McGraw-Hill Encyclopedia of Science & Technology.

Hongjun Yang; Jonathan K. Leland

Bibliography. A. J. Bard (ed.), *Electroanalytical Chemistry*, vol. 10, 1977; A. J. Bard and L. R. Faulkner, *Electrochemical Methods*, 1980; G. F. Blackburn et al., Electrochemiluminescence detection for development of immunoassays and DNA probe assays for clinical diagnosis, *Clin. Chem.*, 37:1534–1539, 1991; J. K. Leland and M. J. Powell, Electrogenerated chemiluminescence: An oxidative-reduction type ECL reaction sequence using tripropyl amine, *J. Electrochem. Soc.*, 137:3127–3131, 1990.

Cholera

Cholera is a bacterial infection that causes severe watery diarrhea, vomiting, and sometimes death. It is usually found in areas with poverty and poor sanitation. In the twentieth century, cholera was confined to southern Asia until 1961, when the seventh pandemic arose on the islands of Celebes in the Malay Archipelago. The pandemic has spread throughout Asia and into Africa, Europe, and Oceania.

Cholera in Latin America. The Americas were spared until late January 1991, when epidemic cholera caused by toxigenic *Vibrio cholerae* 01 appeared in Peru. Emerging nearly simultaneously in several coastal cities, it quickly spread to other urban areas and into the Andes and the Amazon region. The mode of introduction into Peru remains unknown. Possible routes include a person returning from a cholera-affected area or contaminated ballast or sewage from a ship. The epidemic quickly spread to neighboring countries: Ecuador in February, Colombia in March, Chile in April, and Brazil in May 1991. In August 1991, epidemics occurred in Central America. By February 1993, all Latin American countries had reported cholera except Uruguay. As of February 1993, more than 730,000 cases and 6100 deaths had been reported from 21 countries in North, Central, and South America, with more than 100 cases occurring in the United States.

In Latin America, as in other parts of the world, waterborne and foodborne transmission of cholera has been documented. In Latin America, contaminated water has probably caused a majority of cases. The poor condition of many municipal water and sewage systems promotes widespread dissemination of the organism. Water is often not chlorinated, and in many areas illegal connections to major water lines, low water pressure, and intermittent service permit back-siphonage of contaminants into the system. Since running water is often available for only a few hours a day, many families store water in containers in their homes. The water may become contaminated when infected persons dip their hands into it. Cholera has also been transmitted by raw or undercooked seafood and by cooked foods stored at ambient temperature. Contamination of fruits and vegetables can occur where untreated sewage is used to irrigate crops. As in the nineteenth century, epidemic urban cholera is an indicator of an inadequate municipal infrastructure.

Cholera in the United States. An endemic focus of a unique strain of *V. cholerae* 01, distinct from the Latin American strain, was identified in 1973 along the Gulf Coast of the United States. Sixty-five cases were reported during 1973–1990. Sporadic imported cases have been reported among travelers returning to the United States from countries with cholera (41 cases during 1973–1990).

In the 18 months after the introduction of cholera into Latin America, 128 cases of cholera were reported in the United States. Initial cases were in travelers who had consumed undercooked seafood or unboiled water while traveling in Latin America. A second wave of cases occurred in two outbreaks in New York and New Jersey. Eleven persons became ill after eating crab brought back in the suitcases of travelers returning from Ecuador. In February 1992, more than 75 United States residents were infected with *V. cholera* 01 during an airline flight from Latin America.

Most United States communities are not at risk for transmission of cholera because they properly treat water supplies and have adequate sewage systems. However, some areas of the country, such as shanty towns along the Mexican border and migrant encampments in southern California, are at risk for cholera because of lack of safe drinking water and sanitation.

Causative organism. *Vibrio cholerae* is a gram-negative bacterium of the family Vibrionaceae. It is a curved bacillus with a polar flagellum. It causes disease by producing a toxin (cholera toxin) in the gut, which makes the intestine secrete large volumes of fluids. Although there are many serogroups, only serogroup 01 has exhibited the ability to cause epidemics. *Vibrio cholerae* 01 is divided into three serotypes—Inaba, Ogawa, and Hikojima—and two biotypes—classical and El Tor. The epidemic illness is caused by cholera toxin producing (toxigenic) strains of the 01 antigenic serogroup. (Nontoxigenic 01 strains of *V. cholerae* can cause diarrhea and sepsis, but they do not cause epidemics.) The incubation period is typically 1 to 3 days, with a range of a few hours to 5 days. The organism is acid-sensitive and does not survive in foods with low pH. Persons with low gastric acidity are at increased risk for cholera infection.

Clinical manifestations. Toxigenic *V. cholerae* 01 causes a broad spectrum of clinical illness. Most infected persons display no symptoms, some have mild or moderate diarrhea, and few (2–5%) develop cholera gravis, with severe watery diarrhea, vomiting, and dehydration. The watery stools are colorless, with small white flecks of mucus, classically described as rice-water stools. Frequently, the vomiting may be severe. Patients may lose up to 1 quart (1 liter) of fluid per hour in the first 24 h and may lose more than 10% of their body weight. In extreme cases, these losses may lead to shock and death in as little as 2 h. Dehydrated patients may have hypotension and rapid pulse. They may be weak, produce no urine, lose skin turgor, and have sunken eyes, dry mucous membranes, and intense thirst. Patients may be drowsy or unconscious. The loss of sodium, potassium, and bicarbonate in cholera stools causes acidosis, renal failure, cardiac arrhythmias, and severe leg cramps.

Diagnosis. Cholera is diagnosed by culturing stool specimens or rectal swabs from ill persons on a special medium (thiosulfate–citrate–bile salts–sucrose agar [TCBS]) not routinely used by clinically laboratories. Culture for vibrios should be requested for clinically suspicious cases, especially for patients who recently returned from areas affected by epidemic cholera, those who recently consumed raw or undercooked seafood, or those who present with severe dehydrating diarrhea (especially adults).

Treatment. If patients with severe cholera are not treated, up to 50% may die. However, with proper treatment less than 1% die. Rapid replacement of lost fluids and electrolytes is the mainstay of therapy. Patients with mild to moderate dehydration can be treated with oral solutions, but those with severe dehydration need intravenous therapy. Antibiotics may be used to decrease the duration and quantity of diarrhea.

Travelers. The risk of travelers from the United States acquiring cholera in an affected area is thought to be low. Most individuals who developed cholera after visiting Latin America had consumed high-risk items such as raw or undercooked seafood or unboiled water. Travelers should follow the precautions described for prevention of travelers' diarrhea: avoid unboiled or untreated water or ice, fish and beverages from street vendors, raw or undercooked fish and shellfish, and salads and raw vegetables. It is usually safe for the traveler to consume cooked foods that are still hot, fruits peeled by the traveler, and bottled water and beverages. The general rule—"Boil it, cook, it, or forget it"—should be applied.

The cholera vaccine currently licensed in the United States is effective in only 50% of recipients, its protection lasts only 3–6 months, and it is not generally recommended. Experimental oral vaccines may be more effective but are not yet licensed. Travelers to areas with cholera who develop watery diarrhea, especially if accompanied by vomiting, should seek immediate medical attention.

For background information SEE CHOLERA; EPIDEMIC; INFECTIOUS DISEASE in the McGraw-Hill Encyclopedia of Science & Technology.

David L. Swerdlow; Allen A. Ries

Bibliography. D. Barua and W. Burrows (eds.), *Cholera*, 1974; Centers for Disease Control, *Health Information for International travel*, DHHS Pub. (CDC) 90–8280, 1990; G. L. Mandell, G. R. Douglas, and J. E. Bennett (eds.), *Principles and Practice of Infectious Diseases*, 3d ed., 1990; A. A. Ries et al., Cholera in Piura, Peru: A modern urban epidemic, *J. Immun. Dis.*, 166:1429–1433, 1992; D. L. Swerdlow et al., Transmission of epidemic cholera in Trujillo, Peru: Lessons for a continent at risk, *Lancet*, 340:28–32, 1992; D. L. Swerdlow and A. R. Ries, Cholera in the Americas: Guidelines for the clinician, *JAMA*, 267:1495–1499, 1992.

Climatic change

The El Niño of 1991–1992 was the first such event since 1986–1987. El Niño (The Child) refers to the warming of surface waters along coastal Peru that occurs periodically around Christmastime. This phenomenon is part of a far larger set of anomalies of the coupled ocean–atmosphere system of the Pacific Basin that constitutes the El Niño Southern Oscillation (ENSO). There are two extremes of ENSO: the warm event (El Niño) and the cold event (La Niña), each of which is accompanied by characteristic large-scale patterns of drought and heavy rainfall, of cold and warmth, in key regions of the tropics and extratropics. Climate teleconnections are weather and climate anomalies that are linked among widely separated locations. These climate teleconnections result from changes in the intensity and primary locations of the atmospheric circulation cells connecting tropical regions and also connecting the tropics with middle latitudes; the jet streams; and middle-latitude storm tracks. In many ways, the 1991–1992 ENSO was typical of previous warm events, and it even bore some resemblance to the very strong ENSO of 1982–1983. Its onset was predicted by models of ENSO some two seasons previously.

ENSO processes. The large-scale oceanic and atmospheric anomalies associated with a warm ENSO event typically evolve in the 6–12 months prior to maturing during the Northern Hemisphere winter, and they gradually wane through the following northern spring and summer. These anomalies can be considered to involve a set of interrelated phenomena that include pressure changes, wind changes, warm surface water, sea-level fluctuations, clouds and rainfall changes, and interactions among some of the anomalies over large distances.

Pressure changes. The high-pressure system in the southeast Pacific weakens, and surface pressures increase (weakening of the surface low) over northern Australia and Indonesia. The reversal in the pressure gradient across the tropical Pacific that is a result of these changes is measured by the monthly anomaly (from the long-term mean) of surface pressure for Tahiti minus that for Darwin, Australia: the Southern Oscillation Index. In the initial stages of a warm event, this index becomes increasingly negative.

Wind changes. The trend toward negative values of the Southern Oscillation Index signals a reduction in the strength of the zonal (east-west) circulation known as the Walker circulation in the tropical Pacific. Low-altitude easterly winds near the Equator are reduced in strength (westerly anomalies develop). Around the peak of a warm event, actual westerly winds may develop on the Equator near and west of the International Date Line (180° longitude).

Warm surface water. These pressure and wind changes promote an eastward movement of the warmest surface water from the western Pacific and Indonesia into the central equatorial Pacific. The increase in sea surface temperature near the Equator is enhanced by the Equatorial Undercurrent, which

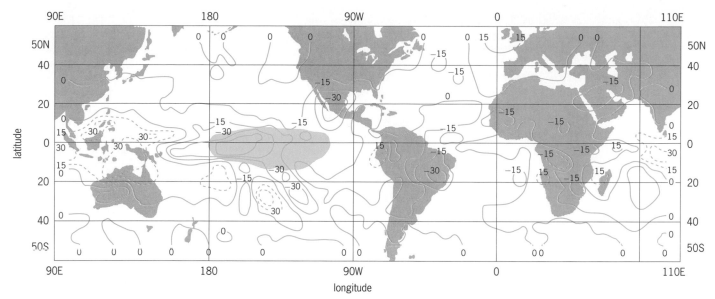

Fig. 1. Patterns of the anomalies of outgoing longwave thermal radiation for January 1992 from satellite information. Positive anomalies are shown as broken lines, negative anomalies as solid lines. Colored area is the tropical Pacific warm-water pool for January 1992. Contour values are given in W/m². (*After U.S. Department of Commerce, NOAA/NWS/NMC/CAC, Climate Diagnostics Bulletin, Near Real-Time Analyses Ocean/Atmosphere: January 1992*)

extends to the surface during a warm event, and by eastward-propagating Kelvin waves. The reduction of wind-induced upwelling along the coast of Peru and expansion of the pool of warmest sea surface temperatures result in the strong increase of the sea surface temperature along coastal South America that defines an El Niño.

Sea-level fluctuations. The upper ocean thermal and atmospheric pressure and wind anomalies are also evident as changes in sea level at coastal stations. Sea levels rise where sea surface temperatures increase in the eastern Pacific Basin, and they fall where sea surface temperatures decrease in the western Pacific Basin.

Cloud and rainfall changes. The eastward migration of the warmest water is accompanied by a relocation of the region of maximum cloudiness and convective rainfall from the Australo-Indonesian region to near the International Date Line. Conversely, cloudiness and rainfall are reduced over the Indian Ocean and the western tropical Pacific. These changes are evident in satellite data of the anomalies of outgoing longwave (thermal) radiation: less than normal over the central Pacific because of the high and cold convective cloud tops; greater than normal over the reduced cloud areas of the western tropical Pacific, as shown in **Fig. 1**.

Interactions. The oceanic and atmospheric anomalies of the tropical Pacific are communicated to the rest of the tropics by interactions among the Walker circulations, and to the extratropics (regions poleward of about 20° latitude) by both increased activity of the meridional (south-north) Hadley cell and interactions of the middle-latitude Rossby waves with the equatorial westerlies. These interactions are favored in the Northern Hemisphere winter when the extratropical westerlies extend toward the anomalous patterns of warm water and convection in the tropics. The

Hadley cell becomes more vigorous in response to the increased convective activity near the International Date Line, helping to intensify the westerly subtropical jet stream. In concert with the intensified jet streams, the Rossby waves influence both the primary regions of frontal cyclogenesis and the locations of heaviest precipitation in the middle latitudes, and they help propagate an ENSO signal to other regions of the Northern Hemisphere extratropics.

Evolution of 1991–1992 El Niño. The 1991–1992 ENSO was broadly representative of the composite picture (average of many warm events) described above. The antecedent conditions were evident in the tropical Pacific by northern hemisphere spring 1991. As early as May, the Southern Oscillation Index recorded its third consecutive month of values exceeding –1.0. The most negative Southern Oscillation Indexes (at or greater than –3.0), recorded in January and March 1992, at the peak of the event, were the lowest monthly Southern Oscillation Indexes since the major warm episode of 1982–1983.

The transition seasons (March, April, May; September, October, November) are particularly critical in the development of a warm event, since both the strongest convection and warmest sea surface temperatures are located near the Equator. Similarly, the latest ENSO intensified and entered its mature stage in October–November 1991, when the sea surface temperature anomalies in the central tropical Pacific increased most strongly, westerly winds developed, and enhanced convection began well east of the International Date Line. By May 1992, the decreasing positive sea surface temperature anomalies in the central equatorial Pacific, and return of the monthly Southern Oscillation Index to near zero, signaled the end of the mature phase and the onset of the decay of the warm event.

Although the 1991–1992 El Niño had many characteristics broadly typical of other warm events, there were departures in the timing of onset of some crucial features. The late development of strongly negative anomalies of the outgoing longwave radiation (enhanced rainfall) in the central tropical Pacific, which occurred in November 1991, contrasts with the positive anomalies of the outgoing longwave radiation existing over Indonesia and the western tropical Pacific from April 1991 onward. In addition, the increases in sea surface temperature in the central equatorial Pacific did not occur until October 1991, even though warming in the eastern tropical Pacific was evident as early as the preceding May. Moreover, unlike the 1982–1983 event, sea surface temperature anomalies in the central and eastern Pacific were substantially weaker, even around the peak of the event (about +2°C or +3.6°F in January 1992). There was also a delay in onset of the strong increases of sea surface temperature along the Peruvian coast. There, the sea surface temperature anomalies were near 0°C (0°F) until October 1991, and reached or exceeded +4°C

(+7.2°F) in March 1992. Thus, the developing stage of the 1991–1992 ENSO event (through September 1991) is probably best characterized as weak, but the mature phase (October through April) was that of a moderate to strong event.

Climatic teleconnections. In many respects the teleconnections to the 1991–1992 El Niño also resemble the general pattern for previous Pacific warm events, and even those of the major ENSO of 1982–1983. For the tropics and subtropics these climate anomalies include (1) heavy rains in northern Peru and Ecuador in the high sun (tropical summer November–February) period; (2) reduced rainfall and drought in southeast Africa in the southern summer; and (3) weak and late monsoon rains over northern and northeastern Australia and drought in Indonesia in the October–April period.

Prominent circulation and climate anomalies occurred in the extratropics that were apparently related to the Pacific warm event. These anomalies included enhanced subtropical jet streams, persistent anticy-

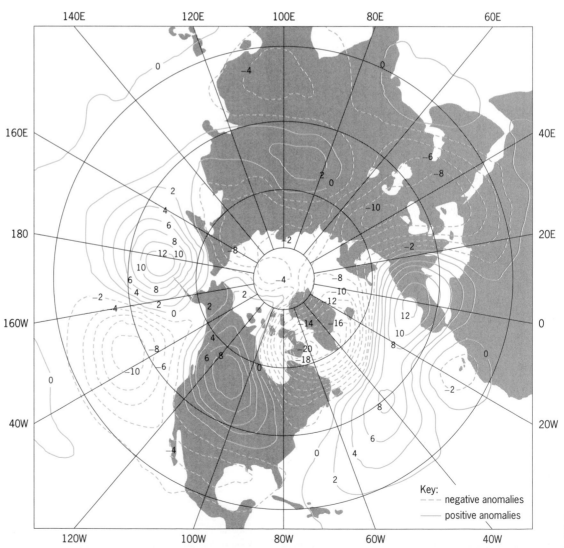

Fig. 2. Patterns of the anomalies of 500 millibars (500 hectopascals) [midtropospheric] geopotential height for the Northern Hemisphere for February 1992. Contour values are given in decameters. (*After U.S. Department of Commerce, NOAA/NWS/ NMC/CAC, Climate Diagnostics Bulletin, Near Real-Time Analyses Ocean/Atmosphere: February 1992*)

clonic blocking, and above-normal rainfall in different regions.

Enhanced subtropical jet stream. An enhanced subtropical jet stream was noted in the western South Pacific in the southern spring of 1991. By early summer, widespread drought was occurring in eastern Australia in response to the strong and persistent subsidence of air over this region.

An enhanced subtropical jet stream in the eastern Pacific and over Mexico was part of a persistent split-flow pattern in the mid- and upper-troposphere over central and eastern portions of the North Pacific during the late winter and spring (**Fig. 2**). A strong blocking ridge of high pressure over western North America was accompanied by a split in the storm track to the west, with branches north into the Gulf of Alaska and south into southern California. The southern track especially was a zone of frequent cyclogenesis. These features produced abnormally warm and dry conditions in the west-central regions of North America, and warm and wet conditions in Mexico, southern California, Arizona, and the Gulf states, especially Texas.

Anticyclonic blocking. Persistent anticyclonic blocking occurred over the eastern North Atlantic and western Europe in Northern winter and early spring (Fig. 2), leading to reduced precipitation and positive temperature anomalies there. Colder-than-normal conditions predominated in the eastern Mediterranean, east of the anomalous high.

Above-normal rainfall. Above-normal rainfall occurred over much of Uruguay, Argentina, and southern Chile in the late southern summer of 1992.

Predictions and implications. In April 1992, statistical and dynamical models of ENSO were predicting the persistence of positive sea surface temperature anomalies in the central and equatorial Pacific through much of the ensuing northern summer. These anomalies often accompany a weakened south Asian monsoon and an enhanced summer rainy season in the southwestern United States and northwestern Mexico. They also tend to be associated with above-average tropical storm and hurricane activity in the central and eastern Pacific.

The irregular occurrence and variable magnitude and persistence of ENSO events complicate the detection of climate trends and longer-term changes, particularly global warming. Thus, while the winter of 1991–1992 was the warmest on record for the coterminous United States, it is not clear to what extent the Pacific warm event was responsible, especially since the past decade was unusually warm. The detection of a clear signal of human-induced global climate change has been further muddied by the eruption of Mount Pinatubo (Philippines) in June 1991. Aside from reducing the ability of satellites to monitor fully the tropical sea surface temperature of the developing ENSO, the injection of large volumes of dust and ash into the stratosphere and their subsequent transport around the globe lowered the global mean surface temperature. That the cooling through early 1992 had been considerably less than predicted, however, probably attests to the impact of the warm ENSO event.

During the northern summer of 1992, colder surface waters were developing in the tropical Pacific. These waters were being monitored for possible signs of a swing to the cold (La Niña) phase of ENSO, similar to that which followed the warm event of 1986–1987.

For background information *SEE* CLIMATE MODELING; CLIMATIC CHANGE; CLIMATOLOGY; TROPICAL METEOROLOGY in the McGraw-Hill Encyclopedia of Science & Technology.

Andrew M. Carleton

Bibliography. M. H. Glantz, R. W. Katz, and N. Nicholls, *Teleconnections Linking Worldwide Climate Anomalies: Scientific Basis and Societal Impact*, 1991; S. G. Philander, *El Niño, La Niña, and the Southern Oscillation*, 1990.

Climatology

For millennia, many of the ills of humanity have been blamed on climate. Droughts, floods, freezes, heat waves, hurricanes, and the like plague societies everywhere.

Climate has usually been viewed as a boundary condition, that is, a condition that people can do little to change. This belief has been captured in the adage that people talk about the weather but cannot do anything about it. To date, no society has been able to weatherproof itself against the vagaries of atmospheric processes, despite constant attempts to do so.

Violent weather is something that both rich and poor nations face, but there is a major difference in the effects. For industrialized countries the impact of climate and weather is, for the most part, an economic matter: a freeze in Florida damages the citrus crop, causing prices in the marketplace for orange juice to increase. However, in the developing world, where food supplies may be at the subsistence level, it can be a matter of survival, since extreme meteorological events can reduce the food supply even further.

Climatic change indicators. Changes in global climate have been viewed at various times as exhibiting both decreases and increases in average temperatures.

Decreasing temperatures. In the early 1970s, following three decades of global temperatures cooler than average, some scientists suggested that the global climate was changing, returning to that of an ice age. Several qualitative, as well as quantitative, indicators were cited in support of this contention: the armadillo (which had migrated as far north as Kansas) had begun to retreat southward; the British growing season had been shortened by 2 weeks; fish previously caught off the northern coast of Iceland were found only off the southern coast; sea ice had drifted equatorward, encroaching on shipping lanes. People began to question the possible consequences of an ice age for contemporary society. United States government agencies such as the Department of State, Central Intelligence Agency, and Department of Agriculture commissioned studies to identify potential impacts of a cooling of the global climate on agricultural production and energy consumption in the nation as well as in the Soviet Union.

Increasing temperatures. By the late 1970s, some scientists had begun to cite new and mounting evidence that the temperature of the Earth's atmosphere was going to increase as a result of emissions of greenhouse gases (such as carbon dioxide, chlorofluorocarbons, methane, and nitrous oxide). The scientific community was asked to identify the possible implications for societies of a global warming.

Climate impacts. The concerns about climatic change that arose in the 1970s sparked the development of a subfield of research referred to as climate impacts. This research focuses on the interactions between the atmosphere and society, and their impacts on each other. Climate was no longer seen as a boundary conditon about which little could be done.

By studying the effects of climate variability on human activities, it might be possible to mitigate (if not avert) some of its worst impacts. Although drought, flood, and freeze cannot be prevented, it is possible to develop strategies to reduce their impacts on society and the environment. A drought, for example, may occur during just one season, but its impact on society may linger for years. The same holds true for the climate-related impacts of other extreme meteorological events, such as tornadoes and hurricanes.

Seasonality. Extreme events are not the only aspects of climate about which researchers are concerned. Seasonality—the natural rhythm of the seasons—has also become a focus of attention. The rhythm of the seasons is something to which every developed and developing society has had to accommodate as its climate-related activities have evolved. An example is the series of devastating droughts in sub-Saharan Africa. These droughts are highly visible and often spectacular, usually forcing governments in the region to operate in a crisis mode. However, such droughts often intensify an already existing, less spectacular, but potentially equally devastating changing of the seasons. People in many parts of the world are most vulnerable just before a harvest. They have depleted their food reserves; and most likely their nutritional status is on the decline, a condition that would be corrected with the fruits of the new harvest. If drought occurs and the harvest is poor, the decline in nutrition will continue unabated.

Ineffective urban planning. The growing number of climate-related impact studies from around the world has shown that society can heighten the impacts of any particular weather event. Urban planning committees have allowed houses to be built in known floodplains (Boulder, Colorado, is a good example) because land is scarce and property near streams and coastlines sells well. Developers, if not prospective homebuyers, knowingly build in flood-prone areas. When a flash flood occurs, its damage will likely be blamed on nature, when in reality the blame rests upon societal decisions that allow houses and other structures to be built in the path of probable flood.

Poor land management. The potential for crop failures can be increased by poor land management as well as by a lack of timely rainfall. For a variety of economic and political reasons, people around the globe have been forced to move in increasing numbers into areas considered marginal for agricultural production, increasing the likelihood that they will be confronted with crop failures. These people often try to grow the same crops they grew in their previous locations, although the new sites may be much drier or wetter. Usually, crops failures are blamed on nature (for example, on a lack of rainfall), but inappropriate human decisions are in many instances responsible for such failures.

Global climate change. Current scientific data suggest that a global warming of a few degrees Celsius could occur in future decades as a result of human activities, such as the burning of fossil fuels, that affect the chemistry of the atmosphere. A global warming will have implications for regional climate regimes around the world. Although much of the existing scientific information on issues of climate change focuses on global averages, most decision makers are interested in the regional and local implications of global warming. There has been considerable interest in assessing climate-related impacts on a regional level by using a variety of research methods such as nested models, historical records, and contemporary analogies. For example, societal responses to the rapid rise in the level of Utah's Great Salt Lake in the mid-1980s could serve as an analog to potential responses to the rise in sea level that scientists expect would accompany global atmospheric warming. In Utah, as a result of an unprecedented 12-ft (4-m) rise in the lake level in a 4-year period, new shoreline and land-use regulations were established, and railroad beddings were elevated. A water pumping station was built to pump excess lake water into the desert. Decision makers in Utah relied on traditional approaches to environmental problems, even when faced with new or unusual conditions.

In 1988, the United Nations General Assembly called for the establishment of an international Intergovernmental Panel on Climate Change (IPCC) in order to assess the current state of the science related to global climate change. The IPCC set up three working groups: working group I focused on the scientific aspects of climate change; II focused on the potential impacts of climate change on ecosystems and on socioeconomic systems; III was given the responsibility for addressing the policy implications of climate change. That climate-related impacts on society were identified as one of the key areas for climate change research attests to their importance to policy makers everywhere. Although the global average temperature might increase, scientists do not yet know how that new average will translate into regional and local climate changes.

In the event of a climate change, there may be regional surprises, that is, counterintuitive impacts. For example, the 1980s have been labeled as the hottest decade in about 100 years of record, registering 6 years of record-setting high temperatures. However, in that same decade, the Florida citrus-growing region was plagued by the largest number of freezes on record. There will doubtless be local and regional impacts that have yet to be reliably identified.

Prospects. In order for societies to better prepare themselves for an uncertain climatic future, impact studies are now required for major developments (such as dams that are expected to be in operation for decades, if not centuries).

The history of each country or region is filled with stories about the impacts of climate-related events: prolonged severe droughts have, at various times, plagued China, India, the former Soviet Untion, Africa, the Americas, Australia, and Indonesia. Floods have caused great damage in Brazil, Peru, Chile, Kenya, India, and Pakistan. The devastating cyclones and floods in low-lying Bangladesh are well known.

In many instances, societies accept climate-related anomalous events and their impacts as acts of nature against which little can be done. Political leaders often resign themselves and their citizens to remaining as passive victims of the vagaries of atmospheric processes. However, recent climate impact research shows that there is no reason for societies to remain passive. Information about climate can be used to mitigate its adverse impacts while taking advantage of its potential benefits. An improved understanding of climate variability and its environmental and societal impacts is an important step toward improving the prospects for economic development in industrialized as well as developing countries.

For background information SEE CLIMATE MODIFICATION; CLIMATE CHANGE; CLIMATE PREDICTION; CLIMATOLOGY; GREENHOUSE EFFECT in the McGraw-Hill Encyclopedia of Science & Technology.

Michael H. Glantz

Bibliography. R. Chambers, R. Longhurst, and A. Pacey (eds.), *Seasonal Dimensions to Rural Poverty*, 1981; D. E. Fisher, *Fire and Ice: The Greenhouse Effect, Ozone Depletion, and Nuclear Winter*, 1990; M. H. Glantz (ed.), *Societal Responses to Regional Climatic Change: Forecasting by Analogy*, 1988; J. T. Houghton, B. A. Callander, and S. K. Varney (eds.), *Climate Change 1992: The Supplementary Report to the IPCC Scientific Assessment*, World Mctcorological Organization/U.N. Environment Programme, 1992; J. T. Houghton, C. J. Jenkins, and J. J. Ephraums (eds.), *Climate Change: The IPCC Scientific Assessment*, World Meteorological Organization/U.N. Environment Programme, 1990; No way to run a desert, *Nat. Geogr.*, pp. 694–719, June 1985.

Coal mining

Methane trapped in coal beds is a vast and virtually untapped source of clean, sulfur-free energy. In the United States, it represents a resource of 400×10^{12} ft^3 (11.3×10^{12} m^3).

The production of gas from coal beds is not destructive to the coal. Miners have not experienced any significant problems in mining through degassed zones. In fact, degassing reduces the potential for gas hazards in underground coal mines.

The origin of coal-bed methane starts with the formation of coal, which acts both as the source of the methane and as its storage reservoir. The unique characteristics of coal make it a prolific producer of gas compared to conventional reservoirs such as porous sandstones.

Origin of coal. Coal is a sedimentary rock formed from accumulated plant debris that is metamorphosed by increases in pressure and temperature. The physical and chemical properties of coal and its gases are influenced by the types of plant material, the original environment of deposition, and the geological history after burial.

The constituents of coal are known as macerals. Different types of plant material (wood, bark, spores, algae, waxes, and so forth) form different macerals, which can be recognized under the microscope. Macerals are analogous to the individual minerals that are found in inorganic rocks. Each type of plant material has specific percentages of carbon, hydrogen, sulfur, nitrogen, oxygen, and ash-forming (noncombustible) components; these contribute to the character of the resulting coal and gases. The degree to which the plant materials decay before and during deposition also affects the process.

Coalification. Burial by increasing thicknesses of inorganic sediments causes the plant material to be metamorphosed, changing its physical and chemical properties. As the pressure and temperature increase, the more volatile, hydrogen-rich components (mostly methane) are released, so that coal becomes progressively richer in carbon. This is the process of coalification. The degree of coalification, or maturity, is expressed in terms of rank, with the material progressing from peat at the time of deposition, to lignite, subbituminous coal, bituminous coal, anthracite, and finally graphite. As rank increases, the percentage of carbon increases and that of hydrogen decreases. Graphite is 100% carbon. The optical reflectance of the macerals is very sensitive to changes in rank, so measurements of reflectance are often used to specify coal rank.

Generation of gas in coal. The coalification process generates large quantities of methane as well as smaller amounts of other gases. Some of this gas is formed by bacterial action in the early stages of diagenesis at temperatures of 70–120°F (21–49°C); the majority is formed by the thermal action during deep burial, beginning at temperatures around 200°F (93°C). This gas is stored both in the coal and, after the gas-expulsion point in the coal has been attained, in the associated sediments. Not all gas migrates or is expelled from the coal seam; significant volumes are retained.

The **illustration** shows the generation of gases at the various stages of coalification. Gas generation begins in lignitic coals and immature subbituminous coals. This early stage includes both the biogenic generation of methane and the expulsion of nitrogen and carbon dioxide. Gas generation increases exponentially, and it reaches a maximum when the coalification process

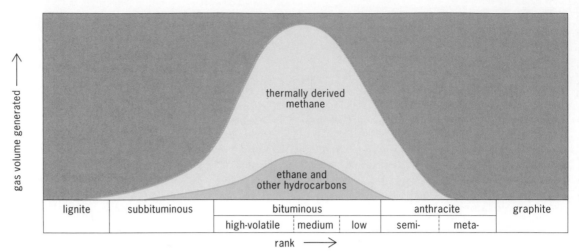

Gas generation in coal. Pressure and heat increase as rank increases. (*Gas Research Institute, Chicago, Illinois*)

reaches the temperatures associated with the bituminous stage. Gas generation then decreases through the anthracite stage and ends around the metaanthracite stage. No gas is generated in the graphite range because no hydrogen remains in the material. Like natural gas from other sources, gas from coal beds can have a range of compositions depending on factors such as source material and rank. In terms of its use as a fuel, the gas produced from coal is similar to conventional natural gas derived from marine rocks rich in organic materials. *SEE FUEL.*

Porosity of coal. The ability of coal to store and release methane is controlled in part by the nature of its porosity. At the beginning of coalification, peat has an interconnected network of very large, water-filled pores. As coalification increases, however, that network collapses; and when bituminous rank is reached, the coal is characterized by a very different type of porosity. The porosity of higher-rank coals is low, usually less than 5%, but their open molecular structure provides very small (10–1-nanometer-sized) micropores into which small molecules such as methane can penetrate. At this scale, gas molecules exhibit weak attractive forces with the solid coal substrate, so the coal has a high apparent "surface area" for gas storage. Consequently, coal can accommodate far greater amounts of gas at any given pressure than can conventional reservoir rocks.

Coal commonly supports a second type of porosity, a network of orthogonal fractures known as cleats. These fractures are interconnected and are much larger than the micropores. The micropores contain the majority of the methane, but the fracture system provides an important pathway that facilitates gas flow.

Coal as a reservoir. Volume for volume, high-rank coals are capable of storing several times as much gas as are porous sandstone reservoirs under similar pressures, because of the extremely high internal surface areas of coals. Gas storage also depends, in part, on pressure and, therefore, on the depth of the reservoir. Shallow coals (at depths to about 1500 ft or 450 m) in the San Juan Basin of New Mexico and Colorado have a storage capacity of about 17:1

standard cubic feet (SCF) per cubic foot (m^3/m^3) of reservoir. A standard cubic foot is the unit for the volume of a gas at standard conditions of pressure (1 atm) and temperature ($0°C$). The deeper coals (to 3000 ft or 900 m) have a storage capacity of about 28:1 SCF/ft^3 of reservoir. For conventional sandstones, the capacities are about 7:1 to 12:1 SCF/ft^3 for shallow and deep reservoirs, respectively.

In coal, methane is stored in two ways. In the system of microporosity, the gas is adsorbed within or onto the molecular structure of the coal substrate or within the micropores. In the system of macroporosity, the gas is stored conventionally as free gas within the fractures or as free gas dissolved in the water within the fractures. The vast majority of the methane is stored in the microporosity systems.

Permeability of coal. Coal has a very large internal surface area; the methane and other gases must diffuse through the micropores to the larger fractures that connect to the well bore. Water in the fractures must also flow to the well bore so that it does not block the gas flow. The gas flow depends on the permeabilities associated with both the micropores and the fracture system. The former is often very low (0.05–0.5 millidarcy), and steps must usually be taken to improve it before gas flow will begin. The most common approach is to reduce the confining pressure in the coal near the well bore by pumping out water; thus many coal-bed methane wells produce large volumes of water along with the gas. Hydraulic fracturing of the coal and special hole completion techniques are also used.

Gas content. The amount of gas in place combined with the degree of permeability determines whether or not it is economically viable to produce coal-bed methane. Therefore, one of the first objectives during drilling should be careful collection of reliable data concerning the gas content of the coal seams. During the early stages of exploration, gas content data are not generally available, but a rough estimation can be made on the basis of rank.

Gas content data are best determined from coal samples taken during the drilling of a well. Several types of coal samples can be collected. The most common sampling method uses conventional cores. A core of the seam is cut as the drill bit penetrates it. The drill string is then quickly withdrawn from the borehole, and the core is carefully removed and placed, usually as 1-ft (0.3-m) sections, into special devices known as desorption canisters. The problem with the conventional coring method is that some gas is lost while the sample is being brought to the surface. It is imperative that the coal be sealed within the canisters as quickly as possible to minimize further loss of gas, and that the time that elapses between the drilling of the core and the sealing of the canister, as well as the reservoir temperature, be recorded.

The simple apparatus used to measure the desorbed gas consists primarily of a canister, a hose, and an inverted water-filled graduated cylinder. The coal is sealed in the canister; as the gas is desorbed, the volume is measured by using the liquid displacement method. The most commonly used method was developed by the U.S. Bureau of Mines; with some modifications it is used throughout the industry. Gas measurements are taken at predetermined times, initially minutes, extended to hours and days, for a duration of up to several months. The volume of gas evolved from the sample is measured as a function of time, temperature, and pressure. The values of the measurement are then plotted on a graph.

The gas volume is calculated according to the equation below, where GIP = gas in place (ft^3 or m^3);

$$\text{GIP} = \text{GC} \times h \times \text{DA} \times p$$

GC = gas content (ft^3/ton or m^3/metric ton); h = net coal thickness (ft or m); DA = drillable area (acres or m^2); and p = coal density (tons/acre-foot or metric tons/m^3)

Reservoir evaluation. The final step in the evaluation of coal-bed methane is the computer modeling of the permeability and flow characteristics within the reservoir. The methods and procedures used for coal-bed methane are similar to those used for conventional oil and gas, but they must be able to model two-phase flow (gas and water) and deal with both fracture porosity and microporosity.

Extraction technology. The first coal-bed methane well was drilled in the latter 1950s in the San Juan Basin of New Mexico. The drilling technology for a coal-bed methane well is virtually identical to that for conventional oil or gas wells. There are, however, several types of special completion methods that may improve the productivity of a coal bed. Recent research in the San Juan Basin suggests that given the right combination of pressure and natural fracturing, open-hole (uncased) techniques tend to be more productive. Under conditions of less favorable fracturing and lower pressures, conventional cased holes that have been hydraulically fractured seem to be more successful.

Ongoing research related to coal-bed methane is focusing on two general themes: methods and techniques to recover the gas more efficiently, and geological controls on coal-bed fracture permeability and development of predictive techniques regarding where

the permeability will be maximized. Interest in coal-bed methane as a fuel source is now spreading around the world. Exploration for and development of this resource is being carried out in Canada, Europe, and Asia, as well as the United States. The exact size and distribution of this energy resource is not yet known. It will take many years before accurate world resource estimates of the coal-bed gas fuel potential and availability can be made; however, the estimates are expected to be very large.

For background information SEE COAL MINING; METHANE in the McGraw-Hill Encyclopedia of Science & Technology.

Dennis J. Nikols

Bibliography. J. E. Fassett (ed.), *Geology and Coalbed Methane Resources of the Northern San Juan Basin, Colorado and New Mexico*, 1988; S. D. Schwochow, D. K. Murray, and M. F. Fahy (eds.), *Coalbed Methane of Western North America*, 1991.

Coenzyme

Pyrroloquinoline quinone (PQQ) was identified in 1979 as an essential cofactor in methylotrophic bacteria. These bacteria are able to grow on methane derivatives (C$_1$ compounds) as their sole source of carbon and metabolic energy. In order to do so, the bacteria must be able to oxidize C$_1$ compounds, as shown in reaction (1), in which formaldehyde (HCOH), hydrogen ions

$$CH_3OH \rightarrow HCOH + 2e^- + 2H^+ \qquad (1)$$

(H$^+$), and electrons (e^-) are produced from methanol (CH$_3$OH).

PQQ structure. The bacterial enzyme that catalyzes reaction (1) is known as methanol dehydrogenase; the cofactor PQQ is required in order for the enzyme to function. The structure of PQQ, in three different oxidation states, is shown in **Fig. 1**. Also shown is the structure of the acetone [(CH$_3$)$_2$CO] adduct of PQQ (Fig. 1d), which was the form of the cofactor first isolated and structurally characterized via x-ray diffraction. The structure of PQQ was independently established by total synthesis. Subsequently, PQQ was shown to be a cofactor in other alcohol dehydrogenases and in glucose dehydrogenase.

Proteins that utilize PQQ as a cofactor are members of a class now known as quinoproteins. PQQ is noncovalently bound to these enzymes; that is, several relatively weak interactions between the protein molecule and the cofactor are responsible for binding. A comparison of the amino acid sequences of enzymes that use PQQ as a cofactor reveals a common (consensus) sequence, (X)$_{14}$-lysine-(X)$_{12}$-proline-(X)$_6$-glycine-(X)$_2$-tyrosine-(X)$_9$, where X is any amino acid. It is possible that this part of the enzyme molecule is involved in binding PQQ. Most quinoprotein dehydrogenases have been shown to require calcium ion (Ca^{2+}) or magnesium ion (Mg^{2+}) for PQQ binding. These metal ions are known to bind strongly to PQQ, suggesting that the PQQ–metal ion complex is the species recognized by the enzyme.

Fig. 1. Pyrroloquinoline quinone structures. (*a*) Pyrroloquinoline quinone (PQQ). (*b*) Semiquinone form. (*c*) Reduced quinol form (PQQH$_2$). (*d*) Acetone adduct at C-5. (*After C. Hartmann and J. P. Klinman, Pyrroloquinoline quinone: A new redox cofactor in eukaryotic enzymes, BioFactors, 1:41–49, 1988*)

PQQ chemistry. The chemistry of PQQ is largely determined by the two adjacent carbonyl (C=O) groups, and the presence of these groups establishes PQQ as an *ortho*-quinone. Consequently, PQQ is well suited to participate in reduction-oxidation (redox) reactions. All three oxidation states of PQQ [quinone (Fig. 1*a*); semiquinone (Fig. 1*b*); quinol, designated PQQH$_2$ (Fig. 1*c*)] may be involved in its biochemical reactions. As implied by the structure of the acetone adduct at C-5 (Fig. 1*d*), the carbonyl groups are not equally reactive toward other molecules. The so-called C-5 carbonyl is more reactive, and it is thought to be the key position in the chemistry of PQQ. Alcohols are believed to form adducts at C-5 as the initial step in their oxidation to aldehydes [an intermediate step in reaction (1)]. It is thought that the alcohol oxidation step consists of a single two-electron transfer, with PQQ being reduced by two electrons to form the quinol (Fig. 1*c*). The subsequent reoxidation of PQQ by nonphysiological electron acceptors, or the physiological partner of the alcohol dehydrogenase, may occur as two one-electron steps involving the semiquinone (Fig. 1*b*) as an intermediate. The PQQ/PQQH$_2$ redox potential (the voltage at which two electrons can be added or removed from the molecule) is relatively high (0.090 V at pH 7).

Biochemical functions. Covalent adducts between substrates and enzymes or enzyme-cofactor complexes are involved in numerous enzyme-catalyzed reactions. The cofactor PQQ combines this type of reactivity with the potential for carrying out both one- and two-electron redox reactions. Hence, it is an extremely versatile cofactor. In part for this reason, and on the basis of a number of suggestive experiments, PQQ or a derivative was proposed to be the cofactor in several other bacterial and eukaryotic enzymes, including some mammalian enzymes. Since there is no known biosynthetic pathway for the synthesis of PQQ in mammals, the presence of PQQ in a mammalian enzyme would imply that it would have to be obtained from the diet or from organisms living within the mammalian host. Thus, the idea that PQQ might be a vitamin was suggested. In fact, PQQ has been shown to be an effective growth stimulant for certain bacteria and for mammalian cells in culture. Recent experiments have shown that the addition of this cofactor to highly purified, chemically defined diets stimulates animal growth. Independent experiments demonstrated that dietary PQQ is absorbed by mammals, principally via the large intestine. PQQ deprivation leads to a variety of disorders in neonatal mice and rats that appear to be related to impaired maturation of connective tissue (**Fig. 2**). PQQ may also have beneficial antioxidant properties. Hence, a reasonable case can be made that PQQ and perhaps related quinones are nutritionally and physiologically important in mammals. The fundamental biochemical basis for the effects of this cofactor in mammals remains uncertain. A major problem is that the natural level of PQQ in mammals appears to be

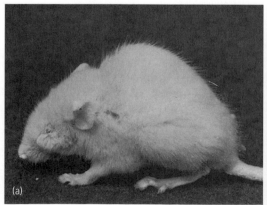

Fig. 2. Appearance of (*a*) PQQ-deprived and (*b*) PQQ-supplemented mice at week 8 after weaning. The PQQ-deprived mice recovered quickly after addition of PQQ to their diet. (*From J. Killgore et al., Nutritional importance of pyrroloquinoline quinone, Science, 245:850–852, 1989*)

so low that it is very difficult to detect PQQ with confidence.

Enzymes. The suggestion that PQQ might be present in mammalian enzymes (and enzymes from other eukaryotes) stimulated a great deal of research. PQQ was proposed to be covalently bound (that is, by strong chemical bonds) in the active sites of three important mammalian enzymes: lysyl oxidase, dopamine-β-hydroxylase, and amine oxidase. All three enzymes also contain copper, which is required for activity. Lysyl oxidase is the key enzyme in the proper development of connective tissue, dopamine-β-hydroxylase is involved in neurotransmitter metabolism, and amine oxidases are widely involved in the metabolism of biogenic amines, which have a variety of physiological roles in higher organisms. Careful and thorough experiments have now established that covalently bound PQQ is not present in dopamine-β-hydroxylase and amine oxidase and is very unlikely to be present in lysyl oxidase. With bound copper, dopamine-β-hydroxylase is fully active and no additional cofactors are required.

A new type of cofactor was discovered in amine oxidases. An active-site fragment derived from bovine plasma amine oxidase was isolated and structurally characterized by nuclear magnetic resonance and mass spectroscopy. The structure deduced from these experiments was the oxidized form of tri-hydroxyphenylalanine (topa), designated topa quinone (**Fig. 3**a). Subsequent work has shown that topa quinone is the cofactor in all copper-containing amine oxidases that have been highly purified.

New quinone cofactors. The cofactor in lysyl oxidase has not yet been conclusively identified. Several other enzymes, from a variety of organisms, that had been claimed to contain PQQ have been shown not to contain it. However, another new quinone cofactor was discovered during this period: tryptophan-tryptophanylquinone (TTQ; Fig. 3b), which is the functional group in the active site of the enzyme methylamine dehydrogenase. As often happens in science, a novel result or hypothesis, although eventually proven erroneous, nevertheless leads to new and unanticipated discoveries.

Both topa quinone and TTQ are similar to PQQ in their reactivity. The enzymatic reactions in which topa quinone and TTQ participate are shown in (2) and (3), respectively. In reaction (2), the R symbolizes a variety of C,H,N-containing groups.

$$RCH_2NH_2 + H_2O + O_2 \rightarrow RCHO + \underset{\text{Hydrogen}}{H_2O_2} + \underset{\text{Ammonia}}{NH_3} \quad (2)$$
$$\text{peroxide}$$

$$H_2O + \underset{\text{Methylamine}}{CH_3NH_2} \rightarrow HCOH + 2e^- + 2H^+ + \underset{\text{Ammonia}}{NH_3} \quad (3)$$

In both reactions, covalent adducts between the quinone and the substrate are believed to be intermediates, similar to the case for methanol dehydrogenase, reaction (1). Further, like PPQ, topa quinone and TTQ can participate in one- and two-electron redox processes. Hence, amine oxidases and amine dehydroge-

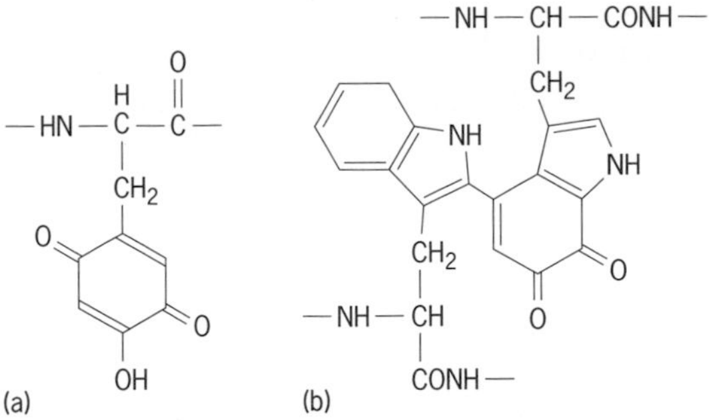

Fig. 3. Quinone cofactors. (*a*) Topa quinone. (*b*) Tryptophan-tryptophanylquinone (TTQ).

nases are also classified as quinoproteins together with the PQQ-requiring enzymes.

Prospects. Quinoproteins are currently of considerable scientific interest, in part because they are examples of a relatively new area of biochemistry concerned with the controlled use of radicals in biological catalysis. One-electron redox processes often involve the production of a radical, which may be simplistically thought of as a molecule with an odd number of electrons. For many years, radicals were considered deleterious to biological systems but are now recognized to serve as critical intermediates in numerous enzyme-catalyzed reactions. For example, an enzyme-bound semiquinone radical has been implicated in the amine oxidase reaction, and tyrosine radicals appear to serve a critical functional role in other enzymes.

Another new and fascinating area of biochemistry has developed as a consequence of investigating the possible roles of PQQ in enzymes. The topa quinone and TTQ cofactors are derived from the amino acids tyrosine and tryptophan, respectively, after the biosynthesis of the polypeptide chain of the enzymes. These cofactors are the first examples of posttranslational modification of protein amino acid residues to produce redox-active cofactors. The basic biochemical mechanisms for this novel process are not yet understood but are under active investigation. It is not known if other enzymes are involved or if the generation of topa quinone and TTQ are examples of unprecedented and sophisticated self-processing. Clearly, the study of PQQ, quinones, and quinoproteins will develop rapidly for the next several years.

For background information SEE AMINO ACIDS; CO-ENZYME; ENZYME; FREE RADICAL; QUINOLINE; QUINONE in the McGraw-Hill Encyclopedia of Science & Technology.

David M. Dooley

Bibliography. V. L. Davidson (ed.), *Principles and Applications of Quinoproteins*, 1992; J. A. Duine, Quinoproteins: Enzymes containing the quinonoid cofactor pyrroloquinoline quinone, topaquinone, or tryptophan-tryptophan quinone, *Eur. J. Biochem.*, 200: 271–284, 1991; J. P. Klinman et al., Status of the cofactor identity in copper oxidative enzymes, *FEBS Lett.*, 282:1–4, 1991; C. R. Smidt, F. M. Steinberg, and R. B. Rucker, Physiologic importance of pyrroloquinoline, *Proc. Soc. Exp. Biol. Med.*, 197:19–25, 1991.

Cognition

A controversial area of psychology is whether females and males have different cognitive abilities, that is, abilities used in acquiring, retaining, and using information. Recent research supports the position that there are some gender-related differences in cognitive abilities, but these differences are multiply determined and their effects are complex. In interpreting research findings, it is important to keep in mind that all conclusions about gender differences are based on statistical findings from large samples. There is great individual variability and gender overlap with regard to cognitive abilities.

The recent research concerning the question of gender differences in cognition has centered on three basic modes—verbal, visual-spatial, and quantitative abilities.

Verbal abilities. Verbal abilities refer to a broad range of language-related activities including spelling, correctly producing fluent speech, using correct grammar, solving verbal analogies, and comprehending complex verbal material. Gender differences do not exist for all verbal tasks, but those found usually show a female advantage. The only exception is the male superiority in making verbal analogies.

Of all gender differences, those involving verbal abilities are the earliest to emerge developmentally. On the average, girls speak at a younger age than boys; they also use more complex grammatical construction (for example, passive voice) at an earlier age. The gender difference is most dramatic at the low end of the verbal abilities distribution, that is, among people with particularly poor verbal abilities. For example, the overwhelming majority of dyslexics, remedial readers, and stutterers are male.

One notable exception to the conclusion that gender differences in verbal abilities tend to favor females involves the verbal portion of the Scholastic Aptitude Test (SAT), the test most frequently used for decisions regarding college admissions. Females scored higher than males on the verbal portion of the SAT in the late 1960s and early 1970s, but their scores relative to males have dropped since 1967, and females now score approximately 12 points lower on this test. The relative loss is probably due to the increased proportion of females now taking this exam (that is, females who took earlier tests were probably selected from the most gifted abilities range, while now the females who take this test also include those who are "average" in abilities) and to a greater emphasis on verbal analogies.

Visual-spatial abilities. Those abilities that are involved in imaging what a figure would look like if it were rotated in space or in discerning relationships among objects are considered to be visual-spatial. These abilities are needed in such professions as architecture, the building trades, dentistry, and chemistry.

Comprehensive reviews of the literature have shown that males, as a group, score higher on tasks that involve visual-spatial abilities, although, as with verbal abilities, there is overlap in the distributions of abilities. The male advantage for visual-spatial tasks

emerges at around 7–8 years of age. The amount of the average gender difference on these tasks varies, with smaller differences on tests of spatial perception and larger differences on tests that require subjects to maintain a spatial representation in memory while mentally transforming it in some way.

Quantitative abilities. The term quantitative abilities is a rubric for a wide variety of mathematics-related tasks. The tasks used to assess quantitative ability include answering computational questions, solving mathematical problems, extrapolating numerical trends, and using quantitative information in novel ways. On the average, girls perform better on computational tasks throughout elementary school, but males consistently score higher on all secondary and postsecondary standardized tests of quantitative ability. For example, males score approximately 50 points higher (SAT scores range from 200–800), on the average, on the mathematics portion of the SAT and have maintained this advantage over the last 25 years.

One possible explanation for these results is that advanced mathematics is highly visual-spatial with topics such as geometry, calculus, and topology requiring the representation and manipulation of visual information in memory—a task that males, in general, perform more accurately and quickly than females.

Reasons for sex-differentiated patterns. It is clear to researchers that there is no single or simple answer to the question of why males and females exhibit differences on cognitive tests. There are biological differences between females and males (other than the obvious genital ones) that are probably involved in cognitive functioning. Several researchers have examined the role of sex hormones (primarily testosterone, estrogen, and progesterone) on cognition. The relative amounts of these hormones that are available during prenatal development play a role in the formation of the developing brain, resulting in differences in the brains of males and females. The most frequently studied locus for sex differences in the brain structures that underlie cognition is the neural fibers in the corpus callosum that connects the two hemispheres. It has been found that females have more interhemispheric fibers and larger fibers. This difference suggests an anatomical basis for the theory that females are more likely to utilize both cerebral hemispheres for language. There are also sex differences in brain areas involved in sexuality, most notably in selected portions of the hypothalamus that are two to three times larger in males (size differences are adjusted for brain and body weight), but these differences probably play no role in cognition.

Sex hormones also play a role in cognition at puberty and in adulthood. For example, a minimal amount of testosterone is needed in males at puberty for the development of good spatial ability. Recent research has shown that some fine-grained cognitive tasks (for example, rapid speech and fine motor movements) vary in females cyclically over the menstrual cycle. Thus, normal variations in sex hormones can influence cognition.

In addition, the sexes are socialized differently from

birth. Boys receive more spatially related toys, such as building blocks, and girls receive more books, dolls, and play-house toys. Furthermore, there appear to be important differences in the overt and covert socialization messages of adults and other children: Girls receive messages encouraging them to be nurturant and supportive, and boys receive messages encouraging them to be assertive and to be leaders. Studies also show that boys receive more encouragement in mathematics and science courses. Pressures to conform to sex role stereotypes are powerful and ubiquitous. They are also difficult to study experimentally.

A psychobiosocial model. The most promising model for integrating biological and psychosocial influences on sex-differentiated development is the psychobiosocial model, which recognizes the reciprocal effects that psychological, biological, and social variables exert on each other. The psychological, biological, and social variables create a seamless web of influence on each individual, so that it is difficult to determine where one type of variable ends and another begins. In the uterus, the brains and genitals of fetuses are formed to reflect either male or female hormones. The sex of the newborn influences how people will react to him or her and the socialization that will be received. Environmental experiences, which are often different for females and males, may alter brain structures and the amount of various hormones that are secreted. There is evidence that brain structures change in response to environmental stimuli. For example, experimental research with nonhuman mammals has shown that enriched environments cause an increase in the weight of the brain and in the complexity and growth of neural structures. Thus, socialization experiences can affect biological structures and functions. Differences in brain structure and hormone concentration also influence the kinds of experiences individuals seek out, and these experiences in turn affect a host of psychological variables, such as self-concept and gender-role identification. In this way, psychological, biological, and social variables act together to influence sex-differentiated patterns of cognitive abilities.

For background information SEE COGNITION; INFORMATION PROCESSING; PSYCHOLINGUISTICS in the McGraw-Hill Encyclopedia of Science & Technology.

Diane F. Halpern

Bibliography. J. Balthazart (ed.), *Hormones, Brain, and Behaviour in Vertebrates: I. Sexual Differentiation, Neuroanatomical Aspects, Neurotransmitters and Neuropeptides,* 1990; A. H. Eagly, *Sex Differences in Social Behavior,* 1987; D. F. Halpern, *Sex Differences in Cognitive Abilities,* 2d ed., 1992.

Combustion

When a reactive fuel is injected onto the lateral surfaces of an airfoil or from the base of a projectile traveling at high speed, combustion can be established. The combustion alters the surrounding pressure field and, if properly controlled, can produce useful forces. The forces can augment the lift, provide attitude con- trol, reduce or cancel the drag, or produce thrust. Fundamental constraints that result from boundary-layer separation limit the magnitude of the forces that can be generated, and this limitation precludes the use of external burning as the primary propulsion system for an accelerating vehicle. Nonetheless, the potential benefits for other applications have spurred the development of external burning technology, from its beginning at the end of World War II to flight tests of candidate systems in 1992.

Early development of external burning. During World War II, solid pyrotechnics were attached to the base of the ammunition used in aircraft-mounted machine guns. External combustion in the base produced the desired luminosity for night fighting. Unfortunately, for economic reasons, only a tiny fraction of the high-speed rounds were fueled. The trajectory of these intermittent base-burning tracers segregated them from the principal mass of the burst. Reduced drag from external burning was the culprit. Consequently, skilled marksmanship was essential to compensate for the offset. Research to explain the phenomenon had to await the end of the hostilities.

Following the war, investigations of wake combustion were made by injecting hydrogen into the base of a 2.5-in.-diameter (10-cm) cone placed in a Mach 1.6 free jet. (Mach number is airstream velocity divided by the local speed of sound; thus $M > 1$ is supersonic.) Similar ground tests were made with solid propellant compositions on the base of 1.57-in. (40-mm) projectiles. These tests provided the data explaining the

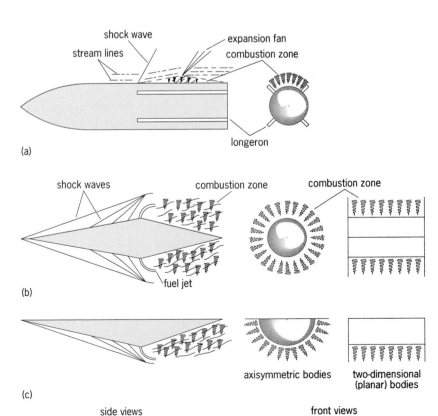

Fig. 1. External burning configurations. (*a*) Attitude controller. (*b*) Thrust generator. (*c*) Thrust and lift generator.

anomalous behavior of the tracer bullets and paved the way for the development of a wide variety of external burning devices.

External burning configurations. Whereas external combustion has been demonstrated on bodies traveling at supersonic speeds, the results from ground tests simulating flight suggest that all of the burning had occurred in subsonic portions of the wake. To obtain its full potential, external combustion would require stabilization of flames in regions of supersonic flow. **Figure 1** shows various concepts for external burning wherein the flow in the combustion zone is supersonic. In effect, external burning raises the temperature, thereby increases the volume of the gas, and causes the streamlines about the body to be deflected, giving a pressure rise similar to that of an aerodynamic flap but with a significantly smaller expansion effect and no drag penalty. There is also a reaction force caused by the injection, which has components in the thrust or lateral directions, depending on the angle of fuel injection.

__Attitude controller.__ The attitude controller for an axisymmetric vehicle (Fig. 1*a*) has injection aft of the center of gravity in any one of four quadrants. (The periphery of the vehicle could alternatively be subdivided into any number of desired segments.) Longitudinal fences separate the quadrants to reduce the dissipation of the positive pressure field through circumferential spillover. The downward force due to external burning leads to positive pitch α and therefore puts the external burning in the leeward zone, so that possibly, at large values of α, adverse conditions for combustion could result. However, if external burning is being used solely to trim the body, an aerodynamically stable vehicle can be designed so that the external burning will always occur in the windward zone. Attitude control systems based on external burning ahead of the center of grav-

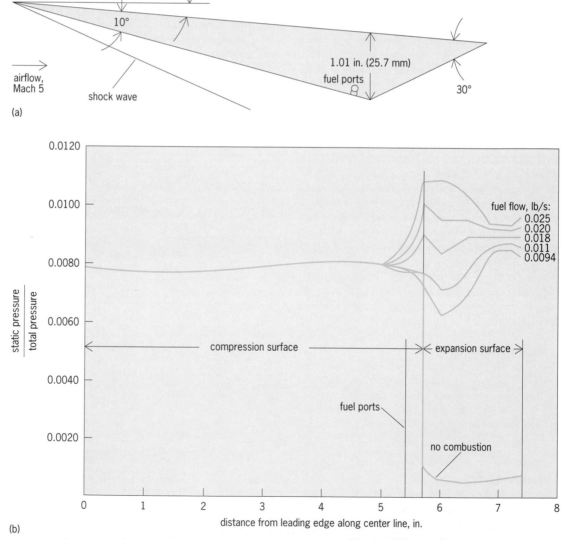

Fig. 2. Results of wind-tunnel tests on a 10–30° airfoil model at Mach 5 and a simulated altitude of 100,000 ft (32.8 km), with external burning of triethyl aluminum fuel. (*a*) Cross section of the airfoil. (*b*) Experimental pressure profiles. 1 in. = 2.54 cm. 1 lb/s = 0.45 kg/s. Total pressure = 78 psia = 5.36 × 10⁵ N/m². Total temperature = 1720°R = 956 K.

ity are conceivable but appear to be less attractive because of the difficulty of confining the positive pressure field to produce an effective pitching moment.

Thrust generator. The thrust-generating device (Fig. 1*b*) could be either the total vehicle or a podded or airfoil engine. At the so-called knee, fuel is added to the air and combustion maintains a positive pressure field on the aft body that is greater than that on the compression surface, thus producing net thrust. However, the specific impulse decreases drastically as the external burning changes from a drag-reducing to a thrust-producing device. Moreover, when more fuel is added above the threshold at which attached flow is sustained, the boundary layer separates just ahead of the injectors, creating more drag on the forebody and a net loss in thrust.

Thrust and lift generator. Figure 1*c* bifurcates the vehicle shown in Fig. 1*b* and therefore provides a combined axial and lateral force capability. The lateral force generated by combustion negates the need for deflection of aerodynamic surfaces and thereby eliminates induced drag, with a corresponding reduction in engine fuel flow. The axial component of force due to the rise in pressure either reduces or cancels drag, and if sufficiently high levels can be obtained, net thrust is produced.

Feasibility of external burning. It is advantageous to establish external combustion immediately following the forebody compression to enhance ignition in the zone of higher static temperature and pressure. Even with this assist it is far more difficult to burn externally than within the confines of an internally ducted engine. Not only are the pressures and temperatures considerably lower in the external burner, but the residence time for completion of heat release is shorter. Moreover, if flame stabilization devices are needed, the drag penalties can be large. Therefore, to generate heat release in external supersonic flows, it has been necessary to use either highly reactive fuels or a flame holder and a fuel with wide flammability limits, namely hydrogen. Both techniques were demonstrated in the late 1950s on configurations of the class shown in Fig. 1*c*.

Experimental pressure profiles. Experimental pressure profiles are shown in **Fig. 2** for tests at Mach 5 and at a simulated altitude of 100,000 ft (32.8 km) using highly reactive triethyl aluminum fuel. The airfoil (Fig. 2*a*) is composed of a 10° forebody wedge and a 30° aft-body wedge, and is at an incidence $\alpha = 4°$ to the flow. The ratio of static to total pressure is 1.9×10^{-3} in the free stream ahead of the model. Compression on the forebody raises the pressure ratio to about 8×10^{-3} (Fig. 2*b*). In the absence of combustion the pressure ratio drops to about 5×10^{-4} on the expansion surface. When fuel is added from ports located just ahead of the knee, the external combustion increases the pressure to levels greater than on the forebody and there is net thrust on the airfoil. Whereas these tests firmly established the feasibility of the external burning concept with burning in super-

sonic flow, the efficiencies were quite low. Moreover, the logistics of handling highly reactive fuels presents such a formidable problem that contemporary applications of external burning have focused on the use of hydrogen fuel with flameholders.

Hydrogen-fueled hypersonic aircraft. Hydrogen-fueled hypersonic ($M > 5$) aircraft provide a unique opportunity for exploiting the benefits of external burning. The engine duct in the air-breathing vehicle operating at hypersonic speeds must be relatively small, leading to a critical thrust deficit as the vehicle accelerates. Some propulsion cycles require fuel flows in excess of that required to produce thrust in the main propulsion system. The hydrogen is stored near its triple point at 24.9°R (13.8 K) and serves as a coolant for both the overall vehicle and in processing the air in the compression phase of the propulsion cycle. At

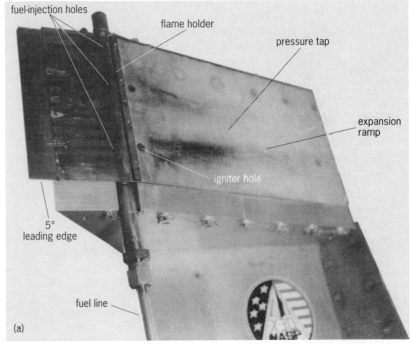

Fig. 3. Flight tests of an external combustion device. (*a*) Model, consisting of 5° leading-edge wedge and 12° expansion ramp. (*b*) U.S. Navy F/A-18 with hydrogen combustion on the wing tip.

high hypersonic speeds, vehicle cooling requires excess hydrogen, and air processing at transonic speeds can also require excess hydrogen. An ideal use of the excess is external burning.

Flight tests. The National Aerospace Plane (NASP) is a hydrogen-fueled hypersonic airplane for which the use of external burning is being considered. It would benefit from the drag reduction that would result from external burning at both transonic and hypersonic speeds. To investigate the feasibility of the external burning concept, flight tests of a model, previously tested in a transonic wind tunnel, were conducted on the wing tip of a U.S. Navy F/A-18 (Hornet). The device comprises a 5° leading-edge wedge and a 12° expansion ramp mounted orthogonal to the wing tip (**Fig. 3**a). The fuel storage and supply system is contained in a separate module, attached to an underwing bomb rack. Fuel is injected from a series of holes located just ahead of a step flameholder. Ignitors are positioned downstream of the flameholder. During the first two test series, stabilized external burning of hydrogen was demonstrated in the transonic speed range, $0.88 < M < 1.05$, at altitudes of 21,000–31,000 ft (6.40–9.45 km). Figure 3b, taken from a chase aircraft, shows the luminous flame zone, which was estimated to be about 3 ft (1 m) in length. Pressure measurements verified the consistency between wind tunnel and flight performance.

For background information SEE AIRCRAFT FUEL; AIRFOIL; BOUNDARY-LAYER FLOW; HYPERSONIC FLIGHT; RAMJET; TRANSONIC FLIGHT in the McGraw-Hill Encyclopedia of Science & Technology.

Frederick S. Billig

Bibliography. G. L. Dugger et al., *Summary Report on External Ramjets*, Johns Hopkins Univ. Appl. Phys. Lab. TG 419, October 1961; E. H. Phillips, Flight tests demonstrate use of external burning as alternative to NASP wind tunnel evaluations, *Aviat. Week Space Technol.*, 137(4):52–53, July 27, 1992; *Proceedings of the 18th International Astronautical Congress*, 1969.

Comfort heating

Most inhabitants of the developed countries spend more than 90% of their time in residences, offices, and other types of buildings. Buildings offer protection from the weather and provide a safe and comfortable environment. Thermal comfort of humans depends on several parameters, including clothing, activity level, and air velocity. During winter months the optimum indoor temperature is 68–75°F (20–24°C) and the optimum relative humidity is approximately 50%. A sedentary person releases about 350 Btu/h or 105 W of heat, which aids space heating in winter but requires additional cooling in summer. Current ventilation standards require 15 ft^3 (0.43 m^3) of outdoor air per minute per person to remove airborne contaminants generated indoors. The indoor environment usually requires mechanical equipment and a control system to provide heating, cooling, and acceptable humidity and airborne contaminant levels. Recent concerns involve

the reduction of both the amount of energy required to heat and cool buildings and the concentration of indoor contaminants.

Energy requirements for space heating. The majority of the energy requirements for space heating result from replacing heat lost through the exterior envelope of the building, and heating and humidifying outdoor air to ventilate the interior.

Insulation. The exterior surfaces of a building must be well insulated. Thermal insulation comprises batts, rigid boards, loose fill, and blown foam. The thicker the insulation, the better the insulation level. Typical United States practice is to specify the R value of insulation. Heat loss through an exterior portion of a building such as a wall or roof is inversely proportional to the R value; thus, a high value denotes good thermal insulation. The R value is the amount of thermal resistance in units of h·ft^2·°F/Btu. Thus, insulation with an R value of 20 indicates a thermal resistance value of 20 h·ft^2·°F/Btu or 3.5 m^2·°C/W. Thermal resistance values for common building materials are listed in the **table**. Typical fibrous insulation is about 20 times better than single-glazed windows. Uninsulated thermal bridges, such as wood studs or steel beams, that pass between inner and outer surfaces of a wall or roof can significantly reduce the effective insulating level. Designs that stagger the bridges or provide thermal breaks can reduce the amount of heat that flows through these elements.

Windows traditionally have been poor thermal insulators; a single-glazed window has an R value of approximately 1.0, and a double-glazed window has an R value of about 2.0. Coatings on the glass surfaces can reduce the amount of infrared radiation that passes between two panes of glass. Better frame designs have reduced the amount of heat that passes through the window frame. Some double-glazed windows now have R values larger than 5.0.

Ventilation. Natural air leakage into and out of buildings can require a substantial amount of energy to heat and humidify the outdoor air to the indoor conditions. This leakage also results in uncomfortable drafts. Warm moist air that leaks out during cold conditions can cause moisture condensation within the structure, resulting in fungal growth and eventual structural deterioration in wood-based products. Many of the cracks in older buildings can be sealed by caulking. Present construction practice is to build homes tight enough that there is a natural leakage rate of about one-half of an air change per hour.

Moisture is the cause of many problems associated with buildings. Large amounts of moisture can be added to the air in a building, particularly in residences. For example, a shower can add moisture at a rate of 2 lb (1 kg) per hour. High levels of indoor moisture can lead to condensation on cold surfaces.

Buildings cannot be sealed too tightly because moisture levels can increase to unhealthy levels and the operation of exhaust appliances such as kitchen-range hoods and clothes dryers can lower air pressure indoors. The pressure reduction can prevent products of combustion from vented equipment (furnaces, water

Thermal resistance values for typical building materials		
Material	h·ft²·°F/Btu	m²·°C/W
5.5-in. (140-mm) fibrous insulation batt	19.0	3.3
5.5-in. (140-mm) fir or pine	5.5	1.0
6-in. (150-mm) concrete	0.9	0.16
8-in. (200-mm) concrete block, hollow core	2.1	0.37
Double-glazed window, wood frame	2.0	0.35

heaters, and fireplaces) from leaving the building, leading to a deadly condition if carbon monoxide is present in the combustion gases. Most large buildings have a mechanical ventilation system with an air-to-air heat exchanger that preheats the outdoor air by utilizing the exhaust air. The use of mechanical ventilation systems is becoming more common in residences.

Heating equipment. Most buildings have their own heating equipment, typically a boiler or a furnace. Heating-equipment efficiencies are now often above 85% and can be as high as 95%. High efficiency is accomplished by cooling the exhaust gases to recover the latent heat of the water vapor and to increase the heat-transfer coefficients in the heat exchangers. The heat pump equipment in some buildings is essentially a refrigeration system that can add heat to the building in winter and remove it in summer. Some buildings in downtown areas and other locations of high building densities use a central heating plant with hot water or steam distributed in a piping network, comprising a district heating system.

Space cooling. Cooling energy requirements are caused primarily by solar energy that is absorbed on exterior surfaces of a building or passes into the building through windows and skylights, internal heat that is generated within the building, and warm outdoor ventilation air that enters the building. Reduction of heat from these sources can be accomplished by several preventative methods. Surface coatings and tinted glass reduce the amount of solar energy that enters through windows. Exterior shading such as overhangs or awnings is effective in reducing the amount of solar energy that enters a building through glazing. Light-colored exterior surfaces absorb less solar radiation than dark-colored surfaces, and therefore produce less heat gain in the building. In thick, massive structures the heat exchange takes many hours, so the magnitude of the heat gains through these structures is very small.

There are several other sources of building heat. People, lights, and equipment within buildings generate heat, creating a significant cooling requirement. The cooling of outdoor ventilation air is also significant, particularly in commercial and institutional buildings that have mechanical ventilation systems. Energy is required to dehumidify air in summer. Moisture brought into the building in the outdoor air and moisture added to the air inside the building must be removed. Moisture stored in building and furnishing materials can create an additional cooling load at certain times of the day.

A typical forced-air heating and cooling system for a small commercial building is shown in **Fig. 1**. Cooling equipment usually consists of a refrigeration device that removes both heat and moisture from the air by achieving a dew point on the heat exchanger that is below the dew point of the air in the building. Most units are vapor compression systems with motor-driven compressors. Large absorption units require a heat source above approximately 200°F (95°C). These units are economical when low-priced heat such as steam is available during the cooling season. Recent regulations on the use of chlorinated fluorocarbons have prompted the search for alternate refrigerants. Chemicals in the methane and ethane families with hydrogen atoms and few if any chlorine atoms in the molecule appear to be promising substitutes. Large central ammonia refrigeration plants or lithium bromide–water absorption systems are other alternatives.

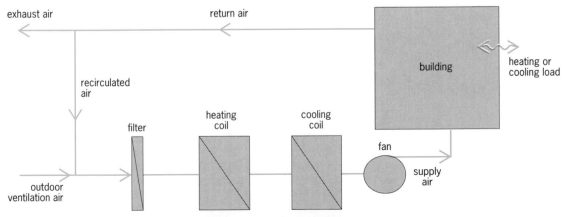

Fig. 1. Simple heating and air-conditioning system for a commercial building.

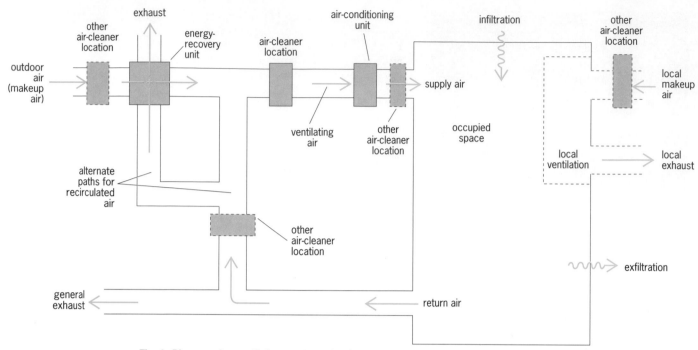

Fig. 2. Diagram of a ventilation system, showing parameters for controlling contaminant concentrations indoors. (*After American Society of Heating, Refrigerating and Air Conditioning Engineers, Ventilation for Acceptable Indoor Air Quality, ASHRAE 62-1989, 1989*)

Indoor air quality. Tighter, more energy-efficient buildings and increased awareness of the health risks associated with exposure to radon, tobacco smoke, gases, and microorganisms have prompted more stringent standards as to the allowable limits of various substances in the indoor air. The American Society of Heating, Refrigerating and Air Conditioning Engineers (ASHRAE) published Standard 62-1989, which prescribes the amount of outdoor air that should be provided to maintain a safe and healthy indoor environment. Allowable levels of various indoor airborne contaminants are not well established. A complicating factor is the different physiological responses of various people to the same level of contaminant.

Indoor concentrations of contaminants can be controlled by three methods. The first involves removing or reducing the source of the contaminants, for example, by providing nonsmoking areas or by eliminating materials that produce gaseous or particulate contamination. The second is the introduction of clean outdoor ventilation air. Both natural air leakage and mechanical ventilation can provide the necessary amount of outdoor air. The third method is cleaning the air with filters or other contaminant-removal devices. **Figure 2** shows various approaches to providing clean air to occupants of a building with a forced-air-conditioning system.

Various methods have been developed to reduce the amount of radon that enters a building from the soil. The recommended method is to exhaust the gas from the lower level of the building by using a low-flow-rate, continuous exhaust system. Particles can be removed effectively with media filters or electrostatic precipitators. Many gases can be removed by activated charcoal beds or by chemical reaction on the surface of granular beds.

Air distribution within rooms is important to provide ventilation air to the occupants. Short circuiting between the supply diffuser and return grille in the ceiling should be avoided. Partitions in large office areas obstruct the flow and can create regions of low air flow and poor ventilation. Open office spaces that have been divided into private office spaces may not have adequate ventilation air distribution because of the original locations of the supply and return grilles and the air-flow rates that are used.

For background information SEE AIR CONDITIONING; CENTRAL HEATING AND COOLING; COMFORT HEATING; DISTRICT HEATING; HEAT INSULATION; HEAT PUMP; RADON; VENTILATION in the McGraw-Hill Encyclopedia of Science & Technology.

Thomas H. Kuehn

Bibliography. *Healthy Buildings: Proceedings of the ASHRAE Indoor Air Quality 91 Conference*, 1991; T. H. Kuehn, Heat and mass transfer in occupied buildings, *Proceedings of the 9th International Heat Transfer Conference*, Jerusalem, vol. 1, 1990; *1989 ASHRAE Handbook of Fundamentals*, 1989.

Communications satellite

Since the INMARSAT consortium began operations in 1982, worldwide traffic between telephones, Telex machines, and facsimile machines, as well as traffic involving ships at sea and offshore platforms, has increased dramatically. To meet this communications growth and to accommodate new services such as voice and data between aircraft in flight and land stations, telephone and data between mobile termi-

Fig. 1. Configuration of *INMARSAT 3* satellite in space. (*Matra Marconi Space Ltd*)

nals, and navigation information to mobile units, INMARSAT planned to inaugurate its third generation of satellites during 1994.

INMARSAT 3 satellite. The satellite (**Fig. 1**) is manufactured by GE Astro, with the communications payload subcontracted to Matra Marconi Space Ltd. This three-axis stabilized spacecraft derives electric power from a 16.7-m (55-ft), Sun-oriented array designed to operate from geosynchronous orbit. Microwave signals in several frequency bands are received and transmitted through parabolic reflectors and other antennas facing the Earth. These antennas can form spot beams that focus the radiated microwave power on particular areas of the Earth. Together with an array of high-power amplifiers, they will allow a large increase in communications capacity. The high power will also accommodate small, inexpensive terminals for maritime, aeronautical, and land mobile use.

INMARSAT's major business at present is voice communications between land and ships. Over 12,000 INMARSAT-A and 1200 INMARSAT-C ship stations are in operation, the former providing voice and Telex and the latter, data services. Voice traffic is growing 40% per year and Telex traffic, 20% per year. With the introduction of *INMARSAT 3*, it is expected that voice and data traffic to commercial aircraft will increase rapidly, especially aircraft on transoceanic flights. Other services, which were not previously available, will be an L-to-L-band channel allowing direct mobile-to-mobile communications, a C-to-C-band channel, and an L-band navigation signal channel. The L-L-band connection will be used for special conditions such as search and rescue operations. The C-C-band channel permits system administrative traffic to be transacted directly between gateway land stations and to bypass the L-band transponders. The navigation payload repeater will enable a new service to INMARSAT users, providing signals that allow users to determine their position and speed.

Communications channelization. Figure 2 is a schematic representation of the mobile communications system. The usual communications connection for any telephone, facsimile machine, message device, or data device originates through the public switched network. A microwave link or a land cable using copper wires or optical fiber strands connects the switching center to the satellite earth station, or shore station. Such a station is located in each of the major countries that are members of INMARSAT, and accesses the satellite following a standardized protocol. The shore stations operate in the C-band frequency range: 6424–6454 MHz in the forward direction (transmitting to the satellite), and 3600–3629

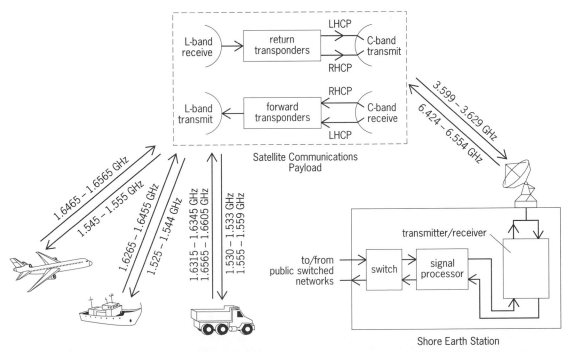

Fig. 2. *INMARSAT 3* mobile communications system. LHCP = left-hand circular polarization; RHCP = right-hand circular polarization.

MHz in the return direction (receiving from the satellite). The forward satellite transponders receive signals in the 6-GHz band, translate them to the L band, and retransmit them to the mobile stations in the frequency range 1525–1559 MHz. The return transmissions from the mobile stations, in the L-band frequency range 1626.5–1660.5 MHz, are received by the return transponders, translated to the C band, and transmitted to shore stations in the frequency range 3600–3629 MHz.

The frequency bands assigned to maritime, land, and aeronautical mobile users, as agreed by all member countries of the International Telecommunications Union (ITU), are severely limited in bandwidth. In order for the satellite to serve these users efficiently, the transponders must be capable of separating the user signals into several channels, arranged as in **Fig. 3**. However, to maintain the power and bandwidth utilization efficiency, with changing mixes of mobile terminal sizes, the channelizing method must be flexible. This flexibility is achieved in *INMARSAT 3* by using narrow-band, bandwidth-switchable, surface-acoustic-wave (SAW) filters in an intermediate-frequency processor, which is part of the transponder. The filters can be switched or tuned to the required bandwidth, shown in Fig. 3, by ground command.

The surface-acoustic-wave filters in the intermediate-frequency processor operate at a nominal frequency of 160 MHz. This operation requires, on the forward link, frequency conversion from 6 GHz to the intermediate frequency and a second conversion to the L band; on the return link, a similar process frequency-shifts the L band to 3.6 GHz. The up and down frequency conversions use a frequency generator source capable of multiple frequency outputs, where each frequency can be controlled by the telemetry, command, and control shore station. The frequency precision and stability must be held to within a few parts per million over a temperature range of 30–40°C (86–104°F).

Since most of the present traffic is maritime, the maritime service is allocated the largest portion of the spectrum, 13.5 MHz with one polarization and 15 MHz in the orthogonal polarization (Fig. 3). The aeronautical service is expected to be the next largest user and is allocated 10 MHz in each polarization. The land mobile service is allocated a maximum of 7 MHz in each polarization, with 3 MHz of it shared with maritime. The C-C-band, L-L-band, and navigation transponders occupy 0.9-MHz, 1-MHz, and 2-MHz bandwidth, respectively.

Spot beams and frequency reuse. The *INMARSAT 3* satellite provides increased communications capacity to the system through technology advances such as higher power transmitters and multiple spot beams. The high-power transmitters on board the satellite are eight times more powerful than those on *INMARSAT 2*. Twenty-two 20-W solid-state amplifiers are arrayed in a linear passive microwave circuit known as a Butler matrix to drive 22 helix-antenna feed elements. The arrangement provides the flexibility to direct the output power of one, several, or all the amplifiers to one output port, or to distribute power equally to each of the output ports. Since each output port generates a narrow beam, the transmitted signals can be the sum of several beams directed to specific portions of the Earth. Thus, the power can be matched to the communications requirements of a particular region.

Two parabolic reflectors, one for receive and the

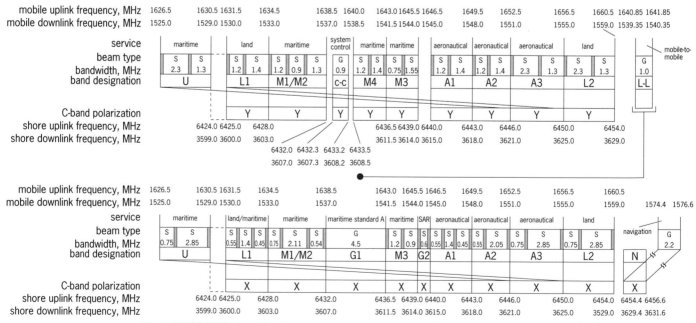

Fig. 3. *INMARSAT 3* channelization plan for transponders. Beam types: G = global beam; S = switchable to any spot beam or the global beam. C-band polarization: X = right-hand circular polarization (RHCP) for uplink, left-hand circular polarization (LHCP) for downlink; Y = LHCP for uplink, RHCP for downlink. SAR = search and rescue. Standard A refers to the type of shipboard terminal currently used on most vessels, a transmitting-receiving terminal using a steerable antenna and a low-noise receiving system.

other for transmit, in conjunction with the receive and transmit feeds and beam-forming networks, form four or five L-band spot beams plus the L-band global beams. As noted above, the transmit feed elements are connected to solid-state power amplifiers, whereas each of the receive feed elements is connected to a low-noise amplifier. The INMARSAT satellites are assigned locations on the geosynchronous orbit to cover the Atlantic Ocean at 26° and 55° west longitude, the Indian Ocean at 63° east longitude, and the Pacific Ocean at 180° east longitude. For optimum coverage of each region with any of the launched satellites, the beam positions and shape are readjustable to a limited extent by ground commands.

The radiated signals at C band and L band are circularly polarized, as shown in Fig. 3 by the designations X and Y for the left- and right-hand senses. The dual polarization applies only to the C-band transmissions to and from the shore stations. The L-band helix-type feeds do not lend themselves to dual polarization. Thus, with two sets of transponders, the dual-polarization capability of the C-band antennas, and the requirement that the L-band multibeam antenna system achieve a beam-to-beam isolation of greater than 18 dB, spectrum use and communications capacity are doubled.

Navigation payload. The U.S. Global Positioning System (GPS) satellites and the Russian GLONASS satellites allow fixed or mobile users to determine their positions to within 10 m (33 ft), and speed with a tolerance of ± 0.1 m/s (± 0.3 ft/s), when four satellites are visible and selective availability (purposeful accuracy degradation) is not activated. The GPS satellites are in an orbit inclined 63° to the Equator at a nominal altitude of 20,000 km (12,400 mi; 12-h orbit) and operate at a frequency of 1575.42 MHz. The GLONASS satellites operate at the same frequency but at an orbit altitude of 19,000 km (11,800 mi) and an inclination of 64.8°. The satellites transmit a highly accurate time signal, derived from on-board atomic clocks, and the satellite's ephemeris, periodically updated by a controlling earth station. The information allows the user's receiver to calculate position by measuring the time difference between the signals received from two or more satellites. The speed is calculated by observing the rate of change of position.

INMARSAT 3 carries a C-L-band transponder that will furnish similar signals, but the ephemeris and time signals are generated at an earth station and relayed through the satellite's global beam at the same frequency band as the GPS. The technique allows changing the information carried by the signals in real time, and it also simplifies the equipment needed on the satellite. The satellite transmission can also broadcast system status information.

The global-coverage navigation beam transmits an effective isotropic radiated power (e.i.r.p.) of 27.5 decibels above 1 watt (dBW). A separate 0.7-m-diameter (2.1-ft) parabolic reflector antenna, on the Earth-facing side of the satellite, yields an edge-of-Earth gain of about 16 decibels above isotropic power (dBi).

The power amplifiers are the same design as those in the communications transponders, yielding 17 W output power. The power assures that the signal at the mobile receiver is the same level as that obtained from the GPS, although the altitude of *INMARSAT 3* is approximately 16,000 km (10,000 mi) higher. To achieve an accuracy comparable to the coarse acquisition GPS signal, which is available to civil (nonmilitary) users, the transponder bandwidth is 2.2 MHz. The composite signal delay through the transponder must be very constant over the range of operating temperature and is specified as 10 nanoseconds. In parallel with the C-band uplink to L-band downlink primary path, a low-power (0-dBW) C-to-C-band link is included. This will be used by control earth stations to adjust uplink timing.

For background information *SEE COMMUNICATIONS SATELLITE; SATELLITE NAVIGATION SYSTEMS; SURFACE-ACOUSTIC-WAVE DEVICES* in the McGraw-Hill Encyclopedia of Science & Technology.

Louis Pollack

Bibliography. American Institute of Aeronautics and Astronautics, *Proceedings of the 14th International Communications Satellite Systems Conference*, part 1, March 1992; S. Egami and M. Kawai, An adaptive multiple beam system concept, *IEEE J. Selected Areas Commun.*, SAC-5(4):630–636, 1987; D. W. Lipke et al., MARISAT: A maritime satellite communications system, *COMSAT Tech. Rev.*, 7(2):351–391, 1977; P. Wood, Mobile satellite services for travellers, *IEEE Commun. Mag.*, 29(11):32–35, 1991.

Computability (physics)

Does light propagate computably? Does heat dissipate computably? In order to answer these and related questions, an understanding of the nature of computability from the theoretical point of view is essential.

The study of the properties of an arbitrary computable process goes back at least to the 1930s, when several mathematical logicians—A. Turing, E. Post, J. Herbrand, K. Gödel, and A. Church—formulated definitions of the term computable function. Although each of these definitions represented a different approach toward the subject, they were all shown to be equivalent. The concept of computability that emerged captures the intuitive notion quite well. Today, it provides the foundation for theoretical work in computer science.

In one respect, all the definitions agree. Both the domain and the range of a computable function can be encoded as subsets of the nonnegative integers N. This will be discussed below.

Turing-machine approach. It is instructive to consider Turing's work in some detail. Turing reasoned that the class of computable functions should consist precisely of those functions—with domain and range included in N—that can be computed by a digital computer. This led him to define the Turing machine, the most general type of digital computer.

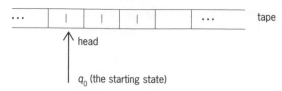

Turing machine at the beginning of the computation discussed in the text.

The Turing machine consists of a read-write head and a tape divided into squares. The head can perform atomic acts of the types given in (1).

$$\text{Print a letter of the alphabet}$$
$$\text{Move one square to the right: } R \qquad (1)$$
$$\text{Move one square to the left: } L$$

It is assumed that the blank B is a letter of the alphabet. Thus, the instruction "Print B" is equivalent to "Erase." In addition, at any given instant, the machine is assumed to be in one of a finite number of internal states, given by (2).

$$q_0 \text{ (the starting state)}$$
$$q_1, \ldots, q_{n-1} \qquad (2)$$
$$q_n \text{ (the final state)}$$

To see how such a machine operates, it is instructive to consider a simple example. Suppose the function f is defined by (3). A Turing machine M adequate for

$$f : x \rightarrow x + 1 \qquad (3)$$

computing f can be designed as follows. The alphabet consists of two symbols, given by (4). The number 0

$$| \text{ (stroke)}$$
$$\qquad (4)$$
$$B \text{ (blank)}$$

is represented by one stroke; the number 1 by two strokes, and so forth.

The computation of $f(2)$ will now be considered. The number 2 is placed on the tape by putting three strokes on consecutive squares. It is assumed that the machine is in state q_0 and that the head is scanning the leftmost square containing a stroke (see **illus.**). The computation is about to begin.

The instructions for the operation of the machine are given in (5). They permit the head to move past

$$q_0|Rq_0 : \quad \text{in state } q_0, \text{ scan '|', move one}$$
$$\qquad \text{square to the right, go into}$$
$$\qquad \text{state } q_0$$
$$q_0B|q_1 : \quad \text{in state } q_0, \text{ scan '}B\text{', print '|',} \qquad (5)$$
$$\qquad \text{go to state } q_1$$
$$q_1|Rq_2 : \quad \text{in state } q_1, \text{ scan '|', move one}$$
$$\qquad \text{square to the right, go into}$$
$$\qquad \text{state } q_2 \text{ (the halting state)}$$

all squares containing a $|$, until it reaches the first blank square. The head then prints a $|$ and halts. Thus, since there are three strokes on the tape at the beginning, the

machine will now have four strokes, indicating that the answer is 3.

The computation for $f(2)$ can be carried out by performing the operations in (6) in the order indicated.

$$q_0|Rq_0$$
$$q_0|Rq_0$$
$$q_0|Rq_0 \qquad (6)$$
$$q_0B|q_1$$
$$q_1|Rq_2$$

Actually, it is possible to go further. It can be shown not only that a Turing machine computes a function whose domain and range are subsets of the nonnegative integers N but also that the whole structure of the machine can be encoded in N by using a device called Gödel numbering. For example, the odd integers can be assigned to the basic symbols $|, B, R, L, q_i$ as in (7).

Symbol	Assignment	
$|$	3	
B	5	(7)
R	7	
L	9	
q_i	$2i + 11 \quad i = 0, 1, \ldots$	

Then sequence (8) can be represented by the se-

$$q_0|Rq_0 \qquad (8)$$

quence of nonnegative integers 11, 3, 7, 11. By using prime factorization this representation can be converted to a nonnegative integer $2^{11}3^35^77^{11}$, which can be denoted by the number r_1. Here, the numbers 2, 3, 5, and 7 are the successive prime numbers. (A positive integer n is prime if $n \neq 1$, and its only divisors are 1 and itself.) Similarly, the second and third instructions can be coded into the numbers r_2 and r_3, respectively. Then the code for the instructions for the machine M is $2^{r_1} \cdot 3^{r_2} \cdot 5^{r_3}$.

In a similar fashion, computation (6) can also be encoded. First, each line of the computation is encoded. Recall that (6) has five lines. If the number associated with the ith line of the computation is s_i, then $2^{s_1} \cdot 3^{s_2} \ldots p_5^{s_5}$ is the number of the computation. (Here p_i is the ith prime in order of magnitude beginning with 2.)

Similar observations can be made in connection with other equivalent definitions of computability. Thus, it is seen that not only is the basic notion of computability discrete but so is the syntax describing this notion.

Two results in the history of computability are worth mentioning. The first states a universal Turing machine exists. A universal Turing machine is a Turing machine that can simulate all other Turing machines. The second, called Church's thesis, was formulated by Church in 1936. It implies that any computable function mapping a subset of N into N can be computed by a Turing machine.

The notion of computability can be, and has been, extended. Thus, for example, there are definitions for a computable real number, a computable complex number, and a computable continuous function of either a real or a complex variable. More generally, the concept of computability on a Banach space is defined. This latter definition actually includes the definition of computability for continuous functions just mentioned. It is these extended definitions that play a role in studying the computability of physical phenomena.

Computability and physical theory. What is the relation between computability and physical theory? This question has been investigated by M. B. Pour-El and J. I. Richards. Two specific examples of this question were given at the beginning of the article, namely the computability or noncomputability of wave propagation and heat dissipation.

Modeling of physical phenomena requires functions of a real or complex variable, rather than functions with domain and range included in N. Hence, the first task is to extend the definition of computability to functions of a real or complex variable. On the basis of this extension, it can be proved that, although heat dissipates computably, waves do not propagate computably. More specifically, wave propagation is governed by the wave equation (9), subject to the initial

$$\frac{\partial^2 u}{\partial x^2} + \frac{\partial^2 u}{\partial y^2} + \frac{\partial^2 u}{\partial z^2} - \frac{\partial^2 u}{\partial t^2} = 0 \qquad (9)$$

conditions of Eqs. (10).

$$u(x, y, z, 0) = f(x, y, z)$$
$$\frac{\partial u}{\partial t}(x, y, z, 0) = g(x, y, z) \qquad (10)$$

It is possible to construct continuous functions of a real variable f and g (with continuous derivatives) so that both of these initial conditions are computable, but the unique solution u of the wave equation, although continuous, is not computable. Physical phenomena can take computable initial data into a noncomputable solution.

There are two ways in which these results have been generalized. The first concerns the equations of classical physics other than the heat and wave equations. The second goes beyond classical physics to theoretical constructs adequate for quantum mechanics. The appropriate model for both is computability on a Banach space. Hilbert space, a special kind of Banach space, plays an important role in quantum theory.

The procedure is as follows. First, the concept of a computability structure on a Banach space is defined axiomatically. It is then shown that the definition is intrinsic. Now, physical phenomena can be associated with linear operators on a Banach space. (For example, on a suitable Banach space, the mapping associated with the wave equation that sends f and g into the solution u is a linear operator.) It is then proved that, under very general conditions, bounded linear operators take computable data into a computable solution and unbounded linear operators do not. The linear operator associated with the heat equation is bounded; the one associated with the wave equation is not. Thus, the results stated earlier concerning heat and waves follow from this general theorem. Of course, this theorem can be, and has been, applied to a host of phenomena of classical physics other than heat dissipation and wave propagation. It has also been applied to the mathematical constructs useful in modeling physical phenomena, such as Fourier series and Fourier transforms.

Quantum theory is concerned with linear operators on Hilbert space, in particular with the so-called self-adjoint operators. Associated with these operators are certain real numbers, the eigenvalues, which are the quantities that are measured. There are two questions dealing with the computability of eigenvalues. Is each eigenvalue a computable real number? (A real number a is computable if there is a Turing machine that can generate a sequence of rationals $\{r_n\}$ so that $|a - r_n| < 1/2^n$.) Is it possible to arrange the eigenvalues in a sequence $\lambda_1, \lambda_2, \lambda_3, \ldots$, so that there is a computer M that, on input n, effectively approximates λ_n? Under very general conditions it can be proved that the answer to the first question is yes, but the answer to the second is no.

AI, the mind-body problem, and physics. The pervasiveness of computers in society—in business, in industry, for research, for recreational games, and in the home—and the speed with which older machines become outdated and are replaced by newer and better ones—suggest a reconsideration of some well-known questions. Will computers become so much better that they are not merely mechanical servants of human beings but intelligent beings themselves? What should the criteria be for judging whether or not a computer is, in fact, intelligent?

The above questions have provided motivation for many specialists in artificial intelligence (AI). There are two subfields of particular interest, robotics and expert systems. Robotics is concerned with the needs of industry for sophisticated devices, that is, machines that can carry out complicated procedures that, at present, require the control or intervention of a human operator. In expert systems, the aim is to code all of the basic knowledge of a specific profession (such as medicine) into a computer package. SEE INTELLIGENT MACHINES.

Some of the accomplishments are quite impressive. For example, there are computers that can play chess so well that they have defeated a grandmaster. Nevertheless, there does not appear to be any device that can be said to simulate human intelligence.

In 1950, Turing wrote a philosophical article on computing machinery and intelligence, in which he presented what is now referred to as the Turing test. The test represents an operational approach to the relation between computers and intelligence. Very roughly, if a machine M interacts with a person so that its behavior is indistinguishable from human behavior, it will be said to be intelligent. More specifically, Turing envisioned a game played between two people and a machine. One of the persons is the interrogator I, who is placed in a room apart from the other two. There is no direct contact between I and either of the others, who are known to I only as X and Y. The interaction between I and either X or Y is limited to questions that I asks one of them and that are answered by the one to whom each question is addressed. The questions are asked and answered in an anonymous way, for example, with a computer as an intermediary. The human responds to questions truthfully, but the machine is programmed to lie so as to persuade the interrogator that it is, indeed, a person. The interrogator's task is to determine which is the machine and which is the person. Turing stated his belief that, by about the year 2000, it would be possible to program computers so that there would not be more than a 70% chance for an average person, I, to make the correct identification after 5 min of questioning. Although this goal will probably not be reached by the end of the century, Turing's work is still of great importance.

Turing's test attempts to determine whether mental processes involve only the execution of a sophisticated algorithm. Perhaps there is a nonalgorithmic element. Much has been written about this controversial question. R. Penrose discusses this and similar matters in a recent book. In brief, he thinks that there is a nonalgorithmic element to mental processes and that present-day physics does not suffice to answer the question. Nevertheless, he believes that physical theory of the future may provide some insight.

The study of the relation between computability, mental processes, and physical theory is only at its beginning. Hopefully, future generations will have much more to say about the topic.

For background information SEE ARTIFICIAL INTELLIGENCE; AUTOMATA THEORY; CONDUCTION (HEAT); EXPERT SYSTEMS; LOGIC; NONRELATIVISTIC QUANTUM THEORY; OPERATOR THEORY; ROBOTICS; WAVE EQUATION in the McGraw-Hill Encyclopedia of Science & Technology.

Marian Boykan Pour-El

Bibliography. R. Penrose, *The Emperor's New Mind*, 1989; M. B. Pour-El and J. I. Richards, *Computability in Analysis and Physics*, 1989; J. R. Searle, Minds, brains and programs, *Behav. Brain Sci.*, 3:417–457, 1980; A. M. Turing, Computing machinery and intelligence, *Mind*, 59:433–460, 1950.

Computer graphics

Many disciplines in science, technology, and medicine generate massive multidimensional data sets. The data sets result from laboratory experiments, satellite observations, or direct numerical simulations carried out with high-performance (massively parallel) computers. Techniques to visualize and quantify this increasing amount of information have become crucial to understanding, analysis, and model formulation. In this article, the visiometric (visualization and quantification) approach will be discussed as applied to some fundamental problems in computational fluid dynamics.

Understanding and visualization. The goal of experiments, observations, and numerical simulations in science and technology is to obtain a quantitative and mathematical understanding of the natural and artificial environment. Examples include direct numerical simulations of fluid and chemical dynamics of the ozone hole (and its health consequences) or of aircraft and automotive flow patterns (for improved design and performance); and satellite or Doppler radar image interpretation for weather prediction. Massive amounts of data are generated or acquired and must be interpreted to obtain a proper understanding to guide design, prediction, and control processes.

Visualization is the process of converting numerical data into a geometric or graphic representation, for example, one-dimensional waveforms, two-dimensional contour diagrams, and three-dimensional solid objects such as a three-dimensional model of the heart from a series of magnetic resonance imaging (MRI) scans. The application of computer graphics techniques to these large data sets is a first step in the visiometric process. Visualization, as an endeavor in science and engineering, has been growing rapidly since the appearance of the National Science Foundation's 1987 report *Visualization in Scientific Computing*, which defined and elaborated the activity in the context of modern computer technology.

To gain an understanding from the visualization, it must be possible to recognize areas of important activity. The persistence of a noted effect (possibly a coherent region of high or low intensity such as the ozone depletion) under variations and perturbations is a signal of the presence of something new. The aim in the different disciplines is to study the evolution and essential dynamics and interactions of these effects and describe them for a finite time period, thus obtaining a partial solution or gaining a new understanding of the original problem. For example, it is possible to track the evolution of a storm front, the change in the ozone levels, or the flow of air over a maneuvering aircraft.

The quantification process includes identifying, classifying, tracking, and projecting evolving amorphous phenomena. Ultimately, it will lead to improved mathematical models of the evolution and interaction of amorphous objects that are observed in time-dependent data sets.

In this article, some of the procedures needed to extract and identify the interacting coherent features (regions, objects, and effects) are highlighted, and are related to ongoing research in computational fluid dynamics. The focus is on the evolution and interaction of coherent scalar, vector, and tensor field structures and topologies.

Visiometrics. Visiometrics involves three principal steps: identification, quantification, and understanding or mathematization.

Identification can be subdivided into visualization, or rendering an image; feature extraction, which involves isolating coherent noted regions; and feature classification, which involves identifying such a region as either a known or a possibly new phenomenon. Examples include eddies, blast waves, rings, tubes, and spikes.

Quantification entails measurements to determine how the regions have changed, where they have moved, and how they interact with other regions. (It

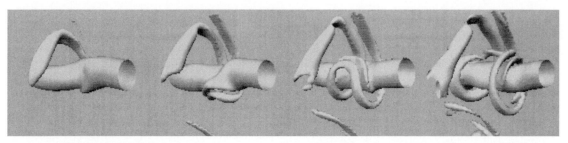

Fig. 1. Four time steps from the three-dimensional simulation of the evolution of two vortex tubes that are initially orthogonal and offset, and of sufficiently different magnitudes to exhibit three-dimensional winding. The isosurfaces of the absolute value of the vorticity vector are shown.

also includes identifying the interactions, such as collisions of waves or galaxies; and mergers, windings, and intensification of vortices in turbulent fluids.)

In the process of understanding or mathematization, the researcher attempts to formulate simpler mathematical models based upon the metrics from the above processes to describe the interactions of observed phenomena. The process also involves juxtaposition, that is, detailed comparisons of numerical simulations with laboratory experiments or field observations.

One of the aims of visiometrics is to develop robust algorithms to aid in the automatic identification and tracking of these multidimensional observed phenomena. Many of the ideas are analogous to those of two-dimensional computer vision, which strives to automatically analyze, recognize, and track objects in motion.

Visualization. The above outline provides a natural progression from data acquisition to data comprehension. The first step, visualization, is pivotal to converting the data into a more understandable form. Some standard visualization techniques include render-

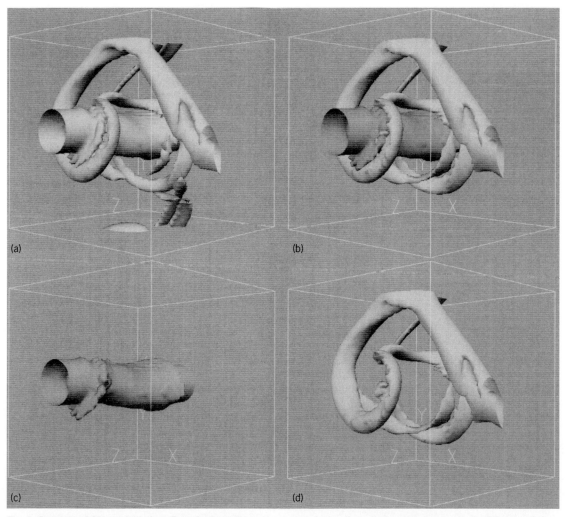

Fig. 2. Example of feature isolation. The winding tubes are separated even though the isosurfaces are connected. (*a*) One large connected region. (*b*) Tubes isolated and shaded differently. (*c*) Inner tube. (*d*) Outer tube.

ing an image by using continuous-tone contour maps (that is, mapping each value to a color and displaying the result), thresholding (separating the field into regions above and below a certain value), displaying isosurfaces (colored surfaces of equal magnitude that are formed from polygonal regions enclosing a thresholded domain; **Figs. 1** and **2**), and drawing icons such as color-coded lines with arrows or tubes to represent vector or tensor fields, respectively (**Fig. 3**).

Quantification. The next step is to quantify the visual representations by isolating, measuring, and tracking features and then classifying these objects and their interactions. Based upon the metrics from this stage, the scientist may be able to formulate a simpler mathematical model describing the simulation.

Feature extraction. To isolate objects in a scalar field, thresholding can be performed in conjunction with connectivity information; that is, objects are distinguished as coherent (connected) components above prescribed magnitudes when certain constraints are satisfied. The process segments the data set into regions that can then be analyzed, classified, and tracked.

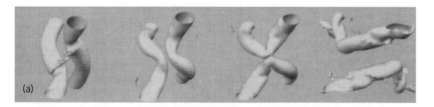

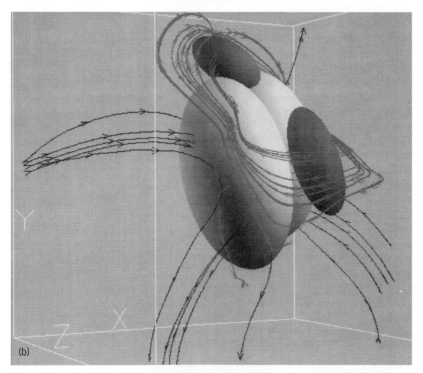

Fig. 3. Three-dimensional simulation of the evolution of two vortex tubes that are initially orthogonal, offset, and of equal strength, so that they display the phenomenon known as bursting reconnection. (*a*) Four time steps from the simulation. (*b*) Juxtaposition of vorticity and vortex-stretching fields. They are both visualized with ellipsoids about extrema and vector field lines threading the ellipsoids. The time shown corresponds to the third frame of *a*. A limit-cycle-like torus emanates from the dark ellipsoids, which represent the vortex-stretching field.

Because evolving objects undergo large deformations during the simulations, classifying them can be difficult. However, this task can be eased by creating abstractions, which are simple forms that quantify the geometric and topological features of space-filling objects. These shapes are elementary and generic to various disciplines. Examples include linelike structures, found in images of two-dimensional projections of jet streams, storm fronts, and galaxy formations; tubelike structures (in three dimensions) such as a tornado, a typhoon, or contrails of flying aircraft; sheetlike structures (interfaces); and ellipsoidlike structures, such as distorted bubbles or regions near extrema of any function. These abstractions can also help in simplifying procedures for tracking objects.

Application to fluid dynamics. The steps outlined above are generic to many disciplines that seek to understand and track coherent amorphous structures. Their applicability to fluid dynamics will be discussed.

The fluid dynamics of liquids, gases, and plasmas is a nonlinear space-time domain of great fundamental and practical interest. It lies at the heart of such diverse fields as aeronautics, ship design, meteorology, oceanography, oil recovery, combustion, and magnetically confined fusion environments and thereby encompasses a large range of experimental environments. These include laboratories (such as wind tunnels and flow chambers), observations (such as satellite and radar imagery), and computations (such as massively parallel supercomputer simulations). In particular, direct numerical simulations can be used to study the underlying interactions that occur in flows dominated by localized concentrations of vorticity (or coherent vortex structures) embedded in a nearly passive, incoherent background (or sea) of intermixed and distributed vorticity. This domain of science and technology is associated with the control of aircraft wakes; the mixing of chemical species and its application to atmospheric and oceanographic species dispersal; the prediction of the arrival time, intensity, and duration of hurricanes, tornadoes, polar vortices and ozone holes; and so forth.

Interactions were studied that eventually lead to the collapse and reconnection of vortex tubes and the accompanying cascade to small-scale vortex structures. Numerical simulations of the three-dimensional (96^3 to 256^3 grids) Navier-Stokes equations were made on high-performance supercomputers. These evolutions begin with two vortex tubes that are initially orthogonal and offset. Different phenomena occur if the strengths of the tubes are unequal (winding) or equal (bursting reconnection). Figures 1 and 3 contain four time steps from the unequal and equal simulations, respectively. The isosurfaces of the magnitude of the vorticity vector are displayed.

In Fig. 1, the weaker vertical tube is wound about the stronger horizontal tube, which undergoes a strong surface deformation. In the last time step, the thresholded isosurfaces are connected in small domains. To aid in analysis, each tube has been isolated by using a maxima-segmentation algorithm (recursively growing the regions from a maxima point). Figure 2 demonstrates the segmentation. Isolating regions is necessary for classification and feature tracking. For example, this technique has been used to track the inner tube of the simulation. The deformation and evolutionary changes can then be determined. Once regions are isolated and identified, more complex measurements can be computed, such as volume and area, moments (which can be represented as ellipses or ellipsoids), and skeletons (medial axes). Furthermore, features can now be juxtaposed, that is, compared to other features and regions from different data sets.

Figure 3a presents four frames from a reconnection process. Figure 3b presents an example of juxtaposition and ellipsoid fitting, namely, comparing two vector fields—vorticity and vortex-stretching. Here, the third frame of the sequence has been enlarged to understand the process of reconnection. The large light ellipsoids represent the magnitude of the vorticity field and were obtained by fitting the isosurfaces with quadratic forms (to highlight the orientation). The arrows on the field lines emanating from these ellipsoids show that the intense vortex regions are antiparallel. Next, the vortex stretching (darker) ellipsoids and corresponding vector fields are juxtaposed, and a toruslike pattern resembling a limit cycle is observed. The torus is a signature of "bursting" reconnection that was discovered by means of this juxtaposition process. This pattern has also been observed in simulations with a simpler (single vortex tube) Biot-Savart model.

For background information SEE COMPUTER GRAPH-ICS; COMPUTER VISION; FLUID-FLOW PRINCIPLES; NU-MERICAL ANALYSIS; SIMULATION; SUPERCOMPUTER in the McGraw-Hill Encyclopedia of Science & Technology.

Norman J. Zabusky; Deborah Silver

Bibliography. F. Bitz and N. J. Zabusky, DAVID and visiometrics: Visualizing, diagnosing and quantifying evolving amorphous objects, *Comput. Phys.*, pp. 603–614, November/December 1990; D. Silver and N. Zabusky, Quantifying visualizations for reduced modeling in nonlinear science: Extracting structures from data sets, *J. Vis. Commun. Image Represent.*, 4(1):46–61, 1993; R. Weinberg et al., *Two and Three Dimensional Visualization Workshop*, SIGGRAPH Course Notes 13, 1989; N. J. Zabusky et al., Visiometrics, juxtapositioning and modeling, *Phys. Today*, pp. 24–31, March 1993.

Computer security

Since information is an important strategic and operational corporate asset, there is a need to have security measures to safeguard sensitive information. In spite of its importance, the issue of security has not been given adequate consideration in the commercial, military, and civil sectors. This oversight is largely due to a lack of commercially available off-the-shelf secure systems and to the high cost of customizing secure systems.

The need for adequate security measures is becoming more and more critical as computers are increasingly relied on for storing information. Without adequate security controls, there is a risk of security breaches that may have disastrous consequences.

Aspects of security. When people think of security, they usually think of computer viruses. Although computer viruses represent an important security threat, they constitute only a small aspect of security. Computer security can be properly defined as consisting of three requirements: confidentiality, integrity, and availability. Confidentiality requires that information be disclosed only to authorized users. Integrity requires that information not be modified improperly. Availability requires that access to information not be improperly denied to authorized users.

Access controls. Most computer systems provide security by controlling modes of privileges to data. These controls are called discretionary access controls since any user who has discretionary access to certain data can pass the data along to other users. Although discretionary access control mechanisms are adequate for preventing unauthorized disclosure of information to most honest users, malicious users who are determined to get access to the data must be restricted by other means. One way to circumvent discretionary access control mechanisms is by using Trojan horse attacks. A Trojan horse is a malicious piece of code that is hidden within a program and leaks information to unauthorized users. As an example, an innocuous-looking sort routine may contain a Trojan horse so that whenever a user invokes the sort routine, in addition to accessing the user's file to be sorted, the sort routine accesses other files of the user and copies them into files belonging to some unauthorized user.

In order to protect the system from the Trojan horse and other direct attacks, the idea of mandatory (or nondiscretionary) access controls together with a protection mechanism called the trusted computing base, for enforcing mandatory access controls, was developed. Mandatory access controls are based on the Bell-LaPadula model, which is stated in terms of subjects and objects. An object is understood to be a data file, a record, or a field within a record. A subject is an active process that can request access to objects. Every object is assigned a classification, and every subject a clearance. Classifications and clearances are collectively referred to as security classes (or security levels). A security class consists of two components: a hierarchical component (usually, top secret, secret, confidential, and unclassified), and a set (possibly empty) of nonhierarchical categories (for example, NATO or nuclear). (Although this discussion is couched within a military context, it can easily be adapted to meet nonmilitary security requirements. In the corporate world, information is similarly labeled as authorized distribution only, company proprietary, and unlimited distribution, for example.) Security classes are partially ordered as follows: given two security classes L_1 and L_2, L_1 is said to dominate L_2 if the hierarchical component of L_1 is greater than or equal to that of L_2 and the categories in L_1 contain those in L_2.

The Bell-LaPadula model imposes two restrictions

on all data accesses: (1) The simple security property: a subject is allowed a read access to an object only if the subject's clearance dominates (in the partial order) the object's classification. (2) The *-property (pronounced "the star property"): a subject is allowed a write access to an object only if the subject's clearance is dominated by the object's classification. These restrictions are intended to ensure that there is no direct flow of information from objects at a high security level to those at a lower security level. Otherwise (since subjects can represent users) a breach of security occurs wherein a user gets access to information for which he or she has not been cleared. Since the above restrictions are mandatory and enforced automatically, the system checks security classes of all reads and writes, thus providing protection from Trojan horses.

The system may not be secure for malicious attacks even if it always enforces the two Bell-LaPadula restrictions correctly. To protect against all malicious attacks, a secure system must guard against not only the direct revelation of data but also violations that do not result in the direct revelation of data yet produce illegal information flows. Covert channels fall into the violations of the latter type. They provide indirect means by which information by subjects at high security classes can be passed down to subjects at lower security classes.

Assurance. Security mechanisms can be implemented with various degrees of assurance. A security mechanism that is easy to subvert has low assurance. As the effort required to circumvent the security mechanism increases, the security mechanism acquires higher assurance.

In *Department of Defense Trusted Computer System Evaluation Criteria*, the U.S. Department of Defense established a metric against which various computer systems can be evaluated for security. It developed a number of levels: A1, B3, B2, B1, C2, C1, and D; and, for each level, it listed a set of requirements that a system must have to achieve that level of security. Briefly, the D level consists of all systems that are not secure enough to qualify for any of the A, B, or C levels. Systems at levels C1 and C2 provide discretionary protection of data, systems at level B1 provide mandatory access controls, and systems at levels B2 or above provide increasing assurance, in particular against covert channels. The level A1, which is the most rigid, requires verified protection of data.

Although these criteria were designed primarily to meet Department of Defense requirements, they also provide a metric for the nonmilitary world. Most commercial systems that implement security would fall into the C1 or D levels. The C2 level requires that decisions to grant or deny access can be made at the level of individual users. In principle, it is straightforward to modify existing systems to meet C2 requirements. This capability has been successfully demonstrated by several operating system vendors; similar projects by database management system vendors are in progress. It is also considered reasonably straightforward to upgrade an existing C2 system to B1. However, upgrading will cause many existing applications to fail because of the enforcement of mandatory access controls. Some vendors are attempting to upgrade existing C2 systems to B2. It is not clear how viable this approach is because B2 imposes modularity requirements on the system architectures. At B3 or A1, it is generally agreed that the system would need to be designed and built from scratch.

Current work. For obvious reasons, the Department of Defense requirements tend to focus on the confidentiality aspect of computer security. However, information integrity is concerned with unauthorized or improper modification of information, such as is caused by the propagation of viruses that attach themselves to executable programs. The commercial world also must deal with the insider threat: authorized users who misuse their privileges to defraud the organization. Many researchers believe that some notion of mandatory access controls is needed, possibly different from the one based on the Bell-LaPadula model, in order to build high-integrity systems. Consensus on the nature of these mandatory access controls has been elusive. There are several efforts under way in the United States, Canada, and the European Community to update the Department of Defense evaluation criteria, and also to extend its scope to the nonclassified government sector and the private sector. In accordance with the U.S. Computer Security Act of 1987, the National Institute of Standards and Technology (NIST) has been charged with developing standards for federal agencies that store unclassified but sensitive information.

For background information SEE COMPUTER SECURITY in the McGraw-Hill Encyclopedia of Science & Technology.

Sushil Jajodia

Bibliography. Department of Defense, National Computer Security Center, *Department of Defense Trusted Computer System Evaluation Criteria*, 1985; M. Gasser, *Building Secure Computer Systems*, 1988; L. J. Hoffman (ed.), *Rogue Programs: Viruses, Worms, and Trojan Horses*, 1990; D. L. Stern, *Preventing Computer Fraud*, 1993; System Security Study Committee, National Research Council, *Computers at Risk*, 1991.

Cosmology

The recent detection of microwave background fluctuations by the *Cosmic Background Explorer* (*COBE*) satellite is a major discovery for astrophysics. Since the initial discovery of the background radiation in 1965, experimentalists have sought to detect fluctuations in the microwave background. The detection of the fluctuations provides important hints about the origin of galaxies and large-scale structure. These fluctuations probably reflect the physical conditions in the early universe. The amplitude of detected fluctuations was somewhat higher than anticipated in the most elegant and popular theory of the 1980s, based on cold dark matter with scale-invariant density fluctuations in a flat universe. Cosmologists are currently struggling to understand whether the *COBE* observations, together with the accumulated observations of the large-scale structure, demand a minor or major modification of the cold-dark-matter-plus-inflation paradigm or the development of a new paradigm. There is a growing sense that the very beginnings of the universe have begun to be understood.

Historical background. The modern era of cosmology began in 1965 with the detection of the homogeneous cosmic background radiation. In the hot-big-bang model, this radiation is the leftover relic heat from the big bang. The Friedmann-Robertson-Walker big bang model assumes that the early universe was homogeneous and isotropic; thus, it predicts that the cosmic background radiation is uniform. Fluctuations in the cosmic background radiation have been observed to be less than those on the surface of a billiard ball.

In the hot-big-bang model, the universe is expanding and adiabatically cooling. Thus, extrapolating the 2.74-K cosmic background radiation photons to the distant past implies that the early universe was very hot and began in an initial singularity of infinite density and temperature.

There are two possible eschatologies in the Friedmann-Robertson-Walker model: If the density of the universe is large enough, the universe is closed and will eventually turn around and collapse in a big crunch. If its density is low, the universe is open and will expand forever. The density of the universe is quantified with the parameter Ω, the ratio of the density of the universe to the critical density. If $\Omega > 1$, the universe is closed. If $\Omega < 1$, the universe is open. Finally, if $\Omega = 1$, the universe is flat.

The most popular way to overcome the metaphysical difficulties of the pure Friedmann-Robertson-Walker cosmology, such as the initial singularity, flatness of the universe, and causally disconnected regions within the horizon, is to incorporate the cosmology into the inflationary model, based on extensions of the standard model of particle physics. The inflationary model predicts that $\Omega = 1$.

One of the major open questions in astrophysics is how the enormous nonlinear density fluctuations called galaxies and clusters formed out of the nearly uniform early universe. Most cosmologists believe that these inhomogeneities grew gravitationally from tiny primordial fluctuations. This model predicts that the cosmic background radiation must have weak fluctuations in temperature generated by the density inhomogeneities. Experimentalists have been searching for these inhomogeneities for more than a quarter-century. In the early 1970s, models of galaxy formation predicted that these fluctuations would be detected at the level of 1 part in 10^3. Experimentalists disproved these theories. A new generation of models that predicted fluctuations an order of magnitude lower were abandoned in the early 1980s, when experiments showed that there were no fluctuations at the level of a few parts in 10^5.

Observations of large-scale structure. Since the early 1980s, observational cosmologists have mapped the three-dimensional distribution of galaxies on scales up to one-tenth the radius of the observable universe. They have identified in the local piece of the universe large-scale coherent structures in the distribution of galaxies called great walls and voids. These structures have been mapped by measurements of the redshifts of thousands of galaxies.

In 1992, results were reported from a three-dimensional survey of galaxies selected from a catalog based on infrared observations by the *Infrared Astronomy Satellite* (*IRAS*). The observations, which complement earlier work based on an optically selected sample, map out large-scale inhomogeneities in the distribution of galaxies. These inhomogeneities are larger than those predicted by most theoretical models for structure formation.

The observations of the redshift distribution of galaxies have been complemented by observations that measure the distance to nearby galaxies on the basis of their intrinsic properties, such as their velocity dispersion or surface brightness. These observations allow the large-scale motions of galaxies to be mapped out. This approach has been applied, based on observations of many of the nearby elliptical galaxies, and evidence has been found that most of the galaxies in the neighborhood of the Milky Way Galaxy are moving toward a so-called great attractor. Recent work suggests that this large-scale motion is coherent on even larger scales. These observations appear to be inconsistent with the simplest version of the cold-dark-matter-plus-inflation model.

Dark matter. Cosmologists quantify these irregularities statistically by measuring the so-called power spectrum, $P(k)$, which is the Fourier transform of the underlying density field. In order to determine $P(k)$, cosmologists must make assumptions about the relationship between the light distribution of galaxies and the underlying mass distribution. This relationship is often described through a bias parameter b, which measures the ratio of the power spectrum of the galaxy distribution to the mass distribution. If light is proportional to mass, then $b = 1$.

The distribution of mass on both galaxy and cluster scales can be inferred from observations of stellar and galactic motions. On the galactic scale, these observations imply that most of the matter in galaxies is not in luminous form. Candidates for this so-called dark matter range from subatomic particles, relics from the hot very early universe, to very low mass stars. On scales of several megaparsecs (1 Mpc = 2×10^{19} mi = 3×10^{19} km), observations of motions of galaxies suggest that if $b = 1$, then $\Omega < 1$. Thus, the inflationary model requires that $b > 1$ on these scales.

Microwave background spectrum. The *COBE* is a satellite of the National Aeronautics and Space Administration (NASA) dedicated to observing the cosmic background radiation. It consists of three instruments, the diffuse infrared background experiment (DIRBE), the far-infrared absolute spectrophotometer (FIRAS), and the differential microwave radiometers (DMR). The DIRBE instrument was designed to detect primordial galaxies. The FIRAS detector was designed to measure the spectrum of the cosmic background radiation. In the hot-big-bang model of cosmology, the microwave spectrum is expected to be thermal. In many alternative models, such as the cold-big-bang model, the spectrum is expected to deviate from the thermal blackbody spectrum. **Figure 1** shows the shape of the spectrum predicted by the hot-big-bang model and the spectrum detected by the FIRAS experiment in 1990. The FIRAS experiment determined that the temperature of the microwave background was 2.735 K above absolute zero to an accuracy of 0.06 K. This thermal spectrum was confirmed by a balloon experiment, also reported in 1990, which obtained virtually the same temperature. The remarkable agreement between the predicted and observed spectrum is a dramatic confirmation of the hot-big-bang model.

Spatial fluctuations. While the FIRAS detector was optimized to measure the absolute spectrum of the background, the DMR detector was optimized to measure the relative variation in flux from point to point on the sky. In April 1992, the detection by the DMR of fluctuations in the microwave background at the level of 1 part in 10^5 was reported (**Fig. 2**).

The DMR was sensitive to microwave fluctuations on angular scales larger than about 8°. The DMR team reported fluctuations over a large range of angular scales, from the smallest scales detectable by the DMR, which correspond to roughly 300 Mpc in a flat universe, up to fluctuations on the scale of the observable universe. These fluctuations were found at several different frequencies in the ratio expected. Detailed analysis by the *COBE* team strongly argued that the fluctuations are not due to dust or hot gas in the Milky Way Galaxy but are primordial fluctuations.

There are several possible physical sources for generating these temperature fluctuations. These fluctuations could reflect small variations in the gravitational potential, in the geometry of space induced by gravitational waves, or in the ratio of the number of photons per electron.

The *COBE* science team interpreted the detection as being due to small fluctuations in the gravitational potential at the surface of last scatter, and used the data to determine the power spectrum of density fluc-

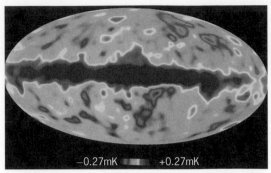

Fig. 2. Map of the fluctuations in the cosmic microwave background recorded by the DMR instrument aboard the *COBE* satellite at a frequency of 53 GHz (5.7 mm wavelength). Scale at bottom indicates correspondence between shades on map and measured temperature. The constant term (2.7 K) and the cosine variation due to the motion of the Milky Way Galaxy have been removed. Galactic coordinates are used, and the central band comes from synchrotron emission in the Milky Way Galaxy. However, the blobs above and below this region, at the level of 1 part in 10^5 of the 2.7-K signal, are believed to represent primordial fluctuations.

tuations. Measurements were expressed in terms of the amplitude and slope of $P(k) \propto k^n$, at very large scales, through Eqs. (1) and (2). Since the simplest version

$$n = 1.1 \pm 0.6 \qquad (1)$$

$$\frac{\Delta T}{T} = 1.1 \times 10^{-5} \qquad (2)$$

of the inflationary scenario predicts a scale-free spectrum, $n = 1$, some scientists viewed this observation as definitive evidence for scale-invariant fluctuations produced by inflation. However, it would probably be a mistake to conclude that these observations mean that the early universe is now understood and that all that remains is to refine engineering details.

Standard cold-dark-matter model. During the 1980s, much effort went into exploring what is now referred to as the standard cold-dark-matter model. This model is a variant of the inflationary scenario: it assumes that the universe is flat ($\Omega = 1$), that the fluctuations are scale-free ($n = 1$), and that the dark matter is made of weakly interacting cold particles. Since the model assumes that $\Omega = 1$, $b \simeq 2$ must be compatible with small-scale galaxy dynamics. Numerical simulations of galaxy formation suggest that this bias could be achieved through a combination of statistical effects and hydrodynamical physics. This model has many attractive features: on the theoretical side, it has no extra parameters and is compatible with the most conservative inflationary models; on the observational side, it is consistent with statistical measurements of the galaxy distribution on scales from 1 to 20 Mpc and many of the properties of galaxies, such as the density profiles of galaxy halos. However, more recent observations of the galaxy distribution on scales larger than 20 Mpc reveal that the universe is clumpier than predicted in the standard cold-dark-matter model. In that model, the dominant source of temperature fluctuations is variations in the gravitational potential. The amplitude of the temperature fluctuations detected by *COBE* exceeds that pre-

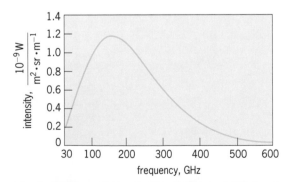

Fig. 1. Spectrum of the cosmic microwave background from the FIRAS instrument aboard the *COBE* satellite. The measured values agree with the thermal spectrum predicted by the hot-big-bang model. (*After J. C. Mather et al., A preliminary measurement of the cosmic microwave background spectrum by the Cosmic Background Explorer (COBE) satellite, Astrophys. J. Lett., 354:L37–L40, 1990*)

dicted by the standard cold-dark-matter model by a factor of 2.

Variations. There has been a burst of activity focusing on minimal variations on this standard model. These variations model the spectrum, the nature of the dark matter, or the parameters of the background Friedmann-Robertson-Walker cosmology.

One possibility is to consider versions of inflation in which the spectrum need not be scale-invariant, that is, $n < 1$. Such versions are called tilted models. This modified spectrum implies higher fluctuations on large scales. A second possibility is to assume that the vacuum energy density is nonzero, implying that the cosmological constant Λ is greater than zero and would modify the expansion rate of the universe. This variation would imply that the universe is older than estimated in the standard cosmology and would help remove the discrepancy between the ages of globular clusters, inferred from stellar evolution theory, and the age of the universe, inferred from measurements of the Hubble constant in the Friedmann-Robertson-Walker cosmology.

A third possibility is to assume that, in addition to the cold dark matter and baryons, a fraction of the density of the universe is composed of a light muon or tau neutrino. This concept is called the mixed-dark-matter model.

All of these models have an extra parameter that can be adjusted to make them compatible with the *COBE* measurements and the large-scale galaxy clustering. In the latter two variations, the fluctuations in the cosmic background radiation are interpreted as being primarily due to potential fluctuations. In the tilted models, both potential fluctuations and gravitational waves contribute to the total cosmic background radiation fluctuations.

Radical alternatives. A more radical alternative is to abandon the notion that gaussian fluctuations from inflation seeded the formation of structure. If the universe underwent a phase transition, fields in different regions of space need not be aligned. The energy associated with the gradients of these fields could have seeded the formation of large-scale structure. There are several variations of this scenario, depending on whether the field energy resides in topological defects, such as domain walls, strings, monopoles, or textures. In these models, the density variations are not imprinted in the first moments of the big bang but are causally generated at later times. These models imply nongaussian fluctuations in both the density field and the microwave background. Improved statistics on the *COBE* maps could test this hypothesis. These models can be, but need not be, embedded into an inflationary scenario.

Besides these two alternatives for structure formation, other speculations are actively being considered. In the primordial isocurvature baryon model, structure grows out of large-scale variations in the ratio of photons to baryons. This model has not yet been incorporated into a whole paradigm and treats all of its parameters as adjustable.

Testing of models. Each of these models makes different detailed predictions for the intensity and spatial variation of the microwave background fluctuations and for the large-scale structure of galaxies. Observations in the coming years will test these models and, it is hoped, enable astrophysicists to determine the origin of galaxies and the large-scale structure.

For background information SEE BIG BANG THEORY; COSMIC BACKGROUND RADIATION; COSMIC STRING; COSMOLOGY; INFLATIONARY UNIVERSE COSMOLOGY; INFRARED ASTRONOMY; RELATIVITY; UNIVERSE in the McGraw-Hill Encyclopedia of Science & Technology.

David N. Spergel; Lev Kofman

Bibliography. M. V. Berry, *Principles of Cosmology and Gravitation*, 1989; E. W. Kolb and M. S. Turner, *The Early Universe*, 1990; E. L. Wright et al., Interpretation of the cosmic microwave background radiation anisotropy detection by the COBE differential microwave radiometer, *Astrophys. J. Lett.*, 396:L13–L18, 1992.

Cryptosporidiosis

Cryptosporidium is one of several genera of protozoa, commonly called coccidia, in the phylum Apicomplexa. Apicomplexa includes protozoa that possess an apical complex used to penetrate cells of the hosts. Except for a few life-cycle stages that transmit infections among hosts, apicomplexans are obligate intracellular parasites (that is, they must enter cells of the hosts in order to undergo development and to multiply). Unlike the tissue-cyst-forming coccidia (such as *Toxoplasma, Sarcocystis,* and *Neospora*), *Cryptosporidium* species require only a single host to complete the life cycle. *Cryptosporidium* infects fish, reptiles, birds, and mammals, causing disease primarily in young and immunosuppressed individuals.

Although human cryptosporidiosis was first reported in 1976, it occasioned little interest until 1982 when severe diarrhea caused by *C. parvum* was reported in patients infected with human immunodeficiency virus (HIV) by the Centers for Disease Control (CDC; Atlanta, Georgia). CDC reported that 14 of 21 patients with acquired immune deficiency syndrome (AIDS) and infected with *C. parvum* had died. By 1986, the CDC reported that 697 of 19,187 AIDS patients were infected with *C. parvum*, and they had higher mortality rates than noninfected AIDS patients. Although *C. parvum* in immunocompetent patients usually causes only a short-term diarrhea that resolves spontaneously, in immunocompromised patients a life-threatening, prolonged, choleralike illness can occur. There is no effective therapy for cryptosporidiosis, and the prognosis is poor for immunocompromised patients, especially those with AIDS.

One of the major reasons that the parasite went unnoticed for so long was the lack of reliable techniques for microscopic diagnosis. In the early 1980s, diagnosis became much easier with the discovery that oocysts of *Cryptosporidium* in stool samples could be detected by using various chemical stains. Since the early 1980s, there has been much interest in the

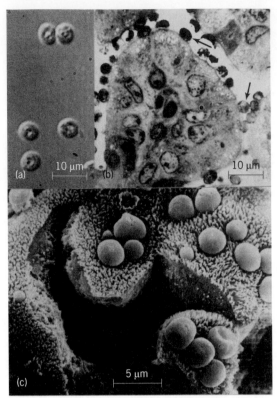

(a) 10 µm (b) 10 µm

(c) 5 µm

Fig. 1. *Cryptosporidium parvum*. (*a*) Live oocysts. (*b*) Parasites (arrows) at the luminal surface of the epithelium of a mouse small intestine. (*c*) Cryptosporidia on the surface of sheep intestinal epithelial cells; note that some of the epithelial cells have sloughed (dark, central area). (*From J. P. Dubey, C. A. Speer, and R. Fayer, Cryptosporidiosis of Man and Animals, CRC Press, 1990*)

epidemiology, diagnosis, treatment, and prevention of cryptosporidiosis.

Species identified. Twenty-one species of *Cryptosporidium* have been named, based primarily on the host the parasite infects, but recent cross-transmission studies have placed the validity of many of these species in doubt. At this point, the following species are considered valid: *C. nasorum* in fish, *C. serpentis* in reptiles, *C. meleagridis* and *C. baileyi* in birds, and *C. muris* and *C. parvum* in mammals with *C. parvum* being infectious to humans. Isolates of *C. parvum* from humans can infect a variety of mammals, and isolates from one mammalian species can infect other mammals.

Life cycle. The life cycle of *Cryptosporidium* is relatively complex and consists of exogenous and endogenous stages involving asexual and sexual reproduction (**Figs. 1** and **2**). Transmission to a new host is accomplished by oocysts, which are environmentally resistant forms of the parasite. Each oocyst contains four sporozoites (an infective stage). When oocysts (Figs. 1*a* and 2*a*) are ingested or inhaled, the sporozoites excyst (escape from the oocyst) through a gap in the wall created by the dissolution of a suture at one end of the oocyst (Fig. 2*b* and *c*). This suture, unique to *Cryptosporidium*, opens in the presence of various enzymes and substances such as trypsin and bile. Sporozoites (Fig. 2*d*) parasitize epithelial cells of

the gastrointestinal or respiratory tract. Interestingly, unlike other coccidia, *Cryptosporidium* species do not penetrate deep into their host cells but remain at the extreme apex or luminal surface of the epithelial cell. By light microscopy, they appear to be attached superficially to their host cell (Fig. 1*b*), but electron microscopy has shown that they are actually intracellular, occupying the cytoplasm within the epithelial cell microvilli (Fig. 1*c*).

The elongate sporozoite makes initial contact with the host cell via its anterior tip and appears to induce the cell membrane of the host-cell microvilli to envelop it so that it resides within a vacuole, called a parasitophorous vacuole. A feeder organelle, unique to *Cryptosporidium*, develops at the interface between all stages of the parasite and the host cell. Although the function of this organelle has not been determined, the organelle appears to substantially increase the surface area between the parasite and host cell, facilitating exchange of materials. After the sporozoite becomes intracellular, it differentiates into a spherical trophozoite (a vegetative stage) with a prominent nucleus. The trophozoite increases in size and undergoes two or three nuclear divisions to form a schizont with four to eight nuclei, which are then incorporated into merozoites that bud at the schizont surface. Merozoites, which are ultrastructurally similar to sporozoites, escape from the host cell, enter other epithelial cells,

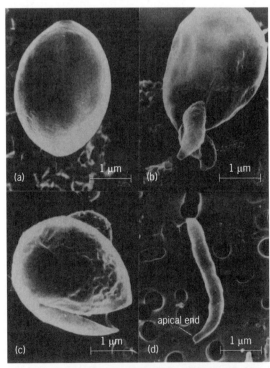

(a) 1 µm (b) 1 µm

(c) 1 µm (d) apical end 1 µm

Fig. 2. Scanning electron micrographs of *Cryptosporidium parvum*. (*a*) Intact oocyst prior to excystation. (*b*) Three sporozoites excysting from oocyst simultaneously via the cleaved suture. (*c*) Empty oocyst. (*d*) Excysted sporozoite. (*From D. W. Reduker, C. A. Speer, and J. A. Blixt, Ultrastructure of Cryptosporidium parvum oocysts and excysting sporozoites as revealed by high resolution scanning electron microscopy, J. Protozool., 32:708–711, 1985, except c from J. P. Dubey, C. A. Speer, and R. Fayer, Cryptosporidiosis of Man and Animals, CRC Press, 1990*)

and undergo a second generation of schizogony. This asexual reproductive cycle is repeated numerous times.

Eventually, some merozoites enter the sexual phase of the life cycle by differentiating into male and female gametocytes, called microgamonts and macrogamonts, respectively. In microgamonts, three nuclear divisions occur to form eight nuclei, each incorporated into a microgamete (equivalent to a male sperm) that develops at the surface and pinches off from the residual body of the microgamont.

Although microgametes are motile and capable of seeking out a macrogamont, just how they move is not known. During fertilization, only the microgamete nucleus and some of its associated microtubules enter the macrogamont where fusion of the two nuclei occurs. The zygote develops into an oocyst with either a thin or thick wall, and sporulation occurs to form four sporozoites. Sporozoites in thin-wall oocysts serve as a means of autoinfection by excysting and infecting other cells of the host, whereas those in thick-wall oocysts are shed in the feces, sputum, or respiratory droplets and serve to transmit the infection to other hosts.

The time interval (prepatent period) between the acquisition of *Cryptosporidium* infection and the shedding of oocysts varies with the host and cryptosporidia species. Prepatent periods vary from 2 to 7 days in cattle, 5 to 10 days in cats, 2 to 14 days in dogs, and 5 to 28 days (mean of 7.2 days) in humans.

Epidemiology. Infection usually occurs during contact with infected individuals or animals and via fecal contamination of water used for drinking or swimming. Some reports implicate cattle or other mammals such as wildlife, zoo animals, and companion animals as zoonotic sources of human infection. Although respiratory cryptosporidiosis occurs frequently in poultry, it is relatively rare in humans and mammals. Respiratory infections probably occur via airborne transmission of the parasite. Arthropods such as flies might also serve as sources of infection, especially in areas with poor sanitation. Fomites (inanimate objects) such as towels, bed linen, and toys may also serve as sources of infection in day-care centers. Contaminated food also seems probable as a source of infection. Although the prevalence of asymptomatic carriers can be relatively high, ranging from 6% in India to 27% in New York City, their epidemiological significance as sources of *Cryptosporidium* infection is not known.

Certain groups are considered to be at relatively high risk of acquiring *Cryptosporidium* infections, including animal handlers, family members and sexual partners of infected individuals, medical or health care workers, veterinarians, farmers, day-care center attendees and employees, immunocompromised individuals, and travelers to highly endemic areas with poor sanitation.

Clinical signs and pathogenesis. The overall effects of cryptosporidiosis are malabsorption and impaired respiration, but just how cryptosporidia cause disease is not known. There is an association between the specific infection site and the clinical features of the disease. Most cryptosporidial infections are found in the gastrointestinal tract and associated organs and respiratory tree. In fish and reptiles, most infections occur in the stomach and rarely in the intestine. In birds, respiratory infections are more common than intestinal infections. In mammals, most infections are intestinal, but immunocompromised mammals might also have infections of the entire length of the gastrointestinal tract and the respiratory tree. The severity of infection can change from subclinical to severe, depending on the species and isolate of the parasite, and the age and immunologic status of the host. Young animals are usually more susceptible to infection and have more severe clinical signs. However, previously unexposed adult humans are routinely susceptible to cryptosporidia infection. In immunocompetent animals and humans, the clinical signs usually last a few days to a few weeks, with recovery occurring spontaneously. Immunodeficient individuals can become chronically and terminally ill.

In mammals and humans, the most prominent signs are voluminous watery diarrhea (with as much as 12–17 liters passed per day), weight loss, dehydration, and abdominal discomfort. Microscopically, infected epithelial cells change from columnar to cuboidal or squamous, and microvilli may be fused or absent at the site of parasite attachment. Infected cells may slough from the epithelium, but this is rare (Fig. 1c).

Immunity and immunoprophylaxis. In general, young immunologically immature humans and animals are more susceptible to the severe effects of cryptosporidiosis than adults. Although there is generally a measurable humoral antibody response to *Cryptosporidium* with the appearance of parasite-specific IgA, IgG, IgE, and IgM antibodies, it appears that immunity is T-cell dependent. Protective immunity might require a T-cell-dependent induction that results in effective humoral and cell-mediated immune responses. Although many antigens have been recognized by antibodies in immune sera and by monoclonal antibodies, the antigens that might provide protection against disease have not been identified. Few attempts have been made to determine which cryptosporidia antigens will stimulate a T-cell response.

Treatment and prevention. More than 100 therapeutic and preventive modalities have been tested against human and animal cryptosporidiosis, but few have shown any appreciable efficacy. Limited success has been obtained with the anticoccidial drugs amprolium, arprinocid, dinitolmide, salinomycin, and sulfaquinoxaline. Recently, dehydroepiandrosterone (DHEA), a drug that appears to stimulate the immune system, provided significant reduction of cryptosporidial infections in immunocompromised hamsters, indicating that it may be an effective prophylactic agent for cryptosporidiosis in immunocompromised humans.

The key to control and prevention of cryptosporidiosis is elimination or reduction of the oocysts in the environment or avoidance of contact with known sources of oocysts. Under favorable conditions such as moderate temperatures and moist conditions, oocysts can remain viable for up to 9 months after leaving the host. Desiccation, freezing, and high temperatures will kill oocysts. The most effective chemical agents appear to be hydrogen peroxide, ozone, Oo-cide (two-phase product producing ammonia), and Exspor (a chlorine dioxide–based cold sterilant).

For background information *SEE ACQUIRED IMMUNE DEFICIENCY SYNDROME (AIDS); CELLULAR IMMUNOLOGY; COCCIDIA; MEDICAL PARASITOLOGY; ZOONOSES* in the McGraw-Hill Encyclopedia of Science & Technology.
                                                    C. A. Speer

Bibliography. C. Ash, The clinical significance of ubiquity, *Parasitol. Today*, 8:181–182, 1992; J. P. Dubey, C. A. Speer, and R. Fayer, *Cryptosporidiosis of Man and Animals*, 1990; K. R. Rasmussen and M. C. Healey, Dehydroepiandrosterone-induced reduction of *Cryptosporidium parvum* infections in aged Syrian golden hamsters, *J. Parasitol.*, 78:554–557, 1992.

Ctenophora

Ctenophores, or comb jellies, are among the most beautiful and voracious of the gelatinous marine zooplankton. Ctenophores of the order Beroida possess a wide mouth and a voluminous stomach that occupies most of the body. Beroid ctenophores are miter- or cucumber-shaped, reaching a length of up to 1 ft (30 cm). They are ravenous predators of other gelatinous zooplankton, particularly other kinds of ctenophores. *Beroë* actively seeks prey by swimming mouth forward, powered by the beating of eight rows of giant ciliary comb plates. The mouth remains closed, with the body streamlined (**Fig. 1**), until prey is encountered. Then the mouth opens suddenly, and the body cavity rapidly expands to gulp in the prey. Engulfment takes only a few seconds. If the prey is too large to be swallowed completely, *Beroë* uses thick compound cilia lining its lips to bite off pieces. These macrocilia bear sharp, toothlike projections at the tips. Macrociliary teeth vary in size and number in different species, probably reflecting different diets and feeding behaviors of various beroids.

Closure mechanism. Recent experiments reveal the mechanism by which *Beroë* keeps its mouth shut until encountering prey and then rapidly opens it. Species with a larger mouth, thinner body wall, and more flattened shape do not keep the mouth closed

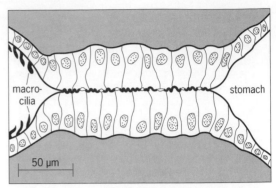

Fig. 2. Close-up view of adherent epithelial strips that encircle the inside walls of the mouth of *Beroë*.

by tonic muscular contraction. The lips remain firmly sealed when all muscular activity is prevented by anesthetics. Examination under a microscope reveals that the lips are fastened by paired strips of adhesive epithelial cells that run around the inside walls of the mouth (**Fig. 2**). By using the electron microscope, it can be seen that the adherent cells are joined together by numerous close contacts between the surface membranes. At these appositions the cell membranes are highly folded and interdigitated. Yet the adjacent membranes conform to each other and run parallel over broad areas, with a uniform separation of about 15 nanometers. The inner cytoplasmic sides of the apposed membranes are coated with dense material and actin filaments, the so-called submembrane cytoskeleton. These features of the adhesive cell junctions provide a stable, interlocking closure mechanism.

The two adherent epithelial strips are structurally identical. Opposite lips are functionally alike as well; that is, they adhere by a symmetric interaction. After surgical bisection of one lip, the two halves stick to each other as firmly as do lips from opposite sides of the mouth.

The molecular mechanism of mouth adhesion in *Beroë* is not yet known. Cell–cell binding may involve oligosaccharide groups of membrane glycoproteins, since lectins, proteins that bind carbohydrates, interfere with lip adhesion. Further research should elucidate the molecular nature and bonding properties of this unusual type of marine tissue adhesion.

Opening mechanism. When the sensory bristles arrayed around the outside of the lips contact prey, mouth opening is triggered. Opening requires contractions of muscles that exert tension on opposite lips. Video-microscopic analysis of the mouth opening shows that the epithelial adhesive strips are not pulled apart all at once but are peeled apart by coordinated muscular activity, starting from the site of initial contact with the prey (**Fig. 3**). *Beroë*, therefore, uses a mechanical advantage to open its mouth: peeling apart two adherent surfaces requires a smaller applied force than separating the surfaces all at once.

The mouth is not opened by muscular effort alone, however. Mouth opening also involves active deadhesion of the epithelial cells themselves. The separated surfaces of adhesive strips from mouths opened by

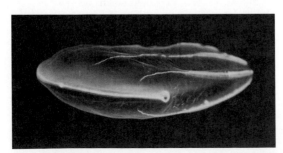

Fig. 1. The ctenophore *Beroë*, its mouth closed, swimming to the left. (*After S. L. Tamm and S. Tamm, Reversible epithelial adhesion closes the mouth of Beroë, a carnivorous marine jelly, Biol. Bull., 181:463–473, 1991*)

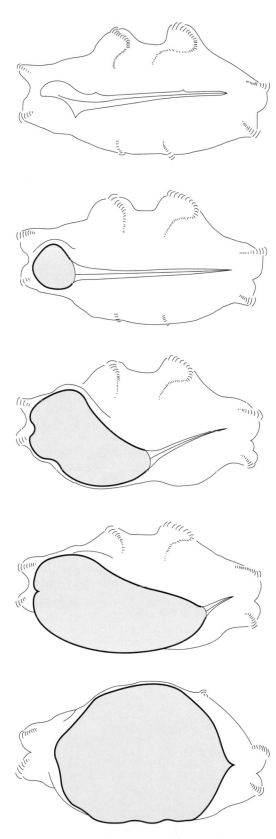

Fig. 3. Video-microscopic analysis of mouth opening of *Beroë* (three times natural size). A progression (top to bottom) over a period of 6.5 s shows that the epithelial adhesive strips are peeled apart by coordinated muscular activity starting from the site of initial contact with food. (*After S. L. Tamm and S. Tamm, Reversible epithelial adhesion closes the mouth of Beroë, a carnivorous marine jelly, Biol. Bull., 181:463–473, 1991*)

contact with food show no trace of the specialized cell junctions. Instead, the cell membranes are now uniformly smooth without any sign of the filamentous actin cytoskeleton. Thus, the interlocking cell junctions found in closed mouths disappear when the mouth is opened in response to prey. This rearrangement of the surface architecture of the epithelial cells likely modifies their binding properties and adhesive interactions. In contrast, the cell–cell appositions do not disassemble when the mouth is forcibly opened with forceps. Instead, the junctions rip apart, leaving remnants on either lip.

Regulation of tissue adhesion. The anatomical pathways responsible for food-induced mouth opening involve the netlike nervous system of the ctenophore. The sensory receptors on the lips makes synaptic contacts with the nerve net, which in turn make synaptic contacts with the lip muscles and cells of the adhesive strips. Food stimuli signal both muscular contractions that peel the lips apart and modifications of the adhesive cell junctions, thereby decreasing their binding.

Immediately after engulfing prey, the mouth closes and the epithelial adhesive strips reseal. The *Beroë* swims slowly away to digest its meal before searching for more prey. The epithelial strips thus adhere only temporarily. Cell–cell bonding rapidly breaks when *Beroë* opens its mouth to ingest prey, and readily reforms after feeding. Tissue adhesion is therefore a reversible process that is regulated by the ctenophore.

Reversible tissue adhesion in *Beroë* shares many structural and functional properties with the transient adhesions made between moving cells of higher animals during embryonic development. In this respect, *Beroë* mouth closure provides unique experimental opportunities for investigating the mechanisms and control of dynamic cell adhesions in higher organisms.

Differences between species. Surprisingly, some species of large-mouthed, thin-walled beroids have paired epithelial adhesive strips that run lengthwise down the center of the stomach cavity instead of around the mouth. These species also have much bigger macrocilia that cover a large area of the stomach. The reason for these differences will remain unknown until more is learned about the feeding behavior and diet of this kind of beroid.

In contrast, beroids with a thicker body wall, a cucumber-shaped body, and a smaller mouth do not possess any epithelial adhesive strips. Neither the lips nor the stomach walls are fastened together in any manner. Evidently the firmer body and smaller oral opening provide sufficient resistance to maintain normal body shape during forward swimming, without the necessity for epithelial adhesion.

For background information SEE BEROIDA; CELLULAR ADHESION; CTENOPHORA in the McGraw-Hill Encyclopedia of Science & Technology.

Sidney L. Tamm

Bibliography. G. A. B. Shelton (ed.), *Electrical Conduction and Behaviour in "Simple" Invertebrates*, 1982; S. Tamm and S. L. Tamm, Actin pegs and ultrastructure of presumed sensory receptors of *Beroë* (Ctenophora), *Cell Tissue Res.*, 264:151–159, 1991; S. L. Tamm and S. Tamm, Jellies with jaws, *Mar. Biol. Lab. Sci.*, 4:28–29, 1990; S. L. Tamm and S. Tamm, Reversible epithelial adhesion closes the mouth of *Beroë*, a carnivorous marine jelly, *Biol. Bull.*, 181:463–473, 1991.

Cutaneous sensation

On cursory examination, it may appear that the human skin serves only to provide a waterproof barrier between the body and its environment and to monitor contact with that environment. Closer examination reveals that the skin has sophisticated capabilities that provide subtle, complex, and vital information often unattainable by the other senses. By using the sense of touch, the roughness of cloth can be instantly graded, minute differences in the masses of weights can be compared, subtle temperature changes can be detected, and the smoothness of a car finish can be judged within micrometers. In addition, the skin may serve as an alternative sensory input channel when sight or hearing is missing or overloaded, for the skin is quite capable of complex pattern perception. For visually disabled persons, text can be read with the fingertips when it is encoded as static embossed patterns (such as Braille) or as dynamic patterns presented electromechanically through vibrotactile reading machines. Similarly, persons who are deaf and read speech (lip-read) can benefit greatly from the information provided through tactile aids that augment vision by sophisticated processing of the speech signal. More recently, the sense of touch has played a significant role in the emerging technology of virtual reality. This technology, through electronic aids such as television goggles, audio headphones, and a sensor-covered glove, allows the user to experience and interact in a computer-generated multisensory three-dimensional world of the programmer's imagination. In such a system, the sense of touch can provide the user with both kinesthetic control and tactile feedback.

Matching stimulus to structure. These capabilities can be fully realized only by choosing a stimulus that takes full advantage of the skin's properties. At the biophysical and physiological levels, these properties are a function of the viscoelastic multilayered character of the 20-ft^2 (2-m^2) skin sheet, the body's largest organ. There are three major layers of skin, the epidermis, the dermis, and subcutaneous sheets of superficial fascia and fat, each made up of unique cellular and morphological components. The epidermis provides a protective function because its outermost layer is composed primarily of dead (keratinized) cells. Deeper, in the dermis, are blood vessels, sweat glands, muscle fibers, hair follicles (in hairy skin), and the neural structures that transduce environmental events into a form that the person can process. The third layer is primarily supportive in function.

An example illustrates the importance of choosing the right stimulus for the skin. Early speaking tubes assumed that simply directing the speech signal (which is made up of vibrating air molecules) at the skin (which responds well to mechanical vibration) would allow deaf persons to "hear." The failure of this system resulted from a poor understanding of at least two properties of the skin. One is biomechanical and is characteristic of energy transmission across any interface: the need for a proper impedance match. Specifically, unless the mechanical properties of the skin are matched by that of the driving signal, much of the energy can be lost. With the speaking tube, the vibrating air molecules do not have sufficient energy to move the layers of skin so as to activate underlying tactile receptors. Impedance matching is a major concern in developing devices specifically designed to present stimuli to the skin.

The other reason for the failure of the speaking tube involves the ability of the neural structures to process the frequencies of vibration it produces. Speech is carried by mechanical vibrations (of the vocal cords, air molecules, and eardrums) ranging in frequency from about 100 to 5000 Hz. Just as the eye is sensitive to only a small part of the electromagnetic spectrum (perceiving neither infrared nor ultraviolet), there is a preferred range of vibration of mechanical energy to which the skin will respond, specifically from about 10 to 500 Hz. The speaking tube was a poor match here also.

Four-channel model. Thus, vibrotactile sensitivity depends on both the biomechanical characteristics of the skin and the distribution and type of the neural structures within it that convert mechanical energy into biological signals for the central nervous system. These structures, or receptors, are primarily found sprinkled throughout the dermis. Recently, a four-channel model of tactile mechanoreception was proposed in which four of these receptors, found in smooth, nonhairy (glabrous) skin, were linked to vibrotactile sensitivity. This model suggests that different structures are responsible for threshold sensation over specific ranges of vibration frequency (**Fig. 1**).

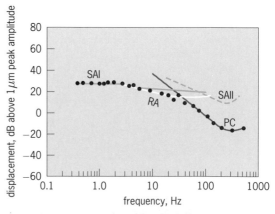

Fig. 1. Vibrotactile sensitivity. The dots represent the threshold for vibration on the palm of the hand as a function of frequency. The curves illustrate the sensitivity of the four cutaneous receptor populations that appear to underlie the overall function. (*After I. R. Summers, ed., Tactile Aids for the Hearing Impaired, Whurr Publishers, 1992*)

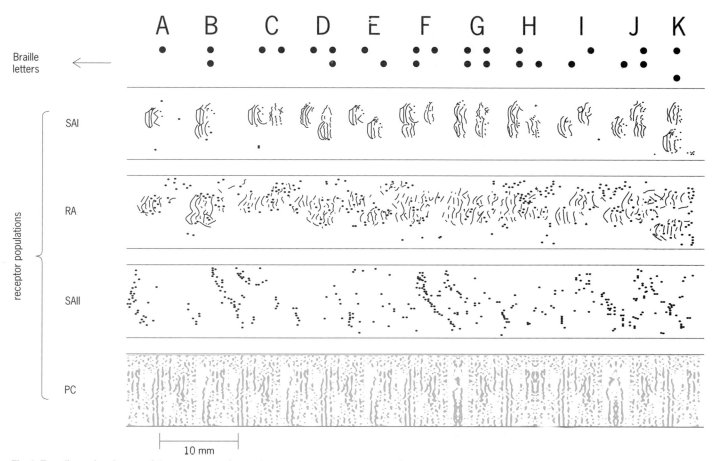

Fig. 2. Two-dimensional maps of the responses of a number of cutaneous receptor populations to coarse embossed patterns (Braille letters). The SAI receptor system clearly translates the spatial aspects of these stimuli best. (*After K. O. Johnson and S. S. Hsiao, Neural mechanisms of tactual form and texture perception, Annu. Rev. Neurosci., 15:227–250, 1992*)

For example, the Pacinian corpuscle (PC) channel serves threshold sensitivity from about 40 to 400 Hz and is most sensitive to 200–300-Hz vibration. In that range, it responds to displacements of less than 0.1 micrometer under optimal conditions.

The three other channels whose threshold sensitivies are graphed in Fig. 1 are the rapidly adapting (RA) channel, presumably associated with Meissner's corpuscles; the first slowly adapting (SAI) channel, presumably subserved by Merkel's disks; and the second slowly adapting (SAII) channel, associated with Ruffini corpuscles. There are similar differences among these channels with regard to their spatial resolution (the ability to distinguish neighboring stimuli), with the PC channel having the poorest spatial acuity for encoding extended two-dimensional stimuli (textures). One reason why the most sensitive system, the Pacinian, shows such poor spatial resolution is that it lies deep in the dermis and receives information from a wide area of the skin.

Hairy skin has characteristics similar to glabrous skin, although somewhat different receptor populations are responsible for the information processed there. Sensitivity varies not only with skin type but also with site. One valid generalization is that sensitivity decreases as the site is moved from the fingertips toward the trunk of the body (with the notable exceptions of the lips and genitalia). There could be as much

as 100:1 difference in vibrotactile sensitivity from the fingertip to the lower back. This variation is directly related to the differences in the density of receptors over the skin's surface and is mirrored in tactile spatial acuity over the body's surface: the fingers are better than other body sites in detecting small gaps between edges. In most of the current tactile aids for hearing-impaired persons, different aspects of the speech signal are encoded by the site of the vibrator and the frequency of vibration. Consequently, in designing an aid, the characteristics of both the site and the signal must be considered.

Response to complex stimuli. In everyday life, however, vibrotactile patterns that are complex combinations of both frequency and texture are typically encountered. Just as it was possible to account for the neural mechanisms responsible for processing simple stimuli, it should be possible to account for how the skin and its receptors appreciate these combinations. Two broad categories of tactile stimuli can be distinguished by their spatial and temporal characteristics. The first involves relatively coarse embossed or raised textures, such as the roughness of cloth or Braille, felt when the fingers contact a surface directly. These stimuli are characterized by low-frequency components, especially in the temporal realm. The second category involves higher-frequency patterns, such as the roughness or texture of paper felt through a pencil while writing, or microtextures such as etched glass. It is to be expected that, depending on the category, the one or more receptor populations whose spatial and temporal properties are best matched will be primarily responsible for encoding the stimulus.

Recently, a considerable amount of research has been conducted on the neural structures and codes underlying the perception of temporally changing two-dimensional information presented to the skin by displays having either high- or low-frequency components. By using these displays to mimic the two categories of normal tactile stimuli described above, researchers have been able to examine the encoding processes of individual cutaneous receptors. For example, **Fig. 2** shows typical responses of the receptor population in the four-channel model to Braille patterns stroked across the finger, a stimulus that falls into the first category. It is clear that the identity of this coarsely textured stimulus is best retained by the SAI receptors, a slowly adapting system sensitive to low vibratory frequencies. The SAI channel also responds well to embossed or raised textures and pressure stimuli, as does the RA channel, and both are well situated for stimulus localization because they are superficial, lying at the dermal-epidermal interface. It is apparent that the PC channel is the poorest channel for representing such coarse stimuli. As discussed above, not only does the PC channel show poor spatial acuity but the design of the corpuscle tunes it to respond optimally to transients rather than to the persistent features that characterize this type of texture exploration. In contrast, the PC channel, because of its exquisite vibratory sensitivity, plays a major role in the perception of the second category of tactile stimuli. These stimuli either vibrate at high frequencies or contain fine spatial textures that can be converted to vibration, and include moving displays from devices such as the Optacon, which translates printed text for blind persons onto a vibrotactile array. Such dynamic patterns were specially designed to appeal to the PC receptors (the Optacon vibrates at 230 Hz), despite the fact that the PC channel provides the poorest spatial resolution. The RA channel also has been found to play an important role in representing these stimuli. The relatively poor spatial acuity of these systems is reflected in data on pattern learning and discrimination: small patterns on the Optacon are harder to learn and read than large ones. Nevertheless, by acting in concert, all of these receptor systems provide a richer view of the tactile world than would otherwise be possible.

For background information SEE BIOMEDICAL ENGINEERING; CUTANEOUS SENSATION; MECHANORECEPTORS; SENSE ORGAN; SOUND in the McGraw-Hill Encyclopedia of Science & Technology.

Roger W. Cholewiak

Bibliography. M. A. Heller and W. Schiff (eds.), *The Psychology of Touch*, 1991; K. O. Johnson and S. S. Hsiao, Neural mechanisms of tactual form and texture perception, *Annu. Rev. Neurosci.*, 15:227–250, 1992; W. Schiff and E. Foulke (eds.), *Tactual Perception: A Sourcebook*, 1982; I. R. Summers (ed.), *Tactile Aids for the Hearing Impaired*, 1992.

Deoxyribonucleic acid (DNA)

Recent research has involved studies of ancient deoxyribonucleic acid as a tool for determining relationships among species, and development of methods for rapid DNA sequencing.

Ancient DNA

Determination of how species are related to each other is possible by observing differences in the sequence of the four bases that make up the DNA strands. These differences result from genetic changes that accumulate in the DNA owing to random processes as well as natural selection. Species that more recently had a common ancestor will have more similarity in DNA sequence than species that shared an ancestor only in the distant past. However, one limitation has been that only the DNA of species alive today can be studied whereas many species are either extinct or rare and therefore available only in museums. Recently developed techniques overcome this problem, permitting retrieval of DNA sequences from remains of animals found at archeological excavations preserved as museum specimens.

Laboratory procedures. The first step in the procedure for isolating DNA from an ancient specimen is to remove a sample; approximately 0.5–1.0 g (0.02–0.04 oz) is generally required. Because the enzymes that start degrading the DNA after death require water to be active, tissues that dried out rapidly after death generally contain the best-preserved DNA. After the tissue is cleaned, a piece is cut out and ground to a fine powder to enlarge the overall surface. The sample is then extracted (for example, by digestion with proteinase), purified (for example, by phenol and chloroform extractions), and concentrated (for example, by centrifugation-driven dialysis). An aliquot of the resulting DNA extract is used as a template in the polymerase chain reaction to amplify a desired DNA fragment. Next, an aliquot of the reaction is loaded on an agarose gel for separation by electrophoresis. On the gel, the target DNA fragment forms a specific band of a known size, which is determined through comparison with markers. The band is cut out and used for DNA sequence determination, where the order of the bases in the DNA is determined by electrophoresis.

Figure 1 is a schematic drawing of the sequence of steps in the polymerase chain reaction. In step 1, the strands of a double-stranded DNA are separated by high temperatures. As a result, the DNA primers, which have been synthesized to match the base sequences flanking the target sequence, can bind to the intended areas of the template DNA strands (step 2). A thermostable DNA polymerase then extends the primers so that two exact copies of the initial template molecules are made (step 3). During the next cycle of denaturation, annealing, and extension, the new as well as old strands act as templates (step 4), yielding four identical copies of the original DNA fragment (step 5). During the process in Fig. 1, repeated cycles of separation of DNA strands by heating, annealing of synthetic DNA primers (which define the region to be amplified) to the extracted DNA, and extension of the primers by a thermostable DNA polymerase result in an exponential growth of the DNA molecules that are to be studied. The sequence of the primers ensures that they anneal only to the sequence of interest and not to other sequences from the organism or from bacteria or fungi that may grow in tissues after death.

An extinct zebra. The first animal from which DNA sequences were determined was the quagga, an extinct African member of the horse family. For this experiment, skin and muscle tissue from a museum specimen 120 years old were used to extract DNA. Two pieces of mitochondrial DNA were sequenced. Such DNA exists outside the cell nucleus in the mitochondria, which produce energy for the cell. Since

each cell contains hundreds or thousands of mitochondrial DNA molecules, it is particularly likely that some will survive for a long time after the death of an animal. Furthermore, the base sequence of mitochondrial DNA accumulates changes faster during evolution than base sequences in the nucleus. Therefore, mitochondrial DNA sequences are very useful for studying how species that shared a common ancestor within the last few million years, such as the zebras, are related to one another.

The sequences of the quagga were compared to those of other members of the horse family and used to reconstruct a phylogenetic tree (**Fig. 2**) that shows how species are related to each other. This tree explains the differences that are observed between the sequences of the different species by postulating the minimum number of mutational events in the past. The tree shows that the quagga was very closely related to the plains zebra and more distantly related to the mountain zebra and other relatives of the horse. The quagga may in fact have been a type of plains zebra. Thus, it may be valid to try to recreate the quagga by breeding experiments where plains zebras are crossed to obtain an offspring that carries the pattern of striation that was typical of the quaggas. Such attempts are under way in southern Africa.

Molecular archeology. In addition to the quagga, DNA has been extracted, amplified, and sequenced from tissues of the extinct marsupial wolf of Australia, giant flightless birds known as moas that once existed on New Zealand, saber-toothed tigers from the La Brea tarpits in Los Angeles, California, and the frozen carcass of a woolly mammoth found in the permafrost of Siberia. These sequences have been used to infer how these extinct animals are related to present ones, and interesting new information about their evolution has emerged. For example, when the mitochondrial DNA sequences from four different species of New Zealand moas were compared to that of the kiwis, which now live in New Zealand, and to that of Australian flightless birds, the kiwis were shown to be more closely related to the Australian birds than to the moas. Therefore, it is not likely that a common ancestor of kiwis and moas became isolated in New Zealand when continental drift caused the landmass to break loose from Antarctica and Australia around 80×10^6 years ago. Rather, the ancestor of the kiwis must have arrived later. Presumably it then was able to fly, and only after its arrival in New Zealand, where no ground-dwelling mammals threatened it, did it lose its capacity to fly.

The possibility of extracting DNA from bones has also been demonstrated. This development is important since often only bones are found by archeologists. Furthermore, dried plant remains often seem to be well preserved. Notably, ribonucleic acid (RNA) as well as DNA exist in corn cobs from Peru and Chile that are up to 4500 years old. Such a quantity of results have been published already that a new area of science has emerged known as molecular archeology.

This new field of study still struggles with many problems. In particular, the DNA in ancient remains is

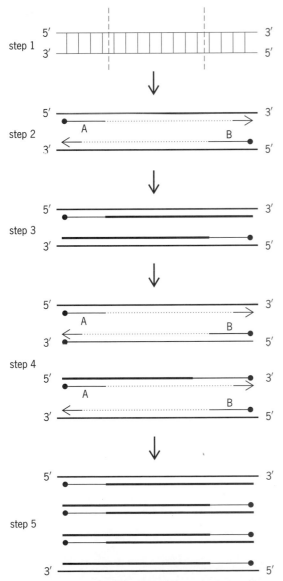

Fig. 1. Schematic drawing of the sequence of steps in the polymerase chain reaction.

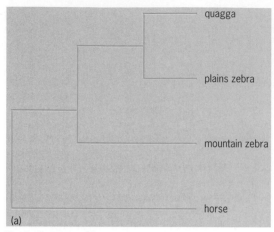

Fig. 2. Quagga and its relatives. (*a*) Phylogenetic tree. (*b*) One of the last living quaggas; its striation distinguishes it from other zebras.

often quite damaged by reactions with water and oxygen over the long periods of time that passed since the death of the organism. This damage makes it impossible to amplify longer DNA fragments. Rather, many short amplifications of just 100–200 bases must be "puzzled together" to reconstruct the longer sequence that existed in the living organism.

Another problem is that the specimens, the extracted material, or the reagents for polymerase chain reaction can be contaminated with traces of modern DNA. Often such contamination stems from miniscule amounts of human DNA that exist in, for example, saliva droplets or perspiration that may have penetrated the specimens during handling by archeologists or museum curators or entered the solutions prepared in the laboratory. A few such molecules can easily obliterate the old molecules in the amplification, since they are less damaged. Therefore, detection of any source of contamination in the experiments is very important. For example, it is always necessary to perform amplifications from control extracts prepared without any tissue samples added. Thus, a check can be made as to whether any modern DNA exists in the chemicals and solutions or whether, for example, dust, which often contains human skin fragments, has entered the extraction. In the case of ancient human remains, it is particularly difficult to distinguish contaminating sequences from ancient ones, since the contaminating sequences most often are human.

Present techniques allow the study of population samples of many animals that have been collected in zoological museums during the twentieth century. Researchers can follow how reductions in population size have resulted in a decrease in the amount of genetic variability in the population. Such studies may assist in understanding how extinction of a species occurs and how living species threatened with extinction can be better protected.

Molecular paleontology. The oldest mammalian sequences so far reported come from a frozen 40,000-year-old mammoth. One fascinating question is whether it will be possible to go back millions of years and retrieve sequences from dinosaurs and other crea-

tures. Probably the best possibility would be to study remains that have been protected from the two agents most likely to destroy DNA over long periods of time: water, which causes bases to come off the backbone of the DNA strands, and oxygen, which chemically modifies the bases as well as the sugar residues in the DNA backbone. One promising source may be amber, made of solidified tree sap. When the sap was still sticky, insects, small animals, or pieces of plants could become embedded in it, and have sometimes been preserved as inclusions that are several million years old. Recently, two research teams have published DNA sequences of insects from amber. Others have recovered sequences from 17-million-year-old plant fossils preserved in clay from Idaho. If these results are reproducible, a new field of research perhaps known as molecular paleontology could emerge. Fascinating possibilities for estimating the rates of changes in genes, as well as for studies of the relationships of ancient animals and plants, could be pursued.

Matthias Höss; Svante Pääbo

High-Speed DNA Sequencing

The Human Genome Initiative is an ambitious international effort to map and sequence the 3×10^9 bases of DNA encoded in the human genome. If the project is successfully completed, the resultant sequence database will be a tool of unparalleled power for biomedical research. However, the project faces a major hurdle in the area of DNA sequencing technology. Existing methods for DNA sequence analysis are one to two orders of magnitude too slow and expensive for successful completion, so new, more effective sequencing technologies must be developed.

An automated DNA sequencing instrument with greatly increased throughput and much smaller sample requirements has been developed. This type of instrument, in conjunction with advances being made in the automation and streamlining of other aspects of the sequencing process, may substantially increase the speed and decrease the cost of DNA sequence analysis.

Sequencing process. The sequencing process is a complex set of procedures of three general types: front end, separation and detection, and back end. The front end is composed of the various molecular biological steps required to obtain the initial DNA fragment to be sequenced, to generate any smaller fragments needed in sufficient quantity, and to perform the necessary sequencing reactions upon those fragments. Next comes the process of separating and detecting, usually by denaturing gel electrophoresis, the DNA fragments produced in the sequencing reactions. Finally, the back end involves the several levels of data processing, correction, and assembly required for converting the short stretches of raw sequence data into a contiguous final sequence encompassing the region of interest.

Given the complexity of the sequencing process and the wide variety of technologies employed, there is no single technology or approach capable of automating the entire procedure. Rather, it is necessary individually to streamline and automate the various aspects of the process and then to combine them to yield a system capable of high throughput. For maximum simplicity, reliability, and ruggedness of the overall process, to as great an extent as possible the various subprocesses should be redesigned to facilitate automation, rather than being automated in the form in which they are commonly practiced.

Separation and detection. Recent work has involved the separation and detection aspect of high-throughput DNA sequencing. The central separation technology in all existing methods for DNA sequencing is gel electrophoresis, usually performed in vertical slab gels and employing either radioactive or fluorescence detection. This electrophoretic separation has extremely high resolution, permitting clear separation of single-stranded DNA fragments differing in length by only one nucleoside subunit out of a total of as many as 600. In order to obtain such high resolution of these relatively short DNA molecules, cross-linked polyacrylamide gels are employed that have an average pore size of the order of 10 nanometers. It is also necessary to include a denaturant, typically urea, in these gels in order to prevent the formation of secondary structures in the DNA molecules as they are moving through the gel. These secondary structures are typically referred to as compressions because their effects on the mobility of the DNA molecules appear as a bunching-up of the DNA bands in the gel; that is, DNA molecules of different lengths have very similar electrophoretic mobilities. These structures are due to the intramolecular base-pairing of the DNA molecule to form, for example, a hairpin structure.

In order for good separation performance to be obtained from the gel, no charged groups can be present in the gel. Charged groups can cause two types of problems. First, they can cause electroosmotic flow, in which the electric field in the gel acting on the mobile counterions creates pressure and bulk flow of solvent in the gel. This flow pattern can distort the migration and shape characteristics of the DNA bands. Second, the charged groups can absorb strongly to the DNA molecules in the gel, causing tailing of the bands and decreased resolution.

The requirement for gel neutrality and the need for denaturants compromise the chemical stability of the gels. The gels are generally employed at an alkaline pH of about 8.5, at which both acrylamide and urea hydrolyze spontaneously at appreciable rates, producing ionic products. Thus, the shelf life of the gel is not long, and commercial preparation of sequencing gels has thus far proven impossible. Rather, these gel mixtures are routinely prepared by the investigator on a laboratory benchtop.

The rate of sequence acquisition in DNA sequencing by denaturing gel electrophoresis is determined by the number of samples that may be analyzed in parallel, and by the speed with which the electrophoretic separation can be performed. Although in principle the speed of the separation might be increased by increasing the electric field, this approach is not possible in conventional electrophoretic systems because of the excess heat generated by the electrophoresis. To address this problem, the electrophoresis system

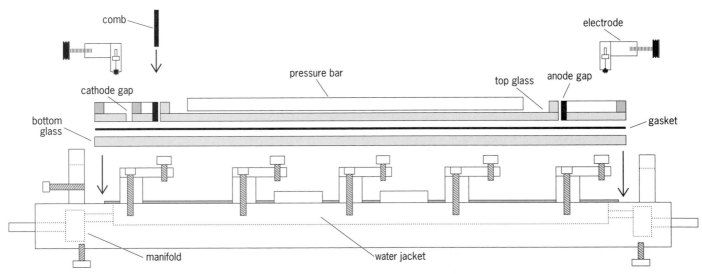

Fig. 3. High-performance electrophoresis cell: side view of the horizontal apparatus base and glass components.

must be redesigned to provide improved heat transfer.

High-performance electrophoresis cell.

A novel high-performance electrophoresis cell known as the HUGE system, for horizontal ultrathin gel electrophoresis (**Fig. 3**), has been constructed. This system provides improved heat transfer because (1) it employs active rather than passive cooling by means of circulating water in contact with one of the lower glass plates; (2) it is horizontal, eliminating the convective temperature gradient found in vertical systems; and (3) the gels are ultrathin (typically 50–75 micrometers), providing an increased surface-to-volume ratio and thereby an increased efficiency of heat transfer. The improved thermal performance of this system is evident from plots of the rise in temperature of the gel as a function of applied power.

In order to employ fluorescence detection with this system, it was necessary to design a sensitive fluorescence detection system (**Fig. 4**). This system employs a sensitive detector consisting of a thermoelectrically cooled charge-coupled-device array and a prism wedge assembly to permit four-wavelength fluorescence detection from multiple samples without the use of any moving parts. At this time it is possible to analyze 18 samples in parallel with a separation time as short as 49 min for fragments up to 410 bases in length. Each sample contains only 100 nanograms of DNA template in a volume of 300 nanoliters. A comparison of the overall throughput of this device with currently available automated DNA sequences is shown in the **table** along with projections for a second-generation device currently under construction.

Several variables play important roles in these gel electrophoretic separations, including temperature, electric field strength, and gel/buffer composition. Optimization of performance involves a trade-off of resolution versus speed involving all of these variables.

Temperature. There are two temperature effects. First, the absolute temperature of the gel affects the electrophoretic mobility of the DNA molecules, and is also important in minimizing secondary structure artifacts in the electrophoresis. The absolute temperature may be adjusted by controlling the temperature of the circulating water bath cooling the bottom gel plate. Second, the temperature gradient across the gel can substantially decrease resolution because of the differential mobility of DNA molecules in different regions of the gel. A gradient as small as 0.05°C (0.09°F) is sufficient to cause substantial band spreading and

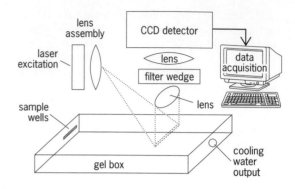

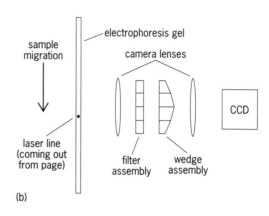

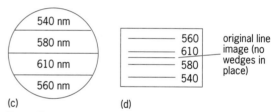

Fig. 4. High-speed automated DNA sequencing. (***a***) **Overview diagram of the instrument.** (***b***) **Side view of the fluorescence detection system.** (***c***) **Filter assembly.** (***d***) **Images on charge-coupled surface.**

loss of resolution. Both the absolute temperature and the temperature gradient may be calculated by solution of the heat conduction equation under appropriate boundary conditions.

Electric field strength. The electric field strength affects both the gel temperature and the gel temperature gradient by means of the increased electrophoresis power dissipated in the gel. Higher field strength also appears to decrease the resolution of the

Throughput of automated DNA sequencers*

System	Samples	Times
ABI 373A	24 samples in parallel	450 bases each in 8 h: 1350 bases/h
Huge system 1 (current)	18 samples in parallel	450 bases each in 52 min: 9300 bases/h
Huge system 2 (projected)	50 samples in parallel	450 bases each in 52 min: 26,000 bases/h

*Electrophoresis time only.

DNA fragments, as has been predicted theoretically and observed experimentally in agarose and acrylamide gels. Similar effects have been observed in these high field separations, as evidenced by a decreased spacing between bands. Both the increased temperature and the spacing effect tend to decrease the resolution of the separations as the electrophoresis power is increased, suggesting that the gains in throughput to be obtained from ever-increasing electric field strength are not open-ended.

Gel composition. The gel composition, and in particular the gel percentage, affects both the speed and the resolution of the separations. Data have been obtained for both 6% and 4% gels. The 6% gels give less rapid separations but are somewhat easier to use and give higher signal-to-noise ratio because the wider peaks permit more data smoothing. The 4% gels give more rapid separations but have thus far been somewhat less reproducible in their performance. Buffer composition and concentration affect the gel conductivity and thus the temperature, as well as bandwidth and spacing. All of these variables are important in performance of the separations.

For background information SEE AMBER; DEOXYRIBONUCLEIC ACID (DNA); ELECTROPHORESIS; NUCLEOPROTEIN; ZEBRA in the McGraw-Hill Encyclopedia of Science & Technology.

Lloyd M. Smith

Bibliography. R. L. Brumley and L. M. Smith, Rapid DNA sequencing by horizontal ultrathin gel electrophoresis, *Nucleic Acids Res.*, 19(15):4121–4126, 1991; E. Golenberg et al., Chloroplast DNA sequence from a Miocene *Magnolia* species, *Nature*, 344:656–658, 1990; R. Higuchi et al., DNA sequences from the quagga, an extinct member of the horse family, *Nature*, 312:282–284, 1984; A. J. Kostichka et al., High speed automated DNA sequencing in ultrathin slab gels, *Bio/Technology*, 10(1)78–81, 1992; S. Pääbo, Molecular cloning of ancient Egyptian mummy DNA, *Nature*, 314:644–645, 1985; L. M. Smith, DNA sequence analysis: Past, present, and future, *Amer. Biotechnol. Lab.*, 7:10–25, 1989.

Digestive system

The understanding of digestive adaptations among mammals that use plants as food has been aided by the recent application of chemical reactor theory to the study of the mammalian digestive system.

Herbivore gut models. There are three basic types of chemical reactors: batch reactors, plug-flow reactors, and stirred-tank reactors.

Batch reactors. Reactors that process reactants (food) in discrete batches and may lie empty for periods between reaction cycles (meals) are known as batch reactors. They are not viable models for herbivores, as the population of microbial symbionts necessary for digestion of plant cell walls would be lost each time the reactor (fermentation chamber) was emptied. However, the model may be applicable to the carnivore stomach, which can retain food for a long time but can be quickly emptied by regurgitation to make way for better-quality food.

Plug-flow reactors. These reactors feature continuous flow of material through a tubular (cylindrical) reaction vessel. There is no mixing of reactants along the axis of the cylinder (although mixing in the radial direction is substantial). Reactant concentrations and reaction rates are maintained at higher levels near the reactor entrance and decline along the length of the vessel. The small intestine most closely meets the description of a plug-flow reactor. Plug-flow reactors are most suited to foods that are easily digested by the animal's own enzymes; this model is therefore applicable to the digestive systems of carnivores, which are dominated by the small intestine.

Stirred-tank reactors. These reactors are characterized by a continuous flow of materials through a usually spherical reaction vessel of minimum volume. The reactor contents are well mixed, but reactant concentration is diluted immediately upon entry into the vessel by material recirculating in the reactor (that is, residues from previous meals). This dilution reduces the reaction rate, but the extent of conversion (digestion) can be high if the flow of material is low enough. These conditions apply where diverticulae of the gut function as fermentation chambers, such as in the forestomach of ruminants (sheep, cattle, antelope) and the cecum distal to the end of the small intestine. In this type of reactor, materials are retained for extended periods for fermentative digestion by resident symbionts, such as bacteria, protozoa, and anaerobic fungi.

Modified plug-flow reactors. When three or more stirred-tank reactors are arranged in series, the reactor system becomes more like a plug-flow reactor. Because there is substantial axial as well as radial mixing, this system is best described as a modified plug-flow reactor. Examples are the forestomach of kangaroos and the proximal colon of large hindgut fermenters such as horses, elephants, rhinoceroses, and marsupial wombats. The modified plug-flow reactor allows higher flow rates of materials, but because retention times are lower the extent of digestion is often less than in a stirred-tank reactor.

Plant components. The optimal configuration of plug-flow, stirred-tank, and modified plug-flow reactors in the digestive system is intimately linked to the natural diet and feeding strategy of the herbivore. Plant material can be divided simply into four major components: cell contents, cell walls, exudates, and secondary compounds.

Cell contents. Soluble organic acids, enzymes, and storage forms of energy such as starch and lipids are cell contents that are readily digested in a plug-flow reactor, such as the small intestine. Fermentation of these constituents leads to loss of energy in the form of methane, hydrogen, and heat. Thus, if the plant material usually eaten by the herbivore is mainly cell contents, restriction of microbial fermentation to the hindgut, distal to the small intestine, is efficient. If cell contents form only a minor part of the plant material eaten, the small loss of energy from fermentation will be inconsequential compared with the advantage to the herbivore of deriving most of its energy from plant cell walls.

Cell walls. Plant cell walls consist of pectin, lignin, hemicelluloses, and cellulose. Although cellulose potentially is completely digestible by microbial cellulases, the rate and extent of attack by microbial symbionts in the herbivore gut is limited by several factors, including the degree of crystallinity of the cellulose; the extent of lignification, cutinization, and silicification of the cell wall; and the surface area (that is, material particle size). Substantial fermentation of cellulose is provided by development of a diverticulum such as in the ruminant forestomach and the cecum of small hindgut fermenters (a stirred-tank reactor), or by increases in the length and cross section of the gut (a modified plug-flow reactor), as in the kangaroo forestomach and the proximal colon of large hindgut fermenters.

Plant exudates. These exudates include saps, nectar, gums, and resins. Although easy to digest in the small intestine, saps and nectar contain mostly water and sugars and are low in protein and minerals. Thus, mammals whose primary source of food is exudates must supplement the diet with food of higher protein content, such as pollen or insects. Mammals that feed on gum often have a large cecum, suggesting that gums are more difficult to digest and microbial fermentation is necessary. This interpretation is consistent with the polymeric structure of some gums.

Plant secondary compounds. Both physical and chemical strategies have evolved as a defense against herbivory. Chemical strategies affect the utilization of both cell walls and cell contents. For example, cytoplasmic proteins of the leaves of many woody plants can be rendered indigestible by complexing with tannins. Lignin and silica, which are used for structural strength, reduce digestibility of plant cell walls. Thus, plant secondary compounds modify the extent to which many types of plant material can be used as food by animals.

Separation of material in the gut. A common strategy among mammalian herbivores to maximize the yield of energy and nutrients from plant material is to separate small from large particles. Large food particles are selectively retained in stirred-tank reactors such as the ruminant forestomach by the passage of only small particles out of the main fermentation area. Thus, cell wall digestibility is maximized, but if the plant material is highly lignified the rate of particle breakdown is so slow and retention in the main fermentation area so long that food intake is inhibited. This bottleneck is alleviated in modified plug-flow reactor guts. In this system there is no restriction on the flow of particles of different sizes. Particles are retained by the outpocketings of the reactor wall (that is, the colon or the kangaroo forestomach) while contractions propel digesta through the reactor.

Most small herbivores feed on plant material of high quality in order to meet their relatively high energy requirements. The option of prolonged reten-tion of large food particles in the gut is not available to them. Instead, many selectively retain fluid (and small particles, including bacteria within the fluid). The separation mechanism is located in the proximal colon, and the fluid and small particles are retained in a large cecum. Thus, the digestive process is effectively concentrated on the most digestible components of the food particles, while the larger, less digestible particles are eliminated from the gut as rapidly as possible. This strategy is employed by rabbits, many herbivorous rodents, and marsupial folivores such as the koala, greater glider, and ringtail possums. In many of these animals, selective retention of fluid and fine particles is coupled with coprophagy (ingestion of feces). This practice further increases the extent of digestion and salvages microbial protein and B vitamins that otherwise would be lost in the feces.

Classification of mammalian herbivores. On the basis of the principal site of microbial fermentation, mammalian herbivores can be grouped into foregut fermenters and hindgut fermenters. The two types of reactors applicable to herbivores (stirred-tank reactors and modified plug-flow reactors) are found in both groups. Among foregut fermenters, most of which are large (> 22 lb or 10 kg), selective retention of large particles maximizes the yield of energy from plant cell walls. This digestive strategy is especially applicable in herbivores with stirred-tank reactors (for example, ruminants), in which food intake can be limited when the plant cell walls are highly lignified. The digestive strategy seems best suited to those environments where food availability is limited, such as deserts. Herbivores with modified plug-flow reactors, such as the kangaroo, can maintain higher flow of food particles that are highly lignified, a strategy best suited to savannas where food availability is usually not limited.

Hindgut fermenters can be subdivided into colon fermenters, in which microbial activity is located primarily in the proximal colon, and cecum fermenters, in which microbial activity is restricted almost entirely to the cecum. Colon fermenters are large, and the proximal colon functions as a modified plug-flow reactor. These herbivores are best suited to savannas where grass quality is often low but availability is not limiting. In contrast, nearly all cecum fermenters are small and must concentrate on higher-quality foods. Most digestion takes place in the small intestine, a plug-flow reactor. The cecum functions as a stirred-tank reactor, but in contrast to the ruminant there is selective retention not of large particles but of fluid and small particles, the most digestible components of the material entering the hindgut. Coprophagy further increases the yield of energy and nutrients to these small herbivores.

For background information *SEE CHEMICAL REACTOR; COLON; DIGESTIVE SYSTEM; HERBIVORY; INTESTINE* in the McGraw-Hill Encyclopedia of Science & Technology.

Ian D. Hume

Bibliography. R. McN. Alexander, Optimization of gut structure and diet for higher vertebrate herbivores, *Phil. Trans. Roy. Soc. London*, B333:249–255, 1991;

I. D. Hume, Optimal digestive strategies in mammalian herbivores, *Physiol. Zool.*, 62:1145–1163, 1989; D. L. Penry and P. A. Jumars, Modeling animal guts as chemical reactors, *Amer. Nat.*, 129:69–96, 1987.

Dinosaur

During the 1920s, world attention was focused on the American Museum of Natural History's Central Asiatic Expeditions to collect dinosaurs from the untapped fossil fields of China and Mongolia. Since that time, central Asia has become one of the richest dinosaur-collecting areas in the world, with a full range of fossil evidence from isolated microsites, larger sites, and bonebeds, including footprints, trackways, isolated teeth and bones, complete skeletons, eggs, nests of eggs, and skin impressions. Exceptionally well-preserved specimens from China and Mongolia have provided some of the best information on the anatomy and evolution of major lineages of dinosaurs.

In the heart of Asia, Upper Triassic and Lower Jurassic rocks of continental origin will eventually yield significant dinosaur finds. Middle and Upper Jurassic beds in northwestern China are rich in dinosaur skeletons, and several new species are in the process of being described. The greatest potential for dinosaur finds in central Asia lies in the extent and variety of Cretaceous sites.

Unlike North America and Europe, Asia has a virtually uninterrupted history of terrestrial dinosaur sites. To date, more than 25% of the known dinosaur species have come from central Asia. In addition, because of the number of well-exposed Mesozoic continental deposits with varied depositional environments, many new species of dinosaurs are expected to be recovered.

Major expeditions. The large, multidisciplinary team approach of the American Museum's expeditions was adopted by later multinational expeditions of Americans, Canadians, Chinese, Mongolians, Poles, Soviets, and Swedes.

American Museum of Natural History. The 1922 expedition found the first evidence of dinosaurs in central Asia, a vast, arid, and then largely unmapped region encompassing parts of China, Mongolia, Kazakhstan, and Uzbekistan. The first site at which the fossil evidence was found, Iren Dabasu, near the modern Chinese city of Erenhot, has since been visited by many major expeditions. It has never become a famous locality because most of the thousands of specimens have never been prepared or described.

The American Museum party continued on into Mongolia, where in 1923 it attracted international attention by discovering the first unquestionable nests of dinosaur eggs at a site now known as Bayn Dzak. Here they also recovered more than 75 skulls of *Protoceratops*, one of the best growth series known for any dinosaur and the first to show sexual dimorphism.

A third site in Mongolia (Anda-Khuduk) produced the first specimens of the early ceratopsian *Psitta-cosaurus*, a dinosaur characteristic of Lower Cretaceous beds across Asia.

In 1990, the American Museum of Natural History initiated a new series of expeditions in Mongolia. A giant lizard capable of eating small dinosaurs as well as the skeletons of several small theropods have already been discovered.

Sino-Swedish and Sino-Canadian expeditions. Sven Hedin's Sino-Swedish expeditions recovered the first Jurassic dinosaur, the long-necked sauropod *Tienshanosaurus*, from Hinjiang in northwestern China. On January 1, 1930, fossils were found in Inner Mongolia near a distinctive pillar of red rock that the team knew as Ulan Tsonchi. The American Museum's expedition had visited the same locality 2 years earlier, and the Sino-Soviet expedition visited in 1960. However, the real importance of this region was realized when, in 1987, a Sino-Canadian expedition found more than 25 skeletons in a single day.

Soviet and Sino-Soviet expeditions. Soviet expeditions into Mongolia in 1946, 1948, and 1949 revisited American Museum localities and discovered new sites, the most notable being the Dragons' Tomb in the Nemegt Valley. The Nemegt has produced an extremely rich Late Cretaceous dinosaur fauna, including well-preserved skeletons of the tyrannosaur *Tarbosaurus* and a giant hadrosaur (*Saurolophus*) that had been previously described from North America. Older dinosaur sites were discovered in the southeastern Gobi of Mongolia in the vicinity of Bayn Shire.

The Soviets collaborated with the Chinese in an ambitious project to collect dinosaurs and other fossils across northern China and central Asia (Kazakhstan, Uzbekistan) over a 5-year period. The Sino-Soviet expeditions got under way in 1959 at the American Museum's Iren Dabasu site. The following year, important new dinosaur sites of Cretaceous age were discovered in the Alashan Desert. The Early Cretaceous site of Maortu produced the iguanodont *Probactrosaurus* and a large theropod. The Upper Cretaceous rocks of Tashuikou yielded a beautiful, virtually complete skeleton of a large ankylosaur. When the political relationship between China and the Soviet Union collapsed in 1960, the Soviets reluctantly returned home. Tons of specimens had been collected, most of which are still in crates in Beijing.

Polish-Mongolian and Soviet-Mongolian expeditions. The Polish-Mongolian Paleontological Expeditions (1964–1971) were extremely successful, revisiting localities worked by Americans and Soviets and pushing into new areas within and to the west of the Nemegt Valley. Small theropods, ornithomimids, tyrannosaurs, sauropods, hadrosaurs, ankylosaurs, and the first good Asian pachycephalosaurs were found. One of the most exciting discoveries was a pair of dinosaurs, *Velociraptor* and *Protoceratops*, that had apparently perished in a death lock.

The longest continuous program in central Asia was the Soviet-Mongolian Paleontological Expeditions (1968–1990), which traversed all of southern Mongolia. Many new dinosaur sites were identified, but the classic collecting sites at Bayn Dzak, Nemegt, and

Bayn Shire continued to produce the best material.

Soviet programs in Kazakhstan and Uzbekistan have been ongoing since the 1920s. Shakh-Shakh, a major locality north of the Aral Sea, produced its first dinosaur skull in 1957. Recently, paleontologists from St. Petersburg and Alma Ata have focused on recovering individual bones from bonebeds and microsites. They have discovered dinosaur faunas that are similar to those from the latest Cretaceous of North America. Large horned dinosaurs of the family Ceratopsidae have been identified for the first time in central Asia.

Sino-Canadian expeditions. The Sino-Canadian Dinosaur Project expeditions (1986–1990), the first to hunt exclusively for dinosaurs, worked in both China and Canada. In Xinjiang, crews collected a new type of large theropod and the neck of the largest sauropod known from Asia. It took four summers to excavate to the end of the neck, where they found the skull, the first known for a mamenchisaur sauropod. In the Ordos Basin, many specimens of *Psittacosaurus* were recovered. In 1988, the most complete skeleton known of a troodontid theropod was uncovered. This find was significant because troodontids have the largest brains among dinosaurs and share many specialized characters with birds, thereby giving clues about the origin of the latter.

The most productive locality, Bayan Mandahu, is 20 mi (32 km) west of the Ulan Tsonchi site. One site revealed a mass death of at least 12 baby ankylosaurs, offering the first evidence that armored dinosaurs were gregarious.

Paleoenvironments. Paleontological, sedimentological, and taphomonic evidence is used to determine what ancient climates and environments were like. Geologists of the 1920s determined that central Asia had been continental since the beginning of the Mesozoic Era, and that many of the best dinosaur-collecting sites were as arid then as they are now. Whereas most regions in the world have preservation biases that favored large dinosaur specimens and destroyed small ones, the arid or semiarid conditions of many localities in central Asia have resulted in preservation of eggs, baby dinosaurs, and complete skeletons of small dinosaurs. An alternative hypothesis suggested that the ancient wind-blown sands were actually dunes on the margins of lakes, and that some of the dinosaurs had been trapped in quicksand.

Sedimentologists from the 1986–1990 Sino-Canadian team tried to resolve which, if either, of these interpretations was correct and whether the Asian dinosaurs had lived in wetland environments like those of most North American sites or in dry regions. Part of the Sino-Canadian objective was to determine the cause of differences between North American and Asian dinosaurs and dinosaur faunas. If the environments had been similar, the differences in anatomy and faunal composition were the result of geographic or temporal separation. However, if the Asian sites had been arid or semiarid, the differences might be correlated with environmental differences.

Sedimentologists with the Sino-Canadian expedition in Asia identified eolian dunes and other features characteristic of a desert. The desert conditions explain the low diversity of species in such Asian sites as Bayn Dzak and Bayan Mandahu, which produce hundreds of dinosaur skeletons, and the generally small size of Asian dinosaurs relative to North American ones. The ancient desert was a stressed environment, and relatively few animals were adapted to living there.

Understanding the environment of places like Bayn Dzak forces a reinterpretation of some of the finds. For example, the pair of dinosaurs *Velociraptor* and *Protoceratops* was once interpreted as being locked in combat when they became trapped in quicksand. It now seems more likely that the *Protoceratops* was attacked and killed by the *Velociraptor* while seeking shelter behind a sand dune during a sandstorm. This pattern of attack is paralleled in the Sahara by hyenas, which are known to attack gazelles under the cover of sandstorms. It appears that the *Protoceratops* locked its jaws on the *Velociraptor*'s arm before dying, resulting in a dual burial in the sand.

The 12 baby ankylosaurs at Bayan Mandahu probably suffered a similar fate. Most of the animals were found oriented in one direction, as if they had aligned their bodies to offer less resistance to a strong wind, like cattle and horses in a rain or snow storm.

The Nemegt sites of Mongolia represent a fluvial environment closer to that of North American sites. The diversity of fossils is much greater than at Bayn Dzak. In fact, there is the same high diversity (35+ species) of dinosaurs at the Nemegt site as at the temporally equivalent Dinosaur Provincial Park in Canada. The faunas of these sites are remarkably similar.

Intercontinental movements. The distribution of dinosaurs, just like those of other animals and plants, are controlled by many factors, including climate and connections between land masses. When dinosaurs appeared, 2.25×10^8 years ago, the continents were still interconnected, allowing plants and animals to spread everywhere, including to Antarctica. But as the land masses pulled apart, the movements of plants and animals were restricted.

The presence of *Dilophosaurus*, a peculiar carnivorous dinosaur with a pair of bony crests on its head, in Arizona and Yunnan (China) shows that these areas were bridged by land during the Early Jurassic. By the Late Jurassic, however, Asian and North American dinosaur faunas seem to have been developing independently. The large carnivorous dinosaurs and sauropods recovered by Sino-Canadian expeditions from the Upper Jurassic rocks of northwestern China are quite different, at the subfamily or family level, from the theropods and sauropods of North America. The differences are still apparent during the Early Cretaceous, when much of central Asia was covered by an extensive series of lakes and *Psittacosaurus* was the most common dinosaur in the region. However, the faunal differences may be attributable to ecological differences rather than continental separation, because at least a few animals seem to be the same in North America and Asia.

Dinosaurs colonized the polar regions and were able

to move between the northern continents across Cretaceous land bridges in the Arctic. Most families of Late Cretaceous dinosaurs found in the Northern Hemisphere have representatives in both Asia and North America.

For background information *SEE CONTINENTAL DRIFT; DINOSAUR; TAPHONOMY* in the McGraw-Hill Encyclopedia of Science & Technology.

Philip J. Currie

Bibliography. P. J. Currie, The Sino-Canadian Dinosaur Project, *Geotimes*, 1991; Z. M. Dong and R. C. Milner, *Dinosaurs of China*, 1988; T. Jerzykiewicz and D. A. Russell, Late Mesozoic stratigraphy and vertebrates of the Gobi Basin, *Cretaceous Res.*, 12:345–377, 1991; D. B. Weishampel, P. Dodson, and H. Osmolska, *The Dinosauria*, 1990.

Drone

Military leaders have always wanted the ability to see beyond the horizon. This need has led to the development of unmanned air vehicles (UAVs), with a payload of cameras or other sensors. Although reconnaissance was the primary goal of this development, innovative civilian applications have focused on placing scientific payloads at altitudes and for time periods not possible for humans.

The early name for these aircraft, remotely piloted vehicles (RPVs), has largely been abandoned because of the often preprogrammed, or autonomous, nature of their flight control. The class of UAV normally associated with the meteorological research of the atmosphere is the high-altitude long-endurance UAV (HALE UAV). The requirements for this class of air vehicle are unusual and unique.

Scientific investigation of the ozone-depleted and carbon dioxide–rich upper atmosphere require data gathering that can be performed only by high-altitude, long-endurance crewless aircraft. With ongoing developments in airfoil design at low Reynolds numbers and transonic Mach numbers, innovative propulsion systems, lightweight and strong composite materials, and miniaturized instrument packages, a family of aerial platforms will be available to help gather the data necessary for accurate global environmental models.

Fig. 1. Boeing Condor UAV. (*Boeing Defense and Space Group*)

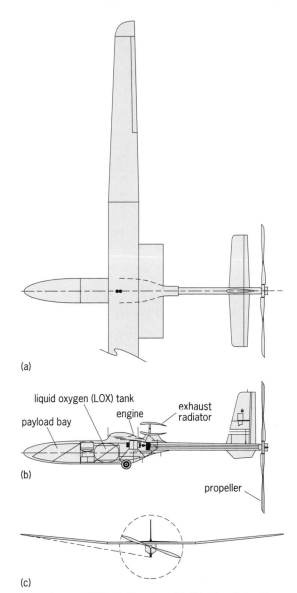

liquid oxygen (LOX) tank

payload bay

engine

exhaust radiator

propeller

Fig. 2. Perseus UAV. (*a*) Plan view. (*b*) Side view (*after News breaks: Perseus tested in closed-cycle mode, Aviat. Week Space Technol., 137(3):16, July 20, 1992*). (*c*) Aft view (*after AIAA Pap. 91–3162, Aircraft Design Systems and Operations Meeting, Baltimore, September, 23–25, 1991*)

Need for meteorological UAVs.

An important factor in understanding the impact of humanity on the global environment is the collection of atmospheric data for accurate modeling of the planetary climate. The accumulation of carbon dioxide and other gases that absorb solar radiation in the troposphere, and the depletion of ozone in the stratosphere, are promoting a measurable and significant change in global atmosphere. Remote measurements made by ground-based sources and space-based platforms such as satellites fail to provide information on the detailed structure of the atmospheric chemistry or dynamics. High-altitude balloons and rockets can make on-site measurements but cannot provide long-term on-station data, nor can they be as easily controlled or maneuvered as low-speed aircraft.

Crewed aircraft have provided most of the atmo-

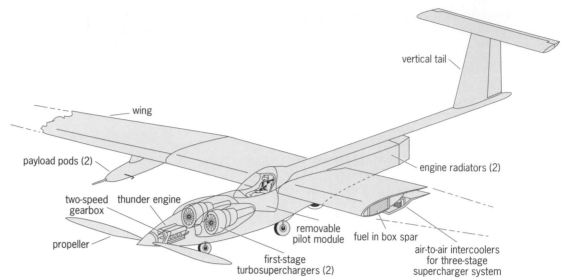

Fig. 3. Conceptual design of NASA stratoplane. (*After A. G. Sim, Modeling, Simulation, and Flight Characteristics of an Aircraft Designed to Fly at 100,000 Feet, NASA TM 104236, 1991*)

spheric data obtained at high altitudes so far. The ER-2 aircraft of the National Aeronautics and Space Administration (NASA), a derivative of the Lockheed U-2 spy plane, has conducted numerous experiments at altitudes up to 70,000 ft (21 km) and with a range of over 3000 nautical miles (5500 km). Antarctic and Arctic overflights have verified a strong correlation between the release of chlorofluorocarbons and ozone depletion. However, the ER-2 cannot reach the higher ozone layers situated in midlatitude or equatorial regions. The need exists for an aircraft capable of a 6000-nmi (11,000-km) cruise at altitudes of 100,000 ft (30 km) or greater.

High-altitude UAV design. Designing an aircraft to fly subsonically at 100,000 ft (30 km) provides unique challenges. The air density at 100,000 ft (30 km) is $1/70$ that at sea level, and the pressure is $1/96$ its sea-level value. Since the speed of sound varies as the square root of temperature, so does the Mach number; aircraft are required to operate at transonic Mach numbers (around 0.7–0.75), even at low flight speeds. Also, the low density drives the operating Reynolds number well below 10^6, with the attendant problems of leading-edge laminar separation bubbles and degraded aerodynamics. Ground-based wind tunnels able to meet these criteria for testing new airfoils are few. Propulsion limits are also a concern. Air-breathing turbine engines lose power almost proportionally to pressure, while advances in supercharged reciprocating engines make them the prime candidates for use with the large (25–30-ft or 7.5–9-m diameter) propellers necessary to maintain efficiency as altitude is increased. Other engine designs under consideration include liquid-oxygen-augmented air-breathing, monopropellant, electric-motor, solar-cell, microwave-beam, and hybrid-cycle engines.

Mini-Sniffer. One of the first high-altitude crewless platforms designed and tested was the NASA Mini-Sniffer, an aircraft developed in 1975 to sense turbulence and to measure atmospheric pollutants at altitudes up to 90,000 ft (27 km). The 22-ft-span (6.7-m), 200-lb (90-kg) aircraft, with swept wings, a pusher 6-ft (1.8-m) propeller, and twin tail booms, was designed to carry a 25-lb (11-kg) air-sampling payload by using a 15-hp (11-kW) hydrazine engine. This air vehicle was also envisioned as an atmospheric sampler above the Martian surface. The average density of the carbon dioxide atmosphere at the Martian surface equals that at altitudes below 100,000 ft (30 km) on Earth. The reduced gravity compared to that of Earth would allow for lower airspeeds, and power requirements would also be reduced. The configuration, unfolding during a parachute descent, would then be released and perform its cruise mission over the Martian surface, conducting aerial, magnetic, and seismic surveys.

Condor. The Boeing Condor UAV (**Fig. 1**) set an altitude record for piston-engine aircraft of 66,980 ft (20,415 m) in 1989, and remained aloft for 2.5 days during flight tests. The 20,000-lb (9000-kg) aircraft carries 60% of its gross weight in fuel and has a 68-ft-long (20.7-m) fuselage and a wingspan of 200 ft (61 m). Construction is of graphite-Kevlar and epoxy sandwich and Nomex honeycomb, producing a lightweight structure with a wing weight of 2 lb/ft^2 (10 kg/m^2). The 36.6-aspect-ratio wing produced a lift-to-drag ratio of 40, a value more commonly associated with sailplanes. The wing operated at lift coefficients up to 1.35 continuously, with 50% laminar flow over its surfaces at Reynolds numbers around 10^6. Two six-cylinder, turbocharged, 175-hp (130-kW), liquid-cooled engines driving 16-ft (5-m) three-bladed propellers powered the aircraft at cruise speeds of over 200 knots (100 m/s). Two flight-control computers flew the vehicle, programmed by 60,000 lines of FORTRAN code. The aircraft was capable of completely autonomous operation, using a strap-down inertial navigation system; Global Positioning System (GPS) use is planned in the future. A microwave landing system was used for guidance during takeoff and landing.

Perseus. The Perseus UAV (**Fig. 2**) is designed to reach an 82,000-ft (25-km) altitude. The Perseus A has a wingspan of 60 ft (18 m), a takeoff weight of 1540 lb (700 kg), and an estimated lift-to-drag ratio of 23.5. The tested power plant is a unique closed-cycle twin-rotor Wankel engine burning liquid oxygen (LOX) and gasoline, using its own exhaust as a working fluid at high altitudes. A central flight computer drives flight surfaces through conventional UAV servos, communication is duplex at 4800 baud over two separate ultrahigh frequencies, and navigation is by GPS. Instruments being developed will sample for chlorine monoxide and bromine monoxide radicals; a diode laser spectrometer will measure methane, nitrous oxide, carbon monoxide, and water vapor traces. Flux radiometers and solar spectrometers will study atmospheric radiation. High-altitude flights were scheduled for 1993, with science missions in 1994.

Stratoplane. Various other platforms are under study to meet the high-altitude environmental-science missions. The NASA Dryden Flight Research Facility has proposed the stratoplane (**Fig. 3**), a configuration with a gross weight of 11,100 lb (5035 kg), a wingspan of 180 ft (55 m), an aspect ratio of 20.9, and a turbocharged gasoline engine developing 600 hp (450 kW). The aircraft could be flown with or without a crew. The mission is to carry a 1000-lb (450-kg) payload over a 3700-nmi (6850-km) range at an altitude of 100,000 ft (30 km). The flight design point of Mach 0.65 at altitude places the wing-chord Reynolds number at 6×10^5; the calibrated airspeed is 47 knots (24 m/s) while the true airspeed will be 382 knots (196 m/s).

Because of the power losses at altitude, the propulsion system designed to fly at 100,000 ft (30 km) is five times heavier than a reciprocating engine for the same horsepower at sea level. Structurally, the stratoplane is expected to be of carbon-fiber sandwich with Nomex honeycomb core, for a wing weight of 1.2 lb/ft^2 (6 kg/m^2). The engine drives a two-bladed propeller, 30 ft (9 m) in diameter. In simulation studies, flight-path-angle and pitch-rate feedback were used to damp the neutrally stable phugoid mode (a "roller-coaster" mode with variations in altitude and airspeed), providing adequate longitudinal-axis response. Bank-angle, roll-rate, and yaw-rate feedback provided acceptable flying qualities about the lateral-directional axes, damping oscillations in the dutch-roll mode (a combination of yaw and roll or banking). Clearly, the thin atmosphere provides a challenge to the designer in aerodynamics, in propulsion, and in stability and control.

For background information SEE AIRFOIL; ATMOSPHERIC OZONE; COMPOSITE MATERIALS; DRONE; RECIPROCATING AIRCRAFT ENGINE; REYNOLDS NUMBER; STRATOSPHERE in the McGraw-Hill Encyclopedia of Science & Technology.

Richard M. Howard

Bibliography. A. Chambers and R. D. Reed, A very high altitude aircraft for global climate research, *Unmanned Sys. Mag.*, 8(3):14–19, Summer 1990; K. Munson, *World Unmanned Aircraft*, 1988; A. G. Sim, *Modeling, Simulation, and Flight Characteristics of an Aircraft Designed to Fly at 100,000 Feet*, NASA TM 104236, 1991.

Echolocation

Echolocation is the process whereby an animal perceives its local environment by the detection of the echoes of calls it has produced specifically for this purpose. Echolocation systems have evolved in several animal groups. Rudimentary systems have been established in fish (catfish), birds (cave swiftlets and oilbirds), insectivores (shrews and tenrecs), some rodents (rats), seals, and a genus of megachiropteran fruit bats (*Rousettus*). By far the most diverse and sophisticated systems have evolved in the microchiropteran bats and odontocete cetaceans. The diversity of echolocation calls, particularly among bats, suggests that there are complex functions for these calls.

Types of calls. Echolocation calls vary widely in structure and can be characterized by using several aspects of this variability. One such parameter is the temporal structure of frequency changes. Calls of animals with rudimentary systems generally comprise a broadband click encompassing many frequencies simultaneously. However, microchiropteran bats have more sophisticated calls with narrow frequency distributions at any particular time. Calls have been generally recognized as having two different types of frequency/time components: constant-frequency components where the frequency remains constant in time or changes only very slowly, and frequency-modulated components, or sweeps, where the frequency changes rapidly in time. Various species combine these components in different ways (see **illus.**).

Functions of call components. Discrimination experiments have revealed that the various components in an echolocation call have different perceptual functions. The bandwidth of a call, that is, the extent of the frequency-modulated sweep, is closely related to ability to locate targets, suggesting that sweeps

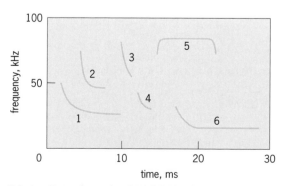

Echolocation pulses of various bat species. 1 = *Lasiurus cinerus*. 2 = *Myotis volans*. 3 = *Pipistrellus hesperus*. 4 = *Myotis thysanodes*. 5 = *Rhinolophus ferrumequinum*. 6 = *Tadarida macrotis*. (*After M. B. Fenton, Echolocation: Implications for the ecology and evolution of bats, Quart. Rev. Biol., 59:33–53, 1984*)

provide useful ranging information. This hypothesis is supported by observations that as bats of many species close in on prey the steepness of the sweep is increased, thus preventing call-echo overlap, and any constant-frequency portion of the calls is eliminated.

Constant-frequency calls are particularly susceptible to Doppler shifts as bats and their prey converge. Some bats that emit constant-frequency calls are able to change the emitted frequency to compensate for the Doppler shift. Hence, the echoes always return in a narrow frequency range. The cochleas of these bats are particularly sensitive to these frequencies. This zone of extreme auditory sensitivity, called the acoustic fovea, allows detection of insects by Doppler-shifted echoes from their fluttering wings. It has been suggested that fluttering wings produce acoustic glints in the echo pattern that can be used by bats to identify prey types. By using constant-frequency calls, bats that can compensate for Doppler shifts can perform complex prey-identification tasks.

Call structure, morphology, and niche.
Since different call components have different functions, it might be anticipated that sympatric bats exploiting different niches of the habitat would vary in the structure of the echolocation calls, while allopatric bats, exploiting similar niches, would have similar calls. A bat that generally feeds by flying very low and picking large insects from leaf surfaces might be expected to have a very different call from a bat living in the same area (sympatric) that feeds only on small insects that are out in the open air. However, a bat living in a different area (allopatric) but using a similar feeding strategy might be expected to have a similar type of call. Moreover, since different niches of the habitat are also likely to influence the flight behavior of bats and therefore aspects of their flight morphology (such as wing size and shape), particular echolocation calls are likely to be consistently associated with certain wing structures.

For a variety of bat species, principal-components analysis of dominant features of wing morphology overlaid with features of echolocation calls has confirmed this association. The first principal component of variation in wing morphology mostly reflects differences in body size. The second component reflects aspect ratio (that is, the width of the wing relative to its span). Large aspect ratios indicate long narrow wings. The third component reflects wing loading (the body mass divided by the wing area). When a large sample of bats is located in the two-dimensional space defined by the second and third components, the bats separate into three distinct groups. In one group are bats that have high values for both the wing-loading and aspect-ratio components. These species have poor turning abilities and because of their high wing loading must fly fast in order to generate lift. Typically, these bats occupy open habitats where there are few obstructions (no clutter). The echolocation calls of these bats are usually long-duration constant-frequency. These bats do not need the constant ranging information that would be provided by frequency-modulated calls to avoid clutter. Additionally, in open habitats, echoes are unlikely to return rapidly, so there is no need of a short call to avoid call–echo overlap. The calls of bats in this first group also tend to be of low frequency, enabling prey detection at long distances (as higher frequency sounds are attenuated faster). However, exact ranging and target localization with such calls is difficult, and thus the bats switch to incorporate frequency-modulated components when they detect prey.

In a second group are the relatively few bats that combine a low value of wing-loading component with a high aspect ratio. The low wing loading enables them to fly slowly in cluttered environments. The echolocation calls of these bats typically combine constant-frequency and frequency-modulated components. The frequency-modulated components reflect the necessity for ranging information in clutter. These bats include those known to use Doppler-shift compensation and acoustic glints to effect complex prey-recognition tasks.

The third group contains the vast majority of bats, with both low values for the wing-loading component, allowing slow flight, and low values for the aspect-ratio component, reflecting short broad wings with high maneuverability. These bats typically feed in dense clutter, and some glean their prey from surfaces at very short range. When these bats echolocate they use steep frequency-modulated calls, reflecting their overriding need for both ranging information in the cluttered environment and target discrimination. The calls are often of very short duration since the close range of objects means that echoes return rapidly, thus preventing overlap of call and echo. A significant proportion of the gleaning bats do not use echolocation to detect prey but rely on passive cues, such as light and sound.

Costs and benefits.
The principal benefit of echolocation is that it allows the animal to exploit the environment, either by revealing prey or allowing movement around obstacles. Although echolocation would appear to have substantial benefits, particularly to animals that are active at night, when insufficient light is available for effective vision, it has evolved in relatively few animals. There are no known echolocating amphibia or reptiles, and the vast majority of birds and mammals do not use echolocation for either prey detection or orientation, indicating that in many situations echolocation may have costs that offset its selective benefit.

Some potential costs of echolocation involve the leak of information on emitting a call. First, the potential prey may detect the echolocator before it can effect capture. Many lepidoptera, for example, can detect echolocating bats and make evasive maneuvers to avoid capture. Second, an echolocating bat may alert conspecific bats about potential sources of prey. Third, the echolocation calls may be used by predators to locate the echolocator.

A different type of cost is the energy needed to produce the call. Measurements on bats have indicated that echolocation is metabolically expensive. A resting bat producing echolocation calls at the same rate as during flight is predicted to expend energy at about 7–12 times the basal metabolic rate. The high energy costs of production may have inhibited the evolution of echolocation in many circumstances. This energy constraint might act in two different ways. First, expending energy at such a high rate for prolonged periods might be beyond an animal's capacity. Second, even if an animal could expend energy at such high rates, the benefits of having an echolocation system might not offset the high energy expenditure.

However, many insectivorous bats have evolved an echolocation system used in flight. In addition, flight itself is extremely costly in energy. Estimates for the energy costs of flight in nonecholocating birds and fruit bats indicate that it costs about 14–16 times basal metabolic rate. If the costs of echolocation are added to those of flight, the total might be as high as 21–28 times basal metabolic rate. If bats can achieve such power outputs or energy gains sufficient to offset these costs, it would seem unlikely that terrestrial animals, which have evolved echolocation systems relatively infrequently, would not have been able to achieve or offset the relatively much lower costs of echolocating when not flying.

Recently, this problem has been resolved by measuring the energy costs of flight in an echolocating bat. By injecting a bat with isotopes of oxygen and hydrogen just prior to a flight session and then examining the differential losses of the two isotopes after the flight, it has been possible to establish the energy costs of flight. The energy costs of flight for an echolocating bat are no different than the costs anticipated for a nonecholocating bat or bird. Echolocating bats achieve this remarkable economy by using one set of muscles to do three jobs. When the flying bat beats its wings, it does so by contraction of the scapular and pectoral muscles. The contraction of these muscles also places pressure on the thoracic cavity, thereby ventilating the lungs with the oxygen necessary to fuel the flight. It also produces a burst of air through the larynx, which modulates the air to form the echolocation call. This reduction of cost explains in part the proliferation of echolocation systems among flying animals and their widespread absence in many terrestrial taxa.

For background information *SEE CHIROPTERA; DOPPLER EFFECT; ECHOLOCATION; SONAR* in the McGraw-Hill Encyclopedia of Science & Technology.

John R. Speakman

Bibliography. M. B. Fenton, Echolocation: Implications for the ecology and evolution of bats, *Quart. Rev. Biol.*, 59:33–53, 1984; U. M. Norberg, *Vertebrate Flight*, 1989; U. M. Norberg and J. M. V. Rayner, Ecological morphology and flight in bats (Mammalia: Chiroptera): Wing adaptations, flight performance, foraging strategy and echolocation, *Philos. Trans. Roy. Soc.*, series B, 316:335–429, 1987; J. R. Speakman and P. A. Racey, No cost of echolocation to bats during flight, *Nature*, 350:421–423, 1991.

Electric distribution systems

Electric power systems are composed of three major components: generation, transmission, and distribution. The electric distribution system (**Fig. 1**) receives power from remote power generation plants through the high-voltage transmission network, and is used to distribute power from the transmission substations to individual customers. It consists mainly of radial feeders, which extend from the substation transformers that convert power to distribution voltage levels to the customer connections. Feeders are constructed in consistent sections of electrically connected equipment that serve customers in a defined geographical area. Adjacent feeders are typically connected by normally open switching devices at tie points to provide alternate service paths for feeder sections as the need occurs. Feeder sections are connected by switching devices at sectionalizing points, which may be opened and closed in conjunction with tie points to alter the feeder configurations. Feeder sections are further segmented by fuses to protect against loss of the entire section in the event of electrical problems. Operation of the distribution system is generally centered on the reconfiguration of feeders to support, for example, isolation of electrical problems, scheduled maintenance, restoration of customer service, and management of equipment loading. Currently, most feeder reconfiguration is performed manually by field crews coordinated through distribution operators. Similarly, distribution operators coordinate many tasks manually by using paper maps, schematic diagrams, books containing facilities information, and wall boards, all representing feeder configurations at the time they were built, with no information describing the current conditions of the feeders until a field crew arrives on site. The application of greater levels of automation to aid distribution operators in more effectively and safely managing the distribution system in both normal and

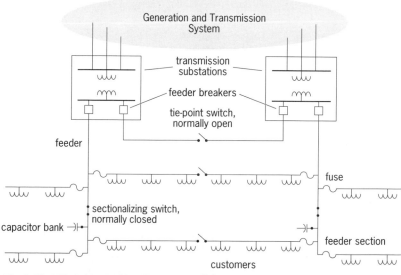

Fig. 1. Simplified electrical distribution system.

emergency situations is emerging as a central issue in meeting the existing and future business needs of electrical utilities.

Forces motivating automation. To account for the significant time required for manual operations activities, the distribution system has generally been designed and operated assuming peak loading conditions. The result has been high utility construction costs with little ability to take advantage of the designed capacity of installed equipment during normal operations. In the past few years, however, growing utility competition, increasing fuel and capital costs, and additional pressure from government rates commissions to improve utilization of existing facilities and reduce customer outage times have emerged as strong forces motivating utilities to establish more efficient operations. These factors, coupled with the recent introduction of more economic computer technologies and solution approaches supporting operations information management, appear to be creating the unique alignment of need, technology, and economy required to make automation of distribution operations a key focus for utilities and vendors in the 1990s.

Forms of distribution automation. Distribution automation can be broadly defined as the implementation of technology to provide improved reliability of service, quality of power delivered, utilization of feeder facilities, and coordination, management, and reporting of distribution operations activities. The main areas of focus are field equipment, substations and feeders, and distribution operations centers.

Field equipment automation. Automation of field equipment corresponds to the implementation of automatic control devices along the extent of a feeder to provide for the protection and regulation of feeder facilities and customer voltage levels. Feeder recloser devices, for example, are frequently used for local protection and restoration. They sense facility overload conditions and automatically open the electrical circuit and reclose after a short period, attempting to clear problems and restore service under electrical fault situations. Other devices, such as voltage regulators and capacitor banks, are used to regulate the feeder voltage within defined limits to help keep electrical energy losses within designed ranges. These high-speed devices operate on individual feeders without human intervention, generally acting independently of other automated equipment on the feeder. Most utilities now have a significant number of these devices deployed on distribution feeders, providing clear benefits in service reliability. Problems that require coordination of multiple devices or feeder reconfiguration are beyond the capabilities of present automated field equipment, generally requiring manual intervention by the distribution operators and field crews.

Substation and feeder automation. This form of automation is focused on the installation of voltage, current, and power monitoring equipment and remote-control switching devices on substation low-voltage equipment, at selected points along the extent of the feeders and, in some cases, at grouped or individual customer load points. Supervisory control and data ac-

quisition (SCADA) systems allow the operator to monitor the present conditions of selected feeder points, manage equipment loading, and reduce coordination time for selected switching activities through these devices. For example, operator monitoring of feeder voltage and remote control of capacitor banks offer capabilities that can result in substantial savings by reducing electrical losses. Installation of monitoring and control equipment is, however, quite expensive, and this expense limits the number of devices feasible for installation and spreads implementation over multiple-year budget cycles. The majority of switches and important points along the feeders will remain unmetered and under manual control through the field crews for some years to come. The access and coordination of information regarding the present configuration of unmetered feeder sections and the correlation with real-time SCADA information is a key aspect of further reducing the time required for operator decision making in managing restoration of service and directing field crews. *SEE PETROLEUM ENGINEERING.*

Distribution information management systems. Comprehensive systems to support automation of distribution operations centers are just beginning to emerge. Generally referred to as distribution information management systems, they establish an information model based on the electrical connectivity of the distribution system feeder equipment. The model is kept up to date with all switching actions performed by the operator through SCADA systems and field crews. Thus, a common information base is available to all distribution operators, and may be coordinated between regional distribution centers. It serves as a base for operations decisions, switching plan development, energization analysis, and reporting in historical, present, and future time frames. These systems offer the capability to integrate many forms of information from related systems (**Fig. 2**)—for example, information on electrical connectivity, derived directly from automated mapping/facilities management (AM/FM) or computer-aided design (CAD) systems, which support the creation and management of electronic feeder maps with corresponding equipment attributes; the graphical display on feeder maps of service problem areas, taken from customer trouble call management systems, which determine the probable cause of customer outages; the graphical display of SCADA metering and selected alarms; and feeder load parameters from customer billing systems. Although not fully proven in practice, these systems fill a void in the automation of distribution systems and begin to more completely address significant benefits, including reduced system restoration times, improved field crew management and safety, better facilities utilization, detailed customer outage recording, and support for condition-based maintenance of facilities.

Implementation and future directions. In general, utility industry trends indicate that distribution system automation will escalate dramatically in the mid-1990s. The implementation of each form of automation will proceed in parallel stages, prioritized according to the individual utility's most press-

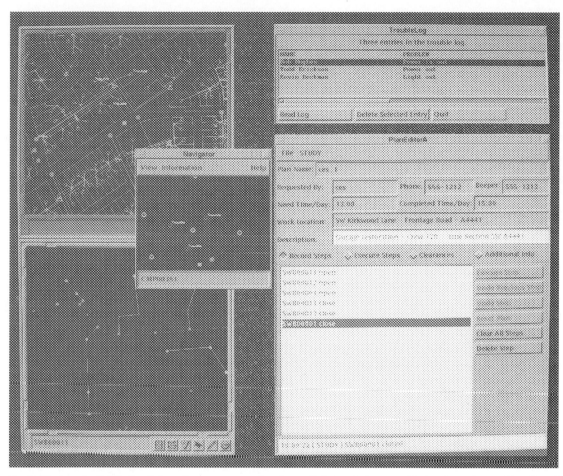

Fig. 2. Display of a distribution information management system, with integrated geographical, schematic, trouble call, and switching plan information.

ing needs and cost effectiveness for the particular situation. To achieve the greatest benefits, the implementations of the various forms of automation need to be directed toward their coordinated use within an integrated distribution automation strategy. In this way, the utility can balance the benefits of the available automation approaches while broadening operations and future implementation options—for example, making use of the improved availability of unmetered feeder configuration information to reduce or delay the required installation of monitor and control equipment. Present utility trends toward implementation of standard company-wide communications networks will further enhance distribution automation options by allowing more timely access to information to support local decision making. Consequently, for example, more intelligent automated field equipment can be designed for high-speed local control and field crew selection of the closest priority customer service outage to reduce response time in emergency restoration situations. Overall, distribution automation will continue to evolve to become an integrated component of a coordinated utility enterprise.

For background information SEE ELECTRIC DISTRIBU-TION SYSTEMS; ELECTRIC POWER SYSTEMS in the McGraw-Hill Encyclopedia of Science & Technology.

Kenneth I. Geisler

Bibliography. K. I. Geisler et al., Model and product based integrated systems for utility operations, *IEEE Computer Applicat. Power*, 5(3):15–20, 1992; IEEE Working Group on Distribution Automation, *IEEE Tutorial on Distribution Automation*, 1988; T. J. Kendrew and J. A. Marks, Automated distribution comes of age, *IEEE Computer Applicat. Power*, 2(1):7–10, January 1989.

Electrical insulation

The efficiency of power transfer from generating sites to load centers increases with voltage, but power is utilized at lower voltages. The transfer of power is most economically performed by a network of transmission (69 kV and higher) lines and distribution (34 kV and lower) lines. Insulators that are used in these lines to mechanically support the conductor and provide the electrical insulation between the conductor and the support structure have traditionally been made from inorganic porcelain and glass. However, since 1967 the use of polymeric materials for outdoor insulation has increased rapidly.

Advantages of polymers. The use of polymers provides several advantages over porcelain and glass, namely, easy installation, light weight, good vandal resistance, novel design concepts, lower over-all cost, and generally better performance under con-taminated conditions. The superior contamination per-formance is due to the lower surface energy of the polymer, which makes the surface water-repellent, as opposed to the hydrophilic or wettable surfaces of porcelain and glass, which have high surface energy. Polymers are used as weathersheds around a fiber-glass rod connected to end fittings, which provide the required mechanical strength (see **illus.**). The weath-ershed is the protective covering on the fiberglass rod that provides the required electrical strength un-der all conditions (wet and dry). The fiberglass rods have better tensile and bending strength than porce-lain and glass. The glass fibers in the rod are bonded with either an epoxy or polyester resin. The weight reduction of a polymer insulator compared with an equivalent porcelain or glass insulator is very signif-icant; therefore, smaller and more compact structures can be used for support. Where improvement in the contamination performance of existing porcelain insu-lators is desired and polymeric replacements are not available or feasible, a long-term solution is to use room-temperature-vulcanized (RTV) silicone rubber as a thin coating (about 10 mils or 0.5 mm thick) over the porcelain.

At present, polymers are used in the majority of cable termination equipment (up to 25-kV class) man-ufactured in the United States, and their use as insu-lators is steadily increasing, although the majority of insulators manufactured are still porcelain and glass. Worldwide service experience of polymeric insulation has been generally favorable.

Aging. A major problem encountered in service with polymeric insulation is aging, which results in a gradual loss of insulating properties and can lead to insulation failure. Aging is the result of perma-nent material changes. The dominant cause of aging is the long-term dry-band arcing, which lasts for a few seconds and repeatedly recurs in the same location. Dry-band arcing is a result of current promoted on the surface of an energized insulator when the surface is moist and contaminated. This current, known as leak-age current, results in ohmic heating of the moisture layer. Because of the nonuniform geometry of the insu-lator weathershed (composed of several areas of large and small cross sections) and nonuniform deposition of contamination, certain parts of the insulator surface dry up faster than others, resulting in the formation of dry regions or bands and consequent nonuniform distribu-tion of voltage across the insulator. The electric stress or the gradient of the voltage across the dry bands can exceed the withstand voltage, resulting in arcs that

jump across the dry band. This phenomenon is known as dry-band arcing, which can result in tracking (car-bonization) and erosion of the material or a flashover across the insulator. Other elements that contribute to aging are ultraviolet radiation from sunlight, moisture, heat, and chemicals.

Composition of polymer insulators. The polymer family is very large. Only a small number of material families are suitable for outdoor insula-tion. Some materials that have been used success-fully for outdoor insulation are silicone rubber (high-temperature-vulcanized), ethylene propylene rubber (EPR), cycloaliphatic epoxy, ethyl vinyl acetate (EVA), modified polyolefins, and polytetrafluoroethylene (PTFE). A polymer is seldom used by itself. Several ingredients are added to the polymer to improve its mechanical, electrical, thermal, and chemical proper-ties for the intended application. For example, inor-ganic fillers such as alumina trihydrate or silica are added to improve the tracking and erosion resistance, zinc oxide or titanium oxide is added to improve the resistance to ultraviolet radiation, and coloring agents, plasticizers, and so forth are added to the base poly-mer. Since the final compound is a composite of many ingredients, large differences in the electrical perfor-mance can exist between different polymer families and also between different compositions in the same material family. Such variation has been well docu-mented both in the field and in the laboratory.

Polymer hydrophobicity. As mentioned above, polymers, unlike ceramics, are low-surface-energy materials, so that water on the surface tends to bead into individual droplets rather than form a con-tinuous film. This property, surface hydrophobicity, is responsible for the superior electrical performance of polymers under contaminated conditions, especially in the initial stage of the insulator life. The surface hy-drophobicity is temporarily or permanently lost during dry-band arcing but can be restored when there is no dry-band arcing. The ability to maintain a hydrophobic surface during dry-band arcing and restore it quickly after it is lost is different for each polymer family and composition. It has been observed that this abil-ity is greater in silicone rubber or silicone-polymer-containing compounds than in nonsilicone compounds. The hydrophobicity recovery is due to the diffusion of low-molecular-weight polymer chains from the bulk to the surface. The diffusion occurs even if the sur-face is contaminated. It is logical to assume that a very thick layer of contamination or material aging can significantly retard or stop the hydrophobicity recovery.

Recent research has shown that even if the poly-mer surface has lost its hydrophobicity the flashover voltage can still be higher than that of porcelain. This capability is attributed to the interaction between the low-molecular-weight polymer chains and the contam-ination, which ensures that the effective water film, which is responsible for the leakage current, is thinner than that formed on a contaminated porcelain surface.

Insulator design. The ultimate performance of the insulator is determined by the weathershed mate-

fiberglass rod   weathershed

end fitting

Cross section of a polymer insulator.

rial, the fiberglass rod, and the design. The interface between the weathershed and the fiberglass rod, and the attachment of the end fitting to the rod, are crucial to insulator performance. Although the hydrophobic property of the weathershed material in the initial stage of the insulator life can be exploited to reduce the leakage distance of polymer insulators, such a reduction can lead to higher leakage current and potential insulator degradation with time. Therefore, present utility practice is to specify that the leakage distance of polymers be the same as for porcelain or glass for the same voltage. (The leakage distance is the total insulating distance on the surface of the insulator between the two metal end fittings.)

Testing and research. Determining the useful life of polymeric insulation is of great importance. Existing standard tests have been developed for porcelain and glass where aging is not a consideration, and they are designed to evaluate the flashover voltage under various conditions in order to determine the amount of insulation. Because of significant differences between ceramics (porcelain and glass) and nonceramics (polymers), new test procedures are needed to evaluate polymers. Comprehensive knowledge of the aging mechanisms is a prerequisite to the development of better materials and test procedures, and this area needs more research.

The mobility of the polymer chains makes the material dynamic, in contrast to the inert nature of porcelain and glass. This dynamic character poses problems in aging research. It is important that the material properties not be affected by sample preparation or by the tests. It is also important that the test procedure be representative of field conditions. For example, to evaluate the flashover voltage by the so-called clean-fog test, the insulator is precontaminated. With silicone rubber insulators, the contamination layer will not stick to the insulator surface unless the surface is chemically altered temporarily. This alteration in itself is an artifact of the test procedure. Moreover, accelerated aging tests require that the insulator be exposed to a rather severe contamination for many hours at a time. Although there may be instances where the field contamination approaches the contamination levels used in these tests, the dry-band activity to which the insulators are continuously subjected represents a much more severe condition than is normally encountered in service. As a result, the degradation mechanisms in these tests could be different from those in service.

Research in polymer aging has benefited significantly from the use of analytical techniques such as infrared spectroscopy for chemical analysis (ESCA), energy-dispersive x-ray analysis (EDX), x-ray diffraction, gel permeation chromatography, and secondary ion mass spectroscopy (SIMS). Yet, the aging mechanisms are not fully understood.

For background information SEE ANALYTICAL CHEMISTRY; DIELECTRIC MATERIALS; ELECTRICAL INSULATION; POLYMER; RUBBER; SILICONE RESINS; SURFACE AND INTERFACIAL CHEMISTRY in the McGraw-Hill Encyclopedia of Science & Technology.

Ravi S. Gorur

Bibliography. R. S. Gorur, High voltage outdoor insulation technology, in C. Leondes (ed.), *Control and Dynamic Systems*, 44(4):131–191, 1991; R. S. Gorur and T. Orbeck, Surface dielectric behavior of polymeric outdoor insulation under HV outdoor conditions, *IEEE Trans. Elec. Insulat.*, 26:1064–1072, 1991.

Electrical utility industry

In 1992, significant changes occurred in the electrical utility industry. The U.S. Congress passed the most comprehensive energy legislation in over a decade. The Environmental Protection Agency (EPA) announced final acid-rain regulations and proposed regulations to control nitrogen oxides. Electrical utility sulfur dioxide emission allowance trading got off to a difficult start amid charges that these allowances permitted pollution. Several major electrical utility mergers, a recent phenomenon in the industry, were completed, and more were proposed. The public's concerns were raised about electric and magnetic fields associated with transmission and distribution facilities with the release of a Swedish study. Finally, Hurricane Andrew did extensive damage to southern Florida and Louisiana.

Energy Policy Act of 1992. After 2 years of debate, passage of the Energy Policy Act eliminated the obstacles to wholesale electricity competition in the Public Utility Holding Company Act of 1935. Both utilities and nonutilities can form exempt wholesale generators, independent power-producing companies that can generate and sell electricity to electrical utilities without triggering the restrictions of the 1935 act. The legislation includes a strong mandate for the Federal Energy Regulatory Commission (FERC) to order use of the transmission system for wholesale electricity sales, but not for retail sales, on a case-by-case basis. Wholesale electricity sales are between an independent power producer or electrical utility and another electrical utility. Retail electricity sales are to the ultimate (retail) customer. The legislation also includes pricing provisions meant to protect the owners of transmission capacity.

A major concern of electrical utilities during the congressional debates on the Energy Policy Act was protection of the reliability of the electric systems if access to the transmission systems was mandated. This concern was mitigated by inclusion of a section on reliability in the energy act that states: "No order may be issued under this section or section 210 [of the Energy Policy Act of 1992] if, after giving consideration to consistently applied regional or national reliability standards, guidelines, or criteria, the [Federal Energy Regulatory] Commission finds that such order would unreasonably impair the continued reliability of electricity systems affected by the order."

The legislation, however, does not authorize regional transmission groups, a subject that was vigorously debated. Although consensus language to form regional transmission groups was agreed upon by most industry and consumer groups, agreement was reached too late to be included in the legislation. Membership in a regional transmission group would be open to all utilities and independent power producers in a given geographic area. The group would provide for coordinated planning and sharing of information to ensure that all known transmission needs are met in an efficient way. In December, the FERC issued a notice of proposed rule making on regional transmission groups.

Sulfur dioxide rules. The EPA announced a package of final rules to cut in half annual emissions of sulfur dioxide (SO_2) from electric power plants, a 1×10^6 ton reduction of 1980 levels by 2010. The package set a permanent national cap on electrical utility sulfur dioxide emissions at just under 9×10^6 tons annually and permits trading of emission allowances. The sulfur dioxide reduction will be accomplished in two phases. Phase I begins in 1995 and affects 110 of the largest power plants at mostly coal-burning utilities in 21 eastern and midwestern states. Phase II begins in 2000. The regulations exempt utility units of 25 megawatts and under from most requirements.

To achieve the sulfur dioxide emission reductions, the EPA will use a market-based allowance trading system. This approach is the centerpiece of the Clean Air Act Amendments of 1990. An allowance gives a utility the authority to emit 1 ton of sulfur dioxide during a given year. The EPA allocates allowances to existing utilities based on a formula using a specified emission rate multiplied by the utility's average fuel consumption during the baseline period of 1985–1987. When the rule is fully implemented in 2010, the aver-age utility will receive allowance allocations for only half the emissions it generated in 1980.

The first sulfur dioxide emission allowance trading involved the sale by Wisconsin Power & Light Company of 10,000 allowances to the Tennessee Valley Authority and 15,000–25,000 to Duquesne Light Company. Although this trade was seen as a positive step by electrical utilities, it was criticized by environmentalists in Wisconsin and Tennessee. Both trades were described in major news accounts as a sale and purchase of pollution rights, allowing the buyer to pollute more. No other public announcements of allowance trades were reported in 1992.

Proposed nitrogen oxide rules. The EPA also announced proposed rules to cut annual emissions of nitrogen oxide (NO_x) from electrical utility power plants by $1.5–2 \times 10^6$ tons annually. This proposal was the first EPA nitrogen oxide reduction program for existing stationary sources of air pollution. Previous control programs were directed at motor vehicles and new industrial sources. Electrical utility boilers using coal, natural gas, and oil account for nearly 60% of the stationary source emissions. As in the sulfur dioxide regulation, the nitrogen oxide regulations exempt new utility units with a nameplate capacity (that is, capacity assigned by the manufacturer at the time of installation) of 25 MW and under from most requirements.

The proposed Phase I standards for nitrogen oxide apply to about 200 coal-fired electrical utility boilers, both tangentially fired and dry-bottom. The proposed standards must be met by January 1995. Applicable emission rates for all other types of coal-fired boilers,

Table 1. United States electric power industry for 1991*

Parameter	Amount	Change from 1990, %
Generating capacity, MW		
Hydroelectric	88,385 (12.0%)	0.97
Fossil-fueled steam	478,435 (64.9%)	0.37
Nuclear steam	110,217 (15.0%)	0.01
Combustion turbine, internal combustion	59,713 (8.1%)	−1.22
TOTAL[†]	736,750	0.25
Transmission, circuit miles		
Alternating current, 230–765 kV	147,107 (236,746 km)	1.56
Direct current, ±250–500 kV	2,426 (3,904 km)	0.00
Noncoincidental demand, [‡] MW	551,320	1.06
Energy production, TWh[§]	2,823.0	0.53
Energy sales, TWh		
Residential	948.7	3.59
Commercial	754.9	2.17
Industrial	929.2	−0.29
Miscellaneous	99.7	2.36
TOTAL	2,732.4	1.80
Revenues, total, 10^9 dollars	183.0	3.68
Capital expenditures, total, 10^6 dollars[¶]	23,370	−3.38
Customers, 10^3		
Residential	97,934	0.96
TOTAL	111,147	0.95
Residential usage, (kWh/customer)/year	9.738	2.42
Residential bill, cents/kWh (average)	7.98	1.92

* After 1992 annual statistical report, *Elec. World*, 206(5):7–12, 1992; North American Electric Reliability Council, *Electricity Supply and Demand 1992–2001*, 1992; and Edison Electric Institute, *Statistical Yearbook of the Electric Utility Industry*, 1991.

† Does not include nonutility capacity available to electrical utilities.

‡ Noncoincident demand is the sum of the peak demands of all individual electrical utilities, regardless of the day and time at which they occurred.

§ 1TWh (terawatt-hour) = 10^{12} watt-hours.

¶ Investor-owned electrical utilities only. Does not include cooperative or public power utilities, estimated at $3.8 billion.

such as cyclone and cell burners, are to be set by January 1, 1997.

The proposed nitrogen oxide regulations allow utilities the flexibility and cost-saving technique of source-emission averaging. They also allow the EPA to grant less stringent emission limits when a plant demonstrates that it cannot meet the proposed standards by using low–nitrogen oxide burner technology. The EPA asked for comment on two options for nitrogen oxide control, use of overfired air on only tangentially fired boilers and use of overfired air on both tangentially fired and wall-fired boilers. As a result, utilities would not know until the summer of 1993 what kind of controls they would have to install. The EPA was 6 months late in issuing its nitrogen oxide regulations, and more delay will result from the request for comments on the use of overfired air. A utility can obtain a compliance deadline extension of 15 months if it can prove that the low–nitrogen oxide burner technology is not available in adequate supply to meet the deadline.

Division by plant and utility type. The division of generating capacity by plant type in 1991, along with other United States electrical utility industry statistics, is given in **Table 1**. The historical breakdown and a forecast of these figures are shown in the **illustration**. Capacity additions in 1992 and a comparison with 1991 capacity are given in **Table 2**.

The electrical utility industry in the United States is pluralistic, divided among investor-owned utilities, cooperatives, municipal utilities, federal agencies such as the Tennessee Valley Authority and the Bonneville Power Administration, and state or public power districts. The division of capacity among these various entities at the end of 1992 was as follows: investor-owned, 573,072 MW (77.4% of the total); cooperatives, 26,453 MW (3.6%); federal, 65,489 MW (8.8%); municipals, 41,126 MW (5.5%); and state and public districts, 34,417 MW (4.7%).

Mergers. Until recently, mergers between electrical utilites were rare. However, increased competition (which was encouraged by federal and state regulators), transmission access demands by distribution utilities and independent power producers (which do not own any transmission facilities), and financial considerations significantly increased merger frequency. Major electrical utility mergers completed in 1992 included Public Service Company of New Hampshire

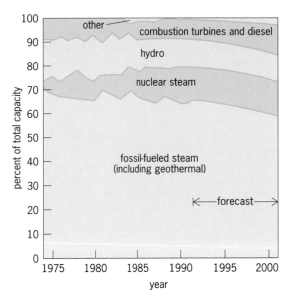

Probable mix of net generating capacity. "Other" includes solar, wind, and other nonconventional generating technologies. (*After North American Electric Reliability Council, Electricity Supply and Demand 1992–2001, 1992*)

into Northeast Utilities; Iowa Public Service Company and Iowa Power, Inc., to form Midwest Power, Inc.; and Kansas Power & Light Company and Kansas Gas & Electric Company to form Western Resources, Inc.

Major mergers proposed in 1992 included Energy Corp's acquisition of Gulf States Utilities Company, which would create the second largest electrical utility in assets in the United States; Cincinnati Gas & Electric Company merger with PSI Energy to form a holding company called CINergy; and the acquisition of El Paso Electric Company by Southwestern Public Service Company.

Electric and magnetic fields. The issue of electric and magnetic fields has captured considerable electrical utility, regulator, and public interest. Most regulators have been taking a cautious approach to electric and magnetic fields as they await the results of additional research.

Early in 1992, the Wisconsin Public Service Commission rejected a proposed 3-year moratorium on new transmission lines, concluding that current research does not show that electric and magnetic fields endanger human health. Later that year, scientists at

Table 2. Generating capacity added in 1992, and comparison of capacity in 1992 and 1991*

	Capacity added in 1992		Capacity, MW		
Plant type	Power, MW	Number of units	1991	1992	1991 total capacity, %
Fossil-fueled steam	590	3	478,435	479,025	64.8
Nuclear steam	0	0	110,217	110,217	14.9
Combustion turbines	1,475	15	55,106	56,581	7.7
Hydroelectric	232	9	88,385	88,617	12.0
Diesel engine	1	1	4,607	4,608	0.6
TOTAL	2,298	28	736,750	739,048	

*Not adjusted for retirements.

three Colorado universities (Universities Consortium on Electromagnetic Fields), after reviewing almost 12,000 published papers, concluded that there is no persuasive evidence directly linking electric and magnetic fields to effects on human health.

The Colorado Supreme Court reversed a lower state court decision preventing Public Service Company of Colorado from upgrading transmission lines because of concerns over electric and magnetic fields and other issues. The transmission lines had been approved by the state's Public Utilities Commission. Near the end of 1992, utilities in Colorado were ordered by the Colorado Public Utilities Commission to minimize exposure to electric and magnetic fields by pursuing prudent avoidance in planning and constructing new transmission lines. Prudent avoidance means striking a balance between the potential health effects of exposure to electric and magnetic fields and the cost and impact of mitigating such exposure.

Litigation over electric and magnetic fields and an administrative law judge's recommendation delayed energizing a 13-mi (21-km), 230-kV transmission line in the service area of Philadelphia Electric Company. The line had been authorized by the state's Public Utility Commission and had been constructed to existing standards. The state administrative law judge, in response to protests, recommended delay in energizing the line until the Public Utility Commission decides whether statewide standards for exposure to electric and magnetic fields should be set. If the Public Utility Commission rules the standards are needed, it is unlikely that the line will be energized before the end of 1993. It was ready for service in June 1992.

Southern California Edison Company has developed computer software to design transmission and distribution lines with reduced electric and magnetic fields. The software, which employs three-dimensional drawings to also design electrical circuits in buildings, substations, and power plants, is being used in the United States and 14 other countries.

Hurricane Andrew. Hurricane Andrew smashed into south Florida with winds of over 200 mi/h (90 m/s) on August 24, 1992. The eye of the storm passed just south of Miami, leaving a trail of devastation in that part of the service area of Florida Power & Light Company (FP&L) and destroying the electricity system of the City of Homestead. It proceeded into the Gulf of Mexico and crossed back onto land in Louisiana west of New Orleans with winds of 140 mi/h (63 m/s) on August 26.

In the service area of FP&L, 1.4×10^6 customers were without electric service after the storm swept through. The storm caused the loss of 472 circuit miles (760 circuit kilometers) of transmission, deenergizing 61 transmission lines serving 110 substations. Forty-three transmission lines serving 58 substations sustained permanent damage. All eight lines from FP&L's Turkey Point plant were damaged, resulting in the loss of off-site power for 4 days. FP&L's damage estimate also includes 8.5×10^6 ft (2.6×10^6 m) of distribution conductor, 18,700 wooden poles, and 16,800 switches.

The Turkey Point site also includes two fossil units

and totals over 2000 MW, 12.6% of FP&L's installed capacity. Unit No. 1 returned to service in March 1993 because of severe damage requiring the stack to be demolished and rebuilt. Unit No. 2 returned to service about 3 weeks after the storm hit. Unit No. 3, a nuclear unit, was being removed from service for refueling at the time of the storm. It was returned to service in December 1992. Unit No. 4 incurred little damage and returned to service late in October 1992. Two other fossil units (Cutler Nos. 5 and 6) in south Florida were damaged and were not returned to service for several weeks.

By the end of September, the City of Homestead electric system demand was still 50% or less of its prehurricane level. In the first few days after the storm, the utility's generating capacity was less than 5% of its previous level.

Other Florida utilities with severe damage from the storm were Florida Keys Electric Cooperative Association, Inc., the Key West Utility Board, and Lee County Electric Cooperative, Inc.

The storm affected six electric systems in Louisiana. Fortunately, most of the 650,000 customers who lost electric service had service restored in 8–16 h; the vast majority had service within 24 h.

In both Florida and Louisiana, utilities and contractors from unaffected areas as far away as North Carolina assisted in the service restoration effort.

Nuclear plant closings. Utilities are pursuing nuclear plant license extension options with the Nuclear Regulatory Commission as well as life-extension options for nuclear plants nearing their 40-year rated life. However, the owners of three nuclear power plants decided to close them before the plants' licenses expired. These decisions are seen as possible precursors of actions by other utility owners of nuclear plants with troubled histories.

Yankee Atomic Electric Company announced that it was closing permanently its 185-MW Rowe, Massachusetts, nuclear plant, citing the lower cost of alternative power in New England as not justifying the expense of modifying and restarting the 31-year-old plant. The plant had been shut down in October 1991 amid charges of reactor container embrittlement.

The 34-year-old, 436-MW San Onofre No. 1 nuclear unit, owned by Southern California Edison Company and San Diego Gas & Electric Company, was shut down late in 1992 as a result of a settlement between the plant owners and the California Public Utilities Commission Division of Rate Payers (CaPUC). The CaPUC order said continued operation of the reactor would be an unreasonable burden on ratepayers because major modifications would be needed by the end of the current fuel cycle in 1992.

Portland General Electric Company decided to close its 16-year-old 1100-MW Trojan nuclear power plant at the end of 1992 instead of waiting until 1996, as originally proposed. Continued operation until 2011 would require spending over $200 million to replace the plant's four steam generators.

Niagara Mohawk Power Corporation (NMPC) decided to study the decommissioning of its troubled

23-year-old, 610-MW Nine Mile No. 1 nuclear plant. NMPC said its economic analysis will study an orderly planned retirement of the unit if the plant has a negative value in 1995.

FCC proposed frequency reallocation. A major controversy arose when the Federal Communications Commission (FCC) proposed reallocating the 1850–2200-MHz (2-GHz) microwave frequency band now used by electrical utilities. The FCC proposed that these frequencies be used by emerging personal communications technologies such as wireless telefacsimiles, telephones, and laptop computers.

Electrical utilities use the 2-GHz frequency band for signaling and data transfer of information about their generation and transmission systems. The utilities maintain that these frequencies are vital to the reliability of the electric systems in the United States. In written testimony to the Senate Committee on Commerce, Science and Technology, the Edison Electric Institute said that the collective investment of the electrical utility industry in point-to-point microwave facilities alone is nearly $800 million. The industry pointed out that communications over the higher frequencies that the FCC proposed for electrical utility use often are susceptible to disruption by adverse weather conditions.

The final FCC order stopped far short of its initial proposal. It adopted a transition framework intended to minimize disruption to the 2-GHz licenses. The FCC agreed not to downgrade to secondary status existing microwave licenses. Most importantly, the FCC will require emerging technology applicants seeking the 2-GHz frequencies held by utilities and others to negotiate with the incumbents during an as yet unspecified transition period, locate alternative frequencies for incumbents that provide reliability at least equal to that of their existing wave bands, and pay all costs for relocating to those frequencies. In addition, public utilities have a special dispensation exempting them from involuntary relocation from existing frequencies. A third Notice of Proposed Rulemaking issued in the fall deals with the timing of the voluntary negotiation period and other aspects of the transition proposal.

The FCC rule mirrors legislation that was sponsored and adopted in the Senate as part of larger appropriations bill. The FCC feared adoption of the Senate-sponsored legislation would set a precedent of congressional management of communications policy. SEE RADIO SPECTRUM ALLOCATIONS; TELEPHONE SERVICE.

Renewable energy. Utilities showed significant interest in renewable energy resources in 1992. Pacific Gas & Electric Company authorized design and construction of a photovoltaic unit to be installed at the end of a distribution feeder. The 500-kW photovoltaic unit will supply a portion of the electricity provided via the feeder, support voltage, and increase the reliability of the feeder.

The use of wind power to generate electricity is now moving outside of California on a much wider scale. Northern States Power Company (NSP) announced a wind-power production goal of 110 MW, with initial operation of the units projected as early as summer

1993. This increase is 100 MW over the 10-MW proposal that NSP currently has before the Minnesota Public Utilities Commission. The utility would obtain the units through competitive bidding. The plants would be turnkey plants owned by the utility or by independent power producers.

In related proposals, four utilities—Portland General Electric Company, PacifiCorp, Idaho Power Company, and Puget Sound Power & Light Company—agreed to build a 50-MW wind farm at Rattlesnake Hills, near Richland in eastern Washington. Construction is scheduled for 1994 with operation in 1996. Bonneville Power Administration solicited 50 MW of wind-power generation and received six proposals totaling 60.5 MW. The 50-MW solicitation is expected to produce, on average, about 38 MW of generating capacity.

New technology. The growing demand for electricity and the increasing difficulty of constructing new facilities require utilities to rely more heavily on their transmission systems. Existing transmission control equipment relies primarily on electromechanical devices, which are relatively slow and require frequent maintenance. Two new electronic control systems to improve the efficiency of transmission systems were put into operation in 1992.

A thyristor-controlled series capacitor bank began operating on a 345-kV line owned by Appalachian Power Company, a subsidiary of American Electric Power Company. This equipment is part of a $20 million facility and related system improvements that can increase the capacity of the 109-mi (175-km) Kanawha River-Funk transmission line. The increased transmission capacity allows the line to handle the higher current flows that occur during a forced outage on a parallel 765-kV transmission line.

The Western Area Power Administration (WAPA) commissioned the first transmission system with variable series compensation. The project, at WAPA's 230-kV Kayenta substation in northeastern Arizona between Shiprock, New Mexico, and Glen Canyon Dam in Arizona, consists of two banks of series capacitors, each rated at 165 Mvar with a single-phase impedance of 55 ohms. One bank is configured for conventional series compensation. The second bank is split: one segment is fixed, the other is adjustable. The equipment permits rapid, continuous compensation control, resulting in quick, continuous, and direct control of transmission-line power flows.

San Diego Gas & Electric Company began a 2-year test of a 200-kW battery storage unit to demonstrate off-peak storage of energy to meet on-peak energy needs. The battery will supply electricity to San Diego Trolley, Inc. The San Diego Metropolitan Transportation Board, a project participant, has the option of purchasing the equipment at the end of the test.

For background information SEE AIR POLLUTION; ELECTRIC POWER GENERATION; ELECTRIC POWER SYSTEMS; ENERGY SOURCES; ENERGY STORAGE; SOLAR ENERGY; STATIC VAR COMPENSATOR; WIND POWER in the McGraw-Hill Encyclopedia of Science & Technology.

Eugene F. Gorzelnik

Bibliography. Edison Electric Institute, *Statistical Yearbook of the Electrical Utility Industry*, 1991; 1992 annual statistical report, *Elec. World*, 206(5):7–12, 1992; North American Electric Reliability Council, *Electricity Supply and Demand 1992–2001*, 1992; North American Electric Reliability Council, *Reliability Assessment 1992–2001*, 1992.

Electrochromic devices

Electrochromic devices are self-contained, hermetically sealed, two-electrode electrolytic cells that include one or more electrochromic materials and an electrolyte. Electrochromic materials are either organic or inorganic substances able to interconvert between two or more colored states upon oxidation or reduction, that is, upon electrolytic loss or gain of electrons. Desirably, the color states available to an electrochromic material include a form where the material is colorless. Consequently, a small bias (1–2 V) across the two electrodes of an electrochromic device induces controlled electrolysis of the electrochromic materials; this electrolysis is perceived by an observer as a change of the transmittance (or reflectance) of the entire device assembly. Electrochromic devices have found a commercial use in antiglare rearview mirrors for automobiles. These so-called smart mirrors include sensors that measure the intensity of the light striking the electrochromic mirror and adjust proportionally the supply of electrical charge to the electrolytic cell. Another potential application of electrochromic devices receiving much attention recently is smart windows. In analogy to smart mirrors, these windows get darker as the level of the ambient light increases. It is expected that electrochromic smart windows in buildings will result in substantial energy savings for interior space heating and air conditioning. Other applications of electrochromic devices under development include light filters of both color and intensity for items such as eyeglasses and camera lenses, interior space dividers, information displays, and erasable optical recording systems.

The phenomenon of electrochromism based on physical phenomena, such as the Franz-Keldish and Stark effects, dates back to 1932. However, since the mid-1970s the major focus of the literature on electrochromism have been on color changes carried out electrochemically, so that today the term electrochromic materials refers to substances as defined above.

Assembly. A typical electrochromic device is a sandwichlike structure of two glass plates and an electrolyte (**Fig. 1**). Each glass plate is coated on the inside with a transparent electrically conducting layer of indium-tin oxide which operates as an electrode. A layer of an electrolyte carries the current inside the cell between the positive (anode) and the negative (cathode) electrodes. The electrolyte can be as simple as a salt (generalized formula M^+X^-) dissolved in a dissociating solvent such as water. Current research is focused on gel and solid electrolytes, because they are easier to confine in the space between the electrodes,

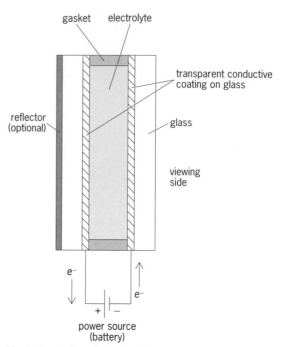

Fig. 1. Electrochromic device. The electrochromic materials can be either dissolved in the electrolyte or coated on the transparent electrodes.

and because they function as laminators, gluing the two glass plates together. Finally, electrochromic mirrors include a reflective coating (such as aluminum) on the outside of one of the glass plates.

The state-of-the-technology devices utilize two electrochromic materials, with complementary properties: the first material is normally in the reduced state (ECM_1^{red}) and undergoes a colorless-to-colored transition upon oxidation [reaction (1)], while the second is

$$ECM_1^{red} \rightarrow ECM_1^{ox} + e^- \qquad (1)$$
$$\text{Colorless} \qquad \text{Colored}$$

normally in the oxidized state (ECM_2^{ox}) and undergoes a similar transition upon reduction [reaction (2)].

$$ECM_2^{ox} + e^- \rightarrow ECM_2^{red} \qquad (2)$$
$$\text{Colorless} \qquad \text{Colored}$$

Reactions (1) and (2) do not show the charge balance of the processes; instead they emphasize in generalized terms that ECM_1 gets colored upon loss, and ECM_2 upon gain, of electrons. ECM_1^{red} and ECM_2^{ox} are selected so that they do not react with each other. Reactions (1) and (2) then are forced by the external power source (Fig. 1), which operates as an electron pump that consumes energy in order to transfer electrons from one electrode to the other. Reaction (1) occurs at the anode and is a source of electrons, while reaction (2) occurs at the cathode and is a sink of electrons. This approach, known as complementary counterelectrode technology, has two distinct advantages. First, the long-term operating stability of the electrochromic cell is greatly enhanced by preventing any electrolytic decomposition of the electrolyte, since both a source and a sink of electrons are provided within the system simultaneously. Second, the reinforcing effect of two electrochromic materials simultaneously changing color enhances the contrast between the colored and colorless states. Depending on the location of the two electrochromic materials within the devices, three main types of devices exist: the solution type, the precipitation type, and the thin-film type.

Solution type. In this device the electrochromic materials are dissolved in the electrolyte and move to the electrodes by diffusion. Electric current forced through the electrode–solution interface is responsible for the electrolysis of the electrochromic materials to their colored forms, which diffuse away from the electrodes back into the bulk electrolyte (**Fig. 2a**). The main advantage of the solution-type electrochromic devices is the variety of materials that

can be used; any material that can be oxidized or reduced and is electrochromic is a potential candidate. In fact, more than 98% of the electrochromic antiglare rearview mirrors for automobiles sold today worldwide are based on solution-type devices. The operation of a solution-type device requires that current be sustained continuously through it, because the color-bearing forms of the two electrochromic materials meet in the bulk solution and annihilate each other back to their respective colorless states. This property increases the power consumption but renders the device self-erasing, thereby providing a fail-safe system that automatically recovers to the colorless state under power-failure conditions. The maximum color intensity of the solution-type device depends

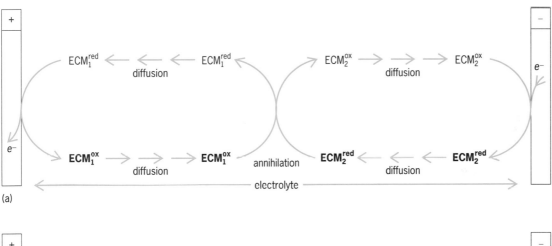

(a)

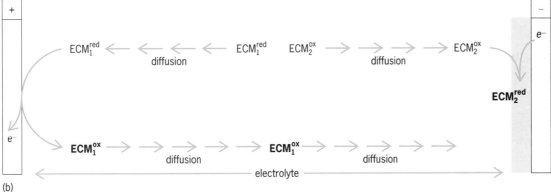

(b)

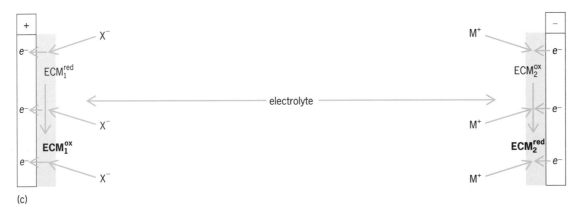

(c)

Fig. 2. Operating principles of electrochromic devices. (*a*) Solution type. (*b*) Precipitation type. (*c*) Thin-film type. The color-bearing forms of electrochromic materials are shown in boldface.

on the concentration of the electrochromic materials, which in turn is limited by their solubility. But, unlike the other types of electrochromic devices, maximum color intensity can be controlled easily by varying the thickness of the layer of the electrolytic solution, typically in the 0.1–5-mm (0.004–0.2-in.) range. A potential drawback of this approach is the relatively slow speed of coloration, which is controlled by diffusion of the electrochromic materials in the bulk electrolyte, a relatively slow process. A typical example of a solution-type electrochromic device employs N,N,N',N'-tetramethyl-p-phenylenediamine (TMPD) as ECM_1^{red}, and N,N'-diheptyl-4,4'-bipyridinium dication (HV^{2+}) as ECM_2^{ox}. In organic solvent-based electrolytes these substances react at the anode [reaction (3)], at the cathode [reaction (4)], and in the bulk electrolyte [reaction (5)].

H₃C \ N —⟨ ⟩— N / CH₃ (TMPD structure)

TMPD, colorless, in solution

$$\left[H_3C, H_3C {>} N -\bigcirc- N {<} CH_3, CH_3 \right]^{\bullet\,+} + e^- \qquad (3)$$

TMPD$^{\bullet\,+}$, blue, in solution

C₇H₁₅—N⁺⟨ ⟩—⟨ ⟩N⁺—C₇H₁₅ + e⁻ ⟶

HV^{2+}, colorless, in solution

$$\left[C_7H_{15}-N\bigcirc-\bigcirc N-C_7H_{15} \right]^{\bullet\,+} \qquad (4)$$

HV$^{\bullet\,+}$, blue, in solution

$$TMPD^{\bullet\,+} + HV^{\bullet\,+} \rightarrow TMPD + HV^{2+} \qquad (5)$$

Precipitation type. In this device the electrochromic materials originally dissolved in the electrolyte, but upon oxidation or reduction at least one of the colored products is electrodeposited onto the corresponding electrode. One of the first examples (1962) of electrolytic cells in light-modulation devices employed the reversible electrodeposition of silver. Another example involves N,N'-diheptyl-4,4'-bipyridinium dication (HV^{2+}) in aqueous electrolytes, where the blue reduction product [reaction (4)] precipitates reversibly as a salt of N,N'-diheptyl-4,4'-bipyridinium monocation radical ($HV^{\bullet+}$).

The precipitation-type electrochromic devices are also diffusion-controlled, but they have lower power requirements than the solution-type devices. Figure 2*b* demonstrates the operating principles of a precipitation-type device in which only one of the two electrochromic materials precipitates on the electrode upon color formation. If the power is disconnected, such devices lose their color slowly, because soluble ECM_1^{ox} diffuses to the opposite electrode, where it reacts with the ECM_2^{red} precipitate to yield colorless ECM_1^{red} and soluble and colorless ECM_2^{ox}.

Thin-film type. In this type, both the electrochromic materials are confined and immobilized as thin-film coatings (thickness between 0.1 and 5 micrometers) on the surfaces of the two electrodes (Fig. 2*c*). In principle, this arrangement has some very desirable properties. For instance, electrode surface confinement of both electrochromic materials is expected to increase the charging (switching) speed of these devices, because coloration no longer depends on electrolysis of electrochromic materials dissolved in the bulk electrolyte. Moreover, physical separation of the two electrochromic materials not only prevents annihilation of the colored forms but also provides an open circuit memory that significantly decreases the average consumption of power. In a sense, a thin-film-type electrochromic device is a rechargeable battery in which the color of the electrodes depends upon the state of charge. Discharging this type of device can be carried out either by reversing the bias or by simply short-circuiting the two electrodes.

Several organic and inorganic materials are available for surface confinement on the electrodes as thin films. A film of tungsten trioxide (WO_3), for instance, is an example of an inorganic electrochromic material that, upon reduction, uptakes H^+ reversibly from an aqueous electrolyte (or Li^+ and Na^+ from a non-aqueous electrolyte) to form a blue tungsten bronze (M_nWO_3) [reaction (6), where $M^+ = H^+$, Li^+, Na^+;

$$WO_3 + ne^- + nM^+ \rightleftharpoons M_nWO_3 \qquad (6)$$
Yellow $\qquad\qquad\qquad$ Blue

and $0 \leq n \leq 1$]. This reaction is analogous to reaction (2). Inorganic materials suitable for thin-film electrochromic devices that are complementary to WO_3 include nickel oxide [NiO; reaction (7)] and Everitt's salt [reaction (8)].

$$NiO + OH^- \rightleftharpoons NiOOH + e^- \qquad (7)$$
Colorless $\qquad\qquad$ Blue-
$\qquad\qquad\qquad\qquad$ black

$$K_4Fe_4[Fe(CN)_6]_3 \rightleftharpoons Fe_4[Fe(CN)_6]_3 + 4K^+ + 4e^- \qquad (8)$$
Everitt's salt $\quad$ Prussian blue
(colorless)

Among organic materials, conductive polymers are currently being evaluated for application in electrochromics. Polymers such as poly(pyrrole), poly(aniline), and poly(thiophene) have all been demonstrated as thin-film electrochromic materials for oxidative coloration. Concurrently, polymeric films derived from diquaternized 4,4'-bipyridine are being studied for reductive coloration in analogy to reaction (4). Although it can be argued that organic polymers are inherently more durable than other surface-confined materials (such as oxides) because they better accommodate the structural changes induced upon oxidation and reduction, colored organic polymers generally show poor stability to sunlight.

Prospects. Electrochromic devices are analogous to liquid-crystal devices in that they do not generate their own light but modulate the ambient light. Unlike liquid-crystal devices, however, electrochromic devices do not require polarizers, allowing for a viewing angle approaching 180°, with the same contrast ratio as black ink on white paper (20:1 or better). Moreover, in liquid-crystal devices the spacing between the electrodes has to be controlled to tight tolerance, a requirement that imposes lateral size limitations. In electrochromic devices, control of that spacing is not important. Other desirable features of electrochromic devices include inherent color, continuous gray scale, and low average power consumption for the thin-film devices. A typical electrochromic device requires 1–2 V, and consumes less than 10 millicoulombs/cm^2 of electrical charge for full coloration. It has also been shown that electrochromic materials can be patterned with a 2–5-μm resolution to form a large number of display elements that can be matrix-addressed. However, even though there is no apparent natural limitation, the best cycling lifetime claimed to date for electrochromic materials is only up to 10–20 × 10^6 cycles, while the cycling lifetime of liquid-crystal devices is of the order of several hundred million cycles. This long lifetime has made liquid-crystal devices a very successful technology in matrix-addressed flat-panel displays.

Electrochromic devices seem better suited than liquid-crystal devices for large-area light modulation applications, such as smart windows, space dividers, and smart mirrors. Lifetime limitations have prevented the utilization of electrochromic materials in continuously refreshed, high-resolution flat-panel displays. Nevertheless, large-area displays that do not need frequent refreshing, such as signs and announcement boards, are a possible application. Several high-resolution applications of electrochromism have been proposed recently, including configurable optical recording devices.

For background information *SEE ELECTROCHEMISTRY; ELECTROCHROMIC DISPLAYS; STARK EFFECT* in the McGraw-Hill Encyclopedia of Science & Technology.
Nicholas Leventis

Bibliography. D. J. Grover (ed.), series of articles in *Displays*, 9:163–206, 1988; N. Leventis and Y. C. Chung, New complementary electrochromic system based on polypyrrole–prussian blue composite, a benzylviologen polymer, and poly(vinylpyrrolidone)/potassium sulfate aqueous electrolyte, *Chem. Mater.*, 4:1415–1422, 1992; S. Matsumoto (ed.), *Electronic Display Devices*, 1990; T. Oi, Electrochromic materials, *Annu. Rev. Mat. Sci.*, 16:185–201, 1986.

Electrode

Recent advances in electrochemistry include modification of electrodes with polymers and inorganic materials, and development of electrodes that incorporate enzyme layers in their structure.

Chemically Modified Electrodes

The attachment of organic monolayers, polymer films, and inorganic solids to electrode surfaces permits combination of a specific chemical function of the overlayer with the electrode's ability to drive oxidation-reduction reactions. This synergistic combination is responsible for new practical technology involving electrocatalysis and electrochromic displays, and it has opened the door to current research in electrochemical analysis, solar energy conversion, and molecular electronics.

Electrode materials. Electrode surfaces are modified by chemisorption or covalent binding of a wide variety of materials, including organic monolayers, ionically or electronically conducting polymers, and various inorganic solids such as clays, zeolites, and semiconductors. Each of these materials imparts a chemical function that can act in concert with the electrode's ability to deliver or take away electrons. **Figure 1** shows an example of a monolayer coating that is bound through gold-sulfur bonds, which are made by simply immersing the electrode in a solution of the appropriate sulfur-bearing molecules. This monolayer contains thiobisethylacetoacetonate (TBEA) molecules that selectively bind copper ions from solution. Fewer than 10^{-10} mole of copper can be sensed electrochemically by reduction through the gold electrode. In this instance the purpose of the modifying layer is to impart selectivity to the electrode so that it will reduce copper and no other ions that, at a bare electrode, would interfere with the analysis.

There are several other strategies for preparing chemically bonded electrode coatings of organic and inorganic materials. One of the most general is to react an oxide electrode surface (for example, superficially oxidized platinum or conductive tin oxide) with molecules or polymers containing silanol groups (—SiOH) in order to form covalent Si-O-metal linkages. Often, simple thermal evaporation or spin coating can be used to prepare films of relatively insoluble materials. Another strategy is to initiate a polymerization reaction electrochemically in a solution of soluble monomers. This technique has been used to grow vinyl polymers on electrodes, as well as electronically conducting polymers such as polypyrrole, which are insoluble once polymerized.

Electrocatalysis. One of the strongest motivations for altering the chemistry of an electrode surface is to speed up the kinetics of an electrochemical reaction. A simple metallic electrode such as graphite, mercury, or platinum can be used to drive electrochemical oxidations and reductions of contacting molecules

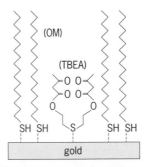

Fig. 1. Modified electrode used for selective sensing of copper ions. The surface monolayer film consists of insulating octadecylmercaptan (OM) and copper-binding molecules of thiobisethylacetoacetonate (TBEA).

or ions; but it often does so in an energetically inefficient manner or without chemical specificity. Electrochemical reactions that involve transfer of more than one electron per molecule, and the simultaneous making and breaking of strong chemical bonds, are usually kinetically sluggish; and they require the application of more energy (per electron) than is stored in the products of the reaction. This extra energy, or overpotential, is simply converted to heat in the process. Avoiding this waste of energy is of great economic importance in large-scale electrochemical reactions.

One such reaction is the electrolysis of brine [aqueous sodium chloride (NaCl)] to produce hydrogen (H), chlorine (Cl), and sodium hydroxide (NaOH) in the chlor-alkali process, which accounts for approximately 3% of the electric power consumed in the United States. Before the advent of modified electrodes, the chloride/chlorine half-cell reaction (1) was carried out

$$2Cl^- \rightarrow Cl_2 + 2e^- \qquad (1)$$

at carbon electrodes, which operated at significant overpotentials and corroded during the reaction, requiring frequent replacement. These electrodes have now been replaced by dimensionally stable (noncorroding) anodes consisting of titanium metal bearing a micrometer-thick coating of ruthenium oxide/titanium oxide solid solution. The electrode coating is electronically conductive and catalytically active for chloride oxidation. Because of cracks and fissures in the coating, the effective surface area of the coating is about 200 times greater than the geometrical area of the electrode. The effect of this high surface area is to reduce the true current density, further lowering the overpotential needed to carry out the oxidation reaction.

Current research involving electrocatalytic coatings is aimed at reducing overpotentials associated with the reduction of oxygen and carbon dioxide. The four-electron reduction of oxygen (O_2) to water [H_2O; reaction (2)] is an essential electrode reaction for air-

$$O_2 + 4e^- + 4H^+ \rightarrow 2H_2O \qquad (2)$$

breathing fuel cells and batteries. Such cells will have very high energy-storage densities; potentially they will have widespread applications in electric vehicles and utility load leveling (provision of supplementary electric power during times of greatest demand). On most electrode surfaces, this reaction requires high overpotentials and is kinetically slow relative to the two-electron reduction of oxygen to hydrogen peroxide (H_2O_2). Since H_2O_2 is a high-energy intermediate, its formation must be avoided in order for these cells to be efficient. While several good electrocatalysts for this reaction have been found, notably metalloporphyrin dimers and highly dispersed noble metals such as platinum and rhodium, at present none are economically viable for large-scale use in fuel cells. The multielectron oxidation of methanol and methane as fuels to carbon dioxide is also of great importance to the development of fuel cells with high energy density that would employ these abundant feedstocks.

Solar energy conversion. Relatively inexpensive electrode coatings capable of converting sunlight directly to electricity may provide an economically competitive alternative to solid-state solar cells based on crystalline and amorphous semiconductors. The design of these coatings (**Fig. 2**) is quite similar to that of the dimensionally stable anodes used in brine electrolysis. A colloidal suspension of titanium dioxide (TiO_2) particles is used to form a porous coating several micrometers thick on a titanium electrode. Sintering under controlled conditions improves electrical contact between these particles while maintaining a very high surface area for adsorption of a light-absorbing dye. While several kinds of dyes (metalloporphyrins, coumarins, cyanometallate complexes) have been tested in this application, the most efficient cells contain ruthenium poly(pyridyl) complexes, which absorb strongly in the visible region of the spectrum and inject electrons efficiently into the TiO_2 layer upon photoexcitation. The electron lost by the complex is replaced by electron transfer from iodide (I^-) at the electrode–solution interface. The iodide is then regenerated by reduction at a transparent tin oxide (SnO_2) counterelectrode. The net effect is to drive current photochemically in a loop through a resistive load. Small prototype cells with solar power conversion efficiencies of approximately 10% have been reported. These cells are composed of relatively inexpensive materials and do not require sophisticated techniques for fabrication. Consequently, they are of potential interest for large-scale applications.

Electrochromic materials. Chemically modified electrodes that change color upon oxidation-reduction are potentially useful in electrochromic displays and as components of so-called smart windows that absorb or transmit sunlight as required for passive heating and cooling of buildings. The most promising materials for these applications are inorganic solids such as tungsten trioxide (WO_3) and certain conducting polymers such as polyisonaphthalene. Both materials can be grown as thin coatings on transparent electrodes such as SnO_2/glass. The display or window consists of the electrochromic layer as well as a thin polymer electrolyte, sandwiched between two sheets

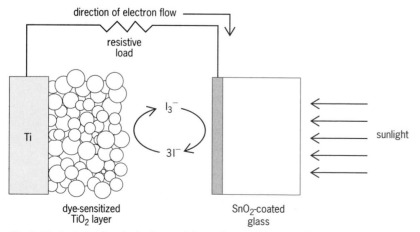

Fig. 2. Photoelectrochemical cell containing a dye-sensitized semiconductor layer of titanium dioxide (TiO_2) grown on a titanium electrode.

of transparent conductor. Tungsten trioxide is a light-yellow solid that is transparent in its oxidized form but acquires an intense blue coloration when it is reduced electrochemically in the presence of protons (H^+) or alkali metal ions (M^+). The cations enter small cavities in the WO_3 structure (**Fig. 3**) to balance the charge that comes from reduction of some of the W^{VI} to W^V. Because the motion of electrons and ions into the solid does not disrupt the bond connectivity of the WO_3 framework, the reduction is reversible electrochemically; displays have been fabricated that can undergo 10^6 coloration-bleaching cycles.

While this consideration is relatively unimportant for smart windows, a challenge remaining for display applications is the fabrication of cells with coloration-bleaching times of 50 milliseconds or less. Polyiso-naphthalene fulfills this requirement but is less stable over many cycles than WO_3. SEE ELECTROCHROMIC DEVICES; SMART MATERIALS.

Molecular electronics. One of the most interesting new research directions in electrode–polymer interfaces is the fabrication of thin-film devices that function in a manner analogous to diodes, transistors, and other elements of electronic circuits. These devices are of great current interest for applications as chemical sensors and, at least in principle, as components of so-called organic computers. Electronically conducting polymers, such as polypyrrole, polyaniline, and polythiophene, as well as their semiconducting relatives such as α-sexithienyl (α-6T; an oligomer of six thiophene molecules), are the essential ingredients of this novel technology. **Figure 4a** shows one of the simplest devices of this kind, an array of three micrometer-size gold wires coated with a layer of conducting polymer. The potential of the middle gate wire is modulated in order to switch the polymer between oxidized (conducting) and reduced (nonconducting) states. By applying a small constant potential difference between the end electrodes (the source and drain), a source-drain current flows, the magnitude of the current being strongly dependent on the potential applied to the gate. Effectively, the device has the same properties as a field-effect transistor (FET). It can also

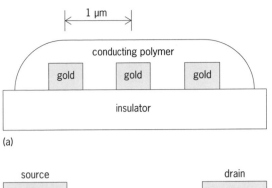

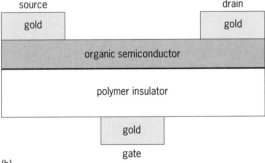

Fig 4. Thin-film devices. (a) Electrochemical conducting polymer transistor. (b) Design of an all-organic field-effect transistor.

be used as a sensor for ions that compensate charge in the oxidation-reduction cycle. Charge compensation is a process in which anions from the solution flow into the solid polymer in order to maintain local electroneutrality.

A serious problem of these electrochemical transistors is their switching speed, which is limited by the motion of counterions in and out of the polymer during oxidation-reduction. Effectively, the switching speed limits device response frequencies to kilohertz, many orders of magnitude slower than conventional solid-state field-effect transistors. A potential solution to this problem is represented by the all-organic thin-film solid-state transistor shown in Fig. 4b. This device consists of a high dielectric insulator (a cyanoethylated polysaccharide) with a coating of the organic semiconductor α-6T and metal gate, source, and drain contacts. Since its operation is similar to conventional silicon field-effect transistors, not requiring counterion motion, it does not suffer the same switching speed limitations as the electrochemical polymer transistor.

Prospects. Since about 1970, chemically modified electrodes have had a revitalizing influence in electrochemical research. As surface chemical techniques for preparing tailored electrochemical interfaces continue to develop, modified electrodes will have an increasing impact on all aspects of electrochemical technology, including chemical sensing, energy conversion, data storage and display, corrosion protection, and microelectronics.   *Thomas E. Mallouk*

Enzyme Electrodes

Enzyme electrodes are amperometric or potentiometric devices that use an enzyme to specifically transduce the concentration of an analyte to a poten-

Fig 3. Structure of tungsten trioxide (WO_3), an active component of electrochromic displays. The structure is composed of corner-sharing WO_6 octahedra and contains large cuboctahedral cavities that can accommodate hydrogen ions (H^+) and alkali metal ions (M^+) upon reduction.

tiometric or amperometric electrode signal. The first enzyme electrode, reported by L. C. Clark in about 1960, was an electrode for glucose that was based on an oxygen/hydrogen peroxide electrode. Since then, the number of applications conceived and the number of analytes detected have greatly expanded. Enzyme electrodes are routinely used by physicians and diabetics to measure blood glucose levels, an essential measurement in diabetes therapy. These glucose enzyme electrodes range from large benchtop analyzers capable of measuring a large number of samples to small hand-held devices that measure glucose in a single droplet of whole blood. Other applications in development include enzyme electrodes for fermentation control, for environmental monitoring, and for use as implantable sensors in emergency rooms, critical care units of hospitals, and diabetes therapy. In addition to glucose electrodes, enzyme electrodes that detect a wide variety of analytes such as lactate, cholesterol, urea, galactose, fructose, and alcohol have been or are being developed.

Enzyme behavior. The enzyme's reaction products or its cofactor must be electroactive in order for the enzyme to be utilized in an enzyme electrode. An example of an enzyme-catalyzed reaction is the oxidation of β-D-glucose by oxygen [reaction (3)]. This

$$\beta\text{-D-glucose} + O_2 \rightarrow \text{gluconolactone} + H_2O_2 \qquad (3)$$

reaction is catalyzed by the enzyme glucose oxidase, a flavoprotein oxidase belonging to a large class of enzymes called donor:acceptor oxidoreductases that also includes dehydrogenases, peroxidases, and monooxygenases. The substrate, β-D-glucose, is oxidized to gluconolactone, which is rapidly hydrolyzed to gluconic acid; and the coreactant, O_2, is reduced to H_2O_2. An amperometric glucose enzyme electrode can measure either the depletion of O_2 at a cathode or the production of H_2O_2 at an anode. A potentiometric glucose electrode measures the pH change resulting from the production of gluconic acid.

The production of H_2O_2 and gluconic acid and the depletion of O_2 are all proportional to the concentration of glucose in the analyte solution. For many enzymes, the relationship between the concentration and the electrode signal is described under defined conditions by a modified form of the Michaelis-Menton equation for enzyme kinetics, an example of which is shown in Eq. (4), where V is the velocity or rate of

$$V = \frac{V_{\max}[S]}{K_m + [S]} \qquad (4)$$

the enzyme-catalyzed reaction, $V_{\max}$ is the maximum rate, $[S]$ is the concentration of substrate, and K_m is a kinetic parameter related to the individual steps of the enzyme-catalyzed reaction and is equal to the substrate concentration at $1/2\ V_{\max}$. For an amperometric enzyme electrode, the concentration dependence of the current directly reflects the kinetic behavior of the enzyme, provided substrate transport is not limiting. The response from a potentiometric enzyme electrode is more complex because potential varies with the logarithm of concentration.

Enzyme immobilization. Enzymes are frequently incorporated into the enzyme electrode by immobilization on the electrode surface. Immobilization is achieved by using a membrane to entrap the enzyme, cross-linking the enzyme on the surface, or covalently binding the enzyme to an activated electrode surface. Many electrode designs use a combination of all three techniques. Enzymes are entrapped by drying the enzyme on the surface, followed by the application of the membrane over the dried film.

Membrane materials that are soluble in organic solvents (for example, cellulose acetate, polyurethane, and Nafion) can be applied by dipping the electrode with the dried enzyme film into a solution containing the membrane material. Membranes are often chemically modified to provide functional groups that can be bound to the enzyme. Cellulose acetate, for instance, has been oxidized to form reactive aldehyde groups. An aldehyde group of the membrane will react with a lysine amino acid of an enzyme to form a Schiff base. Borohydride is used to reduce the Schiff base, forming a stable covalent bond between the enzyme and the membrane. Cross-linking or reticulation refers to using a bifunctional small molecule (for example, glutaraldehyde) to link enzymes to each other, forming a large water-insoluble network on the electrode surface. Because cross-linking can inactivate an enzyme, a protein such as bovine serum albumin is often mutually cross-linked with the enzyme to stabilize it.

Surface activation of an electrode is highly dependent on the electrode material used. Carbon electrodes have been oxidized to form carboxylic acid groups that can be linked to the enzyme by using a carbodiimide coupling agent. Gold and platinum surfaces have been functionalized with reactive primary amines by modifying the surface with mercaptoethylamine and N-[3(trimethoxysilyl)propyl]-ethylenediamine, respectively.

Flavoprotein oxidases. Flavoprotein oxidases have a redox center or multiple centers containing a flavin prosthetic group such as flavin adenine dinucleotide (FAD) or flavin mononucleotide (FMN). During the catalytic cycle, electrons are transferred from the substrate (the electron donor) to the redox center, and from the redox center to the cosubstrate (the electron acceptor). In the case of glucose oxidase, two electrons and two protons are transferred from β-D-glucose to FAD, converting it to FADH$_2$. The two electrons and two protons are then transferred to O_2, creating H_2O_2 and completing the catalytic cycle.

Many enzyme electrodes amperometrically detect the depletion of O_2 or the production of H_2O_2. The O_2 is reduced at -0.4 V or less versus silver/silver chloride (Ag/AgCl), and H_2O_2 is oxidized at $+0.65$ V or less versus Ag/AgCl. Detection of H_2O_2 is the basis for a commercially developed glucose analyzer, one of the first applications of enzyme electrode technology. This type of sensing is, however, highly dependent on the partial pressure of O_2 in the analyte solution. To lessen the dependence on O_2, electrodes have been designed that allow transport of O_2 to the electrode axially and radially while allowing substrate transport only axially. The increased flux of O_2 allows the electrode to operate over a large range of O_2 partial pressures without change in current from fluctuations of the O_2. However, these electrodes cannot operate anaerobically.

Oxygen independence has also been achieved by replacing O_2 in the enzyme-catalyzed reaction with an artificial electron acceptor or mediator. The mediator, a redox couple, must have an electrochemical potential positive of the potential of the flavin prosthetic group for oxidation to occur. In addition, the electrochemistry of the mediator should be fast and reversible. Because the redox centers of many enzymes are surounded by insulating shells of protein or glycoprotein, the mediator must penetrate this shell for electron transfer to take place. Thus, the effectiveness of a mediator is highly dependent on its structure and charge. A mediator is capable of accepting only one electron at a time from the reduced enzyme's redox center. The reduced mediator is then oxidized at the electrode surface, the current from which is proportional to the substrate concentration. The general reactions for a mediated electrode are shown in reactions (5) and (6), using the

$$\beta\text{-}\textsc{d}\text{-glucose} + 2M^o \rightarrow \text{gluconolactone} + M^r + 2H^+ \quad (5)$$

$$M^r \rightarrow M^o + e^- \quad (6)$$

glucose oxidase-catalyzed reaction as example, where M^o is the oxidized mediator and M^r is the reduced mediator.

Small molecule mediators. Small molecules such as ferrocenes and quinones have frequently been used. These enzyme electrodes are often referred to as diffusionally mediated, because mediator transport to within the electron transfer distance of the enzyme's redox center is a diffusion-controlled process. A commercially produced glucose analyzer utilizes 1,1'-dimethylferrocene as the mediator in a single-use enzyme electrode used by diabetics to monitor blood glucose. Components of conducting organic salts such as tetrathiafulvalinium tetracyanoquinomethanide (TTF^+TCNQ^-) are also effective mediators. Enzyme electrodes made with small-molecule mediators are unstable during prolonged use, because the mediator tends to leach from the electrode. Because these mediators can be toxic, mediator leaching is a very important consideration, particularly in biomedical applications, where the implantable glucose sensors and other sensors are sought.

Small molecule mediators can be covalently bound to amino acids of the enzyme to prevent their loss from the electrode. However, if the mediator is not within electron-transfer distance of the enzyme's redox center, electrons will not be transferred to the mediator. If a series of mediators or electron relays are replaced in a path leading to the enzyme's redox center, electrons can be transferred via these relays to the electrode. The incorporation of electron relays into the enzyme's protein is accomplished by partially denaturing the enzyme, covalently attaching electron acceptors to selected amino acids of the enzyme, and then renaturing the enzyme (**Fig. 5**). Alternatively,

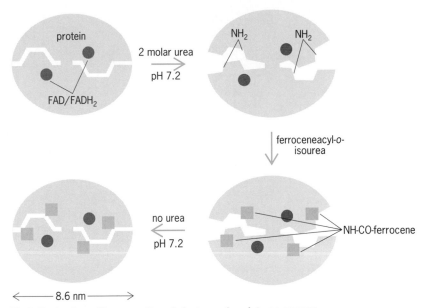

Fig 5. Mechanism of incorporation of electron relays into an enzyme.

mediators have been attached to long flexible chains on the exterior of enzymes. The mediators can then diffuse into and out of the enzyme's protein, with the molecular chain preventing the loss of the mediator.

Polymers. Frequently, large redox polymers and conducting polymers are used in enzyme electrodes to avoid leaching of the mediator. Because they are large, they can be entrapped behind a membrane or cross-linked with an enzyme in a three-dimensional network. Conducting polymers used in enzyme electrodes include polypyrrole and polyaniline, with enzymes typically entrapped in electropolymerized networks of these polymers. However, enzymes have also been covalently bound to functionalized polypyrroles. **Figure 6** shows an example of a redox polymer that has been used in a number of enzyme electrodes. The $Os(bis\text{-bipyridine})_2Cl^{2+/3+}$ redox couple is the electron relay, and covalent linkage to an enzyme is achieved through the ethylamine side chain. When an enzyme, such as glucose oxidase, is mixed with the deprotonated redox polymer and with a cross-linking agent such as poly(ethylene glycol) diglycidyl

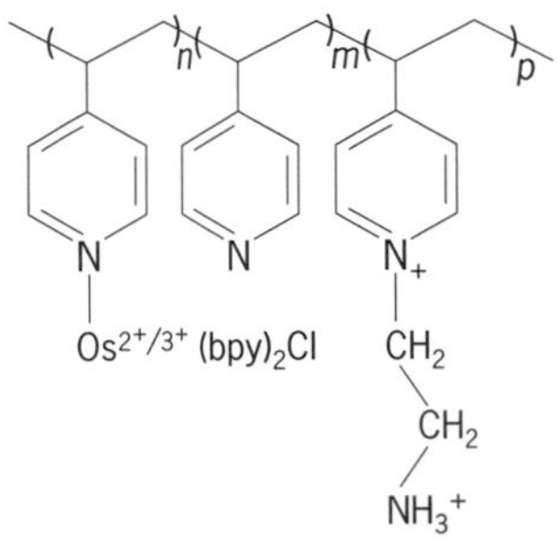

Fig 6. Structure of poly(vinyl pyridine) Os(bpy)$_2$Cl polyamine redox polymer. byp = *bis*-bipyridine, Os = osmium, Cl = chlorine, $n = 1$, $m \approx 4$, $p \approx 1.2$. (*After B. Gregg and A. Heller, A redox-conducting epoxy cement: Synthesis, characterization, and electrocatalytic oxidation of hydroquinone, J. Phys. Chem., 95:5970–5975, 1991*)

ether, a highly permeable three-dimensional network is formed. The redox network immobilizes the enzyme while collecting the current originating in substrate oxidation. Other redox polymers successfully used in enzyme electrodes include poly(vinyl ferrocene), ferrocene-modified polysiloxane, and polyviologens.

Dehydrogenases. Most dehydrogenase enzymes have a redox center in the form of a weakly bound coenzyme, either nicotinamide adenine dinucleotide (NADH) or nicotinamide adenine dinucleotide phosphate (NADPH). Alcohol dehydrogenase, which catalyzes the oxidation of ethanol to acetaldehyde [reaction (7)], is a well-studied example of this type of

$$CH_3CH_2OH + NAD^+ \rightarrow CH_3CHO + NADH + H^+ \quad (7)$$

enzyme. The production of NADH can then be sensed amperometrically. While NADH can be oxidized directly on an electrode surface, dimerization or isomerization can occur. To oxidize NADH to NAD^+, diffusionally mediated enzyme electrodes using dyes such as Meldola Blue or organic conducting salts such as N-methylphenazonium tetracyanoquinomethanide (NMP^+TCNQ^-) have been constructed.

Because the coenzyme is weakly bound to the enzyme, it can diffuse away from the electrode. Typically, NAD(P)H is added to the analyte solution to ensure a sufficient NAD(P)H concentration for catalysis. NADH has also been attached to long polymer chains such as dextran, allowing it to be entrapped behind a membrane along with the enzyme. Long chains with NADH attached have also been covalently bound to enzymes. Perhaps the most significant advance is the development of a genetically engineered dehydrogenase with NADH covalently bound to the enzyme.

Other enzymes. A wide variety of enzymes other than flavoprotein oxidases and dehydrogenases have been used in enzyme electrodes. Other oxidoreductases used in amperometric enzyme electrodes include horseradish peroxidase (used in the detection of hydrogen peroxide and of cyanide) and L-amino acid oxidase. Potentiometric enzyme electrodes for urea, using the enzyme urease, have been widely studied. This electrode measures, at an ion-selective electrode, the change in potential from the production of ammonium ion (NH_4^+) in the enzyme-catalyzed reaction. Ion-selective enzyme electrodes have also measured L-glutamine with glutaminase and L-glutamic acid with glutamate dehydrogenase. Other analytes assayed with potentiometric enzyme electrodes include gluconate, phosphate, amino acids, and acetylcholine.

For background information *SEE ELECTROCHEMISTRY; ELECTRODE; ENZYME; ION-SELECTIVE MEMBRANES AND ELECTRODES; ORGANIC CONDUCTOR; PHOTOCHEMISTRY* in the McGraw-Hill Encyclopedia of Science & Technology.

Michael V. Pishko; Adam Heller

Bibliography. A. Cass (ed.), *Biosensors: A Practical Approach*, 1990; L. R. Faulkner, Chemical microstructures on electrodes, *Chem. Eng. News*, 62:28–45, February 27, 1984; A. Heller, Electrical connection of enzyme redox centers to electrodes, *J. Phys.* *Chem.*, 96:3579–3587, 1992; J. Janata, Chemical sensors, *Anal. Chem.*, 64:196R–219R, 1992; R. W. Murray (ed.), *Molecular Design of Electrode Surfaces*, Techniques of Chemistry, vol. 22, 1992.

Electronic mail

Electronic mail systems have become an increasingly common method of conveying messages between users of computers, both within and between large organizations. The need for a set of common standards for the transmission and reception of electronic mail was foreseen by two international standards bodies, the Consultative Committee in International Telegraphy and Telephony (CCITT) and the International Organization for Standardization (ISO). CCITT developed the X.400 Recommendations, and ISO the Message-Oriented Text Interchange Systems (MOTIS). Both standards are, for all practical purposes, identical.

OSI model. Any electronic communication system is of necessity very complex. To structure and simplify the standardization process, the ISO has established a layered model of such a system. The seven-layer model is known as the Reference Model for Open Systems Interconnection (OSI). *SEE INFORMATION MANAGEMENT.*

The functionality of each layer is defined, as are its interfaces to the adjacent layers. These definitions can be thought of as a set of rules or protocols. Once agreed to, they must be adhered to by both ends of the link for any meaningful communication to take place. The subdivision of the system into seven layers, each with its own protocol and interfaces, allows underlying layers to be implemented with a number of different technologies without affecting the layers above. For instance, the bottom three layers, 1 to 3, could be implemented over a packet switched network (for example, the CCITT X.25 standard) or over an integrated systems digital network (for example, the CCITT I.400 ISDN standards).

Bottom layers 1 to 4 are concerned only with bits and their satisfactory transmission between two end points. Upper layers 5 to 7 are concerned with meaningful messages. Layer 1, the physical layer, defines the actual physical connection over which the data are transmitted. Layer 2, the link layer, defines the way the bits sent over the physical layer between the two ends of the wire are handled as far as such things as error checking are concerned. Layer 3, the network layer, defines the way individual links are connected together to form a circuit between two end points. Layer 4, the transport layer, defines the control and error handling between the two ends of the circuit. Layer 5, the session layer, defines the necessary protocol to enable the computers at the two ends, for instance, to log on to each other. Layer 6, the presentation layer, defines how data items are represented within a session. Layer 7, the application layer, defines the actual user application running over the lower layers. The X.400 Recommendations are a good example of a definition of a layer 7 application.

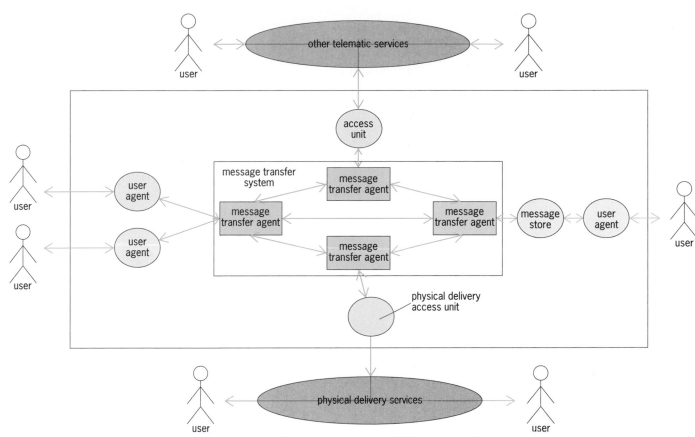

Fig. 1. Message handling system functional model, showing all the components of an X.400 system and how they interact.

X.400 Recommendations. The X.400 Recommendations, then, are not concerned with the lower layers of the OSI model. They simply expect a reliable service to be provided.

X.400 is structured in an object-oriented way. The complete structure is known as the message handling system, and it has two principal objects, the message transfer system and the user agent, and two secondary objects, the message store and the access unit. The message transfer system consists of a number of message transfer agents. **Figure 1** illustrates the relationship between these objects. *SEE OBJECT-ORIENTED PROGRAMMING.*

These objects can be visualized by analogy to the normal postal service. Mail is prepared and put in the mailbox (the user agent). It is taken from the mailbox to the sorting office (the message transfer agent), where the delivery address is examined. If the delivery address is local to this sorting office, the mail item is simply sent out to the appropriate mailbox, or if the recipient has a post office box (message store), the item is deposited there until called for. If the delivery address is not local, the appropriate distant sorting office is selected and the mail item is dispatched to it for onward delivery. A message transfer agent may have many hundreds of user agents or message stores attached to it, or it may have none, simply acting as a transit switching point.

The message transfer agent is implemented on a minicomputer, a mainframe, or even a powerful personal computer dedicated to the task. Message stores reside on the same machine as the message transfer agent. User agents either coexist with the message transfer agent on the minicomputer or mainframe or are implemented as a task on a personal computer. It is for this latter case that the message store exists. Users turn off their personal computers when they go home. If their user agent is not available, their messages may be lost unless they can be stored in the message store until called for. The message transfer agent is expected to be running permanently. The style of user interface to the user agent is not covered in the standard and is entirely implementation dependent (**Fig. 2**). The connection between the user agent or message store and the user terminal is hard-wired, is over a local-area network, or is over the public switched telephone network.

X.400 is essentially a store-and-forward messaging system. That is, a message entering the system will not necessarily be delivered instantaneously. Instead, it may be temporarily stored at some point if, for instance, the required link between two message transfer agents is busy.

The access units are gateways to other non-X.400 networks such as telex. They are required to perform the necessary conversion not only of the data but of the addressing information as well.

X.400 features. The features of an X.400 system include send mail, distribution list, deferred delivery, receive mail, security, delivery notification, non-

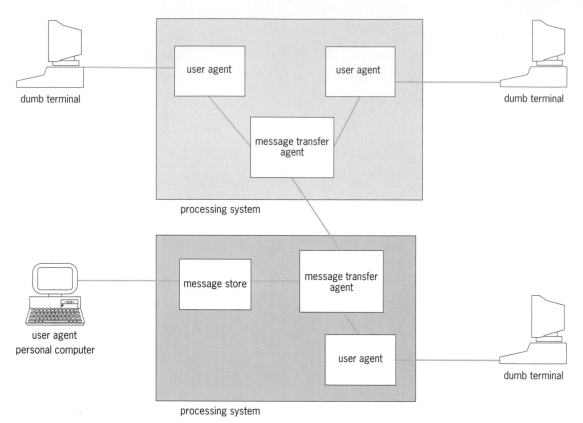

Fig. 2. Two possible implementations of X.400 systems. The processing system could be a mainframe, a minicomputer, or a large personal computer.

delivery notification, receipt notification, nonreceipt notification, autoforwarding, alternate recipient, reply request, conversion, physical delivery, and other message attributes. Prepared files or memos typed online can be sent to one or more recipients (that is, a mailing list). "Carbon copy" and "blind copy" are both supported. A distribution list may be compiled if mail is regularly sent to the same list of recipients. The deferred delivery feature allows a message to be put into the system with the specification that it not be sent out before a certain date or time. A user can also receive mail sent from any other X.400 user anywhere in the world.

A number of perceived security threats can be encountered, including breach of confidentiality, message modification, impersonation, and repudiation.

When mail is sent, the user can request notification of delivery. This notification will be returned to the sender when the message arrives at the destination message transfer agent. If nondelivery notification is requested, a notification will be returned to the user if the mail cannot be delivered. It is quite acceptable to specify both delivery notification and nondelivery notification on the same message.

The sender can request receipt notification, and it will be returned when the recipient actually collects mail from the mailbox. As with delivery and nondelivery notifications, the user can also specify nonreceipt notification, which will be returned when, for instance, a message is autoforwarded to another recipient.

A user can specify that all mail be autoforwarded to someone else. The user can tell the system the reason for autoforwarding, for example, "On holiday until March 10th." This message will be returned to the sender with the nonreceipt notification.

The alternate recipient feature caters for messages that arrive correctly addressed but bearing an unidentifiable recipient name. These messages will normally be delivered to a post room, where someone will have the responsibility of looking at the addressing information and trying to deduce the intended recipient. A number of alternate recipients can be specified either by the sender or the recipient post room. The sender has the option of allowing or preventing this alternate recipient process to take place.

The user can request a reply to a message. The request can specify the date and time by which the reply should be sent, and it can specify the list of recipients to whom the reply should be addressed.

The user can specify whether or not a message is to be translated or converted at any gateway access unit that passes through. A special instance of an access unit provides for delivery of an X.400-originated message to an address with no electronic mail facilities by printing the message and physically delivering it by messenger or the normal postal service.

Other attributes that can be specified by the sender are sensitivity, expiry date, priority, and importance.

For background information SEE ELECTRONIC MAIL; TELEPROCESSING in the McGraw-Hill Encyclopedia of Science & Technology.

Chris F. Wilkinson

Bibliography. Consultative Committee in International Telegraphy and Telephony, *CCITT Recommendation X.200, OSI: Basic Reference Model*, 1984; Consultative Committee in International Telegraphy and Telephony, *CCITT Recommendation X.400, X.400 System & Service Overview*, 1988; International Organization for Standardization, *IS 10021-1, X.400 System & Service Overview*, 1988; C. F. Wilkinson, X.400 electronic mail, *Electr. Commun. Eng. J.*, 3(2):129–136, June 1991.

Electronic navigation systems

Nautical charts on paper are at the heart of marine navigation. However, major changes have been introduced, transforming the static paper chart into a dynamic electronic version that combines the position of the mariner's ship and a radar image of the ship's surroundings into a single display.

Charts and marine navigation. The nautical chart is an official, government-produced map of the shoreline, the depths of water, and the location of buoys, lights, and channels that assist in navigating a chosen route. It serves as an aid to safety and, occasionally, as the legal document that describes an accident at sea. It may be the cause of an accident by falsely showing clear water where underwater hazards are present. The unique legal standing of the nautical chart can be imparted to its electronic descendant by complying with standards being developed by the International Maritime Organization (IMO).

Paper nautical charts show the mariner what hazards lie on the bottom along the ship's route from start to destination. It is up to the mariner to plot the ship's position and determine whether problems will be encountered. Hazards are not only on the bottom; traffic in the harbor is also a threat. It is impractical to plot the track of radar targets on paper, but the cathode-ray-tube monitor of a ship's computer can serve as an electronic plotting surface that contains the chart, the ship's position, and radar targets in the vicinity.

Electronic position finding by the new satellite-based Global Positioning System (GPS) provides a convenient and accurate source of ship location, and the image from a modern digital radar shows the objects on the water.

The electronic chart system currently can be used for navigation in combination with the legal paper chart. There is a large variety of equipment available that ranges from the simplest combination of electronic chart and GPS or older loran up to full-color charts of variable scale, the more accurate form of differential GPS, and an overlay of the radar's video image (see **illus.**). When the IMO standard has been published in final form (expected by 1995) and when national authorities, such as the U.S. Coast Guard, have published rules for its use, the Electronic Chart Display and Information system (ECDIS) can legally supplant the paper chart. *SEE SATELLITE NAVIGATION SYSTEMS.*

The legal ECDIS will be supported by a system of updating the chart by very high frequency (VHF) radio transmissions from the local harbor, or by an international network of satellite radio transmissions by the INMARSAT organization. These satellites will be able to respond to inquiries from ships requesting the latest changes to the charts covering the harbors they are in or about to enter. Alternatively, they may routinely broadcast these changes to all ships within their coverage areas. The local VHF stations will concentrate on the broadcast of temporary warnings concerning hazards or changes in a local harbor. These radio transmissions will automatically change the charts displayed on the ship's computer. *SEE COMMUNICATIONS SATELLITE.*

Any ECDIS system, legal equivalent or not, is free to receive and use the chart updates. Their use will clearly add to the safety of navigation in any harbor covered by these corrections.

Improvements from use of ECDIS. The composite display of chart, ship's present position and previous track, and the image of radar targets is a complete portrayal of the tactical situation facing the vessel. The electronic chart shows water depths in the

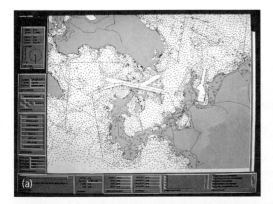 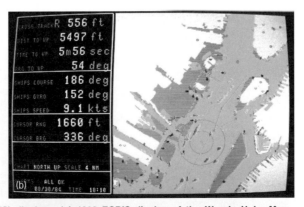

Electronic Chart Display and Information Systems (ECDIS) displays. (*a*) 1992 ECDIS display of the Woods Hole, Massachusetts, area. Resolution is 1728 × 1024 pixels for the entire image. Soundings on the chart cover the whole area, approximately 2.1 × 1.7 nmi (3.9 × 3.1 km). There is no radar image. (*b*) 1984 ECDIS Display, showing New York harbor in the vicinity of Governor's Island. Resolution is 400 × 400 pixels. The distance from top to bottom is 4 nautical miles (7.4 km). Land and water areas and radar images are visible. The ship on which the ECDIS is located is in the center of two concentric circles. The elongated radar image to the left and just above this ship is the *Queen Elizabeth 2*.

form of discrete soundings as well as depth curves. Other hazards, such as wrecks, submarine cables, danger areas, and anchorages, are well marked. Lights, buoys, and channel edges, which are the aids for marking safe routes, are represented electronically. The onboard computer is used to plan a voyage before leaving port; track lines that denote safe routes are drawn on the chart with electronic tools. These track lines are stored for use during the voyage.

The electronic positioning system that provides once-per-second updates of the ship's latitude and longitude is combined with the expected position from the previously stored voyage plan, and errors from this track are computed and displayed. If the vessel is being driven off its planned track by wind or current, it is immediately obvious and corrective action can be taken. If the radar shows a collision possibility with another vessel, course changes can be made and then can be checked against bottom depths on the new heading so as to avoid grounding.

If the radar's map of the surroundings is not aligned with that of the electronic chart there is error in the position computed by the electronic locating system. Shifting the two displays relative to each other so that they match virtually eliminates that error. If the displays cannot be brought into alignment with only minor adjustments, the GPS or loran is malfunctioning.

The display is not limited to just chart, position, and radar. The electronic system can file and display on demand virtually unlimited amounts of supporting information, such as the *List of Lights, Coast Pilot,* or *Sailing Directions,* as well as other nautical publications or items of general information.

The ECDIS display is designed to convey maximum information rapidly and completely. Displaying both harbor traffic and the ship's progress on the background of a chart allows the officer of the watch to verify information obtained from the ship's lookouts and radar and to be sure that the ship's track is over sufficient water depth. If a threatening situation suddenly arises, such as a potential vessel collision, the same display quickly shows whether a turn-away is into safe water. It is an action display rather than one demanding analysis before action can be taken.

Updating the electronic chart by radio transmissions materially adds to safe navigation by advising the ship of the latest situation in the harbor. Even a ship coming into port from sea can be provided with the latest advice and information concerning new hazards in that harbor. The fact that this advice is provided automatically reduces the chances that the change will not be seen by all crew members on the bridge.

Finally, ships equipped with ECDIS will be in a much better position to cooperate with the vessel traffic services (VTS) that are being installed in an increasing number of harbors, both in the United States and abroad. The situation display of ECDIS on the mariner's ship, including radar images of surrounding vessels, is quite similar to the VTS display on shore. A suggested course change sent to a ship will be tested by the ship's officers or pilot against the disposition of ships and tracks as seen on the ship's own ECDIS

display. Since this display and the one on shore will be virtually the same, there is a high probability that the negotiation will be brief and end in accord. *SEE MARINE NAVIGATION.*

Prospects for ECDIS. ECDIS has become a permanently established component of marine navigation systems. The non-legal-equivalent equipment and supporting chart databases can be purchased from many sources in many countries. These systems will continue to grow in number and sophistication, including additional integration of radar, information files, and other ships' instruments and sensors.

The legal-equivalent ECDIS will undergo the most dramatic growth. By the time that international standards have been adopted, new commercial systems will be offered that exploit the features and benefits of ECDIS. Its use is very likely to become mandatory, especially in those harbors and on those ships where hazardous cargoes are being handled. Thereafter its growth will be rapid.

Maritime shipping operations can look forward to an era of reduced frequency of accidental groundings and collisions. Thus, operating costs for shipping companies will be reduced, and the marine and shoreline environment will be better protected.

For background information *SEE ELECTRONIC NAVIGATION SYSTEMS; MARINE NAVIGATION; SATELLITE NAVIGATION SYSTEMS* in the McGraw-Hill Encyclopedia of Science & Technology.

Mortimer Rogoff

Bibliography. M. Rogoff, Electronic charts in the nineties, *J. Inst. Navig.,* 37(4):305–318, Winter 1990–1991.

Electronics

Throughout the twentieth century, the size of active electronic devices has been decreasing and the device count and complexity of electronic systems increasing. Will the pace of development be maintained, and are there ultimate limits to device dimensions? These questions have been posed on numerous occasions since the production of the first integrated circuits in the 1950s. It has been conjectured that molecules are the smallest entities that can display an electronic function. However, although molecules may possess electron distributions, which can be altered and sensed externally, until recently the precise control of individual molecules required to realize even a simple device has seemed an unattainable goal.

Molecular materials are used in electronic devices. Liquid-crystal displays are an important technology, as is the use of molecular semiconductors in photocopiers and facsimile machines. The development of molecular materials for such uses is dependent on the prediction of the macroscopic properties by microscopic models. This area, called molecular materials for electronics, is a vital, integral part of molecular electronics. Molecular-scale electronics, in which the focus is on the properties and use of either individual molecules or small molecular aggregates, is still in its infancy and cannot exist in isolation from the study of molecular materials for electronics.

Molecular materials for electronics.

Major classes of molecular materials used in electronic devices include liquid crystals, conducting polymers, and nonlinear optical materials.

Liquid crystals. Liquid crystals are the prime example of the use of molecular materials in devices and also of the time lag between initial discovery and commercial device production. Liquid crystals have been known for nearly a century, and device applications were patented more than 50 years ago. However, materials suitable for practical liquid-crystal displays were not synthesized until the 1970s. The two most important liquid-crystal phases of rodlike molecules are the nematic and smectic phases (**Fig. 1**). In the nematic phase, the molecules align close to a common direction, the director, but are otherwise disordered. Smectic liquid crystals have planar arrays of molecules with a common director within each layer.

The order in liquid crystals derives from the secondary, nonbonded interactions between molecules. The order is sensitive to molecular structure and external forces, such as electric and magnetic fields and surface forces. This sensitivity is used in the twisted nematic display (Fig. 1c). Rubbed polymer films are used to orient the directors on opposite sides of a thin liquid-crystal film at 90° to one another. The plane of polarization of light traveling normal to the film is rotated as it follows the molecular orientation. An electric field applied via transparent electrodes reorientates the molecular axes normal to the film, eliminating the rotation. Dark-light contrast between areas with and without rotation is obtained by use of an analyzer.

Ferroelectric liquid crystals have a tilted smectic phase and molecules with lateral electric dipoles that align in a ferroelectric phase. Electric-field reorientation occurs by rotation of the director about the normal to the smectic layers, and is much faster than the reorientation of nematic liquid crystals. Since the ferroelectric phase remains ordered, the field is needed only during reorientation. Thus, faster displays with stable switched states and a gray scale can be constructed; these displays could dominate the display market. A variety of high-speed electrooptical switches, filters, and so forth that would be useful in optical communications systems are under development.

Conducting polymers. Polyacetylene (**Fig. 2a**) was the first striking example of a polymer displaying metallic conductivity, that is, a conductivity of about 1000 siemens per centimeter when doped with sodium (Na), lithium (Li), triodide (I_3^-), or arsenic hexafluoride (AsF_6). Sample quality has been improved by the use of soluble precursor polymers and, more recently, by employing carefully aged catalysts. Conductivities as high as 100,000 S cm^{-1} are attainable but involve a temperature dependence unlike that of normal metals because carriers must tunnel between the conductive polymer chains. It is speculated that the ultimate conductivity is in excess of 10^6 S cm^{-1}. (The room-temperature conductivities of copper and silver are about 600,000 S cm^{-1}.)

The physics of semiconducting polymers is dominated by structural relaxation in the vicinity of a charge carrier. New states, termed solitons and polarons, appear in the gap between the valence and conduction bands. In contrast, structural defects reduce the interaction of the carbon atoms in the chain and give a locally wider band gap. Thus, semiconducting polymers have intrinsic states in the band gap but few defect states. Field-effect transistors with polyacetylene as the active layer have switchable absorption below the band edge and a large effective electrooptic coefficient. However, device speed is limited by the low carrier mobility, about 10^{-4} cm^2 V^{-1}s^{-1}. The use of oligomeric materials (substances made of molecules with a small number of identical repeat units) and polymeric substrates gives higher values around 1 cm^2 V^{-1} s^{-1}. The devices can be deformed without degradation in performance, which is close to that of amorphous silicon.

The discovery of electroluminescence in semicon-

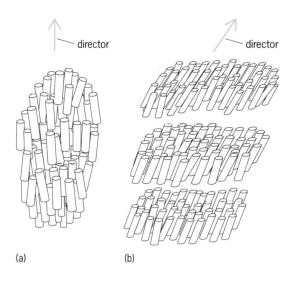

(a)   (b)

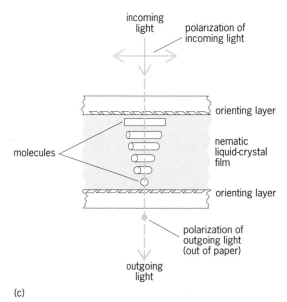

(c)

Fig. 1. Liquid crystals. (*a*) Nematic phase. (*b*) Tilted smectic phase. (*c*) Structure of a simple twisted nematic display.

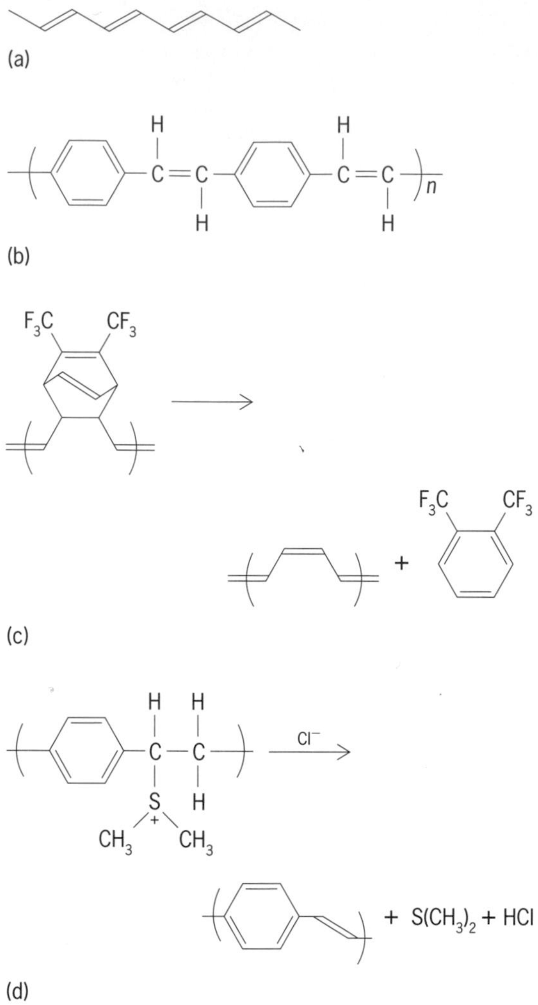

(a)

(b)

(c)

(d)

Fig. 2. Conducting polymers. (a) Polyacetylene. (b) Poly-paraphenylene vinylene. (c) Precursor polymer route to polyacetylene and (d) to polyparaphenylene vinylene.

ducting polymer films has attracted interest because of the ease of coating large-area electrodes. Initial work on polyparaphenylene vinylenes (Fig. 2b) has demonstrated output wavelengths from the blue to the red. The device efficiency is reasonable, but the lifetime is as yet inadequate for commercial use.

Since 1988, progress has been rapid, and there are fascinating prospects for applications as basic studies improve understanding of these materials and how best to fabricate devices.

Nonlinear optical materials. Organic materials have been prominent in the search for better nonlinear optical materials for optoelectronic devices. Large single crystals with high damage thresholds have been produced by using quantum-chemical calculations of molecular nonlinearity as guides for synthesis of new materials. Unfortunately, it is difficult to fabricate either fibers or films as required for integrated optical systems. Glassy polymers, which have low optical loss and are better suited for this purpose, lack the asymmetric structure essential for a nonzero linear electrooptical coefficient. Asymmetry can be introduced into polymers with laterally attached, dipolar, nonlinear optical chromophores by electric fields ap-

plied either with fixed electrodes or a corona discharge. Polarized films relax, but lifetimes of several years at temperatures of 158–176°F (70–80°C) have been reported. Materials suitable for electrooptical devices are available, but further progress may be limited by the perceived size of the market.

On the basis of current theory, conjugated polymers have been studied for all-optical devices. Nonlinear absorption is a problem at wavelengths close to the strong optical absorptions in such materials, and at wavelengths away from resonance the magnitude of the nonlinearity is at the lower limit of usability for devices. Progress has been slow, and practical all-optical devices based on electronic nonlinearities appear some way off.

Molecular-scale electronics. There have been many proposals for the realization of electronic device functions with molecules, but few have been subjected to an experimental test. The advent of techniques capable of atomic resolution imaging of molecules deposited on suitable substrates makes such a test possible. These methods are based on the sensitivity of electron tunneling currents to probe-surface separation (in the scanning tunneling microscope), the repulsive forces between atoms at small separations (in the atomic-force microscope and variants), and the localization of evanescent waves near surfaces (in the near-field optical microscope and variants). These tools also enable individual atoms or molecules to be addressed and interrogated, for example, with electrons in the scanning tunneling microscope, mechanical forces in the atomic-force microscope, and photons in the optical near-field microscope. Thus, molecular-scale science is now possible. Although some results have proved to be due to artifacts, rapid progress has been achieved. Individual molecules have been imaged, spectra recorded, chemical reactions of single molecules effected, and the products observed. These studies will establish whether molecules can be used as electronic devices at the molecular scale.

At the same time, there is a convergence of the disciplines toward the molecular scale. The techniques discussed above have been developed within the framework of physics and engineering wherein nanolithographics has produced low-dimensional metal and semiconductor structures with sizes on the molecular scale. In chemistry, the synthesis of complex structures with strong primary bonds has been mastered, and the self-assembly of complex biological systems is now being mimicked through the concerted production of secondary structures controlled by weaker interactions, such as hydrogen bonding. Since fabrication at the molecular scale, component by component, seems impractical, the ability to direct the self-assembly of the molecular units without external interference appears crucial. Biology provides the proof that molecular-scale electronics is possible in principle through the existence of living systems. Molecular biology focuses on understanding the molecular-scale mechanisms on which life is based, and provides models for the development of simpler synthethic analogs.

Molecular-scale science is in its infancy. The diversity of synthetic and biological molecules makes it difficult to identify particular avenues of research that will lead to applications in the long term. There is a need to pool international as well as interdisciplinary knowledge and resources if progress is to be achieved. Cross-fertilization with the discipline of molecular materials for electronics will follow as molecular-scale science tests the microscopic models used to predict macroscopic properties.

This discussion gives only a hint of the scope of this field. Other important topics include bacteriorhodopsins as materials for optical logic and data storage, deposition by true organic molecular beam epitaxy, ion channels that form living neural networks, analogs of neurons and the visual system in silicon, and molecular rectification in monomolecular films. The challenges are enormous but so are the ultimate prospects for device technology.

For background information SEE ARTIFICIALLY LAYERED STRUCTURES; BIOELECTRONICS; LIQUID CRYSTALS; NONLINEAR OPTICS; ORGANIC CONDUCTOR; POLARON; SCANNING TUNNELING MICROSCOPE; SOLITON in the McGraw-Hill Encyclopedia of Science & Technology.

David Bloor

Bibliography. D. J. Ando and M. G. Pellatt (eds.), *Fine Chemicals for the Electronics Industry II*, Roy. Soc. Chem. Spec. Pub. 88, 1991; D. Bloor, Molecular electronics, *Physica Scripta*, T.39:380–385, 1991; R. A. Hann and D. Bloor (eds.), *Organic Materials for Non-linear Optics II*, Roy. Soc. Chem. Spec. Pub. 91, 1991; H. Kuzmany, M. Mehring, and S. Roth (eds.), *Electronic Properties of Polymers*, 1992.

Electrorheological materials

Electrorheological materials are materials whose physical state and properties of that state are controlled by an imposed electric field. These materials can dramatically change the way work is done by providing a synergistic relationship between work performance and computer technology.

Electrorheology. Rheology is the science of the flow and deformation of matter, that is, the response or resistance of materials to force or stress. At one extreme, an ideally elastic material will deform reversibly to balance the applied stress and then completely recover upon the removal of the stress such that no energy is lost in the process. At the other extreme, an ideally viscous material will undergo an irreversible change in shape (or flow) only in response to a stress; and it will stop flowing when the stress is removed, dissipating all the energy applied to it as heat. An elastic material is characterized by an elastic modulus (stiffness) that is related to its deformation from an applied stress, whereas a viscous material is characterized by its resistance to flow or viscosity. Most real materials are neither perfectly elastic nor viscous but combine both qualities. Such materials are described as viscoelastic; that is, their rheological properties are characterized by complex quantities such as a complex

modulus in which the real part (the storage modulus) characterizes the elastic properties and the complex part (the loss modulus) characterizes the liquid part.

The rheological properties of an electrorheological material (ERM) are very strong functions of the applied electric field. Electrorheological materials that are liquids under ambient (zero field) conditions demonstrate tremendous increases in their resistance to flow (apparent viscosity); under appropriate conditions of stress and field they can solidify (**Fig. 1**). In the solid state both the elastic and loss moduli continue to increase substantially with increasing field. Upon removal of the field, the rheological properties of electrorheological materials return in fractions of milliseconds to their ambient values.

Composition. Electrorheological materials are typically suspensions of fine hygroscopic particles in a hydrophobic, nonelectrically conducting dispersion medium. Particle sizes in the range of 0.1–10 micrometers are common, although much larger particles as well as some macromolecular solutions have demonstrated electrorheological effectiveness. Materials that work well as the dispersed phase include cornstarch, various clays, silica gel, talcum powder, and various polymers. The fluid phase also may consist of a wide range of liquids that have in common the property of high electrical resistivity (high fields may be applied over such fluids without significant currents), and they are hydrophobic. Liquids such as kerosine, mineral oil, toluene, and silicone oil work well, as do many other fluids. In addition, most systems require that significant amounts of water (10–30%) be adsorbed onto the particulate phase. This requirement severely limits the potential use of water-mediated electrorheological materials. The problems include thermal runaway, which

Fig. 1. Solidified electrorheological fluid penetrated by a wooden dowel. The high voltage source is seen in the background.

is a progressively increasing resistive heating of the fluid and associated water loss; relatively high currents and corresponding power consumption; electrolysis; limited use temperatures between 0 and 100°C (32 and 212°F); and irreproducibility of fluid properties because of instability of the adsorbed water.

Recently, particulate phases have been discovered that produce electrorheologically active materials without the need for water, and they exhibit similar properties. Such systems include a number of aluminosilicates, semiconducting polymers, and polyelectrolytes. Although each of these systems has its unique problems, the fact that they exist implies that the electrorheological mechanism can be an inherent part of the molecular structure or chemistry of the particulate phase. Therefore, characterization of these systems should lead to synthesis of materials with enhanced desirable properties of the particulate phase.

Applications. The importance of electrorheological materials is in their application and their ability to compete with conventional methods of vibration isolation, power transmission, and hydraulic control. The criteria for comparison must be based upon economics, reliability, performance, efficiency, and even packaging. It is in these areas that the concept of devices incorporating electrorheological materials seems to hold promise. Such devices are generally much simpler, with far fewer moving parts and less critical tolerances; the control powers can be very small, since control is induced by the field (not the current); and the devices can be controlled directly by microprocessors.

The most appropriate applications include damping (vibration control), torque transmission and control, and hydraulic fluid flow and modulation. These applications are being developed to various degrees, and numerous patents exist for each. None of these applications has yet reached the commercial stage. The potential for implementation depends on the stage of development of electrorheological fluids and the configuration of properties required for specific applications.

Vibration isolation. Isolation devices to control mechanical vibrations are nearest to becoming commercially available. Applications include engine mounts, active suspension devices in automobiles, isolation devices for helicopters, sound damping in submarines, and even earthquake and wind control for bridges and skyscrapers. An example of a possible but simplified design of a shock absorber is shown in **Fig. 2**. Current properties of electrorheological fluids, including yield stresses of around 1 lb/in.[2] (6894 pascals), response speeds of less than a millisecond, and the low powers required, make such damping devices presently feasible. A potentially simple use of electrorheological dampers involves complex or coordinated vibration control, such as coordinating engine vibration control with road vibration. There are some problems associated with present materials. One such problem involves the stability of the fluids, since all the materials are suspensions, which can settle and plug fine orifices. Another problem involves thermal stability at the high temperatures that can be generated

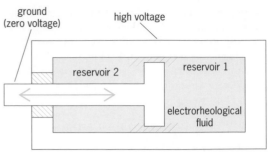

Fig. 2. Diagram of a simplified version of an electrorheological shock absorber. As the piston moves, the electrorheological fluid flows between reservoir 1 and reservoir 2 by way of the narrow gap between the positive and negative electrodes. Varying the voltage difference across the electrodes varies the resistance and thereby the damping characteristics.

locally in the fluids in some device designs. This problem occurs mostly in water-based systems and systems using polymeric particulates (either polyelectrolytes or semiconducting polymers) that are thermally stable only at low temperatures.

Power transmission. Torque transmission or clutching devices are an important application of electrorheological materials for many reasons. Mechanically they are very simple devices (compared to torque converters) with no moving parts and noncritical tolerances; are able to control torque transmission, speed, and lock up (an operation of the clutch in a nonslip mode, when the fluid is solidified), to minimize energy losses; and isolate vibration between the engine and drive train. Devices patented or used for torque transmission are straightforward for the most part (**Fig. 3**), including single or multiple disks alternately attached to the engine and power train, and single or multiple concentric cylinders. The presently available electrorheological fluids permit applications that involve only relatively low levels of load capacity, for example, accessory drives in automobiles or in consumer appliances such as air conditioners. Fluid couplings for automotive drive trains according to present designs and with current fluid strengths of around 1 lb/in.[2] (6894 Pa) will need to be too large to be practical. In-

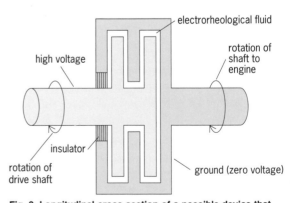

Fig. 3. Longitudinal cross section of a possible device that would employ an electrorheological fluid for torque transmission. When an electric field is applied, the fluid becomes more viscous, transmitting torque from the engine shaft to the drive shaft.

creases of 5–10 times in fluid strength will make such devices practical. Other problems that must be solved but are not insurmountable include fluid durability, containment, effect on devices, and the availability of high-voltage, low-cost power supplies and controls. Unfortunately, relatively little federal and industrial funding is being directed to fluid development.

Hydraulic control. The use of electrorheological materials as the primary fluids in hydraulic systems and circuits is probably the most far-reaching and dramatic technologically. Virtually all large and many small machines—including large aircraft, submarines, heavy industrial machinery, automated industrial operations, and robotic systems—are controlled by hydraulic circuits. Even the previously discussed methods of isolation and torque transmission can involve hydraulic control.

A typical circuit contains a pump that pressurizes the fluid, an actuator that converts the pressurized fluid to be set into motion, and valves that control the flow of the fluid. Valves range from the simple on-off type to very complex servovalves that provide precise control to the actuator. Servovalves not only are expensive and necessitate very precise machining but also can be relatively bulky and require significant amounts of power. In contrast, electrorheological valves can be as simple as electrodes on opposite sides of a flow channel. An electrical field that is imposed across the flow channel can continuously vary the resistance to flow, thus varying the flow rate to the extent that the flow can be completely stopped.

Advantages include valves with no moving parts and potentially higher reliability, much lower cost of production, smaller sizes, lower power consumption, significantly increased speed of operation, and direct computer control. These qualities should permit development of much more complex integrated hydraulic systems that can be housed in smaller packages. At present fluid-performance levels, control of pressures up to 500 lb/in.^{2} (3.5 megapascals) at flow rates of 2 gal/min (8 liters/min) have been demonstrated.

Limiting factors include fluid durability and interaction with the system components, pumping and sealing of the fluids, plating of the fluid particles on the valve surfaces, particle settling, and degradation of the fluid components at the high temperatures that are sometimes generated in the fluids because of shearing in the pump and system. The addition of a heat exchanger in the circuit can alleviate the temperature problem, and use of alternating fields can partially alleviate the plating problem. Electrorheological fluids possessing higher yield stresses must be developed before application as the primary fluid in hydraulic systems.

Prospects. The field of electrorheological materials is still in its infancy. The mode of action of the materials is poorly understood, but the potential for making much better materials is great. The applications presented here are traditional, but others that have not yet been imagined will evolve from the unique properties of these materials in conjunction with computer technology.

For background information *SEE HYDRAULICS; MATE-* *RIALS SCIENCE AND ENGINEERING; RHEOLOGY; TORQUE CONVERTOR; VIBRATION DAMPING* in the McGraw-Hill Encyclopedia of Science & Technology.

Frank E. Filisko

Bibliography. H. Block and J. P. Kelly, Electrorheological fluids, U.S. Patent 4687589, August 18, 1987; H. Block and J. P. Kelly, Electrorheology, *J. Phys. D: Appl. Phys.*, 21:1661–1677, 1988; F. E. Filisko and W. E. Armstrong, Electric field dependent fluids, U.S. Patent 4744914, May 17, 1988; W. M. Winslow, Methods and means for transmitting electrical impulses into mechanical force, U.S. Patent 2417850, March 25, 1947.

Endurance physiology

During locomotion, animals translate the power provided by their muscles into forward propulsion. Running speed is determined by the available power and the mechanical and biochemical efficiency of the muscles. However, as the running duration increases, the level of power output declines exponentially (see **illus.**). Well over 1 horsepower (750 watts) of power can be maintained for bursts lasting only a few seconds, but only a fraction of that power can be maintained for 1 h. Thus, the factors that set peak power for short durations may be very different from those responsible for limiting performance over long durations. Endurance physiology involves adaptations of lung, heart, blood, and skeletal muscle. Perhaps the most effective endurance athlete is the pronghorn antelope, which is described below. Endurance athletes have no special or unique structures, only more of those structures that are responsible for the uptake, delivery, and utilization of oxygen in all mammals.

Speed versus endurance. Two kinds of athletic performance result in very different demands on the muscles, heart, and lungs. Power athletes, such as cheetahs or human sprinters, are limited in performance only by peak power output, which is solely a function of the mass of the working muscles and the contraction velocities of muscle fibers. These athletes have no need for increased oxygen supply to the

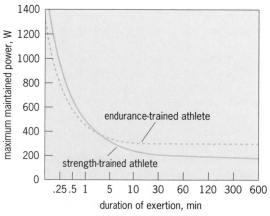

Power output in humans declines exponentially as the duration of exertion increases.

muscles during exercise; in fact, some of the fastest sprinters hold their breath when they run.

However, a different kind of athleticism is involved when sustained high-power output is required over long periods. A foraging hummingbird, a wolf running in a pack, a fleeing pronghorn antelope, and a human marathon runner must all maintain a high power output for long durations. These athletes rely on the delivery of adequate glucose and oxygen to their muscles and the simultaneous removal of heat and carbon dioxide during exercise. Thus, they need sufficient muscle mass to provide the necessary power as well as the ability to refuel the muscles while the muscles are working. The requirement for energy in the muscle fibers is translated into demand for oxygen for the oxidative production of adenosinetriphosphate (ATP), the ultimate source of energy for the muscle.

The performance of endurance athletes may be constrained either by the ability of the muscles to provide sufficient power or, more likely, by the ability of the respiratory and cardiovascular systems to provide adequate glucose and oxygen. The supply of oxygen rests collectively on those structures responsible for its uptake as well as those that deliver it to the tissues. The demand for oxygen is determined by the energetics of the working muscle. That demand is met only when sufficient oxygen is pumped by ventilation into the lungs, moves by diffusion into the blood, is pumped to muscle tissue by the heart, and finally moves by diffusion into the mitochondria within the skeletal muscle tissue. Endurance is largely determined by the amount of oxygen that can be delivered and utilized by these structures.

Human endurance athletes. Human endurance athletes differ from both nonathletes and sprint-trained athletes in the ability to consume and utilize large amounts of oxygen while exercising. Although intuitively one might expect to find that endurance athletes have large lungs, lung volume is actually quite constant among humans. However, once the oxygen passes through the lungs, it must be transported by the blood to the tissue, and endurance athletes have a greater concentration of oxygen-carrying hemoglobin in their blood and larger hearts than either nonathletes or sprinters. In addition, the oxygen must eventually be transported into the working muscle tissue, specifically into the mitochondria, the cell powerhouses that use oxygen in the production of usable energy in the form of ATP. Endurance athletes have many more capillaries that carry blood to muscle mitochondria, which are four to five times more dense than those in untrained individuals. Interestingly, both capillaries and mitochondria are actually less dense in sprinters than they are in untrained individuals. In this regard, the physical attributes of sprint versus endurance athletes are very different.

The best human endurance performance pales in comparison to the top nonhuman athletes. While the fastest humans can cover a 26-mi (42-km) marathon in a little over 2 h, a sled dog can do so in half the time, a race horse in under 1 h, and the pronghorn antelope in just over 30 min.

Pronghorn antelope. The pronghorn antelope (*Antilocapra americana*) evolved in the wide-open prairies of western North America, where its primary predators must have been wolves (endurance athletes) as well as now-extinct cheetahlike cats (power athletes). The combined pressures from both types of predators may have acted through natural selection to produce the unique ability of pronghorn antelopes to maintain unusually high running speeds. They are among the fastest terrestrial animals, with a top speed of 62 mi/h (100 km/h), second only to the cheetah. However, unlike the cheetah, these animals are distance runners, not sprinters. Pronghorn antelopes have been observed to run 7 mi (11 km) in 10 min, averaging over 40 mi/h (65 km/h).

There are two potential explanations for this unusual ability. These animals may have evolved a uniquely high locomotor economy, or low cost of transport. (Cost of transport is the energy needed per unit distance covered per unit body mass, somewhat analogous to gas mileage in an automobile.) Another possibility is that they can process and utilize oxygen at unusually high rates. To distinguish between these two possibilities, the maximum oxygen uptake can be established by running pronghorn antelopes on treadmills and capturing all expired air. The results of oxygen uptake measurements in pronghorn antelopes and many other animals have provided a great deal of information regarding the energetics and biomechanics of locomotion in mammals.

Maximum oxygen consumption in the pronghorn antelope is among the highest recorded for large animals. The maximum weight-specific aerobic capacity in these animals is about equal to that expected for a 0.35 oz (10 g) mouse. Interestingly, locomotor efficiency (or cost of transport) in the pronghorn antelope is apparently unexceptional. The measured rate of oxygen uptake corresponds to that predicted for a mammal of equivalent size to sustain running speeds of 43 mi/h (70 km/h).

Structures responsible for endurance. Maximum oxygen uptake in the pronghorn antelope is nearly five times that of similarly sized domestic goats. Consistent with this observed difference, the lungs of the pronghorn antelope have about five times more tissue for oxygen diffusion into the blood. The heart of the pronghorn antelope is three times larger, and it has twice the blood volume and a 50% greater concentration of hemoglobin in the blood than does the domestic goat. In addition, the total volume of skeletal-muscle mitochondria is much greater than in similarly sized goats. The extremely high aerobic capacity of the pronghorn antelope is thus attributable to enhancement of those structures that exist in all mammals.

For background information SEE BIOLOGICAL OXIDATION; CARDIOVASCULAR SYSTEM; MUSCLE; PRONGHORN; RESPIRATORY SYSTEM in the McGraw-Hill Encyclopedia of Science & Technology.

Stan L. Lindstedt

Bibliography. R. Alexander, Optimization and gaits in the locomotion of vertebrates, *Physiol. Rev.*, 69: 1199–1227, 1989; J. H. Jones and S. L. Lindstedt,

Limits to maximal performance, *Annu. Rev. Physiol.*, vol. 55, 1993; S. L. Lindstedt et al. Running energetics in the pronghorn antelope, *Nature*, 353:748–750, 1991; T. M. McMahon, *Muscles, Reflexes, and Locomotion*, 1984.

Environmental engineering

Municipal solid waste is a discard of society that is generated in the home, workplace, streets, institutions, and places of assembly. If ignored or improperly managed, it can adversely affect public health, damage the environment, and impair the esthetics of the community.

Management. In the United States, management of municipal solid waste had been a responsibility of local and state government until the passage of federal laws such as the 1965 Solid Waste Disposal Act and the 1975 Resource Conservation and Recovery Act (RCRA). These acts, together with air-pollution and water-quality laws, empowered the federal government, through the Environmental Protection Agency (EPA), to promulgate regulations controlling the treatment of municipal solid waste. Federal laws generally establish environmental standards; the states, guided by these standards and by local needs, issue regulations for local governments. The task of defining on-site storage and providing for collection, transport, processing, and disposal service for municipal solid waste continues to be a responsibility of the local government; however, it may be shared with a county or regional agency.

Task definition. Municipal solid waste is defined as essentially household waste and street refuse, but it can include waste from commercial establishments and institutions. However, industrial process wastes, demolition wastes, agricultural wastes, abandoned automobiles, ashes, sewage, and sewage sludge are not defined as municipal solid waste. The definition of municipal solid waste may vary for states and municipalities, but it is generally consistent with this definition.

A 1988 study conducted for the EPA estimated a national average production rate of 3.5 lb (1.6 kg) of municipal solid waste per capita per day, resulting in the generation of 157.7×10^6 tons per year (143.2×10^6 metric tons per year). The study, anticipating population growth, estimated that the annual quantity of municipal solid waste produced in the United States will increase to 192.7×10^6 tons (174.9×10^6 metric tons) by the year 2000.

The study for the EPA also included a breakdown of the components of the current municipal solid waste. This breakdown, based upon national averages, is shown in **Table 1**. It is important to understand that these percentages are average for the United States, and could vary significantly for different localities and population densities. For example, in warm, humid climates such as the southeastern part of the country, yard waste could represent a much higher percentage of municipal solid waste than in arid desert areas such as New Mexico and Arizona.

Those agencies responsible for the management of municipal solid waste must develop a plan for providing a level of service that will protect the public health and the environment as defined by relevant laws, will meet the convenience and esthetic levels as determined by the desires of the community, and can be implemented by the personnel, equipment, and facilities available to the community. The principal processes for treatment and disposal currently used for solid waste management by municipalities are recycling, composting (considered a recycling process in some states), incinerating for the generation of energy, and landfilling. Plans for management of municipal solid waste by urban communities generally include at least two, and often all, of the processes noted. An important aspect of a management plan is a method for the collection of municipal solid waste that must be integrated into the process and disposal methods selected. To conserve resources and reduce the quantity of municipal solid waste that must be disposed of, mandatory recycling, of up to 50% of the municipal solid waste generated by the year 2000, has become a requirement in many states.

There are other processes available for the treatment of solid waste. Pyrolysis is a high-temperature waste degradation process that uses the combustible gases produced from the heating of the waste as fuel for heat generation. The end products of pyrolysis are the combustible gases, a high-viscosity combustible oil, a char, and a noncombustible residue. Although used successfully to convert coal to coke and gas, pyrolysis has not been economically viable as a means of processing solid waste. Some waste-to-energy facilities are designed to process solid waste that has been prepared prior to burning by shredding and removing much of the noncombustibles. The end product of this process, a refuse-derived fuel (RDF), can be incinerated in furnaces designed for the use of refuse-

Table 1. Composition of municipal solid waste, based on national averages*

Component	Weight percent
Yard waste	17.9
Corrugated paper	12.3
Newsprint	8.0
Office paper	3.9
Other paper	16.8
Food wastes	7.9
Ferrous metals	7.0
Aluminum	1.5
Other metals	0.2
Plastics	6.5
Clear glass	4.9
Colored glass	3.3
Wood	3.7
Rubber and leather	2.5
Textiles	1.8
Miscellaneous inorganic materials	1.6
Other	0.2

* After Franklin Associates, Ltd., *Characterization of Municipal Solid Waste in the United States, 1960 to 2000*, March 30, 1988.

Table 2. Average prices for recycled material, in dollars per ton*

Product	1989	1990	1991	1992
Newsprint[†]	(−15)	(−17)	(−22)	(−30)
Office paper	50	60	60	0
Glass: clear	40	30	25	15
green	25	25	20	0
brown	15	15	10	10
Cardboard	60	40	30	0
Tin	15	15	10	10
Aluminum	50	50	50	60
Plastic (polyethylene terephthalate and high-density polyethylene)	N/A	N/A	60	35

*After A. J. Grosso, Executive Director, Delaware County, Pennsylvania, Department of Solid Waste, private communication, May 12, 1992.
[†] The minus figures are the cost per ton for disposing of newsprint.

derived fuel mixed with unprocessed municipal waste prior to incineration, or added, in limited quantities, to coal-fired boilers. This procedure also has had limited success. *SEE FUEL.*

Recycling. Although commercial recycling has been practiced successfully for decades in the United States and is a major factor in handling municipal solid waste in many parts of the world, the newly awakened national interest in recycling is driven by a desire to conserve resources, the public perception that disposal of municipal solid waste in populated areas could be harmful to the environment and public health, and the premise that reducing the amount of waste generated for disposal will reduce the total cost of disposing of it.

Examination of the composition of municipal solid waste, as indicated in Table 1, reveals that approximately 50% of this material could be recycled. In fact, corrugated paper has an impressive recycling record in the commercial waste category; glass, properly separated, was an important recyclable until the increased use of plastic containers; aluminum, because of the low cost to reprocess it into new metal, reportedly is recycled at a rate in excess of 60% of waste product; some yard waste is being composted (and thus recycled) successfully in many communities throughout the country; ferrous scrap metal, including industrial scrap, abandoned vehicles, and discards such as so-called white goods (discarded large enameled kitchen appliances) has a long record of successful recycling by the scrap metal industry; and a limited amount of paper, primarily old newspapers, was recycled profitably for several decades.

In the past, recycling was successful because it was driven by market demand. Recent state requirements that municipalities recycle specified quantities of municipal solid waste have resulted in the accumulation, for some products, of more recyclable material than industry is capable of absorbing. This excess has had a serious impact on the market value of recycled materials. An example of one community's experience with the changing value of recycled material is shown in **Table 2**.

Except for aluminum, the market values of recycled materials have dropped significantly in recent years. In evaluating the impact of recycling on management of municipal solid waste, the desires of the community, with few exceptions, include acceptable service at minimal cost. The costs of recycling include collection of the material, operation of materials recovery facilities, and preparation of the material for market, which could involve treatment of process effluents. Savings result from the reduction in the amount of waste remaining for collection, processing, and disposal, as well as from the offsetting value of the sold recycled materials. The decreasing value of recycled material as experienced by Delaware County, Pennsylvania (Table 2), reveals the difficulty that local governments are encountering in their efforts to meet recycling goals.

Composting. Composting is a process in which organic wastes undergo microbial degradation, yielding a nuisance-free product of potential value as a soil conditioner. Composting has a history of successful application in some parts of the world but poor results in the United States. For compost to be usable as a soil conditioner for agriculture, it must be processed aerobically to kill harmful pathogens. If it is to be used only for horticulture, anaerobic decomposition may be acceptable.

In recent years, as a part of the effort to reduce the quantities of municipal solid waste requiring disposal, the composting of yard waste, particularly leaves, is being encouraged by many states. Homeowners are often requested to operate backyard leaf composters and not to bag grass clippings. It is argued that clipped grass decomposing on the lawn enriches the soil. If, as is true in a number of states, yard-waste compost is acceptable as a recyclable material, the goal of recycling 50% of municipal solid waste may be attainable. However, the value of yard-waste compost has yet to be determined. Furthermore, yard-waste compost may be in competition with sewage-sludge compost, another waste product that is impacted by the perceived shortage of landfill space.

The common static-pile method of composting involves piling leaves in windrows and turning them yearly. This process may take 2 or more years to produce a useful product. However, several mechanical, aerobic processes can produce a usable compost in less than 3 months. These processes are popular in parts of the world where there is a demand for compost to enrich agricultural soils. The limited recent experience with mechanical composting in the United States provides the manager of municipal solid waste with good information about the value of compost so produced. The effectiveness of composting, both of yard waste and other organic fractions of solid waste, could be determined in the next several years.

Waste-to-energy. The use of municipal solid waste as a fuel to generate power has been practiced in the United States for more than 100 years. However, only since the 1970s, with the dramatic increase in fuel costs and the public opposition to landfill siting near populated areas, has generating energy from waste become attractive. Furthermore, the fuel value of municipal solid waste has increased significantly in recent years. In the 1960s, the average heat value of solid waste was approximately 4000 Btu/lb (9.3 megajoules/kg); today the average is approximately 5000 Btu/lb (11.6 MJ/kg). In addition, power plants fueled by solid waste emit considerably less sulfur and nitrogen oxides than most coal- and oil-fired plants generating comparable quantities of electric power. SEE FUEL.

The number of facilities for converting waste into energy is increasing. **Figure 1** depicts the increase in the number of waste-to-energy plants constructed since 1975, with a projection to 1994. The drop in plant start-up rate beginning in 1990 was caused by the introduction of regulations such as mandated levels of recycling that influence the amount of solid waste that communities will have available for incineration. It has been predicted that waste-to-energy plants will continue to increase in number.

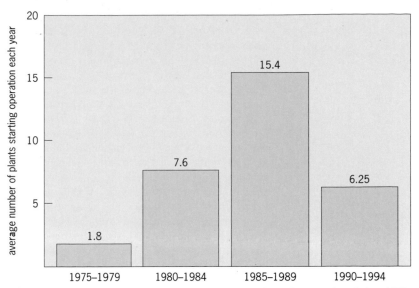

Fig. 1. Chart showing the number of waste-to-energy plants in operation during 1975–1994 (1990–1994 figures are estimated). (*After C. Vansant and M. Hilts, The status of waste-to-energy: What role in integrated waste management?, Sol. Waste Power, pp. 12–19, May/June 1992*)

Sanitary landfilling. Regardless of how successful communities are with processing waste by composting, recycling, or converting it to energy, a significant quantity of nonprocessible waste and treatment residues remains for disposal. The acceptable means of disposing of these materials is by sanitary landfilling.

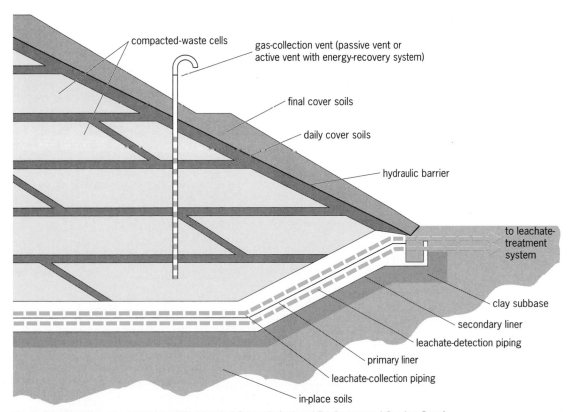

Fig. 2. Cross section of a typical landfill. (*Applied Geotechnical and Environmental Service Corp.*)

Although the controlled dumping of waste has been practiced by humans for thousands of years, the procedure of containing waste in sanitary enclosures each workday, that is, sanitary landfilling, was introduced in the United States in the early 1940s. Since then, there has been little change in the landfill process of dumping and compacting the waste in prepared cells, and controlled covering at the end of each workday. However, there have been significant changes in the preparation of the landfill base and cells, the management and treatment of leachate liquids and generated gases, the placement of the final cover, and the maintenance of the completed sanitary landfill. A cross-sectional view of a typical sanitary landfill based on current design requirements is shown in **Fig. 2**.

Although the landfill in Fig. 2 meets most regulatory requirements, the hydraulic barrier at the top of the completed fill is an aspect of design about which there is a difference of opinion among environmental engineers. The design shown is referred to as a dry tomb, whereas a design with a permeable cover is known as a wet cell. Environmental engineers differ as to the desirability of accelerating decomposition of waste with the addition of moisture, or retarding it by preventing the infiltration of precipitation. This difference, which could significantly affect the long-term management of sanitary landfills, should be resolved with experience.

Prospects. Solid waste management is in a state of flux, and those responsible for providing this service must make decisions, both short- and long-term, that recognize the changing technology, public concern, and economics.

For background information SEE ENVIRONMENTAL ENGINEERING in the McGraw-Hill Encyclopedia of Science & Technology.

Abraham Michaels

Bibliography. Franklin Associates, Ltd., *Characterization of Municipal Solid Waste in the United States, 1960 to 2000*, March 30, 1988; H. I. Hollander (ed.), *Thesaurus on Resource Recovery Terminology*, ASTM Spec. Tech. Pub. 832, 1984; M. D. Hollis, The environmental-health population syndrome, *Environ. Eng.*, pp. 8, 9, 18, and 19, July 1992; C. Vansant and M. Hilts, The status of waste-to-energy: What role in integrated waste management?, *Sol. Waste Power*, pp. 12–19, May/June 1992.

Fault detection

Technical systems include production equipment (chemical plants, steel mills, paper mills, power stations, and so forth); transportation vehicles (ships, airplanes, and motor vehicles); and household appliances (washing machines, air conditioners, and so forth). Although quite varied in purpose, mode of operation, value, and size, these systems share several common properties, such as material or energy conversion and continuous operation. In any of these systems, malfunctions of components may lead to damage of the equipment, degradation of its function or product, jeopardy of its mission, or hazard to humans. While the need to detect and diagnose malfunctions is as old as the construction of such systems, advanced fault detection has been made possible only by the proliferation of the computer. In current terminology, fault detection and diagnosis means a scheme in which a computer (or multiple computers) monitors the technical equipment to signal any malfunction and designate the components responsible for it. The detection and diagnosis of the fault may be followed by automatic actions enabling the system to operate under the particular faulty condition.

Diagnostic concepts. The fault detection and diagnosis activity covers both the basic technical equipment and the actuators and sensors attached to it. In the case of a chemical plant, for example, the basic technical equipment includes the reactors, distillation columns, heat exchangers, compressors, storage tanks, and piping. Typical faults include leaks, plugs, surface fouling, and broken moving parts. The actuators are mostly valves, together with their driving devices (electric motors and hydraulic or pneumatic drives). The sensors are devices measuring the different physical variables in the plant (such as thermocouples, pressure diaphragms, and flow meters). Actuator and sensor fault detection is extremely important because these devices are fault-prone, and a large plant usually employs several thousand of them.

Much recent research and development has been aimed at the on-line detection and diagnosis of faults. Equipment is constantly monitored during its regular operation by a permanently connected computer, and any discrepancy is signaled almost immediately. On-line monitoring is very important for early detection of any component malfunction, before it can lead to substantial equipment failure. In contrast, off-line diagnosis involves the monitoring of the system by a temporarily attached device under special conditions (for example, car diagnostics at a service station).

The diagnostic activity may be broken down into several logical stages. Fault detection is the indication of something going wrong in the system. Fault isolation is the determination of the fault location (the component that malfunctions), while fault identification is the estimation of the magnitude of the fault. On-line systems usually contain a detection and isolation stage; in off-line systems, detection may be superfluous. Fault identification is usually less important (except, for example, in tank-pipe systems where the size of a leak may be of critical significance). While detection, isolation, and identification may be performed sequentially, one stage triggering the next, they may also occur simultaneously as different outputs from the same diagnostic procedure.

Fault detection and isolation can never be performed with absolute certainty, because of circumstances such as noise, disturbances, and model errors. There is always a trade-off between false alarms and missed detections, the proper balance depending on the particular application. In professionally supervised large plants, false alarms are better tolerated and missed detections may be more critical, while in consumer equipment (including cars) the situation is the opposite.

Approaches. A number of different approaches to fault detection and diagnosis may be used, individually or in combination.

Limit checking. In this approach, which is the most widely used, system variables are monitored and compared to preset limits. This technique is simple and appealing, but it is not effective in detecting and isolating faults for two reasons. (1) The monitored variables are system outputs that depend on the inputs; to make allowance for the variations of the latter, the limits need to be chosen conservatively. (2) A single component fault may cause many variables to exceed their limits, so that it may be extremely difficult to determine the source.

Special sensors. These may be applied to perform the limit-checking function (for example, as temperature or pressure limit sensors) or to monitor some special fault-sensitive variable (such as vibration or sound). Such sensors are employed mostly in noncomputerized systems.

Multiple sensors. These may be applied to measure the same system variable, providing physical redundancy. If two sensors disagree (beyond a certain tolerance value), then at least one of them is faulty. A third sensor is needed to isolate the faulty component (and select the accepted measurement value) by majority vote. Multiple sensors may be expensive, and they provide no information about actuator and plant faults.

Frequency analysis. This procedure, in which the Fourier transforms of system variables are determined, may supply useful information about fault conditions. The healthy plant usually has a characteristic spectrum, which will change when faults are present. Particular faults may have their own typical signatures (peaks at specific frequencies) in the spectrum, allowing for their isolation.

Parameter estimation. This procedure utilizes a mathematical model of the monitored system. The parameters of the model are estimated from input and output measurements in a fault-free reference situation. Repeated new estimates are obtained on-line in the course of system operation. Deviations from the reference parameters signify changes in the plant. If the algorithm is so designed that values for the underlying physical parameters are determined from the model, then the faulty component may be isolated.

Consistency checking. This is another way of utilizing the mathematical system model. The fundamental idea is to verify that the observed plant outputs are consistent with those predicted by the model on the basis of the inputs (as shown on the left in **Fig. 1**). Discrepancies indicate a deviation between the model and the plant (parametric faults) or the presence of unobserved variables (additive faults). This concept is also called analytical redundancy, since the model equations describing the relationship among (unlike) plant variables are used in the same way as multiple sensors under physical redundancy. Of course, analytical redundancy hinges on the accuracy of the system model; model errors may cause the same symptoms as faults and lead to false alarms.

Analytical redundancy methods. These usually consist of two stages (Fig. 1). First, residuals are generated. These are quantities representing the inconsistency between the model and the observations. Second, the residuals are subjected to threshold tests and logic analysis to arrive at a detection and isolation decision. The residual generator algorithms rely on a mathematical plant model that may be static or dynamic, the latter in input-output or state-space format. The algorithms usually used are (1) parity equations, that is, the input-output equations in suitably rearranged form, (2) diagnostic observers, and (3) Kalman filters.

Although a single residual is sufficient to detect faults, isolation requires a set of residuals. For better isolation properties, such residual sets are enhanced so that, in response to a particular fault, they either have fixed directional properties or are structured, that is, confined to multidimensional subspaces (**Fig. 2**). Enhanced residuals may be obtained by parity equations or diagnostic observers, both methods leading to identical residual generators. With Kalman filters, residual enhancement is not possible, but the residuals possess certain statistical properties that make testing easier.

Residual generators need to be robust with respect to disturbances and modeling errors. Although decoupling from a limited number of disturbances may be handled within the framework of enhanced residuals, decoupling from modeling errors requires special considerations in the design of the diagnostic algorithm and usually is not completely attainable.

The detection decision, in general, involves determining if the residual vector is of nonzero size. For isolation, with direction-fixed residuals it is necessary to find the fault direction to which the actual residual lies the closest. With structured residual sets, each residual is tested separately, leading to a boolean vector, which is then compared to standard patterns. With Kalman filtering, the actual residual is compared to the residual behaviors evolving under different fault hypotheses.

Application to automobile engines. As an example, the on-line diagnostic approaches used for automobile engines will be briefly described. Although the complexity of this plant is not comparable

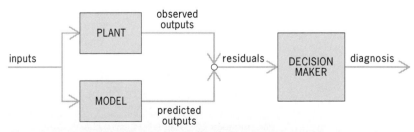

Fig. 1. Two stages of model-based fault detection and isolation. (*After R. Isermann and B. Freyermuth, eds., Proceedings of the IFAC SAFEPROCESS Symposium, Baden-Baden, Germany, September 10–13, 1991, Pergamon Press, 1992***)**

to that of large industrial production facilities, its significance is enormous because of its huge numbers and economic and environmental impact.

Traditionally, a few fundamental variables, such as coolant temperature, oil pressure, and battery voltage, have been monitored by using limit sensors. With the introduction of the on-board microcomputer, the scope of variables subjected to limit checking has been extended, for example, to include the intake manifold pressure. Also, active functional testing is being applied to at least one actuator, the exhaust-gas recirculation valve.

With increasingly stringent air-quality regulations in sight, manufacturers are considering the introduction of more advanced fault detection schemes to cover the components that affect the vehicle's emission control system. One approach under development (**Fig. 3**) uses parity equations to monitor two groups of actuators (fuel injectors and exhaust-gas recirculation valves) and four sensors (throttle position, manifold pressure, engine speed, and exhaust oxygen). The parity equa-

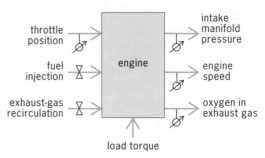

Fig. 3. Automobile engine system with on-board fault detection and diagnosis. (*After R. Isermann and B. Freyermuth, eds., Proceedings of the IFAC SAFEPROCESS Symposium, Baden-Baden, Germany, September 10–13, 1991, Pergamon Press, 1992*)

tions generate residuals that are insensitive to the load torque (a disturbance) and to the vehicle's mass (a model uncertainty). The residuals are also structured for the six faults considered and provide fault codes that, in most cases, uniquely characterize the faulty component.

The successful development of this approach requires the solution of a number of difficult modeling problems. The engine itself is strongly nonlinear; it involves a varying time delay; and the oxygen sensor has a relay-type characteristic with significant temperature dependence. In spite of these difficulties, such systems will probably be deployed on mass-produced automobiles along with more powerful on-board microcomputers.

For background information *SEE AUTOMOTIVE ENGINE; ESTIMATION THEORY; MODEL THEORY; PROCESS CONTROL* in the McGraw-Hill Encyclopedia of Science & Technology.

Janos J. Gertler

Bibliography. J. J. Gertler, Survey of model-based failure detection and isolation in complex plants, *IEEE Contr. Sys. Mag.*, 8(7):3–11, 1988; R. Patton, P. Frank, and R. Clark (eds.), *Fault Diagnosis in Dynamic Systems*, 1989.

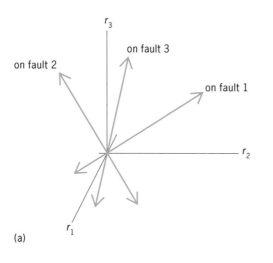

(a)

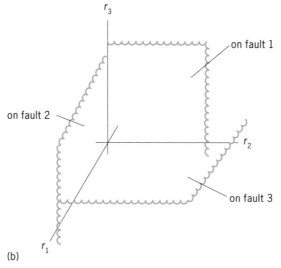

(b)

Fig. 2. Enhanced residual sets in analytical redundancy methods. The coordinates r_1, r_2, and r_3 are elements of the residual set. (*a*) Fixed-direction residuals. (*b*) Structured residuals. (*After S. Dhurjati, ed., Preprints of the IFAC Symposium on On-line Fault Detection, Newark, Delaware, April 22–24, 1992*)

Fescue

Tall fescue (*Festuca arundinacea*) is a cool-season perennial grass widely used for soil conservation, as a turfgrass, and as grazing and hay for livestock. Approximately 3.5×10^7 acres (1.4×10^7 hectares) of this grass were established in the United States before it was discovered, in the mid-1970s, that most tall fescue contains an endophyte, the fungus *Acremonium coenophialum*, that has toxic effects on grazing animals and reduces agronomic performance of the grass.

Historical perspective. Tall fescue, which is native to Europe, was introduced into the United States during the 1800s. A chance observation in 1931 of a tall fescue ecotype that had developed on a mountain hillside in eastern Kentucky led to the release of the variety Kentucky 31 in 1943.

The new tall fescue variety was easy to establish, had a long growing season, produced good forage yields, persisted under a wide range of management conditions, and was found to be widely adaptable in the eastern United States. As a result, from the mid-1940s to the early 1960s there was phenomenal acceptance and widespread planting of Kentucky 31, especially in the lower Midwest and upper South.

However, three distinct disorders of beef cattle, for which it was most frequently planted, were soon identified as being associated with tall fescue. Animals would often develop coats of persistent rough hair, make poor weight gains, and exhibit intolerance to warm weather. These were symptoms of a problem that became known as fescue toxicity.

While fescue toxicity was widespread, two other less frequent but potentially serious problems with cattle were identified. Fescue foot is a gangrenous condition that sometimes occurs on body extremities of cattle, especially the rear feet, during cold weather. Fat necrosis, associated with high levels of nitrogen fertilization, is the accumulation of large globs of hard fat within the abdomens of cattle, a condition that hinders food passage and reproduction.

Moreover, horse producers found that mares grazing tall fescue often had reproductive problems, including difficulty in foaling, delivery of weak foals, and agalactia (poor milk production). Sheep made poor weight gains on the grass. Dairy producers found it to be a particularly poor source of nutrition for lactating animals.

Tall fescue fungus. The only way to determine whether a tall fescue plant is infected with *A. coenophialum* is through laboratory examination of its tissue. The fungus is exclusively seed-transmitted. A sexual stage has never been observed; thus, the fungus cannot move from plant to plant as do other disease organisms. Infected plants occur only as a result of being established from seed containing viable fungus.

Tall fescue samples from thousands of pastures throughout the United States have been tested for this endophyte in recent years, and the overwhelming majority have been found to have 60–100% of the plants infected. This rate of infection can be directly attributed to the establishment of the Kentucky 31 variety from an ecotype that has been found to be over 90% infected. Thus, as this variety was widely disseminated, so too was the endophyte.

Effect on animal performance. Most grazing experiments have shown the daily gains of young beef animals to be decreased by about 0.1 lb (0.045 kg) for each 10% of field infection. When the ambient temperature exceeds 88°F (31°C), gains are even more affected, and visible signs of fescue toxicity become especially pronounced.

Experiments in many locations have revealed that the pregnancy rates, weight gains, and milk production of cows, sheep, and other forage-consuming animals are likely to be dramatically lowered when they consume infected tall fescue. Reproduction is especially likely to be affected when animals are under nutritional stress.

Many of the reproductive problems of horses grazing tall fescue are associated with the endophyte. Evidence indicates it is also responsible for fescue foot, and it has been suggested as the cause of bovine fat necrosis. The problems known to be caused by the endophyte are estimated to cause livestock production losses in the United States of between $500 million and $1 billion annually.

Toxic compounds. The fungus-produced toxins responsible for the disorders of animals grazing on tall fescue have not been clearly identified, but it is believed that one or more alkaloids are primarily responsible. In particular, there is strong evidence implicating ergovaline, an ergot-type alkaloid.

Agronomic characteristics. Knowledge of the relationship between the tall fescue endophyte and animal disorders quickly led to commercial availability of noninfected tall fescue seed. However, tests at various locations have revealed that when the endophyte is not present tall fescue usually displays several undesirable agronomic characteristics.

Traits that can be adversely affected in noninfected plants include germination rate, tiller numbers, seedling dry weights, seedling survival, and seed production. Evapotranspiration, which hastens drought stress, is substantially greater for noninfected plants, which have thinner and wider leaves, have larger stomatal apertures, and exhibit less leaf roll during dry periods.

Pest resistance of noninfected plants is less than for infected plants. Several species of insects prefer or survive better on noninfected fescue. Also, the numbers of several species of plant parasitic nematodes associated with tall fescue roots are typically much higher with noninfected plants.

Because of these differences in stress tolerance and pest resistance, greater stand declines have been shown for noninfected tall fescue in a number of experiments. Also, livestock producers who have planted commercially available noninfected tall fescue seed have often had difficulty establishing or maintaining stands. The differences in agronomic characteristics between infected and noninfected tall fescue are particularly likely to be apparent under stressful environmental conditions.

Impacts and implications. Knowledge of the endophyte and its effects has had several impacts. The tall fescue seed industry has been particularly affected because, although many forage producers want noninfected grass for improved animal performance, those interested in turf or soil conservation prefer the more stress-tolerant infected tall fescue. Thus, seed producers who once could market their seed for either purpose now must choose their market prior to planting a seed field.

Livestock producers who want to establish new fields of tall fescue also face a dilemma. If they plant infected tall fescue, animal performance will be reduced; if they plant noninfected tall fescue, they may have difficulty establishing or maintaining a stand.

Many livestock producers with existing stands of infected tall fescue exercise various management options to reduce the impact of the endophyte on their animals. These measures include keeping infected pastures closely grazed (a practice known to reduce the impact of the toxins), excluding animals from infected fields when the weather is hot or when animals are especially vulnerable to the toxin, and diluting the toxins by providing supplementary feedstuffs or by growing companion species (especially forage legumes) along with the infected grass.

Noninfected tall fescue is less competitive to weeds than infected tall fescue. Thus, weed encroachment in noninfected fields or a slow increase in infection level in fields with both infected and noninfected plants is likely.

Laboratory studies have shown that the reproduction of mice, rats, and rabbits is hurt by consumption of infected tall fescue. Furthermore, infected tall fescue is less palatable to animals. The clear implication is that game animals and other organisms living in the vicinity of infected tall fescue may be impacted by the endophyte, either by direct influence of the toxins on their growth and reproduction or as a result of reduced food intake due to unpalatability.

Prospects. Because tall fescue is an important and widely used species and because the effects of the endophyte are quite significant, research is active in this area. Objectives include clear identification of the toxins associated with the endophyte; identification or development of nontoxic *Acremonium* strains; development of more stress-tolerant noninfected tall fescue varieties; development of livestock treatments to offset the effects of endophyte toxins; and development of management techniques for minimizing toxic effects.

For background information SEE AGRICULTURAL SOIL AND CROP PRACTICES; FESCUE; GRASS CROPS; PLANT PATHOLOGY in the McGraw-Hill Encyclopedia of Science & Technology.

Donald M. Ball; Garry D. Lacefield

Bibliography. D. M. Ball, C. S. Hoveland, and G. D. Lacefield, *Southern Forages*, Potash and Phosphate Institute and Foundation for Agronomic Research, 1991; R. C. Buckner and L. P. Bush (eds.), *Tall Fescue*, Amer. Soc. Agron. Monog. 20, 1979; J. F. Pedersen, G. D. Lacefield, and D. M. Ball, A review of the agronomic characteristics of endophyte-free and endophyte-infected tall fescue, *Appl. Agr. Res.*, 3:188–194, 1990; J. A. Stuedemann and C. S. Hoveland, Fescue endophyte: History and impact on animal agriculture, *J. Prod. Agr.*, 1:39–44, 1988.

Flight

Birds are descendants of carnivorous dinosaurs. The oldest bird, *Archaeopteryx*, appeared about 1.5×10^8 years ago at the close of the Jurassic. *Archaeopteryx* is classified as a bird principally because it had evolved feathers, which are complex epidermal structures unique to living birds. Its skeletal anatomy, however, more closely resembles that of contemporary carnivorous dinosaurs from which it evolved. For many years paleontologists have known that toothed birds with modern avian skeletons, such as *Ichthyornis*, had evolved by the end of the Late Cretaceous, approximately 7×10^7 years ago. The uncertainties in bird

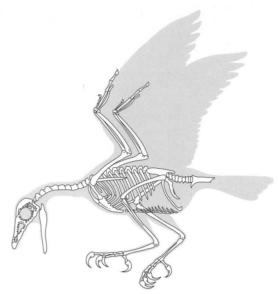

Reconstruction of *Sinornis* in flight, a newly discovered 1.25×10^8-year-old fossil bird from Liaoning Province in northeast China. (*From P. C. Sereno and C. Rao, Early evolution of avian flight and perching: New evidence from the Lower Cretaceous of China, Science, 225:845–848, 1992*)

evolution occurred in the intervening period, when the skeletal changes that allow modern flight performance first appeared. This important 8×10^7-year interval, the first third of avian history, has yielded but few avian remains.

Recently, however, there was a dramatic advance with the discovery of sparrow-sized bird skeletons in northeast China. The new fossil bird, *Sinornis* or Chinese bird, appears to be only $1–1.5 \times 10^7$ years younger than *Archaeopteryx* (see **illus.**). The bones are preserved on slabs of mudstone that once formed the floor of an ancient inland lake. In contrast to the fine-grained ooze that encased *Archaeopteryx* and its feathers, the muddy lake bottom where *Sinornis* came to rest did not preserve impressions of the feathers.

Primitive features of Sinornis. Many features in the skull and skeleton of *Sinornis* attest to its ancient heritage. As in *Archaeopteryx*, the skull had a proportionately short, toothed snout and the hand retained movable fingers with terminal claws that projected from the front margin of the wing. The pelvis, with its pendant and footed pubis, closely resembles that of *Archaeopteryx* and theropod dinosaurs. A series of stomach ribs, or gastralia, covered the ventral surface of the trunk. These primitive features, which are found only in *Archaeopteryx* and dinosaurs and are absent in all other birds, indicate that *Sinornis* represents an early avian offshoot.

Advanced flight function. Other skeletal features related to powered flight suggest that *Sinornis* marks a key point in bird evolution. *Sinornis* lost the long balancing tail of *Archaeopteryx* and theropod dinosaurs, replacing it with a blunt, fused tailbone, the pygostyle. In living birds, the pygostyle supports a fanlike array of tail feathers, or rectrices, which must have been present in *Sinornis*. The stout pygostyle and tail feathers enhance aerial maneuverability and function as a brake during landing.

The trunk in *Sinornis* is proportionately shorter than in *Archaeopteryx* and theropod dinosaurs. The greatly shortened tail and trunk in *Sinornis* shift the center of mass from above the hindlimbs, as in terrestrial precursors, toward the forelimbs, as in modern powered fliers. Thus, hindlimb thrust during locomotion gave way to forelimb thrust during flight in the early evolution of birds.

Many features in the pectoral girdle and forelimb in *Sinornis* attest to its advanced wing design. The shoulder socket, or glenoid, faces laterally rather than posteroventrally as in *Archaeopteryx*, allowing the forelimb to rotate above the vertebral column during the flight stroke as in modern birds. The primary and secondary flight feathers in modern birds attach to the second (middle) digit in the manus, or hand, and one of the forearm bones, the ulna. These bones are correspondingly more robust in *Sinornis* and modern birds than in *Archaeopteryx* and theropod dinosaurs, probably reflecting upon the amount of stress transmitted by the feathers. The wrist joint in *Sinornis* and modern birds is modified to permit the forelimb to fold tightly against the body. In this flexed position the manus collapses against the forearm, allowing the wing to conform to the body contour. In addition, the manus in *Sinornis* is proportionately shorter than that in *Archaeopteryx*. In the latter, the forelimb and manus are particuarly elongated and the terminal unguals are large and recurved. The grasping function of the forelimb in *Archaeopteryx* is maintained despite the presence of a full complement of primary and secondary feathers.

The modifications in the skeleton of *Sinornis* are all related to advanced flight function. The wing could be raised well above the level of the vertebral column during the initial phase of the downstroke, increasing the excursion of the flight stroke. At the beginning of the upstroke, the wing could fold tightly against the body, reducing drag, an important limiting factor in the flight capability of small to medium-sized birds. The small manus in *Sinornis* and most modern birds increases control of the wing tip. Together these adaptations suggest that the flight stroke in *Sinornis* was essentially similar to that in modern birds and much more advanced than that in *Archaeopteryx*. The bony sternum and reinforcing coracoids in *Sinornis* suggest that powerful flight muscles were present. In modern birds, bulky aerobic muscles constitute as much as one-third of body mass and support sustained powered flight during which the supply of oxygen to the muscles that operate the wings is continuously replenished. *Archaeopteryx*, in contrast, was incapable of lifting its forelimb above the vertebral column or flexing its wing tightly, and lacked an ossified sternum for attachment of wing musculature. The long manus and enlarged claws in *Archaeopteryx* suggest that the forelimb retained the grasping function of its theropod ancestors. Nevertheless, it must be assumed that *Archaeopteryx* was capable of primitive flapping flight of

some sort, given the aerodynamic design of its wings and feathers.

Perching capability. In *Archaeopteryx* and theropod dinosaurs, the short first toe, or hallux, is elevated far off the ground, attaching to the inner side of the foot, or pes. This toe did not contact the substrate during locomotion and in theropods is reduced in strength. In *Sinornis*, the first toe articulates at the base of the other toes and has a posterior position. In this configuration, the first toe is capable of flexion in opposition to the other pedal digits. The grasping capability of the pes in *Sinornis* is clearly an adaptation for an arboreal existence. The extremely slender, recurved pedal claws of *Sinornis* are poorly designed for terrestrial locomotion but well adapted for clinging to bark.

Early evolution of flight. The discovery of *Sinornis* reveals that the fundamental skeletal changes involved in modern avian flight function and perching evolved in small-bodied birds only a short time after *Archaeopteryx*. However, *Sinornis* retained primitive characters inherited from theropod ancestors, such as toothed skulls, movable clawed fingers, and stomach ribs. These features, which are not critical to flight or perching functions, were modified long after birds had shifted to an arboreal habit.

Most Mesozoic fossil birds have been discovered in quiet seashore or lagoonal environments, so the perspective of early bird evolution is biased. Perching birds are conspicuously absent among well-known Mesozoic avians, which include the diving bird *Hesperornis* and the shore bird *Ichthyornis*. In contrast, *Sinornis* lived on the wooded margins of a freshwater lake far from the seashore. The new fossils provide a window into the kinds of birds that inhabited ancient inland forests not long after *Archaeopteryx*. These birds, it is now apparent, include small-bodied, advanced fliers that would appear no different, at a glance, from a modern sparrow.

For background information SEE ARCHAEORNITHES; AVES; FEATHER; FLIGHT in the McGraw-Hill Encyclopedia of Science & Technology.

Paul C. Sereno

Bibliography. M. J. Benton (ed.), *The Phylogeny and Classification of the Tetrapods*, 1988; M. K. Hecht et al., *The Beginnings of Birds*, Freunde des Jura-Museums, Eichstätt, 1985; J. H. Ostrom, *Archaeopteryx* and the origin of birds, *Biol. J. Linn. Soc.*, 8:91–182, 1976; P. C. Sereno and C. Rao, Early evolution of avian flight and perching: New evidence from the Lower Cretaceous of China, *Science*, 255:845–848, 1992.

Flow measurement

The movement of a constant-density fluid is described completely if the velocity vector of the fluid, $\mathbf{v} = (v_x, v_y, v_z)$, is known at every point, so that the displacements of all fluid particles can be determined from one instant to the next. This information describes the changes of shape of fluid elements that determine the stresses within the fluid, as well as more global patterns of motion that affect the ability of the flow to transport mass, heat, and momentum. Throughout most of its history, experimental fluid mechanics has used two methods of measurement: probe techniques that measure the local velocity at one location, or at most a few locations, with good accuracy; and flow visualization methods that indicate the overall flow pattern but whose accuracy is seldom sufficient for more than qualitative analysis. Since the early 1980s, new techniques have revolutionized this situation by providing the capability to measure accurately the local fluid velocities at thousands and even millions of locations simultaneously. One technique that has demonstrated widespread utility is particle image velocimetry (PIV), which performs velocity measurement by optically observing large numbers of particles that are suspended in the fluid and move with it.

The result is typified by the vector map in **Fig. 1**, in which the orientation of each vector arrow indicates the direction of the fluid flow at the arrow's base and the length indicates the speed of the arrow's direction. The particle image velocimetry method allows the fluid dynamicist to observe the internal motion of a flow in much the same way that magnetic resonance imaging or computerized tomography allows the medical doctor to observe the interior of a patient.

Principles of PIV. The particle image velocimetry technique is based directly on the definition of velocity as the rate of displacement per unit time. Images of very fine particles are recorded optically at two times, t_1 and t_2, and the displacements of the particles during the time interval $\Delta t = t_2 - t_1$ are found from the

displacements of the images. Letting Δx, Δy, and Δz be the components of particle displacement in the x, y, and z directions, the corresponding components of the velocity vector are estimated from the equations below. These equations are approximations, because

$$v_x \doteq \frac{\Delta x}{\Delta t} \qquad v_y \doteq \frac{\Delta y}{\Delta t} \qquad v_z \doteq \frac{\Delta z}{\Delta t}$$

Δt is not infinitesimally small. However, the errors involved can be made negligible by restricting Δt and hence the displacements to small values such that the particle trajectories during Δt are essentially straight.

To measure the displacements accurately, the particle images and hence the particles themselves must be small. This restriction on particle size also helps to ensure that the particles follow fluid accelerations faithfully without significant lag due to particle inertia. In gaseous flows at normal densities where solid or liquid particles are much heavier than the fluid, particle sizes ranging from 0.5 to 5 micrometers (2×10^{-5} to 2×10^{-4} in.) permit accurate measurements for flow velocities up to several hundred meters per second. The inability of observable particles to follow very rapid acceleration limits particle image velocimetry, roughly speaking, to subsonic flows, although measurements in supersonic flows are possible under favorable conditions. In liquid flows, the increased density of the fluid permits particle sizes between 10 and 100 μm (4×10^{-4} and 4×10^{-3} in.). Naturally occurring aerosols and hydrosols are usable, but often they are replaced by artificially seeded particles whose properties are more uniform.

The duration of the displacement is fixed by illuminating the flow with short pulses of light and

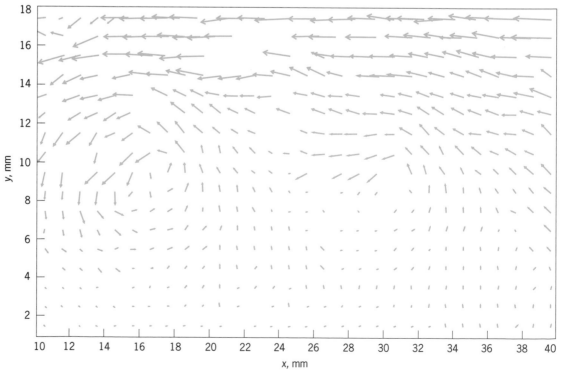

Fig. 1. Velocity vector map of flow along a wall (upper boundary) measured by particle image velocimetry.

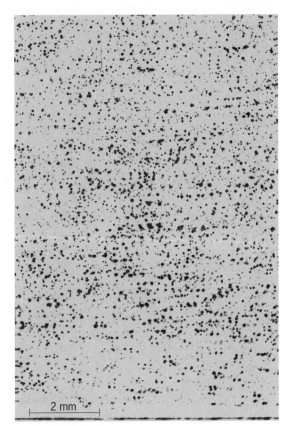

Fig. 2. Double-pulsed particle image photograph of flow over a wall (horizontal line). Flow is from left to right. Image pairs are most easily discerned in the low-speed region close to the wall, where the image displacement is small compared to the particle spacing. The particles are 5-μm solids in water.

recording the images of the particles at t_1 and t_2 with various types of recording media. Pulsed light sources offer two advantages over image shutters. First, they are very fast, making it possible to freeze the images of particles in high-speed flows. Second, pulsed solid-state laser sources concentrate large amounts of light energy into the pulses, making it possible to see very fine particles. For example, the pulses produced by a typical neodymium:yttrium-aluminum-garnet (Nd:YAG) laser are shorter than 10 nanoseconds and contain energies of several hundred millijoules. In low-speed flows, it is possible to use chopped continuous-wave lasers, swept continuous-wave laser beams, or xenon flash lamps.

A typical double-pulsed image is shown in **Fig. 2**, wherein the flow is mainly horizontal, and pairs can be ascertained by visual inspection near the lower boundary. Multiple pulses are sometimes used instead of pairs to make it easier to track the images, but double pulses are adequate if the image processing is done correctly. The principal effect of multiple pulsing is an increase in the number of pairs, which can also be done by increasing the particle concentration.

Images of the particles can be formed by several methods. The most common method is to illuminate the flow with a thin laser light sheet and photograph the particles in the light sheet with a camera at right angles to it. This procedure matches the illuminated region of the flow to the region from which in-focus images of the particles are received. The thickness of the in-focus region, called the depth-of-field, is small when the lens f-number is adjusted to resolve small particles, and hence the light sheet must be thin, typically less than 1 mm (0.04 in.). Use of a light sheet limits measurements of the velocity vector to a two-dimensional domain, and if a single-lens camera is used, the measurements are limited to the velocity components that are in the plane of the light sheet. Figure 2 is an example of such an image. By means of stereoscopic photography with two cameras viewing the light sheet, the full three-dimensional vector can be measured on the planar domain of the sheet.

Extension of these methods to measurement of the three-dimensional vector over a three-dimensional (volumetric) domain has been accomplished in two ways. If the particles are not small, implying low spatial resolution, it is possible to increase the depth of field so that the particles within an entire volume are in focus. Then, conventional photogrammetric methods using multiple cameras can be used to locate each particle. The confusion created by multiple particles normally limits this method to a relatively small number of vectors, of order 1000. High-resolution volumetric recording of fine particles has been achieved recently by using holographic methods.

Image interrogation. The foregoing principles of particle image velocimetry are conceptually simple, but successful implementation tests the limits of technology of lasers and imaging. The process by which the velocity vectors are computed from the recorded images, called interrogation, is much more challenging conceptually, and relies heavily on modern advances in computational power. If the spacing between images is large compared to the displacements between exposures, it is relatively easy to recognize pairs of images by using the principle that the nearest neighboring images probably belong to the same particle. In this case, the interrogation methods become particle tracking procedures in which individual particles are followed. These methods give vectors at random locations, and they fail if the spacing between vectors is made too small.

Velocity vectors on regular, closely spaced grids such as that in Fig. 1 can be measured by increasing the particle concentration so that each small region (interrogation spot) around a grid point contains many particles. Then, the fluid displacement is found by measuring the displacement of the group of particles in the interrogation spot. The motion of the group is found by a procedure in which the pattern of the first-exposure images is shifted in the plane until it matches the pattern of the second-exposure images. Mathematically, the pattern matching is performed by seeking the maximum value of the two-dimensional convolution of the images in the interrogation spot. Such computations are time-consuming, and the speed of the array processors used for these purposes sometimes approaches that of modern supercomputers. There are ongoing efforts to adapt optical correlators to this purpose and to develop fast algorithms that are specifically adapted to the simple character of a particle image.

Planar particle image velocimetry systems typically provide grids of vector data of the order of 100×100 vectors, and holographic systems provide a $100 \times 100 \times 100$ vector cube. The accuracy and resolution of the measurements make it feasible to compute differential quantities such as rate of strain and the vorticity, which are of particular interest in turbulent flows. Particle image velocimetry is also used in large aerodynamic wind tunnels to reduce data acquisition time in complex, unsteady flows such as in internal combustion engine cylinders, and in flows interacting with moving boundaries as in the flow around a propeller.

For background information SEE FLOW MEASURE-MENT; FLUID FLOW; FLUID-FLOW PRINCIPLES; HOLOG-RAPHY; VELOCITY in the McGraw-Hill Encyclopedia of Science & Technology.

Ronald J. Adrian

Bibliography. R. J. Adrian, Particle-imaging techniques for experimental fluid mechanics, *Annu. Rev. Fluid Mech.*, 23:261–304, 1991; T. D. Dudderar, R. Meynart, and P. G. Simpkins, Full field laser metrology for fluid velocity measurement, *Optics Lasers Eng.*, 9:163–200, 1988.

Flower

Recent research on flowers and flowering has included studies of the floral mechanisms that have evolved to attract pollinators, and of the genetic control of floral initiation and development.

Animal-Pollinated Flowers

Few examples of evolution by natural selection are as dramatic as the correspondence of shapes, sizes, and behaviors of flowers with their pollinators. Flowers have evolved many mechanisms for attracting animals, placing pollen on and retrieving it from them, and ensuring that most pollen moves between members of the same species. Research since the early 1980s has significantly improved understanding of how such adaptations originate.

Function of flowers. Bisexual (perfect), angiosperm flowers perform male and female functions in plant reproduction. Female functions involve producing ovules, obtaining pollen, facilitating the fertilization of ovules, developing seeds, and dispersing seeds for germination. The male function is to produce pollen and disperse it to the stigmas of other flowers. Sometimes the male and female functions operate in separate flowers (unisexual flowers) or even separate plants (dioecy), but bisexual flowers are the norm.

Flowers depend on abiotic agents (such as wind) or, more commonly, animals to accomplish pollination. Important adaptive features of flowers promote pollination by advertising to pollinators, rewarding them, facilitating placement of pollen on them, facilitating pickup of pollen from them, and ensuring that pollen gets to conspecific stigmas and comes from conspecific anthers.

Pollinator rewards and advertisements. Plants have evolved a variety of mechanisms for attracting and rewarding animal pollen vectors. Bright colors, unusual shapes, and strong fragrances help advertise the presence of flowers. These features probably promote associative learning by pollinators (that is, association of the reward with the appearance of the flower) and help them locate flowers from long distances. The ultimate reason a pollinator visits a flower, however, is to obtain a reward. The rewarding nature of sweet nectars and nutritious pollen has been recognized for several centuries. Nectar is a primary carbohydrate (energy) source for many insects, birds, bats, and some nonflying mammals (some marsupials and some primates). Pollen is eaten by bees, flies, beetles, and bats. It is also collected by bees as food for their larvae.

Some flowers present sugar-, starch-, protein-, or lipid-rich food bodies as rewards for pollinators. This feature is common in beetle-pollinated flowers and lures potentially destructive beetles away from the sexual parts of the flower.

Several other interesting and widespread reward systems have been discovered. Researchers have found that flowers in a variety of unrelated genera have energy-rich oils (lipids) that bees collect and mix with pollen (usually from other flowers) as food for their larvae. This lipid-pollen reward system has evolved independently in several plant families, including Malpighiaceae, Scrophulariaceae, Orchidaceae, Krameriaceae, Cucurbitaceae, Liliaceae, Solanaceae, and Primulaceae, and is found in both the New and Old World tropics as well as parts of the temperate zones.

Another reward system is the production of floral fragrances that male euglossine bees (Apidae: Euglossini) collect and use as precursors to sex pheromones. This system is restricted to the New World tropics and has evolved independently in the Orchidaceae, Araceae, Euphorbiaceae, Solanaceae, Gesneriaceae, Bignoniaceae, and Haemodoraceae.

A recently described pollinator reward system is the secretion by flowers of triterpene resins that are collected by pollinating bees for use in nest construction. Many tropical and some temperate bees use resins to seal their nests or cement together other construction materials. Members of two unrelated tropical genera, *Dalechampia* (Euphorbiaceae) and *Clusia* (Clusiaceae), attract pollinators by secreting resins.

Pollen-placement mechanisms. Flowers vary widely in mechanisms of pollen placement on pollinators. The simplest mechanism operates in radially symmetrical, open flowers such as the wild rose or buttercup. In these flowers, pollen is released by the opening anther sacs, and pollinators crawl over the anthers while collecting pollen or nectar. Pollen is distributed haphazardly over most of the animal's body, although usually it is denser on the underside. In zygomorphic (bilaterally symmetrical) flowers, the pollinator's approach to the reward is limited, resulting in more precise placement of pollen on the pollinator. Commonly, pollen is placed on some portion of the ventral or dorsal surface of the animal's body. The greatest precision in pollen placement is seen in orchids and the trigger plants (*Stylidium*, Stylidiaceae) of Australia. In most orchids, pollen is united into a single mass, the pollinium, which is released and glued to a precise location on the body when the animal contacts the trigger point. In trigger plants, the pollen is loose, although sticky, and is released from anther sacs at the end of a long motile column. When an insect touches the sensitive trigger point at the base of the column, the column springs forward to dab pollen on the back, side, or ventral surface of the insect, depending on the species of trigger plant. The length of the column also determines the placement of pollen on the insect's body. After the column lashes forward to dab pollen on the pollinator, it recovers to its original position, ready for the arrival of another insect.

Pollen receipt. There has been strong selection for flowers to pick up pollen from pollinators in a fashion coordinated with its placement. The position and shape of style and stigma has evolved to promote the stigma's contacting the pollinator's body in the same place that the anthers contact it. The better the corespondence, the greater the effectiveness in pollen pickup. For example, in trigger plants the receptive stigma occupies the same position on the motile column as was previously occupied by the anther sacs. The column continues to strike the pollinator's body as it did in the male phase, allowing the stigma to pick up pollen from the same spot where it is placed by other flowers in the population.

Adaptations preventing interspecific pollination. It is maladaptive to receive pollen from, or disperse pollen to, flowers of another species for several reasons, such as wasting pollen and therefore producing fewer offspring, and reduced seed set because of insufficient pollination, pollen allelopathy, or stigmatic clogging by alien pollen. When plants with similar flowers and pollinators occur together, there is substantial risk of interspecific pollen flow. Plant adaptations to prevent pollen exchange with similar, usually related species include blooming at different times of the year, opening flowers at different times of day, using different species of pollinators, or placing pollen in different locations on pollinators. The last two strategies involve intricate features of floral morphology; interesting examples are found in *Dalechampia* of the New World tropics and *Stylidium* of Western Australia.

When two or three species of *Dalechampia* occur together, as in most of South and Central America, they generally attract different pollinators. In some cases the sympatric species offer different kinds of rewards. For example, in Amazonian Peru, *D. cissifolia* secretes a resin reward and is pollinated by female resin-collecting bees, whereas the sympatric species, *D. magnoliifolia*, secretes a fragrance reward and is pollinated by male euglossine bees. In other cases sympatric *Dalechampia* species offer different amounts of resin reward, have different-sized blossoms, and are pollinated by species of bees of different sizes.

In Western Australia, up to five or six species of *Stylidium* may occur together. Most of these utilize the same pollinator species (usually bees or bee flies) but place pollen in different areas on the insects. *Stylidium* species have evolved different positions and different lengths of columns in order to place pollen in unique locations (relative to other sympatric species) on their pollinators. For example, at a site near Perth one species places pollen on the top of the pollinator's thorax, a second on the left side of the thorax, a third on the left side of the abdomen, a fourth on the top of the abdomen, and a fifth on the underside of the head and thorax. This diversity promotes pollen getting to the correct species even though the same individual pollinators regularly fly between species.

W. Scott Armbruster

Genetic Analysis of Flowering

Biologists have long speculated on how the initiation of flowering and the development of flowers is controlled. Recently, a combination of genetic and molecular analyses in *Antirrihinum majus* (snapdragon) and *Arabidopsis thaliana* has revealed that many important regulators of flower development are proteins that control gene transcription. Many of these proteins have been evolutionarily conserved in a variety of plant species, and related proteins exist in organisms as diverse as yeast and humans.

Floral organs. Both snapdragon and *Arabidopsis* are indeterminate plants. That is, after the production of a number of vegetative leaves, the plants form a flowering bolt, or inflorescence, which has the potential to form an unlimited number of flowers. The production of leaves and flowers is initiated by the shoot apical meristem, which meristem consists of a group of actively dividing, undifferentiated cells located at the shoot tip. The meristem contributes cells for leaf and flower bud formation, as well as cells to regenerate the apical meristem itself. Whereas apical vegetative and inflorescence meristems are indeterminate, the lateral floral meristems initiated along the inflorescence axis are determinate, and floral meristematic activity ceases after the production of the flower organs.

Flowers formed by snapdragon and *Arabidopsis* have very regular patterns of organ initiation and development. The floral organs are formed in a series of four concentric whorls, with the leaflike sepals in the first or outer whorl, the petals in the second whorl, the pollen-bearing stamens in the third whorl, and the ovule-bearing carpels in the fourth or innermost whorl. The formation of the carpels recruits the last of the undifferentiated floral meristem cells into a differentiated organ.

Floral organ identity. Flower development was the first aspect of the flowering process to be investigated by recent genetic and molecular approaches. Because floral organs are formed in a predictable pattern of concentric whorls, mutants that are defective in this pattern have been used to understand the genetic control of floral organ identity. The key genes identified by these mutations are conserved between snapdragon and *Arabidopsis*.

Three classes of genes are required to specify the identity of all organs in the four whorls of the flower. For the purposes of this review, only the genes of *Arabidopsis* will be discussed. *APETALA 2* gene function is required for the development of sepals and petals (whorls 1 and 2), *APETALA 3* and *PISTILLATA* gene functions are required for the development of petals and stamens (whorls 2 and 3), and *AGAMOUS* gene function is required for the formation of stamens and carpels (whorls 3 and 4) and for the termination of organ initiation within the flower. Mutations in the *APETALA 2*, *APETALA 3*, *PISTILLATA*, and *AGAMOUS* genes result in the production of developmentally transformed organs in the whorls controlled by each gene. For example, unlike the normal pattern of sepal, petal, stamen, carpel found in wild-type flowers, *apetala 2* mutants form flowers that produce carpels instead of sepals in whorl 1, and stamens instead of petals in whorl 2, so that the mature flower has the structure carpel, stamen, stamen, carpel. Thus, the *APETALA 2* gene is required for the identity of sepals in whorl 1 and petals in whorl 2. Because these genes regulate organ identity, they are referred to as homeotic genes. Mutations in these genes disrupt the homeostasis of specific floral organs.

The key to how the floral homeotic genes control flower organ identity lies in understanding how three classes of genes regulate the identity of four whorls of organs. To simplify, the homeotic genes can be classified as controlling organ development in three domains of the floral meristem: Class A (*APETALA 2*) controls whorls 1 and 2 (domain A), class B (*APETALA 3* and *PISTILLATA*) controls whorls 2 and 3 (domain B), and class C (*AGAMOUS*) controls whorls 3 and 4 (domain C). Domain B overlaps part of domain A and part of domain C. Thus, whorl 1 is specified by class A genetic information, whorl 2 by class A and class B genes, whorl 3 by class B and class C genes, and whorl 4 by class C genetic information. A final aspect of this model is that the class A and class C genes act in a mutually antagonistic manner. In the absence of class A gene function, class C gene function expands into whorls 1 and 2. In a class C (*agamous*) mutant, class A (*APETALA 2*) gene function expands into whorls 3 and 4.

Experimental tests of this model have come from both genetic and molecular studies. The first tests were by the construction of multiple mutant lines of *Arabidopsis*. For example, the model predicts that the flowers on a class A/B double mutant would consist of four whorls of carpels, since class C function would occur in all four whorls without class A or B function. This prediction was confirmed when the class A/B double mutant was examined. A very interesting mutant combination was the triple mutant in which defective class A, B, and C genes were combined in the same plant. If these three genes represented major genes required for all floral organ identity, the prediction was that the A/B/C triple mutant would produce "flowers" consisting of only vegetative leaves, as the leaf is considered the developmental ground state for floral organs. Examination of the A/B/C triple mutant confirmed the prediction.

Molecular analysis. The molecular tests of this model for floral organ identity have been made possible by the molecular cloning of several floral homeotic genes from snapdragon and *Arabidopsis*. The floral homeotic genes cloned thus far are related to one another at the level of the amino acid sequence of the encoded proteins. These sequences are very similar to sequences previously identified in two related proteins that regulate the transcription of eukaryotic genes—the MCM1 protein of yeast, and the serum response factor (SRF) of mammalian cells. This region of amino acid similarity with MCM1 and SRF was first recognized with the snapdragon *DEFICIENS* protein and the *Arabidopsis AGAMOUS* protein, and was designated the MADS-box region (from the initial letters of MCM1, *AGAMOUS*, *DEFICIENS*, SRF). The MADS-box genes of snapdragon and *Arabidopsis* represent families of genes, each with about 20 members. Several other floral homeotic genes are MADS-box genes.

A molecular test of the flower development model was performed with the MADS-box gene *AGAMOUS*. *AGAMOUS* is a class C gene of *Arabidopsis*, and thus in wild-type flowers is predicted to function only in whorls 3 and 4. When the pattern of *AGAMOUS* gene transcription was examined, it was indeed found only in the cells of the floral meristem that occur in whorls 3 and 4 and in the differentiating stamens and carpels (whorl 2 and whorl 4 organs). Furthermore, the model predicts that in a class A (*apetala 2*) mutant, the class C (*AGAMOUS*) gene should function in whorls 1 and 2 as well as in 3 and 4. When the pattern of *AGAMOUS* messenger ribonucleic acid (mRNA) was examined in *apetala 2* mutants, it was, as predicted, found in cells of all four whorls of the floral meristem.

Inflorescence development. Recently, progress has been made on understanding the genetic control of inflorescence development in both snapdragon and *Arabidopsis*. Three major genetic functions appear to regulate inflorescence development: one gene controls inflorescence meristem identity, and two genes control floral meristem identity. As with the floral homeotic genes described above, these genetic functions are conserved between snapdragon and *Arabidopsis*. For simplicity, the description of inflorescence development will be limited to the *Arabidopsis* genes.

The *LEAFY* and *APETALA 1* genes of *Arabidopsis* are required for floral meristem identity. In both *leafy* mutants and *apetala 1* mutants, the normally determinate floral meristem fails to form. The *leafy* mutants produce lateral "floral" meristems that are inflorescencelike in that they possess many leaflike organs arranged in an indeterminate pattern of growth. The *apetala 1* mutants form lateral "flowers" in which the sepals are replaced by leaflike organs, petals are absent, and secondary mutant flowers arise from the second whorl position. These secondary flowers in turn form tertiary mutant flowers, and so on. Thus, the *apetala 1* mutant floral meristem displays the indeterminate development characteristic of inflorescence meristems.

Molecular cloning of the *LEAFY* and *APETALA 1* genes has shown that they are likely to encode proteins that act as transcriptional regulators. The *LEAFY* protein is not related to other nonplant proteins; however, the *APETALA 1* protein is another MADS-box protein. Both the *LEAFY* and *APETALA 1* genes are transcriptionally expressed in the newly forming floral meristem prior to the expression of the floral homeotic genes. Neither the *LEAFY* gene nor the *APETALA 1* gene appears to be transcriptionally expressed in the inflorescence meristem.

Inflorescence meristem identity in *Arabidopsis* is regulated by the *TERMINAL FLOWER* gene. The *terminal flower* mutants fail to maintain the indeterminate development potential of the inflorescence meristem and instead form a terminal flower soon after the initiation of the inflorescence bolt. Genetic analysis has shown that the indeterminate growth of lateral "floral" meristem in *leafy* mutants is caused by ectopic *TERMINAL FLOWER* gene function in these lateral structures. Thus, in wild-type *Arabidopsis* plants, it is likely that one of the roles of the *LEAFY* gene product is to regulate *TERMINAL FLOWER* gene function in developing flower meristems.

Prospects. The rapid progress made in understanding the genetic regulation of flowering in snapdragon and *Arabidopsis* is likely to be followed by similar progress with ornamental and agricultural plant species. By using DNA sequence homology to identify similar genes, molecular clones of MADS-box genes and the *LEAFY* gene have been isolated from several plants, including maize, tomato, petunia, and tobacco. Such cloned genes have opened up many new avenues for the analysis and manipulation of the flowering process in these plants.

For background information *SEE FLOWER; GENE ACTION; INFLORESCENCE; REPRODUCTION (PLANT)* in the McGraw-Hill Encyclopedia of Science & Technology.

D. Ry Meeks-Wagner

Bibliography. W. S. Armbruster, Reproductive interactions between sympatric *Dalechampia* species: Are natural assemblages "random" or organized?, *Ecology*, 67:522–533, 1986; W. S. Armbruster, M. E. Edwards, and E. M. Debevec, Pollination ecology and assemblage structure of Western Australian triggerplants (*Stylidium*): Are there any rules?, *Ecology*, 1993; J. L. Bowman, D. R. Smyth, and E. M. Meyerowitz, Genetic interactions among floral homeotic genes of *Arabidopsis, Development*, 112:1–20, 1991; E. S. Coen, The role of homeotic genes in flower development and evolution, *Annu. Rev. Plant Physiol. Plant Mol. Biol.*, 42:241–279, 1991; E. S. Coen and E. M. Meyerowitz, War of the whorls: Genetic interactions controlling flower development, *Nature*, 353:31–37, 1991; L. Real (ed.), *Pollination Biology*, 1983; S. Shannon and D. R. Meeks-Wagner, A mutation in the *Arabidopsis TFL1* gene affects inflorescence meristem development, *Plant Cell*, 3:877–892, 1991; B. B. Simpson and J. L. Neff, Floral rewards: Alternative to pollen and nectar, *Ann. Missouri Bot. Gard.*, 68:301–322, 1981.

Forest fire

Fires have been present during the evolution of much of the Earth's vegetation. The vegetation has shifted in response to long-term climate changes, so that today's species mixes in a particular landscape (that is, a plant community) may reflect groupings that have been stable for only a few thousand years. Fire has probably been associated associated with individual species much longer than the species within plant communities have been associated with one another.

Fire regime. The interactions of fire with various components of an ecosystem depend on the fire regime, that is, the combination of fire frequency, magnitude, extent, seasonality, and synergy with other disturbance factors.

Natural fire frequency, the return interval of fire, can vary from as often as every 2 years in ponderosa pine forests in the southwestern United States to 500 years or more in the temperate rainforests of the Pacific Northwest. Fire magnitude is usually described by fire-line intensity, a measure of heat output per unit length of fire line as it moves across the landscape. Natural fire-line intensities can range from very low surface fires that crawl slowly along the ground (less than 15 Btu·s·ft or 50 kW·m) to raging crown fires (more than 12,000 Btu·s·ft or 50,000 kW·m). The area burned by a fire, that is, the extent of a fire, can be quite large over a short period of time. In the fires of Indonesia in 1982 and northeastern China in 1987, millions of acres were burned. In contrast, the smaller Yellowstone fires of 1988, in the northern Rocky Mountains, are but a continuation of a process that began hundreds of millions of years ago.

Up until the last 20–30 millennia fires in North America were ignited exclusively by natural causes, most significantly by lightning storms, with as many as 1800 in progress at any moment. Later, ignition by humans supplemented or substituted for natural ignition sources.

Plant adaptations. Not surprisingly, some plants have evolved mechanisms to either survive fire or take advantage of environments altered by it. A number of plants possess remarkable adaptations to fire, and sometimes to a particular fire regime. The adaptations include seed-bank strategies, thickened bark, and sprouting mechanisms.

Seed-bank strategies. In seed-bank strategies, seeds are dormant (protected) until an intense fire occurs. Many shrubs, such as ceanothus (*Ceanothus* spp.) or manzanita (*Arctostaphylos* spp.), have hard-coated seeds that will not germinate until the coat is cracked (a process called scarification). The seeds accumulate in the soil year after year, remaining viable for extended periods. Such seed banks may contain as many as 5×10^6 seeds per acre (12×10^6 per hectare). When an intense fire occurs, although many seeds are destroyed others become scarified.

These scarified seeds imbibe water and germinate the next growing season in a competition-free environment.

Another seed-bank strategy involves serotinous (late-opening) cones, found in many species of pine (*Pinus* spp.). Seed remains sealed inside the serotinous cone until there is an intense fire. The cones open within a day of the fire to release seeds, which become buried in the ash and thus are protected from food gatherers.

Thick bark. The thick-bark adaptation functions best in a fire regime of more frequent, less intense fires where the tree crown is not scorched and the bark, a very effective insulator, protects the cambial tissue against heat directed at the tree base. Species such as ponderosa pine (*Pinus ponderosa*) and Douglas-fir (*Pseudotsuga menziesii*), with bark on mature trees up to 4-in. (10-cm) thick, are well insulated from the effects of low-intensity fires. However, at thin places on the bark, enough heat may penetrate to kill the cambium and create a fire scar. By analyzing the fire scars on the annual growth rings of temperate forest species, the record of fires can be reconstructed.

Sprouting. After a fire, sprouting can occur from latent axillary or adventitious buds along the stem, from the root collar, or from the root suckers. Coast redwood (*Sequoia sempervirens*) and many eucalyptus trees (*Eucalyptus* spp.) sprout profusely along burned stems. Root-crown sprouts are common for many shrub and hardwood tree species if the stem has been killed by burning.

Species such as chamise (*Adenostoma fasciculatum*) and some manzanitas and eucalyptuses have a specialized sprouting organ called a lignotuber. This large swelling, surrounding the root–shoot interface at or beneath ground surface, contains food reserves and dormant buds. If the stem is killed, the buds sprout by using food from the lignotuber. Thus, species with lignotubers gain a large advantage over species that must either regenerate from seed or sprout with less available food reserves.

Effects on plant succession. Those species that are best adapted to a particular fire type will be favored for survival. For example, the survival of mature individuals of thick-barked species will be favored by a low-intensity fire, whereas species that either sprout or have protected seed banks will be favored by a high-intensity fire. In response to the complexity of disturbance by fire or other factors, multiple-pathway models for plant succession have replaced the simple successional models. Forest simulation models using computers have enabled the effects on succession of different fire frequencies or intensities to be clearly illustrated.

Effects on soils. Most of the effects of fire on soil are confined to the top few inches of soil, the zone that contains the most active chemical and biological processes and is the entry point for water. The effect of fire on soils is a function of soil heating. Frequent low-intensity fires may have little impact on soil, whereas high-intensity fires may significantly change its physical, chemical, and biological properties. Possible detrimental changes in soil include decreased porosity, increased water repellency, and decreased infiltration, all of which result in increased runoff and erosion. Nitrogen and sulfur, two elements usually limiting for forest growth, can be volatilized by fire. (However, plant species that can fix atmospheric nitrogen through symbiosis with bacteria—for example, legumes and ceanothus shrubs—are often favored by fire.) Beneficial mycorrhizal fungi, which form unions with plant roots and increase the water and nutrient uptake, may be reduced by severe fires.

Other ecosystem effects. Fire has complex interactions with insects and diseases. Historically, fire has controlled major insect and disease problems in dry forests by maintaining low tree density and favoring insect- and disease-resistant species. Effective fire suppression has created forest health problems by increasing the density of insect- and disease-prone tree species. Fire damage can also encourage pest problems. Fire-damaged trees are at increased risk for bark beetle attacks, and fire-scarred stems and roots are often entry points for pathogens that may later kill the tree.

Disturbance by fire and other factors has been associated with an increase in biodiversity. Recurring fire over a landscape creates a mosaic of vegetation types. In the first years after an intense fire, fire-adapted herbs and shrubs, many not represented in the older vegetation patches, will dominate a site. New tree species colonize and dominate intermediate-aged landscape patches. Older landscape patches are dominated by shade-tolerant, fire-sensitive plant species. Wildlife species adapted to each vegetation further increase the biodiversity. Fire maintains a variety of landscape patch ages, thus increasing the total number of plant and animal species in an area.

Effects on productivity by fire are not clearly positive or negative. When forest productivity is defined in terms of tree growth, fire can have some immediate negative effects, caused by stem or foliar damage. Many forests are adapted to survive low to moderate intensity fires, however, and fire can be followed by increased productivity on surviving stems. In drier forests that burn frequently, fire is essential in maintaining productive forests. However, in subalpine forests, fire often has resulted in shrubfields or meadows that have been recolonized by trees very slowly. Understanding the particular fire regime and forest type can help predict effects on forest productivity.

For background information SEE DENDROCHRONOLOGY; ECOLOGICAL SUCCESSION; FOREST ECOSYSTEMS; FOREST FIRE CONTROL; FOREST SOIL in the McGraw-Hill Encyclopedia of Science & Technology.

James K. Agee

Bibliography. T. T. Kozlowski and C. E. Ahlgren, *Fire and Ecosystems*, 1974; J. Walstad, S. R. Radosevich, and D. V. Sandberg, *Natural and Prescribed Fire in Pacific Northwest Forests*, 1990; H. A. Wright and A. W. Bailey, *Fire Ecology: United States and Southern Canada*, 1982.

Fuel

Recent research has yielded advances in development of alternative fuels for vehicles and various refuse-derived fuels for other purposes.

Alternative Fuels for Vehicles

There is an increasing demand for transportation fuels that are attractive from the point of view of environmental quality and energy conservation. Recent research has focused on gaseous fuels, alcohol fuels, reformulated gasoline and diesel fuels, electricity, and hydrogen.

Gaseous fuels. Vehicle emissions from natural gas and propane are expected to be lower and less harmful to the environment than those from conventional gasoline. Both of these fuels are introduced to the engine as a gas under most operating conditions and require minimal fuel enrichment during warm-up. Natural gas and propane reduce cold-start emissions and fuel consumption, and they generate lower emissions of carbon dioxide than gasoline for equivalent energy consumed. These fuels are leaner-burning than gasoline, resulting in lower emissions of hydrocarbon and carbon monoxide. However, because they burn at higher temperatures, emissions of nitrogen oxide are usually higher. To achieve the full environmental and performance benefits of natural gas and propane, vehicles must be designed to run exclusively on gaseous fuels.

Natural gas. Once processed at a gas plant, natural gas is composed of about 97% methane. In its gaseous state at ambient temperatures, natural gas has a very low energy density. To eliminate bulkiness, the fuel is compressed to pressures of 2000–3600 lb/in.^{2} (14,000–25,000 kilopascals), depending on the country and application in question, and stored in cylindrical tanks.

Natural gas has an energy content by weight comparable to that of gasoline and diesel fuel. However, because of its lower storage densities, it has lower energy content per unit volume. Because of the large size and weight of the storage tanks necessary for carrying natural gas supplies equivalent in energy content to gasoline, motorists are required to refuel more frequently than they would with gasoline.

There are some very distinct benefits to using natural gas in vehicles. First, unlike gasoline, natural gas is injected into the engine cylinders in a completely vaporized state, resulting in smoother throttle response, minimized spark-plug fouling, easier cold-weather starting, and reduced engine wear and maintenance. Second, natural gas has an octane rating around 120 (research octane number 130), outperforming even premium gasolines containing antiknock additives. This potential for higher compression ratios results in greater engine efficiency. Third, natural gas has wider flammability limits than either gasoline or diesel fuel; while this may be undesirable from a safety perspective, it results in leaner operating conditions and greater combustion stability.

The physical characteristics of natural gas offer a number of built-in safety advantages in the case of collisions or accidental release. Since natural gas is less dense than air, it disperses immediately upon release, eliminating the possibility of pooling and thus reducing the risk of an explosion. Additionally, natural gas possesses a higher autoignition temperature than either gasoline or diesel fuel, suggesting that a natural gas leak is less likely to result in a fire.

The primary safety concern when using natural gas in vehicles arises from the use of high-pressure storage tanks, which create greater potential for rapid release of gas from the fuel system than is the case with gasoline. This concern has resulted in stringent regulations and requirements for fuel storage cylinders that are stronger and more resistant to damage than standard gasoline tanks. Research is directed at developing lighter, less costly storage cylinders that meet the required safety standards. Advanced methods of adsorbent fuel storage are being examined to allow more fuel to be carried in less space, thereby increasing the achievable driving range.

Acceptance of natural gas as a fuel for vehicles will require optimizing performance while reducing emissions. Monofuel vehicles will have to be specifically designed for natural gas. The potential economic and environmental benefits of the use of these vehicles will have to be balanced against the limited availability of refueling facilities. Recognizing this as a major obstacle to efforts aimed at increasing the demand for natural gas as a vehicle fuel, researchers have developed a safe, vehicle-refueling home appliance.

Although market research has indicated that the home compressor will accelerate the demand for natural gas for vehicles in both the private and small commercial markets, the current compound cost constraints of the refueling appliance and conversion will tend to limit its usage to high-mileage vehicles.

Propane. Propane possesses both a higher octane rating and cleaner combustion characteristics than gasoline. Propane is the main component of liquefied petroleum gas (LPG), which also contains hydrocarbons such as butane and ethane. Auto-propane is a special grade, HD-5, consisting of no less than 90% propane. Unlike gasoline, propane is in a gaseous state at atmospheric pressure. It typically contains about 70% of the energy content of gasoline and 66% of the energy content of diesel fuel (variations occur depending upon the precise grade of gasoline or diesel fuel being considered). However, it burns more efficiently than either of these fuels; therefore conventional engines operating on propane experience only a 5–10% power loss, which can be reduced if the engine is optimized to run on propane.

At ambient temperatures, propane can be liquefied at comparatively low pressures and so, unlike natural gas, sufficient liquid propane can be stored in a relatively lightweight tank to provide a driving range similar to that of gasoline. The vapor pressure of propane varies in relation to the atmospheric temperature. At temperatures of −43°C (−45°F) and colder, the pressure in a fuel tank is zero, the product becomes liquid, and propane fuel systems simply do not operate.

Propane fuel tanks have an excellent safety record. Since the boiling point of propane is lower than that of either diesel fuel or gasoline, precautions have been taken to ensure that the tanks have been tested at pressures as high as 265 lb/in.^{2} (1830 kPa).

Propane's boiling point of $-42°C$ ($-44°F$) versus gasoline's boiling range of $25–210°C$ ($77–410°F$), also has its advantages. Propane's higher volatility allows the fuel to enter the intake manifold of the engine as a gas rather than a liquid, which requires engine heat to vaporize. The fuel is therefore distributed to the cylinders more evenly, improving engine performance. Propane also has a higher octane rating than either premium or regular gasoline (100 versus 92 and 87, respectively), which allows for higher compression ratios and ultimately better engine performance.

Alcohol fuels. In comparison with hydrocarbon-based fuels, the exhaust emissions from vehicles burning low-level alcohol blends (10% alcohol by volume) contain negligible amounts of aromatics and reduced levels of hydrocarbons and carbon monoxide but higher nitrogen oxide content. Further, low-level alcohol blends have a tendency toward increased evaporative emissions, an effect that can be offset somewhat by adjusting the volatility of the base fuel.

Exposure to formaldehyde, which is considered to be carcinogenic, is an important air pollution concern. The aldehyde fraction of unburned fuel, particularly for methanol, is appreciably greater than for hydrocarbon-based fuels; therefore, catalytic converters are needed on methanol vehicles to reduce formaldehyde levels to those produced by gasoline. Use of neat (100%) methanol may not, however, lead to increased ambient formaldehyde levels in the long term. Most of the ambient formaldehyde is formed indirectly in the atmosphere through photochemical reactions involving reactive hydrocarbons, suggesting that neat methanol engines, with their lower hydrocarbon emissions, could actually decrease indirect formaldehyde formation in comparison to gasoline vehicles, possibly offsetting any increase in direct formaldehyde emissions from the exhaust.

The most significant advantage of alcohol fuels over gasoline is their potential to reduce ozone concentrations and to lower levels of carbon monoxide. Another important advantage is their very low emissions of particulates in diesel-engine applications.

Methanol. The appeal of methanol as an alternative to gasoline and diesel fuel stems from its being a relatively inexpensive, clean-burning liquid fuel. Flexible fuel vehicles that can operate on a mixture of gasoline and up to 85% methanol are available.

Methanol is a promising fuel for bus and truck fleets, reducing some of the noxious emissions related to diesel fuel. Methanol-burning engines are being tested at a number of sites in North America for buses and long-haul and garbage trucks.

Because methanol is a clear, colorless toxic liquid, for example, it would require some form of fuel marking (coloring) before it could be widely sold as a transportation fuel. Research suggests that emissions of unburnt methanol would not exceed the lower limits of concern by health administration departments under normal moving-traffic conditions. Improvement of flame visibility and the environmental effects of emissions of methanol and formaldehyde are further areas of ongoing research and development.

Ethanol. Produced from biomass, ethanol is a renewable energy source with potential for increasing use as an alternative fuel. Current technology suggests that the feedstock requirements of an ethanol industry may provide markets for North American grains that are difficult to sell in the domestic and world marketplace.

In comparing the major emissions from methanol-fueled vehicles with those from ethanol-fueled vehicles, ethanol is less toxic than methanol. In addition, acetaldehyde emissions from burning ethanol are less carcinogenic than the formaldehyde emissions from methanol.

With an increasing demand for premium unleaded gasoline, increased octane requirements have been costly to refiners. Oxygenates, oxygen-containing compounds such as ethanol, already recognized for their environmental friendliness, hold promise for meeting these requirements. Although neat ethanol is being tested in both light and heavy vehicles, the current economic viability of ethanol as a neat or near-neat fuel is poor. But, as an oxygenate with a high octane rating, it is a candidate in the North American transportation industry to replace tetraethyl lead.

Reformulated fuels. Conventional gasoline and diesel fuels are complex mixtures of many different chemical compounds. Reformulation simply refers to any significant change in the composition of the gasoline or diesel fuel. Gasoline reformulation can range from mild change, such as backing out some of the butane from the gasoline to reduce volatility, to severe adjustments that change the composition more substantially.

The major current air-quality issues include the increase in ground-level ozone, air toxins, acid rain, and global warming and depletion of stratospheric ozone. Research indicates that ground-level ozone and air toxins can be reduced through reformulating gasoline by adding oxygenates such as methyl tertiary butyl ether (MTBE), reducing the aromatic content, and lowering the 90% boiling point (known as T90), the temperature at which 90% of the fuel is evaporated.

Diesel reformulation has received less attention, and the effects of fuel changes are not yet fully understood. Future reformulated diesel fuels will likely include reductions in sulfur content and increases in cetane number, oxygenates, and additives for durability.

Electricity. The growing interest in electric vehicles arises from their being recognized as emission-free propulsion systems that may reduce noise and air pollution in congested urban centers. The success of these vehicles is ultimately dependent on the success of storage batteries in terms of their weight/energy ratio, their life, the distance traveled between charges, and the recharging time.

Electricity is generated from a wide range of primary energy sources including coal, uranium, oil, natural gas, wood, falling water, wind, and the Sun. While electric vehicles have the potential to reduce emissions in the transportation sector, success in market penetration will result in increased demand for electricity. If the source of electricity at the margin (that is, the excess capacity required to supply the power necessary to meet the demands from the electric vehicle market) is fossil-fuel-fired generating capacity, electric vehicles might simply push emissions farther upstream. This effect could still result in a net reduction in emissions, as it is easier to contain and reduce emissions at a central generating plant than across thousands of automobiles. Similarly, net regional reductions could also result from power generated from hydro or nuclear sources.

A great deal of research by auto manufcturers in the United States and other countries is currently under way in attempts to improve the marketability of electric vehicles. The U.S. Advanced Battery Consortium (ABC), in cooperation with the Department of Energy, is encouraging research in the area of advanced batteries with the objectives of increasing the driving range, improving vehicle performance, and lowering the overall cost of the electric vehicle.

Research is ongoing in the area of advanced fuel-processing systems for reforming fuels such as methanol, ethanol, and natural gas into hydrogen for use in transportation fuel cells. These fuel cells produce electricity and water by electrochemically combining hydrogen and oxygen. Currently, a fuel cell/battery propulsion system is being designed for urban transit buses that will provide higher fuel economy and lower emissions. In these vehicles the battery supplements a phosphoric acid fuel cell to provide additional power.

Also under scrutiny are hybrid vehicles that couple electric vehicles and internal combustion engines using another fuel source such as gasoline, natural gas, propane, methanol, or ethanol. This technology may fulfill environmental concerns until more efficient batteries are developed.

Hydrogen. Hydrogen as a source for vehicle fuel has unique advantages. It does not emit volatile organic compounds, carbon monoxide, particulates, sulfur dioxide (a precursor of acid rain), or carbon dioxide (the principal greenhouse gas). Also, since it can be produced via electrolysis from photovoltaic electricity by using thin-film silicon solar cells, the hydrogen should not be resource-constrained.

However, two major disadvantages are associated with the use of hydrogen. First, combustion of hydrogen results in emissions of nitrogen oxide. Second, the cost of producing hydrogen from natural gas, or by splitting water into hydrogen and oxygen by passing an electric current through it, is prohibitive at this time.

Existing storage mechanisms for hydrogen (compressed gas in heavy tanks at high pressure, as a metal hydride, adsorbed on activated carbon, or as liquid hydrogen at very low temperatures) necessitate significant on-board storage requirements because of the high weight and volume per unit of hydrogen stored. These requirements restrict the amount of hydrogen that can be carried on a vehicle and shorten the driving range.

Hydrogen is generally considered a long-term environmental option, although research into the development of a dual-fuel (hydrogen/gasoline) vehicle with the potential for achieving emission levels similar to those of an electric vehicle in urban centers suggests possible opportunities for greater short-term use. The use of hythane—15 vol % hydrogen and 85 vol % compressed natural gas—has also yielded some favorable emissions results.   *Michelle Heath*

Refuse-Derived Fuel

In a refuse-derived fuel (RDF) facility, preprocessing of solid waste produces a fuel product suitable for combustion in a boiler dedicated for this purpose or for use as a supplement in a fossil-fuel-fired boiler. The various types of refuse-derived fuel have been defined as shown in the **table**.

The prime economic driving force for effective resource recovery is the sale and utilization of energy, because municipal solid waste is composed of approximately 70–75% combustible materials, with the balance being glass, metal, and dirt. The economic justification for resource recovery plants is developed from a combination of tipping fees (charges for the use of plant disposal facilities) and sale of energy (fuel, steam, or electricity), with a minor contribution from sale of recovered materials. Processes for energy recovery from municipal solid waste are generally subdivided into direct combustion of as-received refuse

Types of refuse-derived fuels*

Classification	Description
RDF-1	Wastes used as a fuel in as-discarded form with only bulky wastes removed
RDF-2	Wastes processed to coarse particle size with or without ferrous metal separation
RDF-3	Combustible waste fraction processed to particle sizes, 95% by weight passing a 2-in.-square (5.08-cm-square) mesh screening
RDF-4	Combustible waste fraction processed into powder form, 95% by weight passing 10-mesh (10 openings/in.^{2}) screening
RDF-5	Combustible waste fraction densified (compressed) into the form of pellets, slugs, cubettes, or briquettes
RDF-6	Combustible waste fraction processed into liquid fuels
RDF-7	Combustible waste fraction processed into gaseous fuels

* After American Society for Testing and Materials, *Thesaurus on Research Recovery Terminology*, STP832, 1983.

(RDF-1), known as mass burning, and processing of as-received refuse to separate and reduce the volume of the combustible portion of the total solid-waste stream (RDF-2 through -7). The advantages of processing waste to RDF-2 through -7 include production of a fuel that is more uniform, storable, and transportable.

RDF-1. Combustion of RDF-1 (mass burning) is the most highly developed and commercially proven process available for reducing the volume of municipal solid waste prior to ultimate disposal of the residual material on the land, and also for extracting energy from solid waste. Hundreds of such plants, which incorporate various grate systems and boilers and which differ in details of design, construction, and quality of operation, have been built throughout the world since the mid-1960s. More than 90% of the plants built in the United States since 1965 are still in operation. It is generally agreed that this type of plant currently can be designed and operated with continuous satisfactory service.

RDF-2. At least two basic types of processes, wet and dry, have been used to produce RDF-2, so-called coarse refuse-derived fuel. In the dry process, the material may pass through a bag breaker (flail mill, a type of hammer mill, or other device) before being processed by a horizontal or vertical hammer mill to break apart and reduce in size the incoming solid waste to give a more homogeneous product. After this shredding process, which usually produces a nominal particle size of 4–6 in. (10.2–15.2 cm) measured in any direction, the material may be subjected to the magnetic recovery of ferrous metal before being combusted. Combustion usually takes place in a dedicated boiler utilizing a spreader stoker, the most common type of facility for refuse-derived fuel, other than the mass-burn facility described above.

In the wet process, a hydropulper (an oversized blender) is utilized to produce a solid-waste slurry. A cyclone then separates the light combustible fraction from the heavy fraction that contains glass, metals, and inert materials. Next, the light combustible fraction is dewatered and combusted in a spreader-stroker boiler or fluid-bed furnace. This technology, after being applied in several full-scale facilities and proved not to be cost-effective, is no longer in use.

RDF-3. A fine or fluff refuse-derived fuel is produced by utilizing a horizontal or vertical hammer mill to produce a coarse fuel that subsequently is subjected to magnetic separation to remove the ferrous metal. The coarse refuse-derived fuel then is air-classified to separate the lighter combustible fraction from the heavier noncombustible one. The light combustible fraction undergoes a second shredding operation to reduce its size to about a nominal $1\frac{1}{2}$–2-in. (3.8–5.1-cm) particle size. This fine or fluff refuse-derived fuel can be fed into a new or existing suspension-fired boiler; it can be either combusted alone or cofired, usually with coal. More problems have been encountered in attempts to utilize this technology than with RDF-2, because of the increased equipment required, increased materials handling, and problems related to cofiring

two somewhat dissimilar fuels.

RDF-4. This powdered refuse-derived fuel was produced by a proprietary process and installed in an 1800-ton/day (1600-metric ton/day) plant in Bridgeport, Connecticut. In this process, municipal solid waste first passes through a primary trommel, in which heavier material is separated from light combustible material. The light combustible material is then reduced in size by shredding, and any ferrous metal present is removed by a magnet. Next, the light material is air-classified and conveyed by hot gases to a secondary trommel, in which most of the fine inert material and remaining glass is removed. At the discharge end of the secondary trommel, a chemical embrittling agent (sulfuric acid) is applied to the combustible material. This material is then passed through a ball mill, in which the steel balls are preheated. The combination of heat and embrittling agent causes the cellulosic material to become a powder under the grinding action of the balls. After several years of trial operation, the process was determined not to be cost-effective and the plant was shut down. It was later replaced with a mass-burn facility.

RDF-5. This densified refuse-derived fuel has been produced sporadically over the years at a number of facilities. This material is produced by processing RDF-3 through an extrusion device under high pressure, with or without binding material added, producing pellets, slugs, cubettes, or briquettes. The material produced, if consistent in size, would be easier to feed with crushed coal in a spreader-stoker-fired furnace. A number of trial burns have been conducted in the United States over the years. However, use of this material as a fuel has not generally been accepted.

RDF-6 and RDF-7. Production of RDF-6 (liquid fuels) and RDF-7 (gaseous fuels) may be accomplished either through the proper application of heat (pyrolysis or destructive distillation) or through bioconversion processes. The potential use of bioconversion to convert the cellulosic fraction of municipal solid waste to ethanol has been under study for some time. However, research efforts have not led to development of pilot or full-scale plant applications because of practical process problems involving requirements for construction materials, need for sufficient control of feedstock to ensure resonable yields and quality of end product, and cost effectiveness. Production of methane gas through bioconversion of primarily cellulosic fractions of municipal solid waste in a digester was accomplished in a pilot-scale facility in the 1970s. This process was found to have limited applicability, and the pilot plant was closed. No other facilities have been built since.

In addition, the use of heat (pyrolysis) to produce liquid or gaseous fuels has been actively investigated (particularly in the 1970s) in laboratory, pilot, and full-scale demonstration installations. One demonstration-scale unit, which was intended to produce pyrolytic oil, was built and carried through start-up operation before being shut down in the late 1970s. Problems in the waste-preparation process could not be resolved to the extent that the production process for pyrolytic oil could be effecitvely tested.

Prospects. A number of demonstration and full-scale fuel-gas pyrolysis systems were built and operated from the mid-1970s to the mid-1980s. These installations included a number of facilities in the United States and Europe. After achieving varying limited success in operation, all of them were closed because of an inability to achieve long-term cost-effective operation.

Since the late 1960s, some 40 plants incorporating processes to produce and combust RDF-2 through -7 have been built and placed in operation in the United States. These plants have frequently incorporated some related processes of separation and recovery of materials. Approximately one-third of these 40 plants have subsequently been shut down. The closed plants utilized wet RDF-2 or RDF-3 through -7 technologies.

Refuse-derived fuel has been burned with and without fossil fuel in new dedicated boilers or in existing boilers that were modified to accept the material. Combustion in the boiler takes place partially in suspension and partially on the boiler grate. Numerous trial burns were conducted during 1970–1985 in large utility-type boiler units. Results were generally satisfactory with respect to combusion conditions, but mixed results were experienced with respect to plant operation and maintenance conditions. The major problems included periodic explosions in the initial shredding operation; excessive wear and tear on the equipment, resulting in frequent downtime; and difficulties in storage and retrieval of the refuse-derived fuel. The environmental impacts, and the measures needed to mitigate them, are similar to those required for RDF-1. *SEE ENVIRONMENTAL ENGINEERING.*

For background information *SEE ALCOHOL FUEL; BIOMASS; CETANE NUMBER; LIQUEFIED PETROLEUM GAS (LPG); METHANOL; NATURAL GAS; OCTANE NUMBER* in the McGraw-Hill Encyclopedia of Science & Technology.

Charles O. Velzy

Bibliography. M. Heath, *Alternative Transportation Fuels: Natural Gas, Propane, Methanol, and Ethanol Compared with Gasoline and Diesel,* CERI Study 37, Canadian Energy Research Institute, February 1991; M. Heath, *Conference on Commercializing Clean Fuels for Transportation: Summary,* Canadian Energy Research Institute, 1991; J. L. Jones and S. B. Radding (eds.), *Thermal Conversion of Solid Wastes and Biomass,* ACS Symp. Ser. 130, 1980; J. E. Sinor Consultants, Inc., *The Clean Fuels Report,* vol. 4, no. 1, February 1992; C. O. Velzy, *Energy-from-Waste Incineration,* paper presented at Great Lakes International Solid Waste Management Forum, Lansing, Michigan, March 20, 1990; C. O. Velzy, Incinerator's role in integrated S. W. management, *Managing SW: Options for State Legislative Action,* NCSL Conference, Breckenridge, Colorado, June 13, 1989; P. S. Vesilind and A. E. Rimer, *Unit Operations in Resource Recovery Engineering,* 1981; N. J. Weinstein and R. F. Toro, *Thermal Processing of Municipal Solid Waste for Resource and Energy Recovery,* 1976.

Fullerene

The nearly spherical, all-carbon molecule C_{60} (buckminsterfullerene) is the object of extraordinary interest among chemists, physicists, and material scientists. Although it was first detected in the gas phase in 1985, its recent isolation and purification in large quantities has precipitated intense research activity. Buckminsterfullerene is produced in surprisingly high yield [along with smaller amounts of its congeners (fullerenes) C_{70} and C_{76} and other higher fullerenes] by evaporating graphite in a carbon arc or other source of high temperature. It is readily purified by chromatography. A number of the higher fullerenes have also been isolated, and the structures of some have been determined. A recent report suggests that C_{60} and C_{70} can be isolated in small quantities from natural rocks. Research has elucidated the physical and chemical properties of a number of fullerenes. In addition, some forms in which foreign atoms have been incorporated have been shown to constitute a new class of semiconductors.

Properties

The parent molecule consists of a framework of 60 carbon atoms in a soccer-ball-like arrangement (**Fig. 1**). The general framework for all fullerenes consists of a skeleton of various numbers of unsaturated six-membered rings with 12 five-membered rings. A structure with only six-membered rings would be planar (idealized, graphite is an infinite sheet of this structure). Introduction of five-membered rings brings curvature to the surface; 12 of these rings are required for a closed surface, no matter how many six-membered rings are present. A key structure in fullerenes is the pyraclyene unit [structure (I)] with

(I)

five-membered rings separated by six-membered rings. Since the fusion of two five-membered rings introduces strain, this feature does not appear in fullerenes isolated so far.

Recently, large tubular structures (nanotubes) based on the fullerene geometric framework have been postulated, and they have apparently been observed by electron microscopy. The properties of these nanotubes are of intense interest, but little is known about them.

Only a few of the numerous reports of chemical and physical properties of the fullerenes are summarized in this article. Although much gas- and solid-phase work has appeared, this account is limited to work carried out in the solution phase.

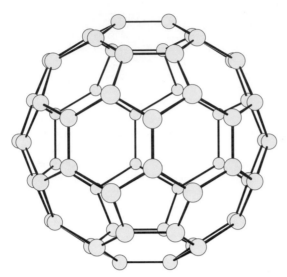

Fig. 1. Structure of C_{60} (buckminsterfullerene).

Physical characterization. The physical characterizations of C_{60} and C_{70} have been most complete; few properties of higher fullerenes [except electronic and nuclear magnetic resonance (NMR) spectra for some] have been reported. An x-ray crystal structure of the nearly spherical C_{60} was extremely difficult to obtain because of its rotation in the crystal, but the problem was finally solved. The vibrational spectra (absorption and Raman) of both C_{60} and C_{70} and numerous derivatives have been reported and extensively analyzed. The ^{13}C-NMR spectra of C_{60} and C_{70} have resonances at rather low frequency compared to aromatic compounds; the spectrum of C_{60} consists of a single sharp line at a chemical shift of 143.2 parts per million, whereas the higher fullerenes have more complex spectra, in keeping with their lower symmetry. The electronic spectra of C_{60} and C_{70} have strong absorptions in the ultraviolet region and much weaker, symmetry-forbidden absorption in the visible region (hence the beautiful purple color of C_{60}). Although C_{60} was originally reported not to fluoresce or phosphoresce, weak fluorescence has been reported recently. Both weak fluorescence and phosphorescence have been observed for C_{70}.

The reaction sequence below shows the paths of the

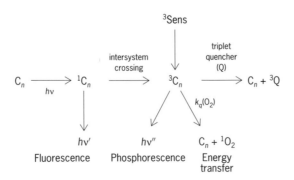

photophysical processes of C_{60} and C_{70}. Absorption of

light by C_{60} and C_{70} produces a very short lived electronically excited singlet state ($^{1}C_{n}$), which converts to a metastable triplet electronic excited state ($^{3}C_{n}$) in nearly quantitative yield. The triplet state lies near 35 kilocalories (146 kilojoules) per mole above the ground state in both C_{60} and C_{70}, as shown by energy transfer from other excited triplet sensitizers (^{3}Sens) and to quenchers (Q) of known triplet energy, and by photothermal techniques. The triplet fullerenes transfer energy readily to oxygen, forming singlet molecular oxygen ($^{1}O_{2}$, a reactive excited state) with high efficiency. In fact, both molecules are excellent sensitizers for production of singlet oxygen, since they are extremely unreactive toward singlet oxygen and quench it only very inefficiently.

The small difference in energy between the excited triplet and the ground singlet state (S-T splitting) in C_{60} and C_{70} [$\sim$ 9 kcal (38 kJ) per mole] is probably a result of the large diameter of the molecule and the resulting small electron-electron repulsion energy. This small S-T splitting, the low value of the fluorescence rate constant, and the expected large spin-orbital interaction in the spherical C_{60} explain why triplet formation is a dominant process in these molecules. Overall, the photophysical behavior of C_{70} is similar to that of C_{60}, but the lower symmetry relaxes the forbiddenness of some absorption and emission processes, explaining its more intense visible absorption and readily observed fluorescence and phosphorescence emissions.

The products of one-electron oxidation and reduction (the radical cation and radical anion, respectively) of C_{60} have been prepared by pulse radiolysis and by photochemical electron transfer, and they have been characterized spectroscopically. A number of products of sequential electrochemical reduction have also been observed. Studies of radical anions, triplet C_{60}, and radical adducts made with electron spin resonance spectroscopy have been reported. The heat of formation of C_{60}, as determined by combustion, is 9.1–9.7 kcal (38–41 kJ) per mole per carbon atom in the crystalline state.

Chemical reactivity. The chemical reactivity of the fullerenes is just beginning to be explored; the production of functional derivatives is in its infancy, and only a few such compounds have been completely characterized. One of the most intriguing properties of C_{60} and C_{70} is that these molecules are extremely electron deficient. They are easily reduced electrochemically (accepting up to six electrons sequentially), but they are very difficult to oxidize. The fullerenes do not behave like aromatic compounds; in fact, their properties are best compared with those of quinones. One consequence of their electron deficiency is that they react with nucleophiles (electron-rich compounds) with surprising ease. In all cases so far, the double bond between two six-membered rings (the pyracylene bond) is the one that reacts. Chemical adducts have been prepared mainly from C_{60}; a monoiridium adduct has also been prepared from C_{70}. Among the C_{60} adducts that have been prepared and characterized thoroughly are derivatives of osmium tetroxide (OsO_{4};

(II)) as well as platinum and iridium.

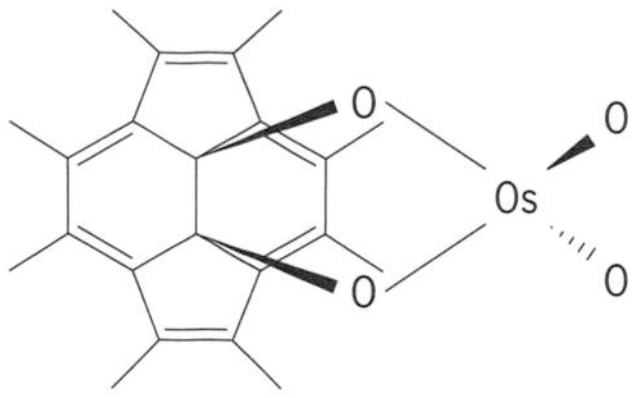

(II)

Diphenylcarbene and other methylene adducts have also been characterized. These adducts have long C—C bonds, suggesting that they are annulenes (III) rather than cyclopropanes (IV).

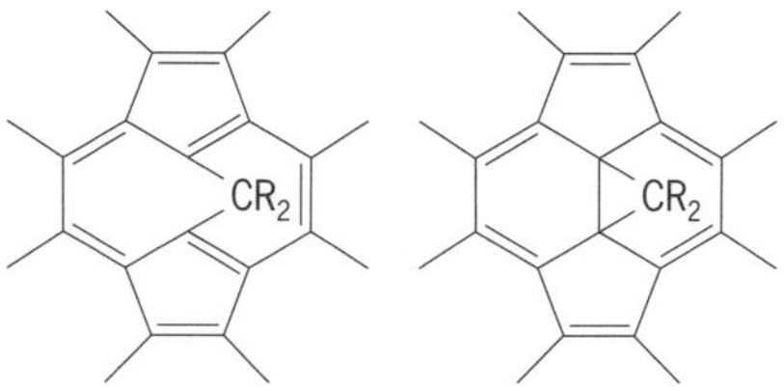

(III)   (IV)

Several adducts of amines and other nucleophiles to C_{60} have been reported, and one has been characterized. In addition, a polymeric derivative in which a metal atom cross-links C_{60} by multiple intermolecular ligand formation has recently been reported. A highly symmetrical bromine-fullerene compound ($C_{60}Br_{24}$) has been prepared, and its x-ray crystal structure has been determined. An epoxide ($C_{60}O$; V) has been

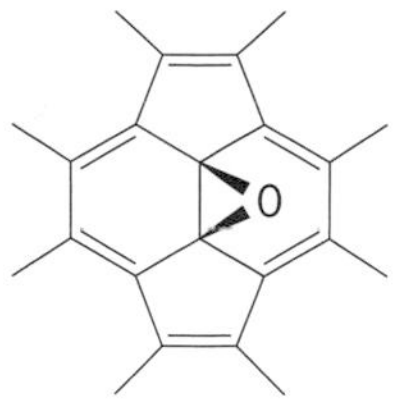

(V)

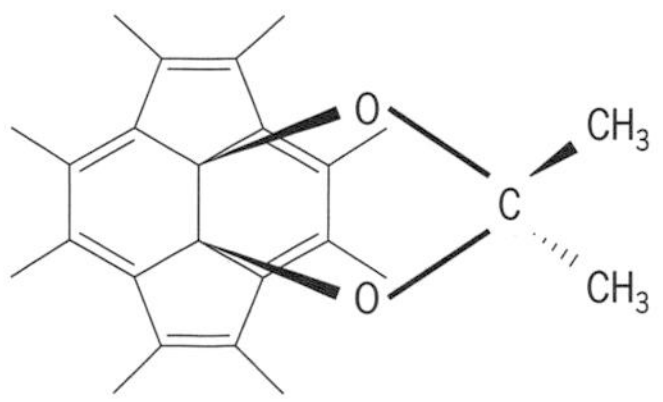

(VI)

reported by two research groups; a small amount is present in soot (as is a small amount of $C_{70}O$), and the epoxide can also be made by a long period of photochemical irradiation under oxygen or by reaction with the highly aggressive oxidizing agent dimethyl dioxirane. In contrast to the methylene adducts, the epoxide

appears to have a three-member ring rather than the open structure. A dioxolane [the acetone derivative of fullerene diol (VI)] is formed along with the epoxide upon oxidation with dimethyl dioxirane.

Free radicals add to fullerenes; the structures of the adducts are not certain, but they are free radicals, which seem to undergo reversible dimerization to form weak carbon-carbon bonds between two adduct fullerenes. A number of products of addition of nucleophiles, reduction, radical addition, and electrophilic substitution have also been reported, but most have not been well characterized (or even isolated, in most cases). Also of great interest is the recent preparation of a number of derivatives with metals inside the ball structure (cage), and of compounds in which carbon has been replaced by boron or nitrogen.

An interesting potential feature of fullerene adducts is inside-outside isomerization. Although it is difficult to conceive of additions to intact fullerenes occurring except with both connecting atoms coming from the outside, such products are not necessarily the most stable. Saturating the nearly planar unsaturated carbons in fullerenes would produce highly strained structures. Recent theoretical studies have shown that the hypothetically fully saturated $C_{60}H_{60}$ with 10 hydrogen atoms inside is 1670 kJ more stable than the all-outside isomer.

The activity in this field is not slackening. The next few years should show a rich organic chemistry of fullerenes and the preparation of many interesting and useful derivatives.   *C. S. Foote*

Superconductors

One of the key interests in superconductivity research is the fabrication of materials that can superconduct at room temperature (300 K or 80°F). A variety of foreign atoms can be incorporated into the fullerene solids to produce doped structures. In the alkali-doped C_{60}, alkali atoms such as sodium (Na), potassium (K), cesium (Cs), and rubidium (Rb) are incorporated in the space between individual molecules. The alkali-doped carbon compounds, $A_{n-x}B_xC_{60}$ (where A and B are alkali atoms and $n, x \leq 3$), form a new class of superconductors.

Solid C_{60} is a semiconductor. However, when doped with the alkali atoms, it becomes a metal and exhibits superconductivity below the critical temperature T_c. The value of T_c varies with the type and concentration of the dopant. The maximum T_c within the fullerene class of superconductors is currently 33 K (−400°F) and is exhibited by the compound

Superconducting transition temperatures (T_c) of alkali-doped C_{60} solids	
Compound*	T_c, K (°F)
K_3C_{60}	19.0 (−425)
Rb_3C_{60}	28.6 (−408)
$RbCs_2C_{60}$	33 (−400)

* K = potassium; Rb = rubidium; Cs = cesium.

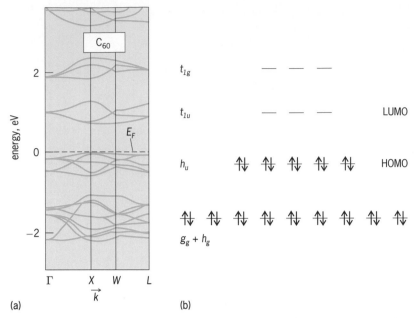

Fig. 2. Correlation between the electron states in the free molecule of C_{60} and those in the solid. (*a*) Electron band structure of solid C_{60} in the face-centered cubic structure. Here $\vec{k}$ denotes the electron wave vector; Γ, X, W, and L denote specific values of $\vec{k}$; and E_F is the Fermi energy. (*b*) Electronic structure (molecular orbitals) for a free C_{60} molecule. The dopant alkali atoms do not change the band structure, but they introduce extra electrons into the unfilled t_{1u} bands, thereby producing a metal. The labels t_{1g}, t_{1u}, h_u, g_g, and h_g describe the symmetry properties of the molecular orbitals.

occupied by the dopant atoms.

Electronic structure. Electrons control most physical properties of a solid. The conduction properties are determined by electrons at the Fermi energy E_F, which is the energy of the highest occupied electronic states in the solid.

There is a close correspondence between the electron states in the free molecule and the solid. The molecular orbitals around E_F are formed by a linear combination of the radially directed p orbitals on the carbon atoms. The highest occupied molecular orbital (HOMO) is a fivefold degenerate orbital with h_u symmetry, while the lowest unoccupied molecular orbital (LUMO) is a threefold t_{1u} orbital, as shown in **Fig. 2**. Since in solid C_{60} the intermolecular distances are relatively large, the interaction between molecules is weak enough that the identity of the individual molecular orbitals is retained in the electronic band structure. The HOMO and the LUMO orbitals are now spread into a band of energy, producing the h_u and the t_{1u} bands.

Solid C_{60} as seen from the band structure is a semiconductor with a band gap of about 1.5 electronvolts as measured by using photoemission and optical absorption. Not being a metal, it cannot superconduct. The simplest way of producing a metal is by doping the solid with alkali atoms. In the alkali-doped C_{60} the single outermost valence electron of the alkali atom is stripped off and is transferred to the t_{1u} orbitals of the carbon molecule. The valence-electron charge-density contour plot of **Fig. 3** shows that there is virtually no valence charge left on the alkali atom. The alkali atom is thus positively charged and the carbon molecule negatively charged, resulting in an ionic compound. Since the t_{1u} band can hold up to six electrons (taking spin into account), it is now partially filled, resulting in a metal unless exactly six electrons are transferred to the carbon cage. If there are six alkali atoms as dopants per C_{60} molecule, this transfer can occur, and the solid becomes a semiconductor again.

Theory of superconductivity. According to the Bardeen-Cooper-Schrieffer (BCS) theory, the superconducting state is formed below T_c by an instability where the electrons near the Fermi surface interact to form pairs with their time-reversed states. Two electrons with opposite momenta and spin ($k\uparrow, -k\downarrow$) can bind together to form a Cooper pair by a weak effective attraction. A Cooper pair has a tendency to form in the presence of other such pairs, but its formation is progressively impeded with increasing temperature. The net result is that superconductivity is destroyed above the transition temperature T_c. For the simple model pair interaction, where the attractive interaction is V for electrons within a characteristic energy $k_B\theta$ from the Fermi energy and zero otherwise, the BCS theory estimates that in the weak-coupling limit, $N(0)V \ll 1$, the value for T_c is given by Eq. (1),

$$T_c = 1.13\theta \exp[-1/N(0)V] \qquad (1)$$

where $N(0)$ is the electron density of states per spin at E_F.

$RbCs_2C_{60}$ (see **table**). One group of researchers has found evidence of superconductivity in $Rb_{2.7}Tl_{2.2}C_{60}$, a similar compound where thallium (Tl) replaces Cs, at as high a T_c as 43 K ($-382°F$), indicating the possibility that the maximum T_c might rise as new dopants are tried and other fullerene compounds are fabricated.

Crystal structure of C_{60}. The C_{60} molecules are produced by first evaporating graphite electrodes in an atmosphere of helium. A carbon plasma is formed that eventually condenses into a graphitic soot on the container walls. The soot contains an assortment of carbon clusters of different sizes, including a few percent of C_{60} molecules, which can be extracted and crystallized into solids by several techniques. One such method involves dissolving the soot in clear benzene and then filtering off and evaporating the resulting red-colored solution. Good-quality single crystals can be produced by heating the soot to about 400°C (750°F) in vacuum and then subliming it onto a substrate.

X-ray diffraction experiments show that at room temperature individual C_{60} molecules are packed in the solid in a face-centered cubic (fcc) structure with a separation of about 1 nanometer between the molecules. The individual molecules rotate freely centered on a fixed lattice site in the crystal. Below 249 K ($-11.5°F$), orientational order develops; however, the molecules flipflop between different equivalent orientations. The rapid flipping stops below the liquid nitrogen temperature (77 K or $-321°F$). The relatively large size of the C_{60} molecules leaves large empty spaces between them that can be

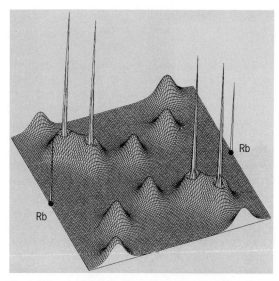

Fig. 3. Valence-electron contour plot of the rubidium (Rb)-doped C_{60}. There is virtually no valence-electron charge density near the rubidium atom. The sole valence electron of rubidium is transferred to the C_{60} molecule partially filling the t_{1u} bands. Spikes indicate atom positions, while the charge maxima indicate accumulation of electrons at midpoints (bond centers) between pairs of nearby carbon atoms.

In a normal metal, the current-carrying electrons move in an uncorrelated fashion and are easily scattered from crystalline imperfections. In the superconducting state, however, the electrons move together in a coherent fashion, making it difficult for the crystalline imperfections to alter their collective motion. Consequently, the electrons experience no resistance, and the solid superconducts.

Phonon-mediated superconductivity. The BCS theory is independent of the mechanism that produces the attractive interaction between a pair of electrons. The most common mechanism is that this attractive interaction occurs via electron coupling to the lattice (electron-phonon coupling). While moving through the solid, the negatively charged electron leaves a deformation trail, inducing a small positive charge in the lattice. This deformation is created somewhat after the electron has passed through the portion of the lattice (retarded interaction), since the electron moves much faster than the lattice can respond. A second electron is then attracted toward the positively charged deformed region, causing an effective attraction between the two electrons. The retarded nature of the interaction is important, because while producing the attractive interaction, it keeps the two electrons sufficiently apart that the undesirable Coulomb repulsion between them is significantly reduced. A large Coulomb repulsion would suppress superconductivity by discouraging electrons from forming bound Cooper pairs.

A powerful version of the BCS theory, general enough to include the retarded interaction, was formulated by G. M. Eliashberg. Based on the Eliashberg equations, W. L. McMillan derived a widely used empirical formula for T_c, which holds if the lattice is not too strongly coupled to the electron ($\lambda \lesssim 1.5$), as

in Eq. (2), where $\langle \omega \rangle$ is an average phonon frequency

$$T_c = \frac{\langle \omega \rangle}{1.20} \exp\left[\frac{-1.04(1 + \lambda)}{\lambda - \mu^*(1 + 0.62\lambda)}\right] \quad (2)$$

of the crystal, μ^* is the effective Coulomb repulsion between electrons on the Fermi surface, and λ is a measure of the effectiveness of the electron-phonon coupling in mediating superconductivity. The last parameter is given by the expression $\lambda = N(0)V_{e-ph}$, where V_{e-ph} is an average coupling of the phonons to an electron near the Fermi surface.

The electron-phonon coupling mechanism is strong enough to explain the superconductivity of the alkali-doped C_{60}. There are several types of phonons: the alkaline phonons, rotations of the C_{60} molecules; the acoustic phonons, the tangential vibrations with carbon atoms moving on the spherical surface of the C_{60} molecule; and the radial vibrations, where the carbon atoms vibrate along the radial direction perpendicular to the sphere surface. Of these, it seems that the tangential vibrations couple most strongly to the electrons. This result is expected since, roughly speaking, the tangential vibrations cause the largest deviation in the carbon-carbon distances on the molecule, thereby maximally affecting the electronic states. The average phonon frequency $\langle \omega \rangle$ has been estimated as 870 cm^{-1}. Band calculations provide the value of $N(0) \approx 10$ states per eV-molecule-spin. The value $V_{e-ph} \sim 50$–80 meV has been estimated by examining how individual phonons affect the electronic energy levels. The magnitude of the electron-phonon interaction parameter then is $\lambda = N(0)V_{e-ph} \approx 0.5$–$0.8$. With an estimated Coulomb parameter of $\mu^* \sim 0.15$, the McMillan formula predicts T_c values of 6 to 40 K (-449 to $-338°$F), in agreement with the observed values for T_c presented in the table.

The separation between the C_{60} molecules can be reduced either by applying external pressure or by choosing smaller alkali atoms. This procedure will increase the width of the t_{1u} bands, reducing thereby the density of states $N(0)$, which is inversely related to the bandwidth. However, the relevant phonon quantities $\langle \omega \rangle$ and V_{e-ph} do not change, as the C_{60} molecules are essentially unaffected. The transition temperature will therefore decrease in both cases, as indicated from the expression for T_c. This decrease is more or less observed, supporting further the electron-phonon mechanism of superconductivity in the fullerenes.

Prospects. New fullerene superconductors will continue to be explored. For instance, it is anticipated that a new class of doped C_{60} solids will be discovered where holes are introduced into the valence bands with appropriate dopant atoms. Also, superconductors based on other fullerene molecules such as C_{70} are possibilities.

For background information SEE BAND THEORY OF SOLIDS; FERMI SURFACE; MOLECULAR ORBITAL THEORY; QUINONE; SOLID-STATE PHYSICS; SPECTROSCOPY; SUPERCONDUCTIVITY; TRIPLET STATE; WORK FUNCTION (ELECTRONICS) in the McGraw-Hill Encyclopedia of Science & Technology.

Sashi Satpathy

Bibliography. F. Diederich and R. L. Whetten, C_{60}: From soot to superconductors, *Angew. Chem.*, 103: 695–697, 1991; G. S. Hammond and V. S. Kuk (eds.), *Fullerenes: Synthesis, Properties, and Chemistry of Large Carbon Clusters*, ACS Adv. Chem. Ser. 481, 1992; A. F. Hebard et al., Superconductivity at 18 K in potassium-doped C_{60}, *Nature*, 350:600–602, 1991; D. R. Huffman, Solid C_{60}, *Phys. Today*, 44:22–29, November 1991; W. Krätschmer et al., Solid C_{60}: A new form of carbon, *Nature*, 347:354–358, 1990; H. W. Kroto, A. W. Allaf, and S. P. Balm, C_{60}: Buckminsterfullerene, *Chem. Rev.*, 91:1213–1235, 1991; H. W. Kroto et al., C_{60}: Buckminsterfullerene, *Nature*, 318:162–163, 1985; S. Saito and A. Oshiyama, Cohesive mechanism and energy bands of solid C_{60}, *Phys. Rev. Lett.*, 66:2637–2640, 1991; S. Satpathy, Electronic structure of the truncated-icosahedral C_{60} cluster, *Chem. Phys. Lett.*, 130:545–550, 1986; S. Satpathy et al., Conduction-band structure of alkali-metal-doped C_{60}, *Phys. Rev. B*, 46:1773–1793, 1992; Special issue on buckminsterfullerenes, *Acc. Chem. Res.*, vol. 25, no. 3, 1992.

Fungi

The geologic history of fungi was not extensively studied because the group was perceived to have an inadequate fossil record. However, interest in these organisms has greatly increased in recent years as their importance in many interactions with plants, animals, and the geologic environment continues to be documented.

Although fossil fungi were some of the first organisms to be studied by paleobiologists, the discipline of paleomycology has been neglected for two principal reasons. One concerns the belief that fungi are fragile and thus poor candidates for fossilization. The second is that most paleobotanists are trained to study vascular plants and thus are not equipped to deal with the life-history complexities and taxonomy of fungi. In recent years, however, paleomycology has emerged as a discipline that is making important contributions both about how the fungal groups evolved and how the ancient fungi interacted with their environment.

Saprophytism. Although the degradation of plant and animal tissues is one of the major activities of fungi today, there are relatively few examples of saprophytism from the fossil record. The earliest documented example of wood-rotting fungi, from the Upper Devonian, includes fungal hyphae in the tracheids and ray cells of the progymnosperm *Callixylon* (**Fig. 1**). Some cells contain specialized fungal spores, while others contain resinous deposits that may have been produced by the plant in response to the invasion of the fungus. As early as there is geologic evidence of terrestrial plants, examples of fungi decomposing plant tissues occur.

In some fossils it is possible to actually see evidence of the fungus in the form of the mycelium, but in others only the symptoms of the fungus are evident (**Fig. 2**).

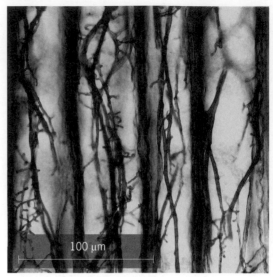

Fig. 1. Fungal hyphae in tracheids of *Callixylon*.

Today there are a number of fungi that attack both hardwood and softwood trees. One is white pocket rot, a type of wood decay characterized by spindle-shaped zones surrounded by sound wood cells. In transverse section the individual pockets are circular and up to 0.14 in. (3.5 mm) in diameter. Some pockets are devoid of cells, while in others the decayed regions consist of white cellulosic tissues in which the lignin is lost from the primary and secondary cell walls. Another form of wood rot that can be documented from the fossil record is white rot. This fungal decay process shows evidence of the simultaneous removal of both cellulose and lignin from the wood cells. The fungal hyphae of white rot are directly associated with the degradation of the cells. In contrast, the fungal hyphae in white pocket rot produce enzymes that diffuse for a considerable distance to form the spindle-shaped pockets. Both of these wood-rot types have been identified as early as the Permian, and in Antarctica they can be traced into the Triassic in several groups of plants. Although yeasts and bacteria are associated with wood-rotting fungi and play an important role in the decay process today, these organisms have not

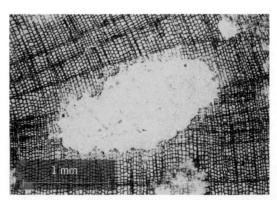

Fig. 2. Cross section of a Triassic woody stem showing effects of white pocket rot fungus.

been identified in fossil woods containing fungi.

Mutualism. Another fungus–plant interaction that can be documented from the fossil record is mutualism, a symbiotic association in which both partners benefit. One form of mutualistic symbiosis between fungi and the roots of terrestrial plants is the mycorrhizal association. Plants benefit from association with mycorrhizae by having an increased ability to withstand drought, high soil temperature, and toxic metals, while the fungus benefits by receiving organic compounds. One example of such an association is the vesicular-arbuscular mycorrhizae (VAM). Vesicular-arbuscular mycorrhizae are characterized by shrublike growths of the fungus, called arbuscules, which are typically present in cells of roots and represent the sites of physiological exchange between the fungus and plant. It has been hypothesized that this type of fungal interaction enabled the early colonization of the Earth by plants by providing the plants with an increased ability to take up nutrients in what must have been poor soils. Some evidence indicates that this type of interaction may have been in existence as early as the Devonian, but to date the best evidence for the arbuscule is found in root cells of a cycad from the Triassic of Antarctica.

Parasitism. The geologic record provides some excellent examples of parasitism, an interaction in which one member of the symbiosis benefits at the expense of the other. This association is more difficult to demonstrate since it is necessary to document both the existence of the fungus and a response by the host. One way in which this type of interaction can be demonstrated in fossils is by identifying the particular type of fungus and relating it to a modern analog. For example, the production of abnormal, proliferated cells in a Carboniferous seed is identical to the host response caused by the fungal parasite *Albugo*. Adding evidence to the existence of this parasitic association in the fossil seed is the presence of certain reproductive cells that are like those of modern *Albugo*.

A much older parasitic interaction between a fungus and plant can be documented from the 4×10^8-year-old Lower Devonian Rhynie chert. One of the plants that lived in this ancient ecosystem was the green alga *Palaeonitella*, a fossil form that is morphologically similar to the modern stonewort *Nitella* or *Chara*. Associated with some of the cells of the alga are several fungi that morphologically resemble living chytridiomycetes, a group of modern microscopic fungi that occur in soil and fresh water. Most living chytrids are saprophytes, but some are parasites on plants, animals, protists, and other fungi. Several stages in the life history of the Devonian chytrids are represented, including flask-shaped zoosporangia that make a definitive identification possible. Zoospores can be seen on the surface of some algal cells, while other cells show that the fungus has penetrated the cell wall. Especially interesting is the response of the algal cells to these fungi. The typical size of a *Palaeonitella* cell is 30–70 micrometers. Cells that have been infected by the chytrids, however, are nearly 300 μm in diameter. The presence of these enormous

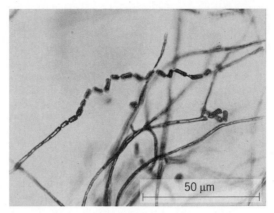

Fig. 3. Chains of specialized fungal spores associated with the leg of an Oligocene spider.

cells intermixed with normal ones is exactly the same host response that has been reported in some modern species of *Chara* attacked by certain types of parasitic chytrids. Thus, in this example of parasitism, stages in the life history of the fungus as well as the response of the host can be identified. The response of the host alga to the fungus has apparently remained unchanged over the 4×10^8-year period.

Interactions with animals. If fungi were associated with the decay process in plants, they were most certainly major saprobes of animal tissues. For example, *Geotrichites* is the name applied to various types of hyphae associated with an arachnid cadaver preserved in amber of probable Oligocene age (**Fig. 3**). In this example the fungus was apparently saprophytic. A more specialized fungus–animal interaction has been reported from Triassic rocks collected in Antarctica. Here the fungus is represented as cylindrical thalli that line the inner surface of what is interpreted as the cuticle of an invertebrate. A specialized cell that anchors the fungus to the insect host is at one end of each thallus; spores that reinfect the host are released from the other end of the thallus. These fungi morphologically resemble modern trichomycetes, a specialized group of zygomycetes that are endosymbionts of various arthropods. The presence of trichomycetes as early as the Triassic suggests that the symbiotic relationship is a very ancient one involving several hundred million years of coevolution between the two groups. As additional information is obtained, it may become possible to determine whether this fungus–animal interaction was initially commensalistic (that is, two organisms living together and sharing a common nutritional base) or pathogenetic.

For background information SEE FOSSIL; FUNGI; PALEOBOTANY in the McGraw-Hill Encyclopedia of Science & Technology.

Thomas N. Taylor

Bibliography. S. P. Stubblefield and T. N. Taylor, Recent advances in palaeomycology, *New Phytol.*, 108:3–25, 1988; T. N. Taylor, Fungal associations in the terrestrial paleoecyosystem, *Trends Ecol. Evol.*, 5:21–25, 1990; T. N. Taylor, W. Remy, and H.

Hass, Parasitism in a 400-million-year-old green alga, *Nature*, 357:493–494, 1992; T. N. Taylor and E. L. Taylor, *The Biology and Evolution of Fossil Plants*, 1993.

Gamma-ray astronomy

Gamma-ray astronomy is the study of high-energy radiation from the cosmos at energies above 100 keV. Extremely diverse astrophysical objects and phenomena emit detectable quantities of radiation in this energy region. These emitters include solar flares, galactic discrete sources typically containing a neutron star or black hole, diffuse gamma rays from the galactic plane, several types of extragalactic sources, a diffuse gamma-ray background that is not well measured, and enigmatic gamma-ray bursts appearing at random with a distance scale that is highly uncertain.

Former progress in gamma-ray astronomy was slow relative to other branches of astronomy. Most observations in gamma-ray astronomy require long exposure times with large, complex instruments above the Earth's atmosphere. Throughout the first two decades of gamma-ray astronomy observations, only limited exposures by instruments on high-altitude balloons or by relatively small satellite experiments could be performed. However, with the launch and operation of two recent spacecraft, the U.S. *Compton Gamma-Ray Observatory* and the C.I.S. *Granat X-Ray/Gamma-Ray Observatory*, the situation has improved dramatically.

Instrumentation. The *Compton Gamma-Ray Observatory*, launched by the space shuttle *Atlantis* on April 5, 1991, was deployed into low Earth orbit. It is the largest scientific spacecraft ever launched by the shuttle and the only multiexperiment spacecraft devoted to gamma-ray astronomy. The 35,000-lb (16,000-kg) spacecraft contains four large instruments designed to cover the entire gamma-ray energy range with a sensitivity approximately 10 times greater than that of previous experiments. The Burst and Transient Source Experiment (BATSE) consists of eight uncollimated, wide-field detectors located on the corners of the spacecraft, permitting a full-sky gamma-ray range from 20 keV to 2 MeV. The Oriented Scintillation Spectrometer Experiment (OSSE) consists of four collimated detectors that are optimized for the study of nuclear gamma rays from discrete sources. The Imaging Compton Telescope (COMPTEL) studies a virtually unexplored region of the gamma-ray spectrum from 1 to 30 MeV. Its detector elements are configured so that an image of the sky can be constructed about 80° in diameter. The Energetic Gamma-Ray Experiment Telescope (EGRET) detects gamma rays above 30 MeV and images a region of the sky about the same size as that of COMPTEL.

The *Compton Observatory* mission performed a

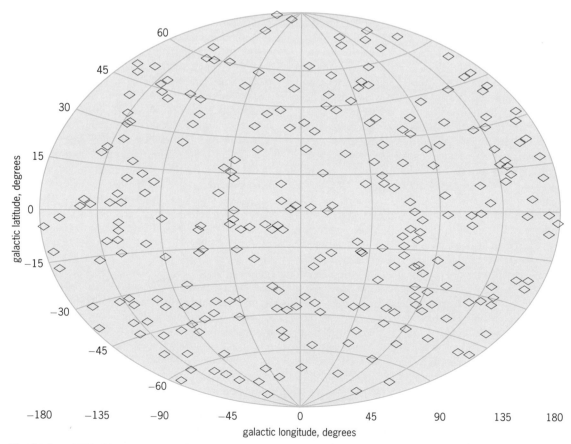

Distribution of 261 gamma-ray bursts observed by the Burst and Transient Source Experiment (BATSE) on the *Compton Gamma-Ray Observatory*.

complete sky survey by the two wide-field imaging instruments on the observatory during the first 2 years of operation. During this period, OSSE performed observations of numerous specific objects and regions, while BATSE provided a full-sky coverage for the detection of gamma-ray bursts and other transient sources. The observatory has operated well since launch, continuously providing its observations of the gamma-ray sky with the expected sensitivity. Several of the most dramatic observations made during the first 2 years of operation are summarized below.

Gamma-ray bursts. Gamma-ray bursts remain one of the most puzzling phenomena in astrophysics. They last from a fraction of a second to several hundred seconds. In spite of two decades of observations, the source objects and emission mechanisms remain unknown. Furthermore, they have never been detected with certainty at other wavelengths. Observations of the distribution of gamma-ray bursts by BATSE on the *Compton Observatory* have provided an unexpected development in the study of these enigmatic objects.

Prior to the BATSE observations, most theorists assumed that the bursts were associated with old neutron stars within the Milky Way Galaxy. It was expected that BATSE would find the bursts, particularly the weaker ones, concentrated in the plane of the Milky Way Galaxy or toward the galactic center. Instead, the gamma-ray bursts observed by BATSE are seen to be isotropically distributed over the sky (see **illus.**). This observation, by itself, seems to indicate that the sources of gamma-ray bursts are uniformly distributed in the three-dimensional space that is observable by BATSE. However, the number of weak bursts seen by BATSE is significantly less than that expected for a homogeneous or uniform distribution, implying that the burst sources are bounded yet isotropically distributed as seen from Earth. The only known astronomical ensemble that fits this description is the universe itself. In this case, the bursts would be at distances comparable to the size of the universe, and their luminosities would be higher than that of any known object, about 10^{51} ergs (10^{44} joules) per burst. There is at present a great deal of speculation as to what type of objects could produce such enormous bursts of gamma rays or what other types of objects at closer distances could produce bursts and match the observed distribution.

Emission from AGNs. Another surprising result from the *Compton Observatory* is EGRET's detection of high-energy gamma-ray emission from over 15 extragalactic objects. These objects, collectively known as active galactic nuclei (AGNs), are classified as quasars, Seyfert galaxies, BL Lac objects, and other galaxies that exhibit radio emission or give other indications of activity from the central, nuclear regions of the galaxy. Several of the quasars found by EGRET are the most distant sources in the universe seen to emit high-energy gamma rays. Also, these sources are variable and their gamma-ray emission does not show any obvious correlation with their radio or optical intensity. These facts indicate that the quasars' emission is beamed and may be produced by highly relativistic beams or jets of particles emitted from the central region. Such jets are observed in the optical and radio emission from active galactic nuclei.

Galactic sources. Within the Milky Way Galaxy, several new objects have been discovered to emit gamma rays. Prior to the *Compton* launch, only two well-known pulsars were observed to emit gamma rays: the Crab and Vela pulsars. Two known radio pulsars have now been added to the list of gamma-ray pulsars. In addition to finding these pulsars, the *Compton* instruments have obtained the most precise observations of the gamma-ray spectra of the Crab and Vela pulsars. These four pulsars and all radio pulsars are powered by the slowing down of the spinning neutron star. Several galactic x-ray binary pulsars were also observed by BATSE and OSSE, providing new data on their periods and spectra in the low-energy gamma-ray region. Unlike the pulsars described above, these x-ray binary pulsars are powered by matter from a binary companion star falling (accreting) onto a rotating neutron star. Transient sources (high-energy sources that appear for only a few weeks or months) have been detected by both the *Compton* and *Granat* observatories. The 1990 *Granat* observation of the transient source Nova Muscae showed a gamma-ray line at about 500 keV for a few days during its outburst. The *Granat Observatory* was also able to provide for the first time images of low-energy gamma-ray sources in the region of the galactic center and measure variability from a number of them. Both the *Granat* and *Compton* observatories succeeded in performing the most precise gamma-ray studies of the galactic black hole candidate Cygnus X-1.

Geminga. Perhaps the most unusual galactic gamma-ray source before *Compton* was an object known as Geminga. It was among the brightest sources in the gamma-ray sky, yet there was no obvious counterpart at other wavelengths. In 1992, a weak x-ray source near the gamma-ray object was discovered to pulse with a period of 237 milliseconds. Soon after, EGRET found gamma-ray pulses with the same period and the identification was made: Geminga is a relatively young, nearby pulsar. Its high intensity is a result of its proximity, but its lack of detectable radio pulses is a mystery.

Radioactive isotopes. OSSE on the *Compton Observatory* also has significant capabilities for detecting gamma-ray lines from radioactive isotopes and from other nuclear processes that may occur in various astrophysical sites. On two occasions in 1991, OSSE observed the remnant of the supernova SN 1987A in the Large Magellanic Cloud. It detected the decaying isotope cobalt-57 (^{57}Co), which was produced in great abundance during the explosion. It was only the second radioactive isotope found in SN 1987A, and its observation will provide data on the formation of elements in these explosions.

Solar flares. Solar flares occasionally produce large fluxes of gamma rays. During June 1991, the Sun produced several very powerful flares that provided opportunities for observation by the *Compton Observatory*. The COMPTEL experiment has the unique capability of detecting neutrons and determining their arrival direction. Its observations of the flare of June 4, 1991, produced the first neutron image of an astronomical object. Another powerful solar flare that month was seen by EGRET to emit high-energy gamma rays many hours after the flare was over. Apparently, high-energy subatomic particles known as pions produced during the flare were trapped in the magnetic field of the flare and decayed at a later time.

The *Compton Observatory*, a pioneer in the young field of gamma-ray astronomy, is expected to continue operation for many years. The *Granat Observatory* has a more limited lifetime and is expected to end operations in 1993 and 1994.

For background information *SEE ASTROPHYSICS, HIGH-ENERGY; GALAXY, EXTERNAL; GAMMA-RAY ASTRONOMY; PULSAR; QUASAR; SATELLITE (ASTRONOMY); SUN; SUPERNOVA* in the McGraw-Hill Encyclopedia of Science & Technology.

Gerald J. Fishman

Bibliography. C. Schrader, N. Gehrels, and B. Dennis (eds.), *The Compton Observatory Science Workshop*, NASA Conf. Pub. 3137, 1992.

Gas turbine

Environmental concerns have focused attention on reducing emissions of nitrogen oxides (NO_x), carbon monoxide (CO), and unburned hydrocarbons from gas turbines, which generate electricity throughout the world. Reduction of NO_x is of particular concern in heavily populated regions, because NO_x contributes to smog, acid rain, and ozone formation in the lower atmosphere. By using catalytic combustion, pollutant levels in the exhaust of gas turbines can be decreased without sacrificing energy efficiency. By comparison with other control techniques in use or under development, catalytic combustion offers the greatest potential for low NO_x emissions.

NO_x formation. The source of energy to rotate a gas turbine is a hot stream of gaseous combustion products created by burning a mixture of natural gas and air. The temperature of the gas stream entering the turbine should be in the range 1100–1250°C (2000–2300°F) for achieving highest efficiency without overheating the blade materials. However, the flame temperature in the combustor must be around 1800°C (3300°F) in order to maintain a stable flame (**Fig. 1***a*). Therefore, the hot combustion products must be diluted with cooler air downstream of the burner to achieve an acceptable turbine inlet temperature.

At the high temperatures encountered in the flame, nitrogen and oxygen molecules in the air combine to form NO_x. In the exhaust from a typical gas turbine combustor, the NO_x level is approximately 200 parts per million. If the combustion of the natural gas–air mixture can be accomplished at a substantially lower flame temperature than usual, NO_x production can be decreased dramatically, from 200 ppm to less than 2 ppm.

NO_x can also be produced from nitrogen-containing compounds present in some fuels. Such compounds are found in some liquid fuels and in coal or coal-derived liquid fuels. The nitrogen in these organonitrogen compounds is converted to NO_x even at relatively low combustion temperatures. Thus, lowering combustion temperatures will not reduce the level of NO_x derived from fuel-bound nitrogen.

Other pollutants. Decreasing the reaction temperature significantly lowers the rate of NO_x production. However, a temperature decrease slows down the rate at which CO and unburned hydrocarbons are eliminated by oxidation. Thus, there must be a balance between slowing the reaction rates in order to control NO_x and driving the oxidation reactions to completion in order to control CO and unburned hydrocarbons. In practice this balance can be achieved by providing a residence time in the combustor sufficiently long to compensate for slower reaction rates, so that burnout of CO and hydrocarbons can be completed.

Catalytic combustion. If the amount of fuel mixed with the air is limited so that the maximum attainable combustion temperature is only 1300°C (2400°F), then the mixture is not flammable; not enough fuel is present for the mixture to sustain a flame, even if an ignition source is present. However, if this mixture is passed over a very reactive catalyst, such as platinum or palladium, the fuel and oxygen in the air will interact with the catalyst surface in an oxidation reaction (Fig. 1*b*). The reaction products will be carbon dioxide (CO_2), water vapor (H_2O), and heat. Although the net result of the overall process is the same as in a combustion flame, the reaction occurs at the surface of the catalyst and does not require the high temperature of a flame in order to be sustained.

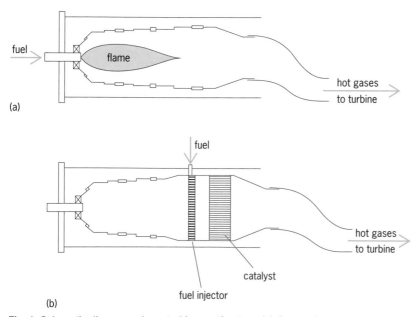

Fig. 1. Schematic diagrams of gas-turbine combustors. (*a*) Conventional combustor; regions near the flame are at temperatures as high as 1800°C (3300°F), resulting in very high NO_x levels. (*b*) Catalytic combustor; maximum temperatures are 1100–1300°C (2000–2400°F), depending on the fuel–air ratio entering the catalyst.

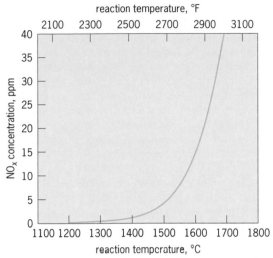

Fig. 2. Graph showing NO$_x$ concentration produced during combustion of methane. Results are for a combustor operated at 150 lb/in.^{2} (1 MPa) pressure and maintained at the indicated temperature for 20 ms.

If the maximum temperature in the catalytic combustion process is kept below 1300°C (2400°F), the NO$_x$ emissions produced will be less than 1 ppm, as shown in **Fig. 2**.

In gas turbines, the catalyst is in the form of a thin coating of fine powder on the interior walls of a honeycomb structure. The powder is commonly made up of extremely small particles of a noble metal, such as platinum or palladium, dispersed within an oxide possessing a high surface area, such as alumina (Al$_2$O$_3$). Oxides of base metals such as chromium or cobalt can also be used as oxidation catalysts, but they are generally less effective. The honeycomb support is made of either ceramic or metal, with 15–60 channels per square centimeter (100–400 channels per square inch). The air–fuel mixture passes through the channels, and the oxidation reaction occurs on the channel walls.

Combustor design. There are three approaches to combustor design: traditional, staged-fuel, and temperature-limited catalyst.

Traditional approach. The most straightforward configuration for a catalytic combustor is shown in **Fig. 3**a. All of the fuel and air are mixed and directed through the catalyst. The fuel–air ratio is such that the adiabatic combustion temperature (the temperature achieved if all the fuel is burned and no heat losses occur) is slightly above that required at the turbine inlet, roughly 1300°C (2400°F). Combustion of the fuel proceeds to completion or near-completion within the catalyst unit, with the gases exiting the catalyst being near the required inlet temperature. If full combustion is not achieved in the catalyst bed, the remaining fuel plus the CO and unburned hydrocarbons are consumed in reactions in the gas phase just downstream.

An advantage of this traditional design is its simplicity. However, as indicated in Fig. 3a, the catalyst and its honeycomb support must be able to withstand the adiabatic combustion temperature of the fuel–air mixture. This requirement rules out the use of metallic supports, which do not retain their physical properties at the high temperatures being used. Ceramic supports, which are suitable in that regard, are also inappropriate because they are subject to failure from thermal stresses induced by the sudden temperature changes that are part of the normal turbine operating cycle. Moreover, typical catalyst formulations deactivate rapidly in the high-temperature environment, because of loss of catalyst surface area from sintering and loss of catalytic components from vaporization.

Staged-fuel approach. This approach (Fig. 3b) limits the maximum catalyst temperature by limiting the amount of fuel supplied to the catalyst. In this way the catalyst temperature can be kept below about 1050°C (1900°F), thus reducing thermal stress. To attain the temperature of 1300°C (2400°F) necessary to drive the turbine at the end of the combustor, additional fuel must be burned. Therefore, fuel is injected into the hot gases that are leaving the catalyst. At temperatures above about 900°C (1650°F), combustion occurs in the gas phase without the need for a catalyst.

The key advantage of the staged-fuel approach is its direct limitation of the maximum catalyst temperature.

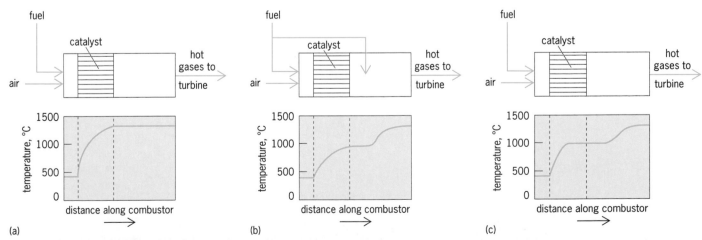

Fig. 3. Approaches to catalytic combustion and the resulting temperature profiles through the combustor. (*a*) Traditional. (*b*) Staged-fuel. (*c*) Temperature-limited catalyst. °F = (°C × 1.8) + 32°.

However, the mechanics of injecting fuel downstream of the catalyst present severe challenges. The temperatures in that region are at or near the limit for any metals that could be used for the injector hardware. The fuel must be mixed into the gas stream uniformly and rapidly, because any nonuniformity causes formation of regions of very high temperatures and consequent production of NO_x. This latter effect may ultimately limit the reduction in NO_x levels that can be achieved.

Temperature-limited catalyst approach. This approach relies on a special design and formulation of the catalyst that limits the maximum temperature that can be attained (Fig. 3c). All of the fuel and air passes through the catalyst, but conversion of the fuel is limited by the catalyst design. Combustion is then completed in the gas-phase region beyond the catalytic reactor.

This approach combines the advantages of mechanical simplicity and low catalyst temperatures. It does not require auxiliary fuel injectors and the associated potential for formation of NO_x in the postcatalyst region. However, a precise design is required to integrate the performance characteristics of the catalyst and the reaction kinetics of the downstream gas phase reaction with the turbine geometry and operating conditions.

Development status. All three concepts for catalytic combustor design have been tested in the laboratory as well as in full-scale, full-pressure (typically 150 lb/in.^{2} or 1 megapascal) gas turbine combustor configurations. In general, the advantages and disadvantages already mentioned have been evident in the test results. The temperature-limited catalyst approach appears to be the most promising. If robust, durable large-scale catalytic combustors can be developed to match the performance achieved in the laboratory, and if legislation continues to dictate further lowering of pollutant levels, it is likely that catalytic combustors will appear on gas-turbine installations in the near future.

For background information SEE AIR POLLUTION; GAS TURBINE; NITROGEN OXIDES in the McGraw-Hill Encyclopedia of Science & Technology.

Ralph A. Dalla Betta; James C. Schlatter; Toru Shoji

Bibliography. S. M. Correa, A review of NO_x formation under gas turbine combustion conditions, *Combust. Sci. Technol.*, 87:329-362, 1993; B. E. Enga and D. T. Thompson, Catalytic combustion applied to gas turbine technology, *Plat. Met. Rev.*, 23:134–141, 1979; L. D. Pfefferle and W. C. Pfefferle, Catalysis in combustion, *Catal. Rev. Sci. Eng.*, 29(2,3):219–267, 1987; R. Prasad, L. A. Kennedy, and E. Ruckenstein, Catalytic combustion, *Catal. Rev. Sci. Eng.*, 26(1):1–58, 1984.

Gene therapy

Gene therapy is a medical intervention that alters the genetic material of living cells. First conceived as a strategy to treat genetic disease, gene therapy has the potential to treat a variety of acquired diseases. To accomplish gene therapy, genetic material is transferred into cells where the new genetic information can be expressed. This process of gene transfer and expression can also be used to mark cells to aid in the understanding of the pathogenesis of disease and to evaluate the effects of other types of therapies. The application of gene therapy to human disease was envisioned by investigators long before technology provided the tools to accomplish the goal. Despite significant obstacles and ethical concerns, gene therapy has now become a reality, propelled to the forefront of medical research by rapid advances in recombinant deoxyribonucleic acid (DNA) technology. Although it should be technically possible to transfer genetic material to any human cell type (including germ-line cells), ethical concerns have limited the focus of human gene therapy to modifying somatic cells.

Ex vivo and in vivo strategies. There are two fundamental strategies for somatic gene therapy, both involving the addition of a normal gene rather than direct replacement or correction of a defective gene. In the ex vivo approach, the new gene is transferred to cells removed from the patient and then the genetically modified cells are reimplanted. The gene transfer is confined to the isolated target cells, and the cells expressing the new gene can be monitored prior to reimplantation. However, not all cells are easily accessible or amenable to ex vivo manipulation. The in vivo strategy of directly modifying cells within the individual is theoretically simpler. Practical concerns relate to the efficiency and safety of in vivo gene delivery.

Vector strategy. The essence of gene therapy is the transfer of a new gene into target cells. Although cells can take up DNA on their own, this process is very inefficient; for practical applications, it is necessary to enhance the efficiency of gene transfer. Vectors, the vehicles used to enhance the efficiency of gene transfer, are classified as nonviral or viral.

Nonviral vectors. In the most common nonviral vectors, genetic information is complexed or conjugated with other molecules. One type of nonviral vector is made by combining DNA and liposomes, charged lipid molecules that readily complex with DNA. The liposome-DNA complex fuses with the cell membrane, thus enhancing the entry of the DNA into the cells. Another type of nonviral vector involves conjugates of DNA with proteins, usually ligands that bind to specific cell-surface receptors, thus facilitating DNA uptake through the ligand-receptor interaction. Compared to viral vectors, nonviral vectors are simple and do not involve the use of infectious agents. However, they are not very efficient in vivo. Further, they do not facilitate the integration of DNA into the target cell genome. They are most useful, therefore, for applications where low level expression of the transferred gene is sufficient or only transient expression of the gene is desired.

Viral vectors. Viral-based vectors capitalize on the ability of viruses to easily transfer their genetic information into target cells. Recombinant retroviral vectors have been most extensively studied and have been used in ex vivo strategies for human gene therapy trials. The retroviral vectors are recombinant ribonucleic acid (RNA) viruses capable of entering cells through specific cell-surface receptors. Once inside a cell, retroviral RNA is converted to DNA, which is inserted into the host cell genome. To utilize retroviruses for gene therapy, the genetic information of the virus is modified to include the new gene and its controlling elements. The recombinant vectors are engineered to infect cells and express the new gene but not to reproduce themselves in the host. In contrast to nonviral vectors, retroviral vectors integrate the recombinant retroviral DNA into the host genome so that the new gene can be passed on to progeny cells. The primary disadvantages of retroviral vectors are the risk that random integration into the host cell genome may result in insertional mutagenesis and uncontrollable proliferation of target cells; the difficulty of producing highly concentrated preparations of the virus; and the requirement for host cell proliferation for successful gene transfer which significantly limits the use of modified retroviruses for in vivo applications.

Recombinant replication-deficient adenoviral vectors are promising alternatives to retroviral vectors, particularly for in vivo applications. The adenovirus is a double-stranded DNA virus that enters cells through specific cell-surface receptors. The adenoviral DNA reaches the nucleus, where it expresses the new gene. In contrast to the retrovirus, it is unlikely that the adenovirus DNA integrates into the host cell genome. However, because host cell proliferation is not required for expression of the new gene carried by the adenovirus, these vectors can be successfully used to transfer genes to a broad range of cell types in vivo. A further advantage of the adenovirus is that highly concentrated preparations can be easily produced, a requirement for efficient gene transfer in vivo. Possible disadvantages of adenoviral vectors include the potential for low level viral gene expression and viral replication; the relatively little information available (compared to recombinant retroviruses) about the safety of adenoviral vectors for human use; and the possible necessity for repetitive administration in order to achieve stable gene expression, given the low likelihood of integration into the host cell genome. If repetitive administration is necessary, the immunologic response of the host to the adenovirus itself may play a role in the success of this strategy. Adenoviral vectors have not yet been used in human trials.

Other viral vectors with potential for human gene therapy include those based upon adeno-associated or herpes simplex viruses.

Review process. Human gene therapy protocols must undergo stringent review before patients can be treated. These reviews focus not only on the scientific basis of the proposed therapy but on the safety of the therapy for the patient and the environment. In the United States, a protocol must first be approved by the local institutional review board and biosafety committee where the therapy will be given. Next, the protocol must be approved by the National Institutes of Health (NIH) Recombinant DNA Advisory Committee, which evaluates the study according to established guidelines. Finally, as with all new drugs, the protocol must be approved by the Federal Drug Administration.

Marker studies. The first human gene transfer study, in 1989, was a cell-marking protocol carried out in cancer patients. The gene for neomycin resistance was used as a marker to track tumor-infiltrating lymphocytes in patients with malignant melanoma. The tumor-infiltrating lymphocytes were isolated from tumors and expanded in tissue culture. A retroviral vector was used to transfer the neomycin resistance gene to a subset of these cells. When the cells were cultured with neomycin, only those cells expressing the newly inserted neomycin resistance gene survived. The genetically marked cells were reinjected into the patient. Marked cells were found in the bloodstream and in tumor samples for several weeks, thus establishing that it was possible to transfer a gene to cells ex vivo, return them to the patient, and subsequently recover the genetically altered cells. Many marking studies have since been approved for diseases including acute liver failure, acquired immune deficiency syndrome (AIDS), leukemia, and several types of cancer.

Disease treatment. The first human gene therapy trial was initiated in 1990 for the treatment of the rare inherited disorder adenosine deaminase deficiency. Adenosine deaminase is an enzyme involved in the purine salvage pathway; adenosine deaminase–deficient T lymphocytes are poisoned by the intracellular accumulation of 2-deoxyadenosine, a toxic metabolite of defective purine metabolism. Thus, an affected individual has a fatal deficiency of both cellular and humoral immunity known as severe combined immunodeficiency. Bone marrow transplantation can be curative, but is limited to those individuals who have matched donors. The only other therapy available, adenosine deaminase conjugated to polyethylene glycol (PEG-ADA), partially corrects the disease when weekly intravenous injections are used.

The second protocol approved for an inherited disease is a treatment of familial hypercholesterolemia, a rare disease in which a lack of normal low density lipoprotein (LDL) receptors in hepatocytes results in elevated LDL cholesterol, premature atherosclerosis, and death from severe coronary artery disease. The Wattanabe rabbit, an animal model for familial hypercholesterolemia, has been used to test the feasibility of gene therapy. A retroviral vector was used to transfer the LDL receptor gene into hepatocytes that had been isolated from the Wattanabe rabbits. The cells were reinstilled into the portal vein of the rabbits with the hope that they would settle in the liver and express the LDL receptor, thus providing the animals with a mechanism for clearing LDL cholesterol. The LDL cholesterol levels of these animals were lowered significantly after gene therapy. Based on these observations, a similar clinical study is ongoing in humans.

The only human protocol approved for direct in vivo gene therapy employs a nonviral vector for the treatment of cancer. In this protocol, DNA is complexed to cationic liposomes and used in vivo to express an allogeneic class I major histocompatibility antigen in the target tumor cells. The hope is that the modified cells will be recognized as foreign by the host immune system and be destroyed.

Cystic fibrosis. Gene therapy has been applied to only rare genetic disorders. Its true promise will be fulfilled when it can be applied to more common diseases such as cystic fibrosis. Cystic fibrosis is a fatal autosomal recessive disorder, with major manifestations in the gastrointestinal and respiratory tracts. The majority of deaths are due to respiratory failure. The disease is associated with defective regulation of cyclic adenosine $3', 5'$-monophosphate acid (cAMP)–mediated chloride secretion in the apical membrane of epithelial cells; consequently, there are abnormalities of the electrical properties of the epithelial cells and the fluid environment that bathes them. The gene responsible for the disease is called the cystic fibrosis transmembrane conductance regulator (CFTR).

Since identification of the CFTR gene, remarkable advances have been made toward understanding the molecular basis of the disease. Studies have demonstrated that the transfer of a normal CFTR gene into a cystic fibrosis epithelial cell will correct the defect in cAMP-mediated chloride secretion. A rational therapeutic approach would be to transfer the normal gene to the respiratory epithelium, where the primary fatal manifestations of the disease occur. However, given the slow proliferation of the airway cells and the difficulty that the geometry of the lung presents for removing cells and reinstillation, the ex vivo retroviral strategy is unlikely to work. Rather, gene therapy for the respiratory manifestations will require a strategy that does not depend upon target cell proliferation and that can be administered directly in vivo.

The recombinant adenoviral vector AdCFTR, containing complementary DNA (cDNA) for CFTR, has been successfully used in vivo to transfer and express the human CFTR in the respiratory epithelium of cotton rats. In addition, the vector corrects the defective chloride secretion of cystic fibrosis epithelial cells and transfers and expresses the CFTR in respiratory epithelium recovered from individuals with cystic fibrosis. Proposed clinical studies in humans will use this type of vector to introduce the cDNA for CFTR directly into the airway epithelium of affected individuals.

For background information SEE GENETIC ENGINEERING; HUMAN GENETICS; IMMUNOTHERAPY in the McGraw-Hill Encyclopedia of Science & Technology.

Melissa Rosenfeld; Ronald G. Crystal

Bibliography. W. F. Anderson, Human gene therapy, *Science*, 256:808–809, 1992; A. D. Miller, Human gene therapy comes of age, *Nature*, 357:455–456, 1992; Recombinant DNA Advisory Committee, Points to consider in human somatic cell therapy and gene therapy, *Hum. Gene Ther.*, 2:251–256, 1991; M. A. Rosenfeld et al., In vivo transfer of the human cystic fibrosis transmembrane conductance regulator gene to the airway epithelium, *Cell*, 68:143–155, 1992.

Geochemical prospecting

The CHIM electrogeochemical prospecting method has only recently attracted western interest as a potential tool in the search for covered mineral deposits. The method has been practiced since about 1970 in the former Soviet Union and since about 1982 in the People's Republic of China. CHIM is an acronym derived from the Russian words *chastichnoe izvlechennye metallov*, meaning partial extraction of metals. In the former Soviet Union it applied principally to exploration for base- and precious-metal deposits. However, the Russian literature states that this method has also been used in the search for vanadium, cobalt, molybdenum, tin, beryllium, rare-earth elements, and even oil and gas.

In 1989 the U.S. Geological Survey started an investigation of the CHIM method, and in 1990 a consortium of Canadian exploration firms funded a field test of CHIM and two other Russian-developed geochemical exploration techniques. A similar field test took place in Australia in 1992.

CHIM method. In principle, the method is quite simple. A group of 20–40 specially designed sampling electrodes (cation collectors; element receivers in Russian terminology) are placed on the ground over the area of interest, and direct current is passed through them for periods of several hours to days. In normal operation, the sampling electrodes are cathodes (negatively charged) and thus collect cations (positively charged ions) from the soil moisture in the vicinity of the electrodes. The current is returned through a single graphite or steel anode that is well removed from the array of cation collectors.

The sampling electrodes are inert plastic cylinders fitted on the bottom with a synthetic parchment membrane through which electrical connection to the earth is made. The cation collectors are filled with reagent-grade acid electrolyte, commonly a 1–4 normal (1–4 N) solution of nitric acid. An inner electrode of chemically pure graphite or titanium is inserted into the acid and used to make connection to a source of electrical power. Typical electrode volumes are in the $3–18$-in.^{3} ($50–300$-cm^3) range.

Cation collection. During operation, cations in the soil moisture move under the force imposed by the applied electric field through the parchment and into the nitric acid. The parchment, or other membrane, must have sufficiently low permeability to minimize electrolyte loss to the earth during the operating cycle. The CHIM method, in effect, constitutes a very weak in-place partial extraction of metals present in the soil. After a run is completed, typically expending $0.1–0.2$ faraday of charge ($2.6–5.2$ ampere-hours), the acid is removed from the cation collectors and analyzed for elements of interest.

Because electrical current conduction within near-surface soils and rocks is entirely by ion migration within the contained water, Faraday's law gives a measure of the number of ions that are collected during a CHIM run. This measure is important because of the constraints it puts on the volume and strength of electrolyte used in the sampling electrodes. In the cation collector, electrochemical reactions at the inner electrode provide a source of hydroxide (OH^-) ions, thus driving the solution to higher pH values. To prevent precipitation of metal hydroxides that can coat and insulate the inner electrode the electrolyte must remain acidic throughout a run. This requirement is accomplished by using an acid electrolyte containing several times the chemical equivalents of the expected charge transfer to be used.

The particular mix of cations that is moved into the collecting electrode is a direct function of the ions present in the soil moisture, their concentration, and the mobility. Ion mobilities in dilute solution are well established. Unfortunately, mobilities in the vadose zone of soils, where most CHIM collection occurs, are poorly known.

Sampling. An important aspect of the CHIM method is the volume of earth that can be sampled. This aspect does not appear to have been addressed directly by Russian investigators. Approximate measures may be obtained by calculating the number of equivalents of ions present in the soil moisture in the vicinity of the electrode. If, for example, 0.1 faraday of charge is transferred at the electrode, ideally the electrode would collect cations from a volume of soil under the electrode containing 0.1 equivalent in the soil moisture. Slow ions would be collected from a smaller soil volume, and faster ions from a corresponding larger volume. By using such simplifying assumptions, a charge transfer of 0.1 faraday (typical of CHIM runs) would sample about 177 ft^3 (5 m^3) in a dry (3% moisture) soil with 20 parts per million sodium chloride in the soil moisture, whereas 0.00177 ft^3 (0.00005 m^3) would be sampled in a wet (30% moisture) saline (2000 ppm) soil. The assumption is that only ions already present in the soil moisture are collected. In practice, such is not the case. Acid loss through the parchment adds hydrogen ions to the soil, which then migrate back to the cation collector under the influence of the electric field.

A necessary consequence of CHIM sampling is that counterions, normally nitrate (NO_3^-) ions for cathode collectors, are pumped out of the collection electrode. These counterions flood the soil in the vicinity of the electrode and change the equilibrium in the soil volume. The importance of the loss of hydrogen ions and the migration of counterions such as NO_3^- needs to be evaluated.

The loss of acid electrolyte to the soil also contributes variability to CHIM data by solubilizing elements and changing their mode of occurrence and thus their potential availability. Many soil constituents such as oxides, carbonates, and organic matter are susceptible to attack in this manner. The magnitude of this problem is not yet well documented.

Power sources. Conventional electric power sources used in CHIM work consist of high-voltage (500–1000-V), high-power (5–20-kW) direct current sources. The dc power is distributed by multiconductor cables to an array of cation collectors. Provision for adjusting individual electrode currents and for mea-

suring the currents is included in the source instrumentation. From 20 to more than 40 electrodes may be deployed at a time. With such equipment, current to each electrode is normally in the range of 0.1–0.5 A. At higher currents, the electrolyte may boil because of resistive heating of the ground beneath the electrodes unless large-contact-area electrodes are used.

A modification to the conventional CHIM power source has been introduced in which individual 12-V batteries are used as current sources at each electrode. In this variation, called APLOCHIM (from the Greek *aplo*, meaning simple), the electrodes are operated at much lower currents, but collection time is normally increased. The use of small battery sources considerably simplifies the logistics of CHIM field operations, making the APLOCHIM technique a more practical field procedure.

Effectiveness and reproducibility. The Russian experience seems to indicate that the CHIM method, in contrast to conventional soil geochemistry, is particularly effective at locating mineral deposits beneath substantial thickness of cover. Presumably the reason is that weak geochemical halos developed through thick cover are more easily identified by the CHIM technique. Recent studies have involved the relationship between CHIM data and various types of conventional soil extractions at sites in Colorado and Arizona. This work showed clearly the close correspondence between CHIM data and weak partial extractions of soils. It also showed clearly the advantage of CHIM or weak partial extractions of soils over the more conventional "total" digestions of soils (in which strong acids are used to dissolve all mineral phases). **Figure 1** shows data comparing CHIM results with

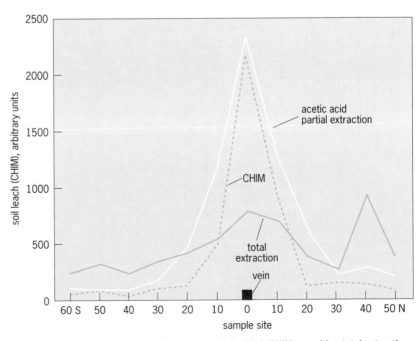

Fig. 1. Comparison of the curves obtained in an 8-h, 0.5-A CHIM run with a total extraction and 0.5 *N* acetic acid partial extraction on 1 gram of soil at the Kokomo mine, Russell Gulch, Colorado. Station spacing is 10 m and the gold vein is at position 0. Note better definition (peaks) provided by CHIM and acetic acid partial extraction.

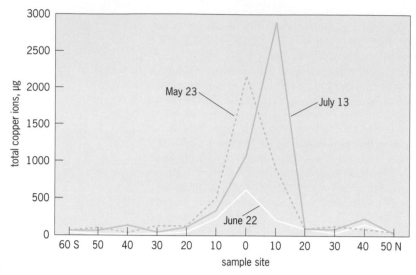

Fig. 2. Comparison of the 8-h CHIM runs at the Kokomo mine, Russell Gulch, Colorado. Traverse is the same as shown in Fig. 1.

soil total digestions and a 0.5 N acetic acid leach for copper at the Kokomo Gold Mine in Colorado. The CHIM extraction was for 8 h at 0.5 A. A 6.6-ft-wide (2-m) quartz vein carrying gold and base metals is directly beneath position 0 and is coverd by 10 ft (3 m) of colluvium. The significantly better definition of the vein system by partial extraction methods, either by weak acetic acid or by CHIM, is evident.

The question of reproducibility has also been addressed by research. **Figure 2** shows copper data from three CHIM runs at the Kokomo mine repeated at about 1-month intervals. Each run was for 8 h at 0.5 A so that an equivalent charge was transferred for each run. The data show high copper concentrations over the vein at position 0 on all runs, but a large decrease in the quantity of copper ions collected on the June 22 run. At the time of the June 22 run the soil in the area was very dry. The reduced quantities of

pathfinder elements were attributed by the researchers to the soil dryness. (A pathfinder element is a relatively mobile element that occurs in close association with an element or commodity being sought and that can be found more easily.)

Studies of the rate of accumulation of ions with time under constant current operation were also made. Typically the accumulation rate of pathfinder elements such as silver or copper falls off significantly after 3–4 h of operation. The causes for this fall-off are not adequately understood but do not appear to be related to the soil properties that exist prior to operation. The effects of this change in the rate of ion accumulation with time are seen when comparing conventional CHIM and APLOCHIM results. At the Kokomo mine, an APLOCHIM run collected quantities of silver, lead, copper, and zinc similar to a conventional run but used about 1/30 of the total charge transfer.

Figure 3 compares zinc values in CHIM and APLOCHIM runs at Johnson Camp in Arizona, a replacement copper-zinc deposit hosted in limestone. Here the CHIM run was for 12 h at 0.5 A, and the APLOCHIM run for 42 h at about 0.05 A. The quantity of pathfinder ions collected was approximately proportional to the charge transfer. This graph also shows the good pattern correlation obtained between CHIM and APLOCHIM results in practice. On this line at Johnson Camp, drilling has confirmed mineralization near the zinc peak at position 95NE buried beneath 100 ft (30 m) of basin fill, and 188 ft (57 m) of barren limestone. No drilling data are available near site 57NE, where the most prominent zinc peak occurs.

Prospects. Research at the Kokomo mine and Johnson Camp has shown that the CHIM method can be an effective tool for identifyng geochemical halos associated with buried mineralization, and that it provides better discrimination of such halos than the soil total digestion techniques now commonly used by industry. It has not yet been demonstrated that CHIM has advantages over carefully selected partial extraction methods. However, it is anticipated that there will be an advantage in areas where the surface halos are weak, so that the larger in-place sampling volume and concentration permitted by the CHIM method will identify halos that cannot be identified by sampling small soil volumes because of limitations of analytical sensitivity.

Much additional research still needs to be done to improve present field practice and understanding of electrochemical processes. Electrode design should be improved to eliminate or minimize effects due to loss of electrolyte through the parchment. Anion collector electrodes need to be developed to permit evaluation of anion collection and anion pathfinders for exploration. Particularly important in this respect is the part that anion complexes may play in the mobility of some metals such as gold. The CHIM method also may play a part in helping to identify the ionic complexes present in the soil.

Although much research remains to be done, the CHIM method is able to identify geochemical halos developed over covered ore deposits. The role of the

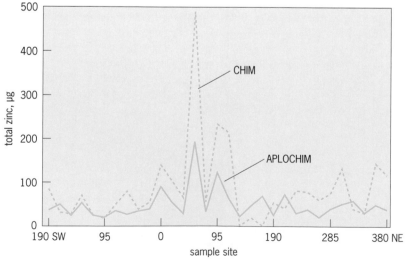

Fig. 3. Comparison of a CHIM run for 12 h at 0.5 A with a APLOCHIM run for 42 h at 0.05 A at the Johnson Camp, Arizona, copper-zinc deposit.

CHIM method in an integrated exploration plan, however, remains to be developed.

For background information *SEE GEOCHEMICAL PROSPECTING; GEOCHEMISTRY; PROSPECTING* in the McGraw-Hill Encyclopedia of Science & Technology.

D. B. Hoover, D. B. Smith

Bibliography. E. Bloomstein, *Selected Translations of the Russian Literature on the Electrogeochemical Sampling Techniques Called CHIM (Chastichnoe Izvlechennye Metallov)*, USGS Open-File Rep. 90-462, 1990; E. E. Good et al. (eds.), *U.S. Geological Survey Research on Mineral Resources: 1991 Program and Abstracts*, U.S. Geol. Circ. 1062, 1991; D. B. Hoover et al., Studies of the CHIM electrogeochemical method in Arizona and Colorado, *Abstracts Society for Mining, Metallurgy, and Exploration Annual Meeting*, Phoenix, Arizona, February 1992.

Geography

Viewed from a distance, a mountain range can be seen to be composed of individual peaks. From a closer vantage point, ridges are revealed on those peaks, and from an even nearer point, those ridges are revealed as being broken into crags and outcroppings. As more detail is exposed at closer views, it becomes evident that the smaller features appear similar to the larger features viewed at a greater distance. Geographers and geologists have long known that photographs of landforms need to include an object of known size—a coin, a person—or else it is difficult to determine the scale of the landforms. That many geographic features, such as mountain ranges, coastlines, and river networks, exhibit this phenomenon of similarity over a range of scales is intuitive to many people.

Fractals. In the mid-1970s, B. B. Mandelbrot coined the term fractal to define that class of objects with noninteger dimensions. A linear fractal function would have a dimension greater than 1 but less than 2, while a fractal surface would have a dimension between 2 and 3. A fractal dimension close to the euclidean dimension (that is, close to 1 for a linear fractal) would represent a relatively smoothly varying function, while a higher fractal dimension would represent a highly irregular function. For example, a sheet of paper can be used to represent a two-dimensional surface. If crumpled into a ball, this highly irregular surface nearly fills space and approximates a three-dimensional object. Thus, its fractal dimension would be nearer to 3 than to its euclidean dimension of 2. Even if the paper were uncrumpled, folds and creases would remain and its fractal dimension would still be greater than its euclidean dimension. Since Mandelbrot's introduction of fractals, techniques have been developed to generate computer images of landscapes, oceanscapes, clouds, and even planets whose fractal dimensions yield scenes that are more realistic than those based on the more traditional euclidean geometry. These techniques combine a set of rules with an element of randomness to create scenes that are nearly independent of scale. When the scaling processes are the same in all euclidean dimensions, the resulting fractal is said to be self-similar. Such is the case, for example, with coastlines and river networks, since the horizontal dimensions have the same scaling, limited by the circumference of the Earth. If the vertical dimension is included (for example, topographic profiles and surfaces), the scaling processes will no longer be the same; scaling in the vertical is controlled by

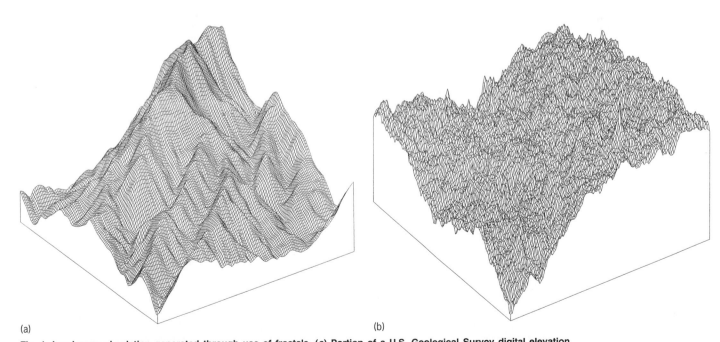

(a)   (b)

Fig. 1. Landscape simulation generated through use of fractals. (*a*) Portion of a U.S. Geological Survey digital elevation model—128 × 128 grid points (3.84 × 3.84 km or 2.38 × 2.38 mi)—for the Fort Douglas (Utah) quadrangle. The fractal dimension of this topographic surface is approximately 2.39, and its elevation range is from 1633 to 2441 m (5358 to 8008 ft). (*b*) Fractal surface having the same dimension as for *a* but generated by using a midpoint displacement method with successive random additions. The mean elevation and standard deviation are also the same as for *a*.

gravity, which limits the maximum relief. Fractals with different scalings are termed self-affine.

Many of the earliest applications of fractals were for the creation of simulated geographies, which remain a subject of continuing effort by geographers, computer scientists, and film makers. However, the processes controlling landscape evolution in nature certainly are not random, and the underlying geologic structure can strongly influence the resulting surface features. Therefore, physical geographers, cartographers, and geologists have little difficulty differentiating between real and simulated landscapes. As a result, scientific applications of fractals to physical geography have been slow to develop and are somewhat limited in number.

Applications of fractals. Fractals have potential for providing information valuable in the fields of geomorphology, climatology, soil science, and remote sensing.

Geomorphology. As might be expected in light of the use of fractals to generate images of more realistic landscapes, most of the geographic applications of fractals have been in the field of geomorphology—the study of landforms and their evolution. While images of landscapes provide sometimes compelling visual evidence that topography is fractal, scientific investigation often reveals the greater complexity of natural landforms. For example, one frequent criticism made by geomorphologists and geologists is the failure of simulated landscapes to properly incorporate stream networks and their effect on the landscape (**Fig. 1**). Even though the geographic surface and the fractal surface have the same fractal dimension, mean elevation, and standard deviation, it is not difficult to

identify which is which. Because no physical constraints (for example, rock strength or gravitational force) have been imposed on the generated surface and because erosional processes have not been taken into account, the generated surface appears much rougher and more jagged than the natural surface. Furthermore, no drainage network can be discerned on the generated surface, while valleys and ridges are clearly seen on the natural surface. Investigation of the differences between simulated landscapes derived from statistically self-affine fractal generators and natural landscapes might lead to a better understanding of the processes at work sculpting the Earth's surface. Moreover, many natural landscapes exhibit more than one fractal dimension over a range of scales and are more properly termed multifractal. This multiplicity may be indicative of different geomorphic processes acting at these various scales. If such is indeed the case, measurement and classification of the fractal dimensions of geomorphic features such as coastlines, mountain ranges, and river networks may allow them to be characterized by the processes controlling their development.

The modeling of geomorphic processes offers another possible use of fractals. Models of stream erosion, for example, often have started with a flat surface, either horizontal or slightly tilted, and allowed erosion to shape the landscape. Because the initial surface is regular, some degree of randomness must be incorporated into the geomorphic processes being modeled to allow erosion to occur at different rates across the landscape. Fractal surfaces, which often are characterized as raw or unmodified by geomorphologists, could be

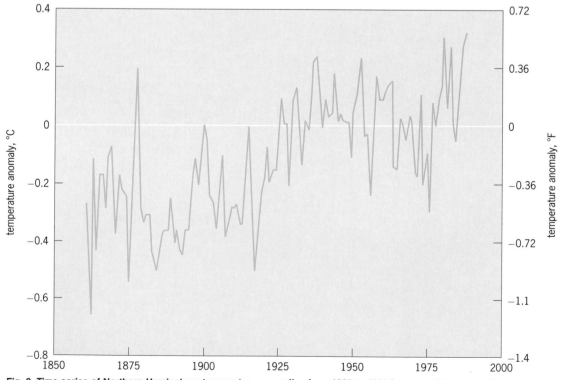

Fig. 2. Time series of Northern Hemisphere temperature anomalies from 1861 to 1988 in terms of departure from the 1951–1980 mean, indicated by the white horizontal line. The fractal dimension of this series is approximately 1.83.

used as the initial, preerosion surface in these models. Thus, the initial surface, not the processes, would contribute the spatial variation necessary to allow erosion, transportation, and deposition to take place at different rates across the landscape. SEE STREAM SYSTEMS.

Climatology. Applications of fractals in climatology have been more limited than those in geomorphology. S. Lovejoy studied clouds and rain areas and found a relationship between the perimeter and area of clouds and rain areas that is fractal. Moreover, he found that the fractal dimension was invariant over several orders of magnitude. This finding indicates that there is no well-defined length scale for clouds and rain areas, since the fractal dimension would differ at scales less than and greater than that length.

Time series of a number of climatic and hydrologic parameters such as temperature, precipitation, and streamflow have been found to exhibit fractal properties. A time series of annual temperature anomalies in the Northern Hemisphere from 1861 to 1988 (**Fig. 2**) exhibits a fractal dimension of 1.83, which is indicative of the highly irregular nature of climatic data. This series also demonstrates why it is difficult to discern evidence of climate change, whether natural or human-induced, from the climate record. Even though inspection of the time series appears to reveal a positive trend in the data, the variation makes it uncertain as to whether this increase is real or is only part of the noise. Fractals also offer a method for analyzing climate that could be used to enhance understanding of both the spatial and the temporal characteristics of climate variability. SEE CLIMATOLOGY.

Soils. Various soil properties such as texture, pH, moisture capacity, and density have been shown to display spatial variation that is fractal, although not necessarily self-similar. This finding has important implications for mapping of soil data, since spatial interpolation of soil properties may not always be appropriate because of sudden changes across boundaries. The same may also be true for studies of vegetation cover, although this possibility has not yet been demonstrated. For example, fractals may be useful in biogeographical studies of landscape patchiness and its effects on wildlife habitat or as a measure of disturbance of ecosystems by humans.

Remote sensing. In remote sensing, fractals may offer a method for classifying images of the Earth's surface. Landforms, vegetation covers, and other surface features may display distinct fractal dimensions that could be used to assist geographers and other users of remotely sensed data in their interpretation of aerial photographs and satellite images. For example, since human-built structures are generally not fractal, techniques might be developed to distinguish those portions of images where humans have influenced the landscape from regions that have not been modified.

Prospects. The use of fractals in geographic research is still in its early stages and has yet to find its full potential. This brief discussion can only hint at the possibilities offered by the study of geographic fractals and the impact they will have on theory and analysis in physical geography.

For background information SEE CLIMATE MODELING; FRACTALS; GEOMORPHOLOGY; REMOTE SENSING in the McGraw-Hill Encyclopedia of Science & Technology.

Clinton M. Rowe

Bibliography. M. Batty, The fractal nature of geography, *Geog. Mag.*, pp. 33–36, May 1992; M. F. Goodchild and D. M. Mark, The fractal nature of geographic phenomena, *Ann. Ass. Amer. Geog.*, 77:265–278, 1987; S. Lovejoy, Area-perimeter relation for rain and cloud areas, *Science*, 216:185–187, 1982; B. B. Mandelbrot, *The Fractal Geometry of Nature*, 1977.

Glass

Only since about 1970 has the existence of the glassy state and the process by which it is usually reached from a supercooled liquid been recognized as a deep problem in theoretical physics. The problem, in which thermodynamics and kinetics seem to be intimately associated, has been one of understanding how a physical system that may exist as a mobile fluid over hundreds of degrees in temperature may suddenly, in the course of a few tens of degrees, change through a region of very high viscosity into a rigid amorphous solid. To tackle this problem, it is first necessary to appreciate the great diversity of behavior exhibited by different liquids as they are cooled from their high-fluidity states.

The best-known example of a glass is that of silicon dioxide (SiO_2), silica glass. Silica glass actually provides one extreme of the spectrum of behaviors of glass-forming liquids, and its thermodynamic properties are almost indistinguishable from those of a normal crystalline solid. As far as can be determined at present, the fundamental thermodynamic property, the heat capacity, of liquid silica is unchanged as the liquid loses its fluidity and settles into the glassy state. In this respect, it is distinguishable from almost all other liquids, which show a pronounced discontinuity, popularly known as the glass transition, as the highly viscous liquid passes into a rigid solid. It is with this transition and its different thermodynamic manifestations that this article is primarily concerned.

Liquid-glass transition. The sudden drop in the heat capacity at a characteristic temperature, called the glass transition temperature, T_g, is best thought of in terms of the popular idea of a glass as a frozen liquid. There exists a temperature at which the motions that distinguish the liquid from the crystalline state no longer occur. At this point in cooling, the contribution to the total heat capacity of those degrees of freedom, associated with the vanishing motions, is lost. Consequently, there is a drop in heat capacity to a value near that of the stable crystal, as is shown in **illus.** *c*. Dealing with overall thermodynamic relationships, the illustration is discussed in detail below.

The drop in heat capacity is, at first sight, very similar to the expectation for a classical second-order phase transition, as described by P. Ehrenfest. However, whereas an Ehrenfest transition is located uniquely in temperature for a fixed value of pressure, the glass transition occurs at a temperature that depends on the rate at which the liquid is cooled. This time dependence is a sure sign that the actual phenomenon is controlled by kinetic factors, although it seems that on a sufficiently long time scale (approaching the lifetime of the universe) the kinetic factors are destined to disappear.

Thermodynamic relationships. The heat capacity, whose decrease signals T_g, is the second derivative of the free energy. To give an overall picture of thermodynamic relationships among crystal, liquid, and glass, their free energies, and also their first two free-energy derivatives, will be examined as functions of temperature. Illustration *a* shows the Gibbs free energy G of liquid and crystal phases of a typical glass-forming substance referenced to their identical free energies at the melting point. The curves show how the instability of the liquid relative to the crystal increases rapidly with decreasing temperature in the temperature interval directly below the melting point T_m, but less

rapidly as temperature decreases further. A broken line indicates the ultimate behavior where the difference between the two must, in principle, become constant. The rate at which the instability of supercooled liquid with respect to crystal builds up reflects the difference in slopes of the curves, which are themselves defined as the entropies S of their respective phases by Eq. (1), where H is the enthalpy and T is the absolute

$$G = H - TS \qquad (1)$$

temperature. The entropy is shown as a function of temperature in illus. *b*, and it may be seen that the difference between liquid and crystal tends to disappear at a temperature designated T_K (after W. Kauzmann, who first recognized the phenomenon). Before this perplexing situation can be reached, however, the liquid inevitably encounters the glass transition at which the rate of entropy change of the liquid becomes the same as that of the crystal because of the decrease in the heat capacity shown in illus. *c*. Thus, the glassy state can retain a nonzero entropy at 0 K, which is one of its best-known properties. The glass transition happens simply because the time for rearranging the particles (the relaxation time τ in illus. *d*) becomes too long at T_g to be detected by the experiment. This situation happens when the viscosity η (which changes very rapidly with temperature) reaches about 10^{13} poises or 10^{12} pascal-seconds (illus. *d*).

Jumps in thermodynamic quantities. Between the case illustrated, in which there is a very large jump in the heat capacity at the glass transition, and silica, in which no such jump can be detected, lie all intermediate cases. The magnitude of the jump seems to be closely related to the structural characteristics of the medium, in particular to the interconnectedness of the particles making up the glassy structure.

In statistical mechanics, the jump in heat capacity ΔC_p at T_g reflects an increase in the mean-square entropy fluctuations as the system gains access to additional degrees of freedom. As the liquid state is entered, there are, in principle, also increases in the mean-square density (or volume) fluctuations, as manifested by a jump $\Delta\kappa$ in the isothermal compressibility. Finally, cross correlations of volume and entropy fluctuations are responsible for a jump $\Delta\alpha$ in the thermal expansion coefficient.

In equilibrium second-order transition thermodynamics, these three quantities are strictly related according to the Prigogine-Defay ratio R, given by Eq. (2). In the case where the changes in C_p, κ, and so

$$R = \frac{\Delta\kappa\Delta C_p}{VT\Delta\alpha^2} = 1 \qquad (2)$$

forth are mediated by kinetic factors, as in the glass transition, the Prigogine-Defay relation becomes inequality (3), unless there is only a single-order param-

$$\frac{\Delta\kappa\Delta C_p}{VT\Delta\alpha^2} > 1 \quad \text{or} \quad \frac{\Delta\kappa}{\Delta\alpha} > \frac{TV\Delta\alpha}{\Delta C_p} \qquad (3)$$

eter controlling the thermodynamic state of the relaxing sytem (which is unknown at present).

The Prigogine-Defay ratio results from the combination of two relationships derived by Ehrenfest for

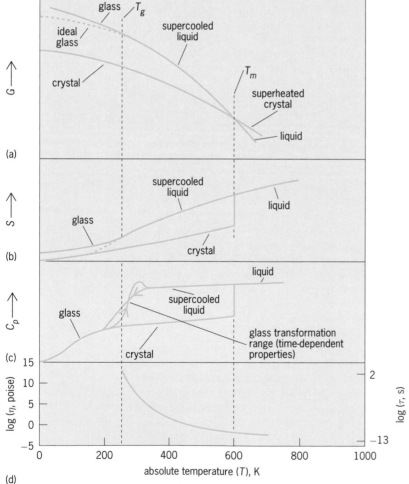

(a)

(b)

(c)

(d)

Dependence of thermodynamic or kinetic quantities on absolute temperature *T* in crystal, liquid, and glass phases of a typical glass-forming substance. (*a*) Gibbs free energy, *G*. (*b*) Entropy, $S = (\Delta G/\Delta T)$ at constant pressure. (*c*) Heat capacity at constant pressure, $C_p = (\Delta^2 G/\Delta T^2)$, showing rapid increase and overshoot for measurements made with increasing *T* (which defines the glass transition, T_g), and hysteresis on cooling. (*d*) Logarithm of viscosity η (in poises) and logarithm of relaxation time τ (in seconds), showing rapid increases at temperatures approaching T_g. 1 poise = 0.1 Pa·s, so log (η, Pa·s) = log (η, poise) −1. °F = (K × 1.8) − 460.

the pressure dependence of a second-order transition at a temperature T_2. These relationships, which have been the subject of a number of experimental checks, are Eq. (4), where the parentheses indicate the value

$$\left(\frac{\Delta T_2}{\Delta P}\right) = \frac{VT\Delta\alpha}{\Delta C_p} \qquad (4)$$

at constant entropy, and Eq. (5), where the parentheses indicate the value at constant volume.

$$\left(\frac{\Delta T_2}{\Delta P}\right) = \frac{\Delta\kappa}{\Delta\alpha} \qquad (5)$$

In glass transition thermodynamics, the validity of the above relationships depends on whether or not the entropy or the volume of the glass is maintained at a constant value as a system passes, at constant scan rate, through the glass transition temperature at different pressures. So far, laboratory experiments suggest that Eq. (5), but not Eq. (4), is obeyed. This result suggests that, to a good approximation, entropy is constant at the glass transition temperature; and in turn, the implication is that it is the entropy that plays some universal role in dictating the relaxation kinetics of viscous liquids. Such a role for entropy is provided in the Adam-Gibbs theory referred to below. These relations have been tested both at the slow heating role of normal calorimetric experiments and on the shorter time scale of ultrasonic experiments.

A large jump in heat capacity at the glass transition is not necessarily accompanied by large jumps in compressibility and expansion coefficient. For instance, there are aqueous systems in which the change in expansion coefficient at the glass transition is effectively zero while the ΔC_p is very large. Conversely, the case of silica is interesting because, although no change in heat capacity is observed at the glass transition, high-temperature light-scattering measurements have revealed a large change in the isothermal compressibility. These observations are typical of the anomalies that may arise when the glass-forming medium has an unusual open structure due to tetrahedral coordination of structural units and preferred bond-angle relationships.

Long time scales. At the moment, theoretical efforts to understand the glass transition are focused in two quite different regions of temperature. For many years, they have been focused on what would happen on long time scales at temperatures below the glass transition. Thus, free-volume models proposed that the glass would tend to reach ideal maximum density for amorphous packing at an ideal glass temperature T_0, if the glass cooled slowly enough. The Adam-Gibbs theory maintains that an ideal glass state of zero excess entropy, $S_{liquid} = S_{crystal}$, as depicted in illus. b, would be reached at a second-order thermodynamic transition in the limit of infinitely slow cooling.

Mode coupling theory. More recently, there has been an upsurge of interest in the possibility that the glass transition, rather than being a kinetic phenomenon largely under the control of a thermodynamic property as suggested above, is tied into the mechanism of diffusive motion in a manner that renders any thermodynamic changes slave to the dynamics.

The theory focuses on liquid dynamics in the domain where packing is loose enough that processes in which particles jump over barriers are not involved. The equations show that, as the system cools, it tends to jam at a sharp dynamical transition not too far below the melting point of most glass-forming liquids. This approach, known as the mode-coupling theory, also leads to predictions about changes in thermodynamic properties at the glass transition, since the jamming automatically causes the extra degrees of freedom of the liquid to drop out. Mode-coupling theory predicts that the dynamical transiton will lie at temperatures well above T_g, where the system can still be studied in the internally equilibrated (metastable) state. This prediction forces proponents of the theory to postulate other mechanisms that allow the system to avoid becoming jammed. At present, there is a division of opinion about which regime of viscous liquids would merit the most theoretical attention.

Secondary relaxations. Within the glassy state are additional processes that have thermodynamic consequences. These processes involve the so-called secondary relaxations, in which a subset of more localized but still cooperative particle motions become frozen, with weak thermodynamic signatures; and the two-level systems that appear to be the tail end of the secondary relaxations and that remain active at temperatures where all other modes of excitation have become inactive or very weak. This regime, which involves heavy-atom tunneling phenomena, is a separate and important subject in glass science.

For background information SEE GIBBS FUNCTION; GLASS; GLASS TRANSITION; PHASE TRANSITIONS; THERMODYNAMIC PRINCIPLES in the McGraw-Hill Encyclopedia of Science & Technology.

C. A. Angell

Bibliography. C. A. Angell, The data gap in solution chemistry and the "ideal" glass transition puzzle, *J. Chem. Educ.*, 47:583–587, 1970; C. A. Angell, Relaxation in liquids, polymers and plastic crystals: Patterns and problems, *J. Non-Cryst. Sol.*, 131–133:13–31, 1991.

Groundwater ecology

Groundwaters, like familiar ecosystems such as forests and lakes, contain a characteristic and specially adapted biota that serves important ecological functions. While serious sampling problems hamper studies of groundwater ecology, enough is known to outline the workings of this diverse biological community.

Groundwater characteristics. Groundwaters, which include all zones beneath the Earth's surface that are saturated with water, differ in many ecologically important ways from surface waters. Groundwaters are dark, so they do not support green plants, the base of food webs in surface-water ecosystems. Consequently, the food used by groundwater communities is transported from the Earth's surface into the aquifers, often after the most readily usable part has been consumed by surface-dwelling organisms.

Except in cave environments, where there may be open channels, groundwater moves very slowly (often less than a meter per day) through tiny pores and cracks in rocks and soils. These pores and cracks form the habitat of most groundwater organisms. The chemistry of groundwater is highly variable, but groundwaters usually contain less oxygen and more carbon dioxide than surface waters as a result of biological metabolism in aquifers and overlying soils. Typically, physical and chemical conditions are less variable seasonally in groundwaters than in surface waters.

Except for caves that can be explored directly, groundwaters are typically sampled by indirect, cumbersome methods. Most often, samples are taken from specially constructed wells. Groundwater organisms collected from ordinary water wells are different from those living in the adjoining aquifer, so these wells usually are unsuitable for scientific studies.

Groundwater biota. Bacteria, protozoa, fungi, and invertebrate and vertebrate animals frequently can be found in groundwaters. Many animals that live in groundwaters are never found in surface waters. Most of these groundwater specialists (stygobionts) were discovered only in the last few decades; many new forms are being found as groundwaters are further explored. Groundwater specialists include species of most kinds of animals known from inland waters, but groundwater crustaceans are especially diversified. The classification of groundwater bacteria is only beginning, but it appears that they are different from those in overlying soils and surface waters.

Groundwater animals are thought to have arisen from several different sources. Some species are closely related to animals now living in soils or surface waters, and probably invaded groundwaters by moving downward and adapting gradually to the peculiarities of the groundwater environment. Remarkably, many species are not at all closely related to species in fresh surface waters but resemble species in ocean beaches or other saltwater habitats. Some of these saltwater species probably entered groundwaters by actively moving along a series of gradually fresher waters, from ocean beaches, to nearshore brackish or fresh-water aquifers or river beaches, and finally into fully fresh groundwaters. More commonly, the ancestors of these species were stranded by falling sea levels into nearshore lakes, beaches, or caves. As these habitats freshened after losing contact with the sea, some species were able to adapt to the changed conditions.

Many groundwater animals are recognizable by their distinctive appearances. Typically, groundwater animals are smaller and more slender than their relatives in surface waters, and they usually are unpigmented and blind. Accompanying these morphological changes is a marked slowing of metabolic rates: groundwater animals have low respiration rates, take a long time to mature, and have long life-spans compared to surface-water animals. Most groundwater animals bear only a few, large eggs. These adaptations presumably aid survival in dark, interstitial habitats where food is scarce.

Distribution. Many groundwater animals occupy small geographic ranges, and the richness of groundwater communities varies widely from place to place. Especially rich communities are found where permeable geological deposits (such as limestone or clean gravel) lie near ancient shorelines where animals could be stranded. For example, the Edwards Aquifer in Texas supports dozens of species of snails, crustaceans, worms, salamanders, and fish that are known from nowhere else in the world. The regions surrounding the Mediterranean Sea, especially the Balkan Peninsula, likewise contain very rich communities of groundwater invertebrates. In contrast, areas that were covered by Pleistocene glaciers have very few specialized groundwater animals. These distributional patterns indicate that groundwater animals move very slowly from their places of origin, and recolonize disturbed habitats with difficulty.

As sampling methods have improved, living organisms have been collected deeper and deeper beneath the Earth surface. Living bacterial communities have now been recovered from hundreds of meters below the surface, and some scientists have speculated that bacterial life may exist as much as 2–3 mi (3–5 km) beneath the Earth's surface. In contrast, animals probably are restricted to much shallower aquifers. In regions of limestone bedrock, where caves are common, or in very clean gravel, animals are found 30–330 ft (10–100 m) below the Earth's surface, but in areas where caves are absent, animals may be scarce more than a meter or two down.

Limiting factors. Factors that limit the distribution of groundwater organisms are poorly known but probably include scarce food supply, harsh physical and chemical conditions, high rates of mortality, and slow dispersal rates. The organic matter available as food usually is sparse and of low quality, either originating from the parent materials of the aquifer or filtering down into the aquifer from the overlying soils or surface waters. Thus, species may be unable to survive in a particular aquifer because the amount or quality of food is too low. The number and size of pore spaces in an aquifer probably controls the biota, both by providing the physical spaces in which the organisms live and by controlling the flow rates of water and waterborne materials. Many kinds of groundwater bacteria and animals require oxygen, which typically becomes scarcer in deeper and more remote aquifers. Other chemical factors of importance include nutrients such as nitrate, sulfate, and phosphate, which are used by certain kinds of bacteria; salinity; and toxic substances, either natural or anthropogenic. Finally, the slow growth rates of groundwater organisms make them very sensitive to predation and environmental disturbance. It is therefore likely that groundwater organisms are absent from near-surface habitats in part because they cannot tolerate the predation and environmental disturbances encountered there.

Functioning of groundwater communities. Little is yet known about the workings of groundwater communities. At the base of the food web are various dissolved and particulate organic materials. In addition, chemically reduced substances such as sulfide, ammonium, and ferrous iron can supply energy to groundwater bacteria. Especially in zones of contact between oxygen-free groundwater (in which these reduced substances may be found) and oxygen-containing groundwater, a substantial part of the food web may be based on inorganic sources of energy. Biological communities may function even in the absence of oxygen: certain bacteria can use nitrate, sulfate, ferric iron, or other chemicals; and many invertebrates can survive for long periods without oxygen.

The importance of ecological interactions such as competition, predation, and mutualism in structuring groundwater communities is not yet known. In cave communities, competition and predation can determine the distribution and abundance of invertebrates.

The activities of groundwater organisms (especially bacteria) can enormously influence chemical and geological conditions in aquifers. Bacteria consume or produce many important substances, including oxygen, carbon dioxide, nitrate, ammonium, sulfate, and sulfide. The carbon dioxide produced increases rates of weathering of rocks and soils, and thereby the hardness of the groundwater and the long-term geological development of the aquifer. Excessive concentrations of bacterially produced nitrate or sulfide in groundwater can make it unsuitable for some uses, including drinking. In addition, bacterial growth might clog the pores in aquifers, thereby impeding water flow, while feeding by invertebrates might reverse this clogging.

Practical problems. Bacteria are often involved in the generation or solution of groundwater-quality problems. In addition to the problems referred to above, iron-oxidizing bacteria sometimes clog the screens of wells with the iron compounds they deposit.

Human activities have contaminated groundwaters with a variety of toxic substances. Some of these substances can be broken down by bacteria under appropriate conditions. A subfield of groundwater ecology called bioremediation is devoted to developing techniques to encourage the growth and activities of appropriate bacteria to clean up contaminated aquifers by biological means.

For background information SEE BACTERIAL PHYS-IOLOGY AND METABOLISM; GROUNDWATER HYDROLOGY; MICROBIAL DEGRADATION in the McGraw-Hill Encyclopedia of Science & Technology.

David L. Strayer

Bibliography. L. Botosaneanu (ed.), *Stygofauna mundi*, 1986; D. Culver, *Cave Life*, 1982; R. A. Freeze and J. A. Cherry, *Groundwater*, 1979.

Heat transfer

A heat pipe typically consists of a sealed container lined with a wicking material. The container is evacuated and backfilled with just enough liquid that the

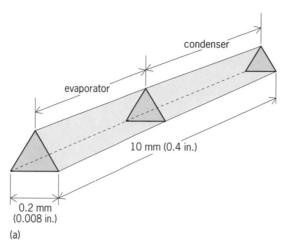

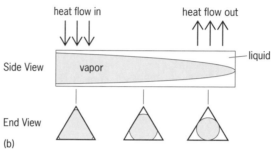

Fig. 1. Micro heat pipe. (*a*) Concept. (*b*) Operation.

wick is wet and only pure liquid and vapor are present. When heat is added to one end of the heat pipe, some of the working fluid vaporizes. The high temperature and corresponding high pressure in this region result in flow of the vapor to the cooler end, where the vapor condenses, giving up its latent heat of vaporization. The capillary forces existing in the wicking structure then pump the liquid back to the evaporator section. This combined vaporization-condensation cycle results in very large heat fluxes, 20–500 times greater than the heat conducted by pure silver.

Many sizes and types of heat pipes have been developed since the 1960s. Recently, significant advances in the fabrication of very small, so-called micro heat pipes have been made. The micro heat pipe concept was formulated in 1984 by T. P. Cotter for use in the thermal control of semiconductor devices. In practical terms, a micro heat pipe is one where the meniscus curvature is approximately equal to the diameter of the heat pipe container. Typically, this type of device is in the form of a small noncircular channel and uses the sharp corner regions instead of a wick to return the working fluid from the condenser to the evaporator, as shown in **Fig. 1**. (A cross section in the form of an equilateral triangle is shown in Fig. 1, but any shape with sharp corners will do.)

As a result of analytical, numerical, and experimental investigations, micro heat pipes have been shown to be viable mechanisms for transferring heat from one location to another or dissipating heat from isolated heat sources over larger areas to reduce the localized heat flux. The potential for use of these devices is enormous, with significant applications in the medical, electronics, space, and aircraft industries already identified.

Applications of micro heat pipes. The initial application identified by Cotter was the use of micro heat pipes approximately 100 micrometers in diameter fabricated directly into silicon wafers to provide a more uniform temperature across the wafer. A considerable amount of work in this area is under way. Additional applications include the development of small individual heat pipes, each approximately 1 millimeter (0.04 in.) in diameter, attached to ceramic chip carriers (**Fig. 2a**) and very small heat pipe arrays to promote the removal of heat from 2-cm-square (0.8-in.-square) chip carriers (Fig. 2b). Heat pipes similar to those in Fig. 2a are currently being utilized in hand-held video cameras, and also to remove heat from laser diodes and other small, localized heat-generating devices.

Applications also include the use of variable-conductance micro heat pipes to kill cancer tissue through either hyperthermia or hypothermia. Micro heat pipes can be inserted into cancerous tumors (**Fig. 3**) and, because these tumors are sensitive to temperature, they can be destroyed by either high or low temperatures, which can be carefully controlled by the noncondens-

able gas reservoir. The major advantage of micro heat pipes is that they provide a method by which heat can be added or removed to maintain the tumor at a precise temperature for a predetermined length of time. More recently, micro heat pipes have been proposed for use in hypersonic aircraft. They can be constructed in the leading edge of hypersonic aircraft wing and can be used to dissipate the heat generated by the high-velocity air passing over the wing and to spread it out over the entire wing surface. Another application is in the thermal control of photovoltaic cells. The efficiency of these cells is very sensitive to the substrate temperature. By using micro heat pipes, it may be possible to reduce the cell temperature and thereby significantly improve the amount of energy that can be produced by these devices.

Investigations of micro heat pipes. To predict either thermal behavior of micro heat pipes during start-up or variations in the evaporator thermal load, a numerical model of their transient behavior has been developed. This numerical model has been used to identify, evaluate, and better understand the phenomena that govern the transient behavior of micro heat pipes as a function of the physical shape, the properties of the working fluid, and the principal dimensions. The modeling results have been compared with the steady-state results from earlier experimental investigation and have been shown to accurately predict the maximum heat-transfer capability. The results of this investigation indicate that, during start-up or rapid transients, the liquid in the sharp-angled corner regions initially flows out of the evaporator. In addition, the wetting angle was found to be one of the most important factors affecting the transport capacity.

An experimental investigation was conducted on several 1–2-mm (0.04–0.08-in.) tapered micro heat pipes similar to the one in Fig. 2a to verify operation, measure performance limits and transient behavior, and determine the accuracy of the previously developed numerical model. Several heat pipes were evaluated under transient conditions, that is, start-up or rapid changes in the thermal load. The experimental data were compared with the results of the previously developed analytical model to determine the accuracy of the model and verify the predicted trends. The results indicated that the transient numerical model is capable of accurately predicting the maximum transport capacity prior to the onset of dry-out, the temperature distribution along the length of the heat pipe, and the temperature difference between locations on the axis of the heat pipe to within 0.3°C (0.5°F). Although the numerical model was found to accurately predict the steady-state behavior, it substantially underestimated the transient response. These early experiments served to verify the micro heat pipe concept and provided the impetus for further efforts to design, fabricate, and construct micro heat pipes on the scale of tens of micrometers.

Investigations of heat-pipe arrays. Following the experimental and analytical-numerical investigations of individual micro heat pipes, a transient three-dimensional numerical model was devel-

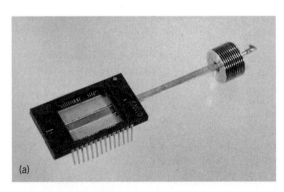

Fig. 2. Micro heat pipe applications. (*a*) Pipe for cooling a ceramic chip carrier (*Itoh Research and Development Corp.*). (*b*) Arrays for cooling high-powered semiconductor chips, shown in various stages of assembly.

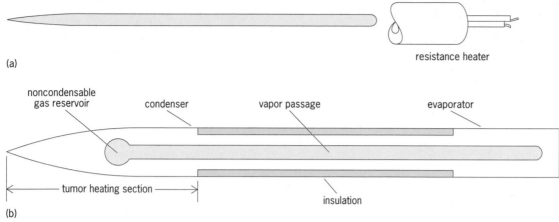

(a)

(b)

Fig. 3. Micro heat pipe designed for use in the nonsurgical treatment of cancer. (*a*) Configuration. (*b*) Enlarged longitudinal ooction.

oped to determine the potential advantages of constructing an array of very small (10–100 micrometers in diameter) heat pipes as an integral part of semiconductor chips. Because of the high effective thermal conductivity, this array functions as a highly efficient heat spreader. The numerical model, when given the physical parameters of the chip and the locations and magnitude of the internal heat generation, was capable of predicting the time-dependent temperature distribution, localized heat flux, and gradients within the chip. The results of this modeling effort indicated that significant reductions in the maximum chip temperature, thermal gradients, and localized heat fluxes could be obtained through the incorporation of arrays of micro heat pipes. Utilizing heat sinks located on the edges of the chip perpendicular to the axis of the heat pipes and an optimized number of heat pipes (12.8% of the cross-sectional area), reductions in the maximum chip temperature of up to 40% were achieved.

This three-dimensional modeling was followed by an experimental investigation to determine how incorporating an array of micro heat pipes directly into silicon wafers would affect the thermal behavior and heat-dissipation characteristics of the wafers. By using an anisotropic etching process, arrays of 19, 29, 39, and 59 parallel channels were fabricated directly into silicon wafers. A clear Pyrex cover plate was then bonded to the top surface of each wafer using an ultraviolet bonding technique to form an array of triangular micro heat pipes each 120 μm wide, 80 μm deep, and 19.7 mm (0.78 in.) long. Once the pipes had been charged with methanol and sealed, the transientand steady-state thermal behavior of the test wafers were measured using an infrared thermal imaging system. The experimental results indicated that incorporating an array of micro heat pipes as an integral part of semiconductor devices could significantly increase the effective thermal conductivity and as a result decrease the temperature gradient across the chip, the maximum chip temperature, and the number and intensity of localized hot spots. The increase in effective thermal conductivity and decrease in the maximum chip temperature were found to be linearly dependent

upon the number of heat pipes in the array and the amount of liquid present. In addition to improvements in the steady-state characteristics, the transient thermal response of the wafers was found to increase significantly with increasing heat-pipe-array density.

For background information SEE HEAT EXCHANGER; HEAT TRANSFER; INTEGRATED CIRCUITS in the McGraw-Hill Encyclopedia of Science & Technology.

G. P. Peterson

Bibliography. B. R. Babin, G. P. Peterson, and D. Wu, Steady-state modeling and testing of a micro heat pipe, *ASME J. Heat Transfer*, 112(3):595–601, 1990; *Proceedings of the 5th International Heat Pipe Conference*, Tsukuba, Japan, 1984; G. P. Peterson, An overview of micro heat pipe research, *Applied Mech. Rev.*, 45(5):175–189, 1992.

Heavy metals, environmental

Heavy-metal contamination of sediments is a primary mechanism by which metals migrate through aquatic systems, damage water-based flora and fauna, and threaten human health. Contaminated sediments result from industrial waste products, the crushing of metal ores, or the adsorption of dissolved metals onto sediment surfaces. Sediments contaminated by metals primarily occur near industrial and mining regions, particularly where sediments contain a large silt, clay, or organic component. Over the long term, contamination levels in sediments generally decrease because of mixing or burial by clean sediments and chemical migration of the metals into surrounding waters or biota. During short-term dispersive events such as floods, however, metal-carrying sediments can move as pulses, leading to contaminant concentrations tens of kilometers downstream that are as high as those at the metal source. Regulation efforts have focused on identifying the concentration at which metals in sediments pose an environmental threat.

Environmental significance. Sediments in river, lake, and marine systems act as long-term stores of heavy metals that can cause environmental damage far from their original source and decades after activities involving processing of heavy metals have ceased. The severity of this problem in the United States is recognized by new proposed amendments to the Clean Water Act for regulating contaminated sediments and by the Environmental Protection Agency's recent effort to develop criteria for maximum permissible metal levels in aquatic sediments.

Aquatic sediments contaminated by metals can directly harm bottom-dwelling organisms that live within or on the sediments. In addition, as these benthic organisms are consumed by aquatic fauna, the metals can bioconcentrate up the food chain, leading to elevated concentrations of metals in higher-level organisms, including humans. In extreme examples, metal contamination of sediments creates sterile environments with no living organisms. In less extreme cases, metal contamination of sediments can hinder restoration efforts by reintroducing toxins to the system long after the overlying waters have been cleaned up. Metal contamination is particularly pernicious because, unlike many organic compounds, the metals do not biodegrade with time.

Metals in sediments that are released to surrounding waters can also cause significant environmental damage. Major factors that can force metals into solution include changes in water salinity, temperature, pH, organic content, and microbial activity. Because metals dissolved in water are more easily assimilated by organisms, the events where sediment-bound metals go into solution can cause immediate and widespread environmental damage. Such events commonly occur during storms, where rapid changes in water chemistry can release metals into solution, kill fish, and contaminate water supplies. At a less extreme level, sediment contamination of water leads to sublethal damage that can affect organism size, health, and reproduction.

Contaminating processes. Sediments contaminated by heavy metals can be produced by three processes. First, contaminated sediments can be produced directly, as with ash from coal-fired power plants or sludges from sewage treatment facilities. Materials of this nature are generally kept on site and contaminate only the immediate disposal area.

Second, materials that naturally contain high metal concentrations are altered so that the metal becomes more available to the environment. This process most commonly occurs when metal-bearing ore is crushed during mining. The increased surface area of the crushed, fine-grained sediments allows them to release their metals to the environment more easily. In addition, the small size of the particles of crushed material allows rivers to transport the contaminated sediments hundreds of kilometers from the original sources.

Fig. 1. Baseline survey of areas where aquatic sediments are probably contaminated by metals. (*After Committee on Contaminated Marine Sediments, National Research Council, eds., Contaminated Marine Sediments: Assessment and Remediation, 1989*)

In the third and most prevalent pathway of contamination, metals that are in solution attach to sediment surfaces. This process can generate metal concentrations in sediments that are orders of magnitude higher than concentrations in the surrounding waters, particularly if the sediments have a high organic or clay content. Under certain chemical equilibria, waters well within federal standards can contaminate sediments to the point where they pose a threat to the environment. Atmospheric deposition and water-carried pollutants can thus contaminate sediments that are tens or even hundreds of kilometers from the original metal source.

Distribution of contaminated sediments. A primary focus of recent sediment-metal research has been to improve the understanding of the distribution of these sediments and the processes controlling their dispersion. This research will provide a clearer understanding of the long-term fate and effects of the sediments and will help determine appropriate techniques for controlling their discharge, dispersion, and cleanup.

National distribution. In a preliminary survey, the Environmental Protection Agency identified over 250 locations in the United States where sediments are probably significantly contaminated by metals from human activities (**Fig. 1**). Metal-contaminated sediments concentrate in industrial regions and mining areas of the West. Figure 1 was developed from a limited database and represents only the tip of the iceberg. Many areas where sediments are badly polluted by metals (including many Superfund sites) are not even shown in the figure.

Downstream distribution. The distribution of metal-bearing sediments along streams depends on the spatial arrangement of the source; on the location, magnitude, and frequency of the hydraulic dispersing processes; and on variation in water and sediment chemistry. Perhaps the simplest case is that of a single point source that places metals into a stream continuously. In this situation, dilution by so-called clean sediments farther downstream and by chemical migration of the metals into other media generally leads to a gradual decrease in metal concentrations (**Fig. 2**).

A point source can release catastrophically, as when a pile of mine tailings collapses during a flood. In this instance, the metal-bearing sediments tend to move as pulses, and concentrations far downstream can be as high as those at the source (Fig. 2). Variations in downstream concentrations during events of this nature largely reflect spatial variations in stream hydraulics, which lead to sorting of the denser metal-bearing sediments from the less dense nonmetalliferous sediments.

Sources of metal-bearing sediment that cover wide areas can result from bedrock naturally rich in metals, nonpoint pollution runoff, point sources of water pollution that contaminate sediments as the dissolved metals flow downstream, or deposition of point source sediments that are remobilized at a later date. For example, sediment eroded from a mine may be deposited over tens of square kilometers in downstream floodplains. The distribution of metals below area sources can be

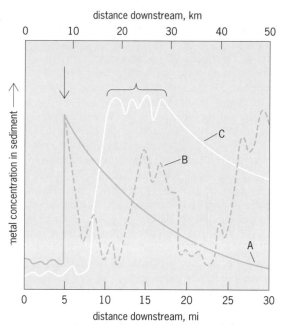

Fig. 2. Graph showing typical spatial variations in metal concentrations in sediments near a metals source. The arrow indicates the location of the point source, and the bracket indicates the extent of the area source of metals. Curves represent continuous dispersion below a point source (A); dispersion that occurs from a one-time catastrophic release such as during a flood (B); and a hypothetical distribution of metals resulting from an area source of contamination (C).

expected to mimic dispersion patterns beneath point sources, except that area sources cover a longer reach of stream and that dilution may occur more slowly because of the larger quantities of contaminated sediment (Fig. 2).

Channel cross sections. Early studies suggested that heavy metals would accumulate in floodplains, perhaps because of the deposition in these environments of fine-grained sediments in which metals tend to accumulate. More recent work, however, indicates that active channel sediments may attain the highest metal concentrations in some settings, perhaps because less dense sediments are flushed away while the more dense metal-bearing sediments are deposited in these areas of relatively high stream velocity. The absence of a clear model describing where peak concentrations can be expected within a channel cross section is one of the major obstacles hindering the development of sampling schemes for regulating and monitoring metal contamination in sediments.

Lakes and marine sediments. Pollution of lake and marine sediments generally results from pollution of the overlying waters in areas where sediment composition and chemistry promote adsorption of metals. Urban lakes and harbors with relatively low pH and sediments consisting of silts, clays, and organic materials are thus classic sites for sediment contamination.

Regulation of contaminated sediments. The wide distribution and severe environmental problems associated with metals in sediments have led to extensive efforts to develop appropriate techniques for regulating the contamination. Most efforts have focused on defining the stage at which metal concentrations in sediments are high enough to cause environmental damage and require mitigative action.

Presently, two broad approaches are being taken to define clean and polluted sediments. The equilibrium partitioning approach defines metal levels in sediments that, under stable chemical conditions, will cause metal concentrations in surrounding waters to exceed amounts permitted by Clean Water Act standards. This approach has the advantages of providing numeric criteria that are applicable on a nationwide basis, and of using existing water quality standards that have a strong scientific basis. The technique does not, however, account for the potential harmful effects of direct ingestion of sediments by benthic organisms, the potential for harmful synergistic effects of multiple metals present at low concentrations, or the possibility of nonequilibria conditions when contaminated sediments can cause the most damage (for example, during storms).

An alternative group of approaches tests the effects of the sediments on living organisms. These effects-based techniques can be used at any site for any mixture of chemicals, and they are not dependent on an assumption of equilibrium. The results of these tests, however, are site-specific, do not discriminate between the effects of different metals in the sediments, and can be thrown off by unmonitored conditions or substances in the sediments.

Most environmental scientists presently recommend using a tiered approach to evelute to what extent the sediments are contaminated. Such an approach consists of a series of steps, each becoming more complex and site-specific. If the sediment passes the test at any point, no further testing is required and the sediment is considered clean. If it fails, further evaluation is required. In the United States the equilibrium partitioning approach may possibly be used on a nationwide basis as a first step for determining if further testing is required. After that point, increasingly specific effects-based techniques would be used.

For background information SEE FOOD WEB; LIMNOLOGY; WATER POLLUTION in the McGraw-Hill Encyclopedia of Science & Technology.

W. Andrew Marcus

Bibliography. Committee on Contaminated Marine Sediments, National Research Council (eds.), *Contaminated Marine Sediments: Assessment and Remediation*, 1989; A. J. Horowitz, *A Primer on Trace Metal-Sediment Chemistry*, USGS Water-Supply Pap. 2277, 1985; W. A. Marcus, Contamination of aquatic sediments: Identification, regulation, and remediation, *Environ. Law Rep.*, 21:10020–10032, 1991; W. A. Marcus, Regulating contaminated sediments: A hydrologic perspective, *Environ. Manag.*, 13:703–713, 1989.

Helicopter

Helicopter design involves a fusion of rotary-wing technology, the fuselage (payload module), the propulsion system, and the control system. The first section of this article focuses on the relationship between weight, power, and rotor size, and also discusses helicopter performance. The second section discusses higher harmonic control, an active method for reducing helicopter vibrations by as much as 90%.

Power Relationships and Requirements for Helicopter Design

The power required by a helicopter comprises three main components: parasite power, induced power, and profile power. There are also other power requirements for cooling, accessories, driving the tail rotor, and so on, but generally, these are secondary and small compared to the main components.

Parasite power. The parasite power is the power required to pull the helicopter fuselage through the air, and does not contribute to lift. When the helicopter is not moving through the air (hovering), the parasite power is zero. The parasite power is equal to the product of the parasite drag and the airspeed V. Since the parasite drag, at a given altitude, varies as the square of the velocity, the parasite power increases rapidly with forward speed, varying as the cube of V.

The parasite drag D of any body can be expressed in terms of an equivalent flat-plate area f such that Eq. (1) is satisfied, where ρ designates the mass density of

$$D = \frac{\rho V^2 f}{2} \qquad (1)$$

the air. The combination $\rho V^2/2$ is usually denoted by q and is called the dynamic pressure. Thus the parasite power is given by Eq. (2).

$$P = \frac{\rho V^3 f}{2} \qquad (2)$$

Induced power. The induced power is the power required by the rotor to produce rotor thrust. That such power is required can be demonstrated by considering a hovering rotor. If the rotor is producing a thrust, there must be an equal and opposite reaction on the air. Thus the air above the rotor is continuously accelerated and moved through the rotor. Initially, this air surrounding the rotor is at rest, but after passing through the rotor it has motion and, hence, kinetic energy. The induced power is equal to the rate at which kinetic energy is added to the flow as the air passes through the rotor.

From momentum principles, the downwash w induced through a hovering rotor is given by Eq. (3).

$$w = \sqrt{\frac{T/A}{2\rho}} \qquad (3)$$

A is referred to as the disk area and equals the area swept by the blades, πR^2, R being the rotor radius. T is the rotor thrust, and the ratio T/A is called the disk loading. The disk loading is the principal parameter that determines the induced power, since the induced power P_i can be shown to equal the product of the rotor thrust and the velocity of the air through the disk, as in Eq. (4). In forward flight, the downwash

$$P_i = Tw \qquad (4)$$

through the rotor takes a different form and is given

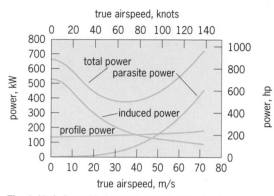

Fig. 1. Variation with forward speed of the total power required by the helicopter to maintain level flight, and of parasite power, induced power, and profile power. Weight = 45,000 N (10,116 lb), rotor diameter = 13 m (42.65 ft), tip speed = 220 m/s (722 ft/s), flat-plate area f = 2 m² (21.5 ft²), σ = 0.003.

approximately by Eq. (5). This expression holds for

$$w = \frac{T/A}{2\rho V} \qquad (5)$$

speeds above approximately 20–30 knots (10–15 m/s), and becomes increasingly accurate as the forward speed increases.

Profile power. The profile power is the power required to overcome the drag of the rotor blades as the rotor rotates, regardless of any lift being produced. If σ denotes the ratio of blade area to disk area (typically σ = 0.07), the profile power can be calculated approximately from Eq. (6). V_T is the tip speed resulting from the rotation of the rotor. Typically, V_T is

$$P_p = \rho A V_T^3 \frac{0.01\sigma}{8} \qquad (6)$$

around 650–750 ft/s (200–230 m/s). The factor 0.01 is typical of the drag coefficient for a section of the rotor blade.

All of the above equations for the power components are in consistent units. For example, in the American Customary System of units, with the velocities expressed in feet per second, areas in square feet, T in pounds, and the air-mass density in slugs per cubic feet, the equations will yield the power components in foot-pounds per second. Dividing by 550 will result in horsepowers. In the International System, the velocity will be in meters per second, the area in square meters, the thrust in newtons, and the air-mass density in kilograms per cubic meter. The resulting power will be in watts.

Variation of required power. The power components depend upon the forward speed of the helicopter in the manner prescribed by the above relationships. The parasite power is zero in hover and increases rapidly with speed at a rate proportional to the cube of the forward speed. The induced power is a maximum in hover and actually decreases with speed at a rate inversely proportional to the speed. The profile power varies only slightly with forward speed. It equals its value in hover increased by the factor $1 + 3(V/V_T)^2$. The total power required by a helicopter, because of the strong dependency of the

induced and parasite powers on speed, will first decrease from that of the hover condition as the forward speed increases. At some speed, typically 15–20 knots (30–40 m/s), the total power will reach a minimum and then will increase as the speed increases (**Fig. 1**).

Performance related to required power. The term trimmed flight means that all of the forces and moments on the helicopter are in balance and the helicopter is not accelerating. Examples include hover and flight at a constant airspeed while climbing, descending, or maintaining a constant altitude. In order to maintain trimmed flight at a constant altitude, the power available from the engine must equal the power required for that flight condition. Any power available from the engine above the required power, known as excess power, can be used either to climb or to accelerate. Since the power required to raise a weight is equal to the product of the weight and the speed at which the weight is being raised, relationship (7) holds. Since

$$\text{Rate of climb} = \frac{\text{excess power}}{\text{weight}} \qquad (7)$$

the power available for a helicopter does not vary significantly with forward speed, it follows that the maximum rate of climb occurs at the speed where the required power for trimmed, level flight is a minimum. Above this speed, the rate of climb will decrease until a speed is reached where the required power and available power are equal. This speed is the maximum speed that the helicopter can obtain in trimmed level flight (**Fig. 2**). In order to fly faster than this, the helicopter must descend, thereby using some of its potential energy to augment the engine power.

As altitude increases, the air-mass density decreases, resulting in two changes that work to adversely affect helicopter performance. The power available from the engine decreases with altitude, and the required power for trimmed level flight increases with altitude. Thus, the maximum rate of climb will decrease as the altitude increases. It is conventional to refer to the rate of

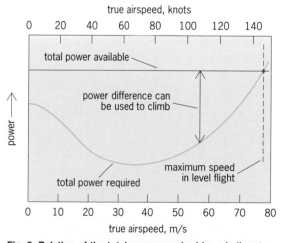

Fig. 2. Relation of the total power required by a helicopter to maintain level flight to the total power available from the engine. The helicopter is able to climb at any forward speed at which the available power is greater than the power required for level flight.

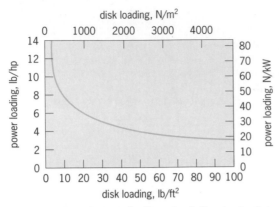

disk loading, N/m²

power loading, lb/hp

disk loading, lb/ft²

power loading, N/kW

Fig. 3. Relation of power loading of a helicopter to disk loading. In calculating this relation, the ideal induced power has been increased 15%, and an increment has been added for profile power.

climb in feet per minute (fpm), and all helicopter rate-of-climb indicators are calibrated in these units. The altitude at which a helicopter (or an airplane) has a rate of climb of 100 fpm (0.508 m/s) is referred to as its service ceiling. The absolute ceiling is that altitude at which the rate of climb is zero.

The absolute hover ceiling for a helicopter is less than that for forward flight. Indeed, if a helicopter is overloaded so that it cannot hover, it may still be capable of forward flight. When the rotor is close to the ground, the downwash w, and hence the induced power, is reduced, allowing the helicopter to lift off a small distance above the ground. The pilot then controls the helicopter to move forward, resulting in a drop in the required power and allowing the helicopter to climb out of the so-called ground effect.

In designing a helicopter, the mission that the helicopter is to perform will determine its overall shape and power requirements. Most mission profiles, whether commercial or military, specify a certain amount of time in hover, a cruising speed and range, and a payload to be carried. The absolute hovering ceiling may also be specified. Because of inherent limitations in rotor aerodynamics, the helicopter configuration is generally limited to operational speeds of approximately 200 knots (100 m/s) or less. Thus, if the mission requires vertical takeoff and landing (VTOL) capability together with a high forward speed, a derivative of the helicopter must be used such as a tilt-rotor airplane or a tilt-wing configuration. In the extreme case where very high forward speeds are desired, the rotor, as a producer of vertical lift, may be replaced by an entirely different type of thrust producer.

Effect of disk loading. For the same structural materials and design stresses, all aircraft tend to scale according to the square-cubed law. If l is a characteristic length, an area is proportional to l^2, and the volume to l^3. If the weight is assumed to be proportional to the volume, the ratio of the weight to the area is proportional to $W^{1/3}$. Therefore, the disk loading for helicopters can be expected to increase with gross weight, as do the wing loadings for fixed-wing

airplanes. The disk loading can also increase because of the mission requirements, as discussed above. Helicopter disk loadings vary from approximately 3 lb/ft² (150 N/m²) for small helicopters to 14 lb/ft² (670 N/m²) for the large Sikorsky CH-53. The V-22, a tilt-rotor airplane, has a disk loading of approximately 21 lb/ft² (1000 N/m²). The disk loadings of tilt-wing aircraft are even higher, and lifting jet engines exceed disk loadings of 1000 lb/ft² (50,000 N/m²).

From the above discussion on induced power, the thrust per unit of power (measured in, for example, pounds per horsepower or newtons per kilowatt) that a rotor can produce is approximately inversely proportional to the downwash. Thus, the ratio T/P, called the power loading, and the disk loading are related by Eq. (8). This expression, of course, con-

$$\frac{T}{P} = \sqrt{\frac{2\rho}{T/A}} \qquad (8)$$

siders only the induced power and neglects the other power requirements such as profile power or tail-rotor power. Nevertheless, it is very descriptive of the installed power in existing helicopters and other VTOL aircraft. The relation is shown graphically in **Fig. 3**. It is seen from this figure that the power required to hover for a smaller helicopter is proportionately less than that required by a larger helicopter because of the lower disk loading of the former.

Barnes W. McCormick

Higher Harmonic Control

Higher harmonic control is an active-control-system concept for helicopters that promises major breakthroughs in such important areas as helicopter vibrations, noise, and even performance. Its potential for vibration reduction was first demonstrated in flight tests of a modified Hughes OH-6A helicopter. The tests were conducted at the U.S. Army's Proving Ground at Yuma, Arizona, in 1982–1984, and they explored the level-flight airspeed regime from hover to 100 knots (51 m/s) as well as transient maneuvers. Up to a 90% reduction in helicopter vibration levels was achieved for the first time, showing that it was feasible to reduce helicopter vibrations to levels comparable to

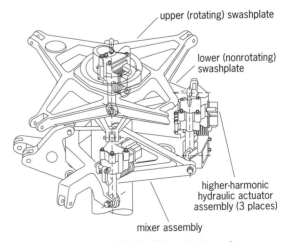

upper (rotating) swashplate

lower (nonrotating) swashplate

higher-harmonic hydraulic actuator assembly (3 places)

mixer assembly

Fig. 4. Higher-harmonic blade-pitch actuator system.

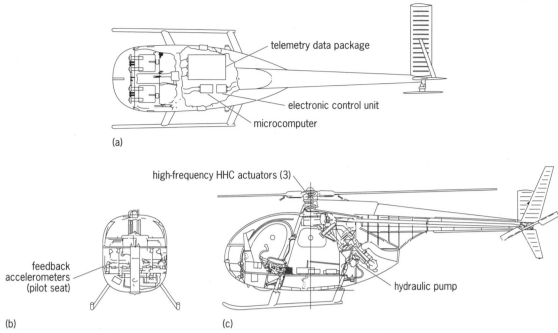

Fig. 5. Installation of a prototype higher harmonic control system on an OH-6A helicopter. (*a*) Plan view. (*b*) Front view. (*c*) Side view.

those of a fixed-wing jet aircraft. Low vibration levels are important in helicopters for crew and passenger comfort and for reducing the structural fatigue of the airframe, the rotor system, and the drive train and its components. Furthermore, for the helicopter as a weapons system, low vibration levels mean improved stability for precision delivery of airborne weapons.

Operating principles. Higher harmonic control is an active control concept, in contrast to the conventional passive means of vibration control, such as vibration absorbers, vibration isolators, bifilar or Frahm absorbers, or nodal beam suspension. A passive vibration-control device treats the vibratory loads after they have been generated, whereas an active vibration-control concept, such as higher harmonic control, alters or reduces the vibratory excitation at its source. In the case of higher harmonic control, the source of the vibrations is the vibratory aerodynamic forces that act on the rotor blades in flight. The rotor system transmits these unwanted vibrations to the airframe. Higher harmonic control applies high-frequency pitch motion to the rotor blades at very small angles (less than 1°) to suppress this aerodynamic excitation.

Thus, higher harmonic control is basically an electronic, computer-controlled vibration suppression system that senses and cancels vibrations in a helicopter airframe by high-frequency pitch motion of the rotor blades at very small blade angles. The means of achieving this major advance were made possible by the rapid evolution of the high-speed, lightweight microcomputer coupled with advances in servoactuator technology.

System components and operations. The primary elements of the active vibration-suppression system are (1) acceleration transducers that sense the vibratory response of the fuselage; (2) a higher-harmonic blade-pitch actuator system (**Fig. 4**); (3) a flight-worthy microcomputer, which incorporates the algorithm or mathematical model for reducing vibrations; and (4) a signal-conditioning system (electronic control unit), which interfaces between the sensors, the microcomputer, and the higher harmonic control actuators.

These and other components of higher harmonic control system are illustrated in **Fig. 5**. Computer-controlled operation of the higher harmonic control is as follows: Triaxial accelerometers mounted beneath the pilot's seat sense vertical, lateral, and longitudinal vibrations there and pass these signals to the electronic control unit, which converts the signals into an electronic format, which in turn can be read by the microcomputer. In the conversion the electronic control unit separates out the primary frequency components of the helicopter vibration at four cycles per rotor revolution (for a four-bladed rotor) and then sends this information to the microcomputer.

The computer, which contains the mathematical model or algorithm for nulling the vibrations, analyzes the input signals from the electronic control unit and determines how much blade feathering is needed to cancel the vibration. This information is sent from the microcomputer to the electronic control unit in computer format. The electronic control unit then closes the loop by converting these values to analog signals at four cycles per revolution that provide the electronic input to the three high-frequency actuators that drive the stationary swashplate at four cycles per rotor revolution in collective motion and in pitch and roll. The result is three-cycle-per-revolution, four-cycle-per-revolution, and five-cycle-per-revolution pitch motion of the rotor blades in the rotating system.

(The swashplates, the heart of a helicopter's control

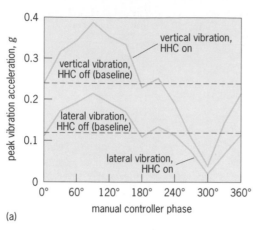

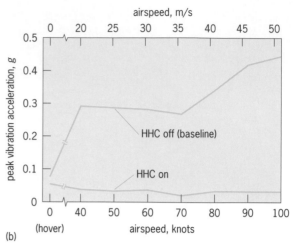

Fig. 6. Peak vibration acceleration of four-cycle-per-rotor-revolution frequency component of helicopter vibration measured at the pilot seat during flight tests of a prototype higher harmonic control (HHC) system on an OH-6A helicopter. (*a*) Variation of vertical and lateral vibrations of an open-loop system with the input phase at an airspeed of 60 knots (31 m/s) and a lateral control input of ±0.33°. (*b*) Variation of vertical vibrations of a closed-loop system with airspeed.

system, are two plates mounted on the main rotor mast that move parallel to each other. Inputs from the pilot's control move the lower, nonrotating plate, and the upper, rotating plate follows this motion. Push rods attached to arms of the rotating swashplate, in turn, convert this motion into feathering motion of the rotor blades. The swashplate has three modes of motion: collective motion; vertical motion on the rotor mast while maintaining the same angle to it, resulting in all blades feathering together; and pitch and roll motion, which refers to angular positioning of the swashplate and imparts once-per-revolution pitch motion to the blades.)

With the prototype (modified OH-6A helicopter) system, the process described above updated itself continuously about every two rotor revolutions (300 ms) to account for changing flight conditions. Flight test results showed that for new designs a higher update rate is desirable for vibration reduction in transient maneuver conditions.

Today engineers are also investigating various alternative methods of blade actuation other than that described. One method, favored by many engineers, applies higher harmonic control motion to a small flap attached to the outboard part of the blade.

Design of prototype system. The OH-6A helicopter selected for the design of the prototype system was one-of-a-kind in the Army inventory in that it had a boosted control system. The standard OH-6A incorporates a mechanical or nonboosted control system. For application of higher harmonic control, an unboosted control system would be unsuitable since it would permit both unwanted feedback to the pilot's controls of the output of the higher harmonic control actuators and corresponding deterioration in the higher harmonic pitching motion of the rotor blades. To preclude this problem, the OH-6A helicopter used for the test program had been modified to incorporate a 1500-lb/in.^{2} (10-megapascal) boost system for the primary controls. An electrically driven pump located in the aft cabin provided the hydraulic source, and

the system was so designed that the aircraft could be flown through a backup mechanical control system in the event of hydraulic failure.

A primary decision in designing the OH-6A higher harmonic control configuration was to accomplish both sensing of vibration and corrective blade feathering in the nonrotating system. With this plan there was no need for a rotating hydraulic manifold and slip-ring assembly. In addition, actuators and tube assemblies were not required to operate in a centrifugal force field.

The concept selected (Fig. 4) replaced existing links with three electrohydraulic servoactuators. The links were located between the control system mixer (a system of bell cranks and linkages that receive input from the motions of the pilot's control rods) and the stationary swashplate assembly. This design had the advantage of keeping the servoactuators in the fixed system, yet locating them as near as possible to the rotor. The individual actuators were about 6 in. (15 cm) long and had a total stroke of ±0.20 in. (5 mm) at 32 Hz (four cycles per rotor revolution for the OH-6A). This design translated into a collective blade-angle authority (that is, maximum angle to which the blades could be moved by the higher harmonic control system) of ±2°, which was about 11% of the OH-6A's total collective blade-angle authority available to the pilot, so that the pilot could easily override the higher harmonic control system if necessary. The 3000-lb/in.^{2} (21-MPa) hydraulic power for the actuators was provided by an engine-mounted pump.

Open-loop flight-test program. The purpose of the open-loop flight-test program (without use of feedback from the sensors) was to obtain basic data on the operation of the prototype higher harmonic control system. The system as designed had a maximum ±2° authority for blade-angle movement. However, most open-loop flight-test points were flown with the manual controller set at an amplitude of about ±0.33° of blade motion.

The program selected for open-loop testing consisted of recording higher harmonic control data at hover and at airspeeds from 40 to 100 knots (21 to 51 m/s) in 10-knot (5-m/s) increments. **Figure 6a** shows typical open-loop test results for a lateral swashplate excitation of $\pm0.33°$ of blade angle at an airspeed of 60 knots (31 m/s). Plotted are vibration accelerations in g's (1 g = 32 ft/s = 9.8 m/s). The variation in vibration level as the input phase to the actuators was varied from zero to 360° in 30° increments is shown. Horizontal lines indicate corresponding baseline (higher harmonic control off) vibration levels.

Figure 6a shows characteristics that were observed at other airspeeds as well. Baseline (higher harmonic control off) vertical vibration levels at the pilot's seat were 0.24 g. As the phase was increased from 0°, the higher harmonic control system initially made the aircraft vibrate more, and a maximum vertical vibration value of 0.38 g was reached at a phase angle of 90°. For 60 knots (31 m/s) indicated airspeed, the maximum occurred at 90° phase angle and the minimum at 300° phase angle on the manual controller. Substantial vibration reduction was achieved at the 300° phase setting in this case. Using only one channel of the manual controller (lateral) with an arbitrary blade-angle input (+0.33°), vertical vibration levels were reduced from 0.24 g to 0.04 g, and lateral vibration levels from 0.12 g to 0.02 g.

Closed-loop flight-test program. The objective of the closed-loop flight-test program was to evaluate the performance of the prototype higher harmonic control system in reducing vibration levels under closed-loop computer control. During this portion of testing, baseline data were taken with higher harmonic control disengaged, and closed-loop data were taken with higher harmonic control engaged for the following flight conditions: hover, 40–100 knots (21–51 m/s) in 10-knot (5-m/s) increments, accelerations from hover to 60 knots (31 m/s), decelerations from 60 knots (31 m/s) to hover, right and left 30° banked turns at 60 knots (31 m/s), and a 1.75-g pull-up at 80 knots (41 m/s).

The success of higher harmonic control in automatically reducing vertical vibration levels at the pilot's seat is shown in Fig. 6b, plotted versus airspeed. All the points shown are for stabilized flight at the respective airspeeds. For reference, a fixed-wing jet aircraft typically has vibration levels of 0.02 g. These data, then, indicate that higher harmonic control results in a jet-smooth ride for the helicopter.

For background information SEE AIRFOIL; CONTROL SYSTEMS; CONVERTIPLANE; HELICOPTER; VERTICAL TAKE-OFF AND LANDING (VTOL) in the McGraw-Hill Encyclopedia of Science & Technology.

E. Roberts Wood

Bibliography. A. Gessow and G. C. Myers, *Aerodynamics of the Helicopter*, 1952, republished 1967; W. Johnson, *Helicopter Theory*, 1980; D. L. Kohlman, *Introduction to V/STOL Airplanes*, 1981; B. W. McCormick, *Aerodynamics of V/STOL Flight*, 1967; R. W. Prouty, *Helicopter Performance, Stability and Control*, 1986; E. R. Wood, Higher harmonic control for the jet smooth ride, *Vertiflite*, 29(4):28–32, May/June 1983; E. R. Wood et al., On developing and flight testing a higher harmonic control system, *J. Amer. Helicopt. Soc.*, 30(1):3–20, January 1985.

Helium

Modern technology makes extensive use of crystal surfaces, which display a rich variety of structural, electronic, magnetic, mechanical, optical, and chemical properties. Surface physics is thus a very broad and active field of scientific research. Helium plays an important role in this field because it is the purest substance in nature and provides opportunities for very clean studies of some basic physical properties of surfaces. Helium crystals exhibit properties that have wide relevance in the whole field of surface physics, but quantum mechanics intervenes in some of their other properties, which are sometimes spectacular.

Roughening transition. A crystal is usually thought of as a solid body with a rather symmetric shape limited by plane facets. However, the existence of facets on the surface of crystals in thermal equilibrium depends on temperature. A crystal surface can be in two different states, a smooth or flat one at low temperature and a rough or rounded one at higher temperature. The change from one to the other is a phase transition that involves the ordering at the crystal surface and is known as the roughening transition. The nature of a crystal surface also depends on its orientation with respect to the main crystal axes, so that each surface in each direction has its own roughening temperature. At zero temperature, crystals are faceted in all directions and, as temperature increases, more and more rounded regions appear between facets, which progressively shrink in size (**Fig. 1**). SEE PHASE TRANSITIONS.

Numerous roughening transitions have been observed on a great variety of crystals, such as those of copper, lead, ice, and plastic, as well as ionic crystals, but the most extensive and quantitative study of this phase transition has been done with helium-4 crystals. The study of helium-4 crystals has recently led to a rather complete understanding of the roughening transition.

Advantages of studying helium. Helium-4 is the main isotope of helium. It exists as a gas under normal conditions and can be liquefied if cooled at 1 atm (101 kilopascals) below 4.2 K ($-269°C$). Helium actually remains liquid down to absolute zero. It crystallizes only under pressure (above 25 atm or 2.5 megapascals) because the attraction between atoms is small and because helium atoms are light so that their quantum zero-point motion is large. Crystals such as those shown in Fig. 1 were grown from the liquid phase by increasing the pressure at low temperature. The crystal surface under study is thus a liquid–solid interface. It can be studied under conditions of extreme purity because it is possible to purify liquid helium-4 down to unique levels. Indeed, when cooled below about 2 K, liquid helium-4 becomes a superfluid,

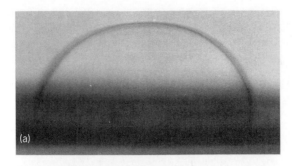

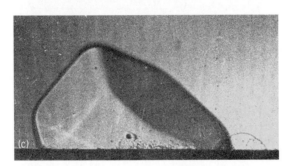

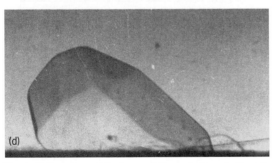

Fig. 1. Helium-4 crystal observed through the window of an optical cryostat at four successive temperatures *T*. (*a*) *T* = 1.3 K. Crystal is completely rounded. (*b*) *T* = 1.1 K. Surface contains one type of facet. (*c*) *T* = 0.4 K. Surface has two different types of facets. (*d*) *T* = 0.35 K. Surface has three types of facets.

meaning that it goes to a quantum state with practically zero viscosity and infinite thermal conductivity. As a consequence, superfluid helium-4 can flow very easily through filters with microscopic pores. In contrast, all chemical impurities solidify above 1 K and cannot flow through such very fine filters. As a result, the only impurity in helium-4 is its light isotope, helium-3. However, adding a counterflow of heat during the filtering can reduce the helium-3 content from the natural concentration of 10^{-7} down to 10^{-12}. No other substance can be prepared with no chemical impurities and such a low level of isotopic impurities. This characteristic is particularly important because the physical properties of surfaces are known to be very sensitive

to the possible adsorption of impurities.

Only two physical parameters are thus relevant in the study of helium crystals, the temperature T and the pressure P. Classical systems are usually difficult to control because their thermal conductivity is finite, so that T is rarely very homogeneous or accurately controlled. On the contrary, at sufficiently low temperatures, most helium phases have a very large thermal conductivity. As already mentioned, the thermal conductivity of superfluid helium-4 is almost infinite, and the pure crystals also have a large thermal conductivity because their quantized accoustical vibrations (phonons) have a very long mean free path between inelastic collisions. Since the pressure is also very homogeneous in a system with zero viscosity, it is possible to study helium crystals under very accurately controlled physical conditions.

Measurements. On a crystal with equilibrium shape, the local curvature of the surface is proportional to the surface stiffness γ, given by Eq. (1), where α is

$$\gamma = \frac{\alpha + \partial^2 \alpha}{\partial \phi^2} \qquad (1)$$

the surface tension or free energy per unit surface area and ϕ is the angle of the normal to the surface. The study of helium-4 crystals showed that the value γ_R of the surface stiffness at the roughening temperature T_R is simply related to T_R through the universal relation (2), where k_B is Boltzmann's constant and a is the

$$k_B T_R = \frac{2}{\pi} \gamma_R a^2 \qquad (2)$$

lattice periodicity along the normal to the surface.

The study of helium-4 crystals also allowed a study of the vanishing of the step energy β at T_R. In the smooth, faceted state below T_R, the crystal surface is flat because it is localized close to one of the lattice planes. There are only a few deviations from this plane, which are terraces limited by steps with atomic height. These steps have a certain free energy β per unit length that is finite at low temperature but vanishes at T_R. The surface becomes rough and loses reference to the lattice planes, because it is invaded by terraces that are easily created by thermal fluctuations since their energy is basically proportional to the vanishing step energy. According to the modern version of the theory of roughening, which was developed from 1976 until about 1987, the step free energy β vanishes exponentially at T_R, as was shown to be correct in studies of helium crystals.

Although measuring step energies is quite difficult, it appeared possible to determine these energies in helium-4 from the measurement of the growth of faceted crystals (**Fig. 2**). Indeed, below T_R, it was shown that the growth rate is limited by the two-dimensional nucleation of terraces, so that the growth rate is an exponential function of the applied force that drives the growth, here a small departure from the solid-liquid equilibrium pressure. From the argument in this exponential, the step free energy was deduced and shown to vanish exponentially as in Eq. (3), where

$$\beta \approx \exp[-c/\sqrt{t}] \qquad (3)$$

$t = 1 - (T/T_R)$ is the reduced temperature.

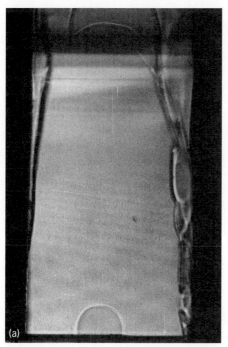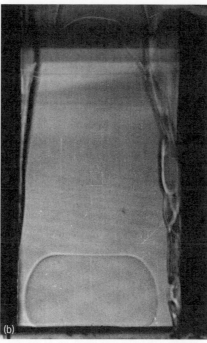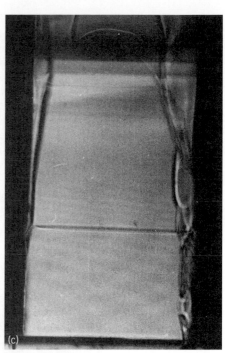

Fig. 2. Evolution of an oriented helium-4 crystal in a transparent cell. (*a*) Oriented crystal is made to appear at the bottom of the cell. (*b*) Horizontal facet develops during growth. (*c*) Horizontal facet occupies the whole cell width. From the growth rate of this facet, the free energy of crystal steps was measured and shown to vanish exponentially at the roughening transition.

Other theoretical predictions have also been checked on helium-4 crystals. They concern the angular dependence of the surface stiffness, as well as the critical variation of the growth rate in the rough, high-temperature phase. The latter study showed that facets disappear not only as a consequence of increasing temperature but also, if the growth rate becomes sufficiently large, as an effect usually referred to as dynamic roughening. Thanks to all these results, it is now widely accepted that the roughening transition is one example of the continuous transitions that belong to the Kosterlitz-Thouless universality class.

Quantum dynamics of rough surfaces. As mentioned above, if T is slightly less than T_R, the growth rate v of a faceted crystal surface varies exponentially with the departure from equilibrium $\Delta\mu$. At even lower temperature, v was observed to be a quadratic function of $\Delta\mu$ because the spiral growth becomes the dominant growth mechanism. However, the growth of rough surfaces above T_R was shown to be proportional to $\Delta\mu$ and thus much faster. Equation (4)

$$v = k\ \Delta\mu \qquad (4)$$

defines the mobility k, which is the inverse surface resistance to the growth.

It is difficult to measure k in classical substances because their growth rate is usually limited by diffusion problems. The crystallization produces heat (the latent heat of crystallization) because the liquid phase and the solid phase have different entropies. This heat needs to be removed by diffusion so that there exists a certain bulk resistance to the growth, which is generally large. In the case of an impure system, the diffusion of impurities can give rise to an even larger growth re-

sistance. Since both liquid and solid helium-4 are very pure and very good thermal conductors where heat can propagate as waves instead of slowly diffusing, the growth of helium crystals is controlled not by bulk diffusion problems but by surface properties described by quantities such as the above-defined coefficient k.

Coefficient k is so large for most helium-4 surfaces below about 0.5 K that helium-4 crystals grow or melt very rapidly under the action of very small values of $\Delta\mu$. As an example, at 0.1 K, if the departure from equilibrium is only 0.2 microbar or 0.02 pascal (to be compared to 25 bars or 2.5 megapascals, the equilibrium pressure), the rough parts of helium crystals grow at 3 m/s (10 ft/s), an enormous value. Usually, crystals rarely grow faster than some micrometers per second, and 0.2 μbar (0.02 Pa) is the tiny pressure difference between two points in liquid helium, resulting from hydrostatic equilibrium, if their heights differ by only 10 μm.

The high value of the mobility k has a spectacular consequence. Crystallization waves propagate on the rough surfaces of helium-4 crystals. These waves are very similar to the capillary or gravitational waves at the surface of a liquid, except that there is no mass motion below the surface; they result from the local growth or melting of the crystal. The appearance of such a crystal, viewed through the windows of an optical cryostat, resembles that of water: the crystal occupies the bottom part of the experimental cell because it is slightly denser than the superfluid above, and tapping gently on the outer wall is enough to excite ripples, which decay slowly in time. The systematic study of these waves provided much interesting information about the growth mechanisms of helium-4 crystals.

The mobility k is large but not infinite because some dissipation occurs during growth. It is mainly due to the scattering by the moving surface of elementary thermal excitations in both phases. This dissipation is small at low temperatures because there are fewer and fewer of these excitations as T tends to zero. Other properties that have also been analyzed concern the release or absorption of entropy during growth, and the resistance to the flow of heat through the crystal surface. All these properties are controlled by the particular nature of thermal excitations in liquid and solid helium-4.

Helium-3 crystals. The lighter isotope, helium-3, can be crystallized at slightly lower temperatures and higher pressures than helium-4, 0.32 K and 29 atm (2.9 MPa), for example. Since its nucleus contains an odd number of particles, it obeys different quantum statistics (Fermi instead of Bose) so that liquid helium-3 is a viscous Fermi liquid instead of a superfluid at temperatures of a fraction of a kelvin. As a consequence, the thermal excitations controlling the crystal growth are Fermi quasiparticles in helium-3 instead of the phonons in helium-4, so that the growth dynamics is much slower. Helium-3 crystals do not display crystallization waves at such temperatures, and actually grow more slowly as T decreases while helium-4 crystals behave in the opposite way. This characteristic illustrates the role of quantum mechanics in the rather peculiar growth processes of helium crystals.

For background information SEE CRYSTAL DEFECTS; CRYSTAL GROWTH; HELIUM; LIQUID HELIUM; PHASE TRANSITIONS; QUANTUM SOLIDS; SURFACE PHYSICS in the McGraw-Hill Encyclopedia of Science & Technology.

Sebastien Balibar

Bibliography. S. Balibar, La transition rugueuse, *La Recherche*, 194:1452–1462, 1987; S. Balibar et al., The growth dynamics of helium crystals, *Physica B*, 163:209–216, 1991; S. Balibar, D. O. Edwards, and W. F. Saam, The effect of heat flow and crystal strain on the surface of helium crystals, *J. Low Temp. Phys.*, 75:119–143, 1991.

Hepatitis

A vaccine affording long-lasting protection against hepatitis A will be widely available in 1993–1994. Hepatitis A is usually a cause of significant morbidity in adults or children over 4 years of age; it is rarely fulminant or fatal, but periodic outbreaks can be quite significant. In a recent outbreak, in Shanghai, People's Republic of China, contaminated shellfish led to more than 350,000 cases.

Potential usefulness and high-risk groups. Hepatitis A occurs worldwide. Paradoxically, outbreaks are most explosive in transitional developing areas where sanitation and water and food hygiene are improving, but subsectors of the population that retain an endemic disease pattern contaminate food or water reaching nonimmune individuals. Underdeveloped areas have few clinically apparent cases, since widespread fecal pollution of food and water leads to almost universal infection in infancy, when the infection is typically subclinical but confers lifelong immunity.

Hygienic waste disposal and water and food supplies free of the hepatitis A virus in developed industrial nations lead to large numbers of nonimmune individuals, who when exposed to the virus are at high risk for the disease. In the United States, the percentage of individuals with evidence of prior hepatitis A virus infection, as determined by blood antibodies, is 38.2%. The percentage of cases that are fatal is highest in older adults, especially when underlying liver disease is present (2.7% in individuals over 49). Immune globulin injection offers 4–6 months of protection against hepatitis A, but such administration to all at-risk individuals on an ongoing basis has generally not been practical or acceptable. A vaccine affording long-lasting protection will therefore be extremely useful. Groups that could benefit from such a vaccine include travelers to endemic areas, persons in recurrently affected areas, day-care-center staff and children, populations in transitional developing nations, food handlers, health workers, gay individuals, and patients with chronic liver disease.

Vaccine development. The development of the marmoset model of hepatitis A virus infection, and the subsequent development of the first inactivated hepatitis A vaccine from virus grown in marmoset liver, led in 1978 to the first demonstration that an inactivated hepatitis A vaccine could protect immunized animals challenged with virulent virus. The first isolation and growth in tissue culture of the virus (virus strain CR326F) from a patient led to the demonstration in 1979 of attenuation, or weakening of the virus, by serial passage in tissue culture, and to studies of different tissue-culture passage levels in primates. A live attenuated vaccine with a very favorable safety and tolerability profile, derived from viral strain CR326F, induced satisfactory blood antibody levels in volunteers after a single dose, but the timing of antibody appearance ranged from one to several months.

Infecting marmosets serially with this strain did not result in the strain's reverting to virulence, but the somewhat long interval to antibody appearance, the difficulty of growing the large amount of virus required, the observed decline of antibody levels over several years, and the potential for loss of potency on exposure to heat led to a decision to use a more productive (faster-growing) further passage level of the strain (called P28). P28 was used in the development of the first killed (formalin-inactivated) hepatitis A virus vaccine that was proven to be protective in volunteers in a field trial. This trial established, at least for the vaccine used, that the appearance of antibodies paralleled protection from disease. The isolation of additional strains of hepatitis A virus has also led to several other candidate formalin-inactivated hepatitis A vaccines at various stages of development. Vaccines derived from parts of the whole virus (subunit vaccines) have, to date, been weakly immunogenic and have not been used in clinical trials.

Vaccine-induced antibodies. The purity of several different vaccines under study differs significantly, and the number of doses needed to induce detectable antibody formation varies from one to four with different vaccines, but all have been found to induce antibodies in blood. Although studies may differ with respect to dose level, regimen, seropositivity cutoff, antibody assay methodology, and age or weight distribution, higher doses have in general yielded the highest antibody levels and the shortest intervals between vaccination and appearance of antibody in most vaccinees. Virtually 100% of children 2–16 years old, and most adults, have detectable antibody within 3–4 weeks after a standard single dose.

In comparison with vaccine-induced antibody levels, levels of antibody provided by a single immune globulin injection are very low. Although these levels become undetectable after 12–20 weeks, they protect from clinically apparent hepatitis A disease.

Thus, anyone with detectable blood antibody can be presumed to be protected from clinical hepatitis A. It is possible, but not proven, that the vaccine can, along with antibody, induce immune memory and therefore long-term protection from clinical hepatitis A. Although exposure of vaccinees to virus may induce a rapid protective memory (anamnestic) response during the 20–40-day incubation period, at present it is best to ensure long-lasting antibody levels by a booster dose of vaccine given 5–6 months after the first appearance of antibodies. Alternatively, immune globulin given simultaneously with the vaccine affords immediate protection, so if immune globulin does not interfere with vaccine responses, such simultaneous administration at two different injection sites might be useful. Hepatitis A vaccine has also been successfully coadministered with hepatitis B vaccine.

For background information *SEE HEPATITIS; VACCINATION* in the McGraw-Hill Encyclopedia of Science & Technology.

David R. Nalin

Bibliography. F. E. Andre, A. Hepburn, and E. D. Hondt, Inactivated candidate vaccines for hepatitis A, *Prog. Med. Virol.* (Base1), 37:72–95, 1990; J. A. Lewis et al., Use of a live, attenuated hepatitis A vaccine to prepare a highly purified, formalin-inactivated hepatitis A vaccine, *Proceedings of the 1990 International Symposium on Viral Hepatitis and Liver Disease*, 1990; P. J. Provost and M. R. Hilleman, An inactivated hepatitis A virus vaccine prepared from infected marmoset liver, *Proc. Soc. Exp. Biol. Med.*, 159:201–203, 1978; A. Werzberger et al., A controlled trial of a formalin-inactivated hepatitis A vaccine in healthy children, *N. Engl. J. Med.*, 527(7):453–457, 1992.

Herbicide

All crops are naturally resistant to some herbicides. However, in selecting herbicides there are often more desirable herbicides for a particular crop than those to which it has natural resistance. Traditional plant-breeding methods have seldom generated a crop variety with sufficient resistance to a new herbicide to warrant commercialization of the new cultivar. Theoretically, plant biotechnology will allow scientists to make any crop resistant to any herbicide. There are already numerous examples of herbicide-resistant crops from biotechnology, many of which have been generated in laboratory studies with objectives other than commercial production. Several herbicide-resistant crops from biotechnology are under commercial development, and some have been introduced to the marketplace. Considerable controversy has arisen over their desirability, although some biotechnological crop herbicide combinations are less environmentally and toxicologically suspect than current crop/herbicide combinations.

Methods. There are many options in producing herbicide-resistant crops from biotechnology. Choices of methods may be influenced by the crop species, the molecular target site of the herbicide, available biotechnological capabilities, and other factors. There are different physiological mechanisms of resistance, and several biotechnological methods can be used to confer these mechanisms.

Plants may be imbued with herbicide resistance by four principal physiological mechanisms: (1) altering the molecular target site of the herbicide so that the herbicide no longer interacts with it; (2) multiplying the number of molecular target sites of the crop plant so that much more herbicide is needed to kill the crop than the weed; (3) providing genes that code for enzymes that degrade or inactivate the herbicide; and (4) altering the plant so that the herbicide does not reach the molecular target site. So far, only the first and third physiological mechanisms have been utilized effectively to generate resistant cultivars.

Two fundamental biotechnological methods for generation of resistant cultivars have been used successfully. The first is selection for resistance, using the herbicide as the selection agent. This approach is similar to the selection process that occurs in agricultural fields sprayed with the same herbicides for several years, eventually leading to herbicide-resistant weed populations. However, the biotechnologist accelerates this process by selecting herbicide-resistant crop cells in cell or tissue cultures. When resistant cells are found, plants are regenerated from tissue cultures of those cells. The second method is to transfer genes for resistance into cells of the crop plant in cell or tissue culture. This process of genetic engineering may involve moving genes from weeds, other crops, or microorganisms into the chosen crop. Successful genetic transformation is followed by plant regeneration from tissue cultures. Regeneration methods have not been developed for all crop species; however, the list of crops for which successful regeneration methods have been developed is growing rapidly.

When genetically engineering a crop to be herbicide-resistant, a gene coding for either a resistant molecular target or a herbicide-degrading enzyme must be available. A potential advantage of the herbicide degradation choice is that the herbicide molecule is less likely to accumulate in edible portions of the crop.

Use. How the availability of herbicide resistant crops from biotechnology will affect management of weed problems in agriculture is a matter of speculation. An argument can be made that such crops will have very little effect, partly because numerous herbicide options are already available and in use for all major crops in developed countries. In some cases, biotechnological crops may offer more desirable herbicide options or may be a tool in reducing evolution of herbicide-resistant weeds by allowing the farmer to rotate herbicides with different mechanisms of action. However, new herbicides to which crops are naturally resistant are continually being introduced, and many of these herbicides may be very competitive with those used with a biotechnological crop. Depending on the crop, the added herbicide option offered by a biotechnological cultivar may be highly advantageous, useful in limited but important situations, or irrelevant. The herbicide to which the biotechnological crop is made resistant and the existing herbicide options for the particular crop are important considerations in predicting the commercial viability of the biotechnological cultivar.

Two examples of herbicide-resistant crops from biotechnology are commercially available. Canola that is resistant to triazine herbicides, introduced in Canada, was the first commercially available biotechnological crop, but it is grown on less than 10% of the canola acreage in Canada. Its limited adoption is due primarily to its 10–15% lower yield than triazine-susceptible canola; unfortunately, the altered gene that confers resistance to triazine herbicides also reduces photosynthetic efficiency. Thus, triazine-resistant canola is used only where there are sufficient weed problems best managed by triazines to justify a sacrifice in crop yield. If secondary genetic effects (pleiotropic effects) of the genes altered in a biotechnological crop are sufficiently deleterious, they may prevent the introduction or success of the crop.

Imidazolinone-resistant corn was introduced in the spring of 1992. It is resistant because a mutant gene codes for an altered form of the branched-chain amino-acid pathway enzyme, acetolactate synthase. Thus far, there are no recognized undesirable pleiotropic effects of this change in the enzyme. How this crop will fare in a very competitive corn herbicide market remains to be seen.

Perhaps the largest biotechnological potential for impacting weed management in major crops involves the use of two nonselective herbicides, glyphosate and glufosinate. These herbicides effectively kill almost all weed and crop species. Thus, in situations in which more than one herbicide is needed to manage a broad spectrum of weed species, a nonselective herbicide might be substituted for two or more selective herbicides.

Many currently used herbicides are applied prior to weed emergence from the soil, as insurance against a possible weed infestation. Herbicides that are sprayed directly onto the foliage of weeds usually require application at specific stages of weed development and are often mildly phytotoxic to crops at levels that kill weeds. Use of foliar applications of nonselective herbicides with biotechnological crops that are highly resistant to these herbicides will allow the use of herbicides only when and where they are needed without concern for crop damage. Such a practice will reduce the amount of herbicides applied to major crops by eliminating the common practice of prophylactic application of herbicides over entire fields before weeds emerge.

For economic reasons, discovery and development of herbicides is oriented toward major crops. For many minor crops, such as horticultural crops, there are few herbicide options available. In some cases, this limitation has resulted in relatively high costs for weed control. Biotechnology has the potential to influence weed control in these crops most dramatically. However, some factors will continue to limit availability of herbicides for both unmodified and herbicide-resistant minor crops. Even when herbicides developed for major crops are found to be ideal for minor crops, the herbicide industry is often reluctant to gain regulatory approval to extend their use to minor crops. The reason is that the cost of approval versus the potential profit may not justify extension of approval to that crop. Also, extension of use to a new crop may increase the calculated potential daily human consumption of the herbicide or its residues, thus jeopardizing use of the herbicide on all crops.

Environmental issues. Environmental concerns about herbicide-resistant crops from biotechnology have been centered on two issues: increased uses of herbicides, and the potential ecological damage due to the crop itself.

Increased herbicide use. Possibly, widespread adoption of biotechnological crops could increase the use of herbicides, thereby increasing contamination of soil, water, and food with herbicides and their breakdown products. However, in some cases such crops will allow replacement of environmentally or toxicologically suspect herbicides with herbicides that are less likely to be of concern. The herbicides used with the biotechnological crops that are under development are generally considered safer to the environment than those that they could replace.

With biotechnological crops, the potential for increased contamination of food with herbicides and their breakdown products is of concern. Because herbicides can be applied directly to the crop at any stage of development without injury, it is possible that in some cases herbicide residues in the edible portions of the crop may be increased. However, use of herbicides is highly regulated, whether with conventional crops or with biotechnological ones. Legal dosages and the stage of crop development beyond which herbicide application is not permitted are set by the registration process. Another concern is that a foreign gene in a transgenic biotechnological crop plant may adversely affect nutritional value or cause the production of an unexpected toxicant via an unpredicted pleiotropic effect. Although improbable, such could occur with any transgenic crop.

Damage caused by the crop. The second major concern is that the biotechnological crop itself may disrupt the ecosystem. If a new gene has been inserted into the crop from an exotic source, the gene could be transmitted to other plant species by interspecific crossing. In most cases, a weed would be the most likely species with which the biotechnological crop plant might interbreed. Thus, the weed would gain a tremendous advantage in the agroecosystem in which the herbicide is used. However, an advantage outside the agricultural ecosystem is highly unlikely. In limited field studies, movement of resistance genes from biotechnological plants to weeds has not been significant. There is no evidence that any of the more than 200 species of weeds that have become resistant to various herbicides acquired their resistance by interspecific crossing with naturally resistant crops for which the herbicides were designed.

However, biotechnological plants could become weeds in other crops for which the same herbicides that they are resistant to are used.

Regulation. In the United States, various aspects of herbicide-resistant crops from biotechnology are regulated by the Environmental Protection Agency (EPA), the Department of Agriculture (USDA), and the Food and Drug Administration (FDA). The EPA regulates the herbicides and their breakdown products, and sets acceptable levels of dietary intake and safe levels in water. The FDA regulates the nutritional quality and presence of any natural toxicants in the food supply. Genetic modification of a crop could alter either factor. The USDA regulates the introduction of new organisms into the environment. Thus, all transgenic organisms, including biotechnological crop plants, are regulated by the USDA.

Economic considerations. The economics of the introduction of a herbicide-resistant crop from biotechnology must be favorable for the herbicide manufacturer, the seed company, and the farmer in order for the crop to be successful. Farmers will adopt such crops if they improve profits by reducing costs or improving the efficacy of weed management. In those cases in which the biotechnological cultivar reduces herbicide damage to the crop, the number of sprayings, or the amount and number of herbicides used, there is potential for lowered cost of weed management. Some farmers might utilize biotechnological cultivars if they provide more environmentally or toxicologically desirable alternatives to currently used weed control methods, and if there are no significant economic disadvantages.

For background information *SEE AGRICULTURAL SCIENCE (PLANT); GENETIC ENGINEERING; HERBICIDE* in the McGraw-Hill Encyclopedia of Science & Technology.

Stephen O. Duke

Bibliography. J. C. Caseley, G. W. Cussans, and R. K. Atkin (eds.), *Herbicide Resistance in Weeds and Crops*, 1991; S. O. Duke et al., *Herbicide-Resistant Crops*, Comments from CAST 1991–1, 1991; R. Goldburg et al., *Biotechnology's Bitter Harvest*, Environmental Defense Fund, 1990; B. J. Mazur and S. C. Falco, The development of herbicide resistant crops, *Annu. Rev. Plant Physiol. Plant Mol. Biol.*, 40:441–470, 1989.

Human intelligence

Theories of intelligence, like most scientific theories, have evolved through a succession of paradigms. The major paradigms have been psychological measurement (often called psychometrics); cognitive psychology, which is concerned with mental processes; the merger of cognitive psychology with contextualism (the interaction of the environment and the mental processes); and biological science, which considers the neural bases of intelligence.

Psychometric theories. Psychometric theories have generally sought to clarify the structure of intelligence. What form does intelligence take, and what are its parts, if any? Such theories have generally been based on and tested by data obtained from paper-and-pencil tests of mental abilities, such as the abilities to make analogies, classifications, and series completions. Underlying the psychometric theories is a psychological model according to which intelligence is a composite of abilities that can be measured by mental tests. This model is often quantified by assuming that each test score is a weighted linear composite of scores on the underlying abilities.

Cognitive theories. Underlying most cognitive approaches is the assumption that intelligence comprises a set of mental representations (such as propositions and images) of information and a set of processes that can operate on these representations. A more intelligent person is assumed to represent information better and, in general, to operate more quickly on these representations than a less intelligent person.

A number of cognitive theories of intelligence have evolved. For example, one theory holds that a critical ability underlying intelligence is that of rapidly retrieving lexical information, such as letter names, from memory. An alternative approach argues for the importance of reasoning processes to intelligence, such as inferring and applying relations between stimuli.

Cognitive-contextual theories. Cognitive-contextual theories deal with the way that cognitive processes operate in various environmental contexts. For example, a theory of multiple intelligences proposed in 1983 argues that there is no single intelligence; rather, intelligences are multiple and include, at a minimum, linguistic, logical-mathematical, spatial, musical, bodily-kinesthetic, interpersonal, and intrapersonal intelligences.

Also taking into account both cognition and context is the triarchic theory of human intelligence, proposed in 1985. Unlike the theory of multiple intelligences, which views the various intelligences as separate and independent, this theory views intelligence as having three integrated and interdependent aspects. The first aspect consists of the cognitive processes and representations that form the core of all thought; the second consists of the application of these cognitive processes to the external world; the third is the integration of the internal and external worlds through experience.

Biological theories. Some theorists have taken a radically different approach, seeking to understand intelligence directly in terms of its biological bases without intervening hypothetical constructs. These so-called reductionists believe that a true understanding of intelligence can result only from the identification of its biological substrates.

Some researchers have investigated types of intellectual performance as related to the hemispheres of the brain from which they originate. For example, it has been found that the left hemisphere is superior in analytical functioning, such as the use of language. The right hemisphere has been found to be superior in many forms of visual and spatial performance, tending to be more synthetic and holistic in its functioning.

Other researchers have used brain-wave recordings to study the correlation between wave patterns and performance either on ability tests or on various kinds of cognitive tasks. Some studies have shown a relationship between certain aspects of electroencephalogram (EEG) waves and scores on a standard psychometric test of intelligence.

More recent research has involved the measurement of blood flow in the brain, which is a fairly direct indicator of functional activity in brain tissue. In such studies, the amount and location of blood flow in the brain is monitored while subjects perform cognitive tasks. For example, it has been found that older adults show decreased blood flow to the brain and that such decreases are more pronounced in some brain areas than in others, notably those areas responsible for close concentration, spontaneous alertness, and the encoding of new information.

Measurement of intelligence. The tradition of mental testing was developed by Alfred Binet and his collaborator, Theodore Simon, in France at the turn of the century. Binet's early test was taken to the United States by a Stanford University psychologist, Lewis Terman, whose version was called the Stanford-Binet test. This test has been revised frequently and continues in use. The Stanford-Binet test and others like it have traditionally yielded at the very least an overall score referred to as an intelligence quotient (IQ). Some tests, such as the Wechsler Adult Intelligence Scale (Revised), yield an overall IQ as well as separate IQs for verbal and performance subtests. An example of a verbal subtest is a vocabulary test, and an example of a performance subtest is a test of picture arranging, in which an examinee arranges a set of pictures into a sequence that tells a comprehensible story.

Intelligence quotient was originally computed as the ratio of mental age to chronological (physical) age, multiplied by 100. Thus, a child of 10 with a mental age of 12 (that is, a performance level of an average 12-year-old on a test) was assigned an IQ of $(12/10) \times 100$, or 120. If the 10-year-old had a mental age of 8, the child's IQ would be $(8/10) \times 100$, or 80. A score of 100, where the mental age equals the chronological age, would be average.

The concept of mental age has fallen into disrepute.

Many tests still yield an IQ, but this figure is most often computed by statistical distributions. The score is assigned on the basis of the percentage of people of a given group expected to have a certain IQ.

Intelligence test scores follow an approximately normal distribution; that is, most people score near the middle of the distribution of scores, and scores drop off fairly rapidly in frequency as one moves in either direction from the center. Only 1 out of 20 scores differs from the average IQ (100) by more than 30 points.

It has been common to associate certain levels of IQ with labels. For example, at the upper end, the label "gifted" is sometimes assigned to people with IQs over a certain score, such as 130. At the lower end, mental retardation has been classified into different degrees depending upon IQ, so that IQs of 70–84 have been classified as borderline retarded, 55–69 as mildly retarded, 40–54 as moderately retarded, 25–39 as severely retarded, and below 25 as profoundly retarded. Such labeling schemes, however, are oversimplifications.

Psychologists now believe that IQ represents only a part of intelligence and that intelligence is only one factor in both retardation and giftedness. Earlier rigid concepts in the field of intelligence measurement, which led to labeling, have had undesirable effects. The growth of a more recent concept, the malleability of intelligence, has also served to discredit labeling.

Malleability of intelligence. Intelligence has historically been conceptualized as a more or less fixed trait. This view perceives intelligence as something people are born with, the function of development being to allow this genetic endowment to express itself. A number of investigators have taken the approach that intelligence is highly heritable, transmitted through the genes. Other investigators believe that intelligence is minimally heritable, if at all. Most authorities take an intermediate position, believing the heritability of intelligence to be about 50%.

Various methods are used to assess the heritability of intelligence. Noteworthy is the study of identical twins reared apart. If it is assumed that when the twins are separated they are randomly distributed across environments (often a dubious assumption), the twins would have in common all of their genes but none of their environment except for chance overlap. As a result, the correlation between their performance on tests of intelligence could provide an estimate of the proportion of variation in test scores due to heredity. Another method of computing the hereditary effect on intelligence involves comparing the relationship between intelligence test scores of identical twins and those of fraternal twins.

It is important to understand that no matter how heritable intelligence is, some aspects of it are still malleable. Intelligence, in the view of many authorities, is not a fixed trait, with its level a foregone conclusion the day an individual is born. To the contrary, a program of training in intellectual skills can increase some aspects of a person's level of intelligence. A main trend for psychologists working in the intelligence field has been to combine testing and training functions in order to enable people to optimize their intelligence.

For background information SEE BEHAVIOR GENETICS; INTELLIGENCE in the McGraw-Hill Encyclopedia of Science & Technology.

Robert J. Sternberg

Bibliography. R. J. Sternberg, *Beyond IQ: A Triarchic Theory of Human Intelligence*, 1985; R. J. Sternberg, *Metaphors of Mind: Conceptions of the Nature of Intelligence*, 1990; R. J. Sternberg, *The Triarchic Mind: A New Theory of Human Intelligence*, 1988; R. J. Sternberg (ed.), *Handbook of Human Intelligence*, 1982.

Human–machine systems

Historically, machines were developed with little concern for the humans who operated them. In most instances, it was expected that the operator would merely adapt to the machine. The more modern approach is to combine the functions of the operator and the machine in such a manner as to optimize the relationship. This approach entails that jobs should be safe, interesting, and challenging and that the required duties should be performed in a reliable, productive, and economical manner.

Machines have become an integral part of most aspects of modern life. With the increase in the number and complexity of the machines being used, more emphasis has been placed on designing machines that are compatible with human capabilities and skills. The design of human–machine systems requires an understanding not only of the limitations of human beings but also of the way machines are built and operated. Researchers in the fields of human-factors engineering and ergonomics have been working to understand how human skills and machine skills can be combined in a manner compatible with improved human–machine performance. One method often used in the design and analysis of human–machine systems is called the information-processing approach, and carefully considers the interface between the machine and the operator. It evaluates the compatibility of the machine display with human sensory-perceptual capabilities, the compatibility of the machine controls with human physical capabilities, and the overall demand the machine places on the cognitive or mental capabilities of the operator.

Information processing. In the design of human–machine systems, humans and machines are often viewed as information processors. Each functions to gather information by a limited set of input or sensory processes. This information is manipulated or processed so as to produce an output of some type through a limited set of output or motor processes. If both the human and the machine are considered in unison, it is apparent that the output produced by the machine is usually the input used by the human; likewise, the output produced by the human is usually the input used by the machine. An example is the relationship between a driver and an automobile. The automobile receives input in the form of manipulations to the steering wheel, accelerator, and brake pedal by the driver. It processes this information and then produces an output in the form of a velocity and direction for the vehicle. In turn, the driver receives as input the velocity and direction of the vehicle and processes this information, usually by comparing it to a desired velocity and direction. The resulting output is in the form of manipulations of the steering wheel, accelerator, and brake pedal.

Although the representation or model of information processing may seem overly simplistic, it provides human-factors engineers with a clearer understanding of how machines must be designed to be compatible with human beings. In particular, the model highlights three areas that machine designers must consider carefully. The first two areas are the interfaces between the machine and the operator. For the operator to perform well, the machine must produce an output that is compatible with the limited input processes of humans. Similarly, the input that the machine receives must be compatible with human output capabilities. The third area to consider is the processing limitations of the human operator. To effectively address these three aspects of the human–machine interface, designers must have a clear understanding of the way in which the human input, output, and processing components function.

Human sensory-perceptual capabilities. The input systems of the human body are known as the sensory and perceptual systems. Sensory processes allow a person to receive information from the environment surrounding the body. These processes include the basic senses of vision, hearing, smell, taste, and touch, plus some less obvious sensory systems such as the vestibular system in the ears, which provides information about the orientation of the body relative to gravity. The term human perception refers to the way in which the information presented by the sensory systems is interpreted and coded so that the cognitive systems can process it.

In the design of most machine systems, human-factors engineers are concerned with providing information on a machine display in a form that can be easily sensed and perceived by the operator. Because displays usually present information visually, it should be presented in a form that can be seen and interpreted easily. Among the factors that are most important in the design of visual displays are the size and shape of the object to be viewed, the brightness or amount of light projected or reflected from it, the brightness contrast (difference in brightness between the object and the background), and the time allowed for viewing. Research in this area has yielded a basic understanding of the way in which all these variables affect sensation and perception as well as the factors most important to other sensory modalities, such as touch and audition.

Human motor capabilities. The output systems used to operate machines are a part of the human physical or motor system. These systems include all the capabilities that permit a person to interact with and change the environment, such as the use of the hands and feet to move about and of the voice to communicate. Machines are usually designed to be directly manipulated by a human's hands and feet, although the use of voice communication control is currently being explored for a number of applications. In designing a machine control suitable for manipulation, consideration must be given to human limitations in the amount of force that can be produced and the accuracy of limb control. **Figure 1** displays several different types of controls used for transmitting different types of information.

At times the input required to operate a piece of machinery exceeds the person's physical or motor capabilities. For example, before the advent of power steering, most people had difficulty in steering very large automobiles. Although it is readily understandable that human strength is limited, it is less obvious that the accuracy of human movement is also a significant factor that must be considered in design. Very few human movements are made exactly as intended. Otherwise, sporting events such as golf, tennis, or baseball would have little challenge. Machines must be designed so that small errors in the control signal do not produce large errors in the action of the machine.

Human cognitive capabilities. For the human–machine system to function smoothly, the op-

erator must be able to reliably produce the proper output, given the available input information provided by the machine. Thus, the designer of a machine also must have a clear understanding of the cognitive or information-processing limitations of human beings. Humans are limited in the amount of information they can process at one time and the extent to which they can store information for later use. Further, humans possess a good deal of information that is unrelated to a task being performed, and this extraneous information can interfere. This information is usually in the form of expectations or prejudices. Based upon their past experience, all operators have expectations as to how machines should work.

Often, expectations are based on past experience with machines that were not designed with human performance considerations in mind. Modern designers must weigh the possible benefits to performance in redesigning these systems against the problems that will result if the machine does not conform to the expectations of the operator. An example is the design of the modern typing keyboard. The traditional QWERTY keyboard was originally designed to slow operators down so that the mechanical typewriter linkage would not jam during operation. Thus, the designers assigned the letters most commonly used in the English language to the slowest, weakest fingers. This linkage system has been outdated for many

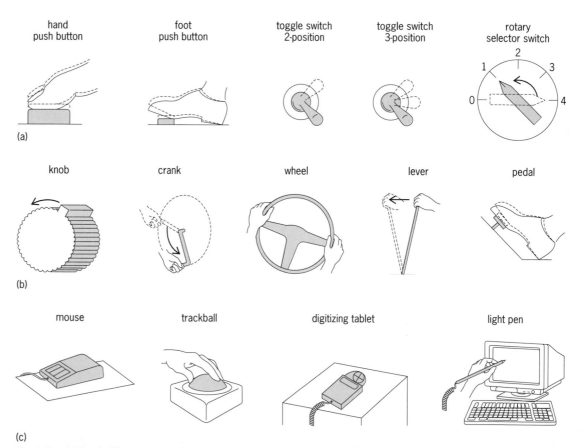

Fig. 1. Examples of different kinds of machine input devices classified by the type of information they transmit best. (*a*) Devices for transmitting discrete information. (*b*) Devices for transmitting traditional continuous information. (*c*) Devices for transmitting cursor-positioning information. (*After M. S. Sanders and E. J. McCormick, Human Factors in Engineering and Design, 6th ed., McGraw-Hill, 1987*)

Fig. 2. Comparison of (*a*) the standard QWERTY keyboard configuration and (*b*) the simplified Dvorak keyboard. (*After M. S. Sanders and E. J. McCormick, Human factors in engineering and design, 6th ed., McGraw-Hill, 1987*)

years, and other key configurations that can make typing much easier and faster have since been developed. One configuration, the Dvorak keyboard, has been recognized as an alternative key configuration by the American National Standards Institute (ANSI; **Fig. 2**). This keyboard layout has been reported to increase typing speed 5–20% with less operator hand fatigue. However, the problems associated with changing to this new keyboard configuration, including the conflict with human expectation, seem to outweigh the benefits; and at present this keyboard is seldom used.

For background information SEE HUMAN-FACTORS ENGINEERING; HUMAN-MACHINE SYSTEMS; TYPEWRITER in the McGraw-Hill Encyclopedia of Science & Technology.

J. C. Woldstad; R. D. Dryden

Bibliography. Eastman Kodak Company, Ergonomics Group, *Ergonomic Design for People at Work*, vols. 1 and 2, 1986; G. Salvendy, *Handbook of Human Factors*, 1987; M. S. Sanders and E. J. McCormick, *Human Factors in Engineering and Design*, 6th ed., 1987; W. E. Woodson, B. Tillman, and P. Tillman, *Human Factors Design Handbook*, 2d ed., 1992.

Immunosuppression

Since the early 1980s, the ability to treat a variety of diseases by transplantation has been greatly aided by the discovery of drugs that block the rejection of the transplanted tissue. Until 1992, little was known of the means by which these drugs paralyzed the immune system, thereby preventing rejection. Research has revealed that these drugs interfere with the process by which information is transferred from the cell membrane, where recognition of foreign tissue occurs, to the nucleus, where a genetic program that initiates transplant rejection is triggered. The events leading to this discovery illus-

trate how genetic studies conducted in lower animals such as yeast are guiding research in many areas of medicine.

Cyclosporin A and its mode of action. The immune system has developed mechanisms to prevent the invasion of foreign tissue, and uses these mechanisms to reject transplanted tissue unless it comes from a genetically identical individual. In the late 1970s, it was found that a drug called cyclosporin A, derived from a soil fungus, paralyzes the immune system and prevents its reaction to foreign tissue. This drug does not diminish the ability of cells to recognize the foreign tissue but blocks the ability of a T cell to react. Since little is known of the intracellular processes necessary for a cell to respond to immunologic stimuli, this drug became a probe of the normal biologic mechanism involved in cell differentiation. In turn, such work helped to unravel the complex mechanism of action of this drug.

Several years after cyclosporin A's discovery, a protein to which cyclosporin A bound very tightly and specifically was found. The molecule, named cyclophilin, was later shown to have the normal function of folding newly synthesized proteins into their natural configuration. This normal function turned out to have little to do with the way that cyclosporin A actually works: a group working on yeast genetics reported compelling evidence that cyclophilin was the definitive receptor for cyclosporin A and that cyclosporin A functioned by forming an inhibitory complex with cyclophilin. The target of the inhibitory complex was found to be a molecule that tightly interacted with this complex. This molecule, calcineurin, had previously been known to be regulated by changes in the concentration of intracellular calcium. Its normal function was to remove a phosphate group from proteins and thereby alter their biologic activity. These test-tube experiments strongly suggested that calcineurin was a target for the actions of the drug; however, evidence in actual cells was essential to verify this information. Verification was provided by experiments that made use of an earlier finding that a specific protein, NF-AT, involved in regulating the activity of the genes that underlie immunologic responses, was completely inhibited by cyclosporin A. By using cultured T cells, calcineurin was shown to be the target of the action of cyclosporin A. FK506, another drug found to block the response of the immune system to transplanted tissue, also blocks the activity of calcineurin. This latter observation, coupled with the fact that effective immunosuppressants have not been found to affect the other steps in the pathway, suggests that perhaps calcineurin is something of a bottleneck in the signal transfer pathway.

These results clearly implicated calcineurin as the target of the drugs cyclosporin A and FK506, but left unresolved the role of calcineurin in transmitting signals essential to the activation of the genetic program leading to immunologic function, in this case graft rejection. Since the protein NF-AT set these changes in motion, clearly it was involved in some way. Subsequently, it was demonstrated that cyclosporin blocked the movement of NF-AT into the nucleus, thereby preventing information sensed at the cell membrane, such as the presence of foreign tissues, from initiating immunologic activation.

Rapamycin and its mode of action. T lymphocytes coordinate the immune response by a two-step process. They are first activated by the transplanted or foreign antigen, resulting in the production of the T-cell growth factor, interleukin-2. In the second step, T cells respond to this growth factor and proliferate, thereby ensuring a sufficient number of cells to reject the transplanted tissue.

A second immunosuppressive drug, rapamycin, has recently been shown to inhibit the ability of interleukin-2 to promote the proliferation of T cells. Here again, the mechanism of action does not interfere with the sensing mechanisms of the cell membranes; it affects the less understood single transmission pathway leading to cell division. One of the most rapidly developing areas of recent research has been the understanding of the mechanism by which information is communicated from the growth factor receptor to the proteins that actually carry out the biologic processes leading to division of a cell. Rapamycin blocks this signal-transfer mechanism in a way that is largely specific for the T lymphocyte. Although little is presently known of its exact site of action, recent efforts have pinpointed it to the cell membrane receptor for the growth factor interleukin-2 and the enzyme p70S6 kinase. The studies suggest that the drug cyclosporin, which blocks the first phase of the immune response, and the drug rapamycin, which blocks the second phase, can be used together at low nontoxic doses. Finally, it is likely that the definition of the mechanism of action of these drugs will lead to the development of much more specific immunosuppressants.

Prospects. The recent developments are being utilized by the pharmaceutical industry in strategies to isolate new and more effective immunosuppressants. The major difficulties with cyclosporin A are that it is not completely specific for T lymphocytes and that it produces adverse effects on the kidney. Because NF-AT is largely restricted to T lymphocytes, it could be the target of new, more specific drugs. If the immune system could be specifically made to tolerate transplanted tissues, transplantation could become a safe, effective, and common means of treating a wide variety of diseases. SEE TRANSPLANTATION BIOLOGY.

For background information SEE IMMUNOSUPPRESSION; TRANSPLANTATION BIOLOGY in the McGraw-Hill Encyclopedia of Science & Technology.

Gerald R. Crabtree

Bibliography. G. R. Crabtree, Contingent genetic regulatory events in T cell activation, *Science*, 243:355–361, 1989; S. L. Schrieber, Chemistry and biology of the immunophilins and their immunosuppressive ligands, *Science*, 251:283–287, 1991; S. L. Schrieber and G. R. Crabtree, The mechanism of action of cyclosporin A and FK506, *Immunol. Today*, 13:136, 1992.

Immunotherapy

Intravenous immune globulin is a concentrated fractionated human blood product derived from plasma pools of several thousand donors. It contains a wide spectrum of antibodies to viral and bacterial pathogens and is of proven benefit in the management of primary and secondary antibody deficiencies, prevention and treatment of certain bacterial and viral infections, and treatment of certain inflammatory and immunoregulatory disorders such as Kawasaki syndrome and immune thrombocytopenic purpura. In addition, intravenous immune globulin has been used in many other disorders with suggestive or equivocal benefit.

Pharmacology. There are now nine different therapeutically equivalent intravenous immune globulins available; all are 5–6% solutions of monomeric IgG globulin, fractionated to remove high-molecular-weight complexes associated with side effects such as flushing, hypotension, and fever. All are sterile and do not contain live viruses, which are removed during fractionation. No viral disease has been transmitted by intravenous immune globulin.

Adverse reactions to intravenous immune globulin are unusual, usually preventable, and easily treated. The most common adverse reactions are nonanaphylactic and occur in about 5% of patients, usually in the first 30 min of administration. Symptoms consist of backache, abdominal pain, headache, chills, fever, and mild nausea. Most reactions can be minimized by pretreatment with aspirin or diphenhydramine and by slowing the infusion. True anaphylactic reactions are extremely rare. Some of these reactions may be due to the presence in the patient of anti-IgA antibodies that react with trace amounts of IgA in the preparations.

Long-term side effects are minimal. A few cases of aseptic meningitis shortly after administration and a few cases of transient or permanent renal insufficiency have been recorded.

Use in immunodeficiencies. About 70% of patients with a primary immunodeficiency (usually congenital and often hereditary) have deficient antibody production; most will need intravenous immune globulin therapy. In addition, about 10–20% of patients with a secondary immunodeficiency (due to immaturity or another illness) have clinically significant antibody deficiency and may also benefit from the therapy.

In primary immunodeficiency, early regimens at low doses of intravenous immune globulin (150–200 mg/kg per month) are sufficient to keep most patients symptom-free. Some immunodeficient patients show improved pulmonary function and decreased symptoms on high doses of intravenous immune globulin (400–600 mg/kg per month).

Patients with secondary antibody deficiency are good candidates for therapy if they have significant hypogammaglobulinemia; or if they have very low levels of natural antibodies to antigens present in all normal subjects (such as *Escherichia coli* antibodies, antistreptolysin-O antibodies, endotoxin antibodies, and isoagglutinins); or if they do not respond to standard antigen (vaccine) challenges.

Hypogammaglobulinemia and antibody deficiency are common in patients with chronic lymphocytic leukemia, multiple myeloma, and other cancers, especially if the disease is advanced. A double-blind study indicated that regular low-dose infusions of intravenous immune globulin every 3 weeks reduced the incidence of bacterial infections in adults with chronic lymphocytic leukemia.

Some pediatric patients develop antibody deficiency associated with proteinuria that occurs in nephrosis, or diarrhea that occurs in protein-losing enteropathy. These patients are candidates for intravenous immune globulin therapy if they have recurrent infection and very low IgG levels.

Premature infants have low levels of maternally derived IgG at birth, develop very low IgG levels in the first months of life, and are very susceptible to infection. Several preliminary studies and large, double-blind, multicenter studies of intravenous immune globulin therapy in premature infants are being completed in the United States to determine if infection can be prevented.

Pediatric acquired immunodeficiency syndrome (AIDS) patients have increased susceptibility to common bacterial and viral infections, poor antibody response to vaccine antigens, and a limited repertoire of antibodies to common pathogens. Intravenous immune globulin significantly reduces the risk of infections in some of the children.

Patients who have had a bone marrow transplant have profound problems with infection in the posttransplant period. Intravenous immune globulin given in the posttransplant period to patients over 20 resulted in decreased infections, decreased graft-versus-host reactions, and improved survival.

Use in bacterial and viral infections. Antibody therapy was important in the history of the treatment of bacterial infections. Immune serum was used successfully in the preantibiotic era to treat *Hemophilus influenzae* meningitis, whooping cough, and meningococcal meningitis and remains an essential part of the treatment of bacterial infections in which toxins are produced.

Intravenous immune globulin, in conjunction with antibiotics, has been advocated in the early management of proven or suspected sepsis of the newborn. Preliminary studies suggest that it is a valuable adjunct to antibiotics in septic premature newborns. Other bacterial infections for which intravenous immune globulin may be of benefit include pseudomonas infections in burns, gram-negative or nosocomial infections in shock or surgery, and refractory staphylococcal infections.

Therapy with special intravenous immune globulin in several viral diseases suggest a role for antibody in the treatment of viral infections. For example, in studies in which intravenous immune globulin enriched with antibodies for cytomegalovirus was used to prevent pneumonia in bone marrow transplant patients, 51% of untreated control patients developed cytomegalovirus infection whereas 35% of patients treated with antibody became infected. Most significantly, the incidence of pneumonia was only 8% in the antibody-treated group compared to 23% in the control group. These results suggest that antibody does not prevent cytomegalovirus infection but may modify the severity of infection and prevent pneumonia in bone marrow transplant patients. In addition to prophylaxis, intravenous immune globulin rich in cytomegalovirus antibodies is of benefit in treating established cytomegalovirus infections.

Preparations of intravenous immune globulin contain antibodies to several common respiratory viral pathogens, including adenovirus, influenza, and parainfluenza. Therapy has been used in some refractory cases. Intravenous immune globulin also contains antibodies to varicella zoster and has been used as a substitute for varicella zoster immunoglobulin in the prevention of chicken pox.

There is one reported cure of pure red-cell aplasia in a 24-year-old man with persistent parvovirus B19 infection. Following intensive intravenous immune globulin therapy, the virus disappeared from the marrow and peripheral blood. When the treatments were discontinued, the patient did not relapse.

Use in immunoregulatory diseases. High-dose intravenous immune globulin therapy has immunosuppressive and anti-inflammatory effects on the immune system that make it a valuable agent in the treatment of several autoimmune or inflammatory disorders. It suppresses antibody synthesis, combines directly with autoimmune antibodies, blocks the uptake of antibody-coated cells in the spleen and liver, and inhibits inflammatory cytokine synthesis.

High-dose intravenous immune globulin therapy has been used extensively in adults and children with both acute and chronic immune thrombocytopenic purpura. Such treatment usually results in prompt increase in platelet counts, equivalent or better than the response to daily prednisone therapy. Intravenous immune globulin had been used in a few cases of autoimmune neutropenia and autoimmune hemolytic anemia. Because of its safety, it is often preferred over steroids or other immunosuppressives, particularly if the patient is immunodeficient or has an ongoing infection. It may also be of benefit in neonatal isoimmune or autoimmune neutropenia.

Intravenous immune globulin is indicated in the treatment of Kawasaki disease, an inflammatory disease of children that often leads to coronary aneurysms. A 1986 controlled study showed that intravenous immune globulin combined with aspirin was superior to aspirin alone in preventing aneurysms, hastening resolution of fever, and lowering the white-blood-cell count and acute-phase reactants.

Intravenous immune globulin has been reported to be beneficial in treating patients with myasthenia gravis associated with anti-acetylcholine-receptor antibodies, patients with Guillain-Barré syndrome, and patients with chronic inflammatory demyelinating polyneuropathy.

Other uses. Intravenous immune globulin has been used in the treatment of allergic diseases such as severe eczema and steroid-dependent asthma; in autoimmune diseases such as dermatomyositis, juvenile rheumatoid arthritis, and lupus erythematosus; and in such miscellaneous disorders as intractable seizures, chronic fatigue syndrome, and recurrent spontaneous abortions.

For background information SEE IMMUNOGLOBULIN; IMMUNOLOGICAL DEFICIENCY; IMMUNOTHERAPY in the McGraw-Hill Encyclopedia of Science & Technology.

E. Richard Stiehm

Bibliography. C. J. Baker et al. and the Multicenter Group for the Study of Immune Globulin in Neonates, Intravenous immune globulin for the prevention of nosocomial infection in low-birth-weight neonates, *N. Engl. J. Med.*, 327:213–219, 1992; R. H. Buckley and R. I. Schiff, The use of intravenous immune globulin in immunodeficiency diseases, *N. Engl. J. Med.*, 325:110–117, 1991; J. M. Dwyer, Manipulating the immune system with immune globulin, *N. Engl. J. Med.*, 326:107–116, 1992; E. R. Stiehm, Recent progress in the use of intravenous immunoglobulins, *Curr. Prob. Pediat.*, 22:335–348, 1992.

Industrial engineering

The rapid introduction of new transportation technology has been significant for economic growth in the United States for the last 200 years. The development of the steamboat gained dominance for the United States in the Atlantic trade in 1850. The large-scale railroad construction in the 1850s through the end of the nineteenth century accelerated westward settlement. The twentieth century was marked by the even more spectacular developments in air and motor vehicle transportation. The development of transportation was accompanied by a large investment in modern infrastructure, most significantly the construction of the 42,000-mi (67,000-km) interstate highway system. These technological developments as well as economic and geographic factors have made the United States the largest investor in and user of transportation technology.

In the future, such dramatic progress is unlikely in the transportation industry, and it will be necessary to use available resources more efficiently. For instance, in recent years many service suppliers and distributors have recognized the importance of efficient distribution strategies to reduce costs of freight transportation, which totaled more than $100 billion in the United States in 1990. In a typical distribution system, servers, based at one or more stations, provide delivery, pickup, or repair and maintenance services to customers at different locations in a given geographic area. A common objective is to find a set of routes for the vehicles that minimizes total fleet operating costs.

This problem, known as the vehicle routing problem, appears in a large variety of applications, such as the distribution of soft drinks, beer, gasoline, and phar-maceuticals, and the pickup and delivery of students by school buses. The wide applications and economic importance of the vehicle routing problem have motivated academic researchers and private and public organizations to consider optimization techniques to improve the efficiency of distribution and transportation systems (Fisher, 1992). Consequently, the problem has been studied extensively.

General routing models. Consider a distribution system with a single depot (such as a warehouse, production center, or school) and n geographically dispersed demand points or customers (such as retailers or bus stops). The demand points are numbered arbitrarily from 1 to n. At each demand point, there are a number of items (such as products or students), known as the demand, that must be brought to the depot by a fleet of vehicles.

Typically, three types of constraints are present. Capacity constraint is an upper bound on the number of items that can be carried by a vehicle; this upper bound is a result of the limited vehicle capacity. Distance (or travel time) constraint is a limit on the total distance (or time) traveled by each vehicle, or a limit on the amount of time an item can be in transit. Time window constraints are prespecified earliest and latest pickup or delivery times for each demand point, and a prespecified earliest and latest time in which vehicles must reach their final destination.

The problem, then, is to design a set of routes such that each route starts and ends at the depot, each item is brought to the depot, no constraint is violated, and total distance traveled is as small as possible. Designing such a set of routes is a generic vehicle routing problem.

Traveling salesperson problem. General routing problems are extremely hard to analyze. One of the most famous problems in optimization, the traveling salesperson problem, is a typical case. In this problem a set of points, cities, or customers has to be visited by a salesperson initially located at one of the points. The task is to design a tour visiting each city and ending at the starting city so as to minimize total distance traveled. The importance of the traveling salesperson problem stems from two facts. First, the problem has numerous applications in many diverse areas ranging from routing in distribution problems to the design of telecommunications networks, as well as in production and manufacturing systems. Second, it serves as a prototype for a class of very difficult combinatorial problems.

Since the traveling salesperson problem has to be solved in any efficient distribution strategy, most research on the vehicle routing problem has been devoted to the development of heuristic algorithms (or simply, heuristics). These algorithms do not necessarily generate the optimal solution (the solution with minimal cost) but tend to find good solutions. These problems are also being attacked by means of neural networks. SEE NEURAL NETWORK.

Capacitated vehicle routine problem. The capacitated vehicle routine problem is a special case of the generic vehicle routing problem in which only the capacity constraint is present. A set of customers dispersed in a given area must be served by vehicles of limited capacity, initially located at a central depot. Associated with each customer is a demand that represents the amount of load that must be delivered to that customer. The objective is to design efficient routes, starting and ending at the central depot, such that each customer receives its demand, the total load delivered by a vehicle does not exceed the vehicle's capacity, and the total distance traveled is as small as possible.

Heuristics for this problem can be classified into three categories: constructive methods, route first/ cluster second methods, and cluster first/route second methods.

Constructive methods. The savings algorithm suggested by G. Clarke and J. W. Wright in 1964 is the most important member of this class. This heuristic is one of the earliest designs for this problem and is the most widely known. The idea of the savings algorithm is very simple: There is one depot and n demand points. Initially a separate vehicle is assigned to each demand point. The total distance traveled by a vehicle that visits demand point i is $2d_i$, where d_i is the distance from the depot to demand point i. Therefore, the total distance traveled in this solution is $2d_1 + 2d_2 + \cdots + 2d_n$.

If two routes are now combined, for example, to service i and j on a single trip (with the same vehicle), the total distance traveled by this vehicle is $d_i + d_{ij} + d_j$, where d_{ij} is the distance between demand points i and j. Thus, the savings obtained from combining demand points i and j, denoted by s_{ij}, is given by the equation below. The larger the savings s_{ij}, the more desirable

$$s_{ij} = 2d_i + 2d_j - (d_i + d_j + d_{ij}) = d_i + d_j - d_{ij}$$

it is to combine demand points i and j. The following savings algorithm is based on this idea.

Step 1: Start with the solution for the capacitated vehicle routing problem in which a separate vehicle serves each demand point.

Step 2: For every pair of demand points i and j, calculate the savings $s_{ij} = d_i + d_j - d_{ij}$. Order the savings in decreasing order. This ordered list is called the savings list.

Step 3: Take the pair of demand points with the largest savings, say i, j, and eliminate it from the savings list. Combine the tour with i in it and the tour with j in it if (1) the combined load of the tours containing i and j does not exceed the vehicle capacity; and (2) both i and j are either the first or last demand point served on the route.

Step 4: Continue with step 3 until the savings list is exhausted.

Route first/cluster second methods. In this class, a traveling salesperson tour is constructed through all the customers (route first), and then the tour is partitioned into segments (cluster second). One vehicle is assigned to each segment and visits the customers according to their appearance on the tour.

Cluster first/route second methods. In this class of heuristics, customers are first clustered into feasible groups to be served by the same vehicle (cluster first), and then efficient routes are designed for each cluster (route second). Heuristics of this class are usually more sophisticated than the previous class, since determining the clusters is often based on a mathematical programming approach, which is outside the scope of this survey.

Recent advances in theory and implementation. In the last few years, a new class of heuristics has emerged. These heuristics take advantage of the structure of the optimal solution in order to guarantee that, for large instances of the problem, the solution provided is close to optimal.

The emergence of this class is a result of the substantial research that has been conducted in recent years on the theoretical analysis of heuristics for the capacitated vehicle routing problem. Almost all the previous literature on this routing problem dealt with empirical analysis of heuristic algorithms, which evaluated the performance of a specific heuristic on a set of standard test problems.

The main advantage of theoretical analysis is that it is possible to identify classes of the capacitated vehicle routing problem for which a specific heuristic is highly efficient or is guaranteed to be inefficient. The major developments in this line of research obtained in the last few years are as follows.

Determination of exact structure of optimal solution value for large-scale problems. Recently the exact cost components that make up the optimal solution of the capacitated vehicle routing problem were identified. This clarification provides new insights into the effects of changes in vehicle capacities, demand sizes, and the shape of the service region on the overall cost of operating the fleet.

Development of new and efficient algorithms for capacitated vehicle routine problem. Knowedge of the structure of the asymptotic optimal cost of the capacitated vehicle routing problem leads to an understanding of the algorithmic structure required to solve large-scale vehicle routing problems. Recently, an efficient algorithm for the capacitated vehicle routing problem has been developed that has the following important property: as the number of demand points increases, the relative error between the solution produced by the algorithm and the optimal solution decreases and in fact goes to zero as the number of customers grows. The algorithm was tested on nine benchmark problems involving real-life instances with 50 to 200 customers. In all cases very good solutions were constructed, and more importantly, in five cases out of the nine, solutions better than or at least as good as those already known were produced.

Complete characterization of a class of heuristics. In recent studies it was proven that heuristics of the route first/cluster second class, which have been the subject of extensive empirical analysis in the last three decades, are guaranteed to be inefficient for the capacitated vehicle routing problem.

An understanding of the importance of taking advantage of the geometry of the area. A very effective method involves a partitioning algorithm that partitions the area where the customers are located into subregions, and then solves a separate problem for each subregion. The effectiveness of this method depends on the partitioning that is carried out.

For background information SEE ALGORITHM; INDUSTRIAL ENGINEERING; OPERATIONS RESEARCH; OPTIMIZATION in the McGraw-Hill Encyclopedia of Science & Technology.

David Simchi-Levi

Bibliography. D. Bienstock, J. Bramel, and D. Simchi-Levi, A probabilistic analysis of tour partitioning heuristics for the capacitated vehicle routing problem with unsplit demands, *Math. Oper. Res.*, 1993; J. Bramel et al., Probabilistic analysis of the capacitated vehicle routing problem with unsplit demands, *Oper. Res.*, 40:1095–1106, 1992; J. Bramel and D. Simchi-Levi, *A Location Based Heuristic for General Routing Problems*, Working Paper, Department of Industrial Engineering and Operations Research, Columbia University, 1992; G. Clarke and J. W. Wright, Scheduling of vehicles from a central depot to a number of delivery points, *Oper. Res.*, 12:568–581, 1964; M. Fisher, *Vehicle Routine*, Working Paper, Department of Decision Sciences, Wharton School, University of Pennsylvania, published in Elsevier Science Handbooks in Operations Research and Management Science, ed. by M. Ball et al., 1992.

Information management

Increasingly, companies are taking the view that information is an asset of the corporation in much the same way that the company's financial resources, personnel, capital equipment, real estate, furniture, and fixtures are assets. Properly employed, assets create additional value with a measurable return on investment. Forward-looking companies carry this view a step further, considering information as a strategic asset that can be leveraged into a competitive advantage in the markets served by the company.

Corporate information management, then, refers to the functions associated with managing the information assets of a corporation. In some respects, the term is interchangeable with management information systems (MIS), although the choice of terminology may reflect the attitude of a company regarding the value of information. Companies using the traditional MIS descriptor are likely to hold the view that MIS represents a cost to be maximized, in contrast to the strategic-asset view with the objective of maximizing return on investments. Another indication may be found in the placement of the function within the corporate hierarchy. The asset view may lead to a chief information officer position, perhaps reporting directly to the chief executive officer, or even a line division as a separate profit center. The cost view may find an MIS position buried several reporting

layers down. The discussions that follow will consider corporate information management in a generic sense, with no distinction between these alternative views.

For the large enterprise with multiple business units and functional elements distributed over a broad geographic area, information management can be an enormously complex task but have a very high potential return on investment. Its successful accomplishment is dependent not only on the diligent application of professional skills and techniques, such as information engineering and systems engineering and integration, but even more on extensive leadership and management abilities strongly rooted in a thorough knowledge of and insight into the business of the enterprise.

Scope. The scope of the corporate information management function may vary between companies. As a minimum, it will usually include the origination or acquisition of data, its storage in databases, its manipulation or processing to produce new (value-added) data and reports via application programs, and the transmission (communication) of the data or resulting reports. While many companies may include, with good reason, the management of voice communications (telephone systems and voice messaging) and even more traditional libraries, this article focuses primarily on the data aspects of corporate information management.

Information versus data. Corporate executives and managers frequently complain of "drowning in a flood of data, but starving for information." There is more than a subtle difference between these two terms. Superficially, information results from the processing of raw data. However, the real issue is getting the right information to the right person at the right time and in a usable form. In this sense, information may be a perishable commodity. Thus, perhaps the most critical issue facing corporate information managers is requirements definition. The best technical solution is of little value if the final product fails to meet the needs of users.

Information engineering. One formal approach to determining requirements is information engineering. By using processes identified variously as business systems planning or information systems (or strategy) planning, information engineering focuses initially on how the organization does its business, identifying the lines from where information originates to where it is needed, all within the context of a model of the organization and its functions. While information management functional personnel are the primary agents in the information engineering process, its success is critically dependent on the active participation of the end users, from the chief executive officer on down through the functional staffs.

An enterprise data dictionary is one of the principal products of the information engineering effort. This dictionary is a complete catalog of the names of the data elements and processes used within the enterprise along with a description of their structure and definition. The dictionary may be part of a more comprehensive encyclopedia, which is typically a formal repository of detailed information about the organization, including the process and data models along with descriptions of the design and implementation of the organization's information systems (applications).

A major advantage of the application of information engineering is that it virtually forces the organization to address the entire spectrum of its information systems requirements, resulting in a functionally integrated set of enterprise systems. However, ad hoc requirements may result in a fragmented set of systems (islands of automation), which at their worst may be incompatible, contain duplicate (perhaps inconsistent) information, and omit critical elements of information.

Systems engineering and integration.

Information engineering thus results in the creation of a logical architecture for an organization's information systems. However, implementation will be within the context of a physical systems architecture. Information engineering ensures functional integration, whereas systems engineering provides for the technical integration of information systems. Together, functional and technical (logical and physical) integration lay the foundation for the integrated enterprise.

Open versus proprietary architectures.

Many critical decisions are involved in establishing a physical systems architecture for an organization. Perhaps the single most important decision is whether the architecture will be based on a single manufacturer's proprietary framework or on a so-called open systems framework. Although the former may be simpler to implement and may offer less technical risk, it is far more restrictive with regard to choices of equipment. An open systems architecture allows for the integration of products from multiple vendors, with resulting increased flexibility in functional performance and the potential for significant cost savings in competitive (multivendor) bidding between equipment suppliers. A significant aspect of increased flexibility is the ability to incorporate new products offering improved performance or previously unavailable functionality. The flexibility to incorporate new products is particularly important in an era of rapid technological advancement, where new technology may become obsolete in as little as 2 years.

Operating systems. An extremely important architectural consideration is the choice of operating systems. This selection may determine the options available in acquiring off-the-shelf software packages. Alternatively, an existing set of software programs (normally representing a major investment) may severely constrain the choice of hardware on which the programs will run. The issue here is whether the operating systems are proprietary to a single vendor's hardware or are vendor-independent (such as UNIX). Most manufacturers make available a UNIX operating system or a variant thereof. Ideally, a vendor-independent operating system will permit true portability of application software between different computers and manufacturers. In some cases, this may be true, while in others, some (often minor) modifications may be required to effect the transition.

Because of the portability it offers, UNIX has become a de facto standard in the industry. The Institute of Electrical and Electronics Engineers (IEEE) has developed a more formal set of standards for operating systems called POSIX. POSIX compliance is an indicator of portability beyond UNIX and derivatives, and can be found from mainframe computers to minicomputers and microcomputers. At the microcomputer level, the dominant operating system by far is MS/DOS, a product of Microsoft. It has also become a de facto standard for microcomputers, permitting almost universal portability of MS/DOS-based software products. UNIX (POSIX) and MS/DOS are typically the principal software components of an open systems environment.

Hardware. Hardware architectural choices are generally more varied and complex. They involve the selection of processor types and sizes, the roles these processors will play, the choices and locations of peripheral equipment (such as storage devices and printers), and ways in which all of these components will communicate with one another.

Most manufacturers have developed their own set of proprietary intercommunications standards. A truly vendor-independent set of standards, however, has been established based on the open systems interconnect (OSI) model. This model comprises seven layers, the physical, data link, network, transport, session, presentation, and application layers. Level 1, the physical layer, is typically a coaxial cable, a twisted pair of wires, or a fiber-optic cable. Each successive layer may be independent of the choices in the level below but utilize its services. Manufacturers and software vendors are free to implement their products for one or more layers, relying on the standards to ensure that their product will correctly interface to and interoperate with products supporting the level below. These standards have resulted in the availability of a rich variety of OSI-based hardware and software products. Numerous local-area network products are available, for instance, to provide network-layer services. *SEE ELECTRONIC MAIL.*

One of the most widely used architectural models is the client/server model. In this model, end users, for example, may employ a personal computer as a client to request data from a corporate database resident at another location on a minicomputer or mainframe (the server). This example could describe how a corporate personnel system makes available information on an employee's eligibility for benefits to a personnel specialist in a field office where that employee works.

Databases. Another major architectural decision is the structures of the databases to be utilized. Most products are based on hierarchical, relational, or object-oriented paradigms. Although the most widely used structure has been relational, increasing attention is being given to object-oriented systems. Information engineering tends to separate data from applications, while object-oriented methods couple the two. (Enhancements to current information-engineering methodologies, however, can be expected to harmonize them with the object-oriented approach.) *SEE OBJECT-ORIENTED PROGRAMMING.*

Software implementation. As discussed above, the information engineering process can produce the complete set of requirements for a corporate personnel system. When all architectural parameters have been determined, the system can be implemented, preferably employing computer-aided software engineering (CASE) tools to automate the process to the maximum degree possible.

In general, the cost and time involved in developing custom software, and the subsequent even greater cost of supporting (maintaining) that software throughout its life cycle, dispose strongly toward the maximum use of off-the-shelf software packages. Only those functions that are truly unique to (and perhaps generate significant competitive advantage for) the organization should be developed specifically for that organization.

For background information SEE DATABASE MANAGEMENT SYSTEMS; INFORMATION SYSTEMS ENGINEERING; LOCAL-AREA NETWORKS; OPERATING SYSTEMS; SOFTWARE ENGINEERING; SYSTEMS ENGINEERING in the McGraw-Hill Encyclopedia of Science & Technology.

Alan B. Salisbury

Bibliography. J. Martin, *Information Engineering*, 1989; A. Simon, *Enterprise Computing*, 1992; T. Wheeler, *Open Systems Handbook*, 1992.

Integrated optics

There is a great deal of interest in arrays of very small lenses with diameters between 20 micrometers and 1 mm. Some applications have been known for many years but have not been fully exploited because suitable lens arrays were not available. Others have arisen as a direct result of developments in technology on a microscopic scale. There now exist a variety of ways of manufacturing lens arrays. Some of these arrays are at the research stage, and others are available commercially.

Manufacture of microlens arrays. One of the earliest techniques of making microlenses was developed by Robert Hooke in 1664. He melted thin strands of glass that assumed spherical shape under the forces of surface tension. A widely used modern variation uses standard photolithographic techniques to generate an array of circular islands of photoresist on a suitable substrate. The islands are then heated in an oven to about 284°F (140°C), at which temperature the resist softens and surface tension draws the islands into the shape of lenses.

For applications requiring a large area or minimum cost, a nickel electroform replica can be made of the photoresist surface and used as a die for embossing onto a plastic sheet or for injection molding. When a particular material, such as silicon, is to be used, the photoresist lenses are first formed on the surface of a substrate of that material and then transferred into the material itself by ion-beam machining.

A totally different approach to the manufacture of microlenses involves manipulating not the shape of the lens but the distribution of refractive index within it. Although various possible forms of these so-called graded-index (GRIN) lenses have been known for a long time, they have become widely available only comparatively recently. There are three basic techniques for achieving a suitable distribution of refractive index: selective exchange of ions in a suitable glass substrate, diffusion polymerization in plastics, and chemical vapor deposition. In ion exchange, the substrate is immersed in a bath of an appropriate molten salt for as long as a week. Metal ions diffuse into the matrix and replace some of the silicon, sodium, or potassium, depending upon the type of glass used. The rate of exchange and the refractive index depend upon the amount of dopant at a given point, so that after a given time there will be a gradation of index from the surface into the bulk of the glass. The process can be speeded up, and to some extent controlled, by applying an electric field in the direction of the diffusion. Similar diffusion effects can be achieved much more rapidly in plastics by immersing a partially polymerized substrate in a bath of a suitable monomer.

Two particularly interesting forms of lens made in this way are the rod lens and the planar microlens, with a three-dimensional distribution of refractive index. As neither of them depends upon refraction at its external surface, these lenses can be cemented directly onto other components, giving rise to greater ruggedness and stability and reducing Fresnel reflection losses. An array of planar lenses with a three-dimensional index distribution can be made by the selective diffusion of the dopant through a mask consisting of a series of circular holes. If the size of the hole and the diffusion parameters are chosen correctly, lenses of good quality can be made with diameters ranging from less than 100 μm up to a few millimeters.

An alternative form of surface-refracting microlens is made by using a photosensitive glass that is exposed to ultraviolet radiation through a mask in the form of opaque dots on a clear background. When the glass is heated, the exposed portion shrinks and squeezes the adjacent material into the form of spherical lenslets. As with the other techniques, it is possible to achieve a performance that is close to the limits set by diffraction.

The diffracting equivalent of a lens is the zone plate, and there is much interest in the manufacture of arrays of miniature surface-relief zone plates and of holographic lenses that work by means of the Bragg effect. These arrays offer the possibility of filling the available area more completely and also offer greater flexibility in design. Their main disadvantage is that their focal length is inversely proportional to wavelength so that they suffer from severe chromatic aberration. Their use is therefore usually restricted to applications with monochromatic light.

Applications of microlens arrays. One of the earliest applications of microlens arrays, proposed in 1908 by Lippmann, is three-dimensional integral photography. Here, the photographic medium is placed at the focal plane of the lens array so that each lens behaves as a miniature camera recording the scene from a different perspective (**Fig. 1**a). When the developed image is viewed through the same lens array, each

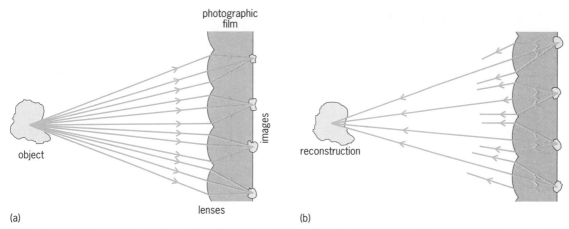

photographic film

images

object

lenses

(a)

reconstruction

(b)

Fig. 1. Integral photography. (*a*) Recording of an object by a series of microcameras with a range of viewpoints. (*b*) Reconstruction, when the recording is illuminated, of an object that has full parallax.

lens behaves as a miniature projector and the object is reconstructed in full parallax (Fig. 1*b*) in a way similar to that of a hologram. The resolution of the microlens array is inferior to that of a hologram, but its advantages are that it can be in full color and that it demands no special lighting or viewing conditions. The microlens array also offers the prospect of three-dimensional television.

A more common use of lens arrays is in photocopying and facsimile machines. If an entire page were to be imaged at one time, a very large lens and a correspondingly large machine would be required. Instead, only a small strip of a page is imaged at a time and the page is scanned (**Fig. 2**). This approach requires the optical system to produce an erect image, which is achieved by what is effectively an array of microscopic telescopes of unit magnification. The images from each telescope coincide to produce so-called integral image, and in this way large areas can be imaged with relatively short object-to-image distance. The most widely used lenses for this application are graded-index rod lenses (Fig. 2*b*), although aligned pairs of surface-relief lenses (Fig. 2*c*) produce the same effect.

The efficiency of detector arrays may be enhanced by positioning a lens array in such a way that for each pixel a lens focuses the light onto the active part of the detector. Light that would have otherwise been lost by falling onto the so-called dead areas occupied by electronic circuitry is now fully used (**Fig. 3**). For infrared detectors, this gain can be put to advantage by reducing the size of the active area and thereby reducing the noise. An acceptable level of noise can then be achieved by Peltier cooling rather than by liquid nitrogen, with a dramatic increase in convenience. A similar improvement in efficiency can be achieved with liquid-crystal displays such as those used in projection television. Not only is the light used more efficiently because it is channeled through the active areas, but it is then focused away from the electronic circuitry, which remains cooler as a result.

The ability of a microlens array to focus light into small areas of a detector chip can be exploited further to enable signal processing to take place on the chip

itself. For example, by comparing the signals from adjacent pixels it is possible to detect edges, motion, or change in an image. This ability reduces the amount of data that has to be transmitted from the chip and mimics the way in which the amacrine cells of the retina compress the data that have to be fed into the optic nerve.

Current developments in electrooptics and digital optics have stimulated a growing trend toward optical systems that consist of arrays of channels operating in parallel and on a miniature scale. Perhaps the simplest application of this type is the coupling between single-mode optical communications fibers in which the lens

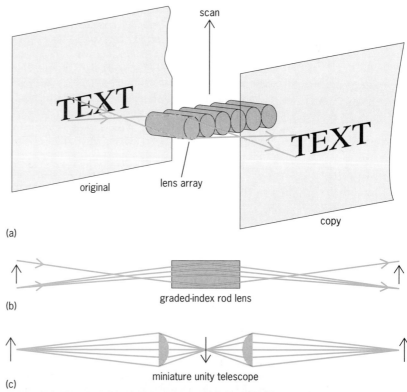

scan

TEXT

TEXT

original   lens array

copy

(a)

graded-index rod lens

(b)

miniature unity telescope

(c)

Fig. 2. Use of a lens array to scan large areas in a photocopier or facsimile machine. (*a*) Forming an erect image by lens array. (*b*) Focusing of rays by a graded-index rod lens. (*c*) Focusing of rays by an aligned pair of lenses forming a miniature unity telescope.

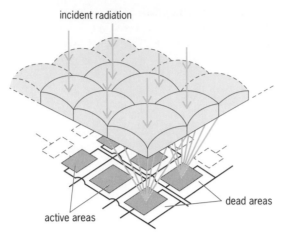

incident radiation

dead areas

active areas

Fig. 3. Microlens array focusing all of the incident light onto the active areas of a detector array, thereby enhancing efficiency.

array is arranged so that the core of each fiber is at the focal point of each lens. The light from each fiber is thereby collimated and then focused by a similar arrangement of lenses onto the second array of fibers.

In high-speed parallel switching networks, optics offers a significant advantage over electronics because beams of light do not interfere with each other whereas electrons traveling in wires do. As circuits become ever more miniature and frequencies become higher, the problem of crosstalk becomes worse, and as a result there is great interest in optical interconnections. In principle, any element of an array can be connected to any element of another array by using a lens that is displaced sideways by just the right amount. By producing arrays of carefully positioned lenses it is possible to make spatially variant optical interconnections (optical-wiring modules) that will achieve such functions as fan-in, perfect shuffle, and cyclic shift. These arrays are likely to be key elements in optical computing and optoelectronic neural networks.

For background information SEE DIFFRACTION; ELECTRONIC DISPLAY; HOLOGRAPHY; INTEGRATED OPTICS; OPTICAL FIBERS; OPTICAL INFORMATION SYSTEMS in the McGraw-Hill Encyclopedia of Science & Technology.

Michael Hutley

Bibliography. M. Hutley (ed.), *Microlens Arrays*, 1991; K. Iga, Y. Kokubun, and M. Oikawa, *Fundamentals of Microoptics*, 1984; E. Wolf (ed.), *Progress in Optics*, vol. 25, 1988.

Intelligent machines

Automation has becomes a key concept in society, and provides improved product quality, productivity, and economic advantages. Automation allows a selected set of operations to be performed repeatedly, without the direct and close supervision of humans. The earliest concepts in automation were directed toward the development of machine hardware that systematically stepped through a sequence of operations. Such an implementation is called hard automation. With the advent of programmable computers, a new idea in automation, called flexible automation, was introduced. In flexible automation, the machine hardware is controlled by a computer (sometimes referred to as a controller), which allows for more flexibility in using the same hardware in manufacturing similar components. Such automation utilizes open-loop control, in which the controller does not monitor how the states of a process are changing. In spite of this limitation, flexible automation has found utility in many industrial applications. This article describes the main ideas behind intelligent robots, which go beyond flexible automation.

Preprogrammed and intelligent robots. Robots are machines. Like other machines, they are designed to accomplish tasks that are important and useful to humans. Machines of the industrial revolution, such as the steam engine, electric motor, automobile, and airplane, made it possible to duplicate and extend various physical capabilities of humans. Within only a century, these machines have become essential elements of human existence in a technological society. Intelligent robots are machines that are designed to duplicate and extend the higher-level perceptual and cognitive capabilities of humans.

Robotic systems can be broadly classified into two groups, one where robots are preprogrammed to perform a specified task, and another where the robot plans and executes its actions intelligently by sensing its environment and analyzing the task requirements. Thus, the actions of both types of robot can be comparable. The main difference lies in whether the actions of a robot result from the detailed sequence of steps provided to the robot before it is asked to perform a task, or whether the robot functions with autonomy by intelligently acquiring and interpreting sensory data, perceiving the nature and changes in its work environment, planning the appropriate course of actions, and executing actions leading to the eventual accomplishment of a task. Robots of the latter type are called intelligent robots. Most robotic systems currently employed in industry are of the preprogrammed variety. Intelligent robots offer potential for improved flexibility, increased versatility, and lower hardware cost (by eliminating the need for the hardware required for precise handling and placement of a workpiece).

Robots of the preprogrammed type have proven quite successful in accomplishing repetitive tasks in structured and static environments. The main limitation of preprogrammed robots results from their inability to adapt to changes and uncertainties in their environments. This limitation needs to be overcome in order to make robotic solutions feasible for a much wider array of applications. Advanced robotic systems will have to be designed to successfully deal with the lack of structure, unexpected events, uncertainties, and complexities associated with their work environments and tasks. These systems will be required to operate with a high level of autonomy and should also be capable of efficient and effective communication with humans. While robots are provided with these capabilities, careful consideration needs to be given to issues of human–machine interaction, error detection and recovery, learning and adaptive behavior, speed, and robustness; and also to various systems engineering issues related to integration, reliability, expandability, and portability.

Integrated robot architecture. The intelligent robot can be viewed as an integrated system having three important capabilities: perception, for perceiving the robot's environment by using an array of sensory devices; action, for performing tasks in the work environment by using mechanical devices; and planning and control of the robot's perception and motor actions. Typically, these capabilities of a robot are provided by four system modules: sensing; perception, planning, and control; motor; and workspace (**Fig. 1***a*). The sensory information acquired in a variety of sensor modalities is analyzed by the perception, planning, and control module of the robot. This module issues commands to the various platforms and effectors of the motor module to perform the appropriate task in the workspace of the robot.

A key idea in developing such complex systems is not to look at these capabilities independently but to examine them in an integrated manner. Such an integrated perspective is most obvious at the system architecture level. Figure 1*b* identifies subsystems and their functions for an integrated robotic system. A number of interesting ideas about integrated robotic architecture and task planning have emerged recently. The operation of two selected integrated autonomous robotic systems is illustrated in **Fig. 2**. Here, industrial robots use their vision, proximity, and force sensors to localize a control panel, to identify and operate various controls (Fig. 2*a*), and to automatically detect a spill and use a vacuum cleaner to clean it up (Fig. 2*b*).

Robot perception. Robotic perception is based upon the sensory inputs acquired from the work environment. Examples of important goals of a perception module are object recognition, object localization, depth determination, and environmental modeling. Perception itself is a rather complex task and requires a range of processing and analysis subtasks, including preprocessing, enhancement, feature extraction, and segmentation at the lower levels, and model-based matching and recognition at the higher levels. In a sense, the computational hierarchy for machine perception provides a road map for the signal-to-symbol transformation inherent in perception.

Sensors are essential to provide the intelligent system with the capabilities that allow an accurate perception of the environment in which the system operates. Without them, the robot environment would have to be precisely structured and static. Depending upon the characteristics of the application and those of the actions required to accomplish the desired objectives, the most appropriate set of sensors can be selected. Some of the important sensor modalities identified in robotics are vision, range, touch, sound, temperature, force, and radiation. Each of these sensory modes requires resolution of many complex issues such as sensor resolution, active versus passive mode, sensor positioning and calibration, processing, and analysis.

The vision modality is often acknowledged as a sensory mode that offers a rich set of scene descriptors in a cost-effective manner. Robot vision systems can be characterized by the types of inputs utilized, outputs generated, and their intended utility. The input consists primarily of digital images. Also utilized is any available prior knowledge about the application domain and the imaging parameters. The output consists of

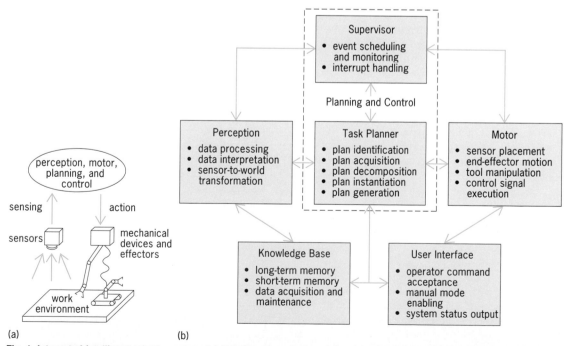

Fig. 1. Integrated intelligent robotic system. (*a*) Relation of system modules. (*b*) Diagram of system architecture, showing subsystems and their functions.

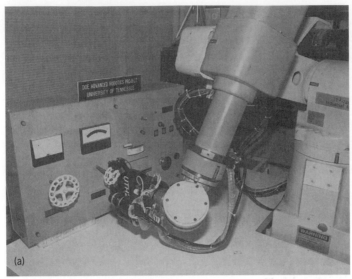

Fig. 2. Operation of autonomous robotic systems where industrial robots use their vision, proximity, and force sensors. (*a*) System to localize a control panel and to identify and operate various controls. (*b*) System to automatically detect a spill and use a vacuum cleaner to clean it up.

high-level descriptors of the imaged scene. Typically, information such as presence or absence of an object, object identity, object location and pose, or object status is desired. The robot vision system gains utility from being part of a robotic system having a separate planning and control module to execute actions in the robotic work environment. Typically, the vision system is incorporated in the dynamic control of the robotic actions, where fast response is a major consideration.

While traditional approaches to machine perception have emphasized the development of generic approaches involving a very detailed and complete analysis of sensory data, the perception module of an intelligent robotic system must satisfy several unique requirements. It must be able to offer cues that are useful in the three-dimensional environment of the robot; it must provide information for real-time operation; and since the robotic system has a definite goal, it needs to provide results that are necessary to accomplish a task (that is, task-specific perception is required). These requirements have inspired researchers to provide fresh insights to the machine perception problem. New computational frameworks for solving depth and shape extraction problems and new sensory systems for real-time range sensing have been introduced, as well as a new paradigm called active perception where a tightly coupled sensory-motor system, used to perceive the robot's environment, is providing promising results.

Robot planning and control. Planning refers to the task of identifying a series of subtasks that need to be performed to accomplish a given goal. It is desirable to specify the goal at a high level without getting encumbered with the numerous low-level details associated with a task. The extent of planning functions depends upon the extent of the knowledge base, which is composed of environmental and behavioral models. This relationship also involves the response time of a system. For example, when a system needs

to perform a task instantaneously after encountering a certain state, there is no time to develop a reasoned plan based upon models. This kind of system performance is known as reflexive, and reactive planner modules are necessary to support it. The tasks that can be handled by reflexive behavior are relatively simple and must be quite closely associated with so-called raw sensory signals (which have not undergone analysis). However, model-based reasoning can be used to derive a series of low-level commands for the perception and motor modules. This kind of planning is called model-based or deliberative, and it can support complex tasks. Naturally, deliberative planners have greater computational requirements and will not be able to respond instantaneously to state changes.

The integrated system, having perception, motor, and planning capabilities, must be properly controlled in order to be of utility. Control functions range from controlling individual joints and links at one extreme to controlling the information and process flow among the subsystems. Advances in control theory have allowed the development of algorithms for controlling robotic structures in real time. Kinematic and dynamic control of multiple-degree-of-freedom manipulators, even those with redundancies and flexibility, are being actively studied with promising results.

Human–machine interface. Intelligent robotic systems are typically considered to be totally autonomous. This conceptualization sometimes stands in the way of developing practical intelligent systems. For many applications, robotic systems are required to perform alongside their human operators. This requirement raises the question of how to get humans and robots to cooperate to accomplish a task.

This observation leads to the view of intelligent systems as integrated human–machine systems. The basic premise of such an approach is to make robots and humans perform complementary tasks. Quite often, the key to success is not to attempt to develop completely autonomous systems but to allow the operator to interact with the robot so that the overall efficiency and productivity is improved. Sensor-based robotics and virtual reality are key technologies in developing such integrated systems. The operator is an essential element of the integrated system and is provided with power displays (for both visual and nonvisual information) for telepresence, and with interactive controls for teleoperation. Thus, the operator, stationed in the safe and virtual work environment, is able to cooperate with the robotic system in accomplishing an assigned task in the most effective manner.

These concepts of an advanced human–machine interface can be illustrated with the example in **Fig. 3**, where a mobile manipulator is utilized to demonstrate the feasibility of a robotic system in handling contaminated filters in a commercial nuclear plant. Currently, this operation is performed by humans, and it accounts for exposure of 20 worker-rads (0.2 worker-gray) per year at this particular site. Successful demonstration of the virtual-world robot simulation in performing this task was necessary to make a case for deployment of the real robot for this application.

Applications and research directions. Given the range of capabilities of intelligent robotic systems, their applications are really limited only by the imagination. Some of the more obvious applications will be discussed. The importance and utility of intelligent robots for operation in hazardous and inhospitable environments such as nuclear plants, space, planetary exploration, and battle fields are well recognized. Intelligent systems have direct applicability in a variety of industrial environments. These systems will definitely help to perform a wide range of inspection, quality-control, maintenance, repair, and assembly tasks. Intelligent robots will also be of much utility in a variety of service-oriented tasks. Systems that can be used for warehouse and office clean-up, mail distribution, and hospital services are being developed. Finally, intelligent robotic systems promise to impact the medical field significantly. Systems that can be used in various surgical procedures are being developed.

Development of intelligent robotic systems required consideration of two different types of tasks. The first task deals with the design and development of individual components required in the system, whereas the second task is related to the proper integration of the individual components to form a complete system. Researchers from multidisciplinary backgrounds have dedicated their efforts to defining and solving important issues underlying the development of such systems. Over the past several years, the main emphasis of research studies has been on the development of individual components that can be utilized in a larger robotic system. These studies have contributed to the development of useful image processing, analysis, and interpretation schemes and various robot control and path-planning algorithms. Research studies with a primary focus on the development of a complete robotic system have been much fewer in number, perhaps because such studies typically require extensive lab-

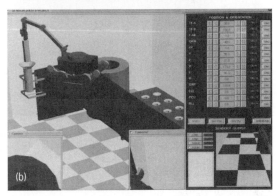

Fig. 3. Mobile manipulator that was utilized to demonstrate the feasibility of a robotic system in handling contaminated filters in a commercial nuclear plant. (*a*) Real robot. (*b*) Simulation and visualization.

oratory resources. Also, whereas both theoretical and experimental approaches were utilized in the research associated with individual component design, the main approach followed in the integrated system development research has been primarily experimental. The complexity of most practical robotic scenarios requires systematic experimental research for system design and performance evaluation.

For background information SEE COMPUTER VISION; HUMAN–MACHINE SYSTEMS; INTELLIGENT MACHINE; ROBOTICS in the McGraw-Hill Encyclopedia of Science & Technology.

Mohan M. Trivedi

Bibliography. *IEEE International Robotics and Automation Conference*, Nice, France, May 1992; P. Maes (ed.), *Designing Autonomous Agents*, 1991; A. Sood et al. (guest eds.), Special issue on intelligent robotic systems and unmanned vehicles, *IEEE Trans. Sys. Man Cybernet.*, vol. 20, no. 6, December 1990; M. Yovits (ed.), *Advances in Computers*, vol. 32, 1991.

Ischemia

Permanent brain damage caused by ischemia (loss of blood flow) and reperfusion (return of blood flow) accompanies disease processes such as stroke and resuscitation from cardiac arrest, and affects close to 200,000 patients in the United States each year. Stroke and cardiac arrest represent different models of damage. In ischemic stroke, only part of the brain is involved, and the ischemia is dense or complete only in the center of the affected area. This area of dense ischemia is surrounded by a penumbral zone in which the blood flow is diminished but not completely lost. In cardiac arrest and resuscitation, the entire brain is subjected to a transient period of complete ischemia followed by reperfusion.

Three major observations have provided the basis for continuing study of the pathophysiologic processes involved in such damage. (1) In transient complete ischemia such as in cardiac arrest, the most prominent structural damage, as seen by electron microscopy, develops during the reperfusion period and not during the ischemic insult itself. (2) Pyramidal neurons in layers III and V of the cortex (involved in thinking and motor activity) and in the CA_1 and CA_4 zones of the hippocampus (a part of the limbic system involved in memory formation) are much more susceptible to damage by ischemia and reperfusion than are neurons in other areas of the brain—a phenomenon known as selective vulnerability. (3) Especially in the selectively vulnerable neurons, there is a generalized and prolonged reduction of the capacity for protein synthesis during reperfusion.

Selective vulnerability. There are currently two major theories that seek to explain selective vulnerability: the excitotoxic amino-acid neurotransmitter theory, which focuses largely on events during ischemia, and the oxygen-free-radical theory, which focuses largely on events during reperfusion. These theories are not mutually exclusive, and considerable evidence supports each of them.

Excitotoxic amino-acid neurotransmitter theory. Although signal transmission along the length of an individual neuron is electrical and involves movements of ions across that neuron's membranes, signal transmission between neurons is not by direct electrical conduction but rather is mediated by chemicals (neurotransmitters) released from one neuron onto the membrane surface receptors of a second neuron. The activated surface receptors then initiate transmembrane ionic movement (that is, electrical conduction) in the second neuron. The excitotoxic amino acid neurotransmitter theory argues that the selectively vulnerable neurons receive an especially high concentration of signals from other neurons, and that these signals are mediated by amino-acid neurotransmitters, such as glutamate. During ischemia, adenosinetriphosphate (ATP) levels in the brain drop to near zero within 4 min, rendering the neuronal membranes unable to maintain the electrical polarization normally accomplished by the partitioning of sodium ions (Na^+) and potassium ions (K^+) across the cells' plasma membrane. The resulting ischemia-induced depolarization is similar to the depolarization that normally accompanies neuronal signaling and results in the release of large amounts of glutamate onto the surface of the cell bodies of the selectively vulnerable neurons. Activated glutamate receptors on the selectively vulnerable neurons then open channels through the membrane that either directly or indirectly result in the influx of calcium ions (Ca^{2+}) into the cell. Overloading the inside of the cell with calcium ions causes the activation of calcium-dependent enzymes, including phospholipases, which damage the cell membrane by removing fatty acids from the lipid components of the membrane. Evidence for this theory includes the demonstration of accumulation of extracellular glutamate in the brain during ischemia, and the ability of excess glutamate to kill cultured neurons.

Oxygen-free-radical theory. Selective vulnerability is seen as a special aspect of reperfusion damage. The theory admits the importance of Ca^{2+} overloading and phospholipase activation during ischemia, but contends that it is not necessary to invoke excitotoxic amino acids to explain the overloading because energy depletion and depolarization themselves will result in Ca^{2+} entry into the cells through voltage-dependent calcium channels, as well as down the 10,000/1 gradient of extracellular/intracellular Ca^{2+} normally maintained by energy-dependent ionic pumps. According to the free-radical theory, the polyunsaturated free fatty acids, especially arachidonic acid, released during ischemia are metabolized by oxidation during reperfusion; in these reactions the oxygen (O_2) is only initially reduced by one electron to form the free-radical superoxide ($\cdot O_2^-$). The excess $\cdot O_2^-$ formed from these reactions reduces insoluble iron ions (Fe^{3+}) normally found in iron storage proteins, such as ferritin, to soluble iron ions (Fe^{2+}). These soluble ions are then available to participate in the formation of other partially reduced oxygen species that are very powerful oxidizers, such as hydroxyl radical ($\cdot OH$). These powerful oxidizers damage the cells' membranes through a set of reactions called lipid peroxidation. In this theory, selective vulnerability is explained by exceptionally high concentrations of iron and the absence of an enzyme, glutathione peroxidase, that efficiently completes the reduction of partially reduced oxygen species without allowing the formation of chemical species such as $\cdot OH$. Evidence in support of this theory includes identification of spin-trapped radicals formed during reperfusion, demonstration of lipid peroxidation products formed during reperfusion, and fluorescence microscopic demonstration that the concentration of lipid peroxidation products is especially high in the selectively vulnerable neurons during reperfusion.

Current therapy and neuronal repair. In spite of the growing understanding of the mechanisms of neuronal damage and death accompanying brain ischemia and reperfusion, effective therapy has remained elusive. Laboratory and clinical trials of drugs that block voltage-dependent calcium channels have shown some (but not dramatic) evidence of neuronal sparing and improved neurologic outcome. Drugs that block glutamate receptors have been largely ineffective in neuronal preservation after complete ischemia, although they appear somewhat more promising in neuronal salvage in the penumbral zone surrounding the area of severe ischemia in stroke models. The effectiveness of glutamate receptor blockers in the penumbral zone is thought to be due to the ongoing occurrence of periodic depolarization in the penumbra. Drugs aimed at inhibiting radical reactions either by inactivating the iron or by directly inactivating the oxygen radicals do appear to reduce biochemical evidence of cellular damage during postischemic reperfusion but show only marginal effects in the preservation of neuronal viability and brain function.

Because of these findings, scientists have begun to focus additional attention on the capacity of neurons to repair membrane damage. Brain neurons are terminally differentiated, that is, they do not reproduce; and the neurons present shortly after birth remain throughout life. Thus, they do not replicate their deoxyribonucleic acid (DNA). Even though there is evidence against significant nuclear or mitochondrial DNA damage during brain ischemia and early reperfusion, the permanent shutdown of DNA replication may be of great importance. Evidence indicates that shutdown is associated with a substantial reduction in the rate of synthesis of new lipids, which would presumably be needed to repair damage to the lipid components of the cell membranes. In this regard, it is interesting that cancer cells, which have their DNA replication machinery inappropriately turned on, are resistant to being killed by oxygen radicals, and in cultured cell lines more differentiated cells are more easily killed by exposure to oxygen radicals.

Both cell replication and the process of differentiation are controlled at the level of the cellular mechanisms governing the cell cycle, and physiologic control of this system is modulated by growth factors, such as insulin, insulinlike growth factor 1 (IGF-1), and nerve growth factor. These growth factors typically bind to protein receptors that span the width of the plasma membrane and have a tyrosine kinase function on the inner surface of the membrane that adds phosphate groups to specific tyrosine amino acids in other proteins. These intracellular tyrosine-phosphorylated proteins exert important regulatory effects on the production and activity of the proteins that control the cell cycle. In the brain the selectively vulnerable neurons have elevated concentrations of receptors for growth factors such as insulin and elevated concentrations of tyrosine phosphorylated proteins in their nuclei. Moreover, there is now evidence that growth factors, such as insulin, nerve growth factor, and IGF-1, protect neurons from dying during postischemic reperfusion. In transgenic mice overexpressing IGF-1, the major effect is an increase in the bulk of myelin, a lipid-rich insulation material arranged to increase conduction velocity in axons. Moreover, insulin is known to stimulate the synthesis of new lipids. Overall, these data suggest that it may be possible to manipulate the machinery for neuronal repair after damage by ischemia and reperfusion.

Although substantial experimental work remains to be done, there is reason to hope that a therapeutic combination that both inhibits the damaging reactions and stimulates immediate neuronal repair processes may soon allow physicians to greatly reduce the extent of brain damage in brain ischemia and reperfusion.

For background information SEE BIOPOTENTIALS AND IONIC CURRENTS; CIRCULATION DISORDERS; HEART DISORDERS; VASCULAR DISORDERS in the McGraw-Hill Encyclopedia of Science & Technology.

Blaine C. White

Bibliography. B. K. Siesjö, Pathophysiology and treatment of focal cerebral ischemia, *J. Neurosurg.*, 77:169–184, 1992; B. C. White, L. I. Grossman, and G. S. Krause, Brain injury by global ischemia and reperfusion: A theoretical perspective on membrane damage and repair, *Neurology*, in press, 1993.

Land plants, origin of

The evolution of land plants (embryophytes) was an extraordinary and far-reaching event in Earth history. Land plants are the basic structural components of terrestrial ecosystems, the primary food source for land animals, and important modulators of climate, atmosphere, and soil chemistry. Land plants are diverse, comprising an estimated 300,000 extant species, but all groups share certain features that strongly suggest a common origin.

Current estimates place the origin of land plants in the Late Ordovician or Early Silurian [450–430 million years ago (m.y.a.)]. The present angiosperm-dominated flora (over 80% of extant species) is the most recent phase of a long history of terrestrial plants, most of which has been dominated by non-flowering groups, many known only through the fossil record. Direct evidence on the nature and origin of vascular plant groups is available from excellent Late Silurian and Devonian macrofossils. The microfossil (spore) record provides additional information on earlier stages of land plant colonization and periods of major floral change.

In addition to information from the fossil record, a better comprehension of phylogenetic relationships in extant groups through the application of microscopic, chemical, and molecular techniques has led to significant advances in understanding land plant origins. Much emphasis is placed now on data analysis through important new developments in systematic theory.

Potential ancestors. Recent work on the origin of land plants has focused on the biology of extant green algae and the relationships between major groups of bryophytes and vascular plants. The macrofossil record has provided information on early vascular plants and their immediate antecedents. Because few macrofossils of early bryophytes are known and there are no unequivocal fossil intermediates between green algae and embryophytes, the nature of early members of these groups has been inferred from the comparative biology of extant species.

A green algal origin for land plants is universally accepted and is based on numerous similarities, including pigments (certain types of chlorophyll *b*), biochemistry, and details of chloroplast and male gamete flagella ultrastructure. Recent advances in understanding green algae systematics strongly suggest that certain members of the charophycean algae (Coleochaetales or Charales) are the living groups most closely related to land plants (see **illus**.). The fossil record is consistent with these general conclusions: The record of green algae considerably predates the earliest evidence of land plants. Furthermore, unequivocal charophycean algae are present in sediments contemporaneous with the earliest land plant macrofossils. A monophyletic origin of land plants is strongly supported by the presence of a similar suite of derived features in all embryophytes.

Relationships between major groups of bryophytes and vascular plants are the subject of much debate. Information from extant and fossil groups supports a monophyletic origin of vascular plants and suggests an early divergence of lycopods and related extinct groups from other major vascular plant lineages. The excellent fossil record of early vascular plants shows that some of these were extremely simple organisms with leafless, dichotomously branching axes. Many morphological and anatomical features appear to be intermediate between those of modern vascular plants and those of nonvascular groups such as mosses. The relationship between bryophytes and vascular plants is distant, but the most convincing hypothesis interprets mosses or hornworts as more closely related to vascular plants than to liverworts.

Structural modifications. A number of important structural modifications related to water economy and transport, support, and dispersal characterize terrestrial plants. Microspores with sporopollenin in the cell wall are characteristic of all land plants and are the primary means of dispersal and colonization in homosporous groups. The structure and chemical composition of the cell wall make microspores resistant to desiccation and ultraviolet light. In addition to their small size and production in large numbers, this feature makes spores ideal units for aerial dispersal.

Water loss and gaseous exchange in the sporophytes of most extant land plants except liverworts are regulated by stomata that control pores in the outer cuticular layer. Cuticles with the outlines of distinctive paired guard cells are known from Late Silurian sediments and are indistinguishable from those in many modern land plants.

Most land plants develop a vascular system for water transport that usually consists of a centrally located, vertically connected system of elongate cells. In many bryophytes and all vascular plants, these cells lose their protoplast, resulting in cell death and the improvement of water flow. Cells of this type in bryophytes (hydroids) are relatively simple, but in vascular plants lignification and distinctive patterns of thickening on the inner cell wall also provide considerable structural

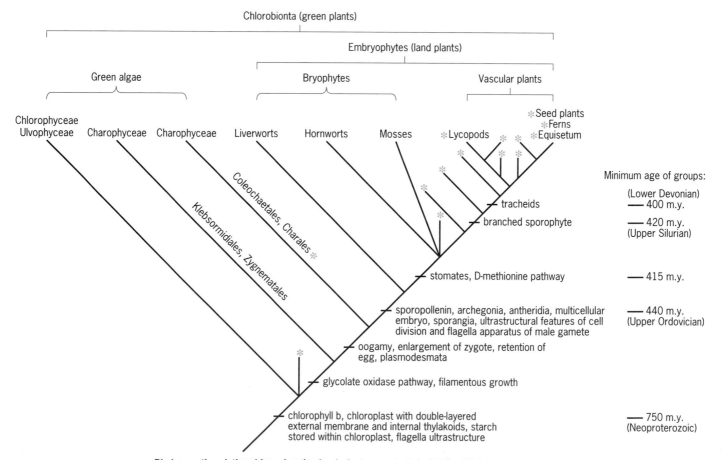

Phylogenetic relationships of major land plant groups and closely related green algae showing the position (*) of known extinct groups (unnamed) and extant groups with a well-known fossil record.

support for the cell (tracheid). The evolution of vascular tissue can be traced through the fossil record, and recent work on early fossils has documented a variety of cell types that are morphologically intermediate between the tracheids and hydroids of extant plant groups.

Gametangia of strikingly similar morphology and function characterize all land plants except the most derived seed plant groups. The archegonium protects the egg from desiccation while allowing fertilization through the passage of sperm down the neck canal. Remarkable details from silicified Early Devonian nonvascular plants indicate that archegonia and antheridia were similar in overall morphology to those in many modern archegoniates.

Earliest terrestrial evidence. The earliest evidence of land plants comes from dispersed spores in sediments of Middle Ordovician age (460–450 m.y.a.). The earliest spores are obligate tetrads, dyads, or monads, which are often enclosed by a delicate membrane. Single trilete spores first appear unequivocally in the Early Silurian (440–430 m.y.a.). Often associated with these dispersed spores are fragments of cuticle with well-delineated cellular patterns and pores, and wefts of microscopic tubes that may be smooth-walled or ornamented and simple or branched. The nature of the organisms producing the cuticles and tubes is still obscure, but the spores may have been produced by land plants at the bryophyte level of organization. Recent work on Ordovician and Silurian spores suggests an increase in diversity of single trilete spores in the Late Silurian and Early Devonian with a concomitant decrease in the earlier obligate tetrads, suggesting a significant change in floral composition during this time.

The earliest unequivocal macrofossil evidence of land plants comes from sediments of Middle Silurian age (430–420 m.y.a.), and these plants appear to be more closely related to vascular plants than to any group of bryophytes.

Life history strategies. All embryophytes have an alternation of generations between multicellular sexual (gametophyte) and asexual (sporophyte) stages. The morphological complexity of gametophytes and sporophytes varies considerably between major embryophyte groups; however, despite this variability certain essential features are consistently present. In all but the most advanced seed plant groups, gametophytes develop archegonia and antheridia; similarly, in all land plant groups, sporophytes produce spores in a sporangium.

Two theories have been proposed to explain the origin of the life cycle in land plants. The antithetic theory proposes that land plants evolved from haplobiontic green algae with a multicellular gametophyte and a unicellular sporophyte. This hypothesis envisages the origin of the sporophyte generation by delay of meiosis and interpolation of somatic cell division in the zygote. The homologous theory proposes that the green algal ancestor had identical multicellular (isomorphic) gametophyte and sporophyte generations. According to this scenario, the major features of the sporophyte

generation were already present in the aquatic ancestor and were subsequently modified by reduction in bryophytes and further elaboration in vascular plants.

Recent phylogenetic studies favor the antithetic theory because the extant green algae most closely related to land plants have haplobiontic life cycles, and the simplicity of the sporophyte generation in bryophytes argues against reduction from more complex types.

Ecology and reproduction. Prior to the extensive development of land plants, terrestrial habitats were very different from modern equivalents, particularly with respect to resources and physical disturbance. In modern terrestrial environments, vascular plants in particular strongly influence the weathering and hydrologic regimes and therefore the patterns of erosion, deposition, water storage, and release of mineral nutrients. They control microclimates, affect macroclimates, and are the main food source for land animals.

One attractive model for the development of terrestrial ecosystems envisions a shift in dominant ecological strategy from resource conservation and mortality replacement in the early stages of land colonization to resource acquisition and competition with the evolution of larger vascular plants. In early terrestrial ecosystems, disturbance was an important factor, favoring organisms with high reproductive capacity (r strategists). However, shortage of critical resources such as mineral nutrients and water would have favored an opposing strategy of resource conservation (k strategists). In other words, early colonizers evolved a strategy intermediate between extreme r and k strategists and similar to that in some extant bryophytes.

The amelioration of terrestrial habitats brought about by the presence of a land flora at the bryophyte level of organization would have led to a slow shift in emphasis away from conservation and replacement strategies to acquisition and competitive preemption of resources. This shift in ecological strategy can be traced through the marked increase in disseminule (microspore, megaspore, seed) size from Silurian through Carboniferous times. Such changes ultimately led to the appearance of a plant canopy, a thick and widespread litter zone, and the development of extensive root and foliage systems. The net effect on the terrestrial ecosystem was an elaboration and differentiation of trophic structure and an increase in productivity at all trophic levels.

For background information SEE EMBRYOBIONTA; PLANT KINGDOM; THALLOBIONTA in the McGraw-Hill Encyclopedia of Science & Technology.

Paul Kenrick

Bibliography. J. Gray and W. Shear, Early life on land, *Amer. Scient.*, 80:444–456, 1992; P. Kenrick and P. R. Crane, Water-conducting cells in early fossil land plants: Implications for the early evolution of tracheophytes, *Bot. Gaz.*, 152:335–356, 1991; T. N. Taylor, The origin of land plants: Some answers, more questions, *Taxon*, 37:805–833, 1988; B. H. Tiffney (ed.), *Geological Factors and the Evolution of Plants*, 1985.

Landscape architecture

Recent developments in landscape architecture in the United States have been marked by growth in the profession and by a renewed focus on the urban habitat. According to the American Society of Landscape Architects, the 1992 estimate of the number of landscape architects in the United States is 34,000, up from 19,000 in 1982. In the United States, most landscape architects practice in metropolitan areas, partly as a reflection of nationwide demography. (The population grew 9.8% from 1980 to 1990, with 90% of this growth occurring in metropolitan areas.) Growth in the profession and in the concern with urban matters, however, can more specifically be traced to recent patterns in the evolution of American cities.

Urban landscape planning. The decades following World War II are commonly regarded as an era of conspicuous consumption fueled by seemingly inexhaustible resources. Cheap energy, the vast areas of undeveloped land, and an abundance of water seemed to make husbanding natural resources unnecessary. Landscape architects trained in land-use planning with an eye to resource conservation were, in many cases, relegated to cosmetic applications such as designing false lakes in deserts or artificial waterfalls in shopping malls.

Eventually, the problems of unchecked development—shapeless urban sprawl, environmental degradation, waste accumulation, abandonment of neighborhoods, resource exhaustion, and traffic gridlock—gave rise to a growing demand for responsible urban planning. Urban design teams today typically include landscape architects along with urban planners, architects, and engineers, as well as sociologists, ecologists, and geographers.

The landscape architect on the urban design team is concerned with all matters that involve land or natural resources, such as building coverage, transportation, infrastructure, and waste management. However, the primary focus is on public outdoor spaces in parks, waterfronts, greenways, and common open space. These areas tend to be ecologically sensitive because they include wetlands, streams, shorelines, steep slopes, and woodlands. Often they are under pressure from competing uses such as recreation, sports, and entertainment, or under stress from surrounding or encroaching development.

Landscape architects view such spaces as parts of larger ecosystems that are tampered with at a community's peril. Disturbed watersheds can result in floods, altered stream beds, erosion, and mud flows. Mismanaged vegetation can result in brush fires; unstable slopes can slide; declining wetlands can lead to water pollution; overdrawing from aquifers can cause contamination of groundwater and ground subsidence. Land development on barrier beaches can expose coastal communities more to storm damage. When such systems are destabilized, a regional chain reaction of damage can ensue. In metropolitan areas, the damage will be particularly severe because of the density of settlement. The landscape architect can help ensure the ecological soundness of these systems through planning and management.

Concepts and methods. The challenge of urban landscape architecture is to balance the ecological aspects with the sociological, esthetic, and infrastructure needs. In pursuing the objective of an attractive, ecologically sound, and user-friendly habitat, urban landscape architects are guided by a number of factors. These factors include the ecological context, ecoregionalism, mixed-use development, and true-cost budgeting.

Ecological context. A city's life support system is its natural environment. The artificial infrastructure must have a benign effect on its host landscape, as the former is dependent on the ecological soundness and sustainability of the latter.

Ecoregionalism. Property lines and the limits of civil land divisions do not always coincide with natural boundaries. Urban development must be viewed in a regional context, regardless of borders, in order to preserve the integrity of ecosystems.

Mixed-use development. Separation of uses defeats the purpose of a city. By mixing compatible uses, communities become viable organisms in balance with their natural surroundings. When a variety of life's activities can take place within a readily traversible perimeter, there are savings in time, energy, and infrastructure capital that would otherwise be devoted to duplication and transport.

True-cost budgeting. True-cost budgeting takes into account the costs of a project's long-term side effects. In addition to costs of development and maintenance, it includes costs of factors such as environmental impact, energy used, and nonrenewable resources consumed.

Sustainable ecological solutions. As professionals with broadly based training ranging from the study of ecological principles to construction technology and the esthetics of design, landscape architects find themselves acting as catalysts on the urban team. While they are conversant with the different vocabularies of planners, engineers, architects, and ecologists, their approach to solving problems is unique.

Landscape architects typically favor sustainable ecological solutions, that is, solutions involving natural processes, over so-called hard solutions. For example, instead of breaking slopes with retaining walls, they favor contouring them to sustainable gradients. Embankments are stabilized by self-sustaining vegetative covers instead of by impervious armor. Instead of forcing water through culverts, streams are maintained in self-stabilizing beds with overflow areas to receive flood water. Erosion control is accomplished through runoff management and velocity control instead of through paving or channelization.

Landscape architects are involved in the reclamation of derelict land in abandoned inner-city zones. Reclamation typically begins with the removal of unsafe building remnants, impervious surfaces, and toxic waste. Materials suitable for recycling are salvaged. Composted organic amendments are introduced into the soil, and low-maintenance, soil-enhancing vegetation (for example, grasses or field legumes) that is sufficiently vigorous to suppress weed emergence is established. The objective is to return the area to ecological health pending disposition in the context of a city plan.

Other declining open spaces, such as neglected urban greenways, are returned to a viable state by reintroducing beneficial biota that may have been wiped out in past indiscriminate applications of control chemicals, such as nonselective pesticides or herbicides, or by toxic discharges, such as from unprotected waste dumps. Soil microorganisms and insect life, for example, form links in the ecological chain and are at risk in urban environments. These little-noticed organisms form the foundation upon which rest the more obvious building blocks of the natural and human-made environment.

With biotechnology, new ecosystems can be created to augment or replace some of the energy-intensive mechanical components of an aging urban infrastructure. For example, landscape architects are developing new wetlands, in suitable terrain, for the biological cleansing of waste water. The high-rate nutrient-consuming capability of selected aquatic vegetation, in combination with soil filtration and microbial action, is harnessed to extract impurities from wastewater that is passed through the wetlands at controlled rates.

New technology. While many landscape architectural techniques are based on age-old principles of managing nature for human benefit, a growing arsenal of new technology and methods is available. These innovations include databases, overlay mapping, ecozoning, and computerization.

Databases. Landscape architects take advantage of a growing computer-accessible database of worldwide information vital to the planning and design process. An example is up-to-date geographical data on population, land use, vegetation, and climate change.

Overlay mapping. Overlay map systems consist of a base map and a series of matched transparent overlays. Each overlay shows one of a number of conditions or resources, such as soil type, vegetation type, and slope gradient. Landscape architects can assess conditions in their study area at a glance by viewing selected combinations of overlays on the base map.

Ecozoning. Ecologically responsive zoning entails regulation of land development based on an area's carrying capacity and natural constraints. The latter includes, among others, topographical, hydrological, and climatological factors. The purpose of this type of regulation is to protect sensitive land from overdevelopment.

Computerization. The hardware and a growing list of software for computer modeling, computer-aided design, computer imaging, animation, and virtual reality systems are being increasingly employed in pursuit of the landscape architect's objectives.

Water resources. Responsible use of natural resources is essential to a city's health and survival. Water is a prime example. Accelerated urban expansion since World War II has caused demand to surpass yield in many areas. Landscape architects exert a marked impact on the planning of urban water systems with potentially high water requirements. Cities in arid climates usually bring in water from afar at considerable expense. Ironically, people in such cities can develop a taste for landscapes alien to the climate, with lawns and plantings that require a large amount of water. Urban streetscapes, parks, golf courses, and office parks are heavy water consumers even in areas where the resource is scarce. In Los Angeles, for example, it is estimated that as much as half the water consumed for domestic purposes is used to sustain home lawns and other ornamental vegetation.

A number of techniques are used by landscape architects to limit the quantities of water used for maintaining vegetation. These techniques include adaptive planting; water-use grouping; delivery targeting; drip-and-soak irrigation; water recycling; use of mulches, wetting agents, and antidessicant coatings; and proportional metering.

Adaptive planning. Adapting a design to the natural vegetation-supporting capability of a site is a traditional method of limiting water consumption to natural availability. Sources of water, climate, soil type, topography, wind exposure, and insulation are taken into account in selecting plant material that is naturally adapted to a particular site. This technique ensures a climate-specific planting. The planting can further be limited to autochthons to make a site regionally compatible.

Water-use grouping. Where landscape objectives cannot be met without supplementary irrigation, plantings can be arranged in groups requiring similar water requirements. This approach avoids overwatering and permits effective zoning of the water delivery system.

Delivery targeting. Where irrigation must be provided, a system is used that will deliver no more water than the amount required by the specified plants and that will deliver water only to these plants and not to the surrounding land. Although this type of irrigation system may be more costly to install than one that provides indiscriminate coverage, there are savings in the reduced quantity of water used.

Drip-and-soak irrigation. This method of irrigation, an example of delivery targeting, delivers metered amounts of water through low-rate emitters directly to the plant. This method avoids water loss due to evaporation and to indiscriminate watering such as in conventional spray systems.

Water recycling. By using recycled water, freshwater consumption can be reduced. Typical sources of recycled water for landscape irrigation are treated wastewater, reclaimed process water, and so-called once-through industrial cooling water. In addition, storm runoff detained in flood control impoundments is a potential source of irrigation water that takes advantage of a surplus and will not deplete a limited source.

Mulches and soil amendments. Organic or synthetic mulch materials (for example, sawmill or agricultural by-products such as shredded bark or crop husks, or synthetics such as plastic sheets) are spread over the soil around and underneath plants to reduce evaporation loss from the soil surface. Soil amendments (for example, composted organic waste or by-products of public works, such as treated sewage sludge) are mixed into soils to enhance their water-retaining capability. Water that would otherwise be lost to evaporation or percolation is thus held longer in the root zone and becomes available for use by plants.

Wetting agents. Wetting agents are water-soluble liquid surfactants that, when added to irrigation water or the soil, reduce surface tension, making it easier for plants to draw water from the soil.

Antidesiccant coatings. Antidesiccants (antitranspirants) are temporary liquid coatings that are sprayed onto plants to reduce water loss through transpiration.

Proportional metering. Computer-controlled metering allows irrigation systems to respond to signals from soil moisture levels and from wind sensors installed near the point of water demand. These systems ensure that water is delivered strictly in proportion to need.

Challenges. The concept of living in harmony with nature while occupying close quarters is as old as the world's earliest cities. The metropolitan areas of the United States—home to some 77.5% of the nation's population—do not, however, always fulfill the vision of great vibrant cities in serene landscapes.

The isolation of urban society from the natural experience diminishes popular understanding of ecological interrelationships and can leave this population open to quick fixes that may result in loss of resources and irreversible damage to the environment. Overcoming indifference to nature, therefore, remains one of the greatest challenges for urban landscape architects.

For background information SEE ECOSYSTEM; LAND-USE PLANNING; LANDSCAPE ARCHITECTURE; WATER CONSERVATION in the McGraw-Hill Encyclopedia of Science & Technology.

Maurice Wrangell

Bibliography. R. B. Gratz, *The Living City*, 1989; P. Langdon and K. McCormick, Reinventing the city, *Landscape Architect.*, 82(5):44–55, May 1992; J. D. Taylor, Take back the water, *Landscape Architect.*, 82(5):52–55, May 1992; R. Walter, L. Arkin, and R. Crenshaw, *Sustainable Cities*, 1992.

Leaf

The anatomy of leaves reflects their most important functions: performing photosynthesis while reducing water loss. Leaves are typically thin, allowing light to penetrate to the photosynthetic mesophyll cells. Stomata open in daylight to admit carbon dioxide (CO_2) for photosynthesis but close at night to reduce evaporation of water. Small veins form a network that both imports the water and dissolved minerals absorbed by the roots and exports the products of photosynthesis to other parts of the plant.

In most plant species, all the photosynthetic cells of the mesophyll are biochemically equivalent and perform C_3 photosynthesis; that is, their first stable product of photosynthesis is a three-carbon compound. Bundle sheath cells, a type of parenchyma cell, surround the veins. Although bundle sheath cells transport water and solutes between mesophyll and vascular tissue, they have small chloroplasts and contribute little to the total photosynthetic activity of the leaf.

All mesophyll cells contain large amounts of RuBP carboxylase, the enzyme that catalyzes the carboxylation of ribulose bisphosphate (RuBP), a five-carbon sugar. RuBp carboxylase also catalyzes the oxidation of the RuBP in a process known as photorespiration. Photorespiration handicaps C_3 plants growing under hot dry conditions. If drought leads to the closing of stomata during the day, the combination of low CO_2 concentration in the intercellular space of the leaf mesophyll and high temperature favors photorespiration, resulting in loss of up to 50% of photosynthetic products.

C_4 photosynthesis. C_4 photosynthesis has evolved as an adaptation to prevent the loss of fixed carbon through photorespiration. In C_4 plants, photosynthesis is a two-step process that requires two types of cells: mesophyll and bundle sheath cells. The first step occurs in mesophyll cells, where CO_2 is combined with the three-carbon phosphoenolpyruvate (PEP) to form a four-carbon compound. These four-carbon compounds diffuse rapidly to adjacent bundle sheath cells, where they are decarboxylated, releasing CO_2 and a three-carbon compound that diffuses back to the mesophyll tissue. The released CO_2 enters the Calvin-Benson cycle via the carboxylase reaction of RuBP carboxylase. This process concentrates CO_2 within the bundle sheath cells, thus boosting the carboxylase reaction of RuBP carboxylase and minimizing photorespiration. Therefore, C_4 photosynthesis loses no carbon through photorespiration. Also, phosphoenolpyruvate carboxylase in C_4 mesophyll cells has a higher affinity for CO_2 than does RuBP carboxylase, thereby allowing carbon fixation to occur at low concentrations of CO_2 in the intercellular space. Thus,

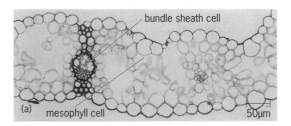

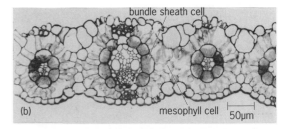

Fig. 1. Cross sections of grass leaves. (*a*) *Bromus tectorum*, a C_3 species. (*b*) *Panicum effusum*, a C_4 species.

photosynthesis occurs throughout the daylight hours, even when the stomata are partly closed to reduce water loss.

C$_4$ leaf anatomy. The C$_4$ pathway of photosynthesis depends on leaf anatomy. Typically, the mesophyll cells have a radiate arrangement around the leaf veins, and almost every mesophyll cell contacts a bundle sheath cell (**Fig. 1**), minimizing the path length for diffusion of four-carbon and three-carbon compounds between the two cell types. The physiological requirement for proximity of mesophyll and bundle sheath cells also demands the close spacing of the leaf veins, a feature that distinguishes the leaf anatomy of C$_4$ species from that of C$_3$ species (Fig. 1). Also, the mesophyll of C$_4$ species has less intercellular space, a feature that may reduce the amount of water evaporating from the moist cell walls of the mesophyll cells while providing enough surface area for the uptake of CO$_2$ into the mesophyll cells.

Although the individual mesophyll cells in C$_4$ plants resemble those of C$_3$ plants, C$_4$ bundle sheath cells are often highly specialized. They are larger than their C$_3$ counterparts, and a greater portion of cell volume is occupied by chloroplasts (**Fig. 2**). Bundle-sheath-cell chloroplasts are large and usually contain starch, unlike C$_4$ mesophyll cell chloroplasts. In the variant of C$_4$ photosynthesis in which the bundle-sheath-cell decarboxylating enzyme is NADP-malic enzyme, the bundle sheath chloroplasts have rudimentary grana, consisting of only a few appressed thylakoids, a feature associated with the lack of photosystem II activity (Fig. 2a).

In many C$_4$ species, the cell wall of the bundle sheath cells is modified by a suberin lamella, a hydrophobic layer similar to that found in cork (Fig. 2). The suberin lamella prevents the leakage of dissolved CO$_2$ from the bundle sheath cells to the intercellular space of the mesophylls, thus contributing to the high concentration of CO$_2$ near the RuBP carboxylase. In C$_4$ species with this cell wall modification, the bundle sheath chloroplasts are located toward the outer wall of the bundle sheath cells, thus minimizing the path length for diffusion between the mesophyll cell and bundle-sheath-cell chloroplasts (Fig. 2a). In those C$_4$ species that lack a suberin lamella, the chloroplasts and mitochondria are located toward the inner side of the bundle sheath cells, a feature that provides the maximum opportunity to recover dissolved CO$_2$ before it leaks into the intercellular space (Fig. 1b).

The adjoining walls of the mesophyll and bundle sheath cells are further modified by the presence of many cytoplasmic channels, called plasmodesmata (Fig. 2b). The density of plasmodesmata is greater between these two cell types than between two mesophyll cells or between bundle sheath cells and cells of the adjacent vascular tissue, indicating that four- and three-carbon compounds move between these two cell types through plasmodesmata rather than across adjoining cell walls. When a suberin lamella is present, it is often thickened around the plasmodesmata, suggesting that it also prevents the leakage of metabolites from the plasmodesmata to the cell walls.

Enzymatic compartmentalization in C$_4$ plants. In addition to the striking structural differences between mesophyll and bundle sheath cells, the two cell types have different complements of photosynthetic enzymes. For example, the mesophyll cells synthesize phosphoenolpyruvate carboxylase and other enzymes required for the formation of the four-carbon acids, while bundle sheath cells synthesize the enzymes required for the decarboxylation of these acids and for the operation of the Calvin-Benson cycle. Enzymatic compartmentalization in mesophyll and bundle sheath cells has been shown by experiments that separate the two cell types and then assay photosynthetic enzyme activity in the two cell preparations. Similar results have been obtained by immunocytochemical methods in which antibodies to enzymes such as RuBP carboxylase are raised in rabbits. The antibodies are concentrated from rabbit blood, labeled with a detection system, and exposed to thin sections of leaf tissue. The labeled antibodies bind with corresponding antigens within the leaf cells, and can be used to show that certain photosynthetic enzymes occur within only one cell type.

Differentiation of cell types. The division of labor between mesophyll and bundle sheath cells provides a good system in which to study the processes of cell differentiation. Although differences occur between mesophyll and bundle sheath cells when leaves are small (0.5 cm or 0.2 in. long), most changes accompanying cell maturation occur as the leaves expand from the apical bud into direct light. Chloroplasts of both the mesophyll and bundle sheath cells have grana when leaves are immature, but only the bundle-sheath-cell chloroplasts lose their grana as the leaves develop. In immature leaves, both cell types synthesize RuBP carboxylase, while in mature leaves only the bundle sheath cells make this enzyme. This pattern of enzyme accumulation results from cell-specific gene expression, as shown by in situ hybridization. In this technique, nucleic acid probes are made that hybridize to specific messenger ribonucleic acids (mRNAs) within tissue sections, making it possible to determine the cell type in which certain genes are being expressed. In corn, the genes for RuBP carboxylase are expressed in both mesophyll and bundle sheath cells in young

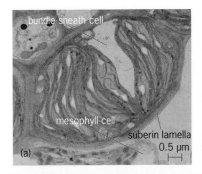

bundle sheath cell

mesophyll cell

suberin lamella
0.5 μm

(a)

suberin lamella

plasmodesma

0.5 μm

(b)

Fig. 2. Transmission electron micrographs of leaves of the C$_4$ grass *Setaria viridis*. (a) Bundle sheath cell showing wall modified by suberin lamella. (b) Cell wall between bundle sheath and mesophyll cell showing thickened suberin lamella at plasmodesmata.

leaves. As the leaf matures, mesophyll cells stop making mRNA for RuBP carboxylase, so that in mature leaves the product of this gene occurs only in bundle sheath cells. At the same time, the mesophyll cells start to express the genes for phosphoenolpyruvate carboxylase.

At least two different factors regulate this cell-specific gene expression: cell position within the leaf and exposure to light. As leaves develop, the immature veins provide positional signals to guide the cell-specific expression of C_4 genes. Evidence to support this idea comes from observations of the husk leaves that surround developing corn cobs. Veins of husk leaves are widely spaced in comparison to the more typical foliage leaves. As the husk leaves develop, only mesophyll cells surrounding the veins differentiate as C_4 mesophyll; the more distant mesophyll cells have features of C_3 mesophyll. Also, if foliage leaves and husk leaves develop under low light conditions, mesophyll and bundle sheath cells are C_3-like. Only after exposure to high light do mesophyll and bundle sheath cells develop their characteristic anatomical and biochemical features.

For background information SEE LEAF; PHOTOSYN-THESIS in the McGraw-Hill Encyclopedia of Science & Technology.

Nancy G. Dengler

Bibliography. N. G. Dengler, R. E. Dengler, and P. W. Hattersley, Comparative bundle sheath and mesophyll differentiation in the leaves of the C_4 grasses *Panicum effusum* and *P. bulbosum, Amer. J. Bot.*, 73:1431–1442, 1986; M. D. Hatch, C_4 photosynthesis: A unique blend of modified biochemistry, anatomy and ultrastructure, *Biochim. Biophys. Acta*, 895:81–106, 1987; T. Nelson and N. G. Dengler, Photosynthetic tissue differentiation in C_4 plants, *Int. J. Plant Sci.*, 153:593, Suppl. 105, 1992; T. Nelson and J. A. Langdale, Developmental genetics of C_4 photosynthesis, *Annu. Rev. Plant Physiol. Plant Mol. Biol.*, 43:25–47, 1992.

Limpet

The majority of grazing herbivores on marine rocky shores are snails, in the phylum Mollusca. These animals eat small algae, such as diatoms, and the young stages of larger algae. This article discusses how one particular snail group, the limpets, forages for algal food, and how recent advances in recording technique have allowed more detailed interpretations of limpet behavior.

With a simple conical shell, limpets comprise a variety of species. Some species breathe through a lung, while others have gills. Some have holes in the top or slits at the edge of the shell through which respiratory water currents are expelled, while others have shells without breaks or apertures. Some are less than 0.4 in. (10 mm) long, while so-called giant species reach 5 in. (130 mm). All, however, feed by using a radula, a long ribbonlike structure covered with teeth. The radula rasps away the algal covering adhering to rock surfaces; and on wave-exposed shores, where limpets are common, the action of the combined radulas of thousands of limpets can be a major factor preventing the growth of seaweeds. Data on limpet foraging are therefore important in understanding the structure of marine intertidal ecosystems.

Feeding tactics. Like most intertidal invertebrate animals, limpets feed not continuously but at specific times. There is a great variety in the times when limpets feed. Some species forage when covered by the tide—the normal pattern for most intertidal animals. Other species feed when the tide is out, usually when the rock surface is still damp. A third group feeds only during the brief period when the water rises or falls over them as the tide floods and ebbs.

Some limpet species feed by foraging apparently at random over the rock surface, but the majority have a home scar, a fixed site to which an individual returns after a foraging trip, often with extraordinary precision. Some species even graze regularly in an area around this home scar, which they defend against other grazers. This territory has a different algal flora from the surrounding rock surface, and is often referred to as a garden.

Feeding studies. One of the species that has been studied intensively is the common limpet of northwest Europe, *Patella vulgata*. This species shows a great variety of feeding patterns and does not seem to have become as stereotyped in its behavior as many other species. It has been observed to feed both under and out of water during the day, and out of water at night. Any one group of individuals seems to maintain a fairly set pattern of activity, so it is assumed that timing is related in some way to selection pressures in the environment—either physical stresses such as desiccation or wave action or biological pressures such as predation or availability of food.

Investigating the timing of foraging trips in an animal that is covered by tides twice a day is not easy. Some workers have used scuba diving, but the frequency and duration of observation by this method is severely limited. The first long-term studies were carried out in Lough Hyne, a marine lough (tidal inlet) in southern Ireland, where tidal rise and fall is limited to about 3 ft (1 m) and it is possible to observe limpet movements by day and night and at high and low water. Even these studies, however, were incomplete because of the difficulties of getting together a team large enough to make continuous observations for a period of weeks.

Automatic recording. Because *P. vulgata* homes precisely to its scar and always rests in the same orientation, it is possible to place a sensor on the rock surface next to the animal to detect the presence of a structure that has been glued to the shell. The most successful technique has involved gluing a magnet to the shell and placing a reed switch with a protective covering on the rock near the end of the magnet. When the magnet is next to the switch, the switch is closed, but as soon as the limpet leaves its scar, the switch opens. The opening and closing of switches can be recorded automatically by connecting them to a data logger or computer at the top of the shore.

Most data loggers and computers have only a small number of input channels, so making recordings from a number of limpets required the design of an appropriate electronic interface to which inputs from a large number of switches could be fed. With the automatic system, recordings at 15-min intervals for many weeks became possible, and this system has been used to record the movement of *P. vulgata*, as well as of some Mediterranean species.

Limpet activity. At Lough Hyne, the tidal amplitude (the total rise and fall of tide) was restricted, thus allowing observation of limpets while the automatic system was working. Because the lough is very close to the sea, it was also possible to use the automatic system on a group of limpets that experienced a normal, high-amplitude tidal regime. Recordings were made over a period of 2 weeks, and allowed a relatively long-term pattern to be followed for the first time.

Four groups were monitored, each with 23 individuals, to see how activity varied with height on the shore as well as with the different tidal regimes inside and outside the lough. Thus, there were high-shore and low-shore groups inside and outside the lough. The activity of low-shore limpets inside the lough is a good starting point for the understanding of how activity changed from day to day (**Fig. 1**).

For the first few days, there was an activity peak in the very early morning, finishing just after dawn. This activity took place out of water, the limpets crawling around on the steep rock faces while they were damp either from the last tide or from falling dew. As spring tides (relatively high-amplitude tides) changed to neap tides (low-amplitude tides), this peak of activity declined, gradually being replaced by another peak in the evening, so that total activity was greater on neap tides than on spring tides. This second peak also occurred while the limpets were out of water.

The activity peak became larger on successive days for about a week and then decreased in size. It did not, however, stay at a fixed time of day but moved forward each day by about 50 min. In this way, the activity peak remained in the period when the limpets were not covered by the tide, even though this period became rather small when the tides changed back from neaps to springs.

In this group of limpets there was very little activity during the day, so that the main feeding excursions were definitely at night. Other groups of limpets, however, do sometimes show activity in the daytime, especially when the rocks are kept damp by spray or wave action.

The pattern of activity in the other three groups of limpets monitored was quite similar to that seen on the low shore inside the lough. The relationship of activity to the spring–neap cycle was different, however, in that at most sites in the other three

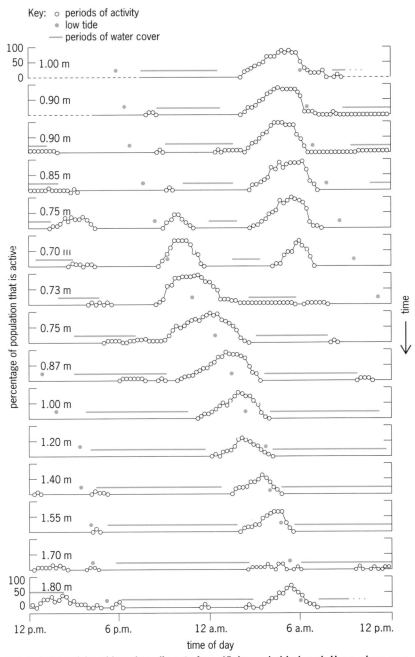

Fig. 1. Daily activity of low-shore limpets for a 15-day period in Lough Hyne, given as a percentage of the population away from or not correctly oriented on the scar. Numbers at left show estimated height of high water: higher numbers are spring tides (above 0.9 m), lower are neaps. (*After C. Little, J. C. Partridge, and L. Teagle, Foraging activity of limpets in normal and abnormal tidal regimes, J. Mar. Biol. Ass. U.K., 71:537–554, 1991*)

groups, activity was maximal at spring tides and minimal at neaps. The low-shore zone within the lough experiences a rather strange tidal cycle, which may explain this difference.

Periodicity of activity. When the records shown in Fig. 1 are subjected to periodicity analysis, three peaks of activity emerge (**Fig. 2**): at 24 h, 12.4 h, and 8.2 h. The 24-h periodicity evidently reflects a diurnal rhythm. The 12.4-h peak is called a circatidal rhythm, related to the periodicity of the tides. The cause of the 8.2-h peak is, as yet, unknown.

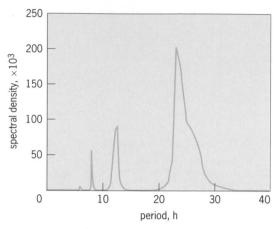

Fig. 2. Activity data for low-shore limpets in Lough Hyne after maximum entropy spectral analysis, which highlights periodicities. (*After C. Little, J. C. Partridge, and L. Teagle, Foraging activity of limpets in normal and abnormal tidal regimes, J. Mar. Biol. Ass. U.K., 71:537–554, 1991*)

For background information SEE LIMPET; MARINE ECOLOGY in the McGraw-Hill Encyclopedia of Science & Technology.

Colin Little

Bibliography. P. Della Santina and G. Chelazzi, Temporal organization of foraging in two Mediterranean limpets, *Patella rustica* L. and *P. coerulea* L., *J. Exp. Mar. Biol. Ecol.*, 153:75–85, 1991; C. Little, Factors governing patterns of foraging activity in littoral marine herbivorous molluscs, *J. Mollusc. Stud.*, 55:273–284, 1989; C. Little, J. C. Partridge, and L. Teagle, Foraging activity of limpets in normal and abnormal tidal regimes, *J. Mar. Biol. Ass. U.K.*, 71:537–554, 1991.

Locomotion (vertebrate)

Recent research has focused on the energetics of locomotion of reptiles and on the design of fish muscles used in swimming.

Locomotor Diversity and Energetics

A large part of the evolutionary success of terrestrial vertebrates can be attributed to locomotor specializations. From a fishlike ancestral condition, tetrapod vertebrates developed a large variety of locomotor appendages and modes, resulting in a diversity nearly as great as that found in the rest of the animal kingdom. Recent study has been directed to the energetics and ecology of vertebrate locomotion, especially to the evolutionary forces that shaped the diversity of existing locomotor types.

The reptile group has played a central role in analyses of locomotor diversity and cost because of its great variety of locomotor methods and structures. Thus, direct comparisons of the energetic and performance effects of various locomotor modifications can be made.

The consequences of the reduction or elimination of limbs constitute one area of interest. Limbs are generally good for running over open ground but can impede locomotion through narrow spaces or in burrows. Frequently, among vertebrates and other groups a wormlike body form has evolved from ancestrally limbed animals.

Direct comparisons of locomotion between lizards and snakes (which are derived from lizards) help in understanding the performance and energy consequences of such evolutionary elimination of limbs. Capacities for burst speed are generally decreased by limblessness. Even though snakes may appear to be very fast, they are generally slower than lizards. A garter snake, for instance, has a peak speed only one-third that of a fence lizard of equal size and body temperature. Endurance, measured as sustainable speed, is approximately the same. However, snakes are far more flexible, because of increased number of body vertebrae. Even within snakes, there appears to be a trade-off between speed and flexibility: constricting snakes can form much tighter coils than other snakes but are not nearly as fast.

Earlier observations suggested that limbless locomotion might be more efficient, that is, less costly in energy expenditure per unit of distance moved. Thus, energy economy might have been an evolutionary factor favoring limblessness. More recent measurements of energy required by a snake to crawl at different speeds on a treadmill found costs to be indistinguishable from those of lizards of the same body size. Indeed, the cost of transport has been found to be a highly uniform function of body size in all terrestrial walkers and runners, including crabs, insects, centipedes, and vertebrates. Hence, limblessness probably evolved not for reasons of increased speed, stamina, or energy economy but for flexibility and locomotion in narrow spaces.

Phylogenetic position. Ancestrally, the reptiles gave rise independently to the birds and mammals, groups characterized by major changes in physiology and morphology, culminating in endothermy. Modern reptiles have been studied as models for the ancestral mammalian and avian conditions, and speculation has centered on the factors responsible for the evolution of endothermy. Certainly, one important component was the thermostability conferred by the high metabolic rates associated with endothermy: body temperatures of mammals and birds are higher and more constant than those of most reptiles. Less obvious are associated increments in work capacity and endurance. A 2.2-lb (1-kg) mammal is capable of sustaining speeds on a treadmill five to ten times those of a similarly sized reptile.

This increased capacity for sustainable work is a direct result of the mammal's greater capacity for maximal oxygen consumption, because the energetic efficiency of locomotion is indistinguishable between the two groups. The greater oxygen consumption capacity of mammals and birds confers the aerobic power to undertake prolonged exertion, such as migration. Reptiles are also capable of intense physical activity, with maximal speeds the same as those of mammals. However, this burst activity is fueled anaerobically and cannot be sustained for more than about 1 min. Thereafter, reptiles rapidly become exhausted and remain fatigued for prolonged periods. During the evolution of birds and mammals, therefore, more than thermoregulatory ability was added. Profound changes also occurred in locomotor stamina and resulting patterns of behavior and activity. These changes may have been important factors, independent of temperature stability, in favoring the evolution of high metabolic rates.

Evolution in mammals and birds greatly modified nearly all aspects of oxygen transport. Lung surface area was increased, and the ventilation apparatus and musculature were considerably modified. Oxygen-carrying capacity of the blood was doubled, and oxygen affinity of the hemoglobin was increased. Cardiovascular structure was greatly modified, including the development of four-chambered hearts with structural separation of oxygenated and deoxygenated blood. The proportional size of organs with high metabolic rates was increased; and the density of mitochondria, the cell organelles in which oxygen is utilized, within these organs was substantially increased. Such a great variety of changes could not have occurred simultaneously, and the evolution of the higher metabolic rates associated with endothermy must have been a long and slow process.

Effect of natural selection and evolution. In addition to retrospective studies on major evolutionary transitions, reptiles are used in studies on the operation of natural selection and evolution on locomotor ability in natural populations. Such studies ask what types of locomotor performance, such as burst speed or endurance, are important to animals in the wild, and how natural selection affects locomotor performance. Preliminary to these measurements is the determination of the susceptibility of the trait to selection.

Three attributes are required for selection to influence distribution of a character: (1) The trait must vary among individuals. (2) This variability must be stable. (3) The trait must be heritable; that is, it must have a genetic basis, and offspring must resemble parents in regard to the trait.

Locomotor performance in reptiles, both speed and endurance, possesses these attributes. For example, among newborn garter snakes from a single population, some individuals are six times faster than others and some have 100 times the endurance (measured as sustained crawling on a treadmill). Under natural conditions, these individual differences are maintained not only from day to day but also over the course of an entire year. Further, differences in performance capacity are heritable. About half of the variability among individuals can be accounted for by genetic effects. For example, a litter of newborn snakes tends to be either fast or slow and to have very distinctive aggressive or defensive behaviors.

Thus, the potential for selection on locomotor capacities is apparently present in reptiles. Studies on garter snakes have found that maximum crawling speed is apparently the most important locomotor mode favored over natural selection. Snakes that are relatively fast have a higher survivorship in their natural populations. These determinations are made by releasing a large number of animals of measured maximal speed and noting which ones survive into the next year. Apparently, the benefits of speed increase with age, because the differential survival becomes easier to detect as the snakes grow older. Differences in endurance capacity are not detectably important in enhancing survival.

The evolution of activity capacities and the behaviors that rely on them continue to be an active area of investigation and speculation. The structural and functional capacities of the vertebrates, in particular, have been greatly influenced by selection for high levels of performance and activity. The earliest vertebrates, the ostracoderm fishes, were actively swimming predators with metabolic and physiological patterns similar to those of modern ectotherms. Because of the low transport costs (energy per distance traveled) involved in swimming, these physiological capacities were adequate for sustained activity in water. Upon emergence onto land, however, the low capacities for oxygen transport and utilization in the primitive tetrapod vertebrates limited the capacity for sustained activity. This condition is evident in modern reptiles, from which much present knowledge has been derived. Only during the evolution of the birds and mammals was sustained oxygen consumption and hence endurance activity again possible in the vertebrate group.   *Albert F. Bennett*

Muscle Design

Animals locomote in a variety of ways, such as flying, jumping, running, and swimming. Different external attributes—wings, legs, and fins—have evolved to perform these activities, and muscle has adapted to power different movements.

Characteristics of muscle. To power locomotion, muscle acts as a linear motor; that is, it shortens and lengthens and generates force along its length. Muscle has particular characteristics, and thus a given muscle tends to work best under a limited set of movements. Muscle is remarkable because its characteristics can vary greatly. Each variant is considered a separate muscle-fiber type. The fact that muscle varies more than any other tissue in the body suggests that it is essential to have the right muscle for a given activity.

Muscle can generate large forces while being held at a constant length (isometric), and it produces smaller forces while shortening. The force generated by a muscle decreases as shortening velocity increases. The mechanical power generated by the muscle (force $\times$ velocity) is extremely important for an activity such as swimming, because mechanical power is necessary to overcome drag while moving through the water. Mechanical power reaches a maximum value at a V/V_{max} of about 0.2 to 0.4 (where V is shortening velocity and V_{max} is the maximum velocity at which the muscle is capable of shortening). Although force generated during active lengthening is larger than during shortening, active lengthening is generally avoided during swimming. The lengthening muscle would absorb mechanical power rather than generate it, thereby reducing the useful power generated by other muscles that are shortening.

V_{max} shows a great deal of variation, ranging almost a thousandfold from the slowest smooth muscle to the fastest skeletal muscle. V_{max} of a muscle is determined by the particular myosin (a muscle protein) present in the fiber type. Thus, even though the shapes of the force-velocity curves are similar for all V_{max}'s (that is, maximum power occurs at 0.2–0.4 V_{max}), the absolute shortening velocity at which this maximum occurs varies tremendously.

Another property of muscles is that the force generated depends on the length over which it is used. At a sarcomere (a microscopic repeating unit of muscle) length of 2–2.2 micrometers, fish muscle generates its maximum force. At shorter or longer lengths, the force declines.

A complex characteristic of muscle is that it must be turned on and off to power movement. Complex molecular processes that require energy are responsible for turning muscle on (activation) and off (relaxation). As with V_{max}, muscle shows a tremendous variation in the rate of activation and relaxation that is controlled by the density of the sarcoplasmic reticulum in the cell. Although it seems useful to rapidly relax the muscle to avoid the muscle's being lengthened while still active, relaxation can be accomplished with less energy if done slowly. Hence, there is likely to be a trade-off between increasing the power output of the muscle and reducing the energy used.

Given the tremendous variation in muscle, it is important to determine how the muscle is adapted to power locomotion. Because the performance of a muscle depends on how it is actually used, its fiber length change and activation pattern during locomotion must be measured and compared to the full range of mechanical properties determined in isolated experiments on the same muscle type.

Fish model. In humans and most other animals, the cellular anatomy of muscles makes it impossible to obtain necessary information. In these animals, the typical muscle is heterogeneous; that is, slow muscle fibers lie next to fast muscle fibers. It is very difficult to identify which fibers are active during particular locomotory activities (that is, electrodes recording from one fiber type will pick up signals from surrounding fibers of different types). In addition, any given bundle of fibers for mechanics measurement will contain more than one type, making it impossible to differentiate the properties of a particular fiber type.

In contrast, muscle fiber types in fish are organized into larger, homogeneous, and anatomically separated regions that are clear to the naked eye. Therefore, researchers can overcome the obstacles involved with other animals. Electrodes can be implanted in the different fiber types, and by the relatively simple technique of electromyography, the swimming activities powered by the different fiber types can be monitored. In addition, bundles of like fibers can be dissected to determine the mechanical properties of a given muscle fiber type. Because of their unique muscle architecture,

fish provide a useful model for muscle function in all animals.

Fish muscular system. Fish perform a variety of motor activities, ranging from slow steady swimming during migrations to a very rapid escape response. The first clue as to how fish are put together so that the muscle can effectively power these different movements comes from electromyographic measurements. Fish use only relatively slow muscle fibers (low V_{max}) during slow steady swimming, but during fast swimming they also use their white muscles (high V_{max}). During the escape response they probably use their white muscles exclusively. Using different fibers to power different activities enables the active muscle to work effectively.

To swim, a fish must bend its backbone, requiring active shortening of the muscle on one side and passive lengthening of the muscle on the other. At low swimming speeds, the red muscle undergoes cyclical sarcomere-length excursions between 1.89 and 2.25 μm, which is the range of lengths where the muscle generates maximum force. The most extreme movement that fish make, the escape response, involves a far greater curvature of the backbone than steady swimming. If the red muscle were powering this movement, it would have to shorten to very low lengths, where low forces and irreversible damage occur. However, fish use the white muscle to power this movement because it is endowed with a fourfold higher gear ratio than red muscle. The basis for the higher gear ratio is the different orientation; the white muscle fibers run in a helical orientation with respect to the long axis of the fish, whereas the red muscle fibers, just beneath the skin, run parallel to the long axis of the fish.

Thus, for a given backbone curvature, the white muscle undergoes only about one-quarter the sarcomere-length excursion of the red muscle and is able to power this most extreme movement while working at lengths that generate near-maximal force. It appears, therefore, that fish are designed so that no matter what the movement, the muscles used generate nearly optimal forces.

Steady swimming and the escape response also require different speeds of movement. The fast movements are accomplished by a combination of the higher V_{max} and higher gear ratio of the white fibers. The V_{max} of carp red muscle is 4.65 muscle lengths per second, and the V_{max} of carp white muscle is 2.5 times higher, 12.8 muscle lengths per second. During steady swimming, the red muscle is used over a range of velocities of about 0.7–1.5 muscle lengths per second, corresponding to a V/V_{max} of 0.17–0.36, where maximum power is generated. At higher swimming speeds (requiring higher V's), the fish recruit their white muscle because the mechanical power output of the red muscle declines.

Red muscle cannot possibly power the escape response. To do so, it would have to shorten at 20 muscle lengths per second, four times its V_{max}. Even if the white muscle were placed in the same orientation occupied by the red, it could not shorten at this rate either, because its V_{max} is only about 13 muscle lengths per second. However, because of its fourfold higher gear ratio, white muscle would need to shorten at only 5 muscle lengths per second, corresponding to a V/V_{max} of about 0.38, at which it generates maximum power.

The white muscle is not used to power slow movements, because it would need to shorten at a V/V_{max} of 0.01–0.03. At this low V/V_{max}, muscle is very inefficient.

Thus, red and white muscle form a two-gear system that powers very different movements. The red muscle powers slow movements, whereas the white muscle powers very fast movements. In both cases, the active muscle works at the appropriate V/V_{max}.

During slow steady swimming, it is not necessary to activate and relax muscle as rapidly as during fast swimming, where the muscle must be relaxed quickly to prevent active lengthening. Experiments in which the muscle is caused to alternately shorten and lengthen while being activated and relaxed (similar to what occurs during swimming) show that the white muscle can relax much faster than the red and operate at much higher cycle frequencies. Thus, slow-relaxing red muscle with a low density of sarcoplasmic reticulum operates best at low frequencies (slow swimming speeds), whereas white muscle with a high density can power very rapid movements.

For background information SEE PISCES; REPTILIA; TETRAPODA; THERMOREGULATION; VERTEBRATA in the McGraw-Hill Encyclopedia of Science & Technology.

Lawrence C. Rome

Bibliography. R. McN. Alexander and G. Goldspink (eds.), *Mechanics and Energetics of Animal Locomotion*, 1977; J. D. Altringham and I. A. Johnston, Modeling muscle power output in a swimming fish, *J. Exp. Biol.*, 148:395–402, 1990; A. F. Bennett, The evolution of activity capacity, *J. Exp. Biol.*, 160:1–23, 1991; A. F. Bennett and R. B. Huey, Studying the evolution of physiological performance, *Oxford Serv. Evol. Biol.*, 7:251–284, 1990; L. C. Rome et al., Why animals have different muscle fibre types, *Nature*, 355:824–827, 1988; L. C. Rome, R. P. Funke, and R. McN. Alexander, The influence of temperature on muscle velocity and sustained performance in swimming carp, *J. Exp. Biol.*, 154:163–178, 1990; L. C. Rome and A. A. Sosnicki, Myofilament overlap in swimming carp: II. Sarcomere length changes during swimming, *Amer. J. Physiol. (Cell Physiol.)*, 260:C289–C296, 1991.

Machine learning

The ability to learn is a fundamental aspect of intelligence, and machine learning has been a topic of artificial intelligence since the 1950s. During the 1980s, interest in machine learning grew rapidly, and it continues to grow in the 1990s. Learning by a computer program is usually defined as the program's making changes that enable it to perform more accurately or efficiently in the future. There are two ways in which a computer system can change: it can acquire new knowledge from external sources, or it can modify itself to exploit its current knowledge more effectively. The bulk of research on machine learning has focused on the first type of change, but both types have been investigated and are surveyed in this article.

Inductive learning. Commonly, a machine learning algorithm is given a collection of examples and counterexamples of some concept, and its task is to infer a general procedure for classifying future examples. This process is called inductive learning. For example, a computer program could be provided with a database of medical records—the age, weight, sex, temperature, blood pressure, and so forth of a clinic's patients, along with a physician's diagnosis for each person. An inductive learning algorithm would produce an automated procedure for predicting the diagnosis of new patients, given their physical characteristics and symptoms. In fact, medical diagnosis is one of the most heavily explored applications of machine learning.

More precisely, an inductive learning program is given examples of the form (x_i, y_i), and it is supposed to learn a function f such that $f(x_i) = y_i$ for all i. Furthermore, the function f should capture the general patterns in the training data, so that f can be applied to predict y values for new, previously unseen x values. Typically, each x_i is a composite description of some object, situation, or event (such as a patient's entering a medical clinic), and each y_i is a simpler description (such as a description of the disease of the patient).

This task is called supervised learning because the y_i values can be thought of as being provided by a supervisor or teacher. When there are only a few possible y_i values, they are often called classes, and the function f can be viewed as assigning each x to an appropriate class. When there are only two possible y_i values, it is often natural to consider the examples as positive and negative examples of some concept (for example, chairs and nonchairs). In this case, f can be viewed as a definition of the concept, and the task is called concept learning.

In most cases, the set of training examples represents only a small sample of the space of all possible $[x, f(x)]$ pairs. Use of a small sample is reasonable, since, for example, physicians are expected to learn how to diagnose diseases without first having seen every possible patient. However, the learning task is ill-posed, because without some other source of constraint, there is no way to know the value of $f(x)$ for an x value that has never before been observed. Thus, learning algorithms are sought that find definitions of f that generalize to new examples, rather than ad hoc definitions of f that apply only to the examples observed during training.

The solution to this quandary is to incorporate some additional constraints into a learning algorithm so that, given a sample of $[x, f(x)]$ pairs, the algorithm can make a reasonable guess concerning the definition of f. These additional constraints are called the inductive bias of the algorithm.

The function learned, f, can be expressed in a large number of ways. If-then rules, predicate logic formulas, mathematical equations, decision trees, and neural networks are among the representations used in machine learning research. One of the most popular and successful inductive learning algorithms is J. R. Quinlan's ID3, which represents f as a decision tree. In a decision tree the nodes contain tests (for example, of patient attributes) and the leaves indicate to which class the patient should be assigned (for example, healthy or sick). The **illustration** shows a simple decision tree for medical diagnosis.

The inductive bias of the algorithm ID3 is to prefer small decision trees. Ideally, it would find the smallest decision tree consistent with the training examples. Unfortunately, such a result is known to be computationally intractable. Hence, ID3 employs a heuristic algorithm based on information theory that constructs a reasonably small decision tree.

Much of the research on neural networks also addresses inductive learning. In addition, since the mid-1980s there has been a sizable amount of theoretical research on inductive learning in a field called computational learning theory. *SEE NEURAL NETWORK.*

Speedup learning. The second kind of learning, in which a system improves its performance by exploiting its current knowledge more effectively, is often called speedup learning or skill acquisition. Most artificial-intelligence problem-solving systems, when presented with the same problem repeatedly, always solve it the same way and in about the same amount of time. It seems shortsighted that they do not adjust their behavior on the basis of their experiences. Speedup-learning systems attempt to restructure their internal knowledge so that if the same or related problems recur they can be solved more rapidly.

Much of the research on speedup learning investigates methods for improving the efficiency of search-based problem-solving systems. Many problems in artificial intelligence can be formulated as a task of choosing a sequence of operators to transform the initial situation into one that is a solution to the problem. The collection of possible situations (states) of a problem solver is called its state space, and the legal operators specify, for each state, which states are directly reachable. Search-based problem solvers look in this space for a path between the initial state and some goal state. An example is solving integrals. The states are possible equations, the operators are the rules of integration and algebra (which rewrite equations into equivalent ones), and the goal states are those equations that do not contain an integral sign.

One way to speed up a search is to learn how to better choose which operator to try next, thus reducing the number of ultimately unfruitful paths considered. Another way to speed up a search is to introduce macro operators that take big steps in the search space. A macro operator is formed by composing a sequence of original operators, and the search algorithm treats the result as a single operator. Thus, the number of operators required to move from the initial state to a goal state is reduced, but also the number of alternatives to consider at each step in the solution can be increased. Returning to the calculus example, integration tables are an extreme form of macro operators. They usually go completely from the initial problem to a solution. However, macro operators need not go directly to a goal state.

An approach called explanation-based learning (EBL) is a popular way to develop a system that performs speedup learning. Again, it is concerned with a problem solver that works by applying operators to transform an initial state (the problem description P) into a final state (the solution S). Many alternative operators can be applied to each state, so the problem solver must conduct a search for a successful sequence of operators. One way to speed up such a problem solver is to retain a cache of previously solved problems and their solutions (that is, pairs of the form $< P, S >$). Each new problem can be checked against this cache to see if it has already been solved.

Unfortunately, this strategy will produce a speedup only if exactly the same problem needs to be solved again in the future. A better approach is to generalize $< P, S >$ so as to drop irrelevant details. New problems can be matched against the generalized description of P, and if a match is found the matching values can be substituted into the generalized description of S to obtain a solution.

To correctly generalize P and S, it is necessary to determine which parts of the problem description P are important for finding the solution S and exactly how details of S depend on details of P. The key insight of explanation-based learning is that a record of the problem-solving steps taken to get from P to S provides the necessary information for generalizing the $< P, S >$ pair. This record is often called an explanation, since it explains how the solution was computed. Explanation-based learning systems analyze this explanation to extract the generalized version of the solution and thereby to learn a new macro operator.

A potential problem with speedup learners is that they can learn too many macro operators. In this case, the time required to select the appropriate one can exceed that required to solve the problem from scratch, using only the original operators. To address this utility problem, learned knowledge must be judged to determine if it is likely to prove useful in the future. If

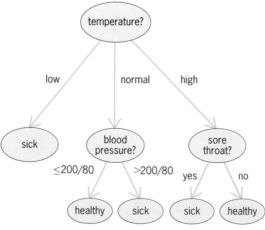

Simple decision tree.

not, it should be discarded. The utility problem may be better understood by considering the example of a cache of all possible multiplications of two six-digit numbers. Rather than looking up solutions in this large cache, it would be faster to simply reapply the rules of multiplication when necessary.

Other types of learning. A number of other forms of machine learning have been investigated, such as unsupervised learning, analogy, case-based learning, genetic algorithms, reinforcement learning, and scientific discovery.

For background information *SEE ARTIFICIAL INTELLIGENCE; DECISION THEORY; EXPERT SYSTEMS; INTELLIGENT MACHINE; NEURAL NETWORK* in the McGraw-Hill Encyclopedia of Science & Technology.

Jude W. Shavlik

Bibliography. J. R. Quinlan, Induction of decision trees, *Mach. Learn.*, 1:81–106, 1986; J. W. Shavlik and T. G. Dieterich (eds.), *Readings in Machine Learning*, 1990.

Magnet

The need for ever higher magnetic fields in many areas of materials research, chemistry, condensed-matter physics, engineering, and biology has provided the impetus for the development of new generations of high-field magnets over the last 50 years. The development of high-field, high-critical-current superconductors has made fields up to 12 teslas relatively inexpensive and accessible to many laboratories. Access to fields above this value, particularly above 20 T, is limited to a few facilities, since the production of such fields requires major investment in power supplies, cooling systems, and resistive magnets. In addition, major magnet development groups and facilities, and provisions for carrying out the best science at these extreme fields, must be available on site.

Science at high magnetic fields. A few examples will provide insight into the importance and significance of research at very high magnetic fields.

Physical studies. In the presence of high magnetic fields, in the range 20–200 T, new phenomena associated with new electronic or magnetic phases are observed and have been studied theoretically. Many more such systems are expected to be found at even higher magnetic fields. Also, many measurements at high fields provide structural, magnetic, and electronic information not available at low fields.

An example that has attracted enormous attention since its discovery in 1980 is the quantum Hall effect. It was observed that in relatively clean, low-carrier-density systems, in sufficiently high magnetic fields (greater than 20 T), and at low temperature, the Hall resistance was quantized in integral units. A great deal of theoretical attention has been focused on this surprising observation, which can be understood in terms of Landau levels, the nearly degenerate orbits allowed by the geometry of the Fermi surface in high magnetic fields. Subsequently, a fractional Hall effect was observed. This effect required fractional excitations,

and has led to new fundamental theories of quantum-mechanical systems in two dimensions.

Also arising from the Landau levels induced by high fields are de Haas–van Alphen and Shubnikov–de Haas oscillations in magnetization and resistivity, respectively, with linearly increasing magnetic field. Data on the oscillations give detailed information on the Fermi surface, effective electron mass, m^*, and anisotropy of the samples studied. At low fields the orbits are large, and scattering destroys the effect except in very pure samples. The ability to do such experiments at very high magnetic fields relaxes this condition and opens the way for exploration of many new systems, such as alloys, that will be of technological as well as fundamental importance. Recent studies of some quasi-one-dimensional organic conductors have revealed a wealth of unusual field-induced phases. At low field and temperature, there is a superconducting phase. At high fields, a field-induced spin-density wave state and a quasi-one-dimensional metallic state appear, and an energy gap apparently opens above about 30 T. There are no available steady-field data above 30 T.

Other systems that have exhibited unusual behavior are the heavy-fermion systems (narrow-band cerium or uranium intermetallic compounds), which exhibit unusual magnetic and electronic transport properties at high magnetic fields. The high-temperature superconductors have very high intrinsic upper critical fields H_{c2}, in some cases greater than 50 T at a temperature of 4 K ($-452°F$). Therefore, to study the important properties of these materials in the low-temperature

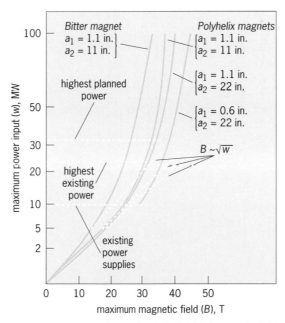

Fig. 1. Measured maximum magnetic field *B* for steady-field (dc) Bitter and polyhelix magnets as a function of maximum power input *w*. Fields calculated from the formula $B \sim \sqrt{w}$ are also shown. All major dc facilities have power supplies of about 10 MW. There are currently no power supplies beyond 24 MW, and no plans beyond 40 MW. a_1 = inner radius; a_2 = outer radius. Magnets with a_1 = 1.1 in. (28 mm) have a bore diameter of 2.0 in. (50 mm); those with a_2 = 0.6 in. (15 mm) have a bore diameter of 1.0 in. (25 mm).

region, very high magnetic fields are essential. In fact, future development of high-field superconducting magnet materials depends critically on the availability of high-magnetic-field test facilities.

Chemical studies. Another area in which high magnetic fields will yield new science and technology is high-resolution nuclear magnetic resonance, which is the single most important tool for structural chemistry and biochemistry. The sensitivity and the chemical shift, which is the important parameter in most structural studies, increase linearly with the field. Because of the homogeneity and stability requirements, the magnets that generate the field must be superconducting. Fourteen-tesla magnets with spectrometers operating at 600 MHz for protons are commercially available and fairly common. The next generation, operating at 750 MHz and 18 T, is scheduled for 1993

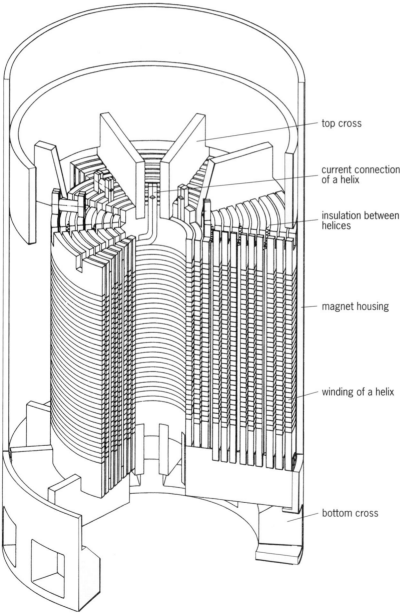

Fig. 2. Diagram of a polyhelix magnet showing the geometry of the helical coils. Cooling water flows at high pressure between the helices.

top cross

current connection of a helix

insulation between helices

magnet housing

winding of a helix

bottom cross

at the National High Magnetic Field Laboratory in Tallahassee, Florida. The superconducting-magnet development group at this laboratory is working toward the development of a 24-T magnet, 1-GHz system. It appears that 24 T is the maximum field attainable with a resolution of 1 part in 10^9 with currently available materials. Superconducting magnets with higher fields must wait for the development of new materials, possibly the high-temperature superconductor, which may be a number of years away.

Other uses of high magnetic fields in chemistry include the modification of chemical reactions in high fields, magnetic alignment of diamagnetic and paramagnetic molecules, and crystal growth and materials processing in high magnetic fields.

Biological studies. An example of the use of high magnetic fields in biological studies is the modification of reactions such as electron transfer and hydrogen atom abstraction, where a pair of paramagnetic (free-radical) species are formed and subsequently react. A strong magnetic field can be used to tune the spin states of the free radicals and thus the reaction products. The technique has recently been applied to studies of the quantum yield in photosynthetic reactions. The study of the motion and response of simple and complex fluids in high fields is being carried out. These studies include both the dynamics of fluid flow and the magnetic orientation, aggregation, and separation of colloids, molecules, and liquid crystal–like ordered fluids. Finally, studies of the effects of high magnetic fields in living systems are important for understanding the interactions of, and in developing safety standards for, steady magnetic fields.

High-field magnets. Magnets in the 20–200-T range are of three basic types: superconducting magnets, steady-field (direct-current or dc) resistive magnets, and pulsed magnets.

Superconducting magnets. These magnets are currently limited to 20–23 T by the critical current and critical field of existing materials. They are not easily swept over large field ranges, and therefore are not useful for studies requiring the measurement of some parameter as a function of field. The most important applications are for high-stability, high-homogeneity applications such as nuclear magnetic resonance and spectroscopic studies.

Steady-field (dc) resistive magnets. These magnets require large power supplies and very high capacity cooling systems. Several design constraints must be recognized. In particular, since the energy stored in the field is proportional to B^2V, for the highest values of the magnetic field B, the volume V of the field must be kept as small as possible consistent with the experimental apparatus. Steady-field magnets are almost all cylindrical solenoids. Typical bore radii are 1.4–4 in. (35–100 mm), which is sufficient for many experiments, including low-temperature experiments. The maximum attainable field is limited by the power supply. Power required versus maximum field is shown in **Fig. 1**.

There are three kinds of high-field dc-resistive magnets. The first magnets to reach fields in the 20-T range

Fig. 3. Pulsed magnet for operation at 55 T. The magnet has a 0.94-in. (24-mm) bore and is cooled by liquid nitrogen to −193°C (−315°F). During the pulse, the temperature rises to near room temperature. The pulse length is 0.02 s, the peak current is about 17,000 A, and the peak voltage is 6900 V. The energy supplied by the capacitor bank power supply is 473,000 J. The magnet is wound with copper wire, with the major strength supplied by a special fiberglass epoxy composite interleaved with the windings and on the outside. The magnet weighs about 6 lb. (2.7 kg). (a) Magnet before being tested. (b) Magnet after being tested to destruction at 62 T. It was blown apart by the magnetic forces on the current in the windings, which place the limit on achievable magnetic fields.

were designed by Francis Bitter. These Bitter magnets consist of stacked circular slotted copper plates with a number of holes through which the cooling water passes. This simple design currently operates up to 20 T with the 10-MW power supply at the Francis Bitter National Magnet Laboratory in Cambridge, Massachusetts, and other laboratories. Designs for 27–30-T Bitter magnets, utilizing the 20-MW supply at Grenoble, France, and the 24-MW supply at the National High Magnetic Field Laboratory are under way.

Another type of resistive magnet is the polyhelix, which is constructed from a series of concentric monolayer coils (**Fig. 2**). It provides more efficient cooling and current distribution, and can be extended beyond the 30-T range with the use of the 24- or 32-MW power supply at the National High Magnetic Field Laboratory.

Hybrid magnets are the third and by far the most sophisticated type of dc magnets. These magnets use a superconducting magnet surrounding a resistive magnet, a Bitter or polyhelix. Hybrid magnets at the Francis Bitter National Magnet Laboratory (1.4-in. or 35-mm bore diameter) and Grenoble (2.0-in. or 50-mm bore diameter) operate routinely up to 30 T. Design is well along for a 45–50-T hybrid at the National High Magnetic Field Laboratory (in collaboration with the Francis Bitter National Magnet Laboratory). This magnet will reach the maximum fields that can be produced with presently available materials. For higher fields new magnet materials must display increased tensile strength while maintaining a conductivity of at least 80–90% that of pure copper. Research on composite materials, such as copper-niobium and copper-silver filamentary microcomposites, are being carried out at a number of laboratories.

Pulsed magnets. These produce the highest fields, since the power and cooling requirements are considerably smaller than for dc magnets. However, the field persists for a very short time, so that only those experiments that can be done on time scales of several milliseconds down to a few microseconds are possible. Luckily, many experiments can be done in pulsed fields, such as studies of spin resonance, the de Haas–van Alphen effect, and magnetoresistance. Long-pulse magnets, with peak fields persisting for a few milliseconds to as long as 100 ms, operate currently at fields in the range 50–65 T (**Fig. 3**). The presently attainable maximum field is 68 T, but it is likely that extension to 70 T for routine experiments will occur very soon. Again, the problem is one of developing new materials with higher tensile strength and conductivity, although clever new designs for currently available composites will probably lead to somewhat higher fields. Pulsed magnets can be constructed relatively simply compared to dc magnets. The development of long-pulse magnets at these very high fields is leading to entirely new research regimes in condensed-matter physics and materials science.

The very highest fields are produced in the laboratory by short pulses, 10–20 μs in duration, with compression of the flux by an explosive charge. The magnet, sample, and probes are destroyed in the process, but fields in the 200-T range have been produced at Los Alamos National Laboratory in New Mexico. A number of important experiments have been done, for example, studies of the Fermi surface by the de Haas–van Alphen method in high-temperature superconductors.

Research facilities. The facilities for research at magnetic fields above 20 T are limited and tend to be concentrated in national centers because of the high cost of power supplies and cooling systems. Major dc facilities exist at the High Magnetic Field Laboratory in Grenoble and at the High Field Laboratory in Sendai, Japan. In the United States, the Francis Bitter National Magnet Laboratory was established in 1960. It served as the only high-field facility in the United States until August 1990, when the National Science Foundation awarded a grant to establish the National High Magnetic Field Laboratory to a consortium formed by Florida State University, the University of Florida, and Los Alamos National Laboratory. The dc magnet facililties are at Florida State University in Tallahassee and the pulsed facilities at Los Alamos.

All of the currently operating major dc facilities have power supplies of about 10 MW (the one at Francis Bitter National Magnet Laboratory is a motor-generator). The Grenoble facility has brought a 20-MW supply into operation. All of the facililties have Bitter or polyhelix magnets with center fields ranging from 20 to 25 T, and 31-T hybrid magnets.

The National High Magnetic Field Laboratory had a 24-MW supply delivered in October 1992. This power supply can easily be extended to 40 MW within 5 years as magnet development projects justify the need. It is the largest power supply in the world, and will probably remain so for some time.

Major pulsed magnet facilities are located in Los Alamos; Amsterdam, Netherlands; several locations in Japan; Oxford, England; Beijing, People's Republic of China; Leuven, Belgium; Toulouse, France; Murray Hill, New Jersey; and Princeton, New Jersey. Essentially, all of these facilities use capacitor energy storage in the range 100–1500 kilojoules, and produce fields of 50–68 T with a few milliseconds' duration.

For background information SEE DE HAAS–VAN ALPHEN EFFECT; ELECTRON-TRANSFER REACTION; FERMI SURFACE; FREE RADICAL; HALL EFFECT; MAGNET; MAGNETIC RESONANCE; PHOTOSYNTHESIS; SUPERCONDUCTIVITY in the McGraw-Hill Encyclopedia of Science & Technology.

William G. Moulton

Bibliography. E. Manousakis et al. (eds.), *Physical Phenomena at High Magnetic Fields: Proceedings*, 1992; Y. Nakagawa and G. Kido (eds.), *Applications of High Magnetic Fields to Materials Science: Proceedings of the Honda International Symposium, Sendai, Japan, August, 1989*, 1990; T. Sekiguchi and S. Shimamoto (eds.), *11th International Conference on Magnet Technology*, 1989; M. N. Wilson, *Superconducting Magnets*, 1983.

Magnetic resonance spectroscopy

Magnetic resonance spectroscopy (MRS), a noninvasive approach to measuring important metabolites in intact living tissues, is based on the principle of nuclear magnetic resonance (NMR). Magnetic resonance spectroscopy has great potential for the safe study of biochemical changes in otherwise inaccessible organs of the human body. It has been studied for several years as a tool for nondestructive biochemical analysis in research laboratories with high-resolution magnets that may not have the capacity for studies of intact living tissues. Even though magnetic resonance spectroscopy

has been established for a much longer period, its clinical application seems to have lagged behind that of magnetic resonance imaging (MRI), perhaps because of its more stringent technical requirements as well as limitations in data on diverse biochemical information needed to interpret findings.

Principles. All atomic nuclei with an odd number of either protons or neutrons have a spin (similar to the Earth's spin around its axis) generating a small magnetic field. In the absence of an external magnetic field, atoms in matter are randomly aligned. In an external magnetic field, they line up along the direction of the field like a compass needle in the Earth's magnetic field. If another field, generated by a radio frequency coil, is applied, the spinning atoms can be "tipped"; if the radio frequency field is now removed, the nuclei return to their original axis (relaxation), inducing a voltage signal that varies with time as it decays (free induction decay). By a mathematical process called Fourier transformation, a frequency spectrum can be extracted from free induction decay in a manner similar to a prism breaking light into the various color components.

Each distinct nuclear species resonates at a unique frequency called the Larmor frequency. Magnetic resonance imaging depends on the principle that the Larmor frequency varies with the strength of the external magnetic field. When a sample to be studied is placed in a linear magnetic-field gradient, the hydrogen protons at any given spatial location will emit a signal whose amplitude is determined by the proton density and whose frequency is determined by the local magnetic field. By repetitive acquisition of these signals, an image of the sample can be constructed. A different principle is exploited by magnetic resonance imaging: the Larmor frequency is also affected by the local magnetic field of the molecular cloud of electrons surrounding the nucleus; this effect is called the chemical shift. Thus, the same element resonates at slightly different frequencies in different chemical compounds, giving rise to a spectrum. The spectrum provides information as to which biochemicals are present by the position of the peak on the horizontal frequency axis. This position is represented as parts per million in relation to a central radio frequency (see **illus.**). The areas under the peaks can represent the quantities of the different metabolites. Magnetic resonance imaging data are presented mostly as ratios of chemical compounds, not as absolute concentrations.

Magnetic resonance imaging emphasizes chemical (spectral) information at the expense of spatial information; magnetic resonance imaging does the opposite. Magnetic resonance imaging utilizes a powerful magnet; similarly, the magnetic resonance spectrometer is essentially a powerful magnet, coupled with a radio-frequency transmitter and receiver coil, a display system, and a computer. To acquire signals from an organ of interest, a surface coil is placed near the area. However, only parts of the body close to the surface are accessible. Several other approaches to localization exist and are identified by a confusing array of acronyms; examples are chemical-shift imaging sequences (CSI), depth-resolved surface coil spectroscopy (DRESS), and image-selected in vivo spectroscopy (ISIS). The majority of these approaches involve some combination of the principles of magnetic resonance imaging and magnetic resonance spectroscopy.

Biochemical aspects. Several elements with intrinsic magnetic spins are naturally present in the body, some in sufficient concentrations to make them observable by magnetic resonance imaging. Elements studied by magnetic resonance imaging include the hydrogen nucleus (proton, or hydrogen-1), the carbon-13 atom, the phosphorus-31 atom, and the fluorine-19 atom. As magnetic resonance imaging has evolved, different nuclei have been deployed to study various aspects of metabolism.

Phosphorus-31. This element has been most widely studied because the ^{31}P nucleus constitutes 100% of all naturally occurring phosphorus nuclei in the body. Further, phosphorus is a constituent of several molecules that have an integral role in physiology. Investigation of phosphorus-31 provides valuable information about bioenergetics, membrane metabolism, and pH of intact organs. Information about energy metabolism is contained in the phosphorus-31 resonances of the high-energy phosphates adenosinetriphosphate (ATP), phosphocreatine (Pcr), and inorganic phosphate (P_i). Adenosinetriphosphate is the main energy currency of the body; any excess is stored as phosphocreatine, which serves as an energy bank. When energy is expended, inorganic phosphate accumulates. Diseases associated with diminished oxygen supply to tissues, anaerobic metabolism, and impaired energy utilization perturb the relative levels of adenosinetriphosphate, phosphocreatine, and inorganic phosphate. The ratio of phosphocreatine to inorganic phosphate has been, therefore, considered a measure of tissue viability.

Phosphorus-31 magnetic resonance spectroscopy also reveals information about membrane phospholipids, which are involved in crucial cellular functions. Phosphomonoesters are the precursors, and phosphodiesters are the degradation products, of membrane phospholipids. By using phosphorus-31 magnetic resonance spectroscopy, the concentration of the phosphomonoester and phosphodiesters can be estimated; these measures reflect membrane turnover and may differ between health and disease states. This method is also practical for noninvasively measuring pH in intact human tissue. The Larmor frequencies of nuclei in relation to an acidic or basic group are sensitive to changes in pH. Therefore, inorganic phosphate can be used as a pH-sensitive probe.

Hydrogen-1. Although hydrogen-1 magnetic resonance spectroscopy has the advantage of high sensitivity, it is limited by the intense background signal from hydrogen-1 in body water, which drowns all other signals. However, computational techniques are available to suppress the water resonance. Another limitation is the relative narrow range of visible chemical shifts, so that signals from different compounds overlap. This problem has been dealt with by

the advent of techniques to edit spectra at the time of acquisition so that different chemical compounds can be selectively detected. Hydrogen-1 magnetic resonance spectroscopy reveals the information about the metabolism of amino acids, neurotransmitters, and their derivatives. The three important signals have been assigned to choline (the precursor of an important neurotransmitter, acetylcholine), creatine (a compound important for energy metabolism), and N-acetyl aspartate. Smaller resonances from amino acids such as glutamate and aspartate are also seen. N-acetyl aspartate, which is primarily an intracellular compound, is considered to reflect the viability of neurons and is reduced in diseases causing nerve cell loss.

Fluorine-19. The sensitivity of ^{19}F is relatively high, making it attractive for magnetic resonance spectroscopy studies. However, there is very little mobile ^{19}F in the body, and it is biologically unimportant. Fluorine-19 magnetic resonance spectroscopy has therefore been performed on artificially introduced chemical compounds containing fluorine, such as blood substitutes and certain medications.

Carbon-13. Only about 0.1% of naturally occurring carbon is ^{13}C, which is magnetic (unlike ^{12}C, the more abundant carbon isotope in the body) and is detectable by magnetic resonance spectroscopy. The main limitation of carbon-13 magnetic resonance spectroscopy is therefore poor sensitivity. However, substrates such as glucose labeled with ^{13}C can be administered orally or by injection in order to study selected areas of metabolism. The advantage of the ^{13}C-labeled studies is that metabolites of the parent compound are simultaneously detected, allowing successive steps in metabolism to be analyzed. Studies with ^{13}C-labeled

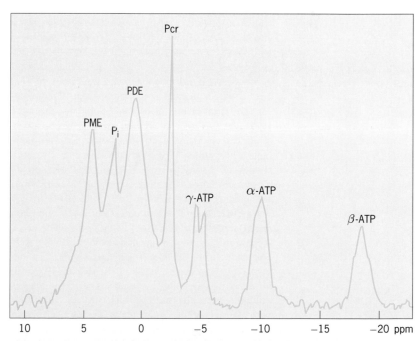

Phosphorus-31 spectrum from a healthy human subject. The spectrum shows resonances from energy metabolites phosphocreatine (Pcr); the gamma, alpha, and beta phosphate groups of adenosinetriphosphate (ATP); and inorganic phosphate (P$_i$). Also seen are metabolic parameters reflecting membrane phospholipid metabolism: phosphomonoesters (PME) and phosphodiesters (PDE).

glucose are now being conducted in disorders of carbohydrate metabolism such as diabetes.

Lithium-7. Lithium is not normally present in the body to any significant extent. However, it is an established treatment for manic-depressive disorders. Lithium-7 magnetic resonance spectroscopy studies have recently been carried out in humans. The brain lithium levels measured appear to correlate well with blood levels.

Medical applications. The noninvasive nature of magnetic resonance spectroscopy has a particular appeal for medical application, both for diagnosis and for monitoring course and treatment response. Disorders of the heart, liver, kidneys, and especially brain have been the focus of much interest. Magnetic resonance spectroscopy potentially provides biochemical information about the physiology of the human body both in health and in disease. An example is the observation that phosphodiesters increase and phosphomonoesters decrease in the brain during normal aging in animals. Similar effects have been demonstrated by phosphorus-31 magnetic resonance spectroscopy studies in healthy elderly humans.

Phosphorus-31 magnetic resonance spectroscopy has been applied for the diagnosis and monitoring of several medical disorders. It has potential value in early detection of rejection of renal and cardiac transplants. Preliminary work suggests that rejection of the transplant is associated with alteration in energy metabolism, that is, reduction in the phosphocreatine:inorganic phosphate ratio. By using phosphorus-31 magnetic resonance spectroscopy, it has been possible to monitor progress of tumors and their response to chemotherapy. The technique has been used in the diagnosis of muscle diseases that are caused by deficiencies of enzymes such as myophosphorylese and phosphofructokinase. An important detectable muscle disease is McArdle's syndrome, a disorder associated with failure of tissue acidification during muscular exercise.

More recently, magnetic resonance spectroscopy has been used for diagnosis and monitoring of neurological and psychiatric disorders. Studies in the early 1980s demonstrated its clinical value in assessing the brain state, and estimating prognosis in newborn infants. Phosphorus-31 magnetic resonance spectroscopy studies of disorders such as congenital cerebral atrophy, propionic acidemia, arginosuccinic acidemia, and meningitis in newborns have revealed decreased phosphocreatine:inorganic phosphate ratios and alterations in pH. Such alterations predict poor outcome for those infants and subsequent neurological abnormalities. Alzheimer's disease is associated with elevations in phosphomonoesters early in the disease, followed by elevations in phosphodiesters. These findings are different from the findings observed in association with normal aging. Findings could thus prove valuable in noninvasive monitoring of dementias. Preliminary studies show that seizures are associated with profound alterations in cerebral metabolism. In schizophrenia, a major psychiatric disorder, decreased phosphomonoester levels and increased phosphodi-

ester levels have been observed, suggesting decreased synthesis and increased breakdown of membrane phospholipids. These findings are similar to those seen with normal aging, and raise the possibility that schizophrenia is associated with a form of premature brain aging. Phosphorus-31 magnetic resonance spectroscopy studies can also prove invaluable in understanding the effects of psychotherapeutic drugs, since they are noninvasive and can be repeated before and after administration of a pharmacological agent.

Several findings of interest have also emerged from hydrogen-1 magnetic resonance spectroscopy studies. Hydrogen-1 magnetic resonance spectroscopy has been studied for its possible value in early detection of brain tumors, which show decreased N-acetyl aspartate. It has been used to detect increased brain phenylalanine in phenylketonuria, a relatively common metabolic disorder causing mental retardation. It has revealed cerebral lactate elevation, suggesting anaerobic metabolism in discrete brain regions of the patients with seizure disorders. Decreases in N-acetyl aspartate have been found in patients with cerebrovascular disease and in multiple sclerosis. Studies of dissected brain tissue have demonstrated decreased N-acetyl aspartate and increased glutamate in brains of Alzheimer patients. An excess of glutamate, a neurotoxic agent, could conceivably lead to nerve cell damage reflected by loss of N-acetyl aspartate. Preliminary studies have shown increases in brain choline, consistent with alterations in neurotransmitter balance in the nervous system in depression. Localized hydrogen-1 magnetic resonance spectroscopy has also been used to monitor brain metabolism following electroconvulsive therapy, an effective treatment for severe depression. It may thus be possible to identify patients at risk for beneficial and adverse effects from this treatment. Lactate infusion, known to induce panic attacks, produces a consistent elevation in brain lactate detectable by hydrogen-1 magnetic resonance spectroscopy. The technique has also been applied to alcoholism research; the alcohol signal can be observed in the human brain by using hydrogen-1 magnetic resonance spectroscopy at levels of legal intoxication (0.1%).

Limitations of method. The main limitation of magnetic resonance spectroscopy is its lack of sensitivity. The current limits of observability are approximately 500 micromoles from 10 cubic centimeters of brain in about 5 min of data acquisition. Thus, in order to obtain meaningful data, magnetic fields must be stronger and more homogeneous to accommodate the need for larger magnet bore size, making the technique more expensive and time-consuming. Only magnetic elements and highly mobile nuclei in solution are visible by magnetic resonance spectroscopy. Subject-related variables such as lack of cooperation and movement artifacts could also affect the ability to interpret studies.

Prospects. Although many data have accumulated regarding clinical applications, magnetic resonance spectroscopy cannot yet be considered an established diagnostic procedure. However, the prospects are promising. The safety of the procedure allows repetition of the studies in response to changes in different clinical states, as well as to medications. A technique known as spectroscopic imaging has been developed to incorporate magnetic resonance spectroscopy data into an image, providing pictures of metabolism. It is also possible to obtain spectra from different elements simultaneously by multinuclear magnetic resonance spectroscopy, and to produce metabolic maps. Newer, more precise methods of localization are being developed. Further research is needed to clarify the specificity of the magnetic resonance spectroscopy findings for diagnoses and their predictive value for course and outcome. Focused and careful research is likely to open a new era in the study of brain disorders, and magnetic resonance spectroscopy has the possibility of becoming a useful clinical investigative procedure.

For background information SEE MAGNETIC RESONANCE; MEDICAL IMAGING; NUCLEAR MAGNETIC RESONANCE in the McGraw-Hill Encyclopedia of Science & Technology.

Matcheri S. Keshavan

Bibliography. P. A. Bottomley, Human in vivo NMR spectroscopy in diagnostic medicine: Clinical tool or research probe?, *Radiology* 170:1–15, 1989; E. B. Cady, *Clinical Magnetic Resonance Spectroscopy*, 1990; M. S. Keshavan, S. Kapur, and J. W. Pettegrew, Magnetic resonance spectroscopy: Potential, pitfalls and promise, *Amer. J. Psychiat.*, 148:976–985, 1991; M. W. Weiner, The promise of magnetic resonance spectroscopy for medical diagnosis, *Investig. Radiol.*, 23:253–261, 1988.

Magnoliophyta

The Magnoliophyta, or angiosperms (flowering plants), are the most diverse and abundant of all the major groups of land plants. They comprise approximately 250,000–300,000 extant species and, with the exception of boreal conifer forests, dominate the vegetation of virtually all terrestrial ecosystems. There is strong evidence that the angiosperms are a monophyletic group (that is, descended from a single common ancestor) based on a suite of diagnostic features that distinguish them from all other groups of seed plants. It is also clear that their initial and most dramatic diversification occurred during the Early Cretaceous, approximately 130–90 million years ago (m.y.a.). The precise time of angiosperm origin remains uncertain. Circumstantial evidence suggests that the lineage leading to angiosperms may have diverged from related groups of seed plants significantly before the Cretaceous, probably during the latest Permian or Triassic, approximately 250–200 m.y.a. Further support for this interpretation comes from a variety of Triassic fossils that show angiospermlike characters. Whether any of the characters of these fossils are homologous with similar features in angiosperms remains to be determined. However, it is clear that the earliest angiosperms significantly postdate most other major lineages of land plants, which can be traced to the Devonian, Mississippian, or Pennsylvanian (400–300 m.y.a.).

Origin of angiosperms. Considerable progress has been made since the early 1960s toward clarifying the phylogenetic origin of angiosperms based on intensive comparative studies designed to resolve the interrelationships among major groups of seed plants. It is now well established that the gymnosperms are not a monophyletic group but merely a heterogeneous assemblage of seed plants, some of which are closely related to angiosperms while others occupy a more distant phylogenetic position. Many important questions still need to be resolved. For example, patterns of relationship among the diverse groups of extinct Mesozoic "seed ferns" are still uncertain because important features of their anatomy and morphology are not yet well understood. Similarly, although the available morphological and molecular data—nucleotide sequences of ribosomal ribonucleic acid (RNA) and chloroplast deoxyribonucleic acid (DNA)—indicate that extant Gnetales, along with fossil Bennettitales and Pentoxylales, are the seed plants most closely related to angiosperms, a very substantial gap still exists between the morphology and biology of these groups and that of Magnoliophyta. That this gap will ever be filled by a continuous series of intermediate forms is unlikely. Nevertheless, the study of phylogenetically informative fossil seed plants may provide a few intermediates and contribute important clues to the origin of critical angiosperm features such as carpels and stamens. Such fossils may also provide insights into the chronological sequence in which the diagnostic features of angiosperms were acquired. Among the more significant recent finds are a variety of small, inconspicuous seed plants that cannot be accommodated readily in existing taxa. One of these plants is now known to have produced the widespread and distinctive fossil pollen grains assigned to the genus *Eucommiidites*, and recent discoveries have shown that these plants share fascinating similarities with angiosperms in having very simple pollen organs and ovules with a double protective covering.

Phylogenetic relationships among basal angiosperms. Improved understanding of the origin of flowering plants depends largely on new paleobotanical discoveries because the four groups of extant gymnosperms (conifers, cycads, *Ginkgo*, and Gnetales) are an unrepresentative sample of the total diversity of the seed plant clade over the last 350 m.y. Clarifying phylogenetic patterns among basal angiosperms presents different challenges because many of the pertinent taxa still have living representatives. Even though many of these extant groups have few living species, which are often highly specialized in their own way, they provide invaluable information from which to infer the structure, biology, and relationships of early angiosperm fossils. Clarifying the early diversification of flowering plants thus requires careful integration of morphological and molecular data from extant plants with paleobotanical evidence of extinct angiosperms.

One of the most significant conceptual advances to emerge in the 1980s in understanding the relationships of basal flowering plants is that the traditional systematic division of angiosperms into monocotyledons and dicotyledons may not provide an accurate reflection of evolutionary relationships. Both morphological and molecular analyses support the recognition of two major monophyletic groups, the monocotyledons and the eudicotyledons (nonmagnoliid dicotyledons), comprising about 23% and 72% of extant angiosperms species, respectively. The remaining 5% of extant angiosperms that have been assigned traditionally to the dicotyledons (as subclass Magnoliidae) comprise a heterogeneous assemblage of taxa, some of which are closely related to monocots while others are more closely related to eudicots. Within the monocots the Araceae or certain Liliaceae seem to be basal, whereas in the eudicots the earliest divergence seems to be into the Ranunculidae and possibly Caryophyllidae, on the one hand, and the Hamamelididae, Rosiidae, Dilleniidae, and Asteridae, on the other.

Relationships among the 20–30 families of extant magnoliid dicotyledons have recently been the focus of several detailed morphological studies as well as phylogenetic investigations based on nucleotide sequences of chloroplast DNA and ribosomal RNA. Analyses based on morphological data support the recognition of four major clades, which together with the eudicots may reflect an early divergence of angiosperms into five primary lineages (see **illus.**). One large group includes the angiosperm families traditionally included in the Magnoliales. Many Magnoliales are characterized by flowers that have numerous parts arranged in

a helix. The second large group, the Laurales, includes several families such as the Monimiaceae (sometimes regarded as including the Lauraceae), together with several other families such as the Austrobaileyaceae and Chloranthaceae, which have often been allied with other groups of magnoliids. Flowers in the Laurales are frequently more simple than those in many Magnoliales and have fewer floral parts, which are typically arranged on a basically trimerous plan (for example, six tepals, nine stamens, one carpel). The third group, the winteroids, includes the predominantly Southern Hemisphere family Winteraceae along with its putative Northern Hemisphere counterparts Illiciaceae and Schisandraceae, while the fourth group, the paleoherbs, includes the monocotyledons along with putatively related magnoliids such as Piperaceae, Sauruaceae, and Aristolochiaceae.

Time of appearance of major angiosperm clades. Despite persisting uncertainty over the interrelationships of major clades of angiosperms, there is clear paleobotanical evidence that all five of the major lineages were already in existence by the Cretaceous, approximately 120–95 m.y.a. Clear evidence of the presence of Magnoliales and winteroids is provided by distinctive fossil pollen grains from early and late Albian rocks, respectively. Probable chloranthoid pollen occurs as early as the Barremian, while pollen and leaves resembling those of monocotyledons are also recorded by approximately the same time. Triaperturate pollen diagnostic of the

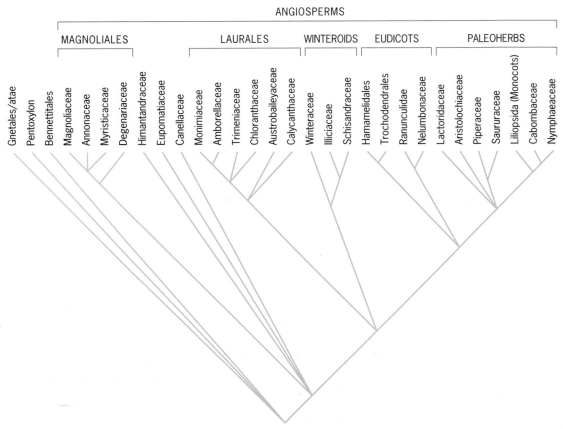

Current hypothesis of relationships among angiosperms and related groups of seed plants.

eudicots is first recorded at around the Barremian-Aptian boundary (about 120 m.y.a.).

These early indications of major angiosperm lineages based on fossil pollen are also supported by the occurrence of flowers, fruits, and other reproductive structures in slightly younger rocks. By around the Lower-Upper Cretaceous boundary (about 100 m.y.a.) the magnolialean clade is represented by large, multiparted, bisexual, magnolialike flowers. The Laurales are known from small, bisexual, trimerous flowers that are virtually identical to those of extant avocado, and the eudicots are known from inflorescences and simple, few-parted unisexual flowers very similar to those of modern sycamore and box trees. Probable chloranthoid and monocot fruits are known from the late Albian–early Cenomanian (about 100 m.y.a.), and unequivocal floral remains of both these groups are known by the Campanian (about 80 m.y.a.). The great diversity in floral form that is now known from the mid-Cretaceous is consistent with the extreme plasticity of floral construction seen among extant magnoliids, and also implies diversity in the nature of interactions with contemporary insect pollinators. Interestingly, while floral structure suggests that pollination in these mid-Cretaceous plants may have been similar to that in their extant relatives, the co-occurring fossil fruits and seeds appear relatively unspecialized for animal dispersal, and may reflect the limited diversity of potential dispersal agents, particularly birds and mammals, during the Cretaceous.

A striking feature of most mid-Cretaceous angiosperms that are currently known from floral remains is the relative ease with which they can be accommodated in extant taxa, even at the level of extant families. This ease of accommodation emphasizes that the angiosperm radiation, like that of many other major clades (such as Metazoa), led rapidly to the establishment of many extant lineages, and also implies that future studies of mid-Cretaceous fossils are more likely to extend the early fossil record of extant taxa than to reveal major angiosperm clades that are now extinct.

Angiosperm diversification. The systematic diversification of angiosperms through the mid-Cretaceous was paralleled by increasing angiosperm abundance as well as profound changes in the floristic composition of terrestrial vegetation. At middle and high paleolatitudes in the Northern Hemisphere, as angiosperms diversified, the abundance and diversity of cycadophytes, as well as many groups of ferns, underwent a marked decline. The conifers were apparently the seed plants least affected as angiosperms radiated, and maintained moderate levels of diversity and abundance throughout the Cretaceous. Similar floristic turnover is seen in comparable areas of the Southern Hemisphere, but at low paleolatitudes (that is, in Cretaceous tropical regions) the pattern of vegetational change shows some important differences.

The vegetation of Cretaceous tropical regions is poorly known, and most of the available information is based on fossil pollen and spores rather than macrofossils. Nevertheless, recent syntheses of the pollen-spore data show clearly that the major increase in angiosperm diversity occurs first in low paleolatitudes (about 20°N–20°S) and then rapidly expands into nonequatorial regions. At low paleolatitudes, in parallel with this increase, ephedroid pollen (characteristic of Gnetales) also expands dramatically in both diversity and abundance. Taken together, the geographic, temporal, and ecological similarities between the mid-Cretaceous diversification of angiosperms and Gnetales suggest that the initial radiation of both groups may have been influenced by similar biological and environmental factors. It is thus of considerable interest that the Gnetales are the group of living gymnosperms that most closely resemble angiosperms in both vegetative and reproductive biology, and also that the mid-Cretaceous was one of the most dramatic intervals of tectonic and climatic change of the last 400 m.y.

With the data currently available, it is impossible to assess the relative importance of environmental change or the evolutionary acquisition of ecologically advantageous innovations in contributing to the angiosperm and gnetalean radiations. However, the parallel increases in diversity and abundance suggest that the causal basis of these patterns is more likely to be associated with features expressed in both groups (such as vessels, reticulate-veined leaves, and perhaps insect pollination) than with features unique to angiosperms. Cretaceous Gnetales and angiosperms may have possessed similar ecological capabilities that predisposed similar responses to environmental change or, alternatively, some innovation may have arisen and spread more or less simultaneously in the two lineages in response to similar selective pressures. Whatever the causal basis of the observed pattern, any features common to both Gnetales and angiosperms must have become less influential during the Late Cretaceous. While angiosperms expanded and continued to diversify in extratropical regions, Gnetales remained restricted to low-latitude areas and rapidly declined to their current low levels of diversity.

For background information SEE MAGNOLIOPHYTA; PALEOBOTANY; PINOPHYTA; PLANT KINGDOM; POLLEN in the McGraw-Hill Encyclopedia of Science & Technology.

Peter R. Crane

Bibliography. S. Blackmore and S. H. Barnes (eds.), *Pollen and Spores: Patterns of Diversification*, 1991; P. R. Crane and S. Blackmore (eds.), *Evolution, Systematics and Fossil History of the Hamamelidae*, vol. 1, 1989; E. M. Friis, W. G. Chaloner, and P. R. Crane (eds.), *The Origins of Angiosperms and Their Biological Consequences*, 1987; P. D. Taylor and G. P. Larwood (eds.), *Major Evolutionary Radiations*, 1990.

Mammalia

The class Mammalia comprises more than 3000 living species, including *Homo sapiens*. Much of the research on mammals concerns their history and evolution and includes the use of phylogenetic trees to indicate relationships among groups (see **illus.**). Research is furthered by continual additions to an already rich fossil

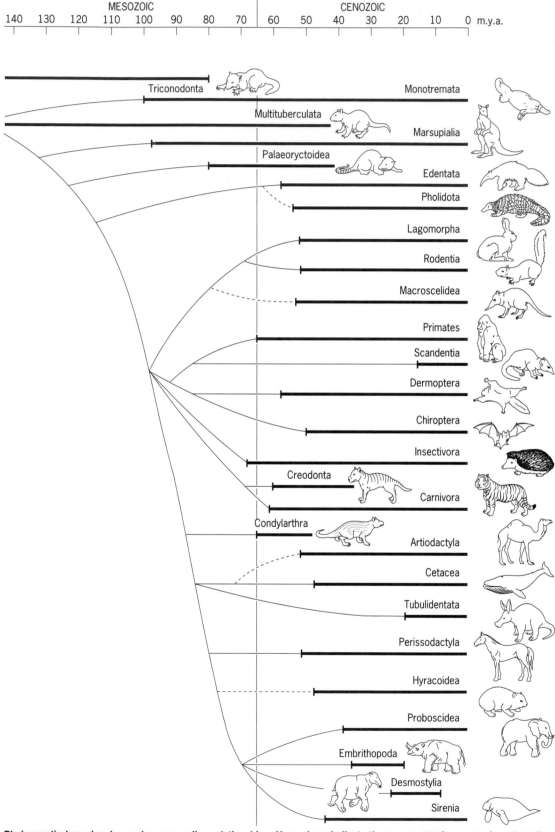

Phylogenetic tree showing major mammalian relationships. Heavy bars indicate the age range of a group, based on first occurrence in the fossil record. Thin solid lines indicate the branching sequence, although the data of actual splitting can only be inferred from relationships of clades and their known ages. Broken lines indicate problematic relationships. (*After M. J. Novacek, Mammalian phylogeny: Shaking the tree, Nature, 356:121–125, 1992*)

record and by a diversity of a scientific evidence— anatomical, physiological, behavioral, and genetic. Although important problems remain, the understanding of mammalian evolution has been greatly expanded.

Contributions from molecular biology provide a comparatively new source of evidence on mammalian evolution. The deoxyribonucleic acid (DNA) base sequences, as identified in selected genes, provide information on relationships independent from that of morphology and the paleontological record. Molecular analysis allows access to an enormous storehouse of information. However, the range of genes thus far studied has been limited. Although DNA has been isolated and amplified in fossil insects dating back 3×10^7 years, the vast majority of fossils do not preserve original DNA. Molecular discoveries in fossil mammals are thus far limited to bone and tissue representing animals much less than 1×10^6 years in age. Data on DNA in mammals are therefore almost exclusively sampled in living form.

Mammalian origins. Mammal fossils 2.2×10^8 years old have been found in Late Triassic faunas. These tiny shrewlike forms show the essential transitions from the therapsid to the mammalian plan. New discoveries of fossil triconodonts have enhanced the understanding of this transition from the premammalian therapsids. The lower jaw of typical mammals is dominated by a single bone, the dentary, which articulates with the squamosal in the skull. The postdentary bones either are lost or are transformed into the quadrate and incus, which along with the stapes form the tiny middle-ear ossicles in mammals. A platelike ledge or palate isolates the nasal cavity from the mouth. This secondary palate allows uninterrupted breathing during prolonged chewing of food. The teeth have also become more complex and variable in structure, allowing for a more efficient means of breaking down kinds of food before swallowing. In addition, the first two vertebrae of the neck are greatly modified to form an atlas-axis complex that enhances mobility of the head.

Major groups. Living mammals are divided into three main groups: monotremes (duckbill platypus and echidna), found only in Australia and New Guinea; pouched marsupials (such as kangaroos, opossums, koalas, and wombats); and diverse placentals, or eutherians (such as whales, bats, primates, carnivores, insectivores, and ungulates). This tripartite division is based on discoveries extending back to the nineteenth century. In addition, lineages of ancient Mesozoic mammals, including the earliest triconodonts and the rodentlike multituberculates, demonstrate the complexity of early mammalian evolution.

New fossil finds document the antiquity of the major Recent groups. These include monotreme fossils recently found in sediments of Cretaceous age in both Australia and South America. Marsupial fossils have been found in very old sediments of South America, Europe, and even North Africa, revealing the widespread range of this group. Exquisitely preserved fossils from the Cretaceous of Mongolia provide important clues to the early evolution of eutherian mammals.

Although distinguished by many morphological features, monotremes, marsupials, and eutherians differ most notably in reproductive structure and mode. The egg-laying monotremes likely retain the primitive mode of vertebrate reproduction. In contrast, marsupials have a short gestation period during which the embryo develops inside the uterus. The poorly developed marsupial young emerge and transfer to the pouch, where they complete their development. Recent experimental work has revealed different adaptations of marsupials to cope with the short internal gestation period.

Eutherians are also called placentals, because of the complex development of extraembryonic membranes that form the nutrient channel, or placenta, between the mother and the fetus. (Marsupials, however, also have a simple placenta.) The placenta and a protective layer (trophoblast) around the developing embryo are innovations that allow prolonged gestation. Recent evidence suggests that the eutherian reproductive mode was derived from a marsupiallike system. However, marsupials show many specializations in the structure and physiology of reproduction that could not be directly ancestral to the eutherian condition.

Marsupial radiation. The ancient history of marsupials coincides with the breakup of the southern continents. It is agreed that all Australian marsupials share a common ancestor exclusive of opossums and other New World marsupials. Fragmentation of the groups probably occurred when Antarctica separated from South America during the Cretaceous. Recently discovered fossil marsupials from early Cenozoic beds in Antarctica seem related to South American forms.

A new theory suggests that an obscure group of South American marsupials represented by a single living genus, *Dromiciops* (the monito del monte), is the nearest relative of all Australian forms. Some, but not all, studies based on deoxyribonucleic acid (DNA) hybridization and immunological cross-reactions support this branching scheme. DNA sequencing studies concerning living and even recently extinct marsupials generally support traditional evidence based on morphology.

Eutherian radiation. Sorting out the major events of early eutherian evolution has proven particularly challenging. The eutherians apparently evolved explosively. More than 20 living orders and several extinct orders are now recognized. The fossil record suggests that many of these orders first appeared, at approximately the same time, more than 6.5×10^7 years ago. Furthermore, orders as distinctive as primates and insectivores show only slight differences among their earliest members. It becomes problematical to separate relatively closely spaced divergences among lineages when such events occurred at very early times. The problem has attracted a range of investigations that draw on skeletal structure, dental microstructure, fetal membrane development, head circulation, brain-visual networks, and protein and gene sequences.

Phylogeny theories include the postulation of a very

early divergence of the South American edentates (sloths, anteaters, and armadillos) from other eutherians. This theory is weakly supported by morphological and molecular evidence. Another suggestion is that the armored pangolins are actually closely related to the edentates, an idea that has proven controversial.

Stronger evidence relates the Rodentia (such as mice, guinea pigs, marmots, and beavers) to the Lagomorpha (rabbits and pikas). Rodentia, the most diverse order of living mammals, is the object of abundant taxonomic and evolutionary research.

Another important eutherian aggregate comprises primates, tree shrews, bats, and flying lemurs. Some morphological evidence suggests that these groups share common ancestry. However, studies of gene sequences do not clearly support this position. Recent debates have concerned the origin of bats. Studies of the nerve connections between the optic region and cortex were used to argue that the nonecholocating fruit bats (Megachiroptera) are closely related to primates and flying lemurs whereas echolocating bats (Microchiroptera) are independently derived. This argument, which would require the dual origin of powered flight in mammals, has not been supported by other evidence. Recent morphological and genetic studies clearly point to a single origin for bats. Data relating primates to flying lemurs, tree shrews, and bats are less clear-cut.

A major sector of the eutherian radiation comprises the diverse, mainly herbivorous lineages known collectively as ungulates. There is evidence that ungulates shared a common origin before diversifying into artiodactyls (such as cows, deer, camels, and pigs); cetaceans (whales); perissodactyls (horses, rhinoceroses, and tapirs); hyraxes; elephants; and sea cows. However, the affinities of many primitive ungulates (including a major South American radiation) remain unclear.

The role of fossil evidence is well illustrated in the case of higher ungulate relationships. For example, the whales and the even-toed artiodactyls seem unlikely relatives by evidence of their living members. However, archaic fossil whales retain vestigial limbs with a digital plan prototypical for artiodactyls. Furthermore, terrestrial fossils belonging to the Mesonychidae show a mosaic of characters that suggest transitions to whales, on the one hand, and to artiodactyls, on the other, from a common ancestor. Studies of mitochondrial gene sequences support some connections between these ungulate groups.

The major outlines of eutherian phylogeny fail to resolve the nearest affinities of several distinctive orders, including carnivores, insectivores, and tubulidentates (aardvarks). Evidence does not suggest a preferred branching scheme for the base of the eutherian phylogeny. Patterns of relationships within many of these orders are, however, much better known.

Prospects. The attempt to map the pathways of descent in mammals has illuminated many issues concerning the evolution of characters and taxa. DNA sequence studies have shown that some genes evolved so rapidly that they are unlikely to provide much evidence on early branching events. At the other extreme, some fossil taxa, despite their early appearance in the record, are markedly specialized and shed little light on deep-seated relationships.

However, in many cases the fossil record does show a very reliable pattern of evolutionary change. Recent analytical studies show a close correlation between the first occurrence date of fossils and the suggested branching sequences based purely on morphological or genetic changes. Correlations are especially strong in horses and other selected ungulates, groups with fairly rich fossil records. Further studies of mammals from paleontological, morphological, and molecular perspectives promise to build on these new insights.

For background information SEE EUTHERIA; MAMMALIA; MARSUPIALIA; MONOTREMATA; THERAPSIDA in the McGraw-Hill Encyclopedia of Science & Technology.

Michael Novacek

Bibliography. R. L. Carroll, *Vertebrate Paleontology and Evolution*, 1988; L. G. Marshall, J. A. Case, and M. O. Woodburne, Phylogenetic relationships of the families of marsupials, in H. H. Genoways (ed.), *Current Mammalogy*, 1990; M. J. Novacek, Mammalian phylogeny: Shaking the tree, *Nature*, 356:121–125, 1992.

Marine navigation

The marine traffic system is largely informal and nondirective. The formal structure, rigorous procedures, directive movement control, real-time precision navigation technology, and system-wide focus characteristic of air-traffic control are not typical of commercial vessel operations. Vessel traffic services (VTS), the maritime alternative to air-traffic control, are increasingly being used to prevent marine accidents, 85% of which are caused by human error. Vessel traffic services provide a more formal structure and process for navigational decision making by captains, mates, and independent marine pilots (local navigation experts) aboard commercial ships, tugs, and ferries. Watchkeepers at vessel traffic centers (VTC) acquire, interpret, and distribute system-wide navigational information to provide a more complete and accurate basis for decision making by pilots. Externally directed control of vessel movements similar to air-traffic control is attempted only under emergency conditions; navigation practices and positioning technologies have not advanced to where shore-based control of vessel maneuvers (steering instructions) is feasible. However, time-space management of vessel movements is sometimes used on a more frequent basis to achieve physical separation, thereby reducing the potential for collisions, rammings, and groundings.

Operating environment. Navigation of commercial vessels in harbors and waterways relies on interdependent decision making by individual units that are not linked together in formal communications network.

In aviation, decision making is shared between the air-traffic controllers and pilots. Controllers coordinate queuing, give clearances for takeoffs and landings, and assign transit paths to aircraft operating under instrument flight rules; pilots fly the planes. Precision guidance along planned air routes takes advantage of real-time electronic navigation systems. Access to airspace is controlled, and physical interactions between aircraft are avoided through vertical and horizontal separation.

In marine transportation, each vessel is operated independently, and virtually all maneuvering decisions are made by its pilot. Self-enforcing rules prescribe procedures for a vessel meeting, overtaking, or crossing the path of another vessel. These rules are less precise when three or more vessels converge, especially in narrow channels where the effect of hydrodynamic forces associated with the interactions of the vessels, channel geometry, and currents is significant. Channels are printed on nautical charts; boundaries may be marked by buoys or aids to navigation on the shore. Vessel positions are estimated by the pilots on the basis of expert local knowledge and radar displays, or the mate may take bearings or radar ranges (distances) to known points and plot them on nautical charts to obtain a position fix. Although feasible, real-time precision electronic navigation systems are mostly in developmental stages rather than in full application. Marine safety authorities overlay vessel traffic services on region-specific operating environments to establish a more consistent framework for decision making while providing improved information sharing to enhance safety performance and efficiency of vessel movements.

VTS operations. Effectiveness of decision making and safety performance are directly related to the quality of navigational information available. Accordingly, the basic operational function of a vessel traffic service is to provide more complete, timely, and accurate information than is normally available to the master, mates, or pilot. Overall, vessel traffic

functions and capabilities vary greatly in the more than 200 facilities worldwide. Staffing, hardware, and specific operations depend on whether the government or private sector installs and operates each system and on the specific objectives: safety, economic efficiency of port facility operations, environmental protection, or protection of waterfront facilities or structures (such as an offshore oil production platform or a critical highway bridge).

Most vessel traffic service systems consist of a manned vessel traffic center (**Fig. 1**) or centers [similar but not completely analogous to airport terminal radar approach control (TRACON) facilities], a voice radio network with dedicated frequencies, radar surveillance, and, sometimes, charted traffic lanes. Visual overlooks, closed-circuit television, and electronic displays of traffic data may also be available. Installations range from very high frequency (VHF) radio through extensive multisensor systems. Remote sensors and multiple traffic centers are linked through microwave or dedicated landline networks. Virtually all vessel traffic service systems are human-resource-intensive. With few exceptions, collection, processing, interpretation, and transmission of traffic and navigational safety information all require manual processing, as does direct intervention (via radio) to prevent collisions, rammings, groundings, and near misses.

VTS technology. Communications equipment augmented by surveillance hardware is the electronic heart of vessel traffic service operations.

Communications. VHF voice radio is standard for marine communications. Local and remote VHF transceivers provide coverage of the VTS service area and are used to communicate routine traffic data and time-sensitive information (such as blockage of a channel by an accident). A single frequency is used for communications in each vessel traffic service sector. A designated frequency is also available for direct (bridge-to-bridge) communications between vessels. Communication channel access is affected by the number and locations of vessels attempting simultaneous communications; increased use of radios during adverse conditions such as reduced visibility in rain, fog, and snow; and sometimes by interference. Electronic data links such as those used by naval forces have not been adapted for marine traffic control, although development of comparable technology for commercial use is in progress.

Surveillance. Independent surveillance (requiring no cooperation from vessels to obtain identifications, positions, trajectories, and velocities) is used to maintain an overview of the activity in the vessel traffic service area. Dependent surveillance, used to provide vessel-specific information, ranges from radio reports of vessel identity through automatic dependent surveillance (ADS) technologies incorporating a vessel's navigation sensors for position and velocity data that are transmitted to the vessel traffic center. Most early vessel traffic service systems adapted shipboard radars for shore-based surveillance rather than developing hardware specifically for vessel traffic service applications; some of this equipment is still

Fig. 1. One of the specially designed vessel traffic control centers in the port of Rotterdam, The Netherlands, which provide unrestricted visual overlooks of critical intersections within the port complex on the Maas River.

Fig. 2. Second-generation display in a vessel traffic center in the port of Rotterdam, The Netherlands. Radar, traffic data bases, and very high frequency (VHF) voice radio transceivers are placed into wraparound consoles. Third-generation displays integrating all information into a single, multicolor display are now entering the market.

ance radar with automatic target acquisition and tracking features, known as ARPA (automatic radar plotting aid), has been installed in some vessel traffic centers. ARPA data are integrated in some systems with video background maps to aid in correlating traffic flows with geographical features. The latest ARPA units use high-light-level raster-scan displays, thereby permitting installation in open rooms with visual overlooks (**Fig. 2**). ARPA units are installed as separate modules or combined with other vessel traffic service sensors into operators' consoles (this closed-system architecture reduces flexibility for modifications and upgrades).

Advanced video displays have been developed specifically for vessel traffic service use or adapted from technology transferred from defense or air-traffic control applications. High-fidelity multicolor visual display systems that integrate and fuse radar data from multiple sensors, closed-circuit television video, vessel-specific data, automated data management, electronic charts, and expert-system decision aids are based on the most advanced available technology. Development of systems that integrate automatic dependent surveillance with differential global positioning system (DGPS) equipment, meteorological and hydrological sensors, digital selective calling (a marine data communications system), and other features is in progress. In the absence of international standards for vessel traffic center displays, there is a proliferation of options in available commercial systems. Portable interactive automatic dependent surveillance systems with integrated electronic charts are in prototype development for coordinated use by marine pilots and shore-based traffic centers. SEE ELECTRONIC NAVIGATION SYSTEMS; SATELLITE NAVIGATION SYSTEMS.

used. Radar information in such systems is presented on plan position indicator displays installed in darkened rooms (to prevent washout). All radar targets are manually acquired, identified, tracked, and correlated with radio and geographic information. Vessel data are recorded on time cards or cardboard strips (as in the original air-traffic control system). Vessel movement is tracked by visual observation, grease-pencil marks on a radar screen, sequencing of transit cards or strips on a card holder, or markers on plot boards.

Advanced technology. Improvements beyond basic vessel traffic service technology include adaptations of advanced shipboard radar technology and development of video displays specifically for vessel traffic service applications. Shipboard collision avoid-

Traffic separation. In some vessel traffic service systems, traffic separation schemes (TSS) are used

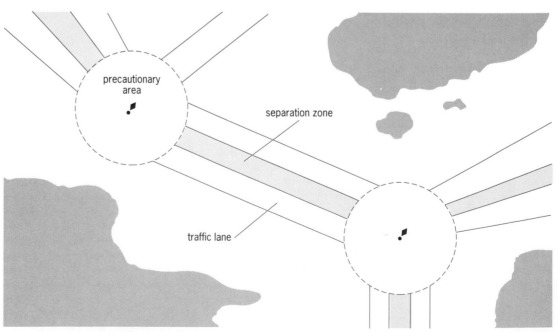

Fig. 3. Major features of a traffic separation scheme (TSS).

to physically separate participating vessels in the two-dimensional horizontal plane (as opposed to three-dimensional vertical and horizontal separation in aviation). A traffic separation scheme (**Fig. 3**) consists of one-way traffic lanes with separation zones between them and precautionary areas at convergence zones (intersections of channels or pilotage routes). Vessels entering precautionary areas are required to proceed with great care. Precautionary areas are often marked in the center with buoys on taut moorings, which remain precisely in position despite changes in currents and winds.

Where narrow channels do not permit use of traffic separation schemes, queuing vessel movement is sometimes used to match traffic flow to physical operating conditions. One-way traffic may be ordered by marine safety authorities (or directly by a vessel traffic service if empowered to do so) to prevent large vessels such as tankers from meeting or overtaking another vessel. Or meetings and overtakings may be timed to occur at the most favorable locations in a waterway. Information needed to support these determinations is acquired by using VHF voice radios and radar surveillance. Informal queuing is also frequently performed by cooperating marine pilots as a routine practice apart from vessel traffic service operations; this practice usually involves foreign flag ships (rather than all vessels that are using the pilotage route) because these ships, whose masters are unfamiliar with local waters, are the primary client base for pilot services in most ports. Separation schemes are also established for approaches to some ports and on some coastal routes offshore to minimize the potential for collisions and environmental damage.

For background information SEE MARINE NAVIGATION; RADAR in the McGraw-Hill Encyclopedia of Science & Technology.

Wayne Young

Bibliography. Canadian Coast Guard, *Proceedings of the 7th International Symposium on Vessel Traffic Services*, Vancouver, B.C., Canada, June 1992; D. Maio et al., *Port Needs Study*, 3 vols., U.S. Coast Guard Rep. DOT-CG-N-01-1.2, August 1991; N. Mizuki, H. Yamanouchi, and Y. Fujii, *The Results of the Third Survey on Vessel Traffic Services in the World*, Electr. Nav. Res. Inst. Pap. 59, Japanese Ministry of Transport, 1989; Nautical Institute, *Pilotage*, 1990.

Metal coatings

Thermal spray is an advanced materials processing tool that is increasingly being used to apply coatings to surfaces, to form composites, and to manufacture new structural materials. Provided it does not decompose, any material can be thermally sprayed. Advances in equipment and materials technology have greatly expanded the range of materials and applications for which thermal spray can be used, and thermal spray has become such an important technology that modern jet aircraft would not be able to operate without it. Today's jet engines each contain over 300 thermally sprayed, coated parts, demonstrating that thermal spray is not only important but commerically viable.

Early developments. Thermal-spray processing began soon after the introduction of the oxy-acetylene torch, between 1890 and 1910. The earliest records of thermal spraying are recorded as far back as 1882 and 1889, when patents were granted to M. V. Schoop in Zurich, Switzerland, for spraying lead. The processes came to be known as metallizing. Wire-combustion spraying of soft metals grew to include hard and wear-resistant metals such as high-carbon and stainless steels. Later it was found that powders could be introduced into the torches' combustion jets and those powders could be melted and deposited onto a base metal. The process name was eventually changed to flame spray, since advances in materials and equipment allowed ceramics as well as metals to be sprayed.

In 1910, Schoop and his coworkers invented a consumable electric wire-arc process that arc-melted and atomized conductive metal wires into droplets and deposited them onto base materials. Meanwhile, applications for thermal spray began to grow, including the coating of bridges, marine structures, and chemical manufacturing equipment. However, not until the introduction of the plasma spray process could ceramic and the higher-melting materials be effectively applied by thermal-spray methods. In the late 1930s a process based on the concept of heating gases within a confined electric arc was introduced. It used the hot, expanding plasma to accelerate, melt, and propel particles onto a surface for coating. When aero-gas turbine engine builders discovered the value of sprayed coatings on engine components, recognition finally came to thermal spray, since it was demonstrated to be a flexible, high-performance coating process. Space reentry studies improved the plasma arc-gas heaters, used in simulating high-velocity, high-temperature surface stagnation conditions of reentry. Improved plasma guns and other thermal-spray devices were developed for improved thermal-spray coating. These devices reliably produced high jet temperatures and velocities, enabling the spraying of ceramics and many other high-temperature materials.

Current status. Thermal spray is a family of processes that use chemical or electrical energy to heat and accelerate small particles, generally smaller than the thickness of a human hair (3–50 micrometers), to form volumes of finely divided molten droplets, rather like an aerosol spray. The particles and droplets are heated and accelerated by expanding gases, which push particle speeds up to over 2000 ft/s (600 m/s), or close to 1300 mi/h (2100 km/h). The combination of the high particle temperatures, where particles are typically either molten or at least softened, and their high velocity causes them to splat (flatten on impact) and conform to the surface of a target (or substrate), placed within the particles' stream. Millions of particles are heated per second by the gas jets to form droplets or softened particles. Coatings are then rapidly built up into layers by particle-by-particle impact onto the target surfaces. The droplets resolidify on impact, build upon each other, and connect to form solid continuous layers. The individual droplet impacts (splats) are very small and thin ($\sim$1–20 μm), so thin that each droplet cools extremely rapidly (millions of degrees per second) to form uniform, very-small-grain-size crystalline materials. In some cases, crystallization in a material can actually be prevented, so that amorphous (glassy) materials, with no grain structure, can be formed.

The various stages of the thermal-spray process include heating of the particles, droplet formation, droplet arrival (occurring at millions of times per second), and subsequent particle deformation, solidification, and interlocking to form a coating. The deposits typically contain some level of open spaces (voids), known as porosity, unmelted or incompletely melted particles, fully melted splats, and varying degrees of other phases, known as inclusions, which are typically produced by oxidation of some of the hot particles traveling in the spray jet due to the mixing with the surrounding air. Since millions of particles are being heated at the same time, many do not experience the same heating; some particles may not be melted at all, thus forming porosity or trapped unmelted particles in the coatings. The degree or amount of these coating features differs depending on what type of thermal-spray process is used, which operating conditions were selected, or which material was sprayed. A thermal-spray coating of a nickel-chromium alloy is shown in the **illustration**. In this micrograph the characteristic layered (splat) structures can be clearly seen. The splats on the coating layers were formed as each droplet impacted the component being coated. Also shown are partially melted particles and dark oxide inclusions, which are often present in thermal-spray coatings made in air.

Processes. Thermal-spray processes that rely on chemical heating as their source of energy are known as combustion-spray processes; they burn a mixture of air or oxygen with a gaseous or liquid fuel (propane or kerosine) to generate a hot, expanding gas jet. Combustion-spray processes use either low-velocity flames (flame spray) or high-velocity flames [high-velocity oxy-fuel (HVOF) spray]. Electrical energy can also be used as the thermal energy source, heating materials either directly or indirectly, in processes known as arc melting or arc gas heating, respectively. The arc heating of consumable, electrically

conducting wires is called wire-arc spray, while arc gas heating raises gases to such high temperatures that the plasma state is reached; thus, these processes are called plasma spray. Plasma is the state beyond gaseous, where gases are heated to such high temperatures (energy levels) that their molecules are separated into their constituent atoms and electrons. In this state the electrons have been removed from these atoms far enough for ionization to occur, and the gases conduct electricity.

Materials are fed into thermal-spray process jets as very fine powder particles, as wires, or in rod form. Hot gases, heated by combustion or by electric discharges, expand and accelerate the stream of particles and droplets, while in wire- or rod-fed processes, high-velocity gas jets melt and break up (atomize) the melting wires or rods. Droplets produced from thermal-spray processes then build up, droplet by droplet, and can be directed to form a coating, a composite material, or a free-standing structure, the last in a process known as spray forming.

Combustion spray. This is the thermal-spray class in which combustible fuels are mixed with an oxidizing agent and are reacted or combusted, subsequently heating and accelerating materials injected into the hot expanding gases. Combustion spraying has been recently divided into two commercially significant technologies. Those that use externally combusting jets, burning unrestricted and at low pressures in ambient air, are considered flame spray. The other processes employ internal combustion at higher pressures such as high-velocity oxy-fuel spray.

Flame-spray processes use externally combusted flames into which powders, wires, or rods are fed. The mixtures burn continuously, and powdered materials or wires are fed into the center of this hot jet from behind the combustion zone. The materials are melted to create a distribution of melted droplets that have been carried through the combustion zone by the expanding gas jet to impact the target surfaces.

High-velocity oxy-fuel spray processes differ from flame spray in that combustion is internal, with the gases being burned at much higher pressures. Specific HVOF process designs vary, but all use oxygen mixed with fuels, such as propylene, hydrogen, propane, or even kerosine. The internally combusted gases are directed into a nozzle or barrel, where powders are mixed with the high-pressure, high-velocity combustion gases that travel. Confined by the nozzles/barrels, the gases and powders exit, supersonically expanding into the ambient atmosphere. The high-pressure combustion jets of the HVOF processes have been shown to improve particle heating uniformity compared to the open lower-pressure flame-spray processes, and they have been shown to be more effective in accelerating the particles and droplets to higher velocities. The better heating and higher particle velocities produce coatings that are denser, with lower oxide content, and more uniform than conventional flame-spray coatings and many plasma-spray coatings.

Wire-arc spray. Another class of thermal spray, wire-arc spray uses two conductive, consumable wires

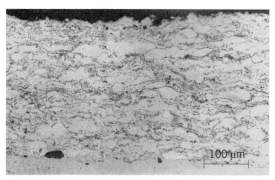

Micrograph of a nickel-chromium-alloy thermal-spray coating made by the high-velocity oxy-fuel (HVOF) spray process.

as electrodes between which a continuous arc melts the advancing wire tips. An atomized gas with a high flow rate breaks up the melting wire tips and carries the dispersed droplets to the substrate. The arc-spray device consists of a nozzle that directs a high-velocity air (or in specialized cases, an inert gas) jet at the wire tips and a wire feed mechanism that positions and advances the wires. Melt rate and the setting of the electrode voltage control the arc gap and affect the droplet size distributions leaving the wire tips, thus affecting the final coating structure. Unlike the other thermal-spray processes, wire-arc spraying uses cool gas jets that break up the molten tips of the arc-melted wires. Very small droplets are formed by the forces of the expanding gas jets, and the small droplets are accelerated to the target to form the thermal-spray deposit. Since the gases are not heated, the components being coated are not heated significantly by the process, an advantage for thermally sensitive base materials. In fact, it has been shown that even thin sheets of paper may be coated by the wire-arc process.

Plasma spray. This is a thermal spray process that uses an electric arc to heat gases to over 20,000 K. At these temperatures, the feed gases (argon, hydrogen, neon, nitrogen, as well as others) disassociate and ionize into their component ions and electrons as energy is pumped into them by the confined arc discharge. This ion-rich plasma state is what gives the process its name and, more importantly, gives the plasma-spray process its thermal capability to melt any material with a stable melting point. The plasma-spray device has a cylindrical tungsten cathode that, with an applied electric field, emits a supply of electrons when heated. The cathode is aligned concentrically inside a water-cooled copper nozzle into which gases are fed. The gases are heated to the plasma state by the electric arc and expanded to form a subsonic or supersonic jet that travels through the nozzle and exits at high velocities. In some cases, these gas-jet velocities may exceed 3000 ft/s (900 m/s), or over 2000 mi/s (3200 km/s). Jet velocities depend on the gun operating power, nozzle shape, gas composition, and gas flow rate. Powders can be introduced at different locations, depending on their melting point. When fed externally, a tube injects a stream of particles radially into the hot gas jet just after the jet exits the nozzle. Powder injection inside the nozzle can also be done; this allows longer particle heating or dwell times for more uniformly melting many materials.

Plasma-spray devices use inert gases to create the hot plasma jets for melting; therefore, the process has more flexibility in terms of the materials and in the type of environment in which the process operates. If operated at low pressures (vacuum) or under inert gases in a chamber or under a shroud, the process has been shown to produce very low levels of oxide inclusion and porosity. This is accomplished by preventing the contact of the heated particles with the oxygen-containing ambient air. The inclusion contents are typically so low that in many cases spray-deposited materials have the same or better properties than the same materials processed by casting or forming. This de-posit characteristic enables the vacuum plasma-spray process to be a spray-forming process that can be used to produce net shapes, which is very advantageous, especially for expensive or difficult-to-form materials (for example, ceramics, refractory metals, and intermetallic compounds).

Characteristics. Depending on the major process classes or on the processing details, the deposit structures and properties may vary. The various processes yield different thermal and velocity histories for the particles. Particle melting and deformation are different in each thermal-spray class, leading to the observed differences in structure and coating performance. The HVOF process typically produces the highest particle speeds, but its gas temperatures limit the maximum particle-melting temperatures. Plasma spraying, on the other hand, has the highest heating potential; however, its particulate velocities, although higher than the conventional flame and wire-arc spray particles, are not as high as those in HVOF spraying. In other cases, accelerated wire-arc spray systems extend the particle velocity range for wire-arc spray, reducing some distinction between processes. Finally, while flame spray has the lowest heating potential and the lowest average particle velocities of all the thermal-spray processes, it finds use in many surfacing applications.

Applications. Thermal-spray processes were originally developed for producing protective surface coatings and for building up undersized surfaces or components degraded by corrosion or wear, such as rotating bearings. Such thermally sprayed coatings were, and still are, used in many applications to modify a material's surface properties, usually to improve its performance or to extend its useful life. Such materials may be subject to wear, oxidation and corrosion, or thermal degradation; and therefore they must be protected. Thermal-spray coatings are used in situations where the service conditions in which the base materials are used are aggressive, for example, in jet or automobile engines, or in corrosive marine environments with their high moisture and salt content. In such environments it is appropriate to coat many base materials so that a less expensive or stronger material can be used underneath. Materials with high contents of aluminum or chromium are more expensive, and they usually reduce a material's strength properties as compared to steels. But when such materials are added as a relatively thin surface layer, they have been shown to protect the underlying base material. In other applications, abrasive or erosive conditions can quickly wear out softer base materials; hence, harder, more wear-resistant surface coatings are needed. Harder materials may require higher carbon, have high carbide or ceramic contents, or have hard phases that grow in the material as they are processed. These types of materials typically can be added to surfaces as coatings that more effectively resist the wearing environment. These materials perform so well that such wear-resistant thermal-spray coatings are the preferred choice in industry.

Heat resistance. Ceramics and refractory metals, such as molybdenum or tungsten, can be added to surfaces to resist heat. Oxide coatings have been used as thermal barrier coatings since their low thermal conductivity actually insulates against heat transfer to the underlying base materials. Thermal barrier coatings have been shown to be useful in enabling the use of base materials with lower heat tolerance or in permitting a reduction of cooling air. Thus, they are now used in many jet engine components that are exposed to high heat. High-melting-point molybdenum and tungsten metals (melting points over 3000°F or 1650°C) typically added to harden a surface are also used to shield lower-melting-temperature base materials from high temperatures in the wall linings of rocket engine nozzles and similar components.

Surface protection. Thermal spray can provide surface protection and surface modification, using a wide range of materials, from polymers to ceramics to superconductors, on a wide range of base materials, from plastics to steel and from high-temperature jet engine materials to titanium alloys. Artificial limbs such as hip joints and knees made from titanium and cobalt base alloys have also been thermally spray-coated. Thermal spraying has been used in optics to coat base materials with oxides that reflect or absorb light, and to provide electrically conductive (metal) coating on insulating (ceramic or plastic) or conductive base materials. Thermally sprayed seals and specialized abradable coatings are used in turbomachinery, pumps, or other rotary devices. These coatings are made from materials that are a composite of metal and polymer or ceramic; they act as a sacrificial layer as one part rubs preferentially into the coating to wear away a sealing path. Solid lubricants such as silver, molybdenum, and molybdenum disulfide deposited as layers that reduce friction on sliding surfaces can also be thermally sprayed or mixed with a harder metal to become a binding matrix in improved sliding-wear coatings. Restoration and refurbishment can also be accomplished, restoring surfaces that may have been worn or mismachined. These restored thermally sprayed surface layers frequently outperform the original surface.

Special catalytic metal or ceramic layers is used to interact chemically with gases or liquids to increase the rates of reactions or to control the products formed. A specialized version of such coatings is used in fabricating the so-called biocompatible ceramics, such as hydroxylapatite, which are sprayed onto artificial titanium hip replacement components. In these applications the thermally sprayed coatings enhance bone attachment and growth. Electromagnetic-shielding metallic coatings such as aluminum, zinc, and tin alloys have been thermally sprayed onto materials that would otherwise be transparent to electromagnetic radiation. These coatings eliminate and shield internal electronics from radio-frequency interference and other types of electromagnetic radiation. Special ceramics with magnetic properties, usually oxides of metals, may also be applied to nonmagnetic surfaces to localize the magnetic properties. Finally, thermal spray has been used for decorative coatings,

applied for their color or texture, to improve the appearance of materials.

Commercial uses. Jet engine manufacturers, automobile makers, electronics companies, industrial machinery producers, chemical plants, and steel/metal-working mills, as well as many others, employ thermal-spray technologies. Widespread use of thermal spray in the aircraft engine/aerospace industry is due to the fact that the jet engine environment is very demanding, and the cost of components used in many of these applications easily justifies the increased cost of the thermal-spray coatings. Thermal spray, including plasma, wire-arc, HVOF, and flame spray, have been found to be more expensive compared to other conventional coating methods, such as painting, dipping, plating, heat treatment, and thermal diffusion. However, the attraction of using thermally sprayed coatings is that many materials could not be applied by any other means. The increased performance of such materials, applied by thermal spraying, typically outweighs any increased processing costs since the materials ensure proper component operation.

Automotive and biomedical use of thermal spraying is increasing rapidly, and major production facilities have been recently committed to the technology. Manufacturers of automotive valve train parts, cylinder and piston parts, hip joints, and high-performance compressors are adopting thermal-spray coating because of the coatings' high performance and flexibility in terms of materials. Thermal-spray coating is also finding application in food processing, textile and paper processing, and chemical and metal-working equipment, as its performance and reliability have increased while its production costs have decreased.

Prospects. While the major role of thermal spray is as an important commercial coating technology, it is now being extended to the production of composites and structural materials. The use of particulate feedstocks with the incremental buildup of the final structures, combined with the rapid solidification of droplets during the formation of the materials, enables thermal spray to be used in forming advanced materials. These advanced materials may consist of fibers, ceramic particulates, or brittle intermetallic materials. Such materials are considered a new family of lightweight, high-performance materials. They have many aerospace applications. Combinations of metals with ceramics (cermets) are easily processed, provided the ceramic materials do not decompose thermally. These materials have unique properties, enabling thermal expansion and thermal or electrical conductivities to be tailored; in addition, they offer increased strength or stiffness. Thermal spray has demonstrated the ability to coat continuous ceramic fibers, putting lightweight, strong intermetallic materials like titanium or nickel aluminides between the fibers to produce a composite material that might be used as the skins for advanced aircraft. The very high thermal-spray processing temperatures, especially those found in plasma spraying, allow the melting and forming of the high-melting-point materials such as molybdenum, tantalum, and tungsten. Such materials are extremely heat-resistant and, except for oxidation, resist chemical attack. Thermal-spray forming of such materials has been demonstrated and holds promise for the commercial production of such advanced composite materials.

For background information SEE CERMET; COMPOSITE MATERIAL; MATERIALS SCIENCE AND ENGINEERING; METAL COATINGS in the McGraw-Hill Encyclopedia of Science & Technology.

Ronald W. Smith

Metallo-carbohedrene

The study of gas-phase clusters, species generally composed of weakly bound assemblies of atoms, molecules, and ions, pervades many fields of science and technology and has undergone an explosive growth since 1980. Within this broad field of research, those systems that assemble into cagelike structures have attracted particular interest. One example is hydrogen-bonded aggregates of water molecules, both pure and in varying combinations with alcohols and ammonia, that form a network structure that can encase central cations and anions (such as the cation H_3O^+, alkali metal ions, and the anion OH^-). Another example is covalently bonded fullerenes, commonly referred to as buckeyballs. In 1992, a new class of molecular clusters was discovered that form comparatively strongly bonded cagelike structures and, moreover, display a heretofore-unknown growth pattern leading to multicage assemblies. These new molecular clusters are composed of particular combinations of carbon and early transiton-metal atoms; to date, they have been observed in systems composed of titanium, vanadium, hafnium, and zirconium. In view of the chemical nature of their bonding, this new class of molecular cluster materials has been termed metallo-carbohedrenes (Met-Cars for short). SEE FULLERENE.

The first member discovered. The metallo-carbohedrene composed of 8 titanium atoms and 12 carbon atoms (Ti_8C_{12}) was the first member of the new class of molecular clusters to be discovered. The identification was made through the use of mass spectrometry, employing several different technical designs and reaction apparatuses. This metallo-carbohedrene was discovered during the course of investigations of reactions of metal atoms and ions in a laser vaporization plasma reactor in which the metal vapor species encounters small hydrocarbons in a gaseous plasma. The plasma is created as a result of ionization processes induced by a laser beam that is tightly focused onto an appropriate metal surface immersed in a stream of helium gas containing the hydrocarbon reactants. This technique not only leads to a wide range of precursor species but also fully dehydrogenates the small hydrocarbons and provides a critical concentration of carbon species that undergo chemical interactions with the metal atoms.

In the usual situation in cluster research, a wide range of cluster species are produced during the course of the ensuing growth processss. However, in the case of the reaction experiments that led to the new discovery, a nearly pure single mass peak arose (**Fig. 1**). Titanium has four times the mass of a carbon atom; therefore, isotopic substitution experiments using deuterated hydrocarbons, where the hydrogen atoms are replaced by deuterium (which has twice the weight of hydrogen), and alternatively using isotopically substituted carbon-13 for the normal carbon-12, were necessary in order to confirm the mass identity of the metallo-carbohedrene. As seen in Fig. 1, the peak at 528 atomic mass units is totally dominant; it is termed supermagic in view of its abnormally enhanced abundance over all other detected species. Investigations conducted with methane, ethane, ethylene, acetylene, propylene, and benzene all led to the same species, namely Ti_8C_{12}. Importantly, no hydrogen atoms are retained in the product.

In order to gain insight into the possible structure of this new cluster material, titration experiments were conducted with ammonia and water vapor. In both cases, the metallo-carbohedrene took up exactly eight molecules of the titrant, showing that the eight titanium atoms, which have d-electrons and readily accommodate ligands (polar groups or molecules), are similarly coordinated and, moreover, are located on the periphery of the cluster structure. A pentagonal dodecahedron structure is the only one that at present is known to be compatible with the requirements imposed by these findings, and it has been proposed as an explanation for the unusually strongly bonded and stable species found in the experiments. As shown in **Fig. 2**, this structure has 12 pentagonal rings, each of which contains two titanium and three carbon atoms. Each of the titanium atoms can bind to three carbon atoms through $Ti-C$ σ-bonds, and each of the carbon atoms may bind to its adjacent carbon through a $C-C$ σ-bond, in addition to bonding two titanium atoms. Those $C-Ti$ and $C-C$ σ-bonds connect all atoms together and form the backbone network of the dodecahedral titanium carbohedrene.

As for the remaining valence electrons, since titanium is viewed as being carbonlike in the model, the π-bonding structure of Ti_8C_{12} is similar to that which might be envisioned for C_{20}. However, in the present case the d-electrons of titanium are evidently involved in the bonding. The metal-carbon double-bonded structure has been widely seen in metal-carbene

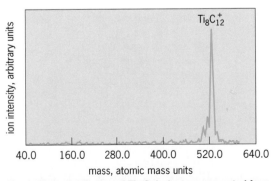

Fig. 1. Mass distribution of $Ti_mC_n^+$ clusters generated from the reactions of titanium with methane. Note the supermagic peak corresponding to $Ti_8C_{12}^+$.

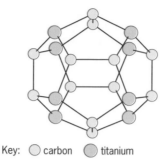

Key: ○ carbon ◉ titanium

Fig. 2. Idealized structure of pentagonal dodecahedron structure proposed to account for the observed stability and ammonia uptake by $Ti_8C_{12}^+$. The eight titanium atoms that appear at the edge of a cubelike arrangement are identically coordinated to three carbon atoms at all equivalent positions.

and metal-carbyne complexes. Since titanium can participate in σ-bonding through other hybridized molecular orbitals, the incorporation of two titanium atoms into the pentagonal ring may somewhat reduce the strain of the ring. The structure and its stability is also borne out by a number of quantum-mechanical calculations, which provide theoretical insight into the structure and stability.

Other members. The new findings of this unusually stable species led to a search for other members of the class. Subsequently, observations of magic peaks corresponding to other M_8C_{12} species were made for cases where M is vanadium (V), zirconium (Zr), or hafnium (Hf). The M_8C_{12} clusters are found to have stable species of cations (positively charged) and anions (negatively charged). Significantly, the stability of the neutral species has been also proven through laser ionization experiments involving the product of the plasma reaction. Titration experiments similar to those performed for the titanium system led to conclusions that the other members are likely to have a pentagonal dodecahedral cagelike structure as well. Similar arguments pertain to the other members' bonding, where it is proposed that the remaining valence electrons can contribute to a delocalized π-electron system throughout the metal-carbon bonded network. In view of the experiments, which reveal the ease of ionizing the neutral species, along with the arguments pertaining to their bonding and theoretical calculations that are consistent with these proposals, it is expected that this class of clusters may be particularly useful as new electronic, optical, and even catalytic materials if bulk amounts can be produced.

Multicage structures. Subsequent investigations of larger members of the zirconium-carbon system revealed evidence for a heretofore-unknown and unique structural growth pattern for these gas-phase clusters. A careful examination of the mass spectrum showed the presence of four distinct groups of clusters, each of which displayed a smooth trend with respect to variations of intensities of the mass peaks with cluster size, followed by termination by a sudden abrupt change in intensity levels, that is, a sudden truncation in the series. The first group of clusters encompassed the peak corresponding to Zr_8C_{12}, designated (8,12).

An abrupt intensity drop occurs after the (8,12), after which there is a second group composed of six peaks. A third group of clusters contains several prominent peaks, as does the fourth group. Truncation of the mass peaks always occurs at cluster sizes corresponding to certain specific metal and carbon compositions, namely (8,12), (14,23), (18,29), and (22,35). These species are characteristic magic numbers, irrespective of the experimental conditions for which they are observed, and are indicative of structures of special stability resulting from cage closings.

Hence, the experimental findings established that a first cage closes at Zr_8C_{12}, as seen for the titanium species first discovered. It is interesting that subsequent cluster growth does not lead to the enlargement of the cage size as it does in the case of most of the clusters (the pure carbon fullerenes and water clusters, for example). Rather, as shown in **Fig. 3**, multicage structures are developed in which adjacent cages, each corresponding to a pentagonal dodecahedron, share common faces; that is, a double cage exists at $Zr_{13}C_{22}$ and $Zr_{14}C_{21}$, a triple cage exists at $Zr_{18}C_{29}$, and a quadruple cage subsequently forms at $Zr_{22}C_{35}$. This unique growth pattern distinguishes the class of metallo-carbohedrenes from the regular metal-doped fullerenes.

Plasma reactor concept. Additional experiments have been conducted that serve to compare and contrast the findings for the early transition-metal–carbon systems with those of other metal-containing carbon species. For example, clusters composed of tungsten, cobalt, tantalum, scandium, and aluminum bound to carbon atoms so far have not been found to assemble into the metallo-carbohedrene structures. The observations to date suggest that those systems that do form metallo-carbohedrenes have dominant

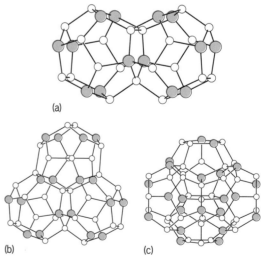

(a)

(b)   (c)

Fig. 3. Multicage metallo-carbohedrene structures. (*a*) A proposed double-cage structure of $M_{14}C_{21}$. When one of the 14 metal sites (solid circles) is occupied by one carbon atom, that is, $M_{13}C_{22}$, the structure is also very stable; two additional carbon atoms (open circles) can form a pentagon through bridging between metal sites on two balls, leading to $M_{14}C_{23}$. (*b*) A proposed triple-cage structure of $M_{18}C_{29}$. (*c*) A proposed quadruple-cage structure.

early precursors in the growth pattern composed of MC_2 or M_2C_3 units. Forming vapors rich in these units may be the key to producing other members of this interesting new class of materials.

In very recent extended studies of the dehydrogenation reactions of other hydrogen-containing small molecules, it has been found that the same plasma reactor technique can be utilized to make a wide variety of materials with varying composition among transition metals bound to nitrides, oxides, and silicides. Elucidating the mechanisms by which these various molecules undergo dehydrogenation and exploring the nature of the new stable and metastable materials that can be produced with the laser plasma reactor concept provide important directions for future research, as does the challenging task of producing metallocarbohedrenes in sufficient quantity to make their use feasible in a wide range of potential applications.

For background information SEE CHEMICAL BONDING; MOLECULAR ORBITAL THEORY; PLASMA PHYSICS in the McGraw-Hill Encyclopedia of Science & Technology.

A. W. Castleman, Jr.

Bibliography. B. C. Guo et al., Metallo-carbohedrenes [$M_8C_{12}^+$($M = V$, Zr, Hf and Ti)]: A class of stable molecular cluster ions, *Science*, 256:515–516, 1992; B. C. Guo, K. P. Kerns, and A. W. Castleman, Jr., $Ti_8C_{12}^+$—Metallo-carbohedrenes: A new class of molecular clusters?, *Science*, 255:1411–1413, 1992; S. Wei et al., Metallo-carbohedrenes as a class of stable neutral clusters: Formation mechanism of M_8C_{12} ($M = $ Ti and V), *J. Phys. Chem.*, 96:4166–4168, 1992; S. Wei et al., Metallo-carbohedrenes: Formation of multi-cage structures, *Science*, 256:818–820, 1992.

Microbial biofilm

Oligotrophic, or low-nutrient, streams usually have very clear water containing few microorganisms. However, the rocks over which these streams flow tend to be covered with biofilms. Biofilms comprise the bacteria, algae, protozoa, and other microorganisms that grow at surfaces, as well as the extracellular substances produced by the microorganisms.

Biofilms form rapidly on synthetic surfaces, such as metal pipes, which are exposed to aqueous environments. Some biofilms are beneficial to human industrial activities. For example, in wastewater treatment, as the water filters through gravel containing biofilms, the bacteria degrade most of the organic material to carbon dioxide and methane. Biofilms have also been used in bioreactors to degrade toxic or environmentally persistent compounds.

Biofilms can also adversely affect certain human activities. Examples include plugging of relief wells, loss of heat transfer in heat-exchange devices, and infection in association with prosthetic devices. Another adverse effect is the enhanced corrosion of metals; such microbiologically influenced corrosion has been recognized for more than 50 years. With advances in technology for studying the biofilms–metal interface,

mechanisms by which biofilms increase the corrosion rates are becoming better understood.

Formation and activity. When a surface is placed in an aqueous environment, dissolved organic material rapidly adsorbs to the surface. Many investigators think this organic conditioning film is necessary for bacterial adhesion to the surface. The bacteria and other microorganisms are transported to the surface via diffusion, gravity, or motility (bacterial swimming). Some species of bacteria are able to sense that they are at a surface and to change their phenotype (physical or chemical attributes). For example, *Vibrio parahaemolyticus* senses that it is at a surface by the restricted rotation of its polar flagella (the organelles used for motility in bulk solutions). Among the changes that this bacterium undergoes is the production of lateral flagella, which are used for swarming motility (movement along a surface). Other bacteria produce an extracellular polysaccharide that helps bind them to surfaces. The initial attachment of the bacteria to surfaces is usually reversible. The bacteria exhibit Brownian motion and can be removed easily by rinsing. Irreversible adhesion usually occurs after the bacteria have produced extracellular organelles or polymers. Mechanical methods, such as brushing, are usually required to remove microorganisms that have become irreversibly adhered.

Once the initial bacteria have bound to the surface, they can utilize nutrients from the bulk solution and from the organic conditioning film for growth. The growth and metabolic activities of the microorganisms change the chemical conditions within the biofilm as well as at the metal–solution interface. Microelectrodes have been used to measure conditions within biofilms, such as pH and dissolved oxygen content. In the absence of light (which stimulates oxygen production by photosynthetic microorganisms), the dissolved oxygen content within the biofilms can be reduced to 0 part per million, whereas the oxygen content in the bulk solution is as high as 8 ppm. Under these conditions of reduced oxygen, anaerobic bacteria (which do not utilize oxygen for metabolism) can flourish. Anaerobic bacteria have been implicated in microbiologically influenced corrosion, since many of them produce corrosive compounds, such as organic acids, as end products of metabolism. Another type of anaerobic bacteria that is commonly found in biofilms and has been implicated in microbiologically influenced corrosion is the sulfate-reducing bacteria. They utilize sulfate as the terminal electron acceptor in oxidative metabolism. The sulfate is reduced to sulfide, which can be particularly corrosive.

Corrosion reactions. In aqueous solutions, freely corroding metal undergoes anodic (oxidation) reactions and cathodic (reduction) reactions. In the anodic reaction, the base metal (M) loses electrons (e^-) and is oxidized to cations (positively charged ions, M^{n+}), as in reaction (1). The electrons released in

$$M \rightarrow M^{n+} + ne^- \qquad (1)$$

the anodic reaction must be consumed in the cathodic reaction, where species such as oxygen (O_2) or protons (H^+) are reduced, as in reactions (2) and (3). An

$$O_2 + 2H_2O + 4e^- \rightarrow 4OH^- \qquad (2)$$

$$2H^+ + 2e^- \rightarrow H_2 \qquad (3)$$

electrochemical potential is associated with the anodic and cathodic reactions. The electrochemical potential of the metal is a mixed potential, with contribution from both reactions. This mixed potential is termed the corrosion potential (E_{corr}), which the corroding metal adopts when placed in solution. No net current is produced at E_{corr}. However, an exchange current between the anodic and cathodic reactions occurs, since electrons are passed between these reactions. This exchange current (or corrosion current; I_{corr}) is directly related to the amount of metal oxidized according to Faraday's law.

Since no net current is produced at E_{corr}, the corrosion rate is usually determined by applying an overpotential to the metal, thus driving the reaction in either the anodic or cathodic direction, with the alternative reaction occurring on a separate inert counterelectrode. The current produced in these experiments is measured with a potentiostat. I_{corr} is then determined by extrapolating I_{corr} at E_{corr} or by using equations derived for this purpose.

Following the anodic reaction, where metal ions are released into the solution, the ions can react with oxygen or water to create insoluble metal oxides or metal hydroxides. Graphs known as Pourbaix diagrams show the metal species that can form at any given pH and potential. The metal oxides form passive films over the base metal and can inhibit further oxidation of the metal. Passive films make some metals, such as stainless steel and titanium, very resistant to corrosion.

A simplified Pourbaix diagram is shown in **illus.** a,

with an idealized anodic polarization curve for iron at pH 5 in illus. b. As the potential of the sample is raised from -0.6 V, the current increases as the base metal is oxidized to soluble ferrous cation (Fe^{2+}). As the potential is raised further, the current decreases as the ferric oxide (Fe_2O_3) inhibits the further corrosion of the metal. The current remains low until the potential reaches 1 V, where the passive film breaks down, and pitting (localized corrosion) ensues. The potential at which the passive film breaks down is termed the pitting potential (E_{pit}).

The presence of ions in solution can inhibit or accelerate corrosion. The most notorious corrosion accelerator is the chloride anion (Cl^-). Chloride reacts with passive films and reduces the pitting potential of the metal. If the pitting potential is reduced to E_{corr}, pitting can occur. Bacteria produce compounds that also can induce pitting.

Study techniques. The study of microbiologically influenced corrosion presents interesting challenges that are not present in the study of chemical corrosion. First, the reactions most relevant to corrosion occur at the metal–biofilm interface. These reactions tend to be more difficult to study than those that occur in the bulk solution. Second, biofilms contain living organisms. Thus, the reactions produced by these organisms can change temporally. Third, biofilms are not uniform over surfaces but tend to be arranged in microcolonies. Therefore, techniques that can detect spatial differences must be utilized.

These difficulties have led to collaborative efforts by scientists from normally disparate disciplines, including microbiologists, electrochemists, and materials scientists. Moreover, new technologies have been utilized. Attenuated-total-reflectance Fourier-transform infrared spectroscopy (ATR-FTIR) has been used to study chemistry at the biofilm–metal interface, and confocal laser microscopy has been used to study

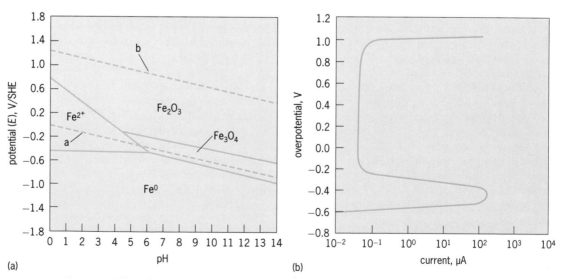

(a)

(b)

Pourbaix diagrams. (a) Simplified diagram for iron in pure water, showing the major insoluble compounds that can form at various levels of pH and potentials. Fe^0 = elemental iron; Fe_2O_3 = ferric oxide; Fe_3O_4 = ferrosoferric oxide; Fe^{2+} = ferrous cation; SHE = standard hydrogen electrode. (b) An idealized anodic polarization curve for iron at pH 5.

chemical reactions occurring within biofilms. Nondestructive electrochemical techniques, such as electrochemical impedance spectroscopy, have been used to study rates and mechanisms of metal corrosion under biofilms. Also, mapping of the potential field has been used to localize the electrochemical reactions. Microelectrode studies have been used to determine local biological effects on surfaces.

Mechanisms of microbiologically influenced corrosion.

Although much study is needed to achieve better understanding of the effects of biofilms on metal corrosion, the proposed mechanisms that have received the most attention are the chemical degradation of passive films by corrosive metabolites produced by bacteria, the physical disruption of the passive film by the biofilm, and the change in the electrochemical potential of the metal induced by the biofilm.

Corrosive metabolites. Corrosive products by bacteria have been implicated in microbiologically influenced corrosion. Anaerobic microenvironments occur in biofilms. Some anaerobic bacteria produce organic acids as end products of fermentative metabolism. The lowering of the pH in localized sites can induce localized corrosion (as seen in the Pourbaix diagram, illus. *a*). Whether this mechanism actually occurs is subject to debate, since the organic acids produced by some anaerobes are rapidly utilized as carbon sources for other bacteria.

Sulfide, produced as an end product by sulfate-reducing bacteria, is probably more important in microbiologically influenced corrosion. Experiments have been performed where the anodic and the cathodic reactions could be spatially separated. The experiments demonstrated that hydrogen sulfide, either biologically produced or added to the solution, could catalyze the anodic reaction of iron or stainless steel, provided oxygen reduction acted as the cathodic reaction. Other experiments demonstrated that biologically produced sulfide lowered the pitting potential of stainless steel, actually having as great an effect as chloride.

Experiments that map anodic and cathodic reactions by using microelectrodes have indicated that most of the bacterial activity is associated with the anodic sites on the metal. These experiments demonstrate that pitting of stainless steel can occur when biofilms containing sulfate-reducing bacteria are present in microcolonies and the remainder of the metal is exposed to aerated solution.

Sulfate-reducing bacteria are often associated with corrosion. Some investigators have performed x-ray diffraction experiments to study deposits associated with microbiologically influenced corrosion. In these deposits they have found certain iron-sulfide minerals, such as mackinawite, which do not occur under normal nonbiological conditions. These crystals have been used as indicators of sulfate-reducing bacteria in microbiologically influenced corrosion.

Disruption of passive films. Physical disruption of passive films by biofilms has been demonstrated. Bacteria that anaerobically reduce ferric iron to the ferrous form have been shown to increase corrosion of steel. It is proposed that the bacteria can reduce the ferric oxide passive film to soluble ferrous ions, thus exposing the base metal to solution.

The extracellular polymers produced by microorganisms have been implicated in microbiologically influenced corrosion. Many polymers produced by bacteria are negatively charged, containing carboxylic acids. The negatively charged polymers bind positively charged metal ions within the passive film and can disrupt the passive film. Experiments with attenuated-total-reflectance Fourier-transform infrared spectroscopy have demonstrated that bacterial polymers, in the absence of bacteria, can bind copper ions and thus can increase the corrosion rates of copper. Similar experiments are being performed to see if bacterial polymers can increase the corrosion rate of stainless steel.

Change in electrochemical potential. The effect of biofilms on the electrochemical potential of metals has received much attention. In studies with naturally flowing seawater, biofilms formed on metal samples, and generally increased the E_{corr} of stainless steels. The potential of the samples was raised to the pitting potential of the steel, resulting in localized corrosion. Analysis of the biofilms by environmental scanning electron microscopy and microelectrodes demonstrated that the rise in potential was due to the presence of photosynthetic microorganisms, which maintained an aerobic environment adjacent to the steel.

Prospects.

Since the research on microbiologically influenced corrosion involves the complex interactions between microorganisms and surfaces, the information developed has potential for counteracting a costly problem. Although much remains to be learned about microbiologically influenced corrosion, progress is being made in identifying the mechanisms involved. Understanding the mechanisms has helped in the identification of microbiologically influenced corrosion (as opposed to chemical corrosion) in the field, and the advances have led to countermeasures. Such application should increase the lifespan of metal structures, even in the presence of organisms that increase the rate of metal degradation.

For background information SEE CONFOCAL MICROSCOPY; CORROSION; FLUORESCENCE MICROSCOPE; FOURIER-TRANSFORM SPECTROSCOPY; MICROBIAL BIOFILM in the McGraw-Hill Encyclopedia of Science & Technology.

Michael Franklin

Bibliography. W. G. Characklis and K. E. Cooksey (eds.) *Biofilms*, 1990; N. J. Dowling, M. W. Mittelman, and J. C. Danko (eds.), *Microbially Influenced Corrosion and Biodeterioration*, 1991; M. Pourbaix, *Atlas of Electrochemical Equilibria in Aqueous Solutions*, 1966; N. Sato, Towards a more fundamental understanding of corrosion processes, *Corrosion*, 45:354–368, 1989; D. C. Savage and M. Fletcher (eds.), *Bacterial Adhesion: Mechanisms and Physiological Significance*, 1985.

Microbial ecology

The science of microbiology underwent a revolution in 1982, when microorganisms that grew at, and even above, 212°F (100°C), the normal boiling point of water, were isolated from shallow marine volcanic vents. About 20 different genera of hyperthermophilic organisms have since been isolated. These organisms can grow at 199°F (90°C) and have an optimal growth temperature of at least 175°F (80°C). Virtually all are classified as Archaea. Formerly known as the Archaebacteria, this group also includes methane-producing organisms and salt-loving or extreme halophilic organisms. On the basis of genetic sequence analyses, Archaea were shown to be closely related to each other but widely separated from Bacteria. The universal phylogenetic tree shows the relationship between the Archaea, Bacteria, and Eucarya, or higher organisms (see **illus.**). Interestingly, the hyperthermophiles are the most ancient organisms within both the Archaea and Bacteria, supporting the notion that the original life forms on Earth evolved at high temperature and suggesting that the environments in which hyperthermophiles are now found are similar to the conditions under which life originated.

Volcanic ecosystems. All of the hyperthermophilic organisms have been isolated from geothermally heated environments, such as marine volcanic vents and the continental hot springs in Iceland, Italy, New Zealand, and the United States. The marine sites are either hot sediments found in coastal waters at depths to 300 ft (100 m) or so, or deep-sea hydrothermal vents as much as 13,120 ft (4000 m) below sea level. The deep-sea vents were first identified in 1977 on the Galápagos Rift off the Pacific coast of Central America. Similar vent sites have been found at deep-sea locations such as the East Pacific Rise, the Mid-Atlantic, the Gulf of California, and the seas of Japan and Polynesia.

Continental and marine geothermal habitats originate in a similar fashion. Surface water percolates deep (as much as 16,400 ft or 5000 m) into the Earth's crust and becomes superheated from the magma chamber, which is brought much closer to the surface as a result of tectonic activity. The water returning to the surface carries dissolved minerals and various gases. On land, this rising water is in the form of hot geysers as well as less spectacular flowing springs. These geysers and springs are frequently highly acidic (pH 1–2) because of the formation of sulfuric acid from the aerobic oxidation of the sulfide-rich water. However, subsurface water not exposed to oxygen, or springs with low sulfide content, are neutral to alkaline (pH 7–9) because of the buffering capacity of sulfide, carbon dioxide, carbonates, and silicates.

Marine vents differ from the continental thermally heated ecosystems in that the rising superheated saline water is more reactive and is extensively mineralized. In addition, the vent waters are highly reducing, anaerobic, and only slightly acidic. The deep-sea-vent waters can emanate at the sea floor as spectacular jets or black smokers. Because the hydrostatic pressure is so high (approximately 100 atmospheres per 3280 ft or 1000 m of depth), the water does not boil and can reach temperatures approaching 752°F (400°C). As this superheated mineralized water meets the cold seawater (ambient is 36°F or 2°C), much of the inorganic material precipitates, resulting in the smoker effect. Microorganisms have not been found in the vent fluids at 752°F (400°C), but they have been isolated both from the spreading smoker plume and from inside the porous mineral sulfide mounds that form the smoker chimney. The deep-sea vents provide remarkable and unique ecological niches, in terms of both temperature and potential nutrients. For example, gases such as methane, hydrogen, and hydrogen sulfide and metals such as iron and manganese are barely detectable in normal seawater (less than 1 nanomolar), but each is present in some vent fluids at concentrations close to 20 millimolar.

Marine hyperthermophiles. The type of hyperthermophilic organisms found in the deep sea reflect the anaerobic, sulfur- and mineral-rich nature of the deep-sea ecosystems. The majority are strictly anaerobic organisms that are obligately dependent upon the reduction of elemental sulfur for optimal growth. They utilize either peptides or carbohydrates to form organic acids, carbon dioxide, and hydrogen gas. The mechanism of elemental sulfur reduction is not clear: several species appear to grow by the fermentation of organic material. Deep-sea hyperthermophilic archaea include species of *Pyrodictium*, *Thermococcus*, and *Pyrococcus*, as well as the hydrogen- and carbon dioxide–utilizing methanogens, *Methanopyrus* and *Methanococcus*. The deep sea is also home for species of the bacterial genus *Thermotoga*. The source of the organic material that these organisms utilize is still a mystery, although one volcanic region, the Guaymas Basin in the Gulf of California, from which several hyperthermophiles have been isolated, is covered by an organically rich sediment almost 1640 ft (500 m) deep.

The immediate vicinity of the deep-sea smoker is sulfide-rich, high-temperature, and free of oxygen.

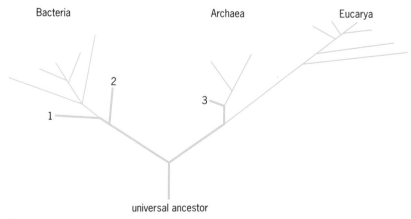

Universal evolutionary tree. The heavy line shows the extent of hyperthermophily, which includes the most ancient Bacteria (1, *Thermotoga*; 2, *Aquifex*) and the most ancient Archaea (3). (*After C. R. Woese et al., Towards a natural system of organisms: Proposal for the domains of Archaea, Bacteria and Eucarya, Proc. Nat. Acad. Sci., 87:4576–4579, 1990*)

Yet, the surrounding seawater which is at 36°F (2°C) and saturated with oxygen can support a variety of mesophilic and aerobic life, including microorganisms and higher life forms such as clams, crabs, and tubeworms. Oxygen-requiring hyperthermophilic organisms have not been isolated from deep-sea environments. The notable exception, the most thermophilic bacterium currently known, is of the genus *Aquifex*, and was isolated from 330 ft (100 m) below sea level off the Icelandic coast.

Fresh-water hyperthermophiles. In contrast to anaerobic and nearly neutral marine vents, continental hot springs can be aerobic and highly acidic. Hence, many of the hyperthermophilic species that have been isolated from fresh-water environments are thermoacidophiles, which use carbon dioxide or sugars as carbon sources and obtain energy for growth by the aerobic oxidation of sulfur to sulfuric acid. The most commonly found species are of the genus *Sulfolobus*. Hyperthermophilic anaerobes can be found in the same habitat, and like many of their deep-sea relatives, they grow by reducing sulfur to sulfide. Hyperthermophilic anaerobes include species of *Acidianus*, which can grow both aerobically, by sulfur oxidation, or anaerobically, by sulfur reduction. Strictly anaerobic hyperthermophiles found in hot springs include species of the genera *Thermoproteus* and *Pyrobaculum*. Hotspring environments also contain methanogens, with species of the genus *Methanothermus* being the hyperthermophilic representatives. Like many of the sulfur-reducing organisms, methanogens use the hydrogen and carbon dioxide in these volcanic ecosystems as carbon and energy sources.

Distribution of hyperthermophiles. All of the marine hyperthermophiles require high salt concentratiuons for growth; hence organisms of the same hyperthermophilic genus have not been isolated from both fresh-water (continental) and marine environments. However, several hyperthermophiles have been isolated from fresh-water springs on three or more continents, and representatives of several marine genera have been found in both shallow and deep-sea vents. Moreover, the volcanic environments in which the hyperthermophiles are found are not infinitely stable: some deep-sea vents have a lifetime perhaps on the order of a few years. Therefore, how are these organisms disseminated?

Viable hyperthermophiles have been found under aerobic conditions and at ambient temperature in the open-sea plume after the eruption of a submarine volcano, suggesting that they may be readily transported throughout the oceans. Their ability to survive under such conditions stems from their being extremely well adapted to their geothermal environments. Some do not grow at temperatures below 175°F (80°C), and most have a minimum growth temperature near 140°F (60°C). Hence, at ambient temperatures they remain dormant, and even the most fastidious of the anaerobes appear to be relatively insensitive to oxygen. Moreover, the growth of deep-sea hyperthermophiles is essentially unaffected by the high hydrosatic pressures of their natural environment. An obligately barophilic hyperthermophile has yet to be isolated.

Upper temperature limit. The upper temperature limit for growth has remained about the same (near 230°F or 110°C) over the last decade, despite the now routine search for organisms that may grow at higher temperatures. However, there are unpublished reports of organisms growing at temperatures in excess of 248°F (120°C). Perhaps the commonly used isolation media do not contain ingredients sufficient to support the metabolisms of superhyperthermophiles; more imaginative substrates may be required. Alternatively, such life forms may not be free-living, but require a solid support, or they may exist as consortia that cannot be separated. A combination of these and other as yet unknown parameters may therefore be required to raise the upper temperature limit to between 266 and 302°F (130 and 150°C). Perhaps such organisms will map on the evolutionary tree even closer to the universal ancestor, and no doubt they will be found within a volcanic ecosystem.

For background information SEE ARCHAEBACTERIA; BACTERIAL PHYSIOLOGY AND METABOLISM; DEEP-SEA BACTERIA; HYDROTHERMAL VENT; MICROBIAL ECOLOGY in the McGraw-Hill Encyclopedia of Science & Technology.

Michael W. W. Adams

Bibliography. M. W. W. Adams and R. M. Kelly (eds.), *Biocatalysis at Extreme Temperatures*, ACS Symp. Ser. 498, 1992; T. D. Brock (ed.), *Thermophiles: General, Molecular and Applied Microbiology*, 1986; S. Burggraf et al., A phylogenetic analysis of *Aquifex pyrophilus*, *Syst. Appl. Microbiol.*, 15:352–356, 1992; H. W. Jannasch, The microbial basis of life at deep sea hydrothermal vents, *Amer. Soc. Microbiol. News*, 55(8):413–416, 1989.

Mitochondria

Mitochondria are energy-producing organelles found within all mammalian cells. The mitochondrion is the only organelle besides the nucleus that contains its own deoxyribonucleic acid (DNA), called mitochondrial DNA (mtDNA). The mtDNA encodes components of the respiratory chain and the oxidative phosphorylation system; however, most proteins located within the mictochondria are products of genes in the nuclear DNA and are imported from the cytoplasm into the mitochondria. Mitochondria, and mtDNA, are inherited exclusively from the mother.

Defects in mitochondrial function are associated with a heterogeneous group of mitochondrial diseases. Because of the dual genetic systems involved (mtDNA and nuclear DNA), heritable errors leading to mitochondrial dysfunction can arise from both genetic compartments. Interestingly, some mitochondrial diseases associated with mutations in mtDNA arise spontaneously, with no apparent inheritance. Recently, there has been a huge increase in knowledge of the genetics of mitochondrial diseases. In addition, there is now evidence that low levels of mtDNA mutations accumulate in normal aging, suggesting that the aging process may be related to defects in mitochondrial function arising from errors in mtDNA.

Maternally inherited disorders. Seven mitochondrial disorders, all maternally inherited, have been reported to be associated with point mutations in mtDNA: (1) Leber's hereditary optic neuropathy is characterized by bilateral loss of vision in the second or third decades, and is more prevalent among men than women. To date, this disorder has been associated with a number of point mutations in mtDNA, all of which reside in structural genes for the mitochondrial respiratory chain. The mutations may be homoplasmic in some families (that is, all mtDNAs are mutated) or heteroplasmic (that is, both normal and mutated mtDNAs are harbored) in others. Specific mutations may act together, perhaps cumulatively, to produce the phenotype of Leber's hereditary optic neuropathy. (2) Myoclonus epilepsy with ragged-red fibers is characterized by myoclonus, ataxia, weakness, generalized seizures, and hearing loss, and by ragged-red fibers in the muscle biopsy (indicating massive mitochondrial proliferation in muscle tissue). This disorder has been found to be associated with a mutation in the gene encoding the transfer ribonucleic acid (tRNA) responsible for the incorporation of lysine into the mtDNA-encoded proteins (tRNA^{Lys}). This mutation results in decreased protein synthesis of mtDNA-encoded polypeptides as well as in synthesis of a few aberrant protein species. (3) Mitochondrial encephalomyopathy with lactic acidosis and strokelike episodes is characterized by seizures, migrainelike headaches, lactic acidosis, episodic vomiting, short stature, and recurrent cerebral strokelike episodes, often causing cortical blindness. This disorder is associated with two different point mutations, both in the gene encoding another tRNA, that required for leucine incorporation (tRNA^{Leu}). The mutations appear to cause a severe decrease in overall mitochondrial protein synthesis, but no aberrant polypeptides have been observed. (4) A maternally inherited adult-onset mitochondrial myopathy and cardiomyopathy is associated with a point mutation in yet another position in the tRNA^{Leu} gene. (5) Another maternally inherited cardiopathy is associated with a point mutation in yet a third tRNA gene that specifies incorporation of isoleucine (tRNA^{Ile}). (6) One mitochondrial disorder is characterized by retinitis pigmentosa, dementia, seizures, ataxia, proximal neurogenic muscle weakness, sensory neuropathy, and developmental delay. It is associated with a heteroplasmic point mutation within the gene encoding subunit 6 of the ATPase complex. (7) Remarkably, when the proportion of the preceding mutation is very high (> 90%), a different clinical phenotype results, namely, Leigh syndrome, a progressive fatal mitochondrial encephalopathy of infancy or childhood.

Other disorders. Three other categories of disorder have been associated with large-scale rearrangements (giant deletions or insertions) of the mitochondrial genome; although these disorders are mutations in mtDNA, none is maternally inherited: (1) Progressive external ophthalmoplegia, or paralysis of the extraocular muscles, is found in two related disorders—a relatively benign ocular myopathy and a frequently fatal

multisystem disorder called Kearns-Sayre syndrome. Morphologically, the progressive external ophthalmoplegia patients display ragged-red fibers in their muscle tissue. Historically, ocular myopathy and Kearns-Sayre syndrome were difficult diseases to classify, because patients were almost all spontaneous, with no apparent genetic component. It is now known that spontaneously occurring giant deletions of mtDNA—up to 10 kilobases of the 16.6-kb mitochondrial genome—are a hallmark of almost all cases of Kearns-Sayre syndrome and about half the cases of ocular myopathy. Besides deletions, duplications of mtDNA have also been identified in Kearns-Sayre syndrome. All patients with such rearrangements have been heteroplasmic, with a large proportion of deleted mtDNAs in the muscle tissue. The size and location of the deletions, and the number of deleted mtDNAs relative to the number of normal mitochondrial genomes, differ among patients, and do not appear to be correlated to the features of the severity of clinical phenotype. (2) Deleted mtDNAs have been identified in Pearson's syndrome, a hematologic disease characterized by pancytopenia, in which the deleted mtDNA population is most pronounced in blood. In both progressive external ophthalmoplegia and Pearson's syndrome, usually only a single species of deleted mtDNA is found in any individual; it is likely that the population of deleted mtDNAs is derived from an initial single mutation event occurring early in oogenesis or embryogenesis. (3) Mendelian inheritance of deleted mtDNAs has also been demonstrated in mitochondrial myopathy with progressive external ophthalmoplegia. This disease is inherited in an autosomal dominant manner. Interestingly, familiar progressive external ophthalmoplegia causes multiple deletions that are apparently generated during the life-span of the individual. The deletions may differ among family members, and different populations of deletions can coexist within the muscle tissue of affected individuals. Multiple deletions have also been observed in patients with multiple symmetrical lipomas, recurrent myoglobinuria, progressive mitochondrial encephalomyopathy, and myoneurogastrointestinal encephalomyopathy.

Mitochondrial mutations and aging. Recently, extremely low levels of deleted mtDNAs have been found to accumulate in tissues of aging normal individuals. It has been estimated that one particular deleted mtDNA species, called the common deletion (also observed in Kearns-Sayre syndrome), is present at a level of 0.1–0.5% in muscle and brain from aged individuals. This accumulation appears to be greatest in tissues that are long-lived, such as muscle and brain; more rapidly dividing tissues (such as liver and spleen) accumulate deleted mtDNAs to a much lesser extent. Other deletions found in Kearns-Sayre syndrome patients have also been found in aged normal individuals. Elevated levels of deleted mtDNAs in brain tissue from Parkinson's disease patients and in heart tissue from patients with cardiac disease have been reported. Taken together, these findings imply that aged tissues may harbor numerous (hundreds or even thousands of) species of deleted mtDNAs; the total number may reach levels high enough to impair oxidative metabolism in aged tissues.

For background information SEE CYTOPLASMIC IN-HERITANCE; HUMAN GENETICS; MITOCHONDRIA; MUTA-TION in the McGraw-Hill Encyclopedia of Science & Technology.

Eric A. Schon

Bibliography. S. DiMauro et al., Mitochondrial encephalomyopathies: Biochemical approach, *Rev. Neurol. (Paris)*, 147:443–449, 1991; C. T. Moraes et al., Mitochondrial diseases: Toward a rational classification, in S. H. Appel (ed.), *Current Neurology*, vol. 11, 1991; E. A. Schon et al., Analysis of giant deletions of human mitochondrial DNA in progressive external ophthalmoplegia, in J. W. Gorrod et al. (eds.), *Molecular Basis of Neurological Disorders and Their Treatment*, 1991: D. C. Wallace, Diseases of the mitochondrial DNA, *Annu. Rev. Biochem.*, 61:1175–1212, 1992.

Molecular evolution

Molecular evolution encompasses two areas of study. The first area examines the evolution of macromolecules such as deoxyribonucleic acid (DNA) and proteins to determine the rate and pattern of change at the molecular level and to identify and understand the factors that govern these changes. The second area, known as molecular phylogenetics, uses DNA or protein sequence data to reconstruct the evolutionary history of organisms. Since DNA is the hereditary material of all living organisms and DNA sequences change in a slow, fairly regular manner, DNA sequence data can be used to infer the evolutionary relationships among organisms and to estimate the time of divergence between species. Since the early 1980s, a large number of genes have been cloned and sequenced because of the dramatic advances in molecular biology techniques, and this rapid accumulation of data has led to much progress in the study of molecular evolution, particularly in the reconstruction of evolutionary history.

Neutral mutation hypothesis. In Darwin's theory of evolution, natural selection is given the dominant role in shaping the genetic makeup of populations and in the process of genetic change between populations or species, although mutation is recognized as the ultimate source of genetic variation. In time, this theory became a dogma in evolutionary biology, and natural selection came to be considered the only force capable of driving the evolutionary process. This view of Darwinism is called selectionism. In this view, other factors, such as mutation and random drift, were thought of as minor contributors at best.

In the late 1960s, the selectionist view was challenged by scientists who proposed that the majority of molecular changes in evolution are not the results of natural selection for advantageous mutations but are due to the random fixation of neutral or nearly neutral mutations. The term neutral refers to a mutation that confers the same fitness to its carriers as does the wild-type gene, or in other words, the mutant type is neither worse nor better than the wild type in terms of fitness. The term nearly neutral refers to a mutation that is not strictly neutral but is very close to being neutral so that its fate is largely determined by chance events rather than natural selection. The neutral mutation hypothesis does not deny that morphological changes in evolution are mainly due to natural selection and that species respond adaptively to environmental changes. Rather, it is concerned only with evolutionary changes at the molecular level and postulates that at this level evolution occurs largely by chance. Because this view is directly opposite to the traditional Darwinian view, it has stimulated one of the great controversies in evolution in the twentieth century.

The essence of the dispute between selectionists and neutralists concerns the distribution of the fitness values of mutations. Selectionists believe that neutral mutations do not exist or occur extremely rarely. Neutralists accept the traditional view that most of nonsynonymous mutations, which are mutations that cause changes in protein sequences, are deleterious and are quickly eliminated from the population so that they do not contribute much to evolution. However, neutralists believe that the majority of nondeleterious mutations are nearly neutral and that the differences between homologous protein sequences from different species are mainly due to the random fixation of such nearly neutral mutations. They believe that the proportion of neutral or nearly neutral mutations is even higher among synonymous mutations, which are mutations that do not cause amino acid changes, and that in noncoding regions [DNA regions that do not code for proteins or any ribonucleic acid (RNA) sequences] most mutations are selectively neutral.

Intense research since the early 1970s has led to a better understanding of the mechanisms of DNA sequence evolution. The majority of evolutionists now accept the neutralist view for evolution in noncoding regions, which can occupy up to more than 90% of the genome in higher organisms such as humans and other mammals but are rare in bacteria. However, whether the neutralist view can largely explain the evolution of coding regions is still much debated. Of course, how coding sequences have evolved is the most important issue in molecular evolution, for these sequences determine the function of an organism. Regardless of whether it will turn out to be true, the neutral mutation hypothesis has had a profound impact on evolutionary thought, and biologists no longer accept natural selection as the only explanation for evolutionary phenomena.

Closest relatives of humans. Traditionally, humans (*Homo sapiens*) were thought to be very different from the apes. The human species was given a family of its own, Hominidae, whereas the African apes (chimpanzee and gorilla) and the Asian ape (orangutan) were placed in a separate family, Pongidae (**Fig. 1***a*). The gibbon, another Asian ape, was classified either separately or with the Pongidae. However, by using a serological test, it was shown that humans, chimpanzees, and gorillas are considerably closer to

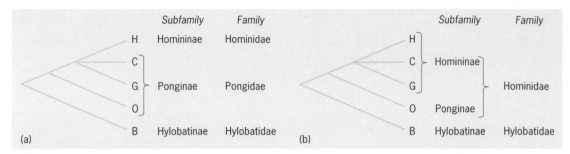

Fig. 1. Two alternative trees and classifications of modern apes and humans. (*a*) Humans in a separate family. (*b*) Humans, chimpanzees, and gorillas in the same subfamily and family. H = humans; C = chimpanzees; G = gorillas; O = orangutans; B = gibbons. (*After W.-H. Li and D. Graur, Fundamentals of Molecular Evolution, Sinauer, 1991*)

one another than any of them is to the orangutan or the gibbon (Fig. 1*b*). In 1967, an immunological method was used to estimate the divergence time between humans and the African apes to be as recent as 5 million years ago (m.y.a), rather than a minimum date of 14 m.y.a. as was commonly thought by paleontologists and anthropologists. The exact date of divergence between humans and the African apes remains to be determined.

Despite many efforts using serological, electrophoretic, or amino acid sequencing methods, no resolution for the evolutionary relationships among humans and the African apes was achieved. Indeed, each of the three possible alternative views shown in **Fig. 2** had been advocated by many authors. DNA sequencing techniques are much more powerful in producing a large amount of sequence data, and the problem of the three views is now commonly believed to have been resolved. The phylogenetic technique employed is to see which of the three alternative branching orders requires the smallest number of nucleotide substitutions to explain the differences among the DNA sequences in the study by using the orangutan as an outgroup (a species that has branched off earlier than the species under study and determines the root of the tree). This technique in known as the maximum parsimony method. Surprisingly, chimpanzees are closer to humans than to gorillas, despite the fact that morphologically chimpanzees look more similar to gorillas than to humans. Thus, the closest relative of humans is the chimpanzee.

Molecular clocks. In the early 1960s, a number of protein sequences became available, and some

biochemists were interested in knowing how protein sequences have evolved with time. A surprising finding was that the hemoglobin sequences from human, cow, rabbit, and horse were roughly equally distant from one another. (The distance between two protein sequences is measured in terms of the number of amino acid differences.) Since these mammalian species were thought to have radiated at about the same time (about 75 m.y.a.), the approximate equality of pairwise distances would suggest that amino acid substitution has proceeded at approximately the same rate in all these mammalian species. It was therefore proposed that the rate of evolution in a macromolecule is approximately constant per year over time and among different evolutionary lineages. This proposal, known as the molecular clock hypothesis, immediately stimulated interest in the use of macromolecules in evolutionary study, because if the hypothesis holds, macromolecules would be extremely useful for dating evolutionary events such as species divergence times. This method would be similar to the dating of fossils by radioactive elements. Moreover, macromolecules would be useful for inferring the relationships among organisms, for the distance between two species would be roughly proportional to their divergence time.

The hypothesis, however, has provoked a great controversy because the clock concept does not fit well with the erratic tempo of morphological evolution. Moreover, it is difficult to imagine why the rate of evolution should be constant per year instead of per generation because the rates of mutation for different organisms appear to be more comparable when measured in terms of generation. Consequently, there has been a strong controversy over whether differences in generation time can have a significant effect on the rate of evolution, or the molecular clock.

Since the early 1970s, numerous studies have been made to test the molecular clock hypothesis, but they have not been able to reject the hypothesis by the use of amino acid sequence data or DNA hybridization data. Therefore, the hypothesis was widely accepted by molecular evolutionists. The rapid accumulation of DNA sequence data has allowed a much closer examination of the hypothesis. Although some scientists still support the molecular clock hypothesis, strong evidence now suggests that no global or universal clock exists. For example, some evidence indicates that the rate of nucleotide substitution may

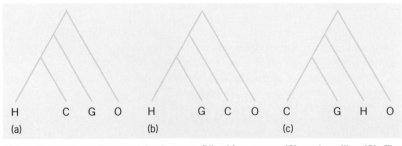

Fig. 2. Three alternative trees for humans (H), chimpanzees (C), and gorillas (G). The orangutan (O) is used as an outgroup to root the tree. (*a*) Chimps are closer to humans than to gorillas. (*b*) Gorillas are closer to humans than to chimps. (*c*) Chimps are closer to gorillas than to humans.

be up to six times higher in rodents than in humans and that the rate in monkey lineages is about 1.5 times higher than that in the human lineage. However, the rate of nucleotide substitution has been found to be nearly the same in the mouse, rat, and hamster lineages. In mammals there seems to exist a local clock for species with very similar generation times, but the molecular clock runs faster for organisms with short generation times. Other lines of evidence negate the existence of a global clock; for example, there is evidence for a sevenfold slower rate of nucleotide substitution in sharks than in mammals.

Classification of guinea pig. The use of DNA or protein sequences in phylogenetic study has sometimes led to big surprises. The guinea pig may be such a case, for current protein sequence data suggest that it may not be a rodent.

Rodentia is the largest mammalian order, consisting of more than half of living mammalian species. It was once divided into the suborders Myomorpha (ratlike rodents), Sciuromorpha (squirrellike rodents), and Hystricomorpha (porcupinelike rodents). The Hystricomorpha includes the New World families of rodents, which are sometimes placed in a suborder, the Caviomorpha (guinea-pig-like rodents), that is independent of the Old World hystricomorph rodents. Today, many rodent taxonomists recognize only two suborders, the Hystricognathi, which includes the New World and Old World hystricomorphs, and the Sciurognathi, which includes the myomorphs and the sciuromorphs.

The guinea pig caught the attention of molecular evolutionists because its insulin is very different from other mammalian insulins. For example, guinea pig insulin differs from both human and mouse insulin by 18 amino acids, whereas the latter two insulins differ by only 4 amino acids. Moreover, many other guinea pig proteins, such as lipoprotein lipase and glucagon, are also quite different from their rat and mouse counterparts. There are two alternative explanations for these observations: either many guinea pig proteins have evolved at high rates, or the guinea pig has been misclassified as a rodent.

To examine the second possibility, 15 proteins were analyzed to see which of the three alternative evolutionary schemes in **Fig. 3** is most plausible. Tree I (Fig. 3a) represents the traditional view that the guinea pig and the myomorphs are sister groups and the primates are an outgroup to them. Tree II (Fig. 3b) clusters the myomorphs and the primates together and puts the guinea pig as an outgroup to myomorphs and primates. Tree III (Fig. 3c) is a third alternative. The parsimony principle was also used to see which tree requires the least number of amino acid changes to explain the differences among the 15 protein sequences. Surprisingly, tree II requires considerably fewer amino acid substitutions than does tree I, which is only as parsimonious as tree III.

The guinea pig is classified as a rodent mainly because it gnaws and has one upper and lower pair of incisors enlarged and ever-growing. The conflict

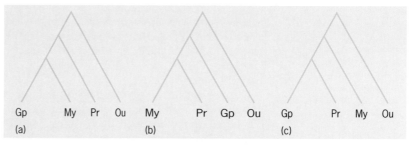

Fig. 3. Three alternative trees for the guinea pig (Gp), the myomorphs (My), and the primates (Pr). (*a*) Traditional view with guinea pig and myomorphs as sister groups and primates as outgroup. (*b*) Myomorphs and primates as sister groups and guinea pigs as outgroup. (*c*) Guinea pig and primates as sister groups and myomorphs as outgroup. The outgroup (Ou) used to root the tree can be either a marsupial or a bird (chicken).

between molecular and morphological data calls for both a careful reexamination of morphological data and further collection of molecular data. If tree II continues to be favored by further molecular data, the guinea pig should no longer be classified as a rodent. In this case, in what would be a major revision of rodent taxonomy, the guinea pig and the other hystricomorphs would have to be separated from the sciuromorphs and myomorphs and put into a new order distinct from the Rodentia.

For background information *SEE MOLECULAR BIOLOGY; MOLECULAR GENETICS; PROTEINS, EVOLUTION OF* in the McGraw-Hill Encyclopedia of Science & Technology.

Wen-Hsiung Li

Bibliography. J. H. Gillespie, *The Causes of Molecular Evolution*, 1991; G. Graur et al., Is the guinea pig a rodent?, *Nature*, 351:649–652, 1991; M. Kimura, *The Neutral Theory of Molecular Evolution*, 1983; W.-H. Li and D. Graur, *Fundamentals of Molecular Evolution*, 1991; A. P. Martin et al., Rates of mitochondrial DNA evolution in sharks are slow compared with mammals, *Nature*, 357:153–155, 1992; C. O'hUigin and W.-H. Li, The molecular clock ticks regularly in murid hamsters, *J. Mol. Evol.*, 35:377–384, 1992; M. Ruvolo et al., Resolution of the African hominoid trichotomy by use of a mitochondrial gene sequence, *Proc. Nat. Acad. Sci. USA*, 88:1570–1574, 1991.

Negative temperature

Nuclear magnetic ordering has been studied extensively at temperatures $T > 0$ in copper and more recently at $T > 0$ and $T < 0$ in silver. These types of experiments have been carried out since the early 1970s. In copper and silver, competition between the dipolar force and the conduction-electron-mediated exchange interaction leads at $T > 0$ to antiferromagnetic spin structures. At negative spin temperatures, ferromagnetic order has been observed in silver. In this metal, the phase transitions occur at 560 picokelvins and -1.9 nanokelvins, respectively. Late in 1992, experiments on rhodium metal produced spin temperatures of 280 pK and -750 pK, the current low-temperature records. The results show that negative temperatures are real, not fictitious, quantities.

Nuclear cooperative phenomena at negative temperatures have been investigated by A. Abragam and his coworkers in dielectric materials such as calcium fluoride (CaF_2) and lithium hydride (LiH). These experiments, however, are limited to ordering by the truncated dipolar force.

Theoretical background. Nuclear spins in metals provide good models to investigate magnetism. The nuclei are well localized, their spins are isolated from the electronic and lattice degrees of freedom at low temperatures, and the interactions between nuclei can be calculated from first principles. Therefore, these systems are particularly suitable for testing theory against experiments. Because the nuclear magneton is small, the critical temperatures for spontaneous magnetic ordering are in the submicrokelvin range.

The hamiltonian of nuclei in silver can be written as in Eq. (1). The dominating spin-spin energy is the

$$\mathcal{H} = \mathcal{H}_{\text{dip}} + \mathcal{H}_{\text{ex}} + \mathcal{H}_Z \qquad (1)$$

nearest-neighbor Ruderman-Kittel exchange interaction $\mathcal{H}_{\text{ex}}$. The dipolar force $\mathcal{H}_{\text{dip}}$ is smaller by a factor of 3, and the Zeeman term $\mathcal{H}_Z$ is proportional to the external magnetic field B.

The energy level diagram for the nuclear spin system of silver is illustrated schematically in **Fig. 1**; because the spin $I = \frac{1}{2}$, there are just two levels, corresponding to nuclear magnetic moments μ parallel and antiparallel to B. The distribution of nuclei among the Zeeman levels is determined by the Boltzmann factor, $\exp[\mu B/(k_B T)]$, where k_B is Boltzmann's constant. At positive temperatures the number of nuclei in the upper level, with μ antiparallel to B, is always smaller than in the lower. When the temperature is increased, more spins jump into the upper band, and at $T = +\infty$ there is an equal number of spins in both levels. At the absolute zero, all nuclei are in the lower energy level. The distribution at $T = +0$ thus corresponds to an energy minimum of the spin system.

Since the energy spectrum of nuclei is limited from above, a population inversion is possible. It can be accomplished by reversing the magnetic field B quickly, in a time $t \ll \tau_2 = 10$ ms, where τ_2 is the spin-spin relaxation time, so that the nuclei do not have a chance to rearrange themselves among the energy levels. This state of inverted nuclear-energy-band populations (Fig. 1) can also be described by the Boltzmann distribution but with a negative absolute temperature. At $T = +0$, an isolated nuclear spin system

has the lowest and, at $T = -0$, the highest possible Helmholtz free energy. Negative temperatures are thus, in fact, hotter than positive ones. In a certain sense, the system passes from positive to negative temperatures via $T = +\infty = -\infty$, without crossing the absolute zero. Therefore, the third law of thermodynamics is not invalidated.

It should be pointed out that negative temperatures are possible only in a system whose energy levels have an upper boundary. In the case of a lattice, for example, a negative temperature would mean that the system has an infinite energy, which is impossible.

In silver, the dominating antiferromagnetic Ruderman-Kittel exchange interaction favors antiparallel alignment of the nuclei and thus leads, in order to minimize the free energy, to nuclear antiferromagnetism when $T > 0$. At $T < 0$, since Helmholtz free energy now must be maximized, the same interactions tend to produce ferromagnetic order. The long-range dipolar force favors formation of ferromagnetic domains.

Experimental procedure. The so-called brute force nuclear cooling method, in cascade, was used in the experiments on silver. During adiabatic demagnetization, only the nuclei are cooled directly; the spin temperature T and the common lattice and conduction electron temperature T_e can differ by several orders of magnitude. There are thus two distinct temperatures in the same specimen at the same time.

To obtain nuclear temperatures in the nano- and picokelvin range, a sophisticated cooling apparatus with two nuclear refrigeration stages was employed. The nuclei of the first stage, made of 20 moles of copper, were first polarized to $p > 0.90$ by using a high initial magnetic field $B_i = 8$ teslas and a precooling temperature $T_i = 15$ millikelvins, obtained by a dilution refrigerator, and then adiabatically demagnetized to a small field $B_f = 100$ milliteslas, whereby a low temperature $T_f = T_i(B_f/B_i) = 200$ microkelvins resulted. At this temperature, the spin-lattice relaxation time τ_1 is 14 h.

Next, starting from this temperature and a 7-T field, a 2-g silver specimen of thin polycrystalline foils, acting as the second nuclear stage, was demagnetized to a low field. The silver nuclei thereby cooled to the nano- or picokelvin range, thermally isolated by the slow spin-lattice relaxation from the conduction electrons, which were thermally anchored to the first nuclear stage and remained at $T_e = 200\ \mu$K.

Finally, the production of negative temperatures was achieved in the nuclear spin system of silver by inverting a 400-microtesla magnetic field in about 1 ms. A special coil system and radiation shields were built to prevent eddy currents, which must have been the problem during earlier, less successful experiments. Nevertheless, the rapid flipping of the magnetic field always resulted in a loss of polarization in the nuclear spin system. The inversion efficiency was about 95% at small polarizations but decreased to 80% for $p > 0.80$. Therefore, the studies at $T < 0$ were limited to negative polarizations up to $p = -0.65$.

One of the difficult problems in these experiments was how to measure the absolute temperature of the

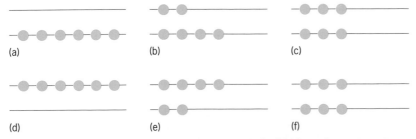

Fig. 1. Energy-level diagrams of silver nuclei in a magnetic field for various temperatures T. (a) $T = +0$. (b) $T > 0$. (c) $T = +\infty$. (d) $T = -0$. (e) $T < 0$. (f) $T = -\infty$.

nuclei. The second law of thermodynamics, $T = \Delta Q/\Delta S$, was used directly; that is, the nuclear spin system was supplied with a small amount of heat ΔQ, and the ensuing entropy increase ΔS was measured. The method is simple in principle but complicated in practice.

Results on silver. The quantity measured in the experiments on silver is the nuclear magnetic resonance (NMR) absorption χ'', recorded by a SQUID. Nuclear polarization was computed from Eq. (2),

$$p = A \int \chi''(\nu) \, d\nu \qquad (2)$$

where ν is the nuclear magnetic resonance sweep frequency and the constant A is determined from measurements near 1 mK. In **Fig. 2a**, the absolute value of the inverse magnetic susceptibility $1/\chi$ of silver, calculated from the Kramers-Krönig relation (3), is

$$\chi = (2/\pi) \int (\chi''/\nu) \, d\nu \qquad (3)$$

plotted as a function of $|T|$ in nanokelvins. At positive temperatures a straight line is obtained with an intercept on the negative side of the temperature axis. This behavior is typical and indicates that silver tends to antiferromagnetic order when $T \to +0$.

At negative temperatures, the intercept is on the positive side of the $|T|$ axis, which shows that when $T \to -0$ the spin system of silver tends to ferromagnetic order; the transition point, however, was not reached in these early experiments. Instead of absorption, as at $T > 0$, the system emits energy when $T < 0$, as is illustrated in the insets of Fig. 2a, where the curves are fits of lorentzian line shapes to the measured spectra.

In later experiments, the ordered phase was reached. Figure 2b displays the static magnetic susceptibility of the nuclear spin system in silver as a function of polarization at negative temperatures. At both zero field and 5 μT, there is first a monotonic increase of χ with $|p|$; this behavior is due to ferromagnetic interactions. Within the scatter of the measured data, the same curve fits the experimental results at both fields. As is typical for dipolar ferromagnetism, the susceptibility does not saturate completely even at high polarizations but, when $|p| > 0.55$, tends to $\chi_{sat} = -1.05$. The saturation tendency of susceptibility was already lost around $B = 10 \, \mu T$, and a reliable determination of the transition point then became impossible.

The crossing of the two lines in Fig. 2b was identified as the transition point to the ferromagnetic state, giving for the critical polarization $p_c = -0.49$. This identification is supported by the value of χ_{sat}, which is close to -1, theoretically expected for the silver specimen. By employing the linear relationship (4)

$$\frac{1}{|p|} - 1 = 0.55 \left(\frac{|T|}{1 \text{ nK}} \right) \qquad (4)$$

between the temperature and inverse polarization, which was found in earlier experiments, $T_c = -1.9 \pm 0.4$ nK was obtained; Monte Carlo calculations predict $T_c = -1.7$ nK. The critical entropy $S_c = 0.82\mathcal{R} \ln 2$ (where $\mathcal{R}$ is the gas constant and $\ln 2$ is the natural

logarithm of 2, whose value is approximately 0.693), computed from p_c, is high when compared with the value of $0.66\mathcal{R} \ln 2$ deduced from the nearest-neighbor Heisenberg model.

The magnetic field versus entropy diagram of silver is shown in **Fig. 3**; the two circles represent the intercept of the two lines in Fig. 2b for $B = 0$ and 5 μT. For $T < 0$, the spin system is ferromagnetic inside and paramagnetic outside the border. The value

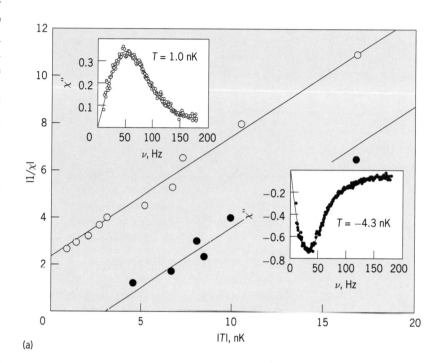

(a)

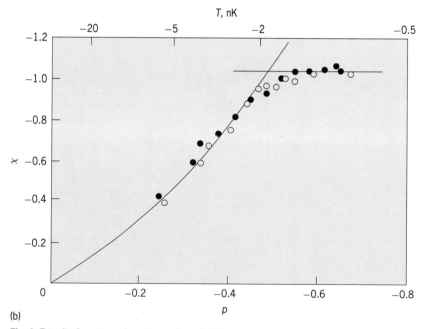

(b)

Fig. 2. Results from experiments on silver. (a) Absolute value of the inverse susceptibility $1/\chi$ versus absolute value of temperature T for nuclei in silver at $T > 0$ (open circles) and at $T < 0$ (solid circles). The insets show measured nuclear magnetic resonance (NMR) spectra (absorption χ'' versus frequency ν) at $T = 1.0$ nK and emission spectrum at $T = -4.3$ nK. The external magnetic field $B = 0$. (b) Static susceptibility χ versus polarization p of silver nuclear spins at $T < 0$ for $B = 0$ (open circles) and $B = 5 \, \mu$T (solid circles).

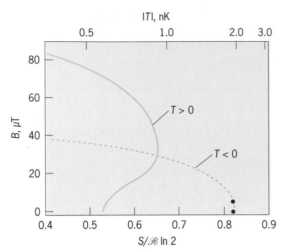

Fig. 3. Phase diagram of nuclear spins in silver metal for positive and negative absolute temperatures in the magnetic field (B) versus entropy (S) plane.

$B_c = 40$ μT was used for fixing the low-entropy end of the transition curve. The shape of the phase boundary was derived from the mean-field theory, with the assumption of a linear relationship between entropy and temperature. The antiferromagnetic phase transition curve at $T > 0$ is included for comparison.

The saturation of susceptibility to $\chi_{sat} \approx -1$ in the ordered state (Fig. 2b) can be explained only by the formation of domains, since otherwise the susceptibility would diverge at T_c. Instead of needles, as at $T > 0$, platelike domains are expected when $T < 0$. Comparison of the saturation values of susceptibility with the mean-field theory and Monte Carlo calculations indicates that in fields parallel to the sample foils there is a multidomain state in silver while in a transverse field a single-domain structure is preferred.

It has been argued that negative temperatures are fictitious quantities because they do not represent true thermal equilibrium in the sample consisting of nuclear spins, conduction electrons, and the lattice. However, the experiments on silver show conclusively that this is not the case. Depending on the sign of the absolute temperature, the same interactions produce ferromagnetic or antiferromagnetic order, depending on whether $T < 0$ or $T > 0$. Besides, true equilibrium, in the strictest sense of the word, hardly ever exists in nature.

For background information SEE ADIABATIC DEMAG-
NETIZATION; ANTIFERROMAGNETISM; DIPOLE-DIPOLE IN-
TERACTION; EXCHANGE INTERACTION; FERROMAGNETISM;
MAGNETIC SUSCEPTIBILITY; NEGATIVE TEMPERATURE; NU-
CLEAR ORIENTATION; THERMODYNAMIC PRINCIPLES; ZEE-
MAN EFFECT in the McGraw-Hill Encyclopedia of Science & Technology.

Olli V. Lounasmaa

Bibliography. A. Abragam and M. Goldman, *Nuclear Magnetism: Order and Disorder*, 1982; P. J. Hakonen et al., Observation of nuclear ferromagnetic ordering in silver at negative nanokelvin temperatures, *Phys. Rev. Lett.*, 68:365–368, 1992; P. Hakonen, O. V. Lounasmaa, and A. Oja, Spontaneous nuclear magnetic ordering in copper and silver at nano- and picokelvin temperatures, *J. Magnetism Magnetic Mater.*, 100:394–412, 1991.

Neural networks

Although applications of neural networks to problems of image and speech recognition have received a great deal of attention, advances are also being made in another application—the role of neural networks in decision problems. Neural networks have been applied to a variety of levels of decision problems, attesting to their flexibility and versatility. Research has yielded great progress in designing and validating the performance of the most popular neural network paradigm, backpropagation. Backpropagation is an algorithm that can be used to find the parameters of a neural network system that enable the system to respond to decision problems in a so-called intelligent way. In addition, there have emerged alternative architectures for neural networks that offer suitable choices when back propagation falters.

Use as decision devices. One reason for the growing popularity of neural networks for decision problems is that knowledge of the rules underlying a process is not required. Instead, neural networks learn mappings (transformations of inputs to outputs) by exposure to a set of training examples.

The low-level decisions addressed by neural networks are typically pattern recognition problems, which use signals generated by sensors. The neural network learns, by way of a set of examples, to associate signals (input vectors) with desired responses. Higher-level decisions are taught to neural networks via training examples generated by human expertise or mathematical analysis. For example, neural networks have been applied to the problem of monitoring a cutting tool from the most basic level of using integrated

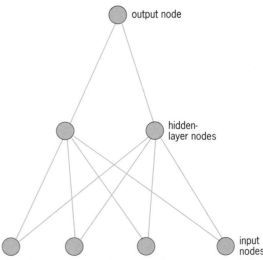

A typical neural network structure for pattern recognition.

sensor signals to determine if a tool is becoming worn, to the higher-level decision of determining cutting parameters to optimize tool life.

A typical pattern recognition neural network appears in the **illustration**. It consists of nodes and connections between them. First-layer nodes simply receive input signals, and output nodes provide the network response to a given input. (In this discussion, only one output node is assumed, but the extension to many is straightforward.) Output nodes and hidden-layer nodes generate responses by passing the net input received through their connections through an activation function. The net input (net_i) to node i is given by Eq. (1),

$$net_i = \sum_k a_k w_{ki} \qquad (1)$$

where k indexes the nodes that are connected to node i, w_{ki} is the weight on the connection between node k and node i, and a_j is the activation or response of node j to the signals that it receives. The activation function is usually a semilinear logistic function such as in Eq. (2).

$$a_j = \left[1 + \exp(-net_j)\right]^{-1} \qquad (2)$$

The activation of each unit at a layer then fans out via the weighted connections to the next layer, thus serving as the input vector to a successive layer. At the topmost layer, the node activations supply the overall system output. Note that the activations of nodes are computed from locally available information. That is, all computation performed at a node requires only information carried by connections from adjacent-layer nodes. Such a structure allows all nodes at a layer to process information simultaneously or in parallel. Its characteristics coarsely model the biological neuron by obeying this locality constraint.

The hidden-layer nodes enable the network to construct complex, nonlinear mappings. The internal parameters—weights on connections—of the neural network adjust in response to the examples of input/output pairs, so that a mapping evolves that enables the network to respond appropriately to new inputs. The mechanism for adapting weights is determined by the specific learning algorithm being used. By far the most popular learning algorithm is backpropagation. For backpropagation, the weight updates derive from the minimization of an error function; error (E) is defined by Eq. (3), where o denotes output node, P is the num-

$$E = \sum_i (d(i) - a_o(i))^2 \qquad (3)$$

ber of patterns in the training set, and $i = 1, \ldots, P$. Thus, the neural network attempts to minimize the error between the desired output, $d(i)$, for pattern i, and the actual network output, $a_o(i)$. The backpropagation weight update [$W_{ki}(\text{new})$] is then given by Eq. (4),

$$W_{ki}(\text{new}) = W_{ki}(\text{old}) - \frac{\alpha \partial E}{\partial W_{ki}} \qquad (4)$$

where α represents a learning rate—the step size.

Essentially, backpropagation successively approximates steepest descent nonlinear optimization. The backpropagation algorithm finds the gradient above in a recursive fashion, backpropagating error through the network. Many passes through the training set are required.

Research in pattern recognition. Adaptive neural networks that perform pattern recognition are used in the majority of current applications, so research in improving these networks has flourished. Backpropagation has historically dominated the choice of learning algorithm for such neural networks.

The performance of the backpropagation network can vary drastically with the number of hidden nodes, which govern the number of weights in the network. Too many hidden nodes can enhance the network's ability to learn training data but can cripple its ability to respond to new cases (that is, to generalize). However, too few hidden nodes can impede its ability to learn the training data at all. The weights actually capture the network's knowledge. Like a statistical model, too many parameters can inhibit the ability to generalize.

Techniques for allowing weights on connections to decay throughout training can allow the network to learn the training problem while preserving generalization ability. Initially, these efforts involved periodic manual pruning of the network's connections, but they have progressed to algorithms that incorporate a penalty term in the error function of the network.

The opposite approach dynamically constructs new nodes during training. For example, an algorithm known as cascade correlation constructs a new node when the training error for a particular configuration has stabilized. All these approaches, whether destructive or constructive, require subjectivity to determine when to prune or add a node or connection or to determine how to weight a penalty term.

Alternative networks. Despite the popularity of backpropagation, its limitations have fueled interest in alternative paradigms. Perhaps its primary drawback is discovering the correct combination of architecture, parameters, and training strategy. Different choices can result in vastly different performance. One of the critical backpropagation decisions is the number of hidden nodes. Another key decision involves the length of the training period and related parameter (learning rate issues). Certain alternative paradigms solve one or more of these problems, and typically they offer much speedier learning processes.

Networks for pattern recognition, whose roots can be traced to studies reported over a decade ago, have resurfaced as an important alternative to backpropagation. These classification networks construct hidden nodes via objective criteria; typically, if a network having a particular configuration misses a classification, an additional node is constructed. Instead of changing weights on connections, parameters defining the region of influence of a node adapt in response to examples. The advantage of such networks lies in their speed and simplicity. These networks are presently restricted to classification tasks, and their accuracy is typically inferior to that achieved by backpropagation.

Networks based on radial basis functions offer more

efficient training than backpropagation, and they can be used for functions such as prediction. Although determination of parameters is again important, the parameters of these networks possess some intuitive meaning, making selection easier.

Selection of training data and training strategy. No matter what paradigm or variant is chosen, the most critical choice made in a neural network application is selection of the training data. Not only the quantity but also the quality of data will affect the network's ability to learn and generalize. That is, if the data chosen do not span the space of the input patterns, it is likely that the network will not learn adequately. Thus, selection of training data has itself become a fertile area for research. In addition, transformations on raw training data can be used to improve performance, and especially generalization, in some cases.

Optimization-oriented networks. Applications of neural networks to different levels of combinatorial decision problems also continue to emerge. The traveling salesperson problem (of finding a minimum distance route through n points) lends itself well to neural network research. At the level of constructing routes, both adaptive and nonadaptive neural networks can evolve solutions. The adaptive systems, like those described above, iteratively update internal parameters in response to problem data. In addition, lateral connectivity among nodes and neighborhood functions allow the network's parameters to actually represent a solution.

Nonadaptive neural systems have fixed weights, but they allow node states to change until the network reaches a stable point of an energy function. This energy function is carefully defined to describe the particular optimization problem at hand, and when it is at its minimum, the neural network should provide a solution that is both good and feasible.

Other combinatorial decision problems to which neural networks have been applied include vehicle routing and scheduling. For these problems, just as with the traveling salesperson problem, the task of finding an optimal solution is mathematically difficult. The solution yielded by the neural network compares to other heuristically generated solutions—not guaranteed to be optimal but obtained rapidly and likely to suffice. The limitations of human cognitive functions preclude effective solutions of large-scale problems in this class.

Often, human experts become involved in a combinatorial decision process such as vehicle routing at the higher decision level of assigning an appropriate algorithm to a difficult problem based on its attributes. To do this assigning, the expert considers characteristics such as number of vehicles and distances and, based on personal experience, assigns an algorithm that has proven efficient and effective for similar routing problems. Researchers have applied neural networks to such problems by compiling a training set consisting of several representative problems and the approach recommended by an expert, and then training a neural network of the pattern recognition type to learn the implicit mapping. SEE INDUSTRIAL ENGINEERING.

For background information SEE ALGORITHM; COMBINATORIAL THEORY; DECISION THEORY; NEURAL NETWORK; OPTIMIZATION in the McGraw-Hill Encyclopedia of Science & Technology.

Laura Burke

Bibliography. J. Hertz, A. Krogh, and R. G. Palmer, *Introduction to the Theory of Neural Computation*, 1991.

Nobel prizes

The Nobel prizes for 1992 included the following awards for scientific disciplines.

Physiology or medicine. Edmond Fischer and Edwin Krebs of the University of Washington, Seattle, were awarded the prize for their discoveries that led to understanding a process critical in regulation of cell proteins.

In experiments on glycogen metabolism in muscle tissue done in the 1950s at the University of Washington, the two biochemists showed that the activity of a glycogen-splitting enzyme called phosphorylase can be turned on and off by the reversible addition of a phosphate group. The phosphate is transferred to the enzyme from adenosinetriphosphate (ATP), converting phosphorylase to an active form, which then breaks down glycogen to release the energy needed for muscle contraction. Removal of the phosphate turns the enzyme off again.

Fischer and Krebs went on to find the enzymes responsible for the addition and removal of phosphate. They isolated the first protein kinase, a class of enzymes that transfer phosphate from ATP to proteins. They also found phosphatase, the enzyme responsible for phosphate removal. Since then, such phosphorylation reactions have been shown to control the activities of hundreds of enzymes and thus regulate many cell processes, such as muscle contraction, immune responses, and cell growth and division. It has been estimated that as much as 1% of the genes in the human genome encode protein kinases.

The work of Fischer and Krebs has stimulated further research into signaling processes that control cellular events, particularly growth control.

Physics. Georges Charpak received the prize for his invention and development of detectors in high-energy physics. He has worked at the European Center for Nuclear Research (CERN) since 1959.

In 1968, Charpak devised the multiwire proportional chamber. At that time, physicists often used photographs taken in bubble chambers to track the numerous charged particles generated in collisions in particle accelerators. However, this detection method was limited because the photographs could be taken only about once a second, which was not fast enough to keep up with the collisions generated by the improved accelerators then being constructed, and because analysis of the photographs was laborious. Much faster electronic detectors were available, but they provided poor spatial resolution.

The multiwire proportional chamber combined high spatial resolution with the speed of existing electronic detectors. This gas-filled chamber contains networks of closely spaced charged wires. A charged particle passing through the chamber ionizes gas atoms, triggering an avalanche of electrons onto the nearest wire. The resulting electrical pulses are amplified and sent directly to a computer, with detection rates of 10^6 particles per second. The many crisscrossing layers of wires provide a three-dimensional record of the particle's path.

An important refinement of the original device was the drift chamber, in which measurement of the time delay between the original collision and the recorded pulse is combined with the known drift rate of the electrons to improve the spatial resolution.

Most current research in high-energy physics depends on detectors that have evolved from Charpak's inventions, and such devices made possible the discoveries of the charmed quark and the W and Z particles. Charpak has also adapted his detectors for use in industry, medicine, and biology.

Chemistry. Rudolph A. Marcus of the California Institute of Technology received the prize for work that provided a mathematical analysis of electron transfer reactions in chemical systems. These are reactions in which at least one electron passes between the reacting atoms or molecules. The processes involving electron transfer are numerous and varied, occurring in both living and nonliving systems. Examples include corrosion of metals, photosynthesis, electroplating, conduction of electricity by certain polymers, and bioluminescence.

The theory of electron transfer reactions, now known as the Marcus theory, provides an explanation of how the overall energy in a system of interacting atoms, molecules, or ions changes and induces an electron to jump from one entity to another. The development of this theory grew out of Marcus's studies during 1956–1965, when he was focusing on the mathematics of oxidation-reduction reactions in solutions in which a single electron is transferred from one entity to another. The theory provides a rigorous explanation of how slight changes in the geometric arrangement of the atoms in the entities in the reaction, or in the surrounding medium, affect the height of the energy barrier that must be overcome in order for the reaction to take place. Enabling prediction of whether a charge-transfer reaction can occur and its rate, the theory has proved valuable in many types of research, ranging from design of biosensors and batteries to elucidation of complex biological processes such as photosynthesis and cellular metabolism.

Marcus's work has stimulated research by other workers in many fields. His present research is focused on electron transfer occurring over a distance on the order of 1 nanometer, a relatively large distance on a molecular level. Such studies hold promise for advances in molecular biology by indicating how energy flows through large molecules, with potential for applications such as devices for improved utilization of solar energy using biomolecules. SEE SURFICIAL AND INTERFACIAL CHEMISTRY.

For background information SEE ELECTRON-TRANSFER REACTION; IONIZATION CHAMBER; OXIDATION-REDUCTION; PARTICLE DETECTOR in the McGraw-Hill Encyclopedia of Science & Technology.

Nutrition

Although controversy remains, much scientific evidence indicates a relationship between dietary fat and degenerative diseases such as cardiovascular disease as well as cancer, obesity, and infectious diseases.

Dietary fat refers to the fraction of foods that is rich in triacylglycerols (formerly called triglycerides). Triacylglycerols are chemically classified as simple lipids, a category that also includes cholesterol, cholesteryl esters, phospholipids, fatty acids, glycolipids, and waxes. Lipids of major concern for humans are triacylglycerols, cholesterol and its esters, and phospholipids.

Triacylgycerol structure. Triacylglycerols consist of glycerol, a three-carbon molecule, to which three fatty acids are attached. **Table 1** gives the names and chemical structures of the fatty acids that are common in lipids of dietary fats.

Fatty acids contain from 4 to 22 carbon atoms and from 0 to 6 double bonds. Fatty acids without double bonds are called saturated; those with double bonds are unsaturated. Dietary triacylglycerols, cholesteryl esters, and phospholipids contain a mixture of saturated and unsaturated fatty acids. Dietary fats rich in unsaturated fatty acids, such as soybean and corn oils, are called unsaturated fats. Fats rich in fatty acids containing two or more double bonds are called polyunsaturated. Fats rich in saturated fatty acids, such as coconut and palm oils and beef tallow, are called saturated. If the polyunsaturated fatty acid contains a double bond that is three carbons from the methyl (CH_3—) end, the molecule is an omega-3 fatty acid. Fish oils are relatively rich in omega-3 acids.

Dietary recommendations. Approximately 37% of calories in the average American diet is provided by lipids (most of which are triacylglycerols), half being derived from animals and half from plants. This 37% of calories comprises 14% from saturated fatty acids, 16% from monounsaturated fatty acids, and 7% from polyunsaturated fatty acids. Studies in the United States recommended that total fat intake be only about 30% of total dietary calories with no more than one-third of this fat coming from saturated fatty acids, at least one-third from monounsaturated fatty acids, and no more than one-third from polyunsaturated fatty acids.

Production of foods with less fat. Food producers and processors are emphasizing the production of raw and processed foods with less fat and saturated fatty acids.

Dairy products. Both the concentration of fat in milk and the fatty acid composition can be altered by selection of dairy cows with more favorable milk composition and by modification of the cows' diet. A variety of dairy foods with lower fat contents have been developed in this way. Technologies are also available, but not yet economical, for removal of most of the cholesterol in dairy foods.

Meat products. A variety of nutritional and management technologies are being used by farmers to produce beef, pork and poultry with less fat. Selective breeding programs and modified animal diets result in meat that contains more monounsaturated and polyunsaturated fatty acids and less saturated fatty acids. The fat content and fatty acid composition of processed meats can be further modified through the use of fat substitutes and water and carbohydrate additions.

Eggs. The cholesterol content rather than the fat content in eggs is of principal concern for humans. An average egg contains around 200 mg of cholesterol. It is recommended that less than 300 mg of cholesterol be consumed per day. The amount of fat and cholesterol in eggs does not respond significantly to changes in genetic makeup or diet of the hen. However, a variety of new technologies can remove most of the cholesterol from eggs and egg products.

Vegetable oils. Although they are liquid at room temperature because of a high proportion of unsaturated fatty acids in the triacylglycerols, vegetable oils may be modified by hydrogenation. This process increases the proportion of saturated fatty acids and thereby increases oxidative stability (that is, decreases the rate of rancidity development). Hydrogenation permits use of vegetable oil as a semisolid food (such as margarine). Increasing concern over saturated fatty acids and over trans fatty acids, which are a specific type of unsaturated fatty acid, that are formed during hydrogenation has caused scientists to seek alternative approaches to change the proportion of different fatty acids in vegetable fats and oils. For example, food products with novel compositions of fatty acids can be developed through biotechnology.

Dietary fat and human health. Genetic makeup and environmental factors such as dietary composition and exposure to virulent viruses, bacteria, and toxins contribute to human disease. Diet is one factor that can be controlled in order to maximize health. Some human diseases and disorders thought to be influenced by diet are cardiovascular disease, obesity, cancer, and infectious diseases.

Cardiovascular disease. Diseases of the cardiovascular system are responsible for about 50% of deaths in the United States. Cardiovascular disease occurs when the blood flow through an artery is impeded by accumulation of a fatty deposit (atherosclerotic plaque) within the lumen of the artery. This accumulation may be associated with formation of a blood clot, which could totally occlude blood flow and lead to stroke, heart attack, or dysfunction.

The rate of development of plaques within arteries, such as the coronary arteries, is correlated positively with blood cholesterol concentration and low-density lipoprotein (LDL) concentration and negatively with high-density lipoprotein (HDL) concentrations. Because the amounts and type of dietary fat can influence the concentrations of total cholesterol, LDL, and HDL, it seems likely that dietary fat influences cardiovascular disease development. In actuality, the blood lipid compositions of individuals respond differently to changes in dietary fat intake. Most studies have shown the impact of dietary cholesterol on blood cholesterol concentration to be minimal. In general, data suggest that with a decrease in cholesterol consumption of 100 mg a decrease of about 5 mg % in total plasma cholesterol may be expected.

The type of dietary fat exerts a greater impact on blood cholesterol concentration than does the amount of fat or cholesterol. Because fats rich in saturated fatty acids seem to be associated positively with blood cholesterol concentration, they are considered a major risk factor for heart disease. Replacement of saturated fats with equal amounts of polyunsaturated fats frequently results in a decrease in blood cholesterol concentration. Saturated fatty acids, however, do not have identical effects on blood cholesterol concentra-

Table 1. Common fatty acids in dietary fats

Symbol	Common name	Structure
Saturated fatty acids*		
$C_{4:0}$	Butyric acid	$CH_3(CH_2)_2COOH$
$C_{6:0}$	Caproic acid	$CH_3(CH_2)_4COOH$
$C_{8:0}$	Caprylic acid	$CH_3(CH_2)_6COOH$
$C_{10:0}$	Capric acid	$CH_3(CH_2)_8COOH$
$C_{12:0}$	Lauric acid	$CH_3(CH_2)_{10}COOH$
$C_{14:0}$	Myristic acid	$CH_3(CH_2)_{12}COOH$
$C_{16:0}$	Palmitic acid	$CH_3(CH_2)_{14}COOH$
$C_{18:0}$	Stearic acid	$CH_3(CH_2)_{16}COOH$
Unsaturated fatty acids [†]		
$C_{16:1}$	Palmitoleic acid	$CH_3(CH_2)_5CH{=}CH(CH_2)_7COOH$
$C_{18:1}$	Oleic acid	$CH_3(CH_2)_7CH{=}CH(CH_2)_7COOH$
$C_{18:2}$	Linoleic acid	$CH_3(CH_2)_4(CH{=}CHCH_2)_2(CH_2)_6COOH$
$C_{18:3}$	Linolenic acid	$CH_3CH_2(CH{=}CHCH_2)_3(CH_2)_6COOH$
$C_{20:4}$	Arachidonic acid	$CH_3(CH_2)_4(CH{=}CHCH_2)_4(CH_2)_2COOH$
$C_{20:5}$	Eicosapentaenoic acid	$CH_3CH_2(CH{=}CHCH_2)_5(CH_2)_2COOH$
$C_{22:6}$	Docosahexanoic acid	$CH_3CH_2(CH{=}CHCH_2)_6CH_2COOH$

*The subscripts indicate the total number of carbons in the fatty acid and the absence of double bonds.
[†]The subscripts indicate the total number of carbons in the fatty acid and the number of double bonds.

Table 2. Fatty acid compostion of selected animal and plant fats*

Fat or oil	Saturated				Monounsaturated		Polyunsaturated	
	$C_{4:0}$–$C_{10:0}$	$C_{12:0}$,$C_{14:0}$,	$C_{16:0}$,$C_{18:0}$	Other	$C_{16:1}$,$C_{18:1}$	Other	$C_{18:2}$,$C_{18:3}$	Other
Animal fat								
Beef tallow	0.1	28.9	21.6	3.0	42.1	1.1	2.8	0.4
Butterfat	9.2	41.0	12.5	2.5	30.1	1.2	3.4	0.1
Chicken fat	—	24.7	6.4	0.3	48.1	0.3	20.2	—
Cod liver oil	—	3.2	3.7	1.2	34.6	14.6	2.4	34.9
Egg yolk	—	26.1	9.9	—	49.9	—	14.7	—
Lard (pork)	0.1	26.4	12.3	0.9	48.2	1.6	10.0	0.5
Mutton tallow	0.2	29.1	24.5	3.6	35.8	1.1	5.3	0.4
Salmon oil	—	15.1	3.8	1.5	42.5	18.5	2.4	12.7
Turkey fat	—	26.0	10.0	0.5	26.5	0.4	21.0	13.5
Plant fat								
Cocoa butter	—	25.9	34.5	1.1	35.6	—	2.9	—
Coconut oil	14.9	74.5	2.5	0.1	6.5	—	1.5	—
Corn oil	—	12.2	2.2	0.1	27.6	—	57.9	—
Olive oil	—	13.7	2.5	0.9	72.3	—	10.6	—
Palm kernel oil	8.2	73.6	2.4	0.1	13.7	—	2.0	—
Palm oil	—	46.5	4.7	0.2	38.9	—	9.7	—
Rapeseed oil	—	3.9	1.9	1.0	64.3	1.0	27.9	—
Soybean oil	—	11.1	4.0	0.4	23.5	—	61.0	—
Sunflower seed oil	—	7.5	4.7	0.4	18.7	—	68.7	—

*Fatty acid composition is expressed as a weight percentage of total fatty acids. Fats and oils are listed from most to least saturated. The subscripts of C indicate the length of the carbon chain of individual fatty acids and the number of double bonds. For example, a 10-carbon chain fatty acid without double bonds is expressed as $C_{10:0}$.

SOURCES: After Van Den Bergh Food Ingredients Group, *Typical Composition and Chemical Constants of Common Edible Fats and Oils*, 1970; and C. Lentner (ed.), *Geigy Scientific Tables*, 8th rev. ed., vol. 1, CIBA-GEIGY Corp., 1981.

tion. For example, lauric ($C_{12:0}$), myristic ($C_{14:0}$), and palmitic ($C_{16:0}$) acids seem to increase blood cholesterol concentrations, whereas stearic acid is neutral in effect (as are the monounsaturated fatty acids known as oleic acids). Fatty acid compositions of several dietary fats derived from animals and plants are presented in **Table 2**. Those fats with the greatest combined concentrations of lauric, myristic, and palmitic acids would be expected to cause the greatest increase in blood cholesterol concentrations.

High blood pressure. Most dietary studies demonstrate that increasing the ratio of polyunsaturated to saturated fatty acids or decreasing the total amount of fat tends to decrease the blood pressure of hypertensive patients. The omega-3 fatty acids decrease blood-platelet aggregation, slightly increase bleeding time, and decrease the tendency of platelets to stick to the vessel wall, and hence decrease chances of thrombosis or blood-clot formation. Thus, changing the amount and proportions of types of fatty acids in the diet should benefit the cardiovascular system.

Obesity. Because a gram of dietary fat contains $2^{1}/_{4}$ times more calories than a gram of carbohydrate or protein, a diet high in fat is a major contributor to obesity. With obesity, the incidence of many diseases such as heart disease and diabetes become greater. Hence, human health can be improved by maintaining ideal body weight. Lowering the amount of fat in the diet facilitates this goal.

Cancer. Along with viruses, radiation, and environmental carcinogens, dietary fat is thought to contribute to cancer development. Several epidemiological studies have demonstrated that dietary fat intake is correlated positively with breast and colon cancer and, to a lesser extent, ovarian, prostate, and pancreatic cancer.

Infectious diseases. Dietary fat can affect the immune system. For example, deficiency of essential fatty acids decreases the immune function of the body, resulting in decreased resistance to and recovery from infectious diseases. However, high intake of fat seems to suppress the immune system. Evidence for a negative relationship of dietary fat to the immune system provides yet another health advantage for decreasing fat intake.

For background information SEE CHOLESTEROL; FAT AND OIL; FAT AND OIL (FOOD); FOOD MANUFACTURING; LIPID; NUTRITION; TRIGLYCERIDE in the McGraw-Hill Encyclopedia of Science & Technology.

Donald C. Beitz

Bibliography. Food and Nutrition Board, Committee on Diet and Health, Commission on Life Sciences, U.S. National Research Council, *Diet and Health Implications for Reducing Chronic Disease Risk*, 1989; G. J. Nelson (ed.), *Health Effects of Dietary Fatty Acids*, American Oil Chemists' Society, 1991; U.S. Department of Agriculture and U.S. Department of Health and Human Services, *Nutrition and Your Health: Dietary Guidelines for Americans*, 3d ed., Home Garden Bull. 232, November 1990; U.S. Department of Health and Human Services, *The Surgeon General's Report on Nutrition and Health*, DHHS (PHS) Pub. 88-50210, 1988.

Object-oriented programming

Object-oriented programming takes a data-centered approach to software development, with a primary emphasis on modeling the semantics of data that are to be manipulated by a system and a secondary emphasis on the processes used to implement the models. A software system is viewed as composed of a set of data objects that communicate through messages. The key ingredients of the approach are the use of data abstraction and encapsulation through classes and the use of inheritance to provide design and code reuse. The approach has become popular because it leads to software systems that can solve complex problems and are easier to construct and maintain than systems that are constructed with traditional procedural programming.

Abstraction and encapsulation. The central concepts of object-oriented programming are abstraction and encapsulation. Conventional programming languages generally support these concepts for built-in data types but not for user-defined ones. A built-in data type such as integer is an abstraction of a mathematical type. The idealized type is modeled by a set of operations and a set of possible values. The programmer who uses a built-in type does not need to know the internal form that represents variables of the type or the details of how the built-in operations on variables of the type are performed. The programmer is satisfied as long as the semantic model of the type is preserved. The language encapsulates the built-in type by allowing access to the type only through the operations provided.

Object-oriented programming allows the definition of classes that provide abstraction and encapsulation for user-defined types. Such types are termed abstract data types.

Classes. Traditional procedural programming in languages such as COBOL, Pascal, and C allows programmers to construct data structures but provides very little protection of the data within those structures. For instance, in Pascal a programmer can declare a Point data type as in example (1). Subroutines

```
TYPE
  Point = RECORD
    x : real;                    (1)
    y : real
  END;
```

that use variables of type point are allowed to access the fields of the Point record directly, as in example (2).

```
FUNCTION DistanceFromOrigin(aPoint
        : Point) : real
BEGIN
  DistanceFromOrigin :=           (2)
    sqrt((aPoint.x * aPoint.x)
        + (aPoint.y * aPoint.y));
END;
```

The ability to directly access fields of a record can cause many parts of a program to depend on the implementation structure of the type. For instance, a programmer may decide that it is more appropriate to represent a point as a pair of polar coordinates, as in example (3). The information represented by the

```
TYPE
  Point = RECORD
    angle : radians;             (3)
    radius : real
  END;
```

Point type is preserved, but the change may cause modifications to a large number of program segments because the data structure is not encapsulated.

In the object-oriented approach, data types are encapsulated so that the only way to access or change the state of a variable is through operations that are defined with the type of the variable. The encapsulated type is referred to as a class, and the operations of the type are referred to as methods. An instance of a class is termed an object instance, or simply an object. The individual data fields of an object instance are referred to as instance variables.

Example (4) illustrates a Point class with a simple set of operations.

```
CLASS Point =
  METHODS
    FUNCTION x : real;
    FUNCTION y : real;
    FUNCTION angle : radians;
    FUNCTION radius : real;
    FUNCTION distance
            (aPoint : Point ) :
            real;
                                   (4)
    PROCEDURE setCartesian
            (x, y : real);
    PROCEDURE setPolar
            (angle : radians;
             radius : real);
    PROCEDURE scale(factor : real);
    PROCEDURE rotate
            (angle : radians);
  END;
```

The declaration for a class describes the methods for the class but does not implement them. The implementation for each method is supplied separately. The data fields (or instance variables) of a class can be described separately from the class interface since the fields will not be directly accessible to clients of the class. The information needs only to be available to the method implementation.

To invoke an operation on an object instance the programmer sends a message to the object. The language support determines the class of the object that receives a message, matches the message to a method defined in the class, and performs a subroutine call to that method. The process of matching a message to a method for a particular class is termed method resolution. The data contained in the object instance that receives a message are automatically made available to the method implementation.

Inheritance. The major addition to abstract data types that is made by object-oriented programming is inheritance. Inheritance allows a programmer to define a new class simply by stating how it differs from an existing one. The new class is said to inherit from the original and is referred to as a subclass of the original class. The original class is referred to as the superclass of the new class.

The data and methods of the superclass are automatically included in the subclass. The subclass may add new data fields and methods and may change the implementation of inherited methods. When a subclass changes the implementation of an inherited method, instances of the superclass and subclass will have an operation in common but the implementation of the operation will differ.

Example (5) illustrates a class that inherits from the

```
CLASS 3D_Point INHERITS FROM Point
  = METHODS
    FUNCTION z : real;
    FUNCTION distance
            (aPoint : 3D_Point) :
            real;

    PROCEDURE setCartesian
            (x, y : real);
    PROCEDURE setPolar              (5)
            (angle : radians;
            radius : real);
    PROCEDURE set3D(x, y, z :
            real);

    PROCEDURE rotateAboutX
            (angle : radians);
    PROCEDURE rotateAboutY
            (angle : radians);
  END;
```

Point class, defined above. The 3D_Point class extends the Point class by providing a function that retrieves the z-dimension value, a method to set the three-dimensional value, and additional rotation procedures. It also specializes the Point class by providing new implementations for the distance, setCartesian, and setPolar methods.

In some object-oriented languages a subclass may inherit from multiple superclasses. This feature is termed multiple inheritance. When a class inherits from multiple superclasses, it receives all the data and methods of each.

Programming languages. The object-oriented languages in most widespread use are Smalltalk and C++.

Smalltalk. This language presents a very uniform object model in which even elementary data types are modeled as classes. The basic operation of the language is sending a message to an object. To add two integers the programmer sends the add message to the first integer with the second integer as the argument of the message. The operation produces a new object instance that contains the result of adding the two numbers. Variables have no declared type and may refer to an instance of any class at run time. The binding of a message to a method imple-

mentation occurs entirely at run time, providing the great flexibility in programming. However, it can also result in run-time errors if a class is found not to have a method implementation for a message that has been received. Smalltalk supports only single inheritance.

Smalltalk is an interpreted language. A method is translated into a sequence of byte codes for a stack-oriented interpreter known as the Smalltalk virtual machine. The language itself is generally not distinguished from the set of class libraries and the programming environment that accompany the language. The programming environment provides an editor, class browsers, and a set of support tools for Smalltalk programming. It also provides an integrated debugger and object-instance inspector. Smalltalk programs are generally constructed and viewed only with the tools of the programming environment. The class library that accompanies the language provides classes for basic types such as integer and for advanced data types such as set and queue, and classes that support graphical user interfaces. Smalltalk provides automatic memory management through garbage collection.

_C++._ This extension to C is a compiled, statically typed language. Basic types such as int and char are modeled as in C. Multiple inheritance is supported for classes, and built-in operators can be overloaded to operate on user-defined classes. Variables must be declared with types, and the compiler performs static type checking. Static type checking ensures that there is a method implementation for every message and allows method resolution for some messages at compile time. Because of compile-time binding and lack of automatic memory management, C++ has less run-time overhead than Smalltalk. Declaration of the Point and 3D_Point classes in C++ are presented in example (6).

```
class Point {
  public:
    float x(void);
    float y(void);
    radians angle(void);
    float radius(void);

    float distance(Point &aPoint);

    void setCartesian
        (float x, float y);
    void setPolar
        (radians angle, float
        radius);
    void scale(float factor);
    void rotate(radians angle);

  private:
    float x_value;
    float y_value;
};

class 3D_Point : Point {          (6)
```

```
public:
  float z(void);

  float distance(Point &aPoint);

  void setCartesian
      (float x, float y);
  void setPolar
      (radians angle, float
       radius);
  void set3D
      (float x, float y, float
       z);

  void rotateAboutX(radians
      angle);
  void rotateAboutY(radians
      angle);

private:
  float z_value;
};
```

Other languages. Common LISP Object System (CLOS), Eiffel, and Objective-C are object-oriented languages with significant user bases. Ada and Modula-2 support abstract data types but not inheritance.

For background information SEE ABSTRACT DATA TYPES; PROGRAMMING LANGUAGES in the McGraw-Hill Encyclopedia of Science & Technology.

John J. Shilling

Bibliography. A. Goldberg and D. Robson, *Small-talk-80: The Language*, 1989; S. B. Lippman, *C++ Primer*, 1989; B. Meyer, *Object-Oriented Software Construction*, 1988; J. Rumbaugh et al., *Object-Oriented Modeling and Design*, 1991.

Organometallic chemistry

Hydrocarbons such as petroleum and methane represent one of the world's most important chemical resources. Unsaturated hydrocarbons (alkenes and alkynes), such as ethylene and propylene, are the precursors used for production of many commercially important organic chemicals and chemical intermediates. However, the more plentiful saturated hydrocarbons (alkanes), such as methane, propane, and the octanes, are used mostly as fuels. The lack of reactivity of alkanes in any reaction other than combustion has hindered their use in major chemical processes. Since 1980, however, experimental research with organometallic compounds has shown that alkanes can be every bit as reactive as their unsaturated counterparts, suggesting that chemical processes may be developed to use these more abundant hydrocarbons.

Carbon-hydrogen cleavage. A number of new metal-based homogeneous catalytic systems based on transition metals have been shown to facilitate the activation of carbon-hydrogen (C—H) bonds. Some of these have promise in developing new catalytic processes for hydrocarbon oxidation and functionalization. The lack of reactivity of alkanes no longer poses a problem in terms of chemical reactivity.

Organometallic compounds are molecular species that usually contain one or two metal atoms covalently bonded to several smaller discrete molecules (ligands), forming species known as complexes. These ligands serve to alter the metal's chemical and physical properties, such as its oxidation potential and its solubility. By careful choice of these ligands, the metal can be induced to react with the strong, unreactive C—H bonds of an alkane under mild conditions. This initial cleavage of the C—H bond is believed to be the first (and most difficult) step in the conversion of alkanes to useful organic compounds. Several types of complexes have been found to be successful in this type of cleavage.

Electron-rich metals. One way that C—H bonds have been cleaved involves the formation of a vacant site on a metal in a low oxidation state that contains electron-donating ligands. The vacant site provides room for the incoming hydrocarbon, and the electron-rich metal readily gives up its electrons in a process called oxidative addition. The net reaction results in the use of two electrons from the C—H bond plus two electrons from the metal to form two metal-carbon bonds, each with two electrons. The vacant site is most easily generated by irradiation of the precursor complex with ultraviolet light, a technique that makes the ligands fall off easily at ambient temperatures. If the irradiation is carried out in a pure hydrocarbon solvent, the metal oxidatively adds to the solvent, resulting in the formation of a complex with metal-hydrogen and metal-carbon bonds. The C—H bond has been completely cleaved and further chemistry can now be envisioned with the fragments attached to the metal.

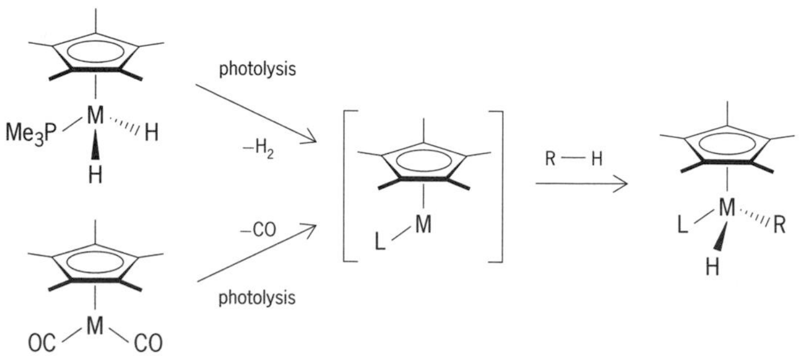

(1)

Reaction (1) shows several specific metal complexes that will undergo this type of oxidative addition chemistry with saturated and unsaturated hydrocarbons such as methane, ethane, hexane, and benzene. In reaction (1), M = rhodium (Rh) or iridium (Ir); Me = methyl group (CH_3); Me_3P = trimethylphosphine; L = PMe_3 or carbonyl (C═O); and R = CH_3, C_2H_5 (ethyl), C_6H_{13} (hexyl), or C_6H_5 (phenyl). With linear alkanes such as hexane, there is a preference of cleavage of the less hindered primary (terminal) C─H bonds at the end of the molecule over the more hindered secondary (internal) C─H bonds of the molecule. Benzene and other aromatic C─H bonds are the easiest to activate and react preferentially.

In one application based on this reactivity, a complex has been used to convert alkanes into alkenes plus hydrogen by using light energy. The complex chlorocarbonyl bis-(trimethylphosphine)rhodium [$RhCl(CO)(PMe_3)$] reacts with alkanes to give many thousands of turnovers of alkenes upon photolysis. The reaction scheme in **Fig. 1** shows how this transformation is believed to occur. Beginning with this complex at the top left, the fundamental step involving activation of the C─H bond can be seen in the catalytic cycle following the loss of carbon monoxide (CO). The loss of a hydrogen atom on the carbon that is adjacent to the one bound to the metal is a common reaction in organometallic chemistry, and it leads to the production of the alkene product still bound to the rhodium. The olefin is then released from the complex. Finally, hydrogen is displaced by reaction with CO to regenerate the initial metal complex. Light energy is required in this cycle to remove CO from the metal.

Another way in which hydrocarbons have been transformed into organic products uses complexes containing isonitrile ligands. These ligands are similar to carbon monoxide in that they can insert into metal-carbon bonds, and consequently oxidative addition of the C─H bond can lead to derivatized hydrocarbons by a process similar to that described above. A catalytic cycle is shown in **Fig. 2** that converts the unsaturated hydrocarbon benzene into a functionalized product by using iron (Fe) and rhodium catalysts.

Electron-poor metals. Another class of organometallic complexes that cleave C─H bonds involves metals that are electron-poor rather than electron-rich. In these systems, one metal-carbon bond is usually exchanged for the C─H bond of an alkane as shown in reaction (2), where Lu = lutetium. The transition metals on the left portion of the periodic table or the lanthanide metals commonly react in this fashion. The exchange occurs by way of a 4-center metathesis reaction in which the metal, hydrogen, and two carbons all

interchange bonds simultaneously. A catalyst for the insertion of an olefin into the C─H bond of pyridine

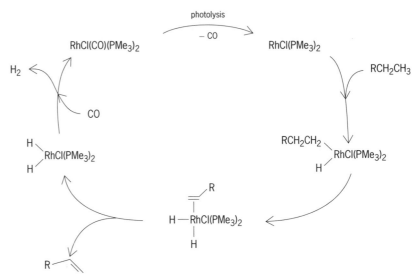

Fig. 1. Reaction scheme showing the transformation of an alkane to an alkene.

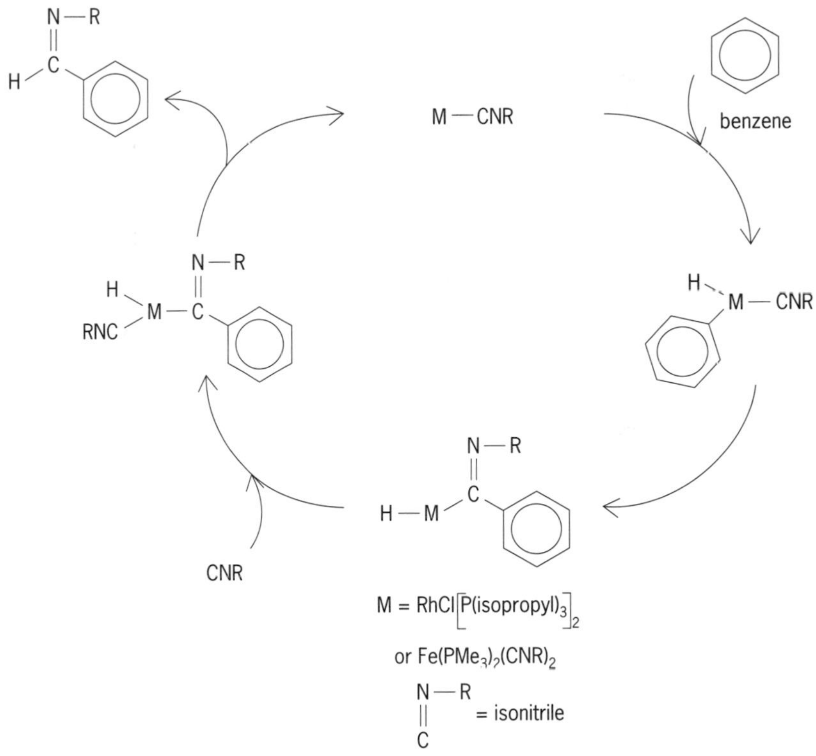

Fig. 2. Reaction scheme showing the catalytic cycle in which benzene is converted to a functionalized product.

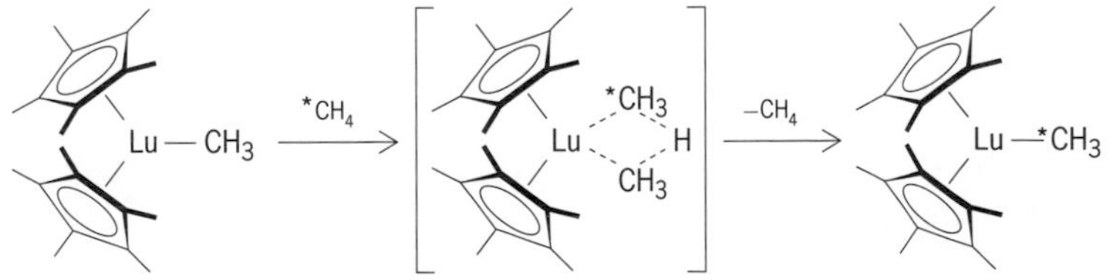

has been developed with this chemistry, as shown in reaction (3), where Zr = zirconium.

$$R + \underset{}{\diagdown} R \xrightarrow[\text{catalytic}]{} R \diagdown \qquad (3)$$

Metal porphyrins. One other class of organometallic complexes that has shown promise in hydrocarbon activation involves metal porphyrins. Formation of oxo derivatives (M=O) of the metal results in a reactive intermediate that can oxidize alkanes to alcohols and ketones. Perhaps the most successful of these complexes is the polyhalogenated porphyrin with the structure below (where Br = bromine and F = fluorine); it is utilized in reaction (4). The presence of

the halogens prevents oxidation of the porphyrin rather than the alkane, thereby ensuring a long life for the catalyst. The oxidation shown in reaction (4) proceeds

$$\underset{\text{Alkane}}{R-H} + O_2 \xrightarrow[\substack{200 \text{ psi } (1.4 \text{ MPa}) \\ 80°C (176°F) \\ 3 \text{ h}}]{\substack{\text{Polyhalogenated} \\ \text{porphyrin catalyst}}} \underset{\text{Alcohol}}{R-OH} \qquad (4)$$

with more than 20,000 turnovers (molecules of product per molecule of catalyst), giving more than 84% alcohol. Other oxidation systems based on porphyrins have used either peroxides or iodosobenzene as the oxidant. A major advantage of this catalytic system is that readily available dioxygen is used as the oxidant. Further improvements in this type of catalyst will be directed at higher catalyst reaction rates and greater selectivity for alcohol products.

For background information SEE CATALYSIS; COORDINATION COMPLEXES; HOMOGENEOUS CATALYSIS; ORGANOMETALLIC CHEMISTRY; TRANSITION ELEMENTS in the McGraw-Hill Encyclopedia of Science & Technology.

William D. Jones

Bibliography. J. A. Davies et al. (eds.), *Selective Hydrocarbon Activation*, 1990; C. L. Hill (ed.), *Activation and Functionalization of Alkanes*, 1989; A. E. Shilov, *Activation of Saturated Hydrocarbons by Transition Metal Complexes*, 1984.

Pain

What patients have often feared most about a surgical procedure is the pain they will experience afterward. This fear was not always unfounded since traditionally patients were not provided with adequate pain relief after surgery. Medical reasons for inadequate pain relief treatment included fear of overdose, fear of addiction, lack of knowledge of the mechanisms of pain, and lack of available techniques for pain relief. Recent advances, however, have resulted in techniques that promise effective pain relief after surgery.

Many of these new techniques involve new ways of delivering established drugs. Typically, the medications are morphine or other narcotics. The fact that morphine has remained the mainstay of pain relief since its first recorded use about 4000 years ago is a testament to its pain-relieving capabilities.

Acute-pain services. One of the newest concepts to have revolutionized the treatment of pain in the postsurgical period is the acute-pain service. Acute pain is short-lived, lasting days to weeks, as opposed to chronic pain, which lasts for months or years. The acute-pain service consists of a group of physicians and other professionals specifically trained in, and dedicated to, the relief of acute pain after surgery. These teams generally include one or more anesthesiologists, a nurse, and often a pharmacist. The team visits the patient in the postanesthesia care unit and initiates an appropriate therapy for the pain. Factors that influence the team's decision about which type of therapy to use include the patient's past medical history; the type, duration, and location of surgery; and the patient's age. The patient is also asked to rate the pain. The rating is often done on a scale from 0 to 10, where 0 represents no pain and 10 the worst pain imaginable. The pain score helps the team to determine the dose of pain reliever required. The pain score is reassessed at frequent intervals, and adjustments in pain therapy are made accordingly. After the initial contact in the postanesthesia care unit, the acute-pain service team typically visits the patient several times each day to assess pain, side effects, and the patient's satisfaction with therapy. This routine generally continues for several days after surgery.

Drug delivery systems. In addition to the traditional postsurgical injection of a pain-killer, other techniques are possible, such as patient-controlled analgesia, spinal blocks, and the patch technique.

Patient-controlled analgesia. Traditionally, pain after surgery was treated with injections of pain-killers at intervals of several hours. Standard doses of these medications were often used for all patients. However, there is a tremendous variation in the amount of pain-killer required by various individuals. A so-called standard dose might be too little for one patient, yet too much for another. Thus, it was very difficult to determine what dose of medicine was appropriate for each patient. This dilemma resulted in the idea of allowing patients to administer their own pain-killers as they felt necessary, and eventually in the practice of patient-controlled analgesia.

Patient-controlled analgesia is rapidly growing in use, and is one of the most common treatments recommended by acute-pain specialists. Typically, the patient can administer pain-killer (usually a narcotic such as morphine) via a computer-controlled pump that releases small preset doses into an intravenous line; the patient controls the pump by pushing a button. The pump is programmed with a type and dose of pain-killer individualized for the patient. It is also programmed with a safety mechanism to avoid overdosing, a so-called lock-out period: generally a 6–10-min period during which no pain medication will be released, even if the patient pushes the button. The dose record is recorded automatically, so that when the acute-pain team visits the patient, it can ascertain how many times the patient has pushed the button as well as how many doses have been received. Adjustments can be made accordingly.

Spinal catheters. Another new therapy for alleviating pain after surgery entails administering pain-killers directly into the spinal area. A narcotic, often in combination with local anesthetics, is infused either into the spinal fluid in the lower back or, more commonly, into the epidural space (the space outside the spinal fluid sac) via a tiny tube. This technique was introduced several decades ago after the discovery of specific receptors in the spinal cord for drugs such as morphine.

Side effects of morphine and other narcotics as traditionally administered include nausea, itchiness, sleepiness, depressed breathing, and decreased ability to urinate. However, it was discovered that delivery of drugs directly into the spinal area provides more effective pain relief with fewer side effects than intramuscular shots. In addition to the side effects from the narcotic, numbness and weakness can occur when local anesthetics are used. However, the spinal technique is now safely and effectively used to treat pain after surgery, particularly after major surgery on the lungs, large blood vessels, or abdominal and pelvic organs.

Patch technique. Another new method of treating pain utilizes a patch filled with the narcotic fentanyl, which is applied to the skin. Thereby the pain-killer is released into the circulatory system in a small, continuous dose over a period of about 3 days.

Nonnarcotic pain-killers. A class of pain-killers called nonnarcotics consists of drugs such as aspirin and Tylenol and the nonsteroidal anti-inflammatory drugs. The latter group of drugs, which includes ibuprofen, can provide excellent analgesia without some of the side effects of narcotics. These drugs have their own group of problems, however, including stomach irritation and upset, and an increased tendency toward bleeding. In addition, most of them were not available in injectable form, so they could not be taken until the patient was well enough to take liquids by mouth. Recently, ketorolac, an injectable form of nonsteroidal anti-inflammatory drug, was introduced.

Nondrug treatments. Drugs are not the only way to treat pain after surgery. Other methods include techniques that numb or stimulate the nerves, hypnosis, relaxation therapy, and patient education.

Numbing techniques. In some cases, temporary numbing of the nerves that supply the area of pain can provide very effective relief. This technique is most easily carried out when the nerves are very accessible. For example, when surgery of the arm or hand is performed, local anesthetics can be infused or injected into the armpit, where the nerves supplying the arm are located. Relief of pain after operations on the chest can be effected by injecting local anesthetics into the intercostal nerves supplying the chest wall. These regional analgesic techniques have the benefit of affecting only the area of pain, with almost no effects on the rest of the body. Nerve blocks are usually carried out with local anesthetics that last a few hours. One technique, called cryoanalgesia, involves freezing the nerve with a probe and can produce pain relief lasting for months.

Stimulation techniques. Acupuncture is a stimulation technique that has been used for thousands of years to relieve all types of pain. Another technique, transcutaneous electrical nerve stimulation (TENS), is carried out with a small battery-operated device wired to adhesive pads that are attached to the skin over the area of pain. Transcutaneous electrical nerve stimulation seems to work, at least in part, by stimulating nerves so that they release the body's own narcoticlike substances in the spinal cord, thereby preventing the transmission of pain impulses via nerves to the brain. An advantage of transcutaneous electrical nerve stimulation is that it produces almost no side effects; a disadvantage is that it helps only a select number of patients.

Hypnosis and relaxation therapy. These approaches have been demonstrated to provide safe and effective relief of pain, although they generally require a period of training for the patient before surgery, and like the stimulation techniques, seem to work in only a select group of patients.

Patient education. One of the most useful tools is patient education before surgery. Several studies have indicated that patients who are told what to expect after surgery require far less pain medication than those who have no such preparation. Therefore, a question-and-discussion session with the surgeon and anesthesiologist prior to any surgical procedure is very important.

For background information SEE ANALGESIC; ENDORPHINS; NARCOTIC; PAIN in the McGraw-Hill Encyclopedia of Science & Technology.

Carol A. Warfield

Bibliography. L. B. Ready et al., Development of an anesthesiology-based postoperative pain management service, *Anesthesiology*, 68:100–106, 1988; C. A. Warfield, Management of postoperative pain, *Hosp. Prac.*, 24(5A): 53–59, 1989.

Paleonutrition

From a nutritional standpoint, the development of agriculture and the domestication of animals may be considered two of the worst mistakes humans ever made.

Humanity first evolved approximately 5×10^6 years ago on the plains of east Africa. The first humans were nomadic scavengers, or hunters and gatherers, and throughout more than 99% of human existence people lived this way. Protein came from fish, fowl, some plant foods, bone marrow, insects of all kinds, and small reptiles and mammals. Later when humans developed better hunting techniques and weapons, protein was also obtained from larger animals. Plant foods included a wide variety of berries, roots, tubers, leaves, nuts, seeds, and fruits—foods that are high in fiber and rich in vitamins and minerals. Diets were low in salt, fat, and sugar. Modern nutritionists have called this assortment the perfect human diet.

Transition to agriculture. Humankind's transition from nomadic hunters and gatherers to farmers and ranchers began about 10,000 years ago in the Middle East. Malnutrition began to occur soon after agricultural groups turned away from diets that centered on gathering a wide variety of fresh fruits, vegetables, tubers, and nuts, especially when the subsequent diet was made up almost entirely of barley and wheat products. Also, consumption of milk, eggs, cheese, butter, and fatty meats from penned animals added high levels of fat to the human diet. Salt and sugars were added to make bland foods such as barley and wheat taste better. These dietary changes increased the nutritional stress on the human body and caused, among other things, a sharp rise in dental caries.

Analysis of bones and teeth. Archeologists analyze the dietary and nutritional effects of the transition to agriculture through a variety of techniques. Common methods of analysis include examining evidence of disease or abnormalities in human bone and teeth.

An important marker of a dietary shift from hunting and gathering to agriculture is dental pathology, because nutrition affects tooth enamel formation. During times of nutritional stress, ameloblasts, which are the enamel-producing cells, may not form in the normal direction. The result is an abnormal enamel matrix composition called Wilson bands. If ameloblast formation is halted for a short time, the thickness of the enamel is affected, resulting in depressed areas in the enamel called linear enamel hypoplasias. Wilson bands and linear enamel hypoplasias can be observed on a tooth surface or in a tooth cross section with a microscope. A limitation of using Wilson bands and linear enamel hypoplasias to determine nutritional quality of the diet is that these types of pathologies can be produced by other stressors, such as weaning or nonnutritionally related disease.

Caries can be seen on a tooth with the naked eye. Caries result from a disease in which the calcified regions of the tooth are disintegrated and eroded through microbial action. Caries have been observed to increase dramatically in agricultural populations. However, such increases have also been observed in certain hunter and gatherer societies that have an extremely high carbohydrate diet.

The frequency of Wilson bands and linear enamel hypoplasias in a large prehistoric sample has been observed to increase from the Late Woodland (mainly hunter and gatherer) to the Mississippian (agricultural) period in Illinois. As the people in this area relied more heavily on corn and less on animal protein, their diet became less nutritious, and they experienced a decreased resistance to infectious disease. The people who survived the infectious diseases showed abnormalities in the development of bones and teeth, including the formation of Wilson bands and linear enamel hypoplasias.

A high frequency of dental pathology has also been observed in the prehistoric Indus Valley civilization of southern Asia. Archeologists have observed that when the population in this region adapted to a full-scale agricultural economy, malnutrition and infection increased. The prevalence of dental disease in the area is high: 72.2% of the sampled teeth contained linear enamel hypoplasias, and 6.8% of the population exhibited caries.

Analysis of parasites. Another method for assessing paleonutrition involves the analysis of parasites from prehistoric human feces (coprolites). A coprolite sample is slowly broken down in a mild detergent solution. The sample is then screened through a micrometer mesh so that the large particles (seeds, bone, fur, and fiber) are caught in the screen and the small particles (pollen, phytoliths, and parasites) pass through. A subsample of the fluid that passes through the screen is analyzed for parasites. The fluid is placed in a vial with a solution of acetic formalin alcohol, which prevents fungal and bacterial growth. After the fluid settles, the upper portion is examined under the microscope for parasites.

Analysis of coprolites has revealed that hunter and gatherer populations were relatively free from parasites and that as subsistence shifted to intensive agriculture populations became increasingly infested by a variety of debilitating parasites. The only parasite that has been observed in coprolite samples of North American hunters and gatherers is *Enterobius vermicularis* (pinworm). This human-specific parasite is not a major health and nutritional risk. In contrast, prehistoric agricultural populations have been observed to be infected with at least 10 types of parasites, many of which are severely debilitating and potentially fatal.

Agricultural populations are more likely to be infected with parasites because of their life-style. Agriculturalists live in large populations and do not move around as much as hunters and gatherers. A large stable community provides a large reservoir for infection, and the proximity of people makes reinfection more likely. Agricultural communities usually have large buildups of garbage and human and animal waste. Because of poor sanitation methods, agricultural living places are favorable breeding grounds for parasites. Other human activities that encourage the spread of parasites include crop irrigational techniques, animal domestication, and grain storage. In contrast, hunters and gatherers maintain small populations, are mobile, and do not have the other problems associated with growing crops.

Protein residue analysis. A newer method for determining paleonutrition involves analyzing coprolites for protein residues. This method allows archeologists to identify the exact types of food items eaten. However, a limitation of this method is that a coprolite sample will only reflect short-term dietary intake. The method involves immunological analysis of tiny amounts of protein through crossover electrophoresis. The unknown protein residue from the coprolites is placed in agarose gel with known antiserum from different plants and animals. The agarose gel is then placed in an electrophoresis tank with a buffer. The electrophoretic action causes the protein antigens to move toward the antibody, which is not affected by the electrical action. The solution containing the unknown protein residue and the matching plant or animal species antiserum forms a precipitate that is easily identifiable when stained. The samples with a precipitate indicate that the matching plant or animal was eaten by the person who deposited the coprolite.

Other analytic techniques. Blood residue analysis can be conducted on a variety of archeological materials, including coprolites, lithic materials, ground stone, and soils. Few studies have been done so far, but the method may be very useful in the future for determining paleonutrition. Other chemical techniques are being developed for the analysis of steroids, wax, molecular compounds, and deoxyribonucleic acid (DNA) in coprolite samples.

For background information SEE ANTHROPOLOGY; ARCHEOLOGY; MALNUTRITION; NUTRITION; PHYSICAL ANTHROPOLOGY in the McGraw-Hill Encyclopedia of Science & Technology.

Kristin D. Sobolik; Vaughn M. Bryant, Jr.

Bibliography. B. Kooyman, M. E. Newman, and H. Ceri, Verifying the reliability of blood residue analysis on archaeological tools, *J. Archaeol. Sci.*, 19(3):265–269, 1992; J. R. Lukacs, Dental paleopathology and agricultural intensification in South Asia: New evidence from Bronze Age Harappa, *Amer. J. Phys. Anthropol.*, 87:133–150, 1992; K. J. Reinhard, Archaeoparasitology in North America, *Amer. J. Phys. Anthropol.*, 82:145–163, 1990; J. C. Rose, K. W. Condon, and A. H. Goodman, Diet and dentition: Developmental disturbances, in R. I. Gilbert, Jr., and J. H. Mielke (eds.), *The Analysis of Prehistoric Diets*, 1985.

Perception

Although the world is rich with motion, the fact that people and other animals move around did not have much impact on discussions of human perception until James Gibson placed self-motion at the center of his theory of ecological psychology. Gibson was tackling what may be the most difficult issue in perception: understanding how humans see the world in the round and not from any particular point of view. He recognized two forms of structure in the optic array: perspective structure, which changes with every point of observation, and invariant structure, which reflects the spatial layout of objects within the environment. Self-motion cleaves these two forms, informing the perceiver at the same time about self-locomotion (changes in perspective) and about the rigid arrangement of environmental surfaces (constants of motion). Recently, these ideas have motivated investigations into the perception of mechanical systems in which a similar structural duality pairs kinematics and dynamics.

Kinematics and dynamics. Motion may be generally described at two levels, comprising a kinematic description of the changes that motion induces in the optic array and a dynamical description that gives a causal account of the motion. Anytime motion is perceived, there is a potential cleaving between these two components of structure. In this sense, kinematics is akin to perspective structure; it is the outward face of a motion event. The dynamics is related to invariant structure in two senses: it is an underlying form that is revealed only through change, and it causally supports the kinematics in the same way that rigid environmental surfaces support perspective structure.

This observation raises the psychological question of whether the dynamics in a motion event is perceptible as a form of invariance. A relative question—whether humans perceive motion events in the round—is particularly relevant because the answer is intimately tied to the phenomenology of everyday experience. Perception of motion in the round involves an awareness of the physical situation that gives the motion its character. There have been numerous studies documenting and calibrating people's awareness of such dynamical quantities as mass ratio, mass, and elasticity. However, the issue is whether awareness of dynamics is tied to anything like a dynamical invariant.

Recently, two viewpoints have been advanced to account for both the existence and quality of the awareness of dynamical quantities. In one view, awareness of dynamics arises from perceiving invariants of motion. The perceived quantities are interpreted as having a one-to-one relationship with real physical quantities such as mass and elasticity. A second view is that although people may have a rich experience of the dynamics, awareness of dynamics arises from constructions generated in order to make sense of the world. These constructions, although tutored by experience, may or may not hold in a given physical situation. These two views have been subjected to experimental analysis primarily in the domain of the perception of mass ratio specified by a two-body collision.

Perceptual problems. Collision experiments have revealed that people perceptually organize collision events by using two heuristics: (1) If an object exits from a collision with a much greater speed, it is less massive. (2) If an object ricochets in a collision, it is less massive. The clearest evidence for heuristic usage is obtained from simulations of real collisions where one object exits with a greater speed and the other ricochets. In such simulations, subjects are completely confident that one of the objects is much more massive but disagree about which one it is. Evidently these heuristics do not relate in an accurate way to the mass ratio that appears in the formal equations describing the event and that is the true dynamical invariant. However, heuristics are not purposeless; they represent the extraction of dynamical information, even if that information is not aligned with distal physical quantities.

Some motion events are patently not perceived in the round; people do not have any awareness of the dynamics, nor do they seem to have any awareness beyond noticing what the object does, that is, they have a kinematic description.

In purely mechanical systems, those not involving electricity and magnetism, the transition between dynamical awareness and lack of perception of dynamics appears to occur with the introduction of rotation. A spinning top falls by precessing, and is an enduring visual delight. A top that is not spinning falls as expected, and this event is neither special nor captivating. A number of reasons may account for this perception. In the first place, rotating objects do have a more complex physics. For instance, there are quantities such as vector cross products that point in directions not obviously related to those implied by the kinematics. That rotation is more complicated than translation would not, however, prevent people from applying heuristics about rotating objects. Yet, there seems to be a glaring absence of rotation heuristics that people readily draw upon. In response to queries as to what holds a top up or why it is easier to balance on a moving bicycle, there is a wide range of answers. Laboratory studies have shown that people cannot distinguish natural from unnatural rotational motions, and are prepared to accept almost anything as possible so long as the object does not stop or reverse direction. People interpret objects stopping or reversing direction as intervention, and they expect to see the agent.

Other accounts of the strangeness of rotation have to do with perceptual learning. One possibility is that people have less experience with rotating objects and so have had less opportunity to learn about rotation. Alternatively, it is arguable that rotation has little environmental consequence and therefore learning does not arise as a real issue. These arguments are not compelling because neither accounts for the nearly complete perceptual opacity that seems to attend rotation. Furthermore, these arguments suggest that certain individuals (such as lathe operators, physics instructors, bicyclists, ice skaters, divers, and gymnasts) might, by virtue of specialized training, be able to perceive rotation dynamics or have sets of rotation heuristics. However, when such people are brought into the laboratory, they invariably behave as naive subjects.

The core reasons for the absence of heuristics about rotating systems and the general level of incomprehension that attends them may have more to do with basic issues in perception and attention. A key observation is that the human visual system is set up to deal with complex arrays of translational motion. Translation is known to be processed in parallel (all object motions are scanned simultaneously), and this ability leads to a grouping principle: objects that share a common translation direction are grouped as a single coherent entity. This principle is referred to as the law of common fate. However, rotation does not lead to grouping on the basis of direction. An array of rotating objects does not constitute a group, and visual search experiments have verified that rotation is processed serially (each object motion is scanned one at a time). The differences that have been found in visual search and in the formation of perceived groups reveal a fundamental limitation in attentional resources. It is not possible to attend simultaneously to more than one rotation axis, perhaps because an axis of rotation demands its own frame of reference and different frames of reference are perceptually exclusive.

The conflict between frames of reference can be experienced by considering the perceptual problem depicted in the **illustration**. If asked how many times the wheel will spin around as it rolls along the line, most people will report that five or six revolutions are required. In fact, only two revolutions are needed to traverse the line. Perceptually, there are two frames of reference here. One is induced by the line; it is an environmental frame. This frame does not specify an origin; it could be anywhere. The second frame is induced when imagining the wheel rolling. This frame has an origin at the center of the wheel, which serves as an axis of rotation. In physics parlance, this frame of reference is body-centered. The perceptual result of having two frames is that the wheel's translational motion seems to be one thing, its rotation another, and what these two have to do with each other is not obvious. This problem can be contrasted with the problem of simply having to decide how many horizontal displacements of the wheel are required to cover the line. It is easy to see that about six displacements are required. The displacement problem can be visualized within the single environmental frame.

The manifest absence of dynamical awareness in the context of rotation may be linked to a perceptual rigidity that attends everyday interactions with the world. The dynamical significance of rotation may not be appreciated because people maintain themselves in

Perceptual problem illustrating the difficulty in combining rotational and translational frames of reference.

an environmental frame of reference. In this frame, people are mostly aware of where objects are going, not how they are moving relative to a body-centered coordinate structure. Furthermore, limitations in the allocation of attention may make it difficult to synthesize the coupling between translational and rotational frames of reference. Thus, the attentional narrowing to a single frame of reference is incompatible with the requirements for extracting dynamical information from rotation.

For background information SEE INFORMATION PRO-CESSING; KINETICS (CLASSICAL MECHANICS); PERCEPTION in the McGraw-Hill Encyclopedia of Science & Technology.

David L. Gilden

Bibliography. J. J. Gibson, *The Ecological Approach to Visual Perception*, 1979; D. L. Gilden, On the origins of dynamical awareness, *Psychol. Rev.*, 98:554–568, 1991; D. L. Gilden and D. R. Proffitt, Understanding collision dynamics, *J. Exp. Psychol. Human Perform. Percept.*, 15:372–383, 1989; M. K. Kaiser et al., The influence of animation on dynamical judgments: Informing all of the people some of the time, *J. Exp. Psychol. Human Perform. Percept.*, 18:669–690, 1992.

Periodic table

The periodic table is a list of elements (atoms) ordered along horizontal rows according to atomic number (the number of electrons in an atom and also the charge on its nucleus). The rows are arranged so that elements with nearly the same chemical properties occur in the same column (group) and each row ends with a noble gas (closed-shell element that is generally inert). One of the forms of the table in use today is shown in **Fig. 1**. The heavy line separates metals from nonmetals. For chemists, the position of atoms in the periodic table provides the single most powerful guide for classifying the expected properties of molecules and solids made from these particular atoms.

The origin of the periodic table was explained in the 1920s in terms of the basic physical laws (quantum mechanics) obeyed by the electrons of an atom. Thus, the horizontal rows in the periodic table correspond to the shell number, n, and groups correspond to a particular electronic configuration designated by the number and type of electrons in its outermost shell. These electrons govern chemical properties and are known as valence electrons. For example, the electronic configuration for nitrogen is $2s^2 2p^3$, where $n = 2$ is the row number and also the electronic shell number; s and p are quantum-mechanical indices, and the superscripts give the number of s-type and p-type electrons present in the valence shell. The indices s, p, d, and f also designate blocks in the table (Fig. 1). When a group in the table is descended, n increases in steps but the type and number of valence electrons remain the same; for example, in group 15, N (nitrogen) = $2s^2 2p^3$, P (phosphorus) = $3s^2 3p^3$, As (arsenic) = $4s^2 4p^3$, Sb (antimony) = $5s^2 5p^3$, and Bi (bismuth) = $6s^2 6p^3$.

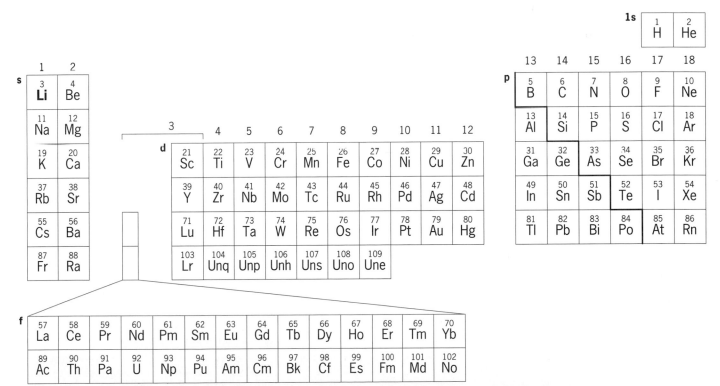

Fig. 1. Contemporary form of a two-dimensional periodic table. The atomic numbers are listed above the symbols identifying each element. (*After J. Emsley, The Elements, Oxford University Press, 1989*)

Configuration energy. Recently, it has been found that additional information from the physical laws of atoms can be incorporated into the periodic table and greatly enhance its organizing capability. The new addition, configuration energy (CE), adds a third dimension to the periodic table (**Fig. 2**). The quantity configuration energy is defined in terms of the ionization energy (I), the energy required to remove an electron from an atom, for example, N $\rightarrow$ N$^+$. Configuration energy is given by the formula below,

$$CE = \frac{aI_{ns} + bI_{np}}{a + b}$$

where a and b are the number of s and p valence electrons. The formula gives the energy of an average valence electron for the atom in question. For N, CE = $(2I_{2s} + 3I_{2p})/5$, where I_{2s} is the energy required to remove a $2s$ electron from N, N($2s^2 2p^3$) $\rightarrow$ N$^+$($2s 2p^3$), with electronvolts as units. For the d-block transition elements, $I_{np} \rightarrow I_{(n-1)d}$, and b represents valence region occupancy. Values for the configuration energy of selected atoms are given in **Fig. 3**.

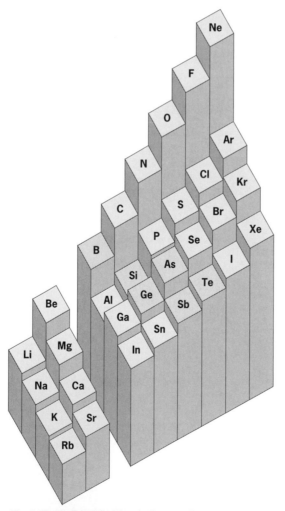

Fig. 2. Three-dimensional periodic table. Only *s* and *p* blocks are displayed. (*After L. C. Allen, Electronegativity is the average one-electron energy of the valence-shell electrons in ground-state free atoms, J. Amer. Chem. Soc., 111:9003–9014, 1989*)

Besides enhancing the organizing capability of the periodic table, the concept of configuration energy also explains many long-standing puzzles about the table itself. It explains the existence of the metalloid band of elements (light-gray atoms in Fig. 2; configuration energy is nearly constant in this band) and why these elements divide the metals from the nonmetals. Elements possessing configuration energies with magnitudes greater than those of the metalloids are nonmetals; elements with lower configuration energies are metals. The topography of the three-dimensional table of Fig. 2 graphically brings out this division. Likewise, the decreasing magnitudes of configuration energies down a group in Fig. 2 demonstrate and quantify the long-known fact that the elements in a group do not have exactly the same chemistry. For example, the chemistry of carbon is quite different from that of tin, even though both are in group 14. This differentiation is called metallization.

Chemical bonding. Configuration energy also quantifies the chemical bonding designations, covalent (C), ionic (I), and metallic (M), which have always been associated with the periodic table but have not been adequately defined. An example of this role for configuration energy is shown in **Fig. 4**, where the elements of the $n = 3$ row and their binary combinations are displayed around a triangle. ΔCE is the difference in configuration energy between a pair of atoms, and it tells the relative polarity or ionic character of a chemical bond. This property is important in the design of new high-technology materials. The values of $\langle\Delta$CE$\rangle$ on the right side of Fig. 3 are the average of the ΔCE values for each pair of atoms along one of the horizontal broken lines through the M-I-C triangle. The numbers show the strong rise in ionic character between pairs of atoms as the ionic bonding limit, NaCl (sodium chloride), for this collection of $n = 3$ atoms is approached. In Fig. 3 the compounds at the points of the triangle are well known, but near the center of the legs of the diagram novel features become apparent. In the middle of a row [Si (silicon) along the horizontal axis] are the metalloids, and going into the triangle there is the 3-5 semiconductor AlP (aluminum phosphide). Close to the center of the I-C leg are polymeric materials, AlF$_3$ (aluminum fluoride) and AlCl$_3$ (aluminum chloride). Close to the center of the M-I leg are the Zintl phases, that is, semiconductors with a metallic sheen and ceramic-type brittleness. These semiconductors have yet to find a commercial application, but constructing triangles like that in Fig. 3 helps focus on new combinations of elements that may spark fruitful investigations. Successive legs can be added to the periphery of the triangle by using elements and binary combinations from the higher n rows; then it is found that along the I-C and M-I legs the middle regions themselves have unexpected, yet meaningful, diagonal patterns that suggest interesting new relationships between solids.

Other properties. Another property of the periodic table is resolved by configuration energy: the limitation of the fluoride oxidation states in the elements N, O (oxygen), F (fluorine), Cl (chlorine), Br (bromine), He (helium), Ne (neon), Ar (argon), and Kr (krypton). These nine atoms cannot reach a fluoride oxidation state equal to the number of their valence electrons, as can all other representative atoms (atoms whose valence shell is made up of s and p electrons only). They have the highest nine configuration energies in the periodic table; some fraction of their valence electrons have ionization energies too high to engage in chemical bonding. For example, IF_7 (iodine heptafluoride) exists but not ClF_7 (chlorine heptafluoride).

Similarly, it had previously been believed that atomic radius was an independent property, but again (for representative atoms) radius can be expressed as a simple inverse function of configuration energy. Likewise, the multiple bonds found for representative atoms of the $n = 2$ row but much less frequently for $n = 3$, 4, 5, or 6 can be shown to follow directly from the high values of configuration energy for $n = 2$.

Significance. Why does the periodic table work so well at organizing the properties of the 15×10^6 known molecules and solids? After all, it is only an array of free atoms. The first reason is easy and straightforward: molecules and solids are made up of atoms that are only very slightly perturbed when they link together; therefore, they largely retain their identity. Another aspect of this question concerns the intrinsically simple nature of atomic electronic structure. Even though the motion of electrons around a nucleus is a highly complex, many-particle swirl requiring the largest digital computers to determine it accurately, to a very good approximation each electron can be individually assigned a type, s, p, d, or f, and a shell number, n. The ability to label electrons individually by these indices is the second reason why the periodic table works so well. On a grander scale, the claim can be made that it would not be possible to have the science of chemistry—or any insight at all into the matter out of which everything on Earth is made—were it not for these two fundamental simplicities.

Because of the intense attention devoted to the periodic table in high school and college first-year chemistry courses, the table has become a cultural icon and a symbol of science for a large part of society.

The number of new synthetic compounds that appear daily (many with very unusual properties) has been increasing at a higher rate than ever before; and in general, there are no well-developed sophisticated theories that can immediately explain how the new compounds fit in with all of the other known compounds or how to predict other new ones with desired properties. Often, the periodic table is the only available organizing scheme that can keep up with the fast-paced change that is taking place in chemistry.

To most professionals, the lack of numerical or analytic connection between the periodic table and the quantum-mechanical concepts and computational models that dominate contemporary explanations and predictions on the structure and reactivity of molecules and solids has reduced its usefulness. These models and concepts are largely built around molecular-orbital-energy-level diagrams and the corresponding

$n = 1$		$n = 4$		$n = 5$		$n = 6$	
H	13.61	K	4.34	Rb	4.18	Cs	3.89
He	24.59	Ca	6.11	Sr	5.70	Ba	5.21
$n = 2$		Sc	6.8	Y	5.9		
		Ti	7.4	Zr	6.6		
Li	5.39	V	8.1	Nb	7.4		
Be	9.32	Cr	8.6	Mo	8.2		
B	12.13	Mn	9.2	Tc	9.0		
C	15.05	Fe	9.9	Ru	9.8		
N	18.13	Co	10.4	Rh	10.6		
O	21.36	Ni	11.0	Pd	11.3		
F	24.80	Cu	10.8	Ag	11.7	Hg	10.44
Ne	28.31	Zn	9.39	Cd	8.99		
$n = 3$							
Na	5.14						
Mg	7.65						
Al	9.54	Ga	10.39	In	9.79		
Si	11.33	Ge	11.80	Sn	10.79		
P	13.33	As	13.08	Sb	11.74		
S	15.31	Se	14.34	Te	12.76		
Cl	16.97	Br	15.88	I	13.95		
Ar	19.17	Kr	17.54	Xe	15.27		

Fig. 3. Configuration energies (in electronvolts) of selected elements.

energy bands and density-of-state patterns in solids. Fortunately, configuration energy, introduced as a new dimension of the periodic table, is just the average atomic energy level, and simultaneously the average density of states, for the atoms out of which the molecular-orbit-energy-level diagrams and energy bands in solids are constructed, thereby tying the periodic table directly to present-day research techniques.

Traditionally, the best auxiliary aid in the use of the periodic table has been the long-recognized chemical concept of electronegativity. However, this quantity

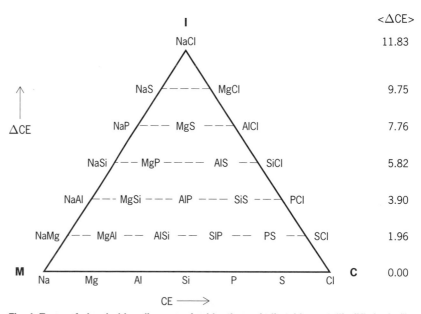

Fig. 4. Types of chemical bonding organized by the periodic table: metallic (M), ionic (I), and covalent (C). Illustrated by $n = 3$ elements. Quantification of bonding type achieved by configuration energy, CE, and by the difference in CE between pairs of atoms, ΔCE. $\langle\Delta$CE$\rangle$ scale is discussed in the text. (After L. C. Allen, Extension and completion of the periodic table, J. Amer. Chem. Soc., 114:1510–1511, 1992)

has lacked a unique definition (and, therefore, an accurate set of values), and the nagging question of whether or not it can be measured, even in principle, remains. Configuration energy can now be identified as fulfilling the role long ascribed to electronegativity, and as suggested above, it has the additional property of strongly correlating with the density of atomic states (that is, the relative spacings of the atomic energy levels). The synergistic combination of configuration energy with the periodic table of Fig. 1 gives rise to Fig. 2 and creates a research tool especially appropriate to today's needs.

For background information *SEE CHEMICAL BONDING; ELECTRON CONFIGURATION; ELECTRONEGATIVITY; IONIZATION POTENTIAL; PERIODIC TABLE* in the McGraw-Hill Encyclopedia of Science & Technology.

Leland C. Allen

Bibliography. L. C. Allen, Electronegativity definition, *J. Amer. Chem. Soc.*, 111:9003–9014, 1989; L. C. Allen, Extension and completion of the periodic table, *J. Amer. Chem. Soc.*, 114:1510–1511, 1992; P. W. Atkins and J. A. Beran, *General Chemistry*, 2d ed., 1992; R. J. Puddephatt and P. K. Monaghan, *The Periodic Table of the Elements*, 2d ed., 1986.

Petrographic image analysis

Scientists have used imaging to study many aspects of the world from the surface of planets and stars to the eye structure of insects. This technology has also been adapted to improving understanding of the storage and flow of hydrocarbons in subsurface rocks.

A major aim of petrographic image analysis is to derive physically relevant microstructural information from analysis of porosity exposed in an intersecting plane (commonly a petrographic thin section). Although developed in the context of permeable sedimentary rocks, petrographic image analysis can be applied to other porous media, such as ceramics or concrete, as well as to multiphase solids.

Porosity in sedimentary rocks (including aquifers and hydrocarbon reservoirs) exists as a three-dimensional network of pores connected by pore throats. Relative sizes and frequency distribution of pores and pore throats govern the way that fluids move within rocks as defined by permeability and relative permeability (the movement of one or more fluids in the presence of others, for example, oil and water). These same parameters also affect physical measurements made on rocks, such as resistivity and acoustical properties, important to evaluation of subsurface reservoirs by the well logging engineer. These properties cannot be fundamentally understood without knowledge of the porous microstructure. Essential microstructural information can be obtained by deriving three-dimensional relationships through analysis of planar sections.

Image acquisition. Petrographic image analysis requires digital images as well as image processing software. Commonly, the pores are impregnated with pigmented epoxy or a low-melting-point alloy to aid in discriminating them from minerals.

Care must be taken that the image represents porosity intersected at the plane of section and that porosity located below that plane is excluded. Images are initially acquired by using tools such as a video camera or backscatter device with a scanning electron microscope. The digitized image represents the field of view produced by a microscope. Digitized images consist of an array of pixels (picture elements). Each pixel is associated with a gray-level value representing the brightness of the pixel. For porosity analysis, the objective is to produce a binary image wherein all pixels representing the rock matrix are set to an intensity of, for instance, unity; and all pixels representing porosity are set to a value of zero. The porosity exposed in the plane of section is discontinuous, with some porosity elements (porels) having simple geometries, and others being more complex.

A single binary image is generally not representative of the section, given the relatively small field of view required to yield adequate resolution. Many images (at least 20 in the case of sandstones) uniformly distributed over the plane must be obtained from a standard thin section (about 1 cm^2 or 0.155 in.^{2}) in order to balance resolution with coverage. The optimal magnification is commonly that used for conventional petrographic assessment.

Low-resolution images taken at very low power (40–100×) are useful to quantify fabric and structure when small-scale resolution of the pore wall is not needed.

Image processing. Scaling from two dimensions to three dimensions is a complex problem. Interpretation of data obtained from each section is based on acceptance of a set of assumptions concerning a degree of randomness in orientation and location of three-dimensional elements in the solid and the relative homogeneity of the solid. Under these assumptions the ratio of pore pixels to the total is equal to pure volume porosity. However, the value of optically determined porosity is commonly less than that of physically measured porosity because of the lack of capability for resolving microporosity. Optically derived porosity, or total optical porosity, and the difference between it and physically measured porosity are useful.

The length of the total pore perimeter per unit area is proportional to the surface area per unit volume; however, it must be understood that this value will be less than the physically based value because of the resolution limit.

Further analysis of the binary image requires determination of sizes and shapes of porels. Porels are patches of porosity consisting of a collection of interconnected pore pixels. A porel may represent a section through a single pore or may contain several pores bridged by sections of connecting throats. In poorly consolidated sandstones the porosity may be so interconnected that a single lacelike porel may cover a large portion of the field of view.

Pore throats occupy narrow portions of pores, and thus they are not commonly exposed in section. However, poorly consolidated sandstones characteristically contain pore throats that are relatively large (that is, throats 40 micrometers in diameter connected to pores that are around 100 μm in diameter), and throats connecting such pores are comonly visible in section. As relative throat size decreases, pores become progressively disconnected and isolated in the plane of section, with the result that porels become smaller. Consequently, there is a positive correlation between the number of large porels and the logarithm of permeability.

Many other variables derived from porels are strongly correlated with the logarithm of permeability. Statistical relationships are useful in that they demonstrate that information in the plane of section is relevant to flow properties. Because of the poor representation of pore throats in the plane of section, such correlations imply a relationship between pores and throat sizes. An understanding of the fundamental relationships between pores and throats requires relating thin-section information to capillary pressure curves. This analysis can be done most clearly by first deriving the three-dimensional sizes and shapes of pores from data concerning sizes and shapes of porels in the plane of section, and then determining which pores are penetrated by a nonwetting phase with increasing injection pressure.

Any three-dimensional object of a fixed size and shape will, if randomly distributed through a volume, generate a characteristic distribution of sizes and shapes on the plane of section. Such a distribution is known as the size/shape spectrum of the object. A given volume may contain many sets of three-dimensional objects, each set composed of objects of similar size and shape. Such a mixture yields a size/shape spectrum in the plane that is the sum of the spectra arising from each class of object. The size/shape spectrum in the plane is always more variable than the related size/shape spectrum in three dimensions. Derivation of the three-dimensional distributions from the planar size/shape spectra thus has two benefits: the statistics are directly related to the fundamental features, and the data are inherently less variable.

Pore types. Pores may exist as a single size/shape population, or several pore types may coexist. Determination of the true situation involves analysis of the size/shape spectrum associated with porels in the plane of section. This analysis requires a measurement scheme sensitive to the nuances of a wide range of irregular geometries and a set of special pattern-recognition algorithms. An image-processing procedure called erosion-dilation differencing produces a spectrum for each porel that measures the number of pixels associated with roughness elements of progressively larger size. The sum of porel spectra through all digitized views from a section produces the erosion-dilation spectrum of a sample. The spectrum may represent a single subpopulation of pores or a composite. The type of spectrum can be determined from simultaneous analysis of a set of related samples.

Given a set of spectra from a collection of related samples (for example, taken from the same core), it is possible to classify the porel data (two-dimensional)

into pore types (three-dimensional) by using two algorithms in a procedure termed polytopic vector analysis. The first algorithm uses a modified factor analysis to determine the number of subpopulations (pore types). The second algorithm derives the size, shape, and relative proportion of each pore type. The validity of this derivation has been physically verified by comparison with nuclear magnetic resonance (NMR) T_1 relaxation spectra. The T_1 term describes the influence of the enclosing matrix on the precession of an atom with an odd number of protons (usually hydrogen).

Relationships to petrophysics. Sandstone and carbonate reservoirs rarely contain more than six pore types. The proportion of porosity accessed by pore throats of various size can be determined by capillary pressure tests, where pressure is inversely related to throat size. Variation in pore-type abundance among samples can be easily correlated with changes in the mercury capillary pressure, with each pore type commonly dominating a specific pressure increment. This relationship can be quantified by using multiple regression procedures, which determine the proportion of each pore type that is invaded in each pressure increment (see **illus**.). A relationship between pore type and throat size indicates a strong tendency for flow circuits to consist of pores of the same type, that is, the porous medium is structured even at small scale.

Structure can be a consequence of layering arising

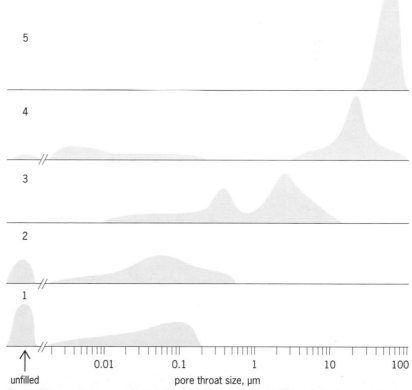

Proportion of volume of each pore type 1–5 from the Cutbank Sandstone, accessed by pore throats of varying size. Each distribution is a frequency distribution whose area is equal to unity. (*After R. Ehrlich et al., Petrography and reservoir physics, III. Physical models for permeability and formation factor, AAPG Bull., 75:1579–1592, 1991*)

from changes in grain size. However, image analysis involving two-dimensional Fourier transform of low-magnification images shows that such structures exist even in well-sorted sandstones and are composed of regions of dense packing that are isolated in a matrix of loose (flawed) packing. The loose packed domains contain large pores connected by large throats. This domainal structure is present in all particle aggregates. The sizes and shapes of each type of domain are largely a function of depositional processes.

Carbonate reservoir rocks typically exhibit a strong relationship between pore type and throat size, even in the case of strongly recrystallized dolomites where all traces of depositional fabric have been erased. Thus, the existence of thermodynamic controls on the relationship is implied.

Permeability modeling. Pores are petrophysically relevant because of the relationship between pore type and throat size. In consequence, variation in flow properties among samples is related to variations in abundance in pore types. This relationship can be made explicit by modeling flow as a function of flow circuits that vary with respect to pore type. The variation in permeability among closely spaced volumes largely arises from the fact that the sample volumes are small with respect to the structural heterogeneities, even in seemingly homogeneous rocks. In sandstones, this structure consists of domains characterized by differences in grain size or packing.

In each sample, the numbers of pores of each type per unit of cross-sectional area can be calculated and, given the relationship between pore type and throat size, a mean throat size can be assigned to pores of each type. These values are sufficient to model a portion of the permeability tensor by using the Hagen-Poiseville capillary tube model.

Most permeability measurements are taken parallel to bedding, while most thin sections are cut perpendicular to bedding. Flow can be envisioned as traveling along circuits whose numbers are defined by the numbers of pores of each type per cross-sectional area. Packing ensures that each circuit is relatively nontortuous. All of these factors contribute to the effectiveness of the Hagen-Poiseville model. The output of such a model consists of the amount of permeability contributed by each pore type. Electrical conductivity of a brine-saturated sample can be modeled similarly. Modeling resistivity (the inverse of conductivity) shows that the parameter known as Archie's cementation exponent, which relates porosity and resistivity, is the ratio of pore area to throat area, measured on an intersecting plane.

These results and others involving multiphase fluid systems demonstrate that reservoir description, previously time-consuming and, at best, only semiquantitative, can be as precise and accurate as physical measurements. With present equipment, about 50 thin sections (1000 images) can be processed per day. Derivation of pore types allows such data to be easily related to previous petrographic classifications. Often, pore facies relate more closely to discharge of fluids in single- and multiple-phase situations than do lithofacies based on depositional environment. Because size/shape frequency distributions are additive, representative samples taken along a core can be combined and, by using the permeability models, a single permeability at reservoir simulator scales (scales that are orders of magnitude greater than the volume of measured samples) can be calculated. Thus, the problem of determining the correct sort of average to calculate among permeability measurements taken at small scale can be circumvented.

For background information *SEE PETROGRAPHY* in the McGraw-Hill Encyclopedia of Science & Technology.

Robert Ehrlich

Bibliography. R. Ehrlich et al., Petrographic image analysis: I. Analysis of reservoir pore complexes, *J. Sed. Pet.*, 54:1365–1376, 1984; R. Ehrlich et al., Petrography and reservoir physics: I. Objective classification of reservoir porosity, *AAPG Bull.*, 75:1547–1562, 1991; R. Ehrlich et al., Petrography and reservoir physics: III. Physical models for permeability and formation factor, *AAPG Bull.*, 75:1579–1592, 1991; J. B. Ferm et al., Petrographic image analysis and petrophysics: Analysis of crystalline carbonates from the Permian Basin, West Texas, *Carbonates and Evaporites,* vol. 8, no. 2, 1993; C. A. McCreesh et al., Petrography and reservoir physics: II. Relating thin section porosity to capillary pressure, the association between pore type and throat size, *AAPG Bull.*, 75:1563–1578, 1991.

Petroleum engineering

Systems for supervisory control and data acquisition (SCADA) have continued to evolve to address the increased levels of automation required by the oil and gas industry. The SCADA systems are applied at various points along the chain comprising oil and gas exploration, production, transportation, refining, storage, and distribution. However, the most pervasive use of SCADA systems is in the pipeline and production areas. SCADA systems are used to improve operating efficiency, reduce labor costs, enhance safety, and reduce the threat of environmental mishaps associated with operation of oil and gas facilities.

Definition of SCADA. The acronym SCADA has been defined in varied and sometimes confusing ways. Differences in the literal interpretation of the acronym itself, evolutionary changes in SCADA technology, and outright misuse of the term have combined to create a somewhat elastic definition. However, in the broadest sense in terms of the technological/market context, SCADA can be defined as an industrial measurement and control system consisting of a central host or master [usually called a master station or master terminal unit (MTU)], one or more field-data-gathering and control units or remotes [usually called remote stations or remote terminal units (RTUs)], and a collection of standard or custom software used to monitor and control remotely located field-data elements, predominantly through open-loop control and long-distance communications.

Open-loop control (that is, a control system that routinely requires the intervention of a human operator in the control decision process) and long-distance communications (that is, involving distances usually measured in miles rather than feet) are the two elements generally considered to be fundamental to the definition of SCADA. Although once considered to inherently exclude their reciprocals (closed-loop control and short-distance communications, respectively), modern SCADA systems now frequently include such elements. Thus, the contemporary definition of SCADA emphasizes open-loop control and long-distance communications but does not require their exclusive use. This definition reflects not only the key technological elements but the equipment elements as well.

Key elements of SCADA systems. The three fundamental building blocks of SCADA systems are the master terminal unit, the remote terminal unit, and software (including associated services).

Master terminal unit. This central (or host) automation platform is most often based on some type of general-purpose computer, minicomputer, microcomputer, workstation, or a combination thereof, in concert with various peripherals for the gathering, manipulation, and presentation of field data to an operator, who acts as a supervisor and initiates control actions when necessary.

Remote terminal unit. This device for remote data acquisition and control is most often based on a microprocessor or microcomputer designed to gather, manipulate, and transfer field data to or from a remotely located master terminal unit over a long-distance communication medium. Remote terminal units may be either special-purpose or general-purpose input-output devices for the implementation of data acquisition, control, calculation, manipulation, and communication functions.

Software and services. SCADA software comprises a set of programs that provide logical linkage, both within and between elements of master and remote terminal units, in order to facilitate required monitoring, control, and communications functions. SCADA services include system engineering, installation, commissioning, training, documentation, and related maintenance and support functions.

Applications and configurations. More than 1000 pipeline and production companies are engaged in the production, gathering, transportation, storage, and distribution of oil, gas, and refined products. SCADA applications are most frequently associated with pipelines and production facilities.

Oil and gas facilities that make extensive use of SCADA include crude oil pipelines, natural gas pipelines, refined products pipelines, crude oil production fields, and natural gas production fields. Typical pipeline applications include gathering lines, transmission lines, and distribution lines, while production SCADA systems are frequently employed to automate production wells, injection wells, and test manifolds, as well as associated storage and delivery facilities.

A pipeline SCADA system typically consists of a central (host) computer platform that communicates via telephone, radio, or satellite communication links with remote terminal units located at pump/compressor stations, block/check valves, metering points, and other key points. A SCADA system for a small pipeline may include a single personal computer or workstation at the master terminal unit and just a few relatively small remote terminal units, while the system for a major cross-country pipeline may involve multiple distributed master terminal units and 50 or more remote terminal units at main line and booster stations, delivery terminals, tank farms, and other important monitoring/control points. SCADA systems for pipelines are generally expanded by the addition of remote terminal units at new pump/compressor stations as capacity (throughput) of the pipeline increases. *SEE PIPELINE.*

Production SCADA systems have become significantly standardized in recent years, often starting with a master control unit based on a personal computer or workstation and communicating with perhaps a dozen remote terminal units. These initial installations are often referred to as pilot systems. The remote terminal units are usually designed for an optimal mixture of input-output points, packaging, and computational capabilities to suit the application. Remote terminal units used in gas production applications have characteristics that are different from those of oil production units; onshore remote terminal units have characteristics different from those of offshore units. The specific design is determined by the applications. These pilot systems are expanded by the addition of similar or identical remote terminal units at an average rate of

Typical performance criteria for pipeline and production SCADA systems in oil and gas applications		
Performance criterion	Pipeline segment	Production segment
Database points	1500–15,000	500–5000
Number of remote terminal units	10–100	5–500
Communication channels	1–4	1–4
Field update time	Seconds	Minutes or hours
Calculations required at remote terminal unit	Often	Some
Relay logic required at remote terminal unit	Often	Some
Graphical user interface required	Often	Some

Fig. 1. Typical workstation-based SCADA master station. (*Fisher Controls International*)

50–100 units per year, until the field is fully automated, usually within 3–5 years.

Performance requirements for oil and gas SCADA systems vary widely from application to application and even from company to company. The **table** gives an overview of typical performance characteristics for pipeline and production SCADA systems.

Technology. The technology of products used for SCADA systems has evolved from hard-wired master control consoles communicating with so-called dumb (that is, hard-wired) remote terminal units via dedicated circuits in the 1970s to minicomputer-based platforms communicating with microprocessor-based remote terminal units over mixed media links in the 1980s. In the 1990s, distributed open-system architecture allows users to create sophisticated hybrid system configurations that employ various types of remote terminal units and mixed communication media on a fully integrated basis.

Although overall system performance is improved dramatically with each successive generation of equipment, the rudimentary characteristics of SCADA sys-

tems have remained relatively intact. Master terminal units and remote terminal units remain the fundamental system building blocks, but software and support services have become the most critical element of SCADA from the perspective of technology and the life cycle of the system.

Master stations. The typical configuration of a SCADA master station has undergone fundamental changes as personal computers and workstations have become more powerful and less costly. Since the early 1970s, configurations of master terminal units based on dual-redundant processors and some type of automatic failover mechanism have been most popular for all but entry-level systems. A dual-redundant configuration uses two computers connected to the same set of field devices in a way that allows either computer to be on-line at any given time. However, the limitations of the data access and expansion in terms of traditional centralized database architecture have been resolved by the more modern SCADA master stations, usually involving a distributed network of personal computers and/or workstations (**Fig. 1**). Modern entry-level systems are frequently based on distributed architecture, although a single personal computer is usually sufficient to address initial database requirements for systems with fewer than 5000 points.

Remote stations. Many commercial products include remote terminal units with varying capabilities. Until recently, such units for the oil and gas industry were characterized by customer application software, special packaging, and unique performance capabilities. The requirements of being able to function in the frequently harsh environments of the oil and gas industry or to interface with specialized oil field apparatus had limited the applications of generic remote terminal units or related devices. However, the advent of microprocessors has permitted greater standardization of remote terminal units without sacrificing the features and performance characteristics needed to satisfy oil and gas applications. An example is the solar-powered radio-linked remote terminal unit shown in **Fig. 2**. This

Fig. 2. SCADA master station with programmable controller interface (center). (*Square D Company*)

unit is designed for automated production of natural gas.

Although conventional remote terminal units produced by traditional SCADA suppliers still predominate, programmable logic controllers are finding more applications and will continue to displace conventional oil and gas remote terminal units, especially in the pipeline segment. However, remote terminal units will still be used extensively for applications involving complex calculations, such as flow calculations. Although contemporary programmable logic controllers are much more robust than earlier models (which were notably weak in computational applications), many users consider them to be inferior to conventional remote terminal units in that respect.

Prospects. Increased levels of automation throughout the energy industry in the 1990s will be necessitated by regulator requirements. Health, safety, and environmental concerns are being emphasized, while staff reductions continue throughout the industry. It is anticipated that SCADA systems will provide the means for meeting stringent regulations with fewer people and at the lowest possible cost.

Some of the factors that may contribute to increased use of oil and gas SCADA systems in the future include larger, more complex operations brought about by mergers and acquisitions of oil and gas companies; the desire to extend other aspects of automation, such as electronic billing, into the real-time computing environment; increasingly stringent health, safety, and environmental mandates for more comprehensive surveillance and control of pipelines and production facilities; and a general trend toward improved and more economical SCADA performance as the amount of available personnel diminishes.

Changes in the oil and gas market infrastructure will affect the automation of pipeline and production facilities in substantially different ways. Operators of pipelines will move more aggressively to take advantage of open-system alternatives, so that fewer complete systems will be purchased. By contrast, production operations will favor complete systems because of the declines in the long-standing practice of using internal staff to support master station hardware and software.

For background information *SEE COMPUTER; CONTROL SYSTEMS; DATA COMMUNICATIONS; DISTRIBUTED SYSTEMS (COMPUTERS); PIPELINE; PROGRAMMABLE CONTROLLERS* in the McGraw-Hill Encyclopedia of Science & Technology.

Michael A. Marullo

Bibliography. Marullo, *Challenges and Opportunities in the Oil and Gas Energy Sector (1990–1995),* SCADAVIEWS Mar. Res. Rep. SCV-1001, TECH-MARC, August 1992; M. A. Marullo, *Challenges and Opportunities in the SCADA Marketplace (1990–1995),* SCADAVIEWS Mar. Res. Rep. SCV-1000, TECH-MARC, March 1992; M. A. Marullo, *Supplier Profiles Dictionary of SCADA Products, Systems and Services,* TECH-MARC, November 1991; *Proceedings of the Energy Telecommunications and Electrical Association,* March 29–April 1, 1992.

Phase transitions

The roughening transition signals a change of state of a crystal–fluid interface. Typically, at low temperatures such an interface displays facets, as seen on common minerals such as quartz or pyrite. As the temperature T is increased, portions of the interface become rounded and facets shrink. Eventually, the size of a facet decreases to zero at its roughening temperature T_R, and the crystal becomes locally rounded, resembling a portion of a water droplet. Inequivalent facets have different roughening temperatures, and full rounding takes place via a sequence of roughening transitions. **Figure 1** shows both rough (rounded) and faceted (flat) portions of lead crystals.

The above scenario assumes full thermodynamic equilibrium at all times. For typical macroscopic crystals, the achievement of such equilibrium, as regards surface properties, requires geologic times (for example, of the order of 10^9 years for a 1-mm^3 copper crystal at a temperature of 500 K or 440°F). Crystal growth often strongly emphasizes facets, and the growth shape is frozen in when growth ceases, as illustrated by typical samples observed in museums of mineralogy. There has not yet been time for these samples to assume their equilibrium shapes. Equilibration is enhanced by facilitating mass and heat transfer. This can be done by using very small crystals (micrometers in size) as in Fig. 1, or by using helium-4, where the flow of matter and heat are very rapid through the superfluid in contact with the crystal and extremely high degrees of purity can be achieved. *SEE HELIUM.*

Experiments. Small metallic crystals such as the lead crystals in Fig. 1 are prepared by first evaporating a thin film of the solid metal onto a substrate such as graphite. When melted, the film breaks up, forming small droplets on the substrate. The droplets are quenched (frozen) and then annealed (for tens of hours) to allow the equilibrium shape to be achieved through surface diffusion. Observations on static shapes are carried out with scanning electron microscopes.

Helium-4 crystals are grown by pressurizing the superfluid contained in a low-temperature cell. Reducing the pressure rapidly melts the crystals. Remarkably,

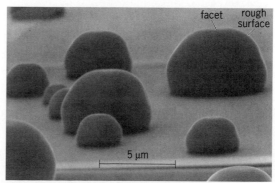

Fig. 1. Electron micrograph of lead crystals on a graphite substrate. (*Photograph by J. C. Heyraud*)

melting and freezing occur so readily that melting-freezing waves, like capillary waves on water, are easily observed on rough surfaces of the crystal at low temperatures. Measurements of both static shapes and growth are made with laser interferometry.

Microscopic picture. Facets are due to the localization of a crystal–fluid interface by the underlying lattice structure of the crystal. At a roughening transition the interface becomes unbound from the crystal structure. It becomes deformable and wanders readily from its average position, resembling a liquid–gas interface. The microscopic changes are illustrated in **Fig. 2**, which shows snapshots of a computer simulation of roughening.

Shape, energy, and universality. In the following discussion, $z(\mathbf{r})$ is the height of the crystal surface above the point $\mathbf{r} = (x, y)$ in the x-y plane. The shape function $z(\mathbf{r})$ is determined by minimizing the surface energy of the crystal at fixed volume. It is a remarkable result of this minimization that $z(\mathbf{r})$ is in fact a surface free-energy function for the crystal. Singularities in $z(\mathbf{r})$ that occur at phase transitions such as roughening are then mirrored in the shape.

If a phase transition is continuous (second or higher order), it is characterized by a correlation length $\xi(T)$, which diverges at the transition. Near the transition, where $\xi(T)$ is large compared to interatomic distances, the essential physics is determined solely by phenomena at the scale $\xi(T)$, and details at shorter length scales, such as lattice structure or interatomic potentials, are irrelevant. There are thus aspects of roughening that are universal in that they are independent of the particular material, be it gold or helium, that is being studied. Among these aspects is the curvature of the crystal at the point on the crystal surface where the facet vanishes at the roughening temperature T_R. Specifically, the geometric mean of the curvature, the gaussian curvature κ, at this point is predicted to be given by Eq. (1), where Δp

$$\kappa = \frac{a^2 \, \Delta p}{\pi k_B T_R} \tag{1}$$

is the pressure difference across the interface, a is the distance between equivalent crystal planes parallel to the facet, and k_B is Boltzmann's constant. This result has been verified by measurements on helium-4.

The edge of a facet is an obvious singularity in the shape, and hence in the surface free energy, signaling a phase transition. If the slope is discontinuous at the edge, the transition is first-order, while if the slope is continuous, there is a diverging correlation length and a universal exponent s characterizing the dependence of the surface height $z(x)$ on the coordinate x parallel to the facet. Explicitly, Eq. (2) is satisfied, where b is a

$$z(x) = -b(x - x_0)^s \qquad x \geq x_0 \tag{2}$$

constant and x_0 is the position of the facet edge. An illustration is **Fig. 3**, which depicts the thermal evolution of a part of a crystal profile. Theory predicts that $s = {}^3/_2$. The experimental evidence in measurements on both helium-4 and small metal crystals is consistent

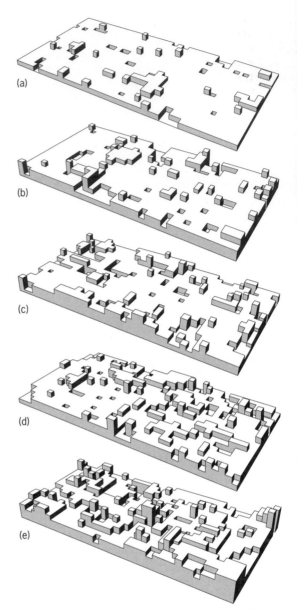

Fig. 2. Snapshots of computer simulations of roughening at temperatures T near the roughening temperature T_R. (*a*) $T = 0.91 \ T_R$. (*b*) $T = 0.95 \ T_R$. (*c*) $T = T_R$. (*d*) $T = 1.05 \ T_R$. (*e*) $T = 1.11 \ T_R$. (*After J. B. Blakely, ed., Surface Physics of Materials, vol. 1, Academic Press, 1975*)

with values of s between 1.5 and 2.

The size of a facet is proportional to the step free energy β, the free energy per unit length required to create a step on an interface. If β is zero, terraces and valleys, whose boundaries are steps, are readily created by thermal fluctuations, and the surface is rough.

The roughening transition is an example of a Kosterlitz-Thouless transition. The superfluid transition in thin liquid helium-4 films and the superconducting transition in metal films are additional examples. The facet-edge transition of Eq. (2) is known as a Pokrovski-Talapov transition, and it is also observed in commensurate-incommensurate transitions in solid films.

An interesting contrast to the case of regular crystals is that of quasicrystals. Though quasicrystals lack the regular periodic structure of crystals, they are nevertheless not believed to roughen at any temperature below their melting temperatures. This surprising result follows from the facts that quasiperiodicity emerges from periodicity in the limiting case that a becomes infinite, and that T_R is proportional to a^2.

Theory. The essential features of the roughening transition emerge from the study of a simple continuum model (as well as other models) for a solid–fluid interface. Universality guarantees that the model will apply to real crystals. Again, the height of the interface is designated $z(\mathbf{r})$, where $\mathbf{r} = (x, y)$ is the coordinate vector in the plane of the interface. The model is defined by an energy functional H given by Eq. (3),

$$H = \int d\mathbf{r} \left[\frac{\alpha_0}{2} |\nabla z(\mathbf{r})|^2 - v_0 \cos\left(\frac{2\pi z(\mathbf{r})}{a}\right) \right] \quad (3)$$

where α_0 plays the role of a microscopic surface stillness and the cosine potential of strength v_0 simulates the preference for integer values of z in terms of the lattice constant a. Renormalization group techniques of statistical mechanics are used to derive results within this model. These results emerge as statistical averages over states of the system, where each state is weighted by a Boltzmann factor, $\exp(-H/k_B T)$.

Of particular interest in the description of roughening is the correlation function, defined by Eq. (4),

$$g(r) = \langle [z(\mathbf{r}) - z(0)]\rangle^2 \quad (4)$$

where the angular brackets indicate the statistical average. The function $g(r)$ gives the average wandering of the interface over a distance r. At low temperature T, where the wandering is negligible and $z(\mathbf{r})$

is close to $z(0)$ for all r, $g(r) \approx 0$. Above T_R, the wandering is important, the periodic potential becomes irrelevant, averaging to zero, and, for large values of r, $g(r)$ is given by Eq. (5). Here $\alpha(T)$ is the macro-

$$g(r) = \frac{k_B T}{\pi \alpha(T)} \ln \frac{r}{a} \quad (5)$$

scopic surface stiffness relating the pressure difference to the mean curvature κ_m of the interface by Laplace's equation (6). This result for $g(r)$, which is also true for

$$\Delta p = \kappa_m \alpha \quad (6)$$

a liquid–gas interface where α is just the surface tension, shows that the interface wanders arbitrarily far from its location at $r = 0$. For $T < T_R$, $g(r)$ is given by Eq. (7), showing that the surface behaves as

$$g(r) = \begin{cases} \dfrac{k_B T}{\pi \alpha_0} \ln \dfrac{r}{a} & \text{for } a \ll r < \xi \\ \dfrac{k_B T}{\pi \alpha_0} \ln \dfrac{\xi}{a} & \text{for } a \ll \xi < r \end{cases} \quad (7)$$

a rough surface over distances r small compared to a correlation length $\xi(T)$, and that it is localized by the periodic lattice potential for larger distances.

At temperatures near T_R, $\xi(T)$ has the universal temperature dependence given by Eq. (8), in which

$$\xi = \xi_0 e^{c/\sqrt{t}} \qquad t \equiv \frac{T_R - T}{T_R} \quad (8)$$

c and ξ_0 are nonuniversal constants depending on details such as the potential strength v_0. This temperature dependence is related to that of the step free energy β by relation (9). The unusual temperature

$$\beta \sim \frac{k_B T}{\xi} \sim e^{-c/\sqrt{t}} \quad (9)$$

dependence of β (β has an essential singularity at $t = 0$) given by Eq. (9) has been quantitatively verified in experiments on the growth of helium-4 crystals. Facet sizes are proportional to β, and though observation of such sizes is difficult near T_R, experiments on both helium-4 and the α-phase of silver sulfide (α Ag$_2$S), a superionic conductor, are consistent with Eq. (9).

Dynamical roughening. Crystalline interfaces are roughened not only by thermal agitation but also by rapid growth. Growth of a flat crystal surface at velocity v is related to the driving force, an overpressure ΔP, by Eq. (10), where the growth rate coeffi-

$$v = k(T, \Delta P)\Delta P \quad (10)$$

ficient $k(T, \Delta P)$ may depend strongly on ΔP. For a faceted interface, at temperatures near but below T_R, growth is a consequence of thermal nucleation of terraces, which then grow laterally by accretion of atoms to complete layers. For this process, $k(T, \Delta P)$ vanishes exponentially with ΔP. At temperatures above T_R, growth is simply by accretion of atoms, as terraces are readily formed by thermal fluctuations, and $k(T, \Delta P)$ is independent of ΔP. In the limit $\Delta P \to 0$, then, $k(T, \Delta P)$ is discontinuous, jumping from zero below T_R to a nonzero value above T_R. For nonzero values of ΔP, this discontinuity is rounded by dy-

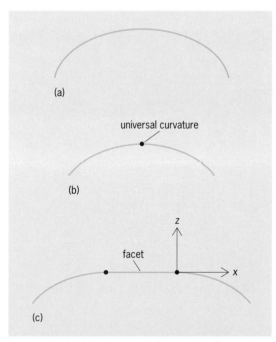

Fig. 3. Thermal evolution of a part of a crystal profile. (a) $T > T_R$. (b) $T = T_R$. (c) $T < T_R$. A facet appears. Beyond the facet edge the surface height z is proportional to $-(x^s)$, where x is the distance beyond the facet edge and $s = {}^3/_2$ is a universal exponent. (*After C. Jayaprakash and W. F. Saam, Thermal evolution of crystal shapes: The fcc crystal, Phys. Rev. B, 30:3916–3928, 1984*)

namical roughening. The rounding occurs when ΔP is sufficiently large that the energy required to nucleate terraces is smaller than $k_B T$, in which case terraces are present in large numbers, as they are for $T > T_R$, and the surface is rough.

A detailed theory of surfaces roughened either thermally or dynamically has been constructed by using a renormalization group treatment of a dynamical equation based on the model of Eq. (3). Good agreement is achieved with experiments on the growth of helium-4 crystals.

For background information SEE CRITICAL PHENOMENA; CRYSTAL GROWTH; LIQUID HELIUM; PHASE TRANSITIONS; QUASICRYSTAL; RENORMALIZATION; STATISTICAL MECHANICS in the McGraw-Hill Encyclopedia of Science & Technology.

William F. Saam

Bibliography. S. Balibar et al., The growth dynamics of helium crystals, *Physica B*, 169:209–216, 1991; C. Godrèche (ed.), *Solids Far from Equilibrium*, 1992; W. Schommers and P. van Blanckenhagen (eds.), *Structure and Dynamics of Surfaces II*, 1987; R. Vanselow and R. Howe (eds.), *Chemistry and Physics of Solid Surfaces VII*, 1988.

Phylogenetic analysis

Biologists who estimate the genealogical relationships of species by using phylogenetic analysis to summarize data are called phylogenetic systematists or cladists. Other major schools of systematics include evolutionary taxonomy, phenetics or numerical taxonomy, and transformed cladistics. Since about 1970, there has been considerable debate concerning the validity of the assumptions made by the various schools.

Comparisons between ingroups and outgroups. Phylogenetic analysis begins when particular taxa (species or groups of species) are chosen for study. These taxa should represent all the descen-

dants of a single ancestral species, as well as can be determined. Then, at least two other taxa (referred to as the outgroups) outside the study group (or ingroup) but related to it as closely as possible are chosen so that comparisons between the groups can be made. Once the ingroup and outgroup taxa are selected, the type of characters that will be examined to elucidate relationships among the taxa are chosen. The characters may be anatomical, behavioral, biochemical, or a combination of these. Each character used in the analysis must be variable among taxa, but preferably not variable within a single taxon. Thus, each species can be assigned a single digit character state (usually 0 or 1) that represents the character as it is observed in members of that taxon. There is no intrinsic limit or end point to data collection, because there is an almost infinite variety of characteristics that can be observed for any set of taxa. Once a character-by-taxon matrix has been generated (**Tables 1** and **2**), the search for hypotheses of relationships among the study taxa can begin.

This search is almost always performed by computer algorithm. With any of the several phylogenetic analysis programs available, the search for relationships consists of two parts. First, the direction (polarity) of evolutionary transformation for each character must be determined, most commonly by comparing the character states of the ingroup and outgroup. If a character state is found both in the outgroup and in some portion of the ingroup, the one in the outgroup represents the evolutionarily primitive state, with the ingroup alternate being derived from it. In Table 1, the primitive state is assigned 0 and the derived state, 1. Only shared derived character states delineate groups within the ingroup. Next, nested sets of groups that are supported by these shared derived character states are searched for by the computer. Prior to the use of computers, this part of the analysis could quickly become prohibitively difficult to perform manually. Computers have greatly enhanced systematists' ability to handle

Table 1. Matrix of anatomical data collected for the phylogenetic analysis of six sunfish species and two outgroup species (white crappie and largemouth bass)

Taxon	Character*									
	1	2	3	4	5	6	7	8	9	10
White crappie	0	0	0	0	0	0	0	0	0	0
Largemouth bass	1	0	0	0	0	0	0	0	0	0
Green sunfish	1	1	0	0	0	0	0	0	0	0
Bluegill	1	1	1	1	1	2	0	0	0	0
Redear sunfish	1	1	1	1	1	2	1	1	0	0
Pumpkinseed	1	1	1	1	1	2	1	1	1	0
Longear sunfish	1	1	1	1	0	1	0	0	1	1
Orangespotted sunfish	1	1	1	1	0	1	0	0	1	1

*Character 1: 0 = more than 3 anal fin spines; 1 = 3 anal fin spines.
Character 2: 0 = no spot on edge of gill flap (earspot); 1 = earspot present.
Character 3: 0 = body elongate; 1 = body shortened, almost disk-shaped.
Character 4: 0 = longest dorsal fin spine shorter than mouth; 1 = longest dorsal fin spine longer than mouth.
Character 5: 0 = pectoral fins short, round; 1 = pectoral fins long, pointed.
Character 6: 0 = mouth large; 1 = mouth moderately sized; 2 = mouth small.
Character 7: 0 = earspot without red margin; 1 = earspot with red margin.
Character 8: 0 = narrow pharyngeal arches; 1 = broad pharyngeal arches.
Character 9: 0 = male breeding color pattern drab; 1 = male breeding color pattern with sharply contrasting red spots on sides.
Character 10: 0 = short earflaps; 1 = long, flexible earflaps.

Table 2. Matrix of hypothetical DNA sequence data for a phylogenetic analysis of six sunfish species and two outgroup species (white crappie and largemouth bass)

Taxon	Nucleotide position*																			
	34	65	73	89	102	126	142	143	144	178	181	184	192	213	229	243	264	269	275	291
White crappie	a	g	g	t	t	g	a	t	t	c	c	t	g	g	c	a	g	t	c	a
Largemouth bass	t	g	g	t	t	g	a	t	t	c	a	t	c	g	c	a	a	t	t	a
Green sunfish	t	g	g	t	t	g	a	t	t	g	a	g	c	g	c	a	a	t	t	g
Bluegill	t	a	c	t	g	a	g	t	t	g	a	g	a	g	t	a	a	c	c	g
Redear sunfish	t	a	c	t	g	t	g	t	a	g	a	g	a	g	a	a	a	c	t	g
Pumpkinseed	t	a	c	t	g	a	g	c	a	g	a	g	a	g	t	a	a	c	t	g
Longear sunfish	t	c	g	g	t	t	a	g	t	g	a	g	c	c	a	c	t	c	t	g
Orangespotted sunfish	t	c	g	g	t	t	a	g	t	g	a	g	c	c	a	c	t	c	t	g

*Character codes are as follows: a = adenosine, g = guanine, t = thymine, c = cytosine. Transversions are changes from a purine (a, g) to a pyrimidine (c, t) or vice versa. Transitions are changes from one purine to the other or from one pyrimidine to the other.

complex problems because millions of comparisons can be made quickly.

Hypothesis of relationships. The result of the phylogenetic analysis is a hypothesis of relationships, usually represented by a tree diagram (cladogram) that summarizes the data used (**Fig. 1**). This summary may suggest that initial assumptions concerning the direction of evolutionary change were incorrect. Additionally, it will almost certainly show that some character state changes support different hypotheses of relationships than do others. Such conflict is termed homoplasy and is hypothesized to be the result of either parallel or convergent evolution. The set of relationships that is most highly corroborated, however, is accepted because it is the most parsimonious. That is, it requires the fewest ad hoc explanations of character state changes.

Sometimes phylogenetic analysis results in two or more hypotheses of relationships that are equally well supported by the data. This type of ambiguity may stem from too little data, or from speciation of taxa in the ingroup being so rapid that change in the observed characters did not keep pace with divergence. This latter possibility leads most systematists to suggest that other types of characters that might be more readily changed through time, and therefore might reflect rapid evolution, should be added to the analysis. An analysis that yields conflicting results may still represent a marked increase in knowledge of the relationships of the ingroup.

Two analyses of the same taxa might produce different hypotheses of relationships for those taxa. For instance, analysis of the molecular data in Table 2 results in the relationships shown in **Fig. 2** rather than those in Fig. 1. Because phylogenetic analysis is a data-summary technique, these differences may be due to poor observation, to poorly described or delineated characters, or simply to different character types responding to different evolutionary pressures. The solution is to critically reexamine all the data, combine the information of the multiple studies, and analyze them to arrive at a synthesis. If the data from Tables 1 and 2 are analyzed together, only one hypothesis results, and it portrays the relationships in Fig. 1. Systematists cannot fully know the genealogical history of the taxa they investigate,

but they try to derive increasingly accurate approximations of that history. For this reason, taxonomy and classifications must change as new data accumulate.

Philosophical issues. Virtually all arguments among systematists have concerned assumptions that are made during phylogenetic analyses. Discussion centers on issues concerning the nature of species, homology, polarity, character equality, and biological classification. Arguments often include the discussion of various computer algorithms, because each algorithm includes or optimizes different assumptions. Phylogenetic analysis has a rich philosophical framework that has been used to justify a consistent, specific methodology.

Nature of species. The metaphysical nature of species has troubled systematists. Phylogeneticists believe that species are real regardless of the ability or inability to detect them. At the other extreme, pheneticists argue that species are unknowable and hence a categorization imposed on reality by scientists. These schools of thought differ as to whether species are diagnosed (detected by investigations) or defined (invented by thoughts). The scientific orientation hence can greatly affect the choice of objects or taxa that are included in studies of relationships and the predictions that are made concerning the evolutionary fate of these objects or taxa.

Homology. Systematists must make the initial as-

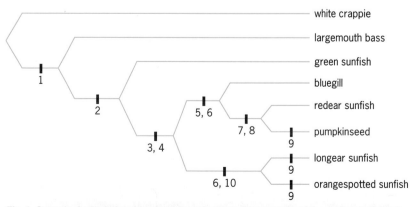

Fig. 1. Genealogical relationships based on phylogenetic analysis of the anatomical data in Table 1. Numbered bars indicate the placement of the change from the primitive condition to the derived condition for each character.

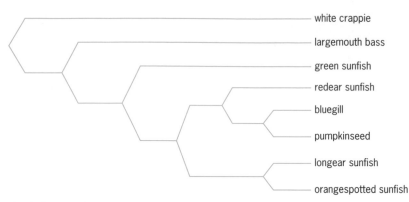

Fig. 2. Genealogical relationships based on phylogenetic analysis of the molecular data in Table 2.

sumption that character states observed in extant taxa are really the same (that is, they are homologs), and that the same character occurred in an extinct ancestor, although it has been transformed through time in lineages descended from that ancestor. This assumption is usually examined through detailed comparisons of the character states. For example, during ontogenetic development of the more recent character state in a taxon, that character may have passed through a stage that resembled its primitive condition—hence the adage, "Ontogeny recapitulates phylogeny." Similarity of position, similarity of composition, and presence of one or more intermediates in a transformation series can be used as auxiliary criteria to reexamine postulates of homology. Systematists also assume that all observed instances of the character state are the result of shared inheritance from a single ancestor. The phylogenetic analysis itself tests both the assumption of homology and the assumption of shared inheritance. When all the data are considered together, the assumptions might be proven incorrect, with the characters being convergent or evolved in parallel (for example, character 9 in Fig. 1). Biologists often argue about specific instances of postulated homologs, leading to reexamination of specimens and collection of more data to add to the analysis. Phylogenetic analysis can therefore be seen as the final arbiter of postulates of homology.

Polarity. The polarity of a character is the direction of evolutionary change from primitive to derived. Evaluation of this transformation may be achieved by using comparisons between outgroups and ingroups, early and later developmental stages, early and later fossils, and hypotheses based on morphology. Most phylogeneticists think that comparison of the characters found in outgroups with those of the ingroup is the superior method because it requires the fewest assumptions about developmental sequences. Transformed cladists hold the opinion that comparison of the character states found in earlier developmental stages with those found in later developmental stages is superior, because they do not have to rely on unobserved assumptions of evolutionary history. Most systematists believe that this issue will be resolved when one or the other school's results are shown to have withstood the test of time.

Character equality. Numerical taxonomists were the first to strictly adhere to the precept that all characters should be presumed equal in their ability to reflect relationships. This idea has also been accepted by phylogeneticists. Differential character weighting remains distasteful to many systematists because they fear the return of idiosyncratic, authoritarian pronouncements about relationships, rather than objective, testable and repeatable analyses. However, with the increase in phylogenetic analyses, biologists and biochemists have begun to argue that all characters are not equal. For example, gains of new characters seem potentially more informative than losses, which could more easily be convergent or in parallel. In addition, the gain of a structurally and functionally complex character seems to be more informative than the gain of a simple character. Similarly, biochemists argue that there are good reasons for weighting transversions in a deoxyribonucleic acid (DNA) sequence as more informative than transitions because the latter occur more frequently and are therefore more likely to show repeated changes at the same site. Therefore, transversions should reflect longer periods of time in evolutionary history. Another reason for differential weighting of biochemical characters is that the third position of a codon has the least influence on codon meaning and is therefore thought to be less reliable through time. Many phylogeneticists now consider differentially weighting their characters.

Parallelism and convergence. Some biologists have suggested that phylogenetic analysis is incapable of dealing with convergent evolution of characters or their parallel acquisition. They have also asserted that in the presence of widespread parallelism or convergence, phylogenetic analysis will produce a false estimate of genealogical relationships. These arguments have had less weight recently for three reasons. First, phylogenetic analysis itself can be used to detect instances of parallel or convergent evolution. Second, tests of phylogenetic analysis using randomized data have shown that true relationships need be reflected by as little as 10% of the data for the technique to produce the correct estimates of those relationships. Furthermore, data that reflect parallel or convergent evolution must covary across all taxa before the analysis is affected. Third, empirically, phylogeneticists have not found instances of widespread parallelism or convergence.

Biological classification. Phylogenetic systematists believe that biological classifications should correspond exactly to the genealogical relationships estimated by phylogenetic analysis, so as to assign the greatest amount of information to the classification. At the other extreme, evolutionary taxonomists believe that biological classifications should reflect an assessment of evolutionary divergence in terms of degree of advancement. Phylogeneticists object to the subjectivity and loss of information concerning genealogy in the classification of evolutionary taxonomists. However, evolutionary taxonomists find the classifications of phylogeneticists lacking in information on degree of advancement. For example, a phylogeneticist would include birds with snakes, lizards, crocodiles, and turtles within the Class Reptilia (because they represent all the descendants of a single ancestor) whereas an evolutionary taxonomist would assign birds a higher rank. In addition, the phylogeneticist sees nested sets within Class Reptilia that the evolutionary taxonomist does not. A classification based on the phylogenetic analysis allows retrieval of the information about genealogical relationships among the organisms from the classification; in an evolutionary taxonomist's classification, this information is irretrievable from the classification itself.

Prospects. There is real predictive power to the estimates of genealogical relationships produced by phylogenetic analysis. For example, although a certain strain of wheat might not be resistant to a particular disease, a related species might be. Phylogenetic analysis could be used to predict which resistant species is the closest relative to the nonresistant species and therefore the most likely donor of the resistant gene for genetic engineering. Knowledge of phylogenetic relationships among mosquitoes has shown that a pesticide program was ineffectual because the insects in the area were distantly related to the species that was known to be susceptible to the pesticide. The results of phylogenetic analysis potentially can be used to predict the evolutionary stability of a particular species by examining the diversity and rate of change through time of the lineage to which it belongs. This information might be used to assess the susceptibility of a species to environmental or community changes. Conservation biologists are only just beginning to utilize phylogenetic systematics in this way, but the results promise a means of choosing which species they should actively work to save when budgets and resources are limited.

For background information *SEE ANIMAL EVOLUTION; ANIMAL SYSTEMATICS; PHYLOGENY; TAXONOMY* in the McGraw-Hill Encyclopedia of Science & Technology.

Kate Shaw

Bibliography. P. Ax, *The Phylogenetic System: The Systematization of Organisms on the Basis of Their Phylogenesis*, 1987; W. Hennig, *Phylogenetic Systematics*, 1966; E. O. Wiley et al., *The Compleat Cladist: A Primer of Phylogenetic Procedures*, Univ. Kan. Mus. Nat. Hist. Spec. Pub. 19, October 1991.

Pipeline

Due to recent advances in pipeline technology, solids are increasingly being transported by pipeline over long distances. Pipelines that transport solids are referred to as freight pipelines. The use of underground pipelines to transport solids not only reduces freight costs but also offers a wide range of environmental and social benefits because of the reduced use of trucks and trains that cause air pollution, noise, traffic congestion, accidents, and damage to highway and rail infrastructures. The general types of freight pipelines are pneumatic pipelines, slurry pipelines, and capsule pipelines. The transport of solids by such pipelines is called pneumotransport, hydrotransport, and capsule transport, respectively.

Pneumatic pipelines. Pneumatic pipelines, which use air to transport solids through pipes, are a common means of bulk-materials transport over short distances, usually not more than a few hundred yards or meters. The solids normally are fine particles so as to facilitate suspension by air during transport. Materials transported include grain, cement, coal, and hundreds of other minerals and industrial products. Sometimes, large particles such as ice cubes, fish, and processed chicken are transported pneumatically.

Pneumatic pipelines are either positive-pressure or negative-pressure. The negative-pressure system works like a vacuum cleaner. Solids are sucked into the pipeline intake by a negative pressure (that is, pressure lower than atmospheric) maintained in the pipeline. The negative pressure is generated by locating the blower (fan) near the outlet of the pipe. In a positive-pressure system, the blower is located near the pipeline intake so that a positive pressure (that is, pressure higher than atmospheric) is generated in the pipeline. Solids are forced into the pipeline by either gravity or a forced feeder, which can be a hopper with a rotating paddle wheel (rotary vanes) on its bottom.

The negative-pressure pneumatic pipeline has simple intake systems; it can have many suction inlets powered by a single blower located at a common outlet. A common application of the negative-pressure system is the refuse-collection system used in modern apartment complexes. Refuse enters the system from inlets located in each apartment and sometimes at street level. After being sucked into the pipe, the refuse is concentrated at a common point for either processing or further transport by truck or other means to a processing or disposal site. The positive-pressure pneumatic pipeline needs a more complex intake. A single blower at the intake can blow solids into various pipes and outlets.

In a negative-pressure pneumatic pipeline system, the maximum pressure change generated by the blower along the pipe is limited to less than 1 atm (10^5 pascals). Consequently, such systems can transport solids over only short distances. Solids can be transported to much greater distances with positive-pressure systems, which can generate many times the atmospheric pressure, especially if blowers are staged.

In both positive- and negative-pressure systems, a rather high speed, of the order of 66 ft/s (20 m/s), must be maintained in the pipe to cause solid suspension during transport. Because of the high velocity, pneumatic pipelines are energy-intensive.

The main advantage of pneumatic pipelines over hydraulic (slurry) pipelines is the use of air instead of water. Not only is air free and available everywhere on Earth, but also it does not wet the cargo. These advantages explain the widespread use of pneumatic pipelines for bulk-materials transport and handling. The principal disadvantage of pneumatic pipelines is their high energy consumption, which limits their use to short distances. The power required is proportional to pipeline length. Another disadvantage is the abrasion of both the materials and the pipe during transport. Finally, because of the separation and accumulation

Fig. 1. Pumping station of the Black Mesa Coal Slurry Pipeline, the longest slurry pipeline in the world. Such slurry pumps are designed to withstand the abrasive action of the slurry. (*Williams Technologies, Inc.*)

of electric charges (static electricity) in pneumatic pipelines, such systems may catch fire or even explode when used to transport combustible materials such as grain or coal. Extreme care must be exercised in design, operation, and maintenance to ensure safety.

Slurry pipelines. Slurry pipelines transport solid-liquid mixtures. The most common liquid in a slurry pipeline is water. Occasionally, other liquids such as petroleum are used to transport solids such as coal in slurry form.

Slurry pipelines using water have been used for more than a century to transport mine wastes and mine tailings for disposal. Such pipelines are usually from a few hundred yards or meters to a few miles or kilometers in length. In recent years, many long slurry pipelines have been constructed in various parts of the world to transport minerals such as coal, copper concentrate, and iron concentrate. They sometimes exceed 100 mi (160 km) in length. The longest existing slurry pipeline, built in 1970, is the Black Mesa Coal Slurry Pipeline from Black Mesa, Arizona, to southern Nevada, over a distance of 273 mi (440 km) [**Fig. 1**]. It uses an 18-in.-diameter (457-mm) pipe and transports 4.8×10^6 short tons (4.0×10^6 metric tons) of coal per year.

Slurry pipelines can be subdivided into coarse-slurry and fine-slurry types. A coarse-slurry pipeline transports large-size solid particles, upt to a few inches (over 100 mm) in size. A fine-slurry pipeline transports small-size particles, of the order of 0.01 in. (0.254 mm) or finer. Because suspension of coarse particles requires a higher velocity, coarse-slurry pipelines are more energy-intensive and more abrasive to pipelines and pumps. Consequently, they are used for only short distances, usually not more than a few miles or kilometers. Coarse-slurry pipelines are widely used in dredging, transport of construction materials (sand and gravel), and lifting of hydraulically cut minerals from underground mines (hydrohoist). An interesting potential future application is for deep-sea mining. A coarse-slurry pipeline system has been studied to

transport manganese nodules collected on the bottom of the ocean to a surface ship.

Slurry pipelines require the use of special pumps that can withstand abrasion (Fig. 1). For short slurry pipelines without booster stations or with only a few booster stations short distances apart, the pump pressure or head required is low, and special centrifugal pumps can be used. For long slurry pipelines with booster stations spaced at large distances, the pressure drop along the pipeline between two neighboring stations is high so that a high pump pressure is needed, and positive-displacement slurry pumps are more appropriate.

Capsule pipelines. A capsule pipeline transports solids inside a large container or solids made into a capsule shape or form. It is the most versatile type of freight pipeline, capable of transporting practically any solid or material that can be fitted or shaped. The fluid medium for transporting capsules can be either gas or liquid. A gas (usually air) is used in a pneumatic capsule pipeline (PCP), and a liquid (usually water) is used in a hydraulic capsule pipeline (HCP).

Pneumatic capsule pipelines. Because air is very light as compared to many solids transported, the air in a pneumatic capsule pipeline may provide insufficient lift force to suspend the capsules moving through the pipe. Consequently, such pipelines must use capsules on wheels rolling through the pipe, unless the cargo is very light. **Figure 2** shows the capsule loading station of a large pneumatic capsule pipeline system in Japan used to transport limestone from a quarry to a cement plant. The pipeline has been in operation since 1983 at very high (98%) availability. The system requires only one operator per shift. Several similar large pneumatic capsule pipeline systems were built in the former Soviet Union for transporting rocks and other materials.

Current use of pneumatic capsule pipelines in the United States is limited to small and short systems. The most common uses are by drive-in banks to transfer cash and other papers between the teller and customer, and by large hospitals to transport blood samples and

Fig. 2. Capsule loading station of the Somitomo Pneumatic Capsule Pipeline in Japan. Crushed limestone is loaded on open-top capsules. The capsules are then propelled through the pipe to a cement plant 2 mi (3.2 km) away.

other materials between rooms or between buildings. A large wheeled pneumatic capsule pipeline system is being marketed by Tubexpress, Inc. in New Jersey to transport coal and other materials over long distances.

Hydraulic capsule pipelines. The water in a hydraulic capsule pipeline produces sufficient lift force to suspend capsules with a density close to that of water. Thus, the capsules do not need wheels; they are plain cargo cylinders or cargo-carrying cylinders.

Hydraulic capsule transport is a technology that has not yet been fully developed and hence is not used commercially. Industrial nations, especially the United States and Japan, have intensive research programs to develop hydraulic capsule pipelines for commercial use. The research in the United States is carried out mostly at the Capsule Pipeline Research Center at the University of Missouri, Columbia. The Center's current focus is to develop the coal log pipeline (CLP), a special type of hydraulic capsule pipeline for transporting coal. In this pipeline, coal is compressed into capsule shapes (logs), which are transported by water in the pipeline over long distances. This method uses less than one-third of the water and transports twice as much coal in comparison to a coal-slurry pipeline of the same diameter.

For background information SEE PIPELINE in the McGraw-Hill Encyclopedia of Science & Technology.

Henry Liu

Bibliography. H. Liu, Hydraulic capsule pipeline, *J. Pipelines*, 1(1):11–23, 1981; H. Liu and G. F. Round, *Freight Pipelines*, 1990; D. Mills, Keeping pneumatic delivery up to speed, *Chem. Eng.*, 97(6):94–105, June 1990; T. L. Thompson and T. C. Aude, Slurry pipelines: Design, research and experience, *J. Pipelines*, 1(1):25–44, 1981.

Plant movements

Plant cells determine which way is up and which way is down by perceiving the direction of gravity. The vector of gravity runs along the radius of the Earth and points toward the Earth's center. For 100 years scientists believed that plant cells detect the direction of gravity by sensing the falling of heavy starch grains onto the bottom of a cell. Recently, the starch grain, or statolith, model has lost some of its appeal since it has been determined that the kinetic energy imparted by the falling starch grain is one-millionth the amount necessary to trigger even the most minute cellular reaction. Moreover, a number of plants that do not have falling starch grains, either naturally or through mutations, perceive gravity. The gravitational pressure model provides an alternative explanation. In this model, starch grains act as ballast to increase the weight of the whole protoplast. Thus, their presence, but not their falling, is important for gravity sensing.

Perception of gravity in Chara. To understand how cells perceive gravity, experiments have been carried out on *Chara*, an algal pond weed with cells longer than 1 in. (2.5 cm) and no falling starch grains. These large cells utilize cytoplasmic streaming, a method of intracellular mixing that allows the nutrients made in one part of the cell to reach all other parts of the cell in a reasonable time. In cytoplasmic streaming, the contents of the cell move down one side and up the other side like a conveyor belt. The motor that provides the force for streaming depends on the interaction between actin and myosin, the same proteins that provide the force for muscle contraction in the human body.

Even though the internodal cells of *Chara* have no falling starch grains, they respond to gravity. Gravity causes the cytoplasm to stream down approximately 10% faster than it streams up. In a horizontal cell, there is no polarity of cytoplasmic streaming, with the cytoplasm streaming at the same rate in both directions. In a vertical cell, the cytoplasm does not accumulate in the bottom of the cell because the fast downward stream is thinner than the slow upward stream. It appeared that this response to gravity was merely physical and did not require the perception of gravity, until it was accidentally discovered that neutral red, a stain used to color the vacuole, caused the cytoplasm to stream up faster than it streamed down.

Polarity in cytoplasmic streaming. Neutral red causes a reversal in the cell's perception of gravity because it causes a change in the movement of ions across the cell membrane and consequently alters the normal electrical potential of the cell membrane. Thus, it seems likely that the gravireceptor in normal cells also causes a change in the movement of ions across the cell membrane and that this change in turn induces a polarity in the streaming cytoplasm. Calcium ion (Ca^{2+}) is the most important ion involved in inducing a polarity in the cytoplasmic streaming, since removing Ca^{2+} from the external medium causes the cytoplasm to stream up faster than it streams down, and a normal response to gravity occurs only when the concentration of external Ca^{2+} is greater than 1 micromolar (10^{-3} mol·m^{-3}). Interestingly, blocking the movement of Ca^{2+} ions into the cells with drugs ordinarily used to prevent the movement of Ca^{2+} into the cardiac muscles of heart patients prevents the gravity-induced polarity of cytoplasmic streaming. Thus, the gravireceptor either is a Ca^{2+} channel or is closely associated with Ca^{2+} channels.

In order to determine where the gravireceptor resides in the cells, a technique called cellular ligation was used to separate one end of the cell from the other in an attempt to alter the density of the two cell halves. However, the effect of altered cell density on gravity sensing could not be determined since the two cell halves, each with its own rotating cytoplasmic stream, were unable to detect gravity following ligation. The indication was that the gravity sensing system had two parts, one in each end of the cell, and the two parts had to be in communication in order to detect the direction of gravity. The way in which the two cell ends communicate is still unknown.

Gravity receptor. It seemed likely that the receptor sensing the direction of gravity was a protein. Therefore, various regions of the cell were irradiated with a microbeam of high-intensity ultraviolet light to inactivate any gravity-sensing proteins. The ultraviolet microbeam prevented the gravity-induced polarity of cytoplasmic streaming when either end, but not the middle, of the cell was irradiated, indicating that proteins, restricted to the two ends of the cell, act as the gravireceptor.

These experiments led to the gravitational pressure theory, according to which the whole weight of the protoplast acts as the gravity sensor. This theory assumes that in a vertical cell the protoplast falls, like a water-filled balloon, a minute distance within the rigid extracellular matrix that surrounds it. The protoplast, however, is tethered by proteins to the extracellular matrix at the top and bottom of the cell. Thus, its falling within the extracellular matrix causes a stretching of the proteins at the interface of the plasma membrane and extracellular matrix at the top of the cell and a squeezing of the proteins on the plasma membrane–extracellular matrix interface at the bottom of the cell. The cell is able to discern which way is up and which way is down by differentially responding to the stretched and squeezed proteins, respectively. When a vertical cell is placed in a solution denser than the protoplast, the cell responds to gravity as if up were down, and down were up; consequently, the cytoplasm streams up faster than it streams down.

The proteins in a horizontal cell can be artificially stretched on one end and squeezed on the other end by applying a hydrostatic pressure to one end of the cell. In this case, the cytoplasm always streams faster away from the end where the proteins are stretched, and slower toward the end where the proteins are squeezed. The hydrostatic pressure–induced polarity of cytoplasmic streaming requires external Ca^{2+}, is inhibited by Ca^{2+} channel blockers, and requires that the two ends of the cell be intact or undamaged by ultraviolet irradiation. The indication is that hydrostatic pressure mimics gravitational pressure in inducing a polarity of cytoplasmic streaming. Such evidence strongly supports the idea that the stretching and squeezing of proteins in the plasma membrane–extracellular matrix interface is sufficient to signal to the cell which way is up and which way is down.

Integrins. In animals, a class of proteins that tether the plasma membrane to the extracellular matrix is becoming well understood, since they are necessary for the migration of cancer cells throughout the body as well as for cell migration during normal development. These proteins are called integrins. Many integrins present in the plasma membrane link to proteins in the extracellular matrix by recognizing a sequence of three amino acids (arginine–glycine–aspartic acid) in the extracellular matrix protein. This attachment provides the friction necessary for the actin- and myosin-dependent cell migration. When migrating animal cells are treated with a small peptide containing the arginine–glycine–aspartic acid sequence, they can no longer move because the peptide competes with the normal binding of integrins to the extracellular matrix protein. Integrins may act as the gravireceptor protein in characean cells since treating *Chara* cells with a peptide containing arginine–glycine–aspartic acid also inhibits the

gravity-induced polarity of cytoplasmic streaming.

Higher plants. The falling of the protoplast releases sufficient energy to activate or inactivate the Ca^{2+} channels that lead to a polarity of cytoplasmic streaming. However, in calculating the energy released by the falling of the protoplast, it becomes clear that the energy released is directly proportional to the volume of the cell.

Since the volume of characean cells is almost 10^6 times greater than the volume of typical higher plant cells, the question arises whether the proposed mechanism is energetically possible in higher plant cells. It appears that the mechanism is energetically possible because higher plant cells are filled with dense starch grains that add to the density of the protoplast and cause sufficient stretching and squeezing of the integrin proteins at the top and bottom of the cell that act as the gravireceptors. If the starch grains themselves were the primary gravity sensors in higher plants, starchless mutants should be unable to sense gravity. If starch grains only act as ballast and add density to the protoplasm, their contribution to gravisensing should depend on the proportion that they contribute to the total weight of the protoplasm (approximate 50–75%). Thus, if the gravitational pressure model is true, the mutants should be able to sense gravity 25–50% as well as the wild type does. Experiments show that the starchless mutants do sense gravity 25–50% as well as the wild type, indicating that the gravitational pressure theory is a better explanation for gravity sensing in higher plants than the classical statolith theory. Since human cells do not have falling starch grains but do have attachments between the plasma membrane and the extracellular matrix, they may, like *Chara*, detect the direction of gravity by sensing gravitational pressure.

For background information SEE CELL MOTILITY; CELL PERMEABILITY; PLANT CELL; PLANT MOVEMENTS in the McGraw-Hill Encyclopedia of Science & Technology.

Randy Wayne

Bibliography. M. L. Evans et al., How roots respond to gravity, *Sci. Amer.*, 255:112–119, 1986; M. P. Staves et al., Hydrostatic pressure mimics gravitational pressure in characean cells, *Protoplasma*, 168:141–152, 1992; R. Wayne et al., The contribution of the extracellular matrix to gravisensing in characean cells, *J. Cell. Sci.*, 101:611–623, 1992; R. Wayne et al., Gravity-dependent polarity of cytoplasmic streaming in *Nitellopsis, Protoplasma*, 155:43–57, 1990.

Pluto

The edge-on orbital alignment of Pluto's satellite Charon that occurred during the latter half of the 1980s, commonly known as the mutual-event season, produced a wealth of new data that are continuing to be analyzed by planetary astronomers. Adding to knowledge of the system are the results from the initial observations by the Hubble Space Telescope, and from ground-based observations utilizing improvements in infrared detector technology.

Physical parameters. Continued analysis of mutual-event data has produced little change in any of the physical parameters for the Pluto-Charon system. The diameters for Pluto and Charon are still estimated to be 1430 mi (2300 km) and 735 mi (1185 km), with a derived mean density of 2.03 g/cm^3. The analysis of multiple chords from the 1988 occultation of a star by Pluto has yielded a slightly larger diameter for Pluto, although the result is somewhat dependent on the assumed atmospheric model. Similarly, a new analysis of a stellar occultation by Charon in 1980 has placed a lower limit of 745 mi (1200 km) on its diameter.

Although the mean density for the system has been known for some time, the individual densities of Pluto and Charon were not known. The first attempt to determine the individual masses (hence densities) was recently made by measuring the wobble of Pluto and Charon about their common center of mass (the barycenter) by using a series of high-resolution images taken by the Hubble Space Telescope. The result suggests that Charon is substantially less dense than Pluto, indicating a much less rocky, more ice-rich composition. This new information is now being incorporated into models of the system's formation, which are expected to shed new light on how the system became a binary.

Imaging. With the launch of the Hubble Space Telescope, high-resolution images of planets could be obtained free of the blurring effects of the Earth's atmosphere. Pluto and Charon were among the first planets to be imaged with the telescope (see **illus.**). These images, surpassing the best ground-based images obtained to date, clearly show Pluto and Charon as separate objects. The initial results from Space Telescope imaging generally confirm the orbital radius as derived from ground-based imaging, although the inclination of Charon's orbit appears to be about 2° less than the value derived from both ground-based imaging and the analysis of mutual-event photometry. Unfortunately, the spherical aberration discovered in the telescope's optics following deployment prevents the disk of Pluto from being meaningfully resolved at this time. As a result, indirect techniques must still be used to provide a picture of the surface.

Surface appearance. By using data from the 6-year-long mutual-event season, astronomers are producing maps that show the distribution of bright and dark material over each of the occulted hemispheres. However, because the system is completely tidally locked, the same hemispheres perpetually face each other, so that maps of the nonfacing hemispheres cannot be made in the same way. The emerging picture of the planet shows two bright polar caps, with the southern one being brighter, and a dark equatorial belt, with one hemisphere having a significantly larger or darker region than the other.

Some structure in the albedo distribution of Charon may also be present, but the mapping efforts are still too preliminary to make any definitive statements about the appearance of Charon's surface.

Surface composition. Recent advances in infrared detector technology have permitted Pluto's infrared spectrum to be measured with much higher spectral resolution and accuracy than was previously possible. The new data have revealed absorption features of molecular nitrogen (N_2) and carbon monoxide (CO), both in solid form (ice). Previous work had detected the strong absorption features of methane (CH_4), so Pluto's surface appears to consist of a mixture of at least these three ices. However, the bulk density of the planet implies that a large fraction of its interior is rocky material. Whether any of this rocky material is exposed on the surface is unknown, although the darker regions of the surface may represent exposed rocky material.

Atmosphere. The existence of an atmosphere around Pluto was firmly established in 1988, when the planet occulted a star. The composition of the atmosphere could not be uniquely determined from these data. Methane gas was presumed to be present, if not dominant, in the atmosphere, because methane had been detected spectroscopically. As mentioned above, nitrogen and carbon monoxide have recently been detected as well. The vapor pressure of nitrogen is the highest of the three gases, followed by carbon monoxide, so if vapor pressure equilibrium is assumed, then (like the Earth's) Pluto's dominant atmospheric gas would be nitrogen.

Conflicting interpretations of the occultation data have led to some uncertainty regarding the structure of Pluto's atmosphere and the planet's size. On the one hand, if the atmosphere is isothermal, the existence of a layer of haze must be hypothesized to reproduce a distinctive feature in the occultation data. Alternatively, the feature could also be produced by a clear atmosphere with a thermal gradient. If a haze layer is present, the mutual-event diameter is overestimating the true size of Pluto, because the true surface is hidden at the limb by the optically thick haze. On the other hand, a thermal gradient would be likely if the atmosphere contains enough methane, a gas that readily absorbs solar infrared radiation.

The presence of an extremely thin atmsphere around Charon has been suggested following a new analysis of occultation data obtained in 1980, when

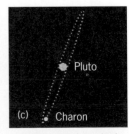

Comparison of images of Pluto-Charon system with ground-based and space-based telescopes. (*a*.) Ground-based image taken with Canada-France-Hawaii telescope in Hawaii. This image is the best achieved to date. (*b*.) Image taken with the faint object camera of the Hubble Space Telescope. At the time of observation, Charon was near its maximum apparent angular separation of about 0.9 arc-second. Charon is the fainter object in the lower left. (*c*.) Charon's orbit around Pluto as viewed from the Earth, showing the position of Charon at the time of observation. (*NASA/ESA*)

Charon passed in front of a star. The data are insufficient to make any significant conclusions about the properties of this proposed atmosphere. Some astronomers are skeptical of both the data and the proposed atmosphere.

Satellites. Pluto is so far from the Sun and the other planets that the region of gravitational stability around Pluto is quite large, extending to about 100 times the separation between Pluto and Charon. Astronomers have recently imaged this region to look for the possible presence of other small, distant satellites, given that multiple satellites are quite common among the outer planets. No satellites were found down to a level of about 10^6 times fainter than the faintest object visible to the unaided human eye at distances greater than six times the Pluto-Charon separation, and six times fainter than that at distances ten times that separation. The region closer to Pluto could not be studied because of the glare of the planet. If a hypothetical second satellite has a surface reflectivity comparable to Charon, the data rule out objects larger than about 70 mi (115 km) and 30 mi (45 km) in diameter for these two regions, respectively. Unlike the other outer planets, but like the Earth, Pluto would appear to have just a single, large satellite.

Origin. Numerical integrations of Pluto's orbit back in time for almost 10^8 years show that Pluto is in a two-to-three resonance with Neptune. That is, for every two orbits that Pluto makes of the Sun, Neptune makes three. This resonance prevents Pluto from closely approaching Neptune. In fact, Pluto comes closer to Uranus than it does to Neptune. This result significantly reduces the probability that Pluto and Charon were once satellites of Neptune, as had been suggested previously. Instead, the independent-formation theory continues to remain in favor with most astronomers.

Charon is so large and massive relative to Pluto that it contains a significant amount of the total angular momentum of the system. If the two bodies were coalesced into a single body, the resulting object would be rotationally unstable. Conversely, it appears unlikely that Pluto and Charon were formed by spontaneous fission of a single, rapidly rotating body. A formation scenario has been suggested in which the majority of the angular momentum was imparted by a small number of large colliding planetesimals early in the history of the solar system. A collision with a large planetesimal could be responsible for both the high obliquity of the Pluto-Charon system and the rotational instability that led to the formation of a binary system.

Future studies. The National Aeronautics and Space Administration (NASA) is currently studying options for a spacecraft flyby of Pluto early in the next century. In addition to being the only planet not yet visited by a spacecraft, Pluto has been given special attention because, as it recedes from the Sun (perihelion passage was in 1989), travel time to the planet increases. The atmosphere will cool and begin to "snow" onto the surface, thereby decreasing the amount of atmospheric bulk that can be studied. An increasing amount of the surface around a pole will be in

shadow for the next several decades during southern-hemisphere winter. As a result, arrival times prior to the year 2020 are considered essential for meeting the scientific objectives of the proposed mission.

Because the rotational period of the planet (6.4 days) is long compared to the duration of a spacecraft flyby, a pair of spacecraft are necessary to achieve global, high-resolution imaging of the entire illuminated surfaces of Pluto and Charon; a twin spacecraft mission is currently envisioned. Arrival times could be staggered by as much as a year so that the second flyby could benefit from the knowledge gained from the first. With a sufficiently energetic launch, possible with currently available boosters, flight times to Pluto of about 7 years could be realized, even without the benefit of a Jupiter gravity assist.

For background information SEE PLANETARY PHYSICS; PLUTO; SATELLITE ASTRONOMY; SOLAR SYSTEM; SPACE PROBE in the McGraw-Hill Encyclopedia of Science & Technology.

David J. Tholen

Bibliography. M. W. Buie et al., Albedo maps of Pluto and Charon: Initial mutual event results, *Icarus*, 97:211–227, 1992; J. L. Elliot and L. A. Young, Limits on the radius and a possible atmosphere of Charon from its 1980 stellar occultation, *Icarus*, 89:244–254, 1991; S. A. Stern et al., A search for distant satellites of Pluto, *Icarus*, 94:246–249, 1991; G. Tancredi and J. A. Fernández, The angular momentum of the Pluto-Charon system: Considerations about its origin, *Icarus*, 93:298–315, 1991.

Precambrian

The Cambrian Period began with an explosive radiation of coelomate animals some 540 million years ago (m.y.a.). Although this event marks the beginning of the Phanerozoic Eon (literally, the age of visible life), it took place relatively late in evolutionary history. Phylogenies based on the comparative biology of living organisms (see **illus.**) indicate that the diversification of the coelomate phyla must have been preceded by the evolution of architecturally simpler animals, which were proceeded by a major radiation of protists. Even the diversification of protists appears to have occurred fairly late in evolutionary history. Paleontological and geochemical data confirm the evolutionary history inferred from comparative biology. Together, studies of comparative biology and the geological record have begun to clarify the nature and timing of major evolutionary events during the long Precambrian Era.

Coelomate radiation. Coelomate invertebrates can be divided into two major evolutionary groups: protostomes and deuterostomes. Protostomes include the arthropods, annelids, mollusks, and a number of less diverse phyla. Deuterostomes comprise echinoderms, hemichordates, and chordates. Fossils in Lower Cambrian rocks document the rapid diversification of both groups. Trace fossils complement this record, showing a dramatic increase in the diversity and density of tracks, trails, and burrows.

Comparative morphology and molecular phylogeny indicate that the coelomate radiation is rooted in an earlier episode of animal evolution during which sponges, cnidarians, and the bilaterally symmetrical ancestors of the coelomate phyla diverged. Fossils deposited near the end of the Proterozoic Eon (2500–540 m.y.a.) confirm this inference. A widely distributed assemblage of soft-bodied animals, generally known as the Ediacaran fauna after occurrences in the Ediacara Hills of South Australia, occurs in rocks as old as about 575 m.y. There is considerable debate about the biological relationships of some Ediacaran fossils, but most researchers agree that cnidarians constitute a major proportion of the fauna. Sponges are also present, as are simple bilaterally symmetrical animals, represented in the trace-fossil record and, possibly, as body fossils. The Ediacaran fauna also contains several taxa whose relationships to extant phyla are problematic.

Protist radiation. Despite a century of study, evolutionary relationships among eukaryotes remained uncertain until recently, when molecular methods were applied. In particular, an interspecific comparison of nucleotide sequences in ribonucleic acid (RNA) molecules from the small subunit of the ribosome was made. Phylogenies based on these molecular data corroborate insights gleaned from ultrastructural and biochemical studies.

These phylogenies suggest that while the eukaryotic domain is ancient, most of the diversity found among living eukaryotes originated during a rapid evolutionary burst late in the history of the group. The red,

green, and chromophyte algae differentiated during this event, as did the fungi, animals, and a heterogeneous group made up of the ciliates, dinoflagellates, and plasmodia. Paleontological data suggest that this rapid burst of protistan evolution occurred about 1000 m.y.a. Neoproterozoic (1000–540 m.y.a.) rocks contain the first identifiable fossils of multicellular green, red, and, perhaps, chromophyte algae; fossils of unicellular protists also document morphological diversification during this period. A practical consequence of this radiation is that protistan microfossils can be used in the biostratigraphic correlation of Neoproterozoic sedimentary successions. Biomarker molecules, that is, specific organic compounds synthesized by specific groups of organisms and preserved in sedimentary rocks, also suggest a marked Neoproterozoic increase in the abundance and diversity of protists.

Appearance of metazoan fossils. If the clade today represented by the animals diverged from other eukaryotic groups 1000 m.y.a., it is not clear why metazoan fossils do not appear in the fossil record for another 400 m.y. One possible explanation is that the genetic controls necessary for metazoan development evolved long after the divergence of the clade. This may be the case, but fossils of seaweeds show that at least some measure of developmental control evolved in algae long before the Ediacaran event. An alternative explanation is that an environmental barrier, specifically a low concentration of atmospheric oxygen, inhibited the evolution of macroscopic animals until about 580 m.y.a. This hypothesis draws support from the geological record. The Neoproterozoic Era

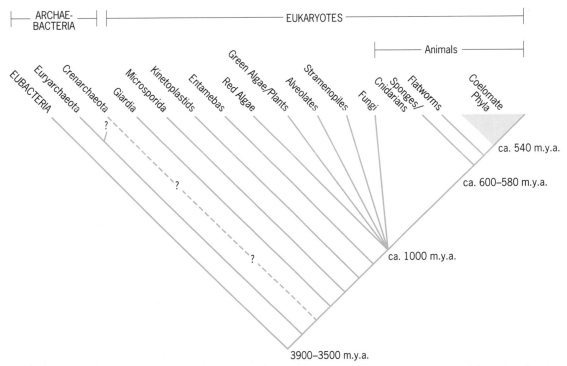

Evolutionary relationships among major groups of living organisms. The alveolates are the chromophyte algae and their heterotrophic relatives. The stramenopiles comprise the dinoflagellates, ciliates, and plasmodia. Dates indicate estimated times of divergence as inferred from the geological record. Question marks and broken lines indicate alternative interpretations of relationships of Crenarchaeota to other organisms.

was a time of pronounced tectonic, climatic, and biogeochemical change; geological data and biogeochemical models suggest that a physiologically significant increase in oxygen levels may well have resulted. A satisfactory understanding of late Proterozoic evolution requires that the biological record be rooted in both a broader phylogenetic context and a framework of geological development.

Early eukaryotic evolution. The relatively late diversification of higher protists raises two additional questions. What biological or environmental events might have initiated this radiation, and how much farther back in time can eukaryotic organisms be traced?

The second question is easier to address. The direct fossil record of eukaryotic organisms goes back about 2100 m.y., fixing a minimum date for the origin of the group. Unfortunately, the fossil record for earlier Earth history is poor, and there is no reason to assume that the earliest known eukaryotic fossils approximate the age of the domain. Molecular phylogenies indicate that the three principal domains—the eubacteria, the archaebacteria, and the eukaryotes—diverged early in evolution. Some data suggest that the eukaryotes are a sister group of the archaebacteria that diverged before or shortly after the archaebacteria began to diversify. Thus, geochemical evidence for methanogenic archaebacteria in 2800-m.y.-old rocks can be used to infer an Archean age (greater than 2500 m.y.) of eukaryotic divergence.

Comparative biology suggests that the earliest eukaryotes were anaerobic heterotrophs whose biochemistry in some ways resembled that of bacteria more than it did extant higher protists. A key attribute of emerging eukaryotes was the presence of a dynamic cytoskeletal and membrane system that enabled these cells to engulf external particles and bring them into the cell. Thus, primitive eukaryotes could capture the aerobically respiring purple bacteria that became symbionts and, eventually, mitochondria, as well as the cyanobacteria that evolved into chloroplasts. Phylogenies indicate that mitochondria became established long before the Neoproterozoic protistan radiation, and geochemical data support this view. Atmospheric oxygen levels capable of sustaining aerobic metabolism in single cells appeared no later than 2100 m.y.a., and possibly 2800 m.y.a. or earlier. Therefore, the acquisition of mitochondria did not trigger the great Neoproterozoic protistan radiation.

In contrast, most photosynthetic eukaryotes occur within the groups that differentiated during the Neoproterozoic event. It might therefore be inferred that the acquisiton of chloroplasts provides a key to understanding the radiation. However, other possibilities exist, including the evolution of meiotic sexuality. No obvious environmental change correlates with the radiation. Thus, in contrast to the Ediacaran radiation of large animals, satisfactory biological or environmental explanations for the Neoproterozoic protistan diversification remain to be found.

Prokaryote radiation. Molecular phylogenies suggest that the archaebacteria and eubacteria dif-

ferentiated before or in association with eukaryotes. Phylogenies indicate that the first archaebacteria were anaerobic, thermophilic, and chemoautotrophic. Distinct branches of sulfur-dependent and methanogenic archaebacteria differentiated early; mesophilic and aerobic lineages subsequently evolved within each branch. New discoveries continue to redefine the deepest branches of the eubacterial tree in which, as in the archaebacteria, are found anaerobes, thermophiles, and unusual chemoautotrophs. A major radiation of eubacteria appears to be related to the evolution of photosynthesis, which is widely distributed within the domain. One particular group of photosynthetic bacteria required the ability to use water as an electron donor. Because oxygen is produced as a by-product of this reaction, cyanobacteria changed the course of environmental and biological history. As in the archaebacteria, aerobic metabolism appears to have evolved repeatedly within the eubacterial domain, often within fundamentally photosynthetic lineages.

Geological data constrain the timing of these events. The oldest metamorphosed sedimentary rocks in which deformation is slight are 3500–3400 m.y. old. They contain small unicellular and filamentous microfossils, indicating that life existed at this early date in Earth history. However, the simplicity of preserved morphologies precludes physiological inference. The same successions also contain laminated sedimentary structures called stromatolites. Stromatolites are built by the activity of microbial communities, and their presence in these rocks indicates that phototactic, and presumably photosynthetic, organisms lived in coastal marine environments. Carbon isotope ratios in organic matter and carbonate minerals provide firmer evidence that photosynthesis had already evolved, although whether the cyanobacteria existed at this time is unclear.

Thus, by 3500–3400 m.y.a., the split between the eubacteria and archaebacteria had already occurred, and the photosynthesis-driven radiation of eubacteria had begun. Despite the possible presence of cyanobacteria, atmospheric oxygen remained at or below 1–2% of present-day levels. Geological data indicate that by the late Archean the cyanobacteria had certainly evolved, as had the major branches of the archaebacterial tree. A polyphyletic radiation of aerobic prokaryotes accompanied the late Archean to early Proterozoic development of stable atmospheric oxygen. Both stromatolites and prokaryotic microfossils occur throughout the Proterozoic record.

For background information SEE ANIMAL EVOLUTION; CAMBRIAN; EDIACARAN FAUNA; GEOLOGICAL TIME SCALE; PRECAMBRIAN in the McGraw-Hill Encyclopedia of Science & Technology.

Andrew H. Knoll

Bibliography. H. Baltscheffsky et al. (eds.), *Early Life on Earth*, Nobel Symp. 84, 1993; A. H. Knoll, The early evolution of eukaryotes: A geological perspective, *Science*, 256:622–627, 1992; J. W. Schopf and C. Klein (eds.), *The Proterozoic Biosphere: A Multidisciplinary Study*, 1992; C. R. Woese, Bacterial evolution, *Microbiol. Rev.*, 51:221–227, 1987.

Quality control

Quality control is a collection of methods used to ensure the quality of manufactured products and to improve the processes by which these products are made. In recent years, product quality has emerged as a key determinant of success in modern industry. By emphasizing the importance of quality, manufacturers have been able to achieve significant gains in market share and profitability.

Although issues pertaining to quality have been studied for many years, research in the United States has intensified recently due to increasing foreign competition. Research in quality falls primarily into two areas, quality management and quality engineering. Quality management focuses on organizational issues. Typical of this concept is the phrase total quality management (TQM), a philosophy in which all components of the firm are committed to quality and focus on customer service.

Quality engineering focuses more directly on technical issues involving the product and the manufacturing process. On the product side, designs are refined and specialized, and features are added to better meet the needs of customers. Although refinements in design features may enhance quality and appeal to customers, these features often make the product more expensive. It is through greater emphasis on the manufacturing process that the most impressive gains have been achieved. When the manufacturing process is improved, the typical results are greater product reliability, less scrap and rework, and fewer customer complaints and returns. By focusing on the process rather than the product, quality can be improved and cost reduced simultaneously.

Process variation. The goal of a manufacturing process is to make each product exactly as specified in the design, a goal that can never be completely attained. Inevitably, the quality of products will exhibit some variation around the design specification. This variation is the primary culprit in quality problems and leads to excessive scrap and rework, poor reliability, and customer dissatisfaction.

Based on the pioneering work of W. A. Shewhart in the 1920s and 1930s, quality engineers divide process variation into two categories, random variation and variation with assignable cause. Very little can be done about random variation, which accounts for the (hopefully small) variation under normal conditions. Variation with assignable cause refers to deviations from targeted quality resulting from a single, identifiable source. Variations of this type present the quality engineer with an opportunity to improve the process. If the variation can be detected and its cause determined, future process variation can be reduced by eliminating the cause. The systematic identification and removal of sources of process variation leads to continuous improvement in quality.

Statistical and engineering process control. Two common approaches to process control that directly address variation in quality and process are statistical process control and engineering process control. (Engineering process control is usually referred to simply as process control, but the word engineering is included here to avoid confusion.) Statistical process control is commonly found in industries that manufacture discrete parts, such as those producing mechanical or electronic devices. Engineering process control is typically found in process industries such as steel and chemicals. In the past the methods of engineering process control and statistical process control have often been viewed as incompatible. At best, they were applied in a hierarchical fashion in which statistical process control performed an infrequent and remote quality-assurance function. However, recent developments have shown how these methods can be unified so that the power of each approach can be achieved simultaneously.

Engineering process control as applied to product quality relies on two key ingredients, a device that can measure the quality of products as they are produced and a control variable that can be used to make adjustments in product quality. Based on the most recent measurement, the control variable is changed up or down to drive the process output back to the level specified in the design. Thus, the idea is to compensate for disturbances to the process by appropriately adjusting a control variable.

Statistical process control has the same goal as engineering process control, the reduction of process variation. However, the approach taken is quite different. Statistical process control attempts to identify variation with assignable cause, ultimately leading to improved quality by systematically eliminating sources of variation. Statistical process control provides a monitoring function in which the occurrence of unusual or suspicious variation is identified. A statistical model of process variation is used to discriminate between random and significant deviations in process quality. Identifying the source of this variation and eliminating the cause is typically the task of engineers.

An example of this approach is a gold-plating operation. Unplated spoons traveling on an overhead conveyor system are immersed in a solution containing gold ions; a voltage is applied to a spoon, resulting in the depositing of gold on it. The important quality measure in this case is the thickness of the layer of gold deposited. It has been determined (and specified in the design) that gold thickness should be at least 0.1 mm (0.04 in.) in order for the spoon to meet the expectations of the customer. Following the plating operation, spoons are passed through an x-ray device that can, to a reasonable degree of accuracy, measure the thickness of the gold layer. The primary variables that affect thickness are the immersion time of the spoon, the concentration of the gold solution in the bath, and the voltage applied. However, the final thickness is a function of many additional factors, which may be unknown or unpredictable. Examples include variations in the raw material from which the spoon is made, impurities in the chemical bath, dirt on the surface of the spoon, slight changes in the speed of the conveyor, and fluctuations in voltage. Because of the unavoidable variation, the target thickness must be set higher than that specified in the design in order that almost all spoons meet the specification (see **illus.**). If process variation can be reduced, the target can be moved closer to the specified thickness, and a savings would result that is literally worth its weight in gold.

The engineering process control approach in this example is to adjust the voltage based on the thickness of spoons as measured by the x-ray device. In this way, a disturbance causing a change in thickness will quickly be compensated for by adjusting the voltage. Of course, the engineering process controller will adjust the process based on any perceived change in thickness, including random process variation and measurement errors made by the x-ray device. There is also a time delay between the occurrence of a disruption and the corrective adjustments. In this example, the delay is the time it takes a spoon to travel from the bath to the thickness-measuring device.

The statistical process control approach in this example is to monitor gold thickness in order to detect unusual variation. While such a disturbance is detected, engineers are asked to isolate the source of the problem and eliminate its cause. In this endeavor, practitioners of statistical process control may well suggest that engineering process controllers be removed from the process, because they tend to obscure the effect of disturbances, thus making the disturbances more difficult to detect. It is also common practice to shut down the production line until engineers have isolated and solved the problem. While these tactics may seem extreme, they closely reflect practices that have been employed with great success in Japanese manufacturing facilities.

Model-based control. Recent model-based approaches attempt to integrate engineering process control and statistical process control so that the desirable features of each may be achieved. A basic ingredient in these methods is a process model that can forecast outputs as a function of inputs. Deterministic dynamic models that capture the nonrandom aspects of the process have been used for quite some time to improve the performance of engineering process control. The model is inverted and used to back-calculate the adjustment in the input required to bring the output back to target. In some cases, process models can be developed from so-called first principles, which are based on scientific laws from thermodynamics, physics, and so forth. More typically, however, empirical models must be developed through systematic experimentation and data analysis. The importance of experimentation and modeling, and the insights they bring, is only now being fully recognized in industrial processes.

One important use of these models is in the integration of engineering process control and statistical process control. A model is developed that contains information on both the deterministic and statistical behavior of the process. In addition to its role in calculating adjustments for engineering process control, the model can be used to predict process performance without the controller adjustments. In this way, the disturbances are not obscured and are thus easier to detect. When the model reveals deviations from target that exceed those explainable by random variation, a search for an assignable cause can be initiated. Because these models contain more information about the process than the simple statistical models typically used in statistical process control, they are able to detect a greater range of disturbances. Engineers are also beginning to use process models to help isolate the source of disturbances. A process model can determine what effect different disturbances will have on the process. By observing which variables are affected and their behavior, much useful information can be obtained as to the sources of the assignable causes of variation. With this information, causes of variation can be more quickly determined and eliminated. Process experimentation, modeling, control, and improvement will continue to be an exciting and fruitful area of research in the future.

For background information SEE MODEL THEORY; PROCESS CONTROL; QUALITY CONTROL in the McGraw-Hill Encyclopedia of Science & Technology.

Robert H. Storer

Bibliography. T. J. Harris and W. H. Ross, Statistical process control procedures for correlated observations, *Can. J. Chem. Eng.*, 69(1):48–57, 1991; J. F. MacGregor, On-line statistical process control, *Chem. Eng. Prog.*, 84(10):21–31, 1988; D. C. Montgomery, *Introduction to Statistical Quality Control*, 1991.

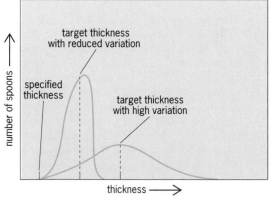

Effects of process variation on target gold thickness.

Quantum interference devices

Electronic fabrication techniques have recently advanced to the point where it is possible to observe, possibly even on a production basis, quantum size effects in the transport of charge carriers in semiconductors. These advances include device patterning techniques such as x-ray and electron-beam lithography, as well as the materials growth techniques of molecular beam epitaxy (MBE), chemical vapor deposition (CVD), and refinements of these two basic techniques. As a result of these advances, several new electronic devices have been designed and fabricated. This new class of devices is known generally as quantum interference devices.

The most important device in solid-state electronics is the transistor. The transistor is basically a three-terminal current valve where the current flow between two of the terminals in the device can be modulated by a voltage applied to the third terminal. Typically, the terminals through which the current flows are named the source and drain (or emitter and collector, depending on the kind of device being discussed) because of the analogy between the transistor and a water valve, and the controlling terminal is called the gate (or base) since it controls the flow of charge carriers. Transistor action can occur as a result of many different physical processes, giving rise to different kinds of transistors. Bipolar junction transistors (BJTs), junction field-effect transistors (JFETs), and metal-insulator semiconductor field-effect transistors (MISFETs) are among the most common. A new set of devices based on quantum interference phenomena associated with the channel current charge carriers is an important recent addition.

Principle of operation. The quantum interference transistor takes advantage of the coherent, wavelike nature of electrons as they propagate through a nanometer-sized device structure. As a result of this coherent propagation, electrons that have taken different paths through the device arrive at the drain (collector) with different phases and interfere, resulting in a high-conductivity channel (in the case of constructive interference) or a low-conductivity channel (in the case of destructive interference). The phase of each electron wave is typically a function of the controlling parameter (such as voltage or magnetic field) for the device, and depends upon the exact physical mechanism governing the transistor action. One such transistor is based upon the Aharonov-Bohm effect. The Aharonov-Bohm transistor essentially relies on a Mach-Zehnder interferometer (two separate electron waveguides that initiate at the source or emitter and terminate at the drain or collector), in which the relative phase differences are controlled by a magnetic field that exists in the region between the two waveguides. This type of transistor effect has been demonstrated in both metals and semiconductors. Another quantum interference transistor, named the resonant tunneling transistor, which will be discussed in detail, is based upon the resonance characteristic of Fabry-Pérot structure.

Resonant tunneling transistor. The resonant tunneling transistor may be thought of as essentially a standing-wave resonator through which electron waves must propagate in order to pass from the transistor source (emitter) to the drain (collector). The mirrors at each end of the resonator are regions of semiconductor material with energy band characteristics sufficiently different from the primary material of which the device is composed so that a substantial reflection of the electron wave will take place at the material boundary. The electron waves reflected from the two material boundary mirrors interfere in a manner determined by their wavelength (see **illus.**). If

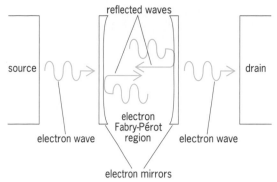

Geometry of a resonant tunneling transistor.

the wavelength of the electron is such that an integral number of half-wavelengths will fit exactly into the space between the two mirrors, a resonance of the resonator is achieved; the electrons pass through the resonator without being reflected by the Fabry-Pérot structure and the conductivity of the channel is very high. If the resonance condition is not achieved, the electron wave is largely reflected by the Fabry-Pérot structure and the channel conductance is low. The electron wavelength is determined by the kinetic energy of the electron as it is injected from the source into the Fabry-Pérot region and, as a result, the channel conductivity of electrons of different kinetic energies can vary widely. Transistor action occurs by changing the average potential of the Fabry-Pérot region of the channel with respect to the source, effectively modulating the kinetic energy of injected electrons and thereby modulating the channel conductivity. This resonant tunneling transistor device has the potential to be very fast because the electrons must propagate through the device ballistically (without scattering) in order for the interference effects to work. Several forms of the resonant tunneling transistor have been investigated. Their differences result primarily from the details of the resonant tunneling structure (Fabry-Pérot in the simplest case) employed in the device. The number, widths, and locations of the tunneling resonances as a function of kinetic energy become important design parameters, and will permit the optimal design of a resonant tunneling transistor for a particular application.

Temperature range. The primary limitation for resonant tunneling transistor devices is the useful temperature range. Due to the fact that the transistor action results from the coherent interference of electron waves, the phase of these waves must be preserved throughout the device. There are several mechanisms by which electrons may be dephased, including scattering from defects and impurities in the device, scattering from other charge carriers in the device, and scattering from phonons (acoustic vibrations). Modern fabrication techniques are good enough that defect and impurity scattering can be minimized, and channel current densities can be kept low enough that carrier-carrier scattering is not important. However, phonon densities increase with temperature, and thus resonant tunneling transistor devices must be operated at extremely low temperatures, of the order of a few kelvins.

Materials. The semiconductor material system from which resonant tunneling devices are typically fabricated is the $GaAs/Al_{1-x}Ga_xAs$ (gallium arsenide/aluminum gallium arsenide) system. This system has several advantages, primary among which are (1) the very small effective mass of electrons in this system, which enhances the quantum size effects that manifest themselves in the device; (2) the very similar lattice constants of GaAs and $Al_{1-x}Ga_xAs$, which result in the growth of an essentially strain-free device structure; and (3) the relatively well understood growth of $GaAs/Al_{1-x}Ga_xAs$, which permits high-quality devices to be fabricated through the use of the molecular-beam epitaxy growth technique. Although molecular beam epitaxy does not readily lend itself to high-volume production, additional techniques such as metallo-organic chemical vapor deposition (MOCVD) are widely used in standard gallium arsenide device technology, and continue to show promise for ballistic resonant tunneling transistor devices as well. The metallo-organic chemical vapor deposition process readily scales to the throughput required for modern production environments.

Applications and advantages. Quantum interference devices including resonant tunneling transistors will have their primary application in very fast switching circuits. The required switching speeds of transistors are currently being pushed to the physical limits for collision-dominated diffusive technology, and the device size is being reduced to the point where quantum effects are becoming observable. As a result, it seems natural to take advantage of the quantum interference effects where possible.

Another advantage of the quantum interference technology is that, because of the absence of lattice-heating collisions, energy loss to the lattice and substrate is minimized, a fact that ultimately could result in more efficient, less lossy devices. However, this advantage is tempered by the fact that a proper efficiency calculation must include the energy used to maintain the device at its operating temperature (which in the case of resonant tunneling transistor devices is equal to or less than 1 K).

A third advantage of resonant tunneling transistor device technology is that it allows circuits to be used with significantly smaller driving sources than are currently required. Present digital technology requires sources with voltages of the order of 3 V or more in order to drive the transistors. There are two primary reasons: threshold voltages or built-in potentials for field-effect-transistor or bipolar-junction transistor devices respectively are of the order of 1 V, and a clear contrast in potential above or below these thresholds is required; and the voltage difference between on and off states must be substantially above the background noise level to assure reliable detection of the output state. Quantum interference devices circumvent these problems. The switching thresholds are entirely a function of the device geometry, and therefore devices may

be designed with switching thresholds anywhere from 0 V to several volts. The noise problem decreases significantly because the operating temperature of the device is so low. Therefore, quantum interference devices will be implemented with lower thresholds, increasing the efficiency of circuits based upon these devices.

Quantum interference devices certainly have the potential to operate faster and with smaller driving sources than transistors used in conventional switching circuits. However, because of the high cost of production resulting from the size and growth tolerance requirements on the fabrication process, probably these devices will be deployed only in areas where very high speed is required and system manufacturing and maintenance costs play only a secondary role. Certain military and space applications are good prospects, and commercial high-speed supercomputing technology may also benefit from these devices.

For background information SEE AHARONOV-BOHM EFFECT; INTEGRATED CIRCUITS; INTERFEROMETRY; SEMICONDUCTOR HETEROSTRUCTURES; TRANSISTOR; TUNNELLING IN SOLIDS in the McGraw-Hill Encyclopedia of Science & Technology.

David R. Andersen

Bibliography. F. Capasso, S. Sen, and A. Y. Cho, Resonant tunneling: Physics, new transistors, and superlattice devices, *Proc. SPIE*, 792:10–17, 1987; H. Ehrenreich and D. Turnbull (eds.), *Solid State Physics: Advances in Research and Applications,* vol. 44, 1991.

Radio spectrum allocations

During February 1992, under the aegis of the International Telecommunication Union (ITU), over 100 countries and international organizations met at the World Administrative Radio Conference (WARC-92). The primary purpose was to reach global agreement as to which radio frequency bands should be reserved for certain new technologies. In addition, WARC-92 sought to develop internationally agreed-upon rules pursuant to which geostationary and nongeostationary satellites could share overlapping swaths of radio frequencies, and to effect a slight expansion of frequencies allocated for shortwave broadcasting in the high-frequency (HF) band and for position-reporting satellites in the very high frequency (VHF) band.

All countries attending WARC-92 signed a treaty committing themselves to implementation of the new frequency allocations over the following 15 years. The treaty also directed that substantial studies of frequency sharing be undertaken in anticipation of certain practical problems involved in placing the new technology in an already crowded frequency spectrum.

WARC-92 will be remembered as the international conference that paved the way for satellite radio broadcasting and for hand-held telephony via satellite. It will also be remembered as the beginning of a spectrum refarming process in which nonmobile technologies now occupying choice frequency bands gave way to mobile communication technologies.

WARC-92 allocations to new technologies

Technology	Allocation*	Geographic limits	Other major limits
Digital audio broad-casting via satellite	1452–1492 MHz (d)	Except in U.S.	None
	2310–2360 MHz (d)	U.S. and India only	None
	2535–2655 MHz (d)	Most of Asia only	None
Portable telephony via satellite	1492–1525 MHz (d)	Western Hemsiphere, except U.S.	None
	1525–1530 MHz (d)	None	None
	1610–1626 MHz (u and d)	None	No interference with navigation satellites
	1675–1710 MHz (u)	Western Hemisphere only	No interference with weather satellites
	1930–1970 MHz (u)	Western Hemisphere only	Secondary
	1970–1980 MHz (u)	U.S. only	After 1996
	1980–2010 MHz (u)	None	After 2005
	2120–2160 MHz (d)	Western Hemisphere only	Secondary
	2160–2170 MHz (d)	U.S. only	After 1996
	2170–2200 MHz (d)	None	After 2005
	2484–2500 MHz (d)	None	None
	2500–2535 MHz (d)	National only	None
Space station and exploration activities	22.55–23.0 GHz	Intersatellite	
	24.45–24.75 GHz	Intersatellite	
	25.25–27.5 GHz	Intersatellite	
	32.0–32.3 GHz	Intersatellite	

*u = uplink; d = downlink.

Allocations to new technologies. New space technologies had become practical since the last frequency-allocating WARC held in 1987. The main technologies were hand-held telephones that work via satellites in low Earth orbit (Mobile Satellite Service or MSS), portable radio receivers for digital audio broadcasts direct from satellites (Broadcasting Satellite Service-Sound or BSS-Sound), and transceivers for astronauts working on space stations or on the Moon (Inter-Satellite Service). The frequencies sought for MSS and BSS-Sound were in the 1–3-gigahertz range, also called the L band (1–2 GHz) and the S band (2–3 GHz). The frequencies sought for space operations were mostly much higher in the spectrum, above 20 GHz. All these frequencies were sought because promoters of the respective technologies believed them to be the most technically suitable, taking into account various constraints that appeared to make yet more technically ideal frequencies unachievable at WARC-92 due to existing patterns of radio spectrum utilization. WARC-92 eventually agreed to the frequency allocations shown in the **table**.

BSS-Sound. In comparative terms, the new radio broadcasting service received a block spectrum nearly twice as large as all previous radio broadcasting allocations combined. Shortwave, AM, and FM radio enjoy about 22 megahertz total bandwidth, whereas the new BSS-Sound service was awarded 40 MHz total bandwidth worldwide, from 1452 to 1492 MHz; 50 MHz additional bandwidth, from 2310 to 2360 MHz, in the United States and India; and 20 MHz additional bandwidth, from 2535 to 2655 MHz, in a dozen, mostly Asian, countries, including the People's Republic of China, Japan, the Commonwealth of Independent States, Pakistan, India, and Thailand.

Mobile Satellite Service. MSS quintupled its previous frequency allocations, increasing from 64 MHz in the bands 1530-1560 MHz and 1626–1660 MHz to over 300 MHz. Not so optimal is the fact that, of the 300 MHz, only 92 MHz will be available worldwide, the rest being available only on a piece-meal basis in different portions of the L and S bands in various areas.

In ITU Region 3 (Asia and Oceania) and ITU Region 1 (Europe and Africa) the MSS now has a total of 197 MHz allocated. This figure includes 64 MHz available before WARC-92 mostly reserved for ships and planes, 35 MHz limited to national-only systems, 60 MHz not available until the year 2005, and 36 MHz available immediately but strewn across 1610–1626 MHz, 1525–1530 MHz, and 2484–2500 MHz. Part of this allocation is usable only on a condition of noninterference with the C.I.S. GLONASS satellite navigation system occupying 1610–1620 MHz.

In ITU Region 2, the Western Hemisphere, there is now 365 MHz of spectrum allocated to MSS in various sizes of blocks across the L and S bands. This bandwidth includes the entire 197 MHz available in ITU Regions 1 and 3, with its same restrictions, plus an additional 80 MHz (100 MHz in the United States) from 2120 to 2160 MHz (2170 MHz in the United States) and from 1930 to 1970 MHz (1980 MHz in the United States), usable only as long as interference is not caused to terrestrial radio relay systems in those bands. An additional 35 MHz, from 1675 to 1710 MHz, is usable only as long as interference is not caused with weather satellites in that band. Also, 33 MHz, from 1492 to 1525 MHz, is usable everywhere in the Western Hemisphere except the United States.

In summary, there was much new bandwidth allocated to MSS at WARC-92, especially in North and South America; but the ability of system operators to use this new bandwidth is dependent on the geographic scope of their service, the noninterference capabilities of their transmitting and receiving apparatus, and, for some of the best spectrum, their willingness to wait until 2005.

WARC-92 also allocated 1000 MHz to MSS from 19.7 to 20.2 GHz and from 29.5 to 30.0 GHz, 20% of which (20.1–20.2 GHz and 29.9–30.0 GHz) was allocated on a primary basis, meaning without any obligation to accept harmful interference from other cochannel systems allocated frequencies in the 20- or 30-GHz range. This WARC-92 allocation is rather anomalous because there is no MSS technology capable of working economically at so high a frequency range. In addition, this frequency range is plagued with very high levels of rain attenuation, which degrades the reliability of MSS links. Nevertheless, and largely at the urging of the National Aeronautics and Space Administration (NASA), the United States space agency, the WARC-92 delegates agreed to make this large addition to MSS in the hope that technological advances would soon render the spectrum useful.

Inter-Satellite Service. Space operations received 3300 MHz of new allocations under the label Inter-Satellite Service. These allocations were at 22.55–23.0, 24.45–24.75, 25.25–27.5, and 32.0–32.3 GHz. Under ITU definitions, a space-walking astronaut with a transceiver, a space station, and a space base on the Moon are all satellites. Hence, communications among these elements of a space operation are called Inter-Satellite Service.

Satellite frequency sharing. WARC-92 sought to agree on how various satellite systems in different orbits could share overlapping bands of frequencies. This issue arises because there are more satellite systems planned than there is MSS bandwidth available for exclusive allocation. Before WARC-92, each of the following countries had announced plans to use approximately 30 MHz of MSS spectrum: the United Kingdom (on behalf of the international organization INMARSAT), Canada, Mexico, France, Japan, C.I.S., Australia, and the United States.

WARC-92 decided that geostationary and nongeostationary satellite systems would have equal access to the new MSS allocations. This changed the status quo, requiring nongeostationary satellite systems to shut off if they caused interference to geostationary satellites. It was agreed that to implement equal access to MSS allocations the satellite systems that would formally announce all of their technical parameters would be entitled to protection from interference caused by later-announced satellite systems. To determine what technically constitutes interference and what technical steps should be taken to avoid it between MSS systems, the ITU decided to commission its technical advisory arm, the International Consultative Committee on Radio (CCIR), to make recommendations. It is believed that between 1992 and 1994 there will be, for the first time, internationally agreed-upon standards for frequency sharing among MSS systems in diverse orbits.

Existing allocation expansion. In effecting some minor additions of spectrum to existing services in the low end of the frequency spectrum,

WARC-92 allocated less than 1 MHz of bandwidth to shortwave broadcasting, at the expense of shortwave point-to-point communications.

About 2 MHz was allocated to position-reporting satellites of a kind similar to the successful COSPAR/SARSAT search-and-rescue satellite system, which has saved more than 1000 lives by pinpointing the location of downed planes and distressed boats. This allocation was made at 138 and 148 MHz in the name of MSS because, unlike COSPAR/SARSAT, MSS systems would not be limited to emergency uses but could be used for general position-reporting purposes. Tracking trucks and their cargo around the globe is a prime intended use for these new low-Earth-orbit VHF-band satellites.

Reallocations. WARC-92 made major changes to the set of internationally agreed-upon frequency allocations. Most of the significant changes were in the 1–3-GHz range, where satellite radio broadcasting and mobile satellite services gained spectrum at the expense of terrestrial radio-relay systems. Many observers see WARC-92 as a harbinger of a major trend in frequency allocations. This trend would displace nonmobile radio services either to much higher frequencies or to nonradio media (for example, fiber optic), so that mobile radio services could implement their absolute need for spectrum at lower frequencies. These changes provide important technical propagation and economic benefits to the mobile radio service, especially when it is provided via satellite.

For background information SEE COMMUNICATIONS SATELLITE; RADIO SPECTRUM ALLOCATIONS in the McGraw-Hill Encyclopedia of Science & Technology.

Martin A. Rothblatt

Bibliography. *Addendum and Corrigendum to the Final Acts of the World Administrative Radio Conference (Malaga-Torremolinos)*, 1992; Federal Communications Commission, *FCC Announced WARC-92 Strategy for Digital Audio Broadcasting*, Public Notice, October 31, 1991; International Telecommunication Union, *Radio Regulations*, vol. 1, 1990; *Proceedings of the Third Bi-Annual Symposium of Satel Conseil: Working with WARC-92*, 1992.

Reproduction (plant)

Relatively complex plants have inhabited land since the Silurian Period [at least 400 million years ago (m.y.a.)]. The fossil record reveals that the most ancient land plants reproduced in a relatively simple fashion and that there has been a pattern of increasing complexity in plant reproduction through time. During the past few years, fossil evidence has been discovered that documents the entire range of reproductive diversity now existing among vascular plants of the modern flora.

Homosporous pteridophytes. During the Middle Devonian all vascular plants were propagated by The first land plants reproduced by a single type of naked spore that was dispersed by wind and water and ultimately germinated to form the gametophyte phase.

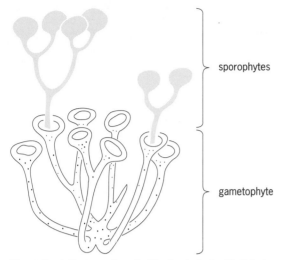

sporophytes

gametophyte

Generalized diagram of a primitive land plant with *Sciadophyton*-type gametophyte and *Cooksonia*-like sporophytes.

Plants with this type of life cycle are referred to as homosporous pteridophytes. As is characteristic of living ground pines (Lycopodiales and Equisetales) and most modern ferns, ancient land plants of the Silurian and Lower Devonian periods were homosporous. In this life cycle, sperm and eggs developed on the same gametophyte. Following fertilization, a new sporophyte grew from a tip of the branched gametophyte, as in living mosses. Therefore, the individual consisted of both the gametophyte and sporophyte phases of the life cycle (see **illus.**). Until recently, all vascular plants of this type were thought to be extinct, but some living pteridophytes recently have been recognized as having a plant body that consists of both gametophyte and sporophyte. Fossil gametophytes of the most ancient land plants are represented by the genus *Sciadophyton*. Fossil sporophytes that were probably produced on such gametophytes include *Cooksonia, Rhynia, Horneophyton,* and *Aglaophyton*.

Heterosporous pteridophytes. Through time, vascular plants acquired progressively more complex methods of reproduction. By the Middle Devonian, some plants produced spores of two sizes, megaspores and microspores. These spores germinated to produce unisexual gametophytes that were smaller but shorter-lived than the gametophytes of homosporous plants. Gametophytes of heterosporous plants also developed within the wall of the spore and therefore were more likely to survive and reproduce than were the unprotected gametophytes of their homosporous ancestors. Because fertilization in heterosporous plants involves sperm and egg from different gametophytes, there is likely to be more genetic diversity than that found in homosporous plants. However, since fertilization in heterosporous plants occurs only if the sperm swim to the egg through a film of water, these plants are largely restricted to moist or aquatic habitats. Living examples of heterosporous pteridophytes include the lycophytes *Selaginella* and *Isoetes* as well as several genera of ferns.

Origin of seed plants. During the Middle Devonian all vascular plants were propagated by naked spores. Homosporous species had restricted genetic variability and vulnerable gametophytes, whereas heterosporous species had restricted habit tolerances because they required environmental water for successful fertilization. However, by the Upper Devonian (about 350 m.y.a.) some heterosporous plants no longer shed naked megaspores. Instead, the megagametophytes and embryos were dispersed while still contained within the protected environment of the intact megasporangium. Fertilization and embryo development also occurred within the megasporangia. As a result, plants of this type were not restricted to the narrow environmental limits of their free-sporing heterosporous ancestors. Somewhat later, the megasporangia of such plants became enclosed within an integument (or seed coat). The resulting structure was a seed. The evolution of seed plants allowed for colonization of a much larger percentage of the land surface, and consequently led to a dramatic increase in the diversity of terrestrial plants during the early part of the Mississippian Period (about 340 m.y.a.).

Advanced seed plants. By the end of the Pennsylvanian Period (325 m.y.a.), several groups of seed plants had evolved, and these in turn dominated much of the land flora. In some groups, fertilization was facilitated by pollen tubes, animals aided in pollination and seed dispersal, and the earliest forms of seed dormancy developed. All of these features are considered to be specializations to ensure greater reproductive efficiency; seed dormancy is of particular significance because it allows groups of seed plants such as conifers to colonize and establish forests in habitats that are only periodically moist enough for successful seed germination and seedling establishment. Throughout the Permian, Triassic, and Jurassic periods (approximately 140–280 m.y.a) several new groups of seed plants evolved, and these were accompanied by the evolution of increasingly complex reproductive mechanisms. Examples are the Glossopteridales, Caytoniales, and Pentoxylales. Fossil flowerlike structures and pollen of Cycadeoidales (=Bennettitales) and Gnetales indicate that the nature of plant–animal interactions in pollination biology and seed dispersal had become more and more sophisticated throughout this range of geological time.

Flowering plants. Undoubtedly, the most remarkable increase in reproductive sophistication is found among the angiosperms (flowering plants), which dominate the land flora today. Although unquestioned fossil evidence for flowering plants does not appear until the beginning of the Cretaceous Period (about 140 m.y.a) the group must have evolved somewhat earlier. Some researchers suggest that flowering plants may have originated as early as the Triassic Period. Flowering plants show a large suite of interrelated features that contribute to their wide distribution in diverse habitats and to their great diversity (250,000–300,000 living species).

Although the relative contributions of each character to the success of the group is difficult to assess, three features are of particular significance: sophisticated plant–animal interrelationships, enclosure of the seeds within the ovary, and herbaceous growth forms. The plant–animal interrelationships provided food or shelter for the animals, and dramatically improved the probability of pollination, fertilization, and seed dispersal in plants. Fossils document a coevolution of interrelated flower types and insect groups. In some instances, the interrelationships are so specialized that neither the plant nor the animal species can survive without the other. Enclosing seeds within an ovary has long been suggested as providing protection, but it is now known to also regulate outcrossing and inbreeding and to increase natural selection for vigorous varieties. Among living seed plants, herbaceous growth forms are found only among the angiosperms. Thus, herbaceous species can grow and reproduce exceedingly rapidly, sometimes in as few as 6 weeks. As a result, many weedy flowering plants can rapidly multiply in number and occupy large areas within a short period of time.

For background information SEE PALEOBOTANY; REPRODUCTION (PLANT) in the McGraw-Hill Encyclopedia of Science & Technology.

Gar W. Rothwell

Bibliography. E. M. Friis, W. G. Chaloner, and P. R. Crane (eds.), *The Origins of Angiosperms and Their Biological Consequences*, 1987; E. M. Gifford and A. S. Foster, *Morphology and Evolution of Vascular Plants*, 1987; W. N. Stewart, *Paleobotany and the Evolution of Plants*, 1983.

Reptilia

The evolution of herbivory among reptiles between 310 and 270 million years ago (m.y.a.) had a critical impact on the history of life on land; it intensified and diversified the interactions among plants and animals, with significant consequences for their evolution as well as ecology. Increasing knowledge of the ecology and evolution of early reptilian herbivores and a better understanding of the physiological and ecological rules that now govern the life of herbivores suggest an explanation for this event and what it meant for the future of life on land.

The chronicle. The oldest fossils representing reptilian herbivores date to Late Carboniferous in the geological chronicle (about 310 m.y.a.), although land plants were abundant 40 m.y. earlier. They occur in small numbers in fossil-bearing rocks from Europe and North America that represent primarily wet, lowland habitats of late Carboniferous and early Permian age (310–270 m.y.a.). There they are associated with an abundance of large carnivorous reptiles, terrestrial and aquatic invertebrates, small insectivorous terrestrial amphibians and reptiles, and predatory aquatic amphibians. In contrast, reptilian herbivores are abundant and diverse in younger sedimentary rocks from western North America, eastern Europe, and southern Africa that date to late Permian (270–250 m.y.a.) and represent relatively dry, subtropical and warm-temperate habitats occupied by relatively open wood and shrubland. In these younger rocks,

their remains are much more abundant than those of large carnivores and of other amphibians and reptiles.

The early herbivores are defined as reptilian by shared primitive features, even though they are only distantly related to turtles, lizards, snakes, and crocodiles. They are recognized as herbivores from the form of their teeth and jaws, which appear more appropriate for cropping and breaking tough plant tissues than for stabbing and tearing flesh. The adults were relatively large with bulky, barrel-shaped bodies, suggesting that they had capacious guts particularly suitable for bacterial fermentation of digestion-resistant plant materials.

Differences in functionally comparable structures, such as teeth, indicate that the critical aptitudes for herbivory evolved independently in at least five distinct groups of early, primitively insectivorous reptiles at least three different times. In addition to these specific aptitudes, early reptilian herbivores share a variety of other characteristics. They have disproportionally small heads, short necks, and short, powerful legs that demonstrate, along with their overall size, that they must have lived on the ground as adults. Other skeletal features indicate that they had relatively low metabolic levels for their size. They required only low rates of energy income and had relatively low capacities for acquisition of food, regulation of disturbance, reproduction, and growth. Thus, they were sluggish, cold-blooded animals comparable physiologically to modern, herbivorous lizards and tortoises.

A large proportion of plant growth in the warm, wet forests occupied by early reptilian herbivores was concentrated in the crowns of the trees, out of reach of the large, squatty, short-necked herbivores. In addition, the threat from large predators would have been severe, and the loss of eggs and hatchlings to small predatory reptiles and amphibians great. The abundance of herbivorous reptiles would thus have depended on the presence of dense patches of shrubby plants at forest edges to provide food and hiding places. The success of these reptiles would have varied with the abundance and size of such patches.

Processing plants. Exploitation of plant tissues is difficult because the greater proportion consists of fiber-rich materials that resist digestion and tend to be low in essential nutrients, particularly protein. Cellulose, the principal constituent of high-fiber tissues, presents a particular challenge, since it is essentially indigestible for animals. In addition, its presence in the cell walls of plants limits access to the digestible, nutritious components within the cells.

Capacities for processing cellulose are largely a function of its consumption by symbiotic bacteria living within the herbivore's gut. These microorganisms ferment cellulose, releasing volatile fatty acids and producing digestible carbohydrates for their own growth. They also convert nitrogenous wastes into protein and buffer or break down potentially toxic plant compounds. The host animal in turn converts most of the volatile fatty acids into usable carbohydrates, as well as digesting a high proportion of these carbohydrates and proteins along with the bacteria that produced them.

The yield from fermentation varies, first, with the size of the food particles in the gut; second, with the duration of fermentation; and third, with the average amount of material fermented per unit time. The first is a function of the effectiveness of the teeth and jaws in chewing; the second and third, of the size of the gut. *SEE DIGESTIVE SYSTEM.*

Evolution of herbivory. The chronicle reveals a common pattern in the physiology and ecology of early herbivores, one that transcends the genealogical differences among them. The pattern includes features that facilitated cropping and crushing fibrous materials and, in most cases, the fermentive processing of plant cellulose by symbiotic bacteria. It also includes characteristics indicating low metabolic levels, thus minimizing requirements for energy and nutrients. The chronicle suggests habitation of warm, wet-lowland forests where food and refuges were rare, and predation intense. Since this pattern transcends inheritance, it must reflect the operation of the basic rules of herbivory in Carboniferous and Permian terrestrial communities, and those rules must explain the evolution of the pattern.

The evolution of herbivory in early reptiles required a transformation in aptitudes from those requisite for consumption of insects and other small invertebrates to ones appropriate for utilization of cellulose-rich plant tissues. A reasonable deduction is that this evolution began with addition of low-cellulose, high-protein items such as seeds to a diet consisting primarily of invertebrates, a shift recorded only by variation in tooth wear. The next stage was probably a shift from omnivory to primary dependence on those low-cellulose, high-protein items, a shift revealed in functional aptitudes for cropping plant tissues The third stage involved the addition of capacities for fermentation with a relatively larger gut, as indicated by greater size and a disproportionally bulky body. A fourth stage included enhanced aptitudes for cropping and crushing fibrous cellulose-rich tissues along with higher metabolic levels and greater abundance and diversity. Such evolutionary sequences are consistent with the available evidence. Further, since this sequence occurred in a considerable number of different groups at rather different times, it seems to have been a relatively frequent and therefore likely event.

On the other hand, the ecological and evolutionary success of these ventures into herbivory was incomplete. No venture appears to have succeeded before the late Carboniferous, or resulted in abundance or diversity before the late Permian. The pattern of failure as well as success indicates a shortage of resources and limited capacities for acquiring food and reproducing. The supply of low-cellulose, high-protein items was very small in the habitats of the early herbivores. This situation allowed the existence of only a small number of herbivores representing only a few species, and it favored animals and species that could utilize, at least as adults, a wide variety of plant tissues (particularly the relatively abundant, high-fiber, high-cellulose

items) and could maintain and replace themselves at relatively low cost. These requirements explain, in turn, the emphasis in early herbivore evolution on aptitudes for collecting and processing high-fiber tissues and on low metabolic levels. They also explain the correlation of herbivore evolution, abundance, and variation with the proportion of woodland habitats where sites rich in food were larger and more widely distributed.

Exposure to intense predation in the face of severe food shortages reinforced these evolutionary tendencies. Predation must have been a critical determinant for body size since large size reduces the risks of attack. Large size in association with adult longevity also facilitated production of a large number of eggs and young, even where the possible investment in reproduction was small.

Finally, early herbivorous reptiles had only relatively limited capacities for searching and collecting food and for mechanical processing of cellulose-rich tissues. These factors reinforced the impact of the overall shortage of food and refuges and of intense predation. The effect would have been particularly severe on smaller animals, which would have had relatively high rates of expenditure and would also have had relatively low capacities for regulation of disturbance.

This explanation for the pattern of early reptilian herbivory through the early Permian is tested and confirmed by the late Permian chronicle. The shift to predominantly drier, open woodland habitats provided a larger supply of food, which could support a larger number and variety of herbivores. This shift brought predator abundance into balance with herbivores, reducing the constraint set by predation. Finally, the appearance of improved mechanisms of food collection and mechanical processing in some groups of herbivores allowed smaller body sizes and higher levels of metabolism. With these changes herbivorous reptiles assumed the primary role in the transfer of energy and nutrients in terrestrial communities.

For background information *SEE HERBIVORY; PALEO-ECOLOGY; PALEONTOLOGY; REPTILIA* in the McGraw-Hill Encyclopedia of Science & Technology.

Richard Beerbower

Bibliography. R. Beerbower, *Looking Backward: Life on Land*, 1993; A. K. Behrensmeyer et al., *Terrestrial Ecosystems through Time: Evolutionary Paleoecology of Terrestrial Plants and Animals*, 1992.

Rice

Rice (*Oryza sativa*) is the staple food for almost half of the world's population. During 1991, rice occupied 3.6×10^8 acres (1.47×10^8 hectares) of land with a production of 5.7×10^8 tons (5.18×10^8 metric tons). To keep pace with population growth, an additional 1.1×10^8 tons (1×10^8 metric tons) will be needed in the year 2000. Rice production will have to maintain the 3% average annual growth rate achieved over the last 20 years. New varieties will be needed for each agroclimatic zone of rice cultivation. Some of the desired features of future varieties are superior grain quality, higher yield potential, enhanced resistance to pests and diseases, and greater tolerance for such stresses as drought, cold, and nutrient deficiencies. Biotechnology is perhaps the most important new resource in meeting this challenge.

Rice biotechnology techniques come from plant tissue culture and molecular biology. Two tissue culture techniques, embryo rescue and anther culture, have already made important contributions. Embryo rescue enables breeders to attempt wide crosses between varieties that could not be hybridized before, while anther culture allows breeding lines to be stabilized faster. Molecular techniques help to accelerate traditional breeding programs through gene tagging and the streamlining of germ plasm management, and to assess population dynamics in pests and pathogens through deoxyribonucleic acid (DNA) fingerprinting. However, biotechnology's most novel contribution will probably be in adding alien genes to the rice gene pool through genetic engineering.

Embryo rescue. Many wild species of the genus *Oryza* fail to hybridize with cultivated rice because of differences in chromosome number or genetic constitution. Fertilization may occur, but the embryo is aborted. Until recently, this phenomenon prevented the transfer into cultivated rice of many useful traits such as disease and pest resistance and tolerance of abiotic stresses. It is now possible, however, to rescue hybrid embryos that contain extra chromosomes and maintain them through several cycles of backcrossing and tissue culture until their chromosome number declines to that of cultivated rice [diploid set (2N) = 24] and fertility is restored. The 12 chromosomes of the haploid set (N) of these new lines closely resemble the chromosomes of the cultivated parent but have small regions replaced by segments of the wild genome through random crossover events.

Anther culture. Repeated self-fertilization, or selfing, of a hybrid can also generate new varieties. At least five cycles of selfing are required to produce stable lines. In geographical areas where only one generation per year is possible, anther culture is advantageous because homozygous plants can be produced in two generations.

In this technique anthers are removed from the hybrid floret, sterilized, and placed on a suitable culture medium. A haploid callus develops from the male gametophyte and spontaneously diploidizes to form a dihaploid that is homozygous at every locus. The dihaploid plant's seeds are usually viable and give rise to plants that can be evaluated immediately. Early generation selection for recessive traits is a major advantage of plants derived from anther culture compared with plants obtained by conventional sexual crossing. Cold tolerance is one of the traits introduced into cultivated indica varieties of rice by this method.

Traditional marker-aided selection. Breeders practice traditional marker-aided selection whenever possible. This technique is used when an important trait that is difficult to assess is tightly linked to a trait that is easily measured. For example, a gene for resistance to the insect pest brown plant hopper is close to a gene specifying the color purple for the coleoptile in some traditional rice varieties grown in northeast India. In this case, coleoptile color is a morphological marker aiding the selection of brown plant hopper resistance. Unfortunately, few morphological markers are known and they tend to be specific for particular rice varieties.

Molecular marker-aided selection. The advent of molecular markers has added enormously to the power of marker-aided selection. Isozyme and DNA markers are more common than morphological markers, are available for any cross, and are codominant, that is, both parental markers can be observed in the hybrid. Additionally, the environment does not affect them, and they do not interact with other genes. They are, however, more difficult to measure than morphological markers.

The most commonly used DNA markers are restriction-fragment-length polymorphisms (RFLPs). To measure an RFLP marker in an individual plant, DNA must be extracted from the plant, transferred to a test tube, and cut into characteristic fragments by using enzymes known as restriction endonucleases. Agarose gel electrophoresis and DNA-DNA hybridization are then used to determine the length of the restriction fragment derived from a particular locus of the rice genome. The locus is specified by the cloned segment of rice DNA used as a hybridization probe. The lengths of restriction fragments often differ between varieties because of mutation. This difference can be used as a marker in much the same way as a difference in coleoptile color. The unlimited number of RFLP markers has allowed the saturated coverage of the genome with markers and the construction of an RFLP map of the 12 haploid chromosomes of rice.

Gene tagging. An important application of an RFLP map is to locate genes of agronomic significance relative to the nearest DNA marker. More than 20 genes have already been tagged with DNA markers, including several for resistance to the rice blast fungus, and major genes for resistance to bacterial leaf blight and to the insects white-backed plant hopper and gall midge. Mapping results can be immediately applied to breeding programs.

It is also possible to tag genes without reference to a map. A technique called bulked segregant analysis is particularly powerful for tagging a major gene, and can be used with RFLP markers or with randomly amplified polymorphic DNA markers. It is highly unlikely, however, that the first RFLP or randomly amplified polymorphic DNA marker shown to cosegregate with an important gene will prove to be the closest marker. A map provides a logical framework for finding closer markers.

Germ plasm management. The collection, conservation, characterization, and dissemination of rice germ plasm provide plant breeders with new raw material for their programs. Molecular markers are used to address questions about germ plasm management, including the genetic diversity, classification, and phylogeny of rice accessions.

DNA fingerprinting. Many of the insect pests and viral, bacterial, and fungal pathogens of rice show considerable geographical and temporal diversity. This diversity is usually exhibited in differences in virulence and host range. Breeders, entomologists, and pathologists screen germ plasm to find new donors of resistance. By applying RFLP and randomly amplified polymorphic DNA analysis to the genomes of rice pests and pathogens, scientists are beginning to see patterns of relationships or lineages that are likely to be important to the eventual understanding of how pests and pathogens evolve and how new resistant varieties of rice might best be deployed.

Protoplast transformation. Alien genes extracted from organisms that cannot hybridize with rice can be introduced through genetic engineering. This process also permits the reintroduction of rice genes that have been extracted and modified to give altered properties. Such gene transfers are impossible with conventional breeding methods.

Several laboratories can produce transgenic rice plants via the uptake of DNA into protoplasts (plant cells freed of their cell walls by enzymatic digestion). Rice is the first agronomically important monocot to be transformed from protoplasts to yield viable, fertile plants. Protoplasts prepared from cells from a suspension culture appear to give the best regeneration frequencies, with japonica varieties responding better than indica varieties.

Biolistic transformation of rice. Many alternative protocols of rice transformation have been reported, but they have not yet proved as successful as the protoplast-based methods. The most interesting protocol is the biolistic approach, in which microscopic gold particles coated with DNA are fired at calluses or intact plant explants such as embryos. The particles may be accelerated by gunpowder, compressed helium gas, compressed air, or electric discharge.

The gold particles penetrate several cell layers, coming to rest either in extracellular spaces or inside the cells. In the latter case, DNA deposited inside the cell may become integrated into the chromosomes of the host cell and stably transform the cell. Other cells derived from the transformed cell by subsequent cell divisions are also transgenic. If the descendant lineage of the transformed cell includes cells responsible for production of pollen or eggs, the foreign gene will be transmitted to the next generation. Otherwise, the transformed plant is merely an unproductive chimera.

The biolistic method of transformation is new and needs to be improved with respect to tissue bombardment and the means of ensuring a high frequency of genetic transmission of the foreign gene.

Marker genes and promoters. The first foreign genes expressed in rice plants were bacterial genes conferring antibiotic or herbicide resistance. When the appropriate antibiotic or herbicide was pres-

ent, these selectable marker genes permitted the few transformed cells to grow while nontransformed cells died. Reporter genes, such as the bacterial β-glucuronidase gene, have been used to optimize the transformation protocol and to study the properties of some plant promoters that may eventually be used to control the expression of useful genes in transgenic rice.

A promoter is a segment of DNA that precedes a gene and controls its activity by instructing the enzyme ribonucleic acid (RNA) polymerase as to where, when, and how often to begin synthesis of messenger RNA. Promoters are diverse in their functions. Some allow activation of their gene only in a certain type of cell, at a particular stage of plant development, or in response to a specific external signal. Other promoters allow permanent expression of genes in a wide range of cell types.

Useful foreign genes. Relatively few useful foreign genes have been introduced into rice to date because reliable transformation protocols have only recently emerged. Several genes encoding potentially insecticidal or fungicidal proteins are currently being introduced.

In the early 1980s, scientists showed that tobacco plants could be protected against tobacco mosaic virus (TMV) infection if the plants had been previously transformed with a virus gene that encoded the tobacco mosaic virus coat protein. Although scientists do not fully understand the mechanism of coat protein cross-protection, they have reported it also for coat protein genes of several other positive-strand RNA viruses, including rice stripe virus. A similar approach might be effective in protecting rice against its two major viruses, Asian tungro and American hoja blanca.

Among the selectable marker genes employed with rice is the *bar* gene for resistance to the herbicide phosphinothricin. In principle, this gene could be utilized to develop herbicide-resistant rice varieties for areas where direct seeding leads to competition with weeds. Such varieties, however, are unlikely to be released because cross-pollination might allow the herbicide resistance gene to escape into local populations of weedy and wild rices, thus negating the original strategy. It may be necessary to place such genes in the DNA of the chloroplast, which is not transmitted through pollen. Biolistic transformation of chloroplasts has already been reported for the green alga *Chlamydomonas* and for tobacco. SEE HERBICIDE.

For background information SEE BIOTECHNOLOGY; BREEDING (PLANT); GENETIC ENGINEERING; HERBICIDE; RICE in the McGraw-Hill Encyclopedia of Science & Technology.

John Bennett; Gurdev S. Khush

Bibliography. P. Christou, T. L. Ford, and M. Kofron, Production of transgenic rice (*Oryza sativa*) plants from agronomically important indica and japonica varieties via electric discharge particle acceleration of exogenous DNA into immature zygotic embryos, *BioTechnology*, 9:957–962, 1991; J. E. Leach and F. F. White, Molecular probes for disease diagnosis and monitoring, in G. S. Khush and G. H. Toenniessen (eds.), *Rice Biotechnology*, 1991; S. R. McCouch et al.,

Molecular mapping of rice chromosomes, *Theor. Appl. Genet.*, 76:815–829, 1988; K. Shimamoto et al., Fertile transgenic rice plants regenerated from transformed protoplasts, *Nature*, 338:274–276, 1988.

Root (botany)

Recent research on the root system of vascular plants has focused on the organization and patterns of gene expression in root apical meristems and on genetic analysis of root cell differentiation.

Organization of Root Apical Meristems

Modern techniques for growing plants and new tools of microscopy and molecular biology have enabled biologists to study roots with new vigor.

Levels of organization. The tissues in the roots of flowering plants are organized in a pattern of concentric cylinders (**Fig. 1**). The outer cylinder, the epidermis, consists of various cell types, including epidermal cells and root hairs, which protect the root from pathogen entry and water loss and absorb water and dissolved nutrients. The cortical cylinder is often several cell layers deep and usually consists of only parenchyma cells. These cells are the sites of many fundamental processes, such as storage. The vascular cylinder contains two tissues, xylem and phloem. The xylem consists of water-conducting cells called vessel members and tracheids, supporting cells called fibers, and parenchyma cells. The xylem parenchyma cells accumulate ions from the absorbed water and transport water and ions into the vessel members and tracheids. The phloem consists of sieve-tube members, which conduct sugars; companion cells, which direct the metabolic activity of the sieve-tube members; and parenchyma cells and fibers.

The next lower level of organization within cylinders and sectors comprises individual cell files. Cell files often consist of cells of a single type joined end to end. Adjacent files, which may consist of entirely different cell types, are joined by a common wall. The cells in one file may divide more frequently (and thus be short), whereas an adjacent file may consist of longer cells. The system of coordinated growth ensures that the connected files do not shear away from each other.

Thus, the sum of the processes of cell division and cell elongation in one file keeps pace with that in the adjoining file, even though the ratio of the two processes often differs.

Transition points. For convenience, the tip of the root is considered to be organized into the root cap, the meristem, the elongation region, and the maturation region. The definition of these regions designates where different processes occur. However, such boundaries do not exist, and each cell file, or group of files (such as cell files of the entire cortex), tends to act independently.

The boundaries between the region of the meristem and the region of elongation, therefore, may be different for cells of the cortex as compared to cells of the epidermis or one of the vascular sectors. This phenomenon can be explained by the concept of transition points. This concept suggests that each cell file has developmental switches, such as the point at which cell division is turned off at the basal boundary of the meristem. Thus, transition points exist at the boundary of the elongation region and at the boundary of maturation. In this way, each file acts independently. Exactly what happens at the transition points is not known, but perhaps expression of specific genes turns events on or off in a cell file–specific manner.

The position of transition points is not static; that is, as the rate of growth of a root changes, the position of the transition also changes. For example, as the rate of root growth increases, the maturation point of xylem vessel members moves farther away from the root tip, and as the root growth slows, maturation occurs closer to the root tip. Thus, the root can spatially accommodate cell division and elongation to its cell differentiation events. One observation of this relationship was made by correlating the height of the root meristem (relative position where cell division was terminated in the cortex) and the position of xylem maturation; a close linear relation existed. In roots with tall meristems, maturation of the xylem occurred farther back in the root; in roots with short meristems, maturation occurred closer to the tip.

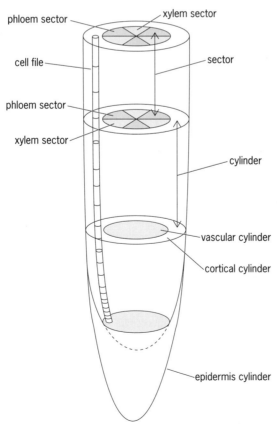

Fig. 1. Levels of root organizations. (*From T. L. Rost, in M. Iqbal, ed., Growth Patterns in Vascular Plants, Dioscorides Press, 1993*)

Types of root apical organization. An important difference between root tips of various species is in the organization of cells in the apex. In one type of organization, called closed, all cell files can be followed directly to a specific tier of initial cells. These initial cells, or histogens, may contribute to different tissues. For example, in the most common monocotyledonous type of root apex there are three initial layers—for the vascular cylinder, for the cortex/epidermis, and for the root cap (**Fig. 2a**). The most common type in dicotyledonous plants also has three layers, but for the vascular cylinder, the cortex, and the epidermis/root cap.

In the second type of organization, cell files do not connect to distinct initial layers but can be followed to a zone of cells lacking special organization. This type of organization, referred to as open since specific cell lineages do not connect to specific initials (Fig. 2b), is found in both monocotyledonous and dicotyledonous plants.

In experiments on root development done in the 1950s, it was found that a group of cells at the tip of the apical meristem divided very infrequently. This quiescent center is composed of cells that progress through the cell cycle slowly. The size of the quiescent center is variable in different plants; in corn this center is composed of approximately 330 cells, and in sunflower it has approximately 80 cells. The function of the quiescent center seems to be related to maintaining the geometry of the root tip. Also, because the cells of the quiescent center are somewhat resistant to certain stress factors, the quiescent center is thought to be a reservoir of fresh cells after acute or prolonged stress.

Gene expression. Many scientists are now studying how genes are regulated in roots and how the expression of genes relates to differentiation in root tissues and cells. Two different methods are generally used in these studies.

One method uses various agents to induce mutations in plant embryos. If the mutation is not lethal, a plant will grow, and researchers study it to determine if the mutation expresses itself in any developmental way. If it does, the mutant plant is allowed to flower, the seeds are collected, and the next generation of mutants is collected. Several root mutants have been observed in different laboratories. These mutants sometimes show deletions or modifications of developmental processes and are useful in identifying the genes responsible.

Another method isolates messenger ribonucleic acid (mRNA) from roots to make complementary deoxyribonucleic acid (cDNA), thus forming a cDNA library. These cDNA strands are inserted into bacteria and replicated. Then molecular biologists match the cDNA to mRNA molecules present in specific root cells. Several genes specifically expressed in roots have been identified. Unfortunately, most genes are expressed in shoots and roots—a predictable observation, because most developmental processes are the same in shoots and roots, only operating in different locations. Breakthrough observations are needed that will correlate

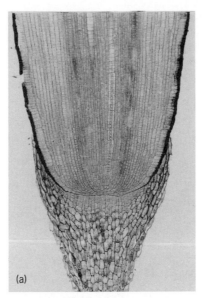

Fig. 2. Types of root apical organization. (a) Corn root tip with closed organization. (b) Pea root tip with open organization.

specific gene expression patterns with the organization levels and transition points of root tips.

Thomas L. Rost

Genetic Analysis of Root Cell Differentiation

Cell differentiation in roots occurs in a highly orchestrated manner in a region near the root apex. As new cells form in the root meristem, they become organized into files. The cells in each file differentiate in form and function from the cells in other files. The mechanisms that control root cell differentiation have not yet been defined. However, recent genetic studies have led to the identification of several genes that control the differentiation of specific root cells.

Many of the genetic studies of root cell differentiation have used a small plant from the mustard family, *Arabidopsis thaliana*. This plant is useful for developmental genetic studies because of its small size, short life cycle, and small genome. Also, the roots of *Arabidopsis* plants have a simple structure: each of the major root cell types (epidermis, cortex, endodermis, and pericycle) is in a single layer around the central vascular region, and the number of cells in two of these layers (the cortex and endodermis) is essentially invariant. These attributes have led to the isolation of many root cell differentiation mutants in *Arabidopsis*. The group of mutants that affect cell differentiation of root hairs has been investigated in the greatest detail.

Root hairs. Root hairs are tubular extensions of epidermal cells. They increase the surface area of the root and are believed to absorb water and nutrients. Root hairs also act as a site for important interactions between the plant and soil-borne microorganisms.

The genetic analysis of root hair differentiation in *Arabidopsis* is facilitated by the fact that root hairs are present on the surface of the root and are easy to observe. Root hairs are single cells that do not divide during cell differentiation. Also, root hairs are not required for the growth of *Arabidopsis* plants, so that root hair mutants of all types, even ones that lack root hairs, can be grown and studied. Mutant study has shown the process of root hair differentiation to be divided into three stages: epidermal cell patterning, root hair initiation, and root hair elongation.

Epidermal cell patterning. Roots of most plants possess two types of epidermal cells: those that form root hairs and those that do not. Although varying in different plant species, the formation of root hairs is usually determined by either cell lineage or cell position. In *Arabidopsis*, a factor controlling epidermal cell type is the position of the cell relative to cells in the underlying cortical layer. Root hairs form only from the epidermal cells that lie over an intercellular space between two cortical cells. The basis for this positional control over epidermal cell fate is not clear. Possibly, the cortical cells produce an inhibitor that is transmitted to epidermal cells located immediately above them to prevent root hair formation. Recently, mutants have been isolated that alter the normal pattern of hair-forming and non-hair-forming epidermal cells. It is therefore likely that the affected genes normally act at an early stage in epidermal cell differentiation to influence the establishment or maintenance of this pattern. These genes are probably members of a group that determines cell fate in *Arabidopsis* roots.

Root hair initiation. The first outward sign of root hair formation is the localized swelling of some of the epidermal cells. This swelling occurs at the end nearest the root apex. One of the *Arabidopsis* genes involved in root hair initiation is called *RHD1*. The epidermal cells of *RHD1* mutant plants produce an abnormally large swelling during the initial formation of the root hairs. At least two other genes in *Arabidopsis* also alter root hair initiation by causing excessive swelling of the epidermal cells when the genes are mutated. As a group, these genes seem to control epidermal cell expansion, since their mutant forms lead to a deregulation of cell expansion. Additional genes are probably required for other aspects of root hair initiation, such as the ability of an epidermal cell to form a swelling at the appropriate location.

Root hair elongation. This phase proceeds by a process known as tip growth and follows the initiation phase. Tip growth is the expansion of cells at one end only (the tip) and generally leads to tubular cells. Secretory vesicles bearing new cell wall components and enzymes involved in cell expansion are directed to the tip of the growing cell. Tip growth also occurs during the development of pollen tubes of plants and the hyphae of fungal cells.

At least three genes in *Arabidopsis*, *RHD2*, *RHD3*, and *RHD4*, are required for the tip growth of root hairs. The *RHD2* gene mutants have stubby hairs that cannot elongate. This gene is normally involved in an early step in tip growth that allows the root hair cells to expand beyond the initial swelling stage. The *RHD3* and *RHD4* genes, however, are required throughout the process. The root hairs of *RHD3* gene mutants are wavy, suggesting that this gene controls the direction of cell expansion during tip growth. Possibly, the

RHD3 gene is required for the directed transport of secretory vesicles to the root hair tip or for the proper orientation of cytoskeletal components that control directional cell expansion. The *RHD4* gene mutants possess bulging root hairs, indicating that this gene normally regulates the degree of cell expansion at the root hair tip. The *RHD4* gene may control the production of secretory vesicles or influence the stability of the growing tip. It is also possible that the *RHD2*, *RHD3*, and *RHD4* genes regulate calcium ion–related processes in the root hairs, since tip growth is known to be influenced by calcium ions.

Prospects. A major challenge in the study of root cell differentiation is understanding the nature of the gene products that have been defined by genetic analysis. A promising approach available in *Arabidopsis* is molecular characterization of the relevant genes, using genetics-based cloning strategies such as insertional mutagenesis or chromosome walking. Such studies will lead to an understanding of the sequence and expression pattern of the gene, as well as the primary structure and cellular location of the gene product.

For background information *SEE APICAL MERISTEM; ROOT (BOTANY)* in the McGraw-Hill Encyclopedia of Science & Technology.

John W. Schiefelbein

Bibliography. F. A. L. Clowes, Development of quiescent centres in root meristems, *New Phytol.*, 57: 85–88, 1958; M. Iqbal (ed.), *Growth Patterns in Vascular Plants*, 1992; N. V. Obrucheva (ed.), *Physiology of Roots*, 1973; J. W. Schiefelbein and P. N. Benfey, The development of plant roots: New approaches to underground problems, *Plant Cell*, 3:1147–1154, 1991; J. W. Schiefelbein and C. Somerville, Genetic control of root hair development in *Arabidopsis thaliana*, *Plant Cell*, 2:235–243, 1990.

Satellite navigation systems

The advent of the Global Positioning System (GPS) has greatly benefited the field of navigation and various other applied fields requiring accurate positioning. With GPS, considerable improvement in positioning accuracies over other widely available systems such as Omega and Loran can now be attained. GPS accuracy is generally considered to be about 50 m (160 ft) for the standard positioning service (SPS), available to all users; and 10 m (33 ft) for the precise positioning service (PPS), available only to authorized users. (Root-mean-square radial error, 1 drms for horizontal position accuracy, is assumed for all accuracy references.) The quest for even finer accuracies than those intended by the original designers in the early days of GPS development led to an accuracy-enhancement concept called differential GPS (DGPS). Accuracy of better than 10 m (33 ft) is now readily available with DGPS, and 1-m (3-ft) accuracy under optimum conditions appears achievable.

The areas of application for which DGPS is potentially useful are many. Numerous applications cannot be satisfied with an SPS or even a PPS solution. Finer accuracies could previously be achieved only with dedicated ground-based microwave positioning systems that are expensive to maintain and have limited coverage. For some of these and many new positioning and guidance applications, DGPS has been capable of providing a much less expensive alternative. In the area of track mapping for highways and railroads, DGPS is used to enhance the available accuracy of the database collected. Automated railroad systems are developing positioning systems with DGPS for vehicle collision avoidance functions, especially in situations involving vehicles on parallel tracks. In the marine arena, the U.S. Coast Guard Service will soon be implementing DGPS for vessel traffic service surveillance and navigation. In aviation, DGPS will provide enhanced accuracy to support traffic surveillance, collision avoidance, and GPS signal integrity assurance functions. DGPS has also been touted to potentially outperform dedicated aircraft landing systems that are more expensive to maintain, although its operational feasibility to do so is not yet verified. *SEE MARINE NAVIGATION.*

DGPS error sources. To obtain a three-dimensional position fix with independent range measurements from at least four satellites, a GPS receiver must synthesize range measurements by timing the shift of the pseudorandom noise code (modulated onto the carrier signal) from transmission to reception. The accuracy of the position fix depends on knowledge of the satellite positions, accounting of all components in the signal propagation paths, and capability of tracking the signal to form the range measurements. Inaccuracies in the broadcast ephemerides produce satellite position errors. Also, ionospheric and tropospheric refraction erroneously lengthens the range measurements. Another source corrupting the signal propagation path is selective availability (SA), a distortion of the transmission timing intentionally induced to control the SPS accuracy level available to nonauthorized users. This encrypted distortion can be entirely removed only by users with access to PPS operation. At the receiver end, both multipath and tracking noise induce measurement uncertainties in precisely locating the pseudorandom code position. Dynamic noise might also arise from model limitations in the receiver's data processing.

The concept of DGPS is predicated on the removal of errors based on strong correlations between those found along separate line-of-sight paths at two locations in proximity (spatial correlation) and at about the same time (temporal correlation). If one of the two locations is a reference station and its position is approximately known, the position of the other (user) location can be derived to a high degree of accuracy by taking advantage of the error observations made by the reference station. These error observations or corrections must be communicated from the reference station to the user's receiver via a separate data link. The accuracy of the user's position thus determined is dependent on the accuracy of the reference station's assumed position.

To attain its refined accuracy, DGPS relies on re-

ductions in the error budget of the stand-alone form of GPS. In the **table**, the error components are significantly reduced for DGPS over stand-alone GPS for the correlated elements. The amount of reduction for each of the correlated elements is principally dependent on spatial and temporal decorrelations that exist between the reference and the user and depend on the reference-to-user spatial separation and correlation data update rates, respectively. Long distances between the user and reference station result in spatial decorrelation, because the ionospheric delays and satellite position errors observed along a significantly different line-of-sight path are no longer highly correlated. To maintain high accuracy over large spatial separations of hundreds of miles, wide-area DGPS methods, discussed below, must be resorted to. Temporal decorrelation arises as a result of time-varying conditions that make the last correction received gradually less accurate until the next correction arrives. To mitigate its effects, a sufficiently high correction update rate is required.

The uncorrelated elements that dominate the DGPS error budget are multipath and dynamic noise. Multipath error becomes spatially decorrelated within even a small fraction of a mile because of its interaction with the surrounding terrain. The reduction of differential multipath can be realized in several ways. First, motion of the user vehicle largely ensures that the signal interaction geometry with the surrounding terrain changes more quickly than if the vehicle were stationary, and the resulting high-frequency error content is easily removed through filtering. Second, the use of complementary information from the tracking of the GPS carrier phase that contains precise relative changes in the pseudorange also allows efficient removal of the multipath errors and tracking noise associated uniquely with the pseudorandom code position. This process is widely known as carrier soothing. (Pseudorange refers to the measured time delay from the transmission to the reception of the signal. This measurement not only has an equivalence in spatial range but contains an element of error due to timing in the measurement process.) Third, if PPS is available, the multipath errors it encounters are reduced because its pseudorandom code

Nominal position error budget comparing stand-alone GPS to DGPS*

	Stand-alone GPS, m	DGPS, m
Correlated elements		
Satellite position	3	0
Ionospheric refraction	5	0
Tropospheric refraction	2	0
Selective availability	33	1
Uncorrelated elements		
Multipath	1	1
Tracking and dynamic noise	2	2
Range root sum square	34	2
Horizontal position error[†]	51	3

* 1 m = 3.3 ft.
[†] The horizontal dilution of precision (HDOP) of 1.5 is used as the scale factor relating the range root sum square to the horizontal position error.

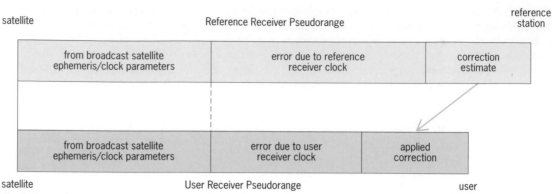

Fig. 1. Correction computation and transmission in DGPS. The pseudorange measured at the reference station includes error components that can be removed by using known information, including reference position. The correction estimate computed is transmitted for application to the user's processing of pseudorange, made at an unknown location.

has a code chip (10^{-7}s, corresponding to a propagation distance of 30 m or 100 ft) ten times smaller than that used with SPS. Dynamic noise is due to the part of a vehicle's motion (specifically, position change) that is uncertain. The reduction of dynamic noise can also be accomplished with carrier smoothing. Information on the vehicle's motion is contained in the relative change of the pseudorange that the continuous carrier phase data accurately convey.

Correction computation and transmission. Even though all the pertinent differential correction information is contained in the raw measurements made by the receiver at the reference station, it is inefficient to transmit these measurements as correction data. Instead, each pseudorange measurement is first stripped of the geometric range of the reference station to the satellite, and the range error due to the satellite clock. These parameters are derived

from broadcast satellite ephemerides. Then, an estimate of the range error due to the receiver clock is removed. Any error in that estimate corrupts all the corrections equally and gets factored out eventually as an additional time error in the user's solution; the user's position is unaffected. What ultimately remains after the pseudorange measurement noise is suppressed by filtering is the part that constitutes the differential error. This error represents the differential correction that is computed separately for each visible satellite and transmitted to users (**Fig. 1**).

The transmitted differential data convey correction and auxiliary data, health statuses, and general almanac information related to the satellites and reference station. The correction information is generally cast in a first-order dynamic model consisting of a pseudorange correction and a range-rate correction. The data can be transmitted in a variety of ways, including cellular

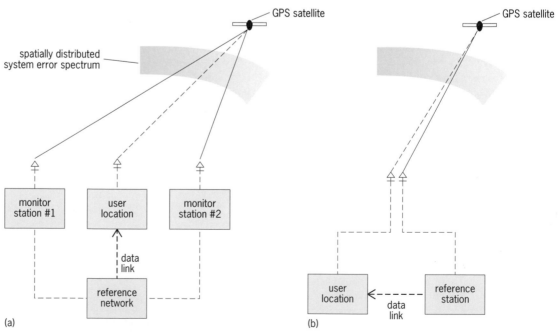

Fig. 2. Comparison of (a) wide-area DGPs and (b) local DGPS.

telephones, radiobeacons, communications satellites, and very high frequency (VHF) data links.

Wide-area DGPS. Spatial decorrelation constrains the operational range of DGPS to well below 160 km (100 mi) without incurring serious accuracy degradation. New developmental work in DGPS is focused on obtaining coverage over larger areas with a minimal number of monitoring stations arranged as a reference network. Data collected from all available stations are processed to create a composite correction solution. This concept is known as wide-area DGPS. In contrast, the single-reference-station approach is now known as local DGPS (**Fig. 2**). Whereas local DGPS monitors system errors along the line-of-sight path to a satellite, wide-area DGPS uses the reference network of monitor stations to observe the system errors of that satellite along several line-of-sight paths. This approach produces a two-dimensional spatial mapping of the errors that can then be interpolated for a user located within the operational domain of the reference network. Thus, users of wide-area DGPS can utilize correction data at locations much farther away from monitoring stations than in the local DGPS case. The observations made at the monitoring stations are processed at a central location to form a composite differential correction that is spatially varying and, hence, dependent on the user's approximate location.

INMARSAT 3 satellites will carry transponder payloads to broadcast wide-area DGPS corrections. The corrections will be broadcast at the GPS L_1 frequency and encoded in GPS-like signals. SEE COMMUNICATIONS SATELLITE.

Recent tests of wide-area DGPS have demonstrated that 2-m (6-ft) accuracy over several hundred miles of separation between reference stations and the user is achievable. Wide-area DGPS would benefit railroad activities, for instance, which cover extensive land areas and require on occasion very precise positioning for verifying a train's correct occupation of one of multiple parallel tracks. Similar situations exist for air and ocean traffic control.

Advanced techniques. Another GPS technique that could potentially achieve major advances in high-accuracy positioning is a differential-carrier-phase method commonly known as kinematic GPS (KGPS), which derives its roots from interferometric GPS terrestrial surveying techniques. KGPS is similar to local DGPS in its use of accuracy enhancement through monitoring of correlated errors at a reference location, but it uses only continuous-carrier-phase and not pseudorange measurements. Although the carrier phase is virtually noiseless, it is incapable of direct ranging. Positioning information is derived instead from estimating its time derivative, the Doppler frequency profile. Several operational constraints are thus imposed on the problem. Most important is an initialization process that can take several minutes of data collection to obtain an accurate solution. Also, if the user is not stationary during initialization, six or seven satellites are needed for optimum operation. Continuous carrier tracking must also be maintained throughout the entire process, imposing fairly severe requirements on the receiver to track the carrier signals reliably and to be able to detect cycle slips when they do occur.

Even with these critical constraints, the promise of decimeter-level accuracy (1 dm = 4 in.) that KGPS affords is leading to intensive research into overcoming them. Although real-time KGPS is being considered as a potential candidate for automatic aircraft landing (autoland), it is at least several years away from operational feasibility. For the present, work on autoland with local DGPS seeks to enhance its accuracy and reliability with the help of other aids such as the radar altimeter and the existing nonautoland instrument landing systems (ILS).

For background information SEE ALTIMETER; INSTRUMENT LANDING SYSTEM (ILS); SATELLITE NAVIGATION SYSTEMS in the McGraw-Hill Encyclopedia of Science & Technology.

Patrick Hwang

Bibliography. P. Y. C. Hwang, Kinematic GPS for differential positioning: Resolving integer ambiguities on the fly, *J. Inst. Navig.*, 38(1):1–16, Spring 1991; C. Kee, B. W. Parkinson, and P. Axelrad, Wide-area differential GPS, *J. Inst. Navig.*, 38(2):123–145, Summer 1991; Radio Technical Commission for Maritime Services Special Committee No. 104, *RTCM Recommended Standards for Differential Navstar GPS Service*, version 2.0, January 1, 1990.

Sensation

Answers to questions of how the brain combines sensations and how it generates the unified experiences of the world are emerging from research that tests people's ability to attend selectively to certain information while ignoring other information. From this research, scientists have isolated a set of rules that the mind uses in processing and combining various sorts of information, from simple sensations to complex meanings.

Perception of wholes. Psychologists often describe human experience as holistic. The world of human experience abounds with wholes; for example, the sounds from instruments in a symphony blend together effortlessly. Separate experiences of sight, sound, and touch meld together, rendering a consistently uniform impression of the world outside the body. Usually, people take for granted the uniformity and coherence of their experiences, but how the brain accomplishes these feats is still unknown.

An appreciation of the brain's accomplishments can be gained by observing the damaged brain. People suffering from certain brain syndromes, such as disjunctive agnosia, experience their world as broken into parts, rather than as blended smoothly into wholes. They might see numerous colors littering a beach but not identify the colors as belonging to beach towels; they might hear someone speaking but not realize that the voice emanates from the person whose lips are moving.

Sensory dimensions. After physical energy reaches the eyes, ears, or skin, just before a perceptual

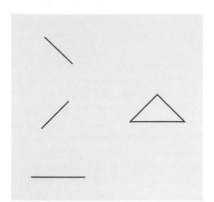

Fig. 1. Shape as an emergent property of the dimension of line orientation. Three lines at different orientations are assembled into a triangle.

experience occurs, the brain reacts vigorously to simple elements, and only later does it react to complex wholes. Certain cells in the primary visual cortex respond to whether the electromagnetic energy is intense or weak and to where it falls on the eyes' retinas. Certain cells in the auditory parts of the brain fire according to whether the mechanical energy is of a high or low frequency. These simple elements are expressed as sensory dimensions. Sound pitch, for example, is one sensory dimension, a specific pitch being one value on that dimension. Human perceptions represent combinations or collages of values on sets of sensory dimensions. The perception of a fire alarm, for example, requires the brain to combine values on the dimensions of pitch, intensity, location, timbre, and duration.

At present, brain-recording techniques have not been sufficiently developed to address complex issues, such as how these combinations occur or whether the brain treats all dimensions equally. In the meantime, scientists are relying on sophisticated, yet inexpensive and noninvasive, techniques that measure how humans perceive stimulation to the senses.

Many of these techniques are based on the concept of selective attention. Subjects are asked to respond to stimulation from one sensory dimension while ignoring stimulation from a different one. Scientists measure how well subjects can attend to the relevant stimulation while ignoring the distracting information. The task is similar to that of a student who tries to complete a homework assignment with the television on.

Dimensions that interact. In a typical selective-attention experiment subjects might be asked to decide whether sounds are loud or soft and to ignore whether they are high- or low-pitched. The experimenter presents each sound to each subject one at a time. The subjects are instructed to press one key if they hear the louder sound and another if they hear the softer sound. The experimenter measures the reaction time of the subjects. In one condition, called baseline, each sound may be loud or soft but every sound has the same pitch. In another, called filtering, each sound may be loud or soft and either high-pitched or low-pitched.

Selective attention is measured by comparing the subjects' average reaction time at baseline with their average reaction time in filtering. If filtering and baseline speeds are equal, selective attention is perfect. That is, concentrating on one dimension is not disrupted by changes along the other dimension. If filtering speed is slower than baseline speed, however, selective attention has failed; the slower speed implies that the brain has combined information from the two dimensions. Such is the case with loudness and pitch. Sensory dimensions such as loudness and pitch are called interacting dimensions. Sensory dimensions that lead to perfect selective attention are called separable dimensions. Color and shape exemplify separable dimensions, since subjects can attend perfectly to shapes (deciding, for example, between circles and squares) without being distracted by the shape's color.

Combining separate dimensions. Of course, even values from separable dimensions are combined eventually by normal brains. For example, in identifying the color of a chair, color information and shape information are joined. This fact implies that results from selective-attention experiments tap a level of brain activity that precedes holistic experience. Moreover, the fact that some dimensions interact while others do not implies that interacting dimensions activate similar brain regions or systems, whereas separable dimensions activate distinct brain regions. Results

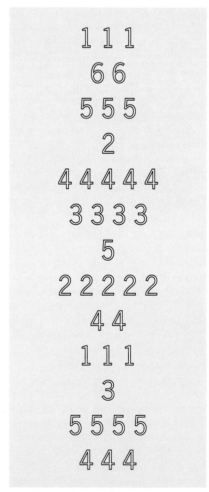

Fig. 2. Dimensions (digit shape and digit quantity) that interact at the level of meaning.

of recent studies on brain recordings support this idea: shape information and color information, for example, are processed by different brain systems.

Dimensions that emerge. Research on selective attention has revealed many other facts about sensory combination. For example, when some sensory dimensions interact, they create new sensory dimensions. It is known that some cells in the primary visual cortex are activated by stimulation from lines in different orientations. Three lines are depicted, each at a different orientation, in **Fig. 1a**. Line orientation is a basic sensory dimension. However, if the three lines are assembled, as in Fig. 1b, a new dimension of shape emerges. Different brain cells are attuned to values along the shape dimension. Shape is a relatively simple dimension, but it is created when the brain combines values on even simpler dimensions.

Dimensions that correspond. Some dimensional interactions suggest very complex brain processes. Much research has explored what happens when values on one dimension into values from another dimension (**Fig. 2**). The number of digits on each line is to be counted as fast as possible from top to bottom. Most people do this relatively slowly; they often make mistakes by naming the digit they are supposed to be counting. When the task is to name the digit on each line while ignoring how many there are, most people execute the task more quickly and more accurately. In general, people's selective attention to the digits succeeds, but their selective attention to the quantity of digits fails. Information in this case interacts at a very high level indeed, namely, at the level of meaning. Moreover, many such high-level interactions are based on past experiences.

Associations between senses. Other relationships are probably unlearned. If asked to decide which shape in **Fig. 3** is called MALOOMA and which is called TAKEETEE, most people will not decide arbitrarily. Humans naturally connect certain sights and sounds. For most people, the staccato-sounding TAKEETEE matches the jagged shape on the left, while the smooth-sounding MALOOMA matches the rounded shape on the right. Such associations indicate how information is combined between senses. It is known, for example, that when deciding whether a color is white or black, subjects are distracted if high- and low-pitched tones are sounded. People tend to asso-ciate white with high pitch and black with low pitch. Apparently, the brain systems that combine values across different sensory dimensions link the values in particular ways.

For background information SEE PSYCHOPHYSICAL METHODS; SENSATION in the McGraw-Hill Encyclopedia of Science & Technology.

Robert D. Melara

Bibliography. M. Livingstone and D. Hubel, Segregation of form, color, movement, and depth: Anatomy, physiology, and perception, *Science*, 240:740–750, 1988; R. D. Melara, Dimensional interaction between color and pitch, *J. Exper. Psychol. Human Percept. Perform.*, 15:69–79, 1989; J. R. Pomerantz, Visual form perception: An overview, in E. C. Schwab and H. C. Nusbaum (eds.), *Pattern Recognition by Humans and Machines: Visual Perception*, 1986; A. M. Treisman, Properties, parts and objects, in K. R. Boff, L. Kaufman, and J. P. Thomas (eds.), *Handbook of Perception and Human Performance*, vol. 2, 1986.

Shrimp

The eyes of marine animals differ in a number of ways from those of terrestrial organisms. Many specializations in the eyes of marine animals are directly attributable to the way that sunlight is distributed in the water column. The eyes of shrimps in the oceanic mesopelagic zone (a region extending over a depth of 990–3300 ft or 300–1000 m) demonstrate subtle adaptations for functioning in the light conditions of the marine world.

In the ocean, light is much more unidirectional than on land. On land, the objects receive lateral illumination from sunlight bounced off reflective surfaces in the environment where in the ocean there is little illumination from the sides and even less from below. In addition, with increasing depth, the intensity of light diminishes rapidly and the full range of colors is not available because light becomes increasingly monochromatic. Objects in the mesopelagic zone are difficult to see, and the eyes of oceanic species have become adapted accordingly.

Underwater light field. Light entering the sea from the atmosphere is refracted according to Snell's law so that light rays become more vertical below the surface. Just below the surface the angular distribution of light is complex, unlikely to be symmetrical, and dependent upon the Sun's elevation and the occurrence of waves. However, with depth the light field becomes increasingly symmetrical with respect to the vertical axis, so that at depths exceeding about 100 ft (30 m) the intensity of light at a particular angle to the vertical is similar in all directions irrespective of the Sun's angle and the surface conditions.

As light penetrates the ocean, radiance levels decline exponentially because of absorption and scattering. Quanta toward the red end of the spectrum (longer wavelengths) are preferentially absorbed by water molecules, whereas those toward the ultraviolet end (shorter wavelengths) are preferentially scat-

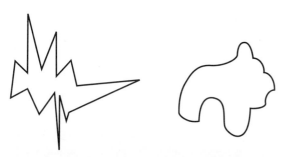

Fig. 3. Two shapes that demonstrate unlearned associations between sight (of the figures) and sound (of the names TAKEETEE and MALOOMA). (*After W. Köhler, Gestalt Psychology, Liveright, pp. 224–225, 1947*)

tered. Consequently, not only does the light decline in intensity with depth, but it becomes progressively restricted to a relatively narrow range of wavelengths extending from about 430 to 530 nanometers, with the wavelength of peak transmission at about 475 nm. Since all sunlight comes from above, it follows that the upwelling light must arise from backscattering. The levels of upwelling radiance is approximately 1/200 of the downwelling radiance because backscattering is a relatively inefficient process. Between these two extremes (vertically downward and upward) there is a characteristic gradient of natural radiance levels that varies with the angle between the vertical and the line of sight.

Reflecting superposition eyes. Shrimps have superposition eyes (**Fig. 1**) that form images upon the receptor cell layer according to the corner reflector principle. Behind each square corneal facet is a mirror box that is twice as long as it is wide. The outer and inner ends of each mirror box are clear; the content of the box is of low, uniform refractive index; and the four sides contain multilayer interference reflector mirrors.

When parallel rays from a distant source enter the eye via neighboring facets, the mirror boxes reflect the light so that the rays are focused on the underlying photoreceptive rhabdoms of the receptor cells. Because of the hemispherical geometry of the eye, most rays do not pass through the center of a box but strike the sides. On average, a ray is reflected first off one side and then off the adjacent side in the process of being redirected onto the receptors (Fig. 1b). Such a corner reflector behaves like a flat mirror at right angles to the plane of the ray. However, like a ship's reflector that irrespective of its orientation redirects radar waves toward the radar source, each mirror box redirects parallel rays to the same point in space even when the eye moves relative to the object being viewed.

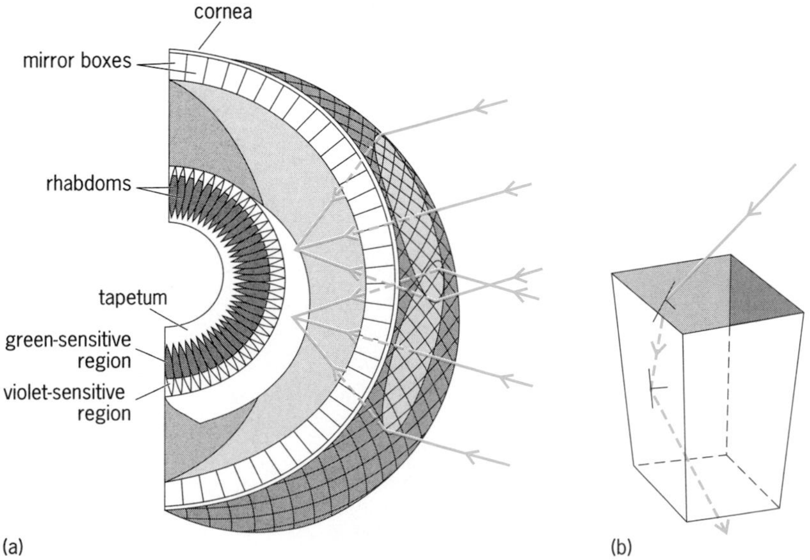

This mirror-box effect enables the eye to form focused images of objects in all parts of the visual field. Thus, light entering many facets in the shrimp eye is focused at a single point on the receptor cell layer, making the eye particularly sensitive.

Graded patterns of tapetal reflection. The eyes of many animals that are active in dim light contain a reflective tapetum behind the receptor cells (Fig. 1a). Such tapeta improve the efficiency of the eye by effectively doubling the light path through the retina and increasing the chances of photon capture by the receptor. Photons not absorbed by the rhodopsin photopigment on their first passage through the receptor are reflected into it for a second chance at absorption.

Mesopelagic shrimps possess tapeta that like a cat's eye reflect bright light toward the observer. Such reflected light, known as eyeshine, has the disadvantage of making the shrimps visible to predators. If the tapetum is very efficient, it will increase sensitivity without producing bright eyeshine when illuminated. However, the tapetum is needed to enhance the shrimps' visual sensitivity in the dim light. Consequently, the brightness of tapetal reflection is determined by the need to see without being seen.

When the eyeshine patterns of a number of species of oplophorid decapod shrimps are examined, certain consistent features emerge. First, eyeshine is graded over the eye with a clear gradient of increasing brightness from the upwardly looking parts of the eye to the downwardly looking parts. In species such as *Systellaspis debilis* these features are particularly clear (**Fig. 2**). This distribution of eyeshine is the converse of the radiance distribution in the sea: those parts of the eye looking downward at the dimmest part of the environment have the most reflective tapetum, thus helping the eye to discriminate dim shapes against a very dim background.

There is also a gradient of reflectivity in the horizontal axis of the eye so that the forward-looking parts of the eye have much brighter eyeshine than the parts that look behind the shrimp. In this case, the region of the eye concerned with prey capture has a better tapetal reflector and therefore increased sensitivity. The reason for such gradients is almost certainly the need to avoid being seen by predators. Upward-looking parts of the eye with an efficient tapetum would be particularly visible to predators above the shrimp because the eye would appear bright against the dim upwelling light of the background. Consequently, in shrimps that live in the upper regions of the mesopelagic zone, such as *Oplophorus spinosus*, the tapetum is absent in the most dorsal regions of the eye but present in graded amounts elsewhere in it. A highly reflective tapetum in downward-looking parts of the eye does not significantly increase the visibility of the shrimp because it reflects only the very dim upwelling light and does not appear as a bright spot against the background. Thus, this part of the eye can have a maximally reflective tapetum without increasing the shrimp's visibility. Horizontal-looking lateral regions of the eye have tapeta of intermediate reflectivity.

Additional evidence that reduced tapetal reflectiv-

(a)   (b)

Fig. 1. Organization of the eye of the shrimp *Systellaspis debilis*. (a) Section through eye. (b) Mirror box. Parallel rays entering different facets are focused onto the retina by mirror boxes behind the curved cornea.

Labels: cornea, mirror boxes, rhabdoms, tapetum, green-sensitive region, violet-sensitive region

Fig. 2. Regional variation in levels of eyeshine brightness in the shrimp *Systellaspis debilis*. (*a*) Dorsal eyeshine is low, (*b*) ventral eyeshine is high, and (*c*) lateral eyeshine is intermediate. Similarly, (*d*) the anterior, forward-looking part of the eye has much brighter eyeshine than (*e*) the posterior, backward-looking region. Since the path of tapetally reflected light is the reverse of that of the incident illuminating light, the area of eyeshine effectively defines the aperture of the system.

ity enables shrimps to avoid being seen by predators comes from the counterintuitive finding that those shrimp species with the brightest eyeshine live nearest to the surface. One examination of different species from increasing depths shows eyeshine levels decreasing. Species near the surface are relatively well illuminated, and the presence of bright eyeshine does not greatly enhance their visibility. In deeper water, eyeshine becomes more important in revealing the shrimps to potential predators. Therefore, for deeper species tapeta are less reflective even though the eyes are consequently less sensitive.

Many of the deep-water species produce flashes of bioluminescent light either from photophores or from regurgitated stomach contents. Highly reflective tapeta are very visible in the presence of such a flash. At about 3300 ft (1000 m) bioluminescence is the main source of light, possibly explaining why a species such as *Acanthephyra purpurea* possesses a very incomplete tapetum with a hole in the center so that no lateral eyeshine is produced. The tapetum of *A. purpurea* is otherwise normally distributed to give sensitive downward and forward vision.

Other graded properties of the eye. The eyes possess other graded properties that can be related to the behavior of sunlight in the water column. The photopigment of the photoreceptor cells is contained within special structures called rhabdoms, with one rhabdom for each facet. In *S. debilis*, each rhabdom consists of a large green-sensitive proximal part and a smaller distal region suspected to have violet sensitivity. Moving from the dorsal upward-looking part of the eye to the ventral downward-looking region, the relative proportions of the two parts of the rhabdom alter so that the distal component changes from contributing about one-tenth of the rhabdom dorsally to contributing about one-third of it ventrally (Fig. 1*a*). This change may be an adaptation to the varying spectral composition of light that occurs with viewing angle. At least in the upper regions of the sea, the backscattered upwelling light is significantly richer in shorter-wavelength light than is the downwelling light. The enlargement of the distal part and reduction of the green-sensitive part of the rhabdom is almost certainly an adaptation to match the photoreceptor cells to the spectral quality of the light in different parts of the visual field.

For background information *SEE DECAPODA (CRUSTACEA); EYE (INVERTEBRATE)* in the McGraw-Hill Encyclopedia of Science & Technology.

Peter M. J. Shelton

Bibliography. E. Gaten, P. M. J. Shelton, and P. J. Herring, Regional morphological variations in the compound eyes of certain mesopelagic shrimps in relation to their habitat, *J. Mar. Biol. Ass. U.K.*, 72:61–75, 1992; P. J. Herring et al. (eds.), *Light and Life in the Sea*, 1990; M. F. Land, Superposition images are formed by reflection in the eyes of some oceanic decapod crustacea, *Nature*, 263:764–765, 1976; P. M. J. Shelton, E. Gaten, and P. J. Herring, Adaptations of tapeta in the eyes of mesopelagic decapod shrimps to match the oceanic irradiance distribution, *J. Mar. Biol. Ass. U.K.*, 72:77–88, 1992.

Silicon

Silicon is the most widely used semiconductor in the computer industry. It is the material of choice for microprocessors and volatile memory (random-access memory, or RAM) chips in virtually all consumer electronics products. Although it has superior electronic characteristics, silicon does not have the ability to give off visible light, and so it cannot be used for optical applications such as light-emitting diodes or flat-panel displays. If optical connections, such as those that are now used for long-range (kilometer) fiber-optic telecommunications, could be incorporated into the small-scale (centimeter) environment of computer chips, the chips would be faster and more efficient. With the 1990 discovery of efficient visible light emission (luminescence) from a porous form of silicon, this goal may be achieved.

Porous silicon production. Luminescent porous silicon can be made by chemically or electrochemically corroding a wafer of crystalline silicon (**Fig. 1**). An electric contact is made to the silicon, which is then immersed in a solution containing a 1:1 volume ratio of 49% aqueous hydrofluoric acid (HF) and ethanol. As electric current is passed, the corrosion process creates submicrometer-size holes that can extend as deep as tens of micrometers into a wafer that is typically 300 μm thick. As the holes expand, they begin to overlap, leaving behind a disordered array of

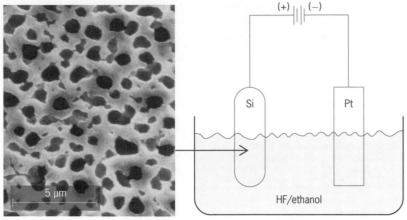

Fig. 1. Apparatus used to make luminescent porous silicon. At left is an electron microscope image of the surface of a silicon wafer after it was electrochemically etched.

silicon wires. The isolated silicon fingers create a very porous network. A layer of luminescent porous silicon can be up to 80% empty space. The resulting spongy material photoluminesces at wavelengths between 530 and 900 nanometers (green to near-infrared) when excited with a blue or green laser. The color and intensity of photoluminescence from a given sample depends on a number of parameters, including semiconductor dopant concentration and type, current density during the etch, duration of the etch, and subsequent chemical treatments. Luminescence can also be induced by passing electric current through the porous silicon sample—a crucial feature if porous silicon is to find practical application in electrooptic circuits.

Possible luminescence mechanisms. The origin of visible luminescence in porous silicon is still controversial. Bulk silicon, with an energy gap of 1.12 electronvolts (1100 nm wavelength) emits light weakly in the near-infrared region of the electromagnetic spectrum. As already mentioned, porous silicon's luminescence peak energy can range into the green (2.3 eV or 540 nm wavelength). Additionally, the photoluminescence from porous silicon is much more intense than expected. Three main theories have surfaced to explain these observations.

The original explanation involves destroying the crystallinity of silicon to create an amorphous layer. Just as graphite and diamond are polymorphs of carbon with profoundly different electronic properties, an amorphous phase of silicon formed via the electrochemical etch may have properties quite distinct from those of crystalline silicon. For example, amorphous silicon can have an energy gap up to 2.0 eV, markedly different from the gap of crystalline silicon, at 1.12 eV. There is evidence for the existence of amorphous silicon in at least some samples of luminescent porous silicon.

The luminescence of porous silicon can also be qualitatively explained by invoking size-dependent quantum confinement effects. This model derives from the familiar particle-in-a-box problem from quantum physics. As an electron is confined to successively smaller boxes, its energy increases and takes on quan-

tized values. In this model, porous silicon is viewed as containing an ensemble of quantum wires of silicon with a distribution of sizes, and the visible photoluminescence then derives from recombination of quantum-confined electron-hole pairs within these wires. Transmission electron microscopy data support the existence of crystalline silicon domains in porous silicon that have dimensions small enough (approximately 5-nm diameters) to be responsible for such phenomena.

The third model postulates that a surface or bulk chemical species is responsible for the luminescence observed in porous silicon. Polysilanes or polysilylenes, which are polymers containing only silicon and hydrogen, can photoluminesce at visible wavelengths. Siloxene, which contains silicon, oxygen, and hydrogen, also has emissive properties that resemble those of luminescent porous silicon. Any of these chemicals, or related ones, may be created in the silicon etching process, and may be the origin of porous silicon's unique properties.

A large amount of research effort is being spent to determine which of the above theories or combination of theories plays a dominant role in determining the luminescence properties of porous silicon. The results of this effort should ultimately determine the feasibility of potential applications for the porous silicon material.

Porous silicon applications. Devices based on luminescent porous silicon have potential applications in electrooptics and as chemical sensors and photodetectors.

Electrooptics. In microelectronic applications, the electronic properties of silicon are the most important. Silicon's ability to transmit or hold electric charge under precisely defined conditions is essential to the operation of capacitor and transistor structures that make up the devices. Silicon does its "thinking" with electrons, and it communicates with other devices and the user with these same electrons, through electrical connections. Electrical signal transmission suffers from several drawbacks. Losses are associated with resistance in the connecting wires, capacitive and inductive charging can reduce and slow down signals, and crosstalk between adjacent wires can increase noise levels and possibly cause erroneous data transmission. These problems can be reduced or eliminated if the information is carried by a light beam rather than by electrons in a wire. Thus, research has focused on incorporating light-generating and transmitting elements into conventional silicon-based devices. Solutions typically involve expensive or complicated processes, such as prefabricating arrays of very small light-emitting diodes from different materials and mechanically placing them on the surface of a silicon chip. Because luminescent porous silicon can be easily fabricated directly on a chip, it has the potential to overtake more elaborate or expensive techniques for converting computer signals into light.

Beyond allowing electronic elements to communicate via light, optical elements have the potential to replace them, resulting in a machine that thinks as well as communicates with light. The first functional optical computer, developed in 1990, performed calculations by using optical beams and switches instead of electricity and transistors. Although still very much at the research level, all-optical computers have the potential to replace their electronic counterparts, because they can operate at much faster speeds. The role, if any, of porous silicon's optical properties in these types of devices is less certain.

Although the fundamental mechanism responsible for the luminescence of porous silicon is still hotly debated, the phenomenon has generated a great deal of interest in silicon-based optoelectronics applications. Optical devices made of silicon would be far more easily integrated, in terms of fabrication and processing, than the currently used light-emitting materials. To integrate optical with electronic components, methods are needed to pattern the light-emitting porous silicon onto conventional silicon wafers and to excite the lu-

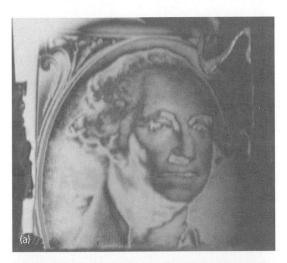

Fig. 2. Photographs of an image of George Washington from a United States $1 bill etched into silicon. (*a*) White-light photograph. (*b*) Fluorescence microscope photograph. The fluorescence was excited with a hand-held ultraviolet lamp and observed through a 455-nm cutoff filter, so that only the red-orange photoluminescence from the porous silicon is seen. (*From V. V. Doan and M. J. Sailor, Luminescent color image generation on porous silicon, Science, 256:1791–1792, June 26, 1992, ⓒ American Association for the Advancement of Science*)

minescent material electrically. A variety of so-called electroluminescent devices made of porous silicon have already been demonstrated, including a porous silicon *np* heterojunction diode that emits light at relatively low voltages. Also, a resistless photolithography technique etches patterns of porous silicon into single-crystal wafers with a lateral resolution of 20 μm.

Specialized etching techniques have been developed that further expand porous silicon's potential as an optoelectronic material. Photochemical etches have been developed that allow the patterning of gray-scale images and diffraction gratings into porous silicon layers. An example of an image etched into porous silicon is shown in **Fig. 2***a*. Colors in the image (which are reproduced here as various shades of gray) arise from optical interference similar to the iridescence observed on the surface of soap films. In the case of Fig. 2*a*, the thin film is the porous silicon layer, and the variations in color arise from variations in the porosity and thickness of this layer. These images also photoluminesce (Fig. 2*b*). The fact that luminescent porous silicon can be generated with such optical uniformity is good for applications such as holographic data storage or computation, and possibly for optical waveguide purposes.

Chemical sensors. The luminescence of porous silicon has been found to be extremely sensitive to adsorbed chemicals. For instance, exposure to ethanol can reduce the intensity of photoluminescence by a factor of 100. The effect is reversible, so evaporation of the chemical restores the original photoluminescence intensity. This sensitivity gives the surface of porous silicon another potential application, as a liquid or gas sensor.

Photodetectors. Porous silicon has been shown to have superior optical response for use in the detection of visible and lower-energy light. In addition, its spongelike surface can absorb up to 97% of light incident upon it. Conventional silicon detectors must be covered with polymer-based antireflection coatings to reduce reflected light losses. Because porous silicon's absorption and emission properties can be readily manipulated at the photochemical and electrochemical preparation steps, porous silicon may also find use in wavelength-specific detector needs.

Porous silicon limitations. Whether porous silicon will actually be used in any of the above-described applications is an open question. Some serious drawbacks with the material have already emerged. First, the luminescence of porous silicon can be affected, usually adversely, by exposure to air, light, and heat. These instabilities can be alleviated somewhat by encasing the material in polymer or glass. A potentially more serious drawback relates to the persistence of luminescence after the excitation source is turned off (photoluminescence decay time). For porous silicon, the photoluminescence decay is long, typically on a microsecond time scale. Although there are indications that the decay times may be shortened, the microsecond switching rate implied by such long decay times is at least three orders of magnitude too slow for computing applications. For optical display applications, the efficiency of electroluminescence is low relative to commercially available light emitters based on gallium arsenide phosphide. This efficiency needs to be improved if light-emitting porous silicon devices are to be competitive.

For background information SEE AMORPHOUS SOLID; ELECTROLUMINESCENCE; INTEGRATED CIRCUITS; INTEGRATED OPTICS; LIGHT-EMITTING DIODE; LUMINESCENCE; NONRELATIVISTIC QUANTUM THEORY; OPTICAL INFORMATION SYSTEMS in the McGraw-Hill Encyclopedia of Science & Technology.

Michael J. Sailor; Corrine L. Curtis; Vincent V. Doan

Bibliography. H. H. Arsenault, T. Szoplik, and B. Macukow (eds.), *Optical Processing and Computing*, 1989; L. T. Canham, Silicon quantum wire array fabrication by electrochemical and chemical dissolution of wafers, *Appl. Phys. Lett.*, 57:1046–1048, 1990; V. V. Doan and M. J. Sailor, Luminescent color image generation on porous silicon, *Science*, 256:1791–1792, 1992; M. J. Sailor and K. L. Kavanagh, Porous silicon: What is responsible for the visible luminescence?, *Adv. Mater.*, 4:432–434, 1992; R. L. Smith and S. D. Collins, Porous silicon formation mechanisms, *J. Appl. Phys.*, 71:R1–R22, 1992.

Smart materials

Smart materials and structures are defined as those that are capable of performing some desired function or functions beyond the normal functions of load bearing and volume encasing. This intentionally broad definition includes most aspects of the technology that are of interest today for applications such as threat sensing, conformal/embedded avionics, active structures, and monitoring structural health. The terms and definitions used when describing this technology are multiple and varied, depending upon the desired application and results. The terms intelligent materials, adaptive structures, active materials, smart skins, structural-integrity monitoring, acoustic suppression, and threat sensing are all related in their description of aspects of this technology, though they differ in choice of adjectives or applications.

In a more specific sense, smart materials and structures can be defined as those that are capable of sensing and reacting to their environment in a predictable and desired manner. This objective is currently achieved through the integration of various elements—such as sensors, actuators, power sources, signal processors, and communications networks—with structural materials.

The goals of smart materials and structures include sensory structures, which sense their external or internal environments; adaptive structures, which sense and react to their environments; smart skins or conformal/embedded avionics; and multifunctional materials. The materials and structures must be integrated with some form of intelligence, either in real time within integrated systems of smart materials and structures and artificial intelligence, or prior to deployment through selection of materials that sense and react to predicted or known environments in desired manners.

Thus, smart materials and structures involve multidisciplinary technology, requiring expertise in a number of areas, including but not limited to structural dynamics, materials, control, avionics, sensors, actuators, processing, and artificial intelligence.

Current approaches. There are currently two primary approaches to achieving smart materials and structures. One approach is the integration of individual and distinct elements such as sensors, actuators, processors, power sources, and communication networks with structural materials. A less developed but more elegant approach is the creation of unified multifunctional materials that combine the desired functions within themselves.

Current technology, in pursuit of the goal of demonstrating feasibility, primarily employs the approach of integration. The first generation of smart materials and structures systems will most likely be achieved through the integration of host structural materials and specialty elements. Yet, this approach is not as attractive as the development of multifunctional materials, and it poses many challenges. The integratable elements are often parasitic to the material and structure. Some workers in the field have suggested embedded elements may themselves be elastic inclusions and the local state in the host material may be altered by the very presence of embedded elements.

There are questions of the interface between the sensor and the host material. Current sensor technology requires much more development for successful applications for smart materials systems for air vehicles. Candidate sensors must be evaluated against criteria that include not only measurand type and sensitivity but also cost, reliability, connectivity, and supportability. The supportability, repairability, and manufacturability of structures with embedded or integrated elements such as sensors evoke much concern among users. In addition, there are questions of the very performance of integrated sensing systems. Long-term performance of fiber-optic sensors has yet to be characterized; it is not yet known if something that works during the first decade of a structure's life will do so in the second.

Many programs and efforts are currently targeted at the development of smart materials and structures technologies. Many industries in the United States have ongoing research in the area, as do numerous universities and government laboratories. These technologies are being developed aggressively not only in the United States; in Japan, discussion of intelligent materials concepts and research into applications have continued since 1987. The focus of possible applications is in the fields of medicine, aerospace, and electronics. Sponsorship for such work has been provided by the Council for Aeronautics, Electronics and Other Advanced Science and Technology of the Japanese government. The program has brought together specialists from diverse fields of materials science; chemical, electrical, and computer engineering; and medicine. International meetings on smart materials and structures technologies have been held in Japan, and there is regular technical exchange between the United States and Japan.

The first full European conference devoted to smart structures and materials, in May 1992, called them one of the cornerstone technologies for the twenty-first century, a field that offers immense benefits in applications ranging from space structures to buildings and bridges to medical implants and even perhaps to toys and domestic goods. Canada is also the scene of active research in the technology. *SEE BIOMIMETIC MATERIAL.*

Sensors. At present, sensory structures are achieved through the integration of advanced sensors with structural materials. These sensors can be either embedded within or attached to the surface of the material. Many factors must be considered when determining which type of integration is to be used. A need for superior supportability, manufacturability, or repairability may dictate attachment over embedment. Certain locations as well as sensor types may dictate how the sensor should be integrated.

Conventional sensors may have applications for sensory structures. Foil strain gages are simple, proven, and inexpensive. However, they are also limited in durability and multiplexability and may best be limited to attachment. Piezoelectric ceramics show potential for use in strain gages. Such materials develop an electrical charge across the poling faces as they are strained, and they are very sensitive strain sensors for local or point measurements.

The sensor currently being worked with most is the fiber-optic-based sensor, which is sensitive, immune to electromagnetic interferences, small (less than 150 micrometers), lightweight (which is good for embeddability), and relatively strong ($> 10^6$ lb/in.^{2} or 3.5×10^{12} pascals). Fiber-optic-based sensors are often functional at high temperatures, and provide the potential for monitoring composite curing and severe operational temperature environments. Such sensors also have the potential for multiplexing, which minimizes cost and connections. Frequency, phase, or polarization modulation is converted to power modulation and detected electrically. Fiber-optic-based sensors have been developed that can detect multiple parameters, including strain, temperature, acoustic emission, and pressure. Such information can be used to provide information on impact and battle damage and delaminations. Many sensors of this type are not commercially available.

Corrosion is an important parameter for sensing on many metallic systems. Techniques for sensing corrosion through the detection of corrosion products, moisture, or environments conducive to corrosion may be developed by using smart materials.

Measurement based on acoustic emission is a promising development. The release of strain energy during deformation and fracture in a material results in acoustic emission. This phenomenon can provide information on remote delaminations and impact damage in structures as well as crack noises.

Actuators. Actuators are components of control systems, and their use must take into account the desired end product and the environment of application. Materials for such mechanisms must be assessed against numerous criteria prior to selection for system application, including but not limited to power requirements, response time, displacement limits, stroke, durability, shock sensitivity, and temperature sensitivity.

There currently exist a number of multifunctional materials that could be utilized for actuators in smart structures. For example, piezoelectric materials, which can also perform as sensors, are being developed for actuator use. Piezoelectric materials convert electrical and mechanical energy, producing electrical charges in response to applied stresses, and displacements in response to applied voltages. Piezoelectric materials produce displacements on the order of a few microinches, often requiring amplification to be useful. Piezoelectric ceramic materials offer benefits in that they can be fabricated in various sizes and shapes; these materials have isotropic electrical, mechanical, and electromechanical properties, and they do not become piezoelectric until they are polarized. They also have a characteristic temperature, the Curie point, above which the material suffers loss of piezoelectric activity. Structural applications of such materials have been demonstrated.

Shape memory alloys, another actuator material, are being worked into smart materials and structures systems. These alloys comprise materials that are capable of changing their stiffness characteristics through thermally induced phase transformation. Heating induces a phase transformation from the austenitic phase, which is crystalline and rigid, to the easily deformable martensitic phase; cooling transforms the material back to the austenitic phase. This effect has been exploited to achieve actuation for flexible structures and acoustic transmission control. Piezoelectric films, electrorheological fluids, magnetostrictive and electrostrictive materials, conductive polymers, and polymeric biomaterials also are potential actuators for smart materials and structures systems. *SEE ELECTRORHEOLOGICAL FLUID.*

Integrated elements. Many efforts are approaching smart materials and structures technologies through the integration of structural materials and specialty elements. These elements are typically either embedded or attached. If embedment is the preferred approach, its effects upon material and structural properties must be characterized and understood, including the long-term durability and fatigue.

Regardless of whether the smart structures system utilizes integrated sensors, actuators, or other elements, some sort of knowledge base and algorithms also must be integrated to provide the intelligence and adaptability of the system. The smart structure must be capable not only of detecting certain states but also of assessing, making decisions, transforming sensed information into models, and reacting to the information in the desired fashion. Other aspects required of a smart structure include knowledge base, algorithms, power supply, and data links.

Applications. Materials that can intrinsically sense and adapt to their environment have many applications to weapon systems. Such materials could be tailored to continually monitor their internal strain, temperature, vibration, damage state, and external threat environment. Structures fabricated from such materials could offer sensory capabilities such as structural integrity or health monitoring, and active capabilities such as vibration, sound radiation, or shape control.

Aircraft structures with capabilities for monitoring structural integrity could result in reduced requirements for inspection, leading to increased availability and increased sortie rates. Some studies have shown that a majority of inspections may be unnecessary, with only a few resulting in maintenance actions. Elimination of unnecessary inspections could significantly reduce supportability costs. Development and use of sensory materials and structures for integrity monitoring in weapon systems for the U.S. Air Force could lead to greater confidence in, and thus more effective utilization of, advanced composite materials, resulting in less overdesign and increased applications.

Accurate component tracking could also be performed through the use of structural monitoring; as a component is exchanged from vehicle to vehicle over its life, exact knowledge of the load spectrum that a particular component has experienced could be used to determine remaining use time allowed for that particular component. Today's aircraft are remaining in the fleet much longer than originally anticipated, and their missions are often changing from those for which they were originally designed. Knowledge of the structural health and remaining life of these aircraft is imperative; the current prediction methods used do not accurately monitor the structures. A complete reassessment of, and improvement upon, the way the Air Force performs its Aircraft Structural Integrity Program could result from monitoring structural health through the use of smart materials and structures. The useful life of aircraft structures could be realized through the application of smart materials and structures for monitoring integrity as well. Other projected yields from this diverse technology include increased payload, range, and maneuverability through active structures; monitoring of composite cure and quality control; and suppression of vibration and flutter through active and adaptive structures. Control of sound radiation from vibrating structures through the use of embedded or bonded piezoelectric actuators is another application for active and adaptive structures.

Spacecraft structures could also greatly benefit from the application of structural health monitoring. Space systems could eventually sense their health in the space environment, and lead to adaptive structures that could sense and compensate for undesired states. Vibration control and shape control for large precision structures are two examples. Significant damping can be achieved through the use of embedded sensors and actuators such as piezoelectric ceramics.

For background information SEE AIR ARMAMENT; MATERIALS SCIENCE AND ENGINEERING; SERVOMECHANISM; SHAPE MEMORY ALLOYS in the McGraw-Hill Encyclopedia of Science & Technology.

Tia Benson Tolle

Bibliography. R. Clark, Control of sound radiation with adaptive structures, *1st Joint U.S./Japan Conference on Adaptive Structures*, November 13, 1990; D. Dehart, Astronautics laboratory smart structures/skins overview, *1st Joint U.S./Japan Conference on Adaptive Structures*, November 13, 1990; C. R. Lawrence, Active member vibration control experiment in a KC-135 reduced gravity environment, *1st Joint U.S./Japan Conference on Adaptive Structures*, November 13, 1990; J. Sirkis, What do embedded optical fibers really measure?, *1st European Conference on Smart Structures and Materials*, Glasgow, May 12, 1992; T. Takagi, A concept of intelligent materials and the current activities of intelligent materials in Japan, *1st European Conference on Smart Structures and Materials*, Glasgow, May 12, 1992.

Snail

Snails, any of the approximately 74,000 species in the class Gastropoda of the phylum Mollusca, are among the most common animals of shallow lake and stream areas. Snails feed on decaying plants or on algae that cover plants or rocks. Some float upside down at the water surface, supported by the surface tension, and feed on floating algae. Snails are also found at considerable depths in many lakes and form the basis of food chains dominated by sport fish. Fresh-water snails can adapt to the short-lived and variable nature of their habitats. Fresh-water streams and lakes are uncertain in an evolutionary sense because they last only for hundreds of years, a relatively short period in comparison to that necessary for natural selection to occur. Ponds are also uncertain in an ecological sense, as they often dry up at unpredictable times, based upon the amount of rainfall or snow the preceding winter. Extensive intraspecific variation in life histories, productivity, morphology, and feeding habits enable fresh-water gastropods to adapt to these uncertain habitats.

Prosobranchia and Pulmonata. Gastropods are the most diverse class of the phylum Mollusca, making up almost three-quarters of the 110,000 or so species of known mollusks. Over 50,000 of these species belong to the mostly marine and fresh-water subclass Prosobranchia, while another 20,000 species belong to the subclass Pulmonata, of which most are terrestrial.

Prosobranch snails. Prosobranch snails, whose marine ancestors invaded fresh waters via estuaries, possess a gill (ctenidium) and a horny (flexible) or calcareous operculum, or trap door, which can be closed to protect the animal. Prosobranch snails are usually dioecious (that is, sexes occur in separate individuals), and males use the enlarged right tentacle as a copulatory organ, or possess a specialized penis or verge, or have no copulatory organ. Prosobranchs lay clutches of a few eggs; or they may brood eggs, which hatch in a fold of the anterior mantle, where development continues, and the offspring are born free-living. Some species are parthenogenetic (that is, females can reproduce without being fertilized), an adaptation for

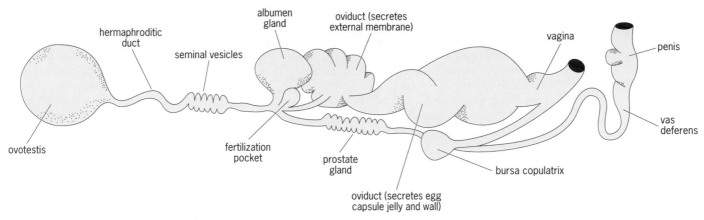

Anatomy of the reproductive system of a hermaphroditic pulmonate snail.

colonizing unpredictable habitats where densities are low and chances of finding mates rare.

Pulmonate snails. Pulmonates, in contrast, have terrestrial ancestors that reinvaded fresh waters, use a modified portion of the mantle cavity as a lung, and lack an operculum. All are hermaphroditic or, more correctly, monoecious (that is, both sexes occur in one individual). The basic components of the pulmonate reproductive system are shown in the **illustration**. Sperm and eggs are produced in the ovotestis and exit via a common hermaphroditic duct. Eggs are fertilized either in the hermaphroditic duct by the same individual's sperm (which has been stored in the seminal vesicles) or in the fertilization pocket by sperm from another individual. The external egg membranes and egg capsule jelly are then secreted in the oviduct. Eggs are laid in gelatinous cases and attached to plants or rocks. Although pulmonates are hermaphrodites, the great majority of species outcross whenever possible. Pulmonates that self-fertilize usually mature at later ages and have lower fecundity.

The adaptive value of hermaphroditism in pulmonates is usually explained in the following way. Pulmonates are slow-moving organisms, and populations often go through seasonal bottlenecks (that is, precipitous declines in density). A monoecious individual under such situations can fertilize its own eggs, whereas a dioecious individual requires an individual of the opposite sex. The chances of finding a mate in such situations are small, providing a selective advantage for hermaphroditism. Because pulmonates are dispersed passively as small juveniles trapped in mud on birds' feet, being monoecious is obviously beneficial. The cost of producing two reproductive systems is evidently less than not being able to reproduce at all in such situations.

Variations in life history. Studies of life-history variation in fresh-water snails are common because these snails are fairly easy to sample, have relatively short life cycles, and are easy to rear in the laboratory. They are also extremely interesting because of the variety of observed life-history patterns. At one end of the spectrum of life-history variation are pulmonate snails that reproduce once in the spring and die; that is, there is complete replacement of genera-

tions. Most pulmonates belong to this group, including species from the genera *Lymnaea, Physa*, and *Aplexa*. A second category in the spectrum includes species in which reproduction occurs in both spring and late summer and in which either both cohorts survive the winter or there is complete replacement of generations. There can be as many as three reproductive intervals, with varying degrees of replacement of generations, in populations in subtropical or tropical environments. Finally, there are populations that can be considered perennial (life cycles of 4 to 5 years) and iteroparous (that is, with repeated periods of reproduction throughout the life cycle); most are prosobranchs.

Clear differences in life-cycle patterns occur between marine and fresh-water snails. Marine snails have enormous fecundity, but individual eggs are extremely small and usually produce a pelagic larval stage. Most marine snails are prosobranchs, and fresh-water prosobranchs have relatively small eggs probably as a result of being their descendants. Pulmonates produce fewer but larger eggs than marine prosobranchs. The loss of the planktonic veliger larval stage and the shortening of the developmental period in all fresh-water snails have been attributed to the more variable physicochemical conditions in fresh water. Pulmonate genera, on the average, reproduce at smaller sizes and earlier ages, produce more eggs, have larger clutch sizes, have greater shell-growth rates, and have shorter life cycles and smaller final shell sizes than the prosobranchs that brood their young and release them as miniature adults. Prosobranchs that brood their young, however, have a long reproductive interval and are iteroparous with relatively small clutch sizes in most cases.

Reproductive effort (that is, the percentage of energy devoted to reproduction) is lower in iteroparous fresh-water snails than in semelparous ones (that is, those that reproduce only once). Presumably, the reduced fecundity and increased parental care found in some prosobranchs have evolved to increase offspring survival. Survival to maturity can be much less than 1% in pulmonate populations but more than 40% in prosobranchs that brood their young.

Prosobranchs are often sexually dimorphic in life-history patterns. Females reach larger sizes and live longer, males usually surviving for only one reproductive season. Possibly, males expend more energy to locate mates in their reproductive season, resulting in lower survivorship.

Causes of variation. The relative importance of environmental and genetic factors in explaining life-history variation in many fresh-water snails has been extensively studied in some groups. A large number of studies has implicated environmental factors such as algal productivity in determining the number of generations per year, growth rates, fecundity, and gastropod secondary production. Other important environmental factors include physicochemical variables such as water hardness and water temperature. For example, populations of pulmonates in Canada often take several seasons to complete their life cycle, while populations in the warmer waters of the midwestern United States have annual life cycles. Abundant predator populations may also favor rapid snail-growth rates, since most crayfish or fish predators feed preferentially on smaller snails.

For example, populations of pulmonates in England in wave-swept habitats have life-history traits characteristic of what ecologists call r-selected populations (having early reproduction and high reproductive output to counteract high mortality rates) in comparison to populations in less harsh habitats. The theory predicts that populations in harsh environments (such as wave-swept shores) should be kept far below their equilibrium abundance by high mortality, and life-history traits that cause rapid population growth (and thus result in large values for the growth rate r) should be selected for. Thus these populations should be genetically different from those in protected areas, where competitive ability is selected because of limiting resources.

However, other studies of life-history variation in mollusks do not agree as well with the predictions of r and K theory. Furthermore, transplant studies, where individuals from separate populations are reared in a common environment, usually indicate that environmental effects on life histories are much more important than genetic differences between populations. For example, populations of one pulmonate species found in more productive ponds lay nine times as many eggs, have an annual versus a biennial reproductive cycle, and reach larger individual sizes.

Genetic polymorphism. Genetic polymorphism has been studied extensively by using gel electrophoresis in terrestrial pulmonates and fresh-water prosobranchs but only rarely in aquatic pulmonates. Such techniques allow determination of the types of alleles at a number of gene loci, and the degree of polymorphism is usually expressed as the percentage of all loci tested, averaged over all individuals, that are heterozygotic (that is, that possess two different alleles). Fresh-water snails have average levels of genetic polymorphism intermediate to terrestrial and marine species. This pattern may be due to the fact that terrestrial snails inhabit microclimates that, because they are patchily distributed, increase chances for low population densities and self-fertilization, resulting in low

levels of genetic polymorphism. Fresh-water snails experience low densities because of seasonal bottlenecks and thus may self-fertilize, potentially reducing levels of genetic variation. Marine environments are much less seasonal, and many marine snails have planktonic larvae, facilitating gene flow even further and thus increasing polymorphism. However, more electrophoretic data are needed for fresh-water pulmonates before stronger genetic comparisons can be made. Such comparisons may also be confounded by systematic differences, because most terrestrial snails are pulmonates whereas fresh-water snails contain both pulmonates and prosobranchs and marine snails are almost all prosobranchs.

For background information SEE GASTROPODA; PROSOBRANCHIA; PULMONATA; SNAIL in the McGraw-Hill Encyclopedia of Science & Technology.

Kenneth M. Brown

Bibliography. K. M. Brown, Intraspecific life history variation in a pond snail: The roles of population divergence and phenotypic plasticity, *Evolution*, 39:387–395, 1985; K. M. Brown and T. D. Richardson, Genetic polymorphism in gastropods: A comparison of methods and habitat scales, *Amer. Malacol. Bull.*, 6:9–17, 1988; P. Calow, The evolution of life-cycle strategies in fresh-water gastropods, *Malacologia*, 17:351–364, 1978; T. A. Crowl and A. P. Covich, Predator-induced life history shifts in a freshwater snail, *Science*, 247:949–951, 1990.

Snow cover

Across the middle and high latitudes of the Northern Hemisphere, the impact of snow cover on humans and the environment is considerable. Snow lying on the ground or on ice influences hydrologic, biologic, chemical, and geologic processes. Snow exerts an impact on activities as diverse as engineering, agriculture, travel, recreation, commerce, and safety. Observational and modeling studies also show snow cover to have an influential role within the global heat budget, chiefly through the snow's effect of increasing surface reflectivity. Global models of human-induced climate change suggest enhanced warming in regions where snow cover is currently seasonal. For this reason, snow cover has been suggested as a useful index for detecting and monitoring such change.

Monitoring. Accurate information on snow cover is essential for understanding details of climate dynamics and climate change. It is also critical that snow observations be as lengthy and geographically extensive as possible. Snow data are gathered from ground sites, aircraft, or satellite platforms. Advantages and liabilities of extracting data from each of these sources are related to their accuracy and coverage. Observations of snow cover from aircraft are limited in both space and time, and they tend to be for specific investigations.

Station observations. Surface-based snow cover data are gathered mainly from observing stations on a once-per-day basis. The general practice is

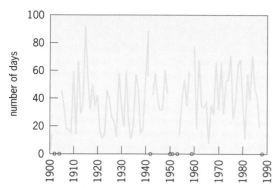

Fig. 1. Days with 7.5 cm (3 in.) or less of snow cover at Fairbury, Nebraska, during the winters (September–May) of 1899/1900 through 1986/1987. Missing years are plotted along the bottom axis. No year with data had a complete absence of snow cover days.

to record the average depth of snow lying on level open ground having a natural surface cover. Current station observations of snow cover are of a sufficient density for climatological study in the lower elevations of the middle latitudes of the Northern Hemisphere. Elsewhere, data are spotty at best.

Efforts are in progress to organize, verify, and analyze long-term station observations of snow cover. **Figure 1** is an example showing the number of days with 7.5 cm (3 in.) or less of snow on the ground between September and the following May at Fairbury, Nebraska, for the winters of 1899/1900 through 1986/1987. Only seven years during this period had insufficient data that prohibited a seasonal summation. The remaining years show considerable variability in duration of the snow cover. The 1920s and 1930s had a number of years with infrequent cover, while years in the late 1960s to middle 1980s frequently had extended cover.

Visible satellite observations. Regional and continental snow extent is gleaned from visible satellite observations of solar radiation reflected off the Earth's surface, and from microwave radiation emitted by the surface. The visible approach has the benefits of being directly interpretable by the human eye and having global imagery with a resolution of one to several kilometers available on a daily basis. Disadvantages of a visible approach include the inability to monitor surface conditions where clouds are present and where dense vegetation precludes reliable observations of the underlying surface. Low solar illumination is not a significant liability, since most high-latitude regions are snow-covered before the diminution of light and remain covered until spring.

For the past several decades the U.S. National Oceanic and Atmospheric Administration (NOAA) has mapped snow cover over Northern Hemisphere lands on a weekly basis. NOAA charts are based on a visual interpretation of photographic copies of visible imagery by trained meteorologists. Accuracy in charting is such that this product is considered suitable for continental-scale climate studies.

Mean snow extent. An analysis of the NOAA visible data since 1972 finds that, on average, some

46.5×10^6 km^2 (17.9×10^6 mi^2) of Eurasia and North America are covered with snow in January, the snowiest month of the year. February is a close second with an average of 46.0×10^6 km^2 (17.8×10^6 mi^2). August has the least cover, averaging 3.9×10^6 km^2 (1.5×10^6 mi^2), most of this being snow on top of the Greenland ice sheet. The annual mean cover is 25.5×10^6 km^2 (9.8×10^6 mi^2). The mean position of the North American snow line for four months of the year is shown in **Figure 2**. The snow season over North America and Eurasia begins in September, when snow cover becomes established over the high Arctic and on lofty mountain peaks. By January, snow over lower elevations extends southward to roughly the 40th to 45th parallels, and dips farther toward the Equator over mountainous regions. In April, the snow at the low elevations retreats across the United States–Canadian border and to about 60°N in Scandinavia and western Russia and 55°N in Siberia. By June, snow lies close to the Arctic coast and in the high mountains; and in mid-July, all lands and even the sea ice over the Arctic Ocean are essentially snow-free. Only the Greenland ice sheet maintains a year-round cover of snow in the Northern Hemisphere. Snow cover is also a permanent feature of the Antarctic ice sheet, and it occurs seasonally on the sea ice surrounding the Antarctic continent, as well as over some of the mountains and highlands of South America, Australia, and Africa.

Snow cover variability. NOAA charts indicate a great deal of the year-to-year variability in snow extent over Northern Hemisphere lands. The snowiest year of the past two decades was 1978, with a mean cover of

Fig. 2. Boundaries separating snow-covered from snow-free ground over North America in October (A), January (B), April (C), and June (D). Boundaries are defined as isolines denoting a 50% frequency of cover.

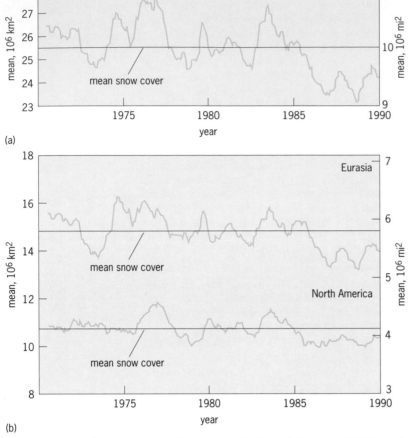

(a)

(b)

Fig. 3. Twelve-month running means of snow cover for the period January 1972 through May 1992. (*a*) Over Northern Hemisphere lands. (*b*) Over Eurasia and North America (including Greenland).

on the order of several tens of kilometers, making a detailed delineation of snow extent difficult, particularly where snow is patchy. It is also difficult to identify shallow or wet snow by using microwaves, and recognition of snow is a problem where vegetation masks the surface. Microwave-derived snow data are available since the late 1970s, although the data are still considered to be experimental in nature. The lack of sufficient ground truth data (data gathered at the surface) on snow depth makes an adequate assessment of the reliability of such microwave estimates uncertain. The focus of research into the microwave monitoring of snow cover remains on the extent of snow. Comparing a hemispheric time series derived from microwave data with NOAA visible observations suggests that microwave estimates of snow extent are between 80 and 90% of the more reliable visible areas in winter and spring and at times as much as 50% lower than visible values in the summer and fall.

Prospects. The critical role that snow cover plays in the global heat budget and the expected impacts of snow feedbacks in human-induced climate change support the continued diligent monitoring of snow cover over continents and sea ice. With the availability and better understanding of data from a variety of satellite and ground sources, and the ability to integrate and examine these data by using geographic information techniques, more accurate and extensive knowledge of snow cover across the globe is within reach. This information, along with expanded retrospective analyses using historical station data, is solidifying the position of snow as one of the key indicators of future change in the climate system.

For background information *SEE CLIMATIC CHANGE; SNOW; SNOW LINE; SNOW SURVEYING* in the McGraw-Hill Encyclopedia of Science & Technology.

David A. Robinson

Bibliography. A. T. C. Chang, J. L. Foster, and D. K. Hall, Satellite sensor estimates of Northern Hemisphere snow volume, *Int. J. Remote Sens.*, 11:167–171, 1990; M. Matson, C. F. Ropelewski, and M. S. Varnadore, *An Atlas of Satellite-Derived Northern Hemispheric Snow Cover Frequency*, NOAA/NESDIS/NWS, 1986; D. A. Robinson and K. F. Dewey, Recent secular variations in the extent of Northern Hemisphere snow cover, *Geophys. Res. Lett.*, 17:1557–1560, 1990; D. A. Robinson and M. G. Hughes, Snow cover and climate change on the Great Plains, *Great Plains Res.*, 1:93–113, 1991.

27.4×10^6 km^2 (10.8×10^6 mi^2); 1990 was the least snowy at 23.2×10^6 km^2 (9.0×10^6 mi^2). Twelve-month running means of continental snow extent best illustrate the periods of above-normal cover that occurred in the late 1970s and mid-1980s (**Fig. 3**). Intervals with lower snow extents include the mid-1970s and early 1980s; however, neither interval approaches the deficit of snow cover observed in recent years. Of the 58 months between August 1987 and May 1992, only five had above-normal snow cover. Spring cover showed pronounced deficits during 1988–1992 in Eurasia and during 1987–1992 in North America; snow extents in these spring seasons were at or below lows established prior to this period. During the same interval, both continents had low seasonal cover in the fall and summer, although frequently neither continent was at or approached record low levels. Winter cover was close to average during 1987–1992.

Microwave satellite observations. Microwave radiation penetrates winter clouds, permitting an unobstructed signal from the Earth's surface to reach a satellite. The discrimination of snow cover is possible mainly because of differences in emissivity between snow-covered and snow-free surfaces. Estimates of the spatial extent as well as the depth or water equivalent of the snowpack are made by using multiple-channel microwave data. Spatial resolution is

Soil ecology

Recent research in nematology emphasizes biological methods of controlling plant-parasitic nematodes and soil insects. New technology is being developed to mass-produce and deliver biological control agents to the soil in an efficient and effective manner. Other new studies involve the genetics and molecular biology of phytoparasitic and bacteria-feeding nematodes.

Nematode prevalence. Some 800 genera and 6000 species of terrestrial nematodes are described in the literature, including 225 genera and 5000 species of plant-parasitic nematodes and 175 genera and 800 species of insect-parasitic nematodes.

Nematodes are the unseen but most numerous multicellular organisms in the soil, ranging 380,000 to 2.8×10^6 per cubic foot ($4–30 \times 10^6/m^2$). In soil under agricultural crops, nematodes often constitute 90% of the invertebrate multicellular fauna. The number of nematode species in a community varies with habitat. Prairie and woodland habitats are generally richer in total numbers of all nematode species than are cultivated fields, prairies being richer than woodlands. The number of nematode species in a cultivated field can be as high as 74, with 16 of these being plant-parasitic. More than 170 plant-parasitic nematode species have been associated with corn, but usually only 3–8 species are found around a plant at any one time.

In agroecosystems, the largest numbers of plant nematodes are found in the top 2.3–7.8 in. (15–20 cm) of soil, but some may be found as deep as 94 in. (240 cm). Vertical migration is largely controlled by temperature, moisture conditions, and root distribution.

Although nematodes are small, they are scattered throughout the soil by the billions, so that they must constitute an important mechanical factor. From the time they are hatched until death, nematodes seem to be in constant motion.

Nematodes are soft-bodied, thin, and usually less than 0.04 in. (1 mm) long. When a few thousand are extracted from the soil and placed in a test tube with a milliliter of water, and the test tube is held to the light, a white cloudlike suspension of nematodes can be seen moving in a serpentine manner. A few nematodes in a like amount of water probably would not be visible. The movement of a nematode in a film of water in the soil would closely resemble that of a rattlesnake gliding across the sand.

Nematodes are in the phylum Nemata. Nematology is the science of plant-parasitic, insect-parasitic, and free-living nematodes, whereas parasitology deals with nematodes that are animal parasites. Nematodes occur all over the world in almost every soil type; in forest, desert, and arctic areas; in plants and animals; and in hot springs, lakes, rivers, and oceans. They are found exploiting every ecological niche.

Nematodes can be both harmful and beneficial. Harmful nematodes feed on roots, stems, and leaves of crop plants, on humans, and on domestic animals. Beneficial nematodes feed on insects, plant-pathogenic fungi, and bacteria, and are important members of the food chain.

Nematodes usually go through both male and female stages, producing hundreds of eggs. In some species, males do not exist, and either the eggs develop without fertilization (parthenogenesis) or the female gonads produce both eggs and sperm (hermaphroditism). Sex ratios are variable, but for most free-living, plant-parasitic, and insect-parasitic nematodes males and females occur in about equal numbers. However, when a population is subjected to environmental stress, more males are produced. When many juvenile mermithid nematodes parasitize a young grasshopper, most if not all become males because of the lack of sufficient food to become the larger female nematode. When only one mermithid nematode parasite is present, it almost always becomes a female nematode.

Nematode life cycles vary from 3 days to a year or more. Nematode juveniles molt four times as they become larger and develop into adult males or females.

Classification. Nematodes are classified as plant-parasitic, insect-parasitic, and free-living.

Plant-parasitic. A major area of nematological study in the United States deals with plant-parasitic nematodes. The wheat seed-gall nematode, discovered in 1743, appears to be the first recorded plant-parasitic nematode. **Figure 1** shows a typical parasitic nematode with feeding stylet. Plant-parastic nematodes feed on the roots of soybean, potato, tomato, cotton, corn, rice, citrus trees, and many other economic crops. In the United States, damage from plant-parasitic nematodes amounts to over $6 billion per year.

Plant-parasitic nematodes cause galls, lesions, discoloration, deformity and, in some cases, tissue death in the penetration and feeding areas of the affected plant. The plants lose vigor and grow more slowly,

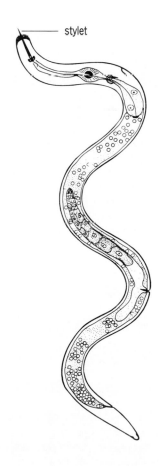

stylet

Fig. 1. A typical plant-parasitic nematode. (*From R. P. Esser, What Is a Nematode?, State of Florida Dep. Agr. Bull., 1981*)

often with a reduction in size and quality of fruits or vegetables. Entire plantings are sometimes destroyed. Some nematodes transmit plant viruses and other plant-disease organisms. Various plant-parasitic nematodes are capable of surviving unfavorable soil conditions by entering a dormant state for long periods, from a few months to 39 years.

Trends in agronomic nematology include the movement away from the use of chemical nematicides as a means of controlling plant-parasitic nematodes. The use of biocontrol, natural products, pheromones, resistant varieties, solarization, and crop rotation is advocated in order to reduce the amount of pesticide entering the groundwater. Extensive research is directed at finding fungal and bacterial soil parasites that can kill troublesome plant-parasitic nematodes. Also, technology is being developed to mass-produce and deliver the parasites of nematodes in a manner suitable for control of these pests.

Insect-parasitic. Nematodes have evolved to parasitize almost every kind of plant and animal. There are species that can kill, sterilize, or otherwise debilitate, by the millions, many different kinds of insects, such as rootworms, mosquitoes, grasshoppers, caterpillars, soil maggots, and wireworms. Insect-parasitic nematodes enter the insect larva by way of the mouth, anus, spiracles, or directly through the integument. These nematodes are found in the insect's body cavities, usually in the abdomen, and they obtain some amino acids and other nourishment directly from the blood of the host.

Insect-parasitic nematodes first appear in the fossil record from Rhine lignite and Baltic amber. Although there were a few scattered references to insect-parasitic nematodes in seventeenth- to nineteenth-century research, the most productive work in the search for biological control organisms for pest-insect control has taken place since the mid-1960s.

Insect-parasitic nematodes, especially steinernematids (**Fig. 2**), are used for biological control of soil insect pests. These nematodes transmit a bacterium that kills the insect quickly, often in fewer than 36 h. Nematodes that carry bacteria that kill insects are known as entomopathogenic. Research to develop new technology for practical application of these parasites is expanding. The infective stages of these nematodes are available commercially. Entomopathogenic nematodes can be propagated by a liquid fermentation process at a cost of less than 10 cents per million.

About 95% of all insects spend some time in the soil, where they are exposed to parasitism by nematodes. The potential for use of such nematodes as biological control agents exists in situations where chemical pesticides are too expensive, are not practical, or are harmful to humans and the environment. Ideal habitats for nematode parasite control of insects are groundwater, streams, and ponds; areas near livestock and other animals or near human dwellings; and soil situations with concentrated irrigation or river bottoms devoted to agricultural production.

Free-living. Other, free-living nematodes feed on bacteria and fungi. Bacteria-feeding nematodes can

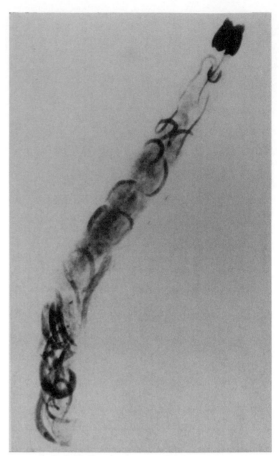

Fig. 2. Dead larval mushroom fly with insect-parasitic nematodes in its body cavity. (*Courtesy of W. R. Nickle*)

consume as much as 50% of the annual production of the microfloral biomass. Nematodes inhabit decaying organic material, and frequently individuals of this group compose the greater portion of nematode populations in soils with high organic content. Rarely is a soil sample processed that does not contain several microbivorus nematode species. Little is known about their role in soil biology, but no doubt they aid in breaking down organic matter and reducing it to plant food. Populations of these saprophagous nematodes fluctuate rapidly. As supplies of organic materials are exhausted, they may almost disappear from a given area within a period of weeks.

Scientists are using the free-living soil nematode *Caenorhabditis elegans* to study how its genes control development and behavior. This genetic process could be applicable to the higher forms of life. *Caenorhabditis elegans*, with its short life cycle, can be easily cultivated and manipulated, and it is small enough to be handled in large numbers. It has few cells, thus facilitating study of lineage and patterns, and it may be amenable to genetic analysis.

For background information SEE INSECT CONTROL, BIOLOGICAL; INSECT PATHOLOGY; NEMATA; NEMATICIDE; SOIL ECOLOGY in the McGraw-Hill Encyclopedia of Science & Technology.

William R. Nickle

Bibliography. R. P. Esser, *What Is a Nematode?*,

State of Florida Dep. Agr. Bull., 1981; W. R. Nickle (ed.), *Manual of Agricultural Nematology*, 1991; W. R. Nickle (ed.), *Plant and Insect Nematodes*, 1984.

Soil erosion

Recent research in soil erosion has involved the topographic parameters associated with the ephemeral erosion, the effects of wind erosion on cropland and the measures for control, and the use of windbreaks for reducing erosion.

Parameters of Ephemeral Erosion

Ephemeral gullies are channels that form in cultivated fields when precipitation exceeds soil infiltration rates. Excess water moves downslope as thin-sheet flow but eventually coalesces into small streams. Increased scouring occurs in these concentrated flows because water velocity is greater. Small channels, or rills, subsequently develop on upper slopes; larger channels form on lower slopes in concave swales that serve as surface drains for relatively larger watersheds. These larger gullies are called ephemeral gullies when they are small enough to permit passage of tillage implements.

Ephemeral channels tend to form in the same location each season, primarily because gully location is strongly controlled by landscape configuration. Unlike ephemeral channels, rills occur at random locations on the slope each season; thus soil is removed from the entire slope, although the magnitude of erosion varies with slope position. Tillage acts as a cut-and-fill process, extending the impact of the gully several meters beyond the ephemeral channel on both sides. Repeated cycles of channel formation and tillage-filling remove a greater volume of topsoil from these areas and can quickly reduce crop yields. Adjacent slopes become steeper, hastening processes of rill and interrill erosion.

Factors influencing gully formation are those that determine (1) precipitation rates, (2) infiltration and water-retaining capacities of the soil, (3) resistivity of the soil to detachment and transport, and (4) transport capacity of overland flow. Recent research has determined how landscape topography influences the occurrence and severity of ephemeral gullies in a given watershed. This dependence occurs over a range of landscape scales.

Topography external to watershed.

Topographic features occurring beyond the watershed boundary influence ephemeral erosion via impacts on the patterns of watershed precipitation and temperature transitions. The potential for erosion becomes greater as number and intensity of rainstorms increase. Occurrence of temperature transitions can increase erosion, particularly in early spring when soil frost just below the surface prevents infiltration of warm rainwater. The thawed surface layer becomes saturated, and soil particles are easily dislodged by concentrated flow. Both regional and local physiography may influence ephemeral erosion.

Regional influence. At the regional scale, the impact of orography on precipitation patterns is well known. Topographic barriers may decrease cyclonic precipitation on the leeward side, owing to drying associated with descending air. A reverse effect can occur when convectional storms develop over mountains and drift over leeward valleys. (Convectional storms are created when air that is warmed at the Earth's surface rises into the cooler upper atmosphere and, upon cooling, forms clouds and precipitation.) For example, a watershed separated from moist, temperate marine air by a mountain barrier will be subject to less severe ephemeral erosion than an identical watershed not so separated. Not only will the number of annual storms be reduced at the drier location, but the climate of the location may also be more continental; winters may be colder, perhaps cold enough that precipitation may fall in frozen form, eliminating the erosion potential in that season entirely.

The position of the watershed with respect to a topographic barrier may also determine whether storms are dominantly cyclonic or convectional in character. Rainfall from convectional storms is of higher intensity than that from cyclonic systems. High-intensity rains generally produce more runoff, because higher rates of precipitation commonly exceed capacities for soil infiltration. In addition, the kinetic energy of larger raindrops is greater, possibly leading to rapid formation of a surface seal that can reduce infiltration by as much as 80%. Watersheds in convectional rainfall areas may experience erosion throughout the warm season. Even when crop cover reaches its maximum, runoff and erosion may result from convectional storms because of high precipitation rates.

Local influence. Local features influence rainfall and temperature regimes between different watersheds. Certain landscape configurations can channel airflow, producing zones of low-level moisture convergence. In these zones, a relatively warm, moist airflow collides with another airmass flowing from a different

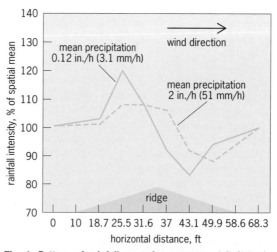

Fig. 1. Pattern of rainfall occurring across a 3-ft (0.9-m) ridge, along the wind path, when wind speed is 14 mi/h (6 m/s) and meteorological rainfall intensity is either 2 in./h (51 mm/h) or 0.12 in./h (3.1 mm/h). 1 ft = 0.3 m.

Table 1. Topographic indices commonly computed by digital terrain models

Code	Name	Definition
S	Slope	Maximum rate of change of elevation of the surface (m/m)
ASP	Aspect	Compass bearing of the maximum downward slope (degrees clockwise from north)
P_F	Profile curvature	Second derivative of arc defined by the intersection of the surface with a vertical plane that passes through slope vector and node (m/m²; positive → convex)
P_L	Planform curvature	Second derivative of arc formed at the surface by a vertical plane perpendicular to slope vector and passing through node (m/m², positive → concave)
A	Upstream contributing area	Upstream area (m²) that contributes flow to the surface point corresponding to each node
A_U	Unit area	Area/unit contour length (m²/m; unit contour length is the size of land surface appraisal unit)

direction and from a different source area, so that locally higher mean rainfall results. Complex relief is thought to reduce the efficiency with which developing storms assimilate latent energy from the atmosphere. As a result, convectional storm activity in complex physiography consists of numerous small raincells that produce less intense rainfall. In smooth terrain, raincells can grow larger and produce more intense precipitation of longer duration. A watershed positioned in depressions or near the lower terminus of canyons or drainages experiences more temperature transitions than one not exposed to drainage of cold air; as a result, incidents of rain on frozen soil increase in the former.

Internal topography of watershed. The physiography of the watershed itself can affect runoff, and hence ephemeral erosion, by influencing spatial distribution of precipitation and infiltration, and by playing a role in controlling runoff or subsurface flow. Within the watershed, microclimates associated with surfaces of different slope and aspect create variable temperature patterns, crop production, and soil properties. Hydrologic response varies accordingly. For example, greater crop or residue cover or greater soil organic matter impedes the formation of surface seals due to raindrop impact; thus infiltration is better maintained, and runoff is reduced. Evidence indicates that interaction between wind and ridge-shaped relief causes unequal distribution of precipitation at the surface. The effect of wind on the pattern of rainfall intensity received over a hill, where rainfall is converted to an equivalent depth received on a level surface is shown in **Fig. 1**. Thus, the configuration and orientation of divides or included ridges within watersheds determine surface precipitation inputs under a given wind regime, and these inputs influence the location and severity of ephemeral channel development. Internal topography primarily controls ephemeral erosion by determining the distribution of soil moisture in the watershed and the erosive power of emergent streams of concentrated flow. An understanding of these processes is essential in order to evaluate the erosion potential inherent in different landscapes.

Digital terrain models. In order to examine spatially dependent processes in landscapes and to de-

velop predictive relationships that are applicable in diverse environments, researchers require a nonpositional method of relating spatial properties within landscapes. In other words, the location in a landscape associated with ephemeral gully formation must not be defined in terms of fixed coordinates, but by parameters that describe erosion potential inherent at the location. Because ephemeral erosion processes are very sensitive to landscape configuration, parameters have been derived from topographic attributes.

Topographic parameters describing each location in a watershed are calculated by using a three-dimensional numerical representation of the watershed surface, the digital elevation model. Commonly, the digital elevation model is given as a series of elevations (Z values) for X and Y coordinates, as defined by the nodes of a uniform grid. A computer program analyzes the digital elevation model and outputs a digital terrain model; it models surface configuration by using topographic indices computed for surface points corresponding to all nonperipheral grid nodes in the digital elevation model. The indices commonly computed for each grid node are listed in **Table 1**. The relationship between curvature parameters and surface configuration is illustrated in **Fig. 2**.

Predicting ephemeral gullies. The associations between simple and combination indices and the occurrence and severity of ephemeral gullies in water-

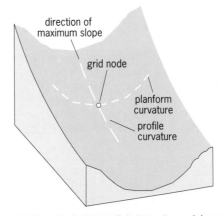

Fig. 2. Relationship between digital-terrain-model curvature indices and surface configuration.

Table 2. Simple and combination topographic parameters and associated indicator or predictive potential

		Hydrologic indicator[‡]	
Code	Name/definition[†]	Channel occurrence (high soil moisture)	Channel severity (erosive power)
S	Slope		+
P_F	Planform curvature	+	+
P_L	Profile curvature	−	
A	Upstream contributing area	+	+
A_U	Unit area	+	
CTI	P_L*A*S		+
PFS	P_F*S	$-/+^§$	
LNAS	$Log(A_U/S)$	+	$-^§$
ABS	A_U*S	−	−

[†] Asterisks indicate that the term in the first column is derived from the product of the indices given, that is, $P_L*A*S = P_L \times A \times S$.

[‡] Plus and minus signs indicate the sign of the correlation.

[§] Relationship may vary depending on watershed character.

sheds are presented in **Table 2**. Recent studies have shown that the presence of ephemeral channels is most strongly related to the topographic parameter of planform curvature (the curvature of the Earth's surface as measured along the contour). Once this factor is accounted for, indices coded as PFS, LNAS, and ABS (see Table 2) provide further explanation of the variability observed with regard to gully positions in watersheds. The severity, or size of ephemeral channels, was also found to be primarily related to planform curvature. Secondary relationships with the indices LNAS, ABS, and CTI have been observed. The nature of the relationships observed between the occurrence and severity of ephemeral gullies and topographic parameters appears to differ between watersheds having contrasting soil properties or other differences. The suggestion is that one, two, or three topographic parameters may not adequately describe ephemeral erosion hazards in various landscapes, and that topographic parameters alone are not adequate to predict the pattern of the erosion that may develop at a given site. Current approaches used to evaluate the ephemeral erosion potential in landscapes have employed topographic parameters or have endeavored to develop physically based mathematical models. The influences of internal topography on the patterns of rainfall, temperature transitions, and soil properties have not been addressed in these efforts, and they need to be included in future designs.   *Rodrick D. Lentz*

Parameters of Wind Erosion

The erosion of soils by wind has always presented a hazard to society. Many human activities can accelerate this basic geomorphological process, but erosion can be controlled with some basic practices. Wind erosion can render land almost useless for traditional agriculture. Only the most rudimentary agricultural systems can operate in a severely eroded region. In addition to damaging the land, wind erosion degrades the environment by generating large dust clouds that obscure the Sun, render traffic extremely hazardous, deteriorate painted surfaces, and damage or destroy plant seedlings and any moving mechanical device. Managing and using available resources is the key to effective and efficient control of wind erosion.

Soil erosion by wind is a subtle process, but the damage to plants can be dramatic. The impact on the soil may not be apparent until irreparable damage has already been done. Damage to plants is immediately apparent, because plants may be cut off at the soil surface; in addition, plants may be sufficiently damaged that the seedlings will die later, or the damage is such that crop quality and yields are severely reduced. Wind erosion has been studied for many years, but only recently have workers begun to understand the complete process and the complexity of trying to accurately measure and model wind erosion in the field. To describe the impact of wind erosion on the soil's ability to produce crops will require many years of additional research, because the impact depends on depth of the soil profile, the crop being grown, and the climate.

Effects on soil. Wind can detach soil particles, roll them along the soil surface, or inject them into the wind stream. Small particles become suspended in the air stream and may be transported hundreds to thousands of kilometers. In the detachment and transport process, the fine material is sorted in a manner similar to the winnowing of grain. The finest particles, 2–100 micrometers in diameter, are suspended in the air stream; the intermediate-size particles, 100–500 μm, are bounced along the soil surface in a process known as saltation; and the largest particles, 500–1000 μm, are moved along at the surface in a process known as soil creep.

As the soil surface continues to erode, it may be subjected to additional abrasion such that nonerodible aggregates are broken into erodible fractions. The fine material and soil organic matter that is lost represents the most productive portion of the soil profile.

In many areas of the United States, wind erosion is most prevalent at the time that crops are established. Most plant seedlings are very susceptible to damage by wind-blown soil particles. In fact, crops such as peppers and carrots can be destroyed when exposed to a 10-min windstorm. Yields of major cash crops like cotton can be reduced 50–75% by a 15-min exposure to blowing sand. The quality of horticultural crops and

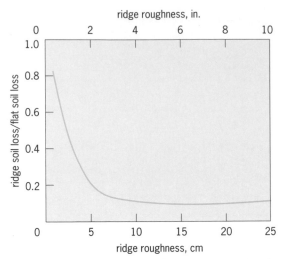

Fig. 3. Relationship between soil ridge roughness (height of the ridge when spacing between ridges is four times the height) and soil loss when the surface is ridged, divided by soil loss when the same soil surface is smooth. 1 mm = 0.04 in. (After D. W. Fryrear, Soil ridges-clods and wind erosion, Trans. Amer. Soc. Agr. Eng., 27(2):445–448, 1984)

tobacco may be reduced when sand particles become embedded in the plant tissue or when the damaged tissue serves as an avenue for attack by pathogens.

The importance of soil loss by wind erosion depends on the depth of the active rooting zone. If the productive soil is deep, like the loess soils of Iowa, the loss of a few millimeters of topsoil may not impact soil productivity. If the soil is shallow, the loss of any amount of topsoil may result in a permanent decline in soil productivity. The shallower the soil, the less the rooting depth; consequently, crops will have less available soil water and nutrients, and crop yields will decline.

Eroded soil material may be transported to the edge of a field or to cities downwind, or may even circle the globe several times before settling to the surface. The deposition zone may be in a house, a fence row, an open field, or a lake.

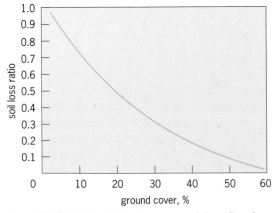

Fig. 4. Relationship between the percent of the soil surface covered with nonerodible material and the soil loss ratio (ratio of soil loss when the soil surface is covered to soil loss when the surface is bare). (After D. W. Fryrear, Soil cover and wind erosion, Trans. Amer. Soc. Agr. Eng., 28(3):781–784, 1985)

Control. The most effective methods for controlling wind erosion use available resources. Those methods involve using nonerodible soil aggregates, roughening the soil surface, maintaining crop residues, reducing the wind velocity at the soil surface by leaving crop stubble standing, or using wind barriers to reduce wind velocity adjacent to the downwind surface.

Soil roughening. Nonerodible soil aggregates occur on the surface of most soils when the appropriate tillage implement is used with adequate soil moisture conditions. Generally, the sandier the soil, the more fragile the soil aggregates. When the soil surface is roughened, a loose erodible particle may move only a few centimeters before larger nonerodible aggregates or a small depression protect the particle. Thus, soil movement stops and erosion by wind is controlled (**Fig. 3**). Ridges 10 cm (4 in.) high and spaced every 40 cm (16 in.) across the field will reduce soil losses by 90%. If tillage is used to control wind erosion, the soil must be tilled again whenever the previous nonerodible aggregates are destroyed by rain or weathering.

Soil cover. Maintaining crop residue on the soil surface can be an effective method for control of wind erosion in regions where crops such as wheat, oats, barley, rye, corn, sorghum, or forages are grown. In semiarid regions, the quantity of residue produced may not be sufficient to cover the soil. Research has shown that a 50% cover of the erodible soil surface will reduce soil losses 95%, and a 30% cover will reduce losses 70% (**Fig. 4**). The cover required depends on the erodibility of the soil. Sandy soils require a higher level of reduction to control erosion than silt loam or clay soils.

Wind barriers. Wind barriers reduce wind velocity to the lee of the barrier. The reduction in wind velocity depends on the porosity and height of the barrier. For most design purposes, the protected zone is assumed to be 10 times the height of the barrier. The actual protected zone depends on the velocity of the wind and the erodibility of the soil surface. The higher the wind velocity and the more erodible the soil, the smaller the protected zone.

Optimum control. For maximum effectiveness, the optimum wind erosion control system uses a combination of all control methods. While one method may provide better control under one set of soil-crop-weather conditions, a different method may be required during drought periods. For example, a wind barrier provides little protection when the wind is parallel to the barrier; but if tillage or crop residues are used in combination with a wind barrier, more effective erosion control is possible over a greater portion of the erosion season.   *D. W. Fryrear*

Wind Barriers

Wind barriers or windbreaks, which include shelterbelts, herbaceous barriers, and constructed barriers such as fences or nets, can effectively reduce wind erosiveness and soil erodibility. Effective windbreak networks are compatible with modern farming operations. Agricultural producers are increasingly turning to herbaceous windbreaks. Many governments offer

Table 3. Tree species commonly used in shelterbelts, their area of use, and their characteristics

Species	Area of use	Characteristics						
		Height	Width	Porosity	Drought hardness	Growth rate	Life-span	Competitiveness
Green ash	North America, Russia, Europe	medium	low	high	medium-high	medium	medium-high	low
Caragana	North America, Russia	low	medium	low	high	high	high	low
Poplar	North America, Russia, China, Europe	high	medium-high	high	low-medium	low-medium	low-medium	medium-high
Eastern redcedar	North America	medium	low-medium	low	medium-high	medium	medium-high	low
Spruce	North America, Russia, Europe	high	low-medium	low	low-medium	high	medium	medium
Scots pine	North America, Russia, Europe	medium-low	medium	medium	high	medium	medium-high	low
Birch	Russia	medium	medium	medium	medium-high	medium	medium	medium
Radiata pine	Australia-New Zealand	medium-high	low	medium-high	medium-low	high	low	medium
Siberian elm	North America, Russia	low	high	medium	high	high	low	high
Acacia	Europe, Australia-New Zealand, Africa, India	low	medium	low	high	high	low	high
Casuarina	Europe, Australia-New Zealand, Africa	high	medium	high	high	high	medium	high
Eucalyptus	North America, China, Australia-New Zealand, India	high	high	high	medium-high	high	medium	medium

assistance to encourage establishment of shelterbelts, reflecting the importance that the general public puts on soil conservation and promotion of environmental diversity.

The threshold velocity to make a wind erosive is about 19–24 km/h (12–15 mi/h). A soil is made erodible by conditions related to management or to intrinsic soil factors such as light texture, poor aggregation, lack of moisture, lack of crop residue, and excessive tillage. Wind erosion is controlled mainly by making soil less erodible through crop residue management and reduced tillage and by rendering the wind less erosive through use of windbreaks. These practices are most effective if they are combined in an overall conservation strategy.

Shelterbelts. Shelterbelts consist of rows of trees or shrubs whose designs vary depending on the purposes for which they are planted. Networks of single-row, single-species shelterbelts have been widely planted in North America and many other regions to control erosion and trap snow. In this type of shelterbelt, an ideal species is tall, narrow, moderately porous, drought-hardy, resistant to insects and disease, fast-growing, long-lived, and noncompetitive with adjacent crops. Since few species possess all these qualities, a wide variety is used in shelterbelts (**Table 3**).

Strips of natural trees are often left as field dividers but serve as effective windbreaks as well. Multirow, multispecies shelterbelts act as dense windbreaks, and they have yielded additional benefits such as providing wildlife habitat, timber, or fuelwood. In some coun-

tries, such as Denmark, the desire to promote environmental diversity has caused designs to evolve from utilitarian, monoculture barriers to true multiple-use barriers of 15 or more species. Herbaceous barriers include rows of perennial or annual crops such as grasses, cereal grains, or sunflowers. They are inexpensive to establish and are especially valuable in semiarid regions, where it is difficult to establish trees. Fences or nets are generally expensive to erect and maintain, but they are sometimes used to protect high-value crops such as fruits or vegetables.

Windbreak characteristics. Figure 5 shows schematically how windbreaks affect the wind. Low-level winds are partially blocked as they move

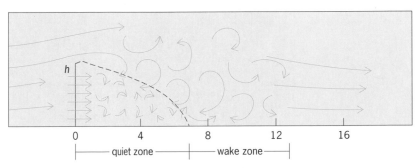

horizontal distance/barrier height

Fig. 5. Schematic representation of turbulence in the wake zone and in the quiet zone behind a model windbreak. The size of the circular arrows is proportional to the energy and scale of the turbulence. h = windbreak height. (*After K. G. McNaughton, Effects of windbreaks on turbulent transport and microclimate, Agr. Ecosys. Environ., 22/23:17–39, 1988*)

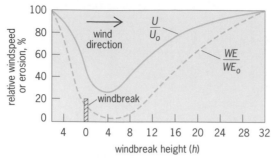

Fig. 6. Ratio of shelter to open-field wind speed (U/U_o) and ratio of wind erosion (WE/WE_o) with all wind speeds above threshold velocity and normal to a 40% porous windbreak. Wind speeds are measured at 0.12h above the surface. h = windbreak height. (*After L. J. Hagen, Windbreak design for optimum wind erosion control, in R. W. Tinus, ed., Shelterbelts on the Great Plains: Proceedings of the Symposium, Denver, Colorado, April 20–22, 1976, Great Plains Agr. Counc. Pub. 78, 1976*)

through the pores in the windbreak. The back pressure on the windward side forces high winds over the barrier. The low-level winds coming through the barrier move at a reduced velocity because of the energy that has been absorbed by the windbreak. The shear stress between the two zones causes the higher winds to be pulled toward the ground, creating a turbulent zone in which the slow-moving and fast-moving air streams mix. Farther from the windbreak, the upper winds return to ground level and restore the original wind regime.

A typical wind profile leeward of a moderately porous barrier is shown in **Fig. 6**. The effectiveness of any windbreak in reducing erosive winds depends on its physical characteristics, including height, length, orientation, and porosity. Wind protection is proportional to barrier height as long as porosity is constant. Therefore, shelterbelts become more effective as they increase in height, and taller species give more protection than shorter species. Windbreaks need to be at least 12 times as long as they are high, since wind that sweeps around the end of a windbreak changes direction because of the lower pressure created in the quiet zone, so that only the land within 60° from the end of the shelterbelt is protected.

Orientation. Windbreak orientation should be perpendicular to the predominant direction of erosive winds. In regions where winds come from more than one direction, shelterbelts should be planted in networks to protect from all directions. For a wind crossing a two-dimensional barrier at an angle θ, the protected distance d_p is given by the equation below,

$$d_p = d_o \times \sin \theta$$

where d_o is the distance protected from a perpendicular wind. Shelterbelts, being three-dimensional, give better protection than two-dimensional windbreaks from such winds, and the effective porosity changes as the angle changes. They even reduce parallel winds significantly.

Porosity. In two-dimensional barriers, windbreak porosity is the open space expressed as a percentage of the entire area. In shelterbelts, porosity is more

difficult to measure and varies from 10% or less for dense multirow barriers to over 80% for single-row shelterbelts in winter. The porosity causes the barrier to be more or less resistant and therefore affects the turbulence and wind speed on the leeward side. Wind erosion is affected mainly by the wind speed, which regains erosive force beyond the zone of turbulence. A porosity of 40% or less is best for good wind erosion protection.

Open-field wind speed. The distance to which a barrier completely stops erosion from a perpendicular wind depends on open-field wind speed. A 20% wind reduction would reduce a 30-km/h (19-mi/h) wind to the threshold velocity, while a 40% reduction would be needed for a 40-km/h (25-mi/h) wind. Therefore, the barrier in Fig. 6 would give total protection to 19h (h is the windbreak height) in the first case but only to 12h in the second case. Beyond the area of total protection, the amount of erosion is proportional to the cube of the wind speed. The result is a wind erosion profile as shown in Fig. 6, indicating significant wind erosion protection to over 20h.

A study done in Saskatchewan, Canada, involved direct measurements of soil movement leeward of shelterbelts. Soil traps were set in three fields in lines perpendicular to shelterbelts when erosive winds were blowing across the shelterbelts. Total erosion protection to 9h was found, and some protection was apparent as far as 20–25h. None of the curves show an equilibrium of wind erosion by the last observation, so shelterbelt benefits may extend even farther. Since shelterbelts normally reduce horizontal wind speed to a distance of 25–30h, soil movement can be expected to reach an equilibrium somewhere beyond this zone.

Advantages. Windbreaks also reduce soil erodibility by increasing crop growth and soil moisture. The growing conditions are improved in shelter; that is, wind damage to crops is reduced, daytime temperatures are higher, nighttime temperatures are lower, and evapotranspiration is reduced. Yields, especially vegetative yields, increase so that sheltered soils are higher in organic matter. The organic matter binds soil aggregates together, rendering them less erodible. These soils maintain their moisture- and nutrient-holding capacity better than eroded soils, and therefore an even greater yield advantage in the sheltered zone develops over time. Snow trapment and reduced evapotranspiration increases soil moisture and reduces soil erodibility in shelter even further, and helps the soil to resist pulverization during tillage.

The biological interaction between shelterbelts, crops, and wind erosion has especially significant economic implications for the producer who grows the type of vegetable crops that leave little vegetation to protect the surface. These crops are also more valuable than cereal grains and are more responsive to shelter influence. Soil erosion control is especially important in many vegetable crops, since abrasion by soil particles and wind can significantly reduce their quality. The interaction of low residue, high value, high shelter response, and increased quality makes shelterbelts or some other type of windbreak essential for vegetable producers.

For background information *SEE EROSION; SOIL; SOIL CONSERVATION* in the McGraw-Hill Encyclopedia of Science & Technology.

John Kort

Bibliography. R. A. Bagnold, *The Physics of Blown Sand and Desert Dunes*, 1943; C. Carlson (ed.), *Soil Conservation: Assessing the National Resources Inventory*, vol. 2, 1986; W. S. Chepil and N. P. Woodruff, The physics of wind erosion and its control, *Adv. Agron.*, 15:211–302, 1963; D. W. Fryrear, Wind erosion: Mechanics, prediction, and control, *Adv. Soil Sci.*, 13:187–199, 1990; D. W. Fryrear et al., Wind erosion: Field measurement and analysis, *Trans. Amer. Soc. Agr. Eng.*, 34(1):155–160, 1991; G. M. Heisler and D. R. DeWalle, Effects of windbreak structure on wind flow, *Agr. Ecosys. Environ.*, 22/23:41–69, 1988; I. D. Moore, G. J. Burch, and D. H. Mackenzie, Topographic effects on the distribution of surface soil water and the location of ephemeral gullies, *Trans. Amer. Soc. Agr. Eng.*, 31(4):1098–1107, 1988; I. D. Moore, R. B. Grayson, and A. R. Ladson, Digital terrain modelling: A review of hydrological, geomorphological, and biological applications, *Hydrolog. Proc.*, 5:3–30, 1991; E. L. Skidmore and L. J. Hagen, Reducing wind erosion with barriers, *Trans. Amer. Soc. Agr. Eng.*, 20(5):911–915, 1977; G. Tibke, Basic principles of wind erosion control, *Agr. Ecosys. Environ.*, 22/23:103–122, 1988; N. P. Woodruff and F. H. Siddoway, A wind erosion equation, *Soil Sci. Soc. Amer. Proc.*, 29:602–608, 1965.

Soil mechanics

Soil generally consists mainly of sand, clay, and silt particles bonded into aggregates. The strength of this bonding and the size of these aggregates depend upon the content of cementing materials such as clay, organic matter, and iron oxides. Every soil that supports plant life is thus structured at the surface to some extent, and the proportion of soil that is aggregated and the sizes of the aggregates vary widely. The larger the aggregates, the better are the macropore characteristics and thus the water-transmission and air-exchange properties of the soil. Therefore, the stability and size distribution of aggregates in a particular soil are important to its productivity. However, aggregate integrity and strength are threatened by such factors as intensive land use and climatic stress. Alternate freezing and thawing, which is typical of the cool seasons of temperate, maritime regions, is a potent aggregate-disintegrating force.

Aggregate breakdown. Reduction in the strength and size of aggregates (aggregate breakdown) results in reduced resistance of the soil to erosion. The reduced resistance potentially can lead to a loss of soil and nutrients from farm or forest land and their transference by rill or interrill flow (sheet flow between rills) to streams, rivers, and estuaries as pollutants.

Aggregate breakdown of the surface soil takes place in four stages. The first stage is characterized by large

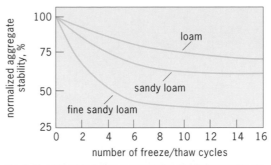

Fig. 1. Relationship between normalized soil aggregate stability and the number of freeze/thaw cycles for three agricultural soils. (*After L. M. Edwards, The effect of alternate freezing and thawing on aggregate stability and aggregate size distribution of some Prince Edward Island soils, J. Soil Sci., 42:193–204, 1991*)

intergranular pore spaces. In the second stage, crust begins to form in small pore spaces near the surface, some microaggregates break off, and the aggregates at the surface flatten. By the third stage, the aggregates are broken down further, and some of the surface crust is washed downward; thus pore size is further decreased. At the fourth stage, the macropores are significantly reduced.

In the laboratory, aggregate breakdown can be studied and expressed as the ability of a selected size of aggregate to maintain its integrity (aggregate stability) while being agitated in water for a fixed time; or it can be defined as the size distribution of aggregated fractions resulting from the breakdown (upon similar agitation) of a selected size of aggregate. Measurements are based on dry mass.

The soil aggregate breakdown characteristics of three agricultural soils (loam, sandy loam, and fine sandy loam) were studied in relation to repeated freezing and thawing (daily freeze/thaw cycles; **Fig. 1**). Aggregate breakdown was examined as aggregate stability and as mean weight diameter, which is a summary measurement of changes in aggregate sizes (that is, aggregate size distribution). In response to an increasing number of freeze/thaw cycles, there was a nonlinear pattern of decline in aggregate stability for all three soils in which there was, generally, an initial, rapid breakdown, which afterward became progressively less rapid and approached a minimum at which there was a leveling-off (its asymptote). **Figure 2** shows a graph in which the data are plotted in a transformed (normalized) state; that is, the data for all three soils have a common origin (zero cycle) at the point 100 on the *y* axis for simplifying comparisons.

Patterns of decrease. The specific patterns of decrease in normalized aggregate stability and normalized mean weight diameter for the three soils were dissimilar. The rate of breakdown (curvature) to the asymptote was significantly gentler for the loam than for either the sandy loam or the fine sandy loam, a result that is attributable to greater macroaggregate strength associated with a higher content of clay and iron oxides in the loam. The extent of breakdown of the fine sandy loam was significantly greater than that of either of the other two soils, attributable to its

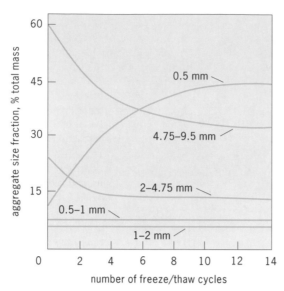

Fig. 2. Relationship between soil aggregate size fraction and the number of freeze/thaw cycles for four size grades averaged over three agricultural soils. 1 mm = 0.039 in. (After L. M. Edwards, The effect of alternate freezing and thawing on aggregate stability and aggregate size distribution of some Prince Edward Island soils, J. Soil Sci., 42:193–204, 1991)

relatively low content of organic matter. The number of freeze/thaw cycles to maximum aggregate breakdown was up to 3.5 times greater for the loam than for the sandy loam or the fine sandy loam. Thus, for example, in the Atlantic region of Canada, where an average of 42 freeze/thaw cycles occur during the cool period, the loam would emerge without severe loss of aggregate structural integrity, in contrast to the sandy loam, which fully breaks down after 15–17 cycles, or the fine sandy loam, after 20–25 cycles.

Although with freeze/thaw substantial changes in the soil water composition cause chemical reactions that lead to aggregate breakdown, by far the greater effect on soil aggregate stability is physical.

Size distribution. As with repeated stress on soil aggregates, the greatest change in size distribution with freeze/thaw cycling occurs in the largest- and smallest-size fractions (Fig. 2). In the above study of aggregate breakdown in three agricultural soils, aggregate sizes from 9.5 to less than 0.5 mm (0.37 to less than 0.02 in.) were examined; it was determined that there was, on average, an overwhelming increase (33%) in the mass of the finest fraction with increasing freeze/thaw cycling, and a simultaneous decrease in the largest and second-largest fractions (28% and 10%, respectively). Intermediate-size fractions showed practically no change. The magnitude of these changes varied with soil type. Whereas the smallest aggregate fraction showed similar percentage increases (with increasing freeze/thaw cycling) in all three soils, the percentage reduction of the largest two fractions decreased in the order fine sandy loam, sandy loam, loam. The loam thus showed a relatively small decrease in macroaggregate content with increasing freeze/thaw cycling. Seventy percent of the mass of the fine sandy loam was composed of the finest fraction

after freeze/thaw cycling, while only 15% of these microaggregates were present in the loam.

The relatively good structural stability of loams has rendered them invaluable for sustained soil productivity in climates such as the Pacific northwest and the Atlantic northeast of North America, where alternating freeze/thaw prevails.

Some studies have shown substantial reduction (up to 50%) in soil cohesion strength after only a single freeze/thaw cycle. By subjecting aggregates to a single freeze/thaw cycle, significant improvement has been made in the methodology for determining aggregate stability.

In the Atlantic northeast, studies of agricultural soils subjected to freeze/thaw cycles have shown increases of over 200% in runoff volume, about 90% in sediment loss in interrill flow, and about 30% in sediment loss in rill flow from bare soil. The erosion resistance of soil experiencing freeze/thaw has, however, been known to drop to 15% of that of nonfrozen soil.

There is evidence as well that through moderate aggregation, freeze/thaw could increase the hydraulic conductivity of poorly aggregated soils under conditions of low soil water content, and improve the structure of near-impervious parent materials. However, the ultimate effect of freeze/thaw on a moist, aggregated soil is breakdown.

Water retention and transmission. The relationship between soil water content and soil aggregate breakdown under the influence of freeze/thaw is cyclical. Research in the northwestern United States has shown that the initial drainage status of a soil, and thus its water content, determines the extent or speed of aggregate breakdown by freeze/thaw. Freeze/thaw is most effective in this process where the aggregates are large and saturated, because of pressures and associated shearing forces exerted by the water in the process of freezing. During freezing, aggregates act as a sink for free subsurface water due to an increased hydraulic potential (progressively nearer the soil surface), which attracts water to the freezing front—a process that virtually ensures the vulnerability of aggregates even in subhumid, temperate climates. The fine aggregates that result from the breakdown of macroaggregates cause reduced macroporosity and, consequently, reduced infiltration.

The adverse effects on the water retention and transmission characteristics of a soil have no chance to be improved before late spring, when reaggregation sets in under conditions of reduced pressure (originally set up by the development of ice lenses) in the soil mass, and under the influence of increased microbial activity in the root zone. Up to 60% microbial mortality and a virtual cessation of the aggregate building that results from microbial activity can be expected upon freeze/thaw. After the cool period, however, no more than moderate reaggregation may be expected where soil management is mechanically harsh (as under intensive cultivation), chemically destabilizing (as with heavy application of univalent cations to the soil), or biologically poor (as with depleted soil organic matter).

The vulnerability of soil to erosion under circumstances of freeze/thaw is expected to differ on the basis of soil type. Researchers in the Atlantic region of Canada, looking at the responses of three soils to freeze/thaw, found significantly greater sediment loss from a fine sandy loam than from a loam or sandy loam, matching the pattern of aggregate breakdown separately observed for similar soils.

Soil aggregate stability may undergo many changes in the field. In a practical sense, however, any climatic or management event that covers the soil near the beginning of the cool period will improve the characteristics of soil heat and water flow, thus minimizing the effects of freezing on the aggregated surface of that soil. There are thus obvious benefits from an early, even, deep snow cover that lasts until spring, a cover crop or crop residue cover, or standing stubble, as has been demonstrated by researchers.

For background information SEE SOIL; SOIL CHEMISTRY; SOIL MECHANICS in the McGraw-Hill Encyclopedia of Science & Technology.

Linnell M. Edwards

Bibliography. L. M. Edwards, The effect of alternate freezing and thawing on aggregate stability and aggregate size distribution of some Prince Edward Island soils, *J. Soil Sci.*, 42:193–204, 1991; P. A. Frame, J. R. Burney, and L. M. Edwards, Laboratory measurement of freeze/thaw, compaction, residue and slope effects on rill erosion, *Can. Agr. Eng.*, 34(2):143–149, 1992; H. Kok and D. K. McCool, Quantifying freeze/thaw induced variability of soil strength, *Trans. Amer. Soc. Agr. Eng.*, 33(2):501–506, 1990; J. L. Walworth, Soil drying and rewetting, or freezing and thawing, affects soil solution composition, *Soil Sci. Soc. Amer. J.*, 56:433–437, 1992.

Soil microbiology

Soils are highly complex ecosystems in which numerous physical, chemical, and biotic factors continuously interact. Changes in any one factor may bring about variations in many others, both directly and indirectly. Many changes result from human activity, some having greater impact than others. One such manipulation of soils occurs at opencut (surface; also called opencast) mine sites. In opencut mining, the ore body (for example, coal) is made accessible by removing the material that lies above it: topsoil, subsoil, and overburden. These three materials are generally stored in large separate mounds in and around the mine site. When the ore body in an area has been worked out, the resulting void is filled by sequentially replacing overburden, subsoil, and topsoil.

Ideally, the land once restored would support vegetation much as before the mining operation began. However, the operations of soil stripping, storage, and respreading have the potential for profoundly changing the parameters that affect soil fertility, such as microbiological status. A typical soil contains 1–5% carbon, 1–3% of which is made up of a living component, or biomass. Over 90% of the biomass of most soils is composed of soil microorganisms (bacteria, fungi, actinomycetes, algae, and protozoa), and the terms soil biomass and soil microbial biomass are often used synonymously. The significance of the soil microbial community is far greater than its contribution to total soil mass (0.01–0.15%), since it represents a highly dynamic, active pool of carbon, responsible for decomposition of organic matter and nutrient cycling. Changes in the size, composition, and activity of the soil microbial community may, therefore, be predicted to have consequences for soil productivity. In the context of opencut mining operations, soil biomass is subjected to major stress during two phases: stockpile construction and subsequent soil storage, and mound dismantling and land restoration.

Mound storage. An opencut mine site may be in a previously worked area or in an undisturbed so-called greenfield site. In either case, topsoil (the surface layer, 4–12 in. or 10–30 cm thick) is stripped off with mechanical diggers and stockpiles are constructed, often around the periphery of the working site, where they act as visual screens and noise baffles. The normal practice is to incorporate existing vegetation in the mound and to reseed the surface with a ryegrass-based mix, to protect against erosion. Dimensions of the mounds are variable, but stockpiles tend to be 17–26 ft (5–8 m) wide at the base, 10–17 ft (3–5 m) wide at the top, 10–20 ft (3–6 m) high, and less than 330 ft (100 m) long.

An inevitable consequence of storing soil in a mound is the increase in the soil mean bulk density as

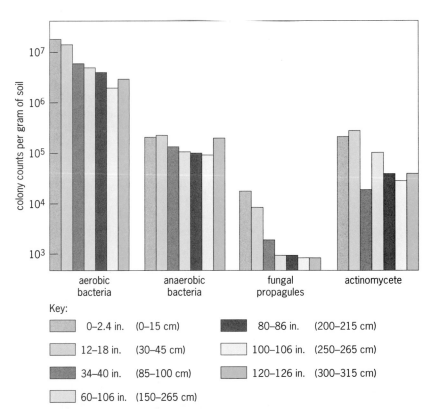

Fig. 1. Variation of microbial numbers (indirect counts) with depth in an 11-year-old topsoil storage mound in north England. The ranges identified in the key refer to the depth from which the soil was sampled.

a result of compaction (a relatively rapid reduction in volume) and consolidation (a more protracted phase) due to the weight of overlying soil and to the effect of traffic. When soil bulk density increases, pore volume decreases and the percentage of micropores (that is, those pores that tend to hold water rather than air) increases. The increase in micropores, coupled with the fact that the gaseous interchange between soil in the bulk of the mound and atmospheric air is very limited, means that microbial metabolism quickly depletes the oxygen in the mound and anaerobic conditions set in. Although many microorganisms are unable to function in the absence of oxygen and either die out or become inactive, others (principally bacteria) can adapt to or may even require anaerobic environments. A number of significant differences exist between the microbiology of aerobic and anaerobic soils. For example, microbial diversity tends to be greater in aerobic soils; energy available from organic matter breakdown is greater under aerobic conditions, so that biomass tends to be smaller under anaerobic conditions; and some products of anaerobic metabolism (such as sulfide and certain organic acids and alcohols) are phytotoxic.

Profound time- and depth-related changes occur in the microbiology of soils in stockpiles. However, because of heterogeneity of mound soils on a macroscale as well as microscale, considerable variability occurs in their biomass, particularly in the earlier phase of storage. Within the first few weeks of mound construction, there is actually a stimulation of soil biomass in general and bacteria in particular. This stimulation arises principally from the mixing of soil and vegetation from the surface, but it also stems from the physical manipulation of soil, which makes accessible to microorganisms organic materials that had previously been protected from attack. This flush of activity tends to be short-lived, and within 6 months to 1 year the surface layer of soil in the mound usually contains both a larger total biomass and greater numbers of all individual microbial groups (wth the exception of anaerobic bacteria) than deep-buried soil. With time, biomass variations due to soil heterogeneity decrease, and differences in the microbial status of surface and deep-buried soil become more pronounced. **Figure 1**
shows data from an 11-year-old soil mound in Cumbria, north England. Total biomass in deeply buried soil is often not too dissimilar to that at the mound surface, but this condition is due to a high percentage of microorganisms at depth being moribund. For example, although fungal propagules may be recovered from deep soil, over 99% of fungal biomass is either dead or nonviable. Only about 1 in 1000 bacteria in the core of storage mounds is capable of growing on solid media, compared to 1 in 100 in surface soil. Of those isolates, most are capable of forming endospores and are probably inactive at depth. The actual delineation of surface and deep-buried soil is somewhat arbitrary, as oxygen concentrations decline steadily with soil depth. However, soil less than 3.3 ft (1 m) in depth can generally be considered as aerobic, and soil below this level, anaerobic. Therefore, in a typical stockpile 50–70% of soil is in the anaerobic zone.

Land restoration. Dismantling a storage mound and spreading topsoil (usually to a depth of 4–6 in. or 10–15 cm) on restored land has a second significant effect on soil microflora. The accessibility of oxygen results in a short-term flush of activity of principally aerobic bacteria (**Table 1**), which utilize organic materials that have accumulated under anaerobic conditions in the mound. Experiments with soils taken from different depths in a mound have confirmed that this flush is restricted to anaerobic zone soils. In contrast, numbers of actinomycetes and fungal propagules actually decline (relative to surface mound soil) in the initial phase of restoration. This situation is, however, transient, and after some months bacterial numbers decline while those of actinomycetes and fungi increase. Concomitant with those changes is a change in the overall composition of soil biomass, with soil fungi accounting for around 50% of total biomass after 2 weeks but more than 95% of total biomass 1 year after restoration. One other important ongoing change is the proportion of total soil carbon that is in the form of biomass carbon, in the region of 1–2% for normal soil ecosystems and less than 1% in stockpile mounds (particularly in the anaerobic zone). The amount of biomass carbon declines further following land restoration (Table 1). Restored land, monitored over a 3-year period, showed continued increases in

Table 1. Comparison of the microbiological status of stockpiled and restored soils*

Soil[†]	Biomass, mg carbon/kg	B/C, %[‡]	Aerobic bacteria, $\times 10^6$	Anaerobic bacteria, $\times 10^3$	Fungal propagules, $\times 10^3$	Actinomycetes, $\times 10^6$
Mound[¶]						
Surface, $n = 4$	225	0.81	26	150	22	3.3
Buried, $n = 8$	142	0.53	7.1	140	11	0.62
Restored						
2 weeks, $n = 16$	160	0.56	230	440	4.8	0.61
26 weeks $n = 9$	90	0.33	190	850	29	1.9
52 weeks, $n = 9$	100	0.39	43	770	31	1.2

* Data are representative mean values from an opencut coal mine in central England.
[†] n = number of soil samples used in the analysis.
[‡] B/C = percentage of total soil carbon in the form of biomass.
[¶] Surface = 0–1.7 ft (0–0.5 m); buried = 3.3–17 ft (1–5 m).

Table 2. Dissolved organic carbon (DOC) and ammonium-nitrogen (NH₄-N) contents of mound soils*

Soil depth, in. (cm)	DOC, μg/g soil	DOC, % total soil C	NH₄-N, μg/g soil	NH₄-N, % total soil N
0–2.4 (0–15)	95	0.35	12	0.5
40–46 (100–115)	237	0.99	154	4.8
80–86 (200–215)	—	—	183	59
120–126 (300-315)	157	1.05	59	3.7

*Data are from a 4-year-old stockpile at an opencut coal mine in central England.

total soil carbon, indicating that a steady-state situation had not been reached within this time.

Environmental implications. Within the anaerobic zone of storage mounds, soils accumulate large quantities of soluble carbon compounds and of ammonium-nitrogen (**Table 2**). The former result from incomplete mineralization of organic matter under anaerobic conditions, and the latter from the fact that ammonium oxidation, unlike ammonium formation, is inhibited by the absence of oxygen. When land is restored, some of the soluble organic matter will be utilized by endemic microorganisms (a major cause of the bacterial flush described above), and some will be leached out in percolating rainwater; the relative proportions depend on climatic as well as soil factors. Metabolizable organic matter entering streams and rivers from water draining restored land will stimulate heterotrophic microbial activity and thereby contribute to oxygen depletion. Ammonium is rapidly oxidized to nitrate by nitrifying bacteria under the aerobic conditions that prevail in freshly restored soils. Interestingly, these obligate aerobes can survive, albeit in reduced numbers, in the anaerobic zone of stockpiles for more than 10 years. Nitrate is a highly labile anion in soil ecosystems, and it is readily lost by the processes of leaching and denitrification. **Figure 2** shows nitrate levels in drainage water from restored land peaking 3–4 weeks after restoration at more than 180 μg/ml, and also nitrate output being lowered by incorporating straw (a high carbon:nitrogen microbial substrate) or by specifically inhibiting nitrifying bacteria with dicyandiamide. Enhanced nitrate levels in aquifers are causing considerable concern, since they contribute to water eutrophication and have been implicated in human health problems. In the absence of leaching, denitrification can lead to significant nitrogen loss from restored soils. In moist soils incubated in the labortory at 65°F (20°C), over 10% of total soil nitrogen has been found to be lost from soil taken from the anaerobic zone of a soil mound within the period of 3 weeks. A high proportion of gaseous nitrogen is lost in the form of nitrous oxide (N_2O) rather than dinitrogen (N_2), a phenomenon that has also been observed when denitrification occurs in soils that contain elevated concentrations of nitrate. Nitrous oxide is significant as a greenhouse gas that contributes to global warming.

Limiting deleterious effects. Clearly, the operations of stripping, stockpiling, and respreading of soils at opencut mine sites adversely affects the soil microbial community. However, deleterious effects can be minimized. Many of the problems deriving from the development of anaerobic conditions in the central bulk of storage mounds can be avoided altogether if soil stripping and restoration are carried out at the same time, the stripped soil being used to restore land without any interim storage period. Cutting down the time that soils spend in storage is a less satisfactory solution. Reducing the size of stockpiles minimizes the percentage of stored soil that becomes anaerobic, and limiting the amount of vegetation buried in mounds slows down the rate of oxygen depletion. When ammonium-rich soil is used in land restoration, it is both financially and environmentally beneficial to limit the amount of nitrogen loss following oxidation to nitrate. Current research indicates that by inhibiting nitrifying bacteria (for example, with dicyandiamide) and stimulating biomass (to capture ammonium ions) a high percentage of nitrogen can be conserved in newly stored soils.

For background information *SEE LAND RECLAMATION; SOIL MICROBIOLOGY; SURFACE MINING* in the McGraw-Hill Encyclopedia of Science & Technology.

D. Barrie Johnson

Bibliography. D. B. Johnson, J. C. Williamson, and A. J. Bailey, Microbiology of soils at opencast coal sites: I. Short- and long-term transformations in stockpiled soils, *J. Soil Sci.*, 41:1–8, 1991; J. C. Williamson and D. B. Johnson, Determination of the activity of soil microbial populations in stored and restored soils at opencast coal sites, *Soil Biol. Biochem.*, 22:671–675, 1990; J. C. Williamson and D. B. Johnson, Microbiology of soils at opencast coal sites: II. Population transformations occurring following land restoration and the influence of ryegrass/fertilizer amendments, *J.*

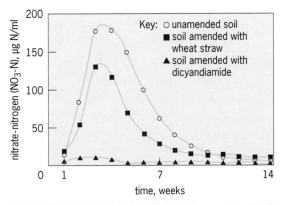

Fig. 2. Nitrate levels in drainage waters from restored soil that had been stored within the anaerobic core of a topsoil stockpile.

Soil Sci., 42:9–15, 1991; J. C. Williamson and D. B. Johnson, Mineralisation of organic matter in topsoils subjected to stockpiling and restoration at opencast coal sites, *Plant Soil*, 128:241–247, 1990.

Solar energy

Among the numerous ways in which the energy from the Sun can be harnessed, two that have been studied in recent years are solar chimneys and solar power satellites.

Solar Chimneys

The solar chimney system is a natural-draft device that uses solar radiation to provide upward momentum to a mass of air, thereby converting the thermal energy into kinetic energy. This kinetic energy is extracted from the air with suitable wind machines. Because of its lack of dependence on natural occurrence of wind, this system is a very attractive development. In effect, the load factor of the wind turbine is increased substantially by providing a more or less regular source of wind. Moreover, since the direction of air movement is fixed, the complicated and expensive tracking mechanism necessary for a standard wind turbine is not needed.

The solar chimney system can be conveniently described and analyzed by examining its three main components: the solar collector or the greenhouse, the chimney, and the wind machine (**Fig. 1**). The greenhouse collects solar radiation and heats the air mass near the ground. The chimney, near the center of the greenhouse, acts as a funnel to increase the velocity of the air. The wind turbine, mounted within the chimney, taps the kinetic energy of the air and converts it into rotational mechanical energy that can be readily converted into electricity. This electricity either can be connected into the main power grid or can be stored conveniently in a useful form. One possible form of storage, which also provides convenient end-point use, is as hydrogren produced by electrolysis. For countries like India, with a large geographical spread, the solar chimney system has been proposed as a way of reliably and inexpensively providing electricity to remote rural areas.

The solar chimney system was originally proposed

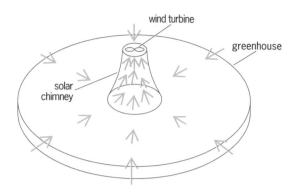

Fig. 1. Overall view of a solar chimney system.

in the 1970s, and by 1982 a pilot plant was operating in Manzanares, Spain, as a joint project between the German government and a Spanish utility. Although the operating experience of the prototype is encouraging, technical and economic improvements are needed in order to make the system more viable and marketable. The **table** gives parameters of the Spanish facility and of hypothetical designs for larger plants.

In the beginning, the solar chimney was conceived as a constant-diameter chimney, and the pilot plant was built according to that design. The next significant step in the improvement of the solar chimney system may come with the design of an optimally shaped chimney, probably of a converging type. A shape of this kind would have two basic advantages: it would act as a funnel to accelerate the flow, and it would be structurally much more stable than the constant-diameter chimney.

Performance parameters. The performance of the solar chimney system may be described by two parameters: the efficiency of conversion of thermal energy to electricity, and the amount of electricity generated at design conditions. The efficiency of the system was found to be proportional to the square of the ratio of the chimney entrance and exit areas. This result is intuitively obvious, since, by continuity, the kinetic energy of the exit airstream must vary as the square of the area ratio: the higher the area ratio, the higher the efficiency.

The effect of the height of chimney entrance above the ground can be explained by examining a simplified description of the operation. The air near the ground is

Parameters of solar chimney systems*

Plant size	0.1 MW[†]	10 MW	100 MW	1000 MW
Chimney height	200 m	400 m	600 m	900 m
Collector diameter	250 m	1780 m	4350 m	9500 m
Collector area	5×10^4 m²	2.5×10^6 m²	1.5×10^7 m²	7.1×10^7 m²
Chimney diameter	8 m	50 m	130 m	400 m
Turbine diameter	6 m	40 m	103 m	7×123 m
Average power with 540 W/m² for 10 h/day	0.05 MW	10 MW	100 MW	1000 MW
Maximum power at 1000 W/m²	0.1 MW	15 MW	150 MW	1470 MW
Average efficiency	0.2%	0.75%	1.25%	1.8%

*1 m = 3.28 ft; 1 m² = 10.76 ft².
[†] Pilot plant in Manzanares, Spain.
SOURCE: L. B. Mullet, The solar chimney—Overall efficiency, design and performance, *Int. J. Ambient Ener.*, B(1):35–40, 1987.

heated by the insolation and is forced to rise because of the buoyant acceleration resulting from the change in density. The buoyant acceleration continues until the flow enters the chimney. Once in the chimney, the flow is basically governed by continuity: what goes in must come out. In order to conserve the mass, the airstream keeps accelerating until it exits the chimney. Thus, the final velocity of the air (or equivalently, the final kinetic energy available to the wind turbine) is governed by two factors: the buoyant acceleration, causing the hot air to enter the chimney with a certain velocity, and the nozzle type of acceleration, causing the air to accelerate further. It can be argued that the higher the chimney entrance above the ground, the higher is the inlet velocity to the chimney and the higher is the power output. Early studies found that the efficiency is in direct proportion to this entrance height. The power output, however, was found to be proportional to the 3/2 power of the entrance height. The extra factor of the square root of the height comes from the dependence of the power output on the mass flow rate through the chimney.

The effect of the temperature rise near the ground is to increase the power generated. Some investigators found that the power generated is proportional to the 3/2 power of the initial temperature rise (to a first-degree approximation), taking into account the variation of kinetic energy (proportional to the temperature rise, to first degree) and the mass-flow rate (proportional to the square root of the temperature rise).

Future research. Four major areas may be outlined for further investigation: thermal and fluid design, wind-turbine design, structural design, and greenhouse design.

Thermal and fluid design. The solar chimney system is essentially a heat engine, working between the heat source provided by the Sun and the atmospheric sink. It deals with high mass-flow rates of air, as well as high velocities. Therefore, losses in the available energy are from two sources: thermal losses due to heat transfer between the hot air inside the chimney and the cold air outside, and entrance, exit, and frictional losses from the airstream.

Because of the high velocity of air within the chimney, the heat-transfer coefficients are expected to be rather high. Therefore, a material with a low thermal conductivity should be used for the chimney construction.

The frictional losses are incurred because of the physical contact of the airstream with solid surfaces. Since the flow is most likely to be turbulent, these losses can be assumed to be negligibly small. The losses due to entrance into and exit from the chimney, however, cannot be neglected. These losses can be minimized by properly designing the chimney envelope so that a smooth streamlining of the airflow is maintained at all mass flow rates. Early simulations of the airflow inside the chimney indicated a possible recirculating zone within the entrance if the envelope is not well designed. This recirculation zone could dissipate a major portion of the available kinetic energy. Thus, the determination of the shape of the chimney envelope is crucial.

Another interesting effect worth investigating is the swirling motion that the airstream is most likely to have. At the scale of the chimney, the Coriolis force due to the rotation of the Earth may not be negligible.

The several commercially available fluid-flow simulation packages can be used for further study. Laboratory-scale testing and corroboration of the theoretical results also need to be carried out.

Wind-turbine design. Since the power that can be extracted from a solar chimney system varies as the cube, and the efficiency as the square, of the air velocity at which the wind turbine operates, it is desirable to operate at as high an exit velocity as possible. Early theoretical studies suggested the possibility of increasing the air velocity to cause significant compressibility effects. However, present-day turbines cannot handle wind speeds above 80–120 km/h (50–75 mi/h). The elimination of the tracking mechanism from the solar-chimney wind turbine, because of the highly regulated wind direction, allows the turbine to withstand higher velocities.

Structural design. There are essentially two major structural paradigms: the compression structure and the tension structure. In compression structures, such as conventional buildings, the load is transmitted to the foundation through successive members. In contrast, a tension structure is suspended from the top, and the members of the structure are in tension. Suggestions for using the tension structure concept for solar chimneys include suspending the chimney from helium balloons. Another alternative would be to make the actual solar chimney from a material that can support only tension, such as tarpaulin, and suspend it from a central mast. Obviously, much can be done in this area, and there is always scope for further improvement and reduction in costs.

Greenhouse design. The greenhouse of the solar chimney system is the solar energy collector and consists of a canopy of a transparent material stretched over the ground on a frame. The air near the ground gets hot and rises through the chimney, creating a low-pressure region in the greenhouse. The cool air from outside is drawn in and in turn gets heated and rises through the chimney. The greenhouse may be viewed as the most crucial item in a solar chimney. Because of its vast spread on the ground, it is the costliest.

The cover of the greenhouse deserves particular attention. It must be made of a transparent material that can withstand continuous exposure to sunlight, and it must be tough enough to accommodate the effects of air movement underneath it. It must also be as cheap as possible, since it may represent as much as half the total expense.

The solar chimney system is a natural convection device in that the driving force (temperature rise) and the system response (mass-flow rate) are intimately coupled. Therefore, a systematic study of the total dynamics of the airflow within the greenhouse is needed. More specifically, the exact dependence of the mass-flow rate on the temperature rise, and ultimately of the available kinetic energy on the solar radiation, needs to be established by employing heat-transfer analyses. An interesting suggestion is to paint the ground under the greenhouse black so as to increase absorptivity. Sensitivity analyses should be carried out to determine the merits of this idea, and also to determine whether it is better to paint the ground fully or partially. Other important factors include the height of the greenhouse (constant or variable), the design of the supporting frame, and the ground area covered.

S. A. Sherif; M. M. Padki

Solar Power Satellites

Renewable energy sources such as solar power have been under consideration for a number of years. Solar collectors on Earth are limited by atmospheric attenuation, cloud blockage, the day–night cycle, and the oblique angle of the Sun's rays with respect to the ground. To circumvent these limitations, P. Glaser proposed in 1968 that large solar collectors, called satellite solar power stations or solar power satellites (SPS), be placed in geostationary orbit 22,300 mi (35,800 km) above the Equator.

In geostationary orbit, a given solar power satellite will always be over the same spot on the Earth's surface. For maximum illumination, it can be oriented so that its collecting surface is perpendicular to the Sun's rays. Sunlight striking the satellite is converted to electricity, and the electricity is converted to microwaves, which are beamed to the Earth's surface. The microwaves are received and converted back to electricity by a rectifying antenna, or rectenna. The rectenna is actually a large array of many small antenna elements. It converts alternating current at microwave frequencies to direct current. This direct current is then converted to 50 or 60 Hz alternating current and distributed to users (**Fig. 2**).

The solar power satellites would be in darkness for only brief periods near each equinox. Since microwaves can penetrate clouds, the rectennas would receive power almost continuously. The satellites would be so large that they would have to be assembled in space. United States government research in the 1970s resulted in a design for a solar power satellite that would measure 3.1×6.2 mi (5×10 km), with a mass of 37,000–56,000 tons (34,000–51,000 metric tons), and would beam up to 5 gigawatts of power to an elliptical rectenna measuring 6.2×8.1 mi (10×13 km). Each solar power satellite could therefore eliminate the need for several conventional power plants. The intensity of the beam at the rectenna would be fairly low, perhaps 148 mW/in.^{2} (23 mW/cm^2) at the center and 6.5 mW/in.^{2} (1 mW/cm^2) at the edge. Since the United States safety standard for human exposure to microwaves is 65 mW/in.^{2} (10 mW/cm^2), the area around the edge of the rectenna, and even some of the area inside the edge, presumably would be safe. However, stricter standards for microwave exposure are in effect in some countries, so the issue of microwave safety is still open.

SPS-to-Earth power transmission. In order to transmit power from solar power satellites to Earth in a safe and efficient manner, the proper microwave frequency must be chosen. The most widely studied frequency for solar power satellite use is 2.45 GHz, which is in a band allocated for industrial, scientific, and medical use. Frequencies from about 1 to 10 GHz can pass through clear air as well as rain and clouds. Microwave beams are electromagnetic waves and are subject to the laws of diffraction. Therefore, the smaller the transmitting antenna, the more the beam will spread out as it reaches the Earth. Transmitting antennas under consideration thus tend to be large, typically 0.6 mi (1 km) wide. They are circular and consist of many small antenna elements. Even such large antennas will result in a transmitted beam that is several miles wide at the Earth's surface. A solar power satellite must therefore transmit several gigawatts of power if it is to be economical. The use of higher frequencies will cause the beam to become more concentrated. Since a more concentrated beam may heat the ionosphere, a large number of lower-power solar power satellites will be necessary, possibly making the individual solar power satellite units more technically and economically feasible. As the frequency of the microwave beam is increased above roughly 15 GHz, atmospheric attenuation becomes higher. However, certain higher frequencies, such as 35 and 94 GHz, are not greatly attenuated by clear air. Such frequencies, known as atmospheric windows, can be attenuated by rainfall, and care must therefore be taken in selecting feasible rectenna sites.

Lasers operating in the infrared or visible portion of the spectrum could also transmit power from solar power satellites to Earth. At the Earth's surface, photovoltaic cells would convert the laser light to electricity. A laser beam would be much narrower than

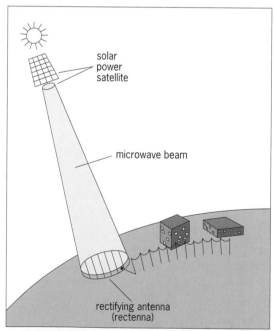

Fig. 2. Model of a solar power satellite system to supply Earth with electricity from outer space. (*After Arthur D. Little Company; M. L. Wald, Solar power from space grows more feasible, The New York Times, p. 49, December 25, 1991*)

solar
power
satellite

microwave beam

rectifying antenna
(rectenna)

a microwave beam. Such a concentrated beam would allow for considerable savings in land, but might be hazardous to the environment and passing aircraft. In addition, clouds can block such a beam. An intense laser beam may be able to punch a hole in the clouds, with possible undesirable environmental effects.

Alternative designs. Although silicon or gallium aluminum arsenide photovoltaic cells have been suggested for the solar power satellite's conversion of the Sun's energy into electricity, other means have been considered. Solar concentrators can be used to heat a fluid, which would drive a turbine to generate electricity. In addition, a hybrid concentrator-photovoltaic system could be employed. In such a system, solar concentrators would focus the Sun's rays onto photovoltaic cells. The energy thus produced would be transmitted to Earth through a microwave or laser beam.

The cost and environmental impact of launching large amounts of material into space are major constraints on solar power satellite construction. The use of lunar materials may alleviate this problem. Launching material from the Moon requires a fraction of the energy needed to launch it from the Earth. Since silicon for solar cells and aluminum for the structure can be mined from the Moon, at least 90% of the mass of the solar power satellite can consist of lunar materials.

Because of the size and development cost of a geostationary solar power satellite, as well as the need to build a lunar infrastructure, solar power satellite proponents hope to prove the concept on a smaller scale before a full-size solar power satellite is constructed. A Japanese proposal, known as SPS 2000, would be prism-shaped, 2625 ft (800 m) long, with 328-ft (100-m) sides. It would consist of several segments that would be launched separately in crewless rockets and assembled in space. The satellite would orbit at an altitude of about 620 mi (1000 km). Since such an orbit has a period of 105 min, the satellite would be within view of its ground station for only a small fraction of its orbit. However, this fraction may be enough to supply electricity to remote regions near the Equator.

Advocates of the solar power satellites maintain that it is one of the cleanest means of producing large amounts of energy currently under study, although there is some controversy over the health and environmental effects of microwaves. Unlike nuclear fusion, no technical breakthroughs are required. However, a great deal of engineering needs to be done to obtain a substantial portion of the world's energy through solar power satellites.

For background information SEE MICROWAVE; SOLAR ENERGY; WIND POWER in the McGraw-Hill Encyclopedia of Science & Technology.

Seth D. Potter

Bibliography. *The Final Proceedings of the Solar Power Satellite Program Review*, Conf-800491, U.S. Department of Energy, Office of Energy Research, Solar Power Satellite Project Division, 1980; P. E. Glaser, Power from the Sun: Its future, *Science*, 162:857–861, 1968; T. B. Morrow, L. R. Marshall, and S. A. Sherif (eds.), *Industrial Applications of Fluid Mechanics*, 1991, ASME FED, vol. 132, 1991; *Proceedings of SPS '91: Power from Space*, Société des Electriciens et des Electroniciens et Société des Ingénieurs et Scientifiques de France, August 1991; *Proceedings of the Seminar on Energy Conservation and Cogeneration Through Renewable Resources*, Indian Institute of Metals, Ranchi Chapter, January 1989; R. Richards, Hot air starts to rise through Spain's solar chimney, *Elec. Rev.,* 210(15):26–27, 1982; *Spirit of Enterprise: The 1984 Rolex Awards*, 1984; M. L. Wald, Solar power from space grows more feasible, *The New York Times*, p. 49, December 25, 1991.

Solar neutrinos

The solar neutrino problem, in which less than half of the expected flux of high-energy solar neutrinos is observed, has been a puzzle for more than 20 years. The solution to the deficit could lie in astrophysics, where the solar processes in which high-energy neutrinos are produced are not understood; or it could lie in new particle physics, where the neutrinos are produced in the correct numbers but disappear or oscillate away before they reach the Earth. In order to understand the source of the problem, it has proven necessary to observe the low-energy p-p solar neutrinos. Recent results from two solar neutrino experiments employing gallium to detect the p-p neutrinos, the Soviet-American Gallium Experiment (SAGE) and the Gallium Experiment (GALLEX), also observe deficits and apparently indicate that new particle physics is at play.

Production. Solar neutrinos are produced during the fusion processes that power the Sun. The vast majority (91%) of the neutrinos are produced by the reaction in which two protons fuse to form deuterium, a positron, and an electron neutrino. The p-p neutrinos produced in this reaction have a very high flux but are low in energy, making them very difficult to detect. The two other neutrino-producing reactions of particular interest generate the so-called beryllium-7 (^{7}Be) and boron-8 (^{8}B) neutrinos. These isotopes, which are produced by nucleosynthesis in the Sun, produce neutrinos by electron capture to form lithium-7 (^{7}Li) and by beta decay to beryllium-8 (^{8}Be), respectively. The ^{7}Be neutrinos are copious, accounting for 7% of the solar neutrino flux, and occur as line sources (monoenergetic) at intermediate energies. The ^{8}B neutrinos are the highest in energy but have a very low flux, accounting for only about 10^{-4} of the total flux.

Detailed calculations using so-called standard solar models are able to predict the fluxes from the different solar neutrino reactions. These models use various inputs (such as the primordial abundance of helium) and then track the development of the Sun as the Sun begins to burn hydrogen and progresses through the various stages of nucleosynthesis. Depending on the data used for inputs, different standard solar models predict most of the observed properties of the Sun accurately. They all predict essentially the same flux for the p-p neutrinos but vary in their predictions of the fluxes of the higher-energy neutrinos.

High-energy measurements. Neutrinos interact so weakly with matter that it is extremely difficult to detect them. Although the fluxes of solar neutrinos are very high, the cross sections are so small that the neutrinos can travel through light-years of solid material before interacting. Thus, very large detectors typically containing hundreds of tons of material are required to observe a few solar neutrinos per week. The detectors must be located deep underground to provide shielding from cosmic rays, which can produce reactions mimicking the neutrinos.

The first experiment to observe solar neutrinos is an ongoing radiochemical experiment at the Homestake Gold Mine in South Dakota in which a tank containing 615 metric tons (678 tons) of perchloroethylene is used. During a 3-month period a few atoms of argon-37 (^{37}A) are produced by inverse beta decay of the solar neutrinos on chlorine-37 (^{37}Cl). At the end of a run, the ^{37}A atoms are collected and transferred into miniature proportional counters, where they are detected as they decay back to ^{37}Cl with a 35-day half-life. This experiment is sensitive to both ^{7}Be and ^{8}B neutrinos. Averaged over 20 years of measurements, it has observed only 28–40% of the expected rate (depending on which standard solar model is used).

In 1988, another experiment, which is located at the Kamiokande mine in Japan, reported measurements that are sensitive only to ^{8}B neutrinos. This experiment consists of a large tank containing 2140 metric tons (2359 tons) of ultrapure water. Solar neutrinos passing through the detector elastically scatter on electrons in the water molecules. The recoil electrons radiate light (Cerenkov radiation) that is detected by an array of 948 large photomultipliers. The results indicate that the flux of ^{8}B neutrinos is only 46–61% of that expected by the standard solar models. Taking into account the experimental uncertainties, the chlorine and Kamiokande experiments agree with each other reasonably well only if the flux of ^{7}Be neutrinos is very far below the predictions of the standard solar models.

Solutions of the problem. Numerous efforts have been made to explain the deficit of ^{8}B neutrinos. The fundamental uncertainty is that the flux of ^{8}B neutrinos depends very strongly on the core temperature of the Sun, T_c, scaling as T_c^{18}. Thus, reducing T_c by only 4–6% reduces the ^{8}B flux by a factor of 2 to 3. Nonstandard solar models have been invented that suppress T_c, but they run into problems with reproducing other measured properties of the Sun. The chlorine and Kamiokande experiments indicate that the ^{7}Be neutrinos are more suppressed than the ^{8}B neutrinos. But the flux of ^{7}Be neutrinos is much less temperature-dependent than ^{8}B, scaling only as T_c^8. Thus, any mechanism that suppresses the ^{8}B flux by reducing T_c will not suppress the ^{7}Be neutrinos sufficiently, leading many to conclude that the solution to the solar neutrino problem cannot be of astrophysical origin. Nonetheless, the extreme energy dependence of the fluxes makes it difficult to completely rule out a solar physics explanation.

The other class of possible explanations for the observed deficit invokes new physics. A large variety of explanations have been proffered that require the existence of either new particles or new properties of the neutrino. Essentially all of these explanations have been ruled out by either astrophysical observations or laboratory measurements. Apparently, the sole remaining possibility involves neutrino oscillations. Neutrinos are known to exist in three forms: electron, muon, and tau neutrinos (plus their antiparticles). Electron neutrinos are produced in common processes, such as nuclear beta decay and nuclear fusion. The muon- and tau-type neutrinos accompany high-energy processes and cannot be produced in the Sun. If neutrinos have a nonzero mass and there is some mixing between the different types, then as neutrinos propagate they can oscillate from one type of neutrino to another. Grand unified field theories (GUTs), which attempt to unify the four known forces in nature, generally predict the existence of neutrino oscillations, and an observation of oscillations would be one of the great triumphs of modern physics.

The solar neutrino detectors are sensitive only to electron-type neutrinos, so if such neutrinos oscillate into muon or tau neutrinos during their journey from the Sun to the Earth, they would not be detected. The long baseline from the Sun to the Earth, the low energy of the neutrinos, and the high densities in the Sun (which can serve to resonantly amplify neutrino oscillations) make the Sun a unique laboratory to search for neutrino oscillations. Both the chlorine and Kamiokande results can be accommodated in a consistent manner, assuming neutrino oscillations by appropriate choices of the neutrino mass and mixing parameters.

Gallium experiments. Gallium has a very low threshold for inverse beta decay and thus provides a means to observe the *p-p* neutrinos. The standard solar models predict that 53% of the signal in a gallium experiment should come from the *p-p* neutrinos, with ^{7}Be providing most of the remaining signal. If it is assumed only that the Sun is in thermal equilibrium, a minimal value of the flux is predicted to be 60% of the standard solar model. If the flux is less than 60%, new physics must be the explanation of the solar neutrino problem.

Two radiochemical gallium solar neutrino experiments, SAGE and GALLEX, are under way. Both experiments chemically extract the few atoms of germanium-71 (^{71}Ge) formed in inverse beta decay by solar neutrinos on gallium-71 (^{71}Ga). SAGE employs 57 metric tons (63 tons) of gallium in the form of the liquid metal (gallium melts at 29.8°C or 85.6°F), while GALLEX uses 30 metric tons (33 tons) of gallium in an aqueous chloride solution. The chemical extraction procedures differ in the two experiments, but both detect the ^{71}Ge by extracting it into miniature proportional counters and observing the decay of the ^{71}Ge atoms back to ^{71}Ga with an 11.4-day half-life.

The scheme of the SAGE experiment is shown in the **illustration**. The metallic gallium is held in tanks in the underground laboratory in the Caucasus mountains in southern Russia. Solar neutrinos should produce about 2.3 atoms of ^{71}Ge per day in 57 metric tons (63 tons) of gallium (assuming the standard solar model flux). The ^{71}Ge atoms are chemically extracted once a month, and their decay is observed for a period of several months in order to accurately determine both the number of ^{71}Ge atoms extracted and the backgrounds in the proportional counters.

SAGE was the first to come on line, and in 1990 measurements were reported from 5 months of running using 30 metric tons (33 tons) of metallic gallium. The initial SAGE results indicated that the flux was only about 15% of that expected. However, because of the low statistics, the accuracy of the measurement could not rule out possible solar physics explanations of the deficit. However, the SAGE scientists concluded that the solar neutrino problem might also apply to the low-energy p-p neutrinos, indicating the existence of new neutrino properties. Given the very low observed flux and the important implications of new physics, considerable effort was expended to demonstrate that everything in the experiment worked as expected and that no ^{71}Ge atoms were unexpectedly lost. Several tests were carried out, including putting a known number of ^{71}Ge atoms in the gallium and extracting and counting them. All tests demonstrated the validity of the techniques, and no possible loss mechanism was found. SAGE then worked to upgrade the experiment of 57 metric tons (63 tons) of gallium and to reduce backgrounds. Data acquisition was started again in spring 1991 on a monthly basis.

GALLEX began operation at about the same time as SAGE. After solving a problem with an unexpected background, the GALLEX scientists began measurements of the solar neutrino flux in summer 1991. Their results after 1 year of data taking showed a positive signal and a flux 63% of that expected. While the present accuracy of the experiment is not sufficient to conclude unambiguously that solar physics cannot explain the deficit, the GALLEX scientists stated that the agreement with these models (standard solar models), even with severe modifications, was marginal.

While there appeared to be some discrepancy between the SAGE and GALLEX results, further data from SAGE using 57 metric tons (63 tons) of gallium proved to observe a signal of about 44% of the standard solar model prediction. Thus, it appears clear that SAGE and GALLEX are observing p-p neutrinos, but they both observe a deficit. Scientists working on both experiments plan to carry out upgrades and to take several more years of data to improve the measurement accuracy. They also plan to produce an artificial chromium-51 (^{51}Cr) neutrino source and expose the gallium to it. This source should produce a known number of ^{71}Ge atoms, which can be used to test the full extraction and counting efficiencies.

Conclusions. The results of all four solar neutrino experiments indicate a deficit of solar neutrinos, and can be explained in a consistent manner by in-

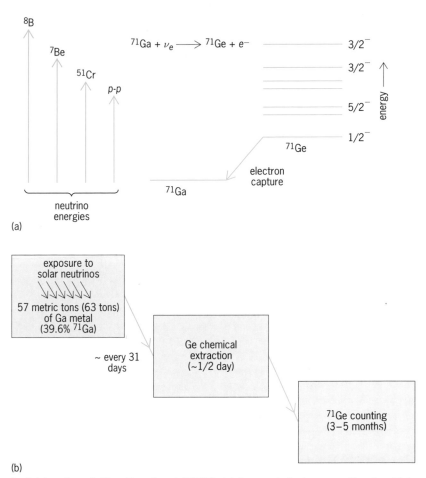

Soviet-American Gallium Experiment (SAGE). (*a*) Inverse beta-decay reaction in which neutrinos produce germanium-71 (^{71}Ge). The subsequent decay of the ^{71}Ge atoms back to gallium-71 (^{71}Ga) by electron capture, with an energy release (Q) of 233 keV and a half-life of 11.4 days, is shown in an energy-level diagram. The symbols on the right of the energy levels are the spins and parities of the levels. (*b*) Operation of the experiment.

voking neutrino oscillations with a fairly constrained set of parameters for the neutrino masses and mixing probabilities. The two gallium experiments, taken together, indicate that the source of the solar neutrino problem is very unlikely to be solar physics. Thus, the solar neutrino experiments may finally have provided the first window showing evidence for new physics predicted by the grand unified field theories.

Future experiments. In order to absolutely demonstrate the existence of neutrino oscillations, it will be necessary to demonstrate that solar neutrinos arrive at the Earth as muon or tau neutrinos. An experiment to do just that, the Sudbury Neutrino Observatory, is under construction in Canada. It will employ 1000 metric tons (1100 tons) of heavy water (D_2O) in a water Cerenkov detector and will be able to measure all types of neutrinos. In addition, two other large solar neutrino projects are under way: SuperKamiokande, a 50,000-metric-ton (55,000-ton) version of the Kamiokande experiment, and BOREXINO, which will use a liquid scintillator to observe the ^{7}Be neutrinos directly. These experiments will have count rates 10 to 100 times higher than the current experiments and should finally be able to resolve the solar neutrino problem.

For background information *SEE GRAND UNIFICATION THEORIES; NEUTRINO; SOLAR NEUTRINOS* in the McGraw-Hill Encyclopedia of Science & Technology.

Thomas J. Bowles

Bibliography. A. I. Abazov et al., Search for neutrinos from the Sun using the reaction ^{71}Ga(ν_e, e^-)^{71}Ge, *Phys. Rev. Lett.*, 67:3332–3335, 1991; P. Anselmann et al., Solar neutrinos observed by GALLEX at Gran Sasso, *Phys. Lett.*, B285:376–397, 1992; J. N. Bahcall, *Neutrino Astrophysics*, 1989.

Solid-state physics

In 1916, H. H. Poole established that the conductivity of lamellae of mica depends exponentially on the electric field E, as in Eq. (1) [where σ_0 and γ_P are

$$\sigma_P = \sigma_0 \exp(\gamma_P E) \tag{1}$$

constants]. In 1938, J. Frenkel proposed a model of high-field conductivity in which it varies as in Eq. (2),

$$\sigma_{PF} = \sigma_0 \exp\left(\gamma_{PF}\sqrt{E}\right) \tag{2}$$

allowing for a reasonable fitting of the experimental data on solid dielectrics. This latter behavior is called the Poole-Frenkel effect. These two laws have since been unified in a theory in which they are derived from the distortion by E of the donor or acceptor potential wells.

Theory of Poole-Frenkel effect. The theory of the bare Poole-Frenkel effect relies on the assumptions that all contacts are ohmic; carrier mobility is weak; the potential wells associated with donors or acceptors are isolated, that is, their density is so small as to make interactions between them negligible; and there is only one impurity level.

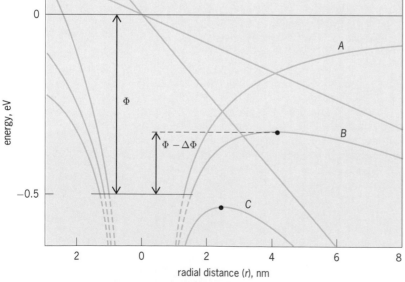

Fig. 1. Coulomb potential well of a donor in the energy gap of a solid dielectric. Curve *A* is potential for electric field *E* = 0. Curves *B* and *C* show lowering of potential at increasing electric fields. The sloping straight lines indicate how much the conduction band edge is shifted by the electric field in these two cases. (*After R. Ongaro and A. Pillonnet, Poole-Frenkel (PF) effect for high field saturation, Rev. Phys. Appl. (France), 24:1085–1095, 1989*)

The theory is semiclassical; that is, E is assumed to distort the potential well of a donor (**Fig. 1**), with depth Φ below the conduction band. (This depth is the zero-field activation energy.) The barrier lowering by E follows the relative depths of the donor and the Fermi level, and can be written as either Eq. (3) or Eq. (4), where β is given by Eq. (5). (Here, e is the

$$\Delta\Phi_{PF}^{(1)} = \beta\sqrt{E} \tag{3}$$

$$\Delta\Phi_{PF}^{(2)} = \frac{\beta}{2}\sqrt{E} \tag{4}$$

$$\beta = \left(\frac{e^3}{\pi\epsilon_0\epsilon}\right)^{1/2} \tag{5}$$

magnitude of the electron charge, ϵ_0 is the permittivity of free space, and ϵ is the relative permittivity.) These cases generally hold in compensated (by acceptors) materials and noncompensated materials, respectively. The enhancement by the field of the probability of electron thermal release, and therefore of the conductivity σ_{PF}, is then taken to be proportional to expression (6). The result is a straight line on a Poole-Frenkel

$$\exp\frac{\Delta\Phi_{PF}}{kT} \tag{6}$$

plot (which plots the logarithm of σ_{PF} against $\sqrt{E}$), with a slope of either $\beta/(kT)$ [slope S1] in the case of compensated materials, or $\beta/(2kT)$ [slope S2] in the case of noncompensated materials. Sometimes, noncoulombic wells are introduced, either as r^{-n} potentials (where r is the radial distance and $1 < n < \infty$), yielding Eq. (7) for the conductivity, or

$$\sigma_{PF} = \sigma_0 \exp\left[\gamma_{AK}E^{n/(n+1)}\right] \tag{7}$$

as screened potentials given by expression (8), whose

$$r^{-1}\exp\frac{-r}{r_0} \tag{8}$$

Poole-Frenkel plot displays a reduced linearity.

Recently, a new formulation of the Poole-Frenkel theory was derived by substituting Fermi-Dirac statistics for the Boltzmann approximation, yielding Eq. (9)

$$\sigma_{PF} \propto 2s\frac{q-1}{q}\left(1+\frac{s}{q}e^{\eta-\alpha}\right)^{-1}$$

$$\times \left[1+\left(1+4s\frac{q-1}{q}\frac{e^{\eta-\alpha}}{\left(\frac{1+s}{q}e^{\eta-\alpha}\right)^2}\right)^{1/2}\right]^{-1} \tag{9}$$

for a compensated material and for any kind of well. Here, $s = N_d/N_c$, $q = N_d/N_a$, $\eta = \Phi/(kT)$, $\alpha = \Delta\Phi_{PF}/kT$ for coulombic wells, and N_a, N_d, and N_c are the densities of acceptors, of donors, and of states in the conduction band. Thus, s is a measure of the density of donors, q is related to the compensation, η is a measure of the depth of the wells, and α is a measure of the barrier lowering by the electric field. The related Poole-Frenkel plot then displays bilinear curves with successive slopes S1 and S2 for Coulomb potentials, followed by a final saturated part (**Fig. 2**). This improvement unifies the two previous formulations of Poole-Frenkel theory, which then appear as limiting cases available for compensated and uncompensated materials, respectively. In a well-compensated material, q is approximately 1, as in the case of curves A and A' in Fig. 2. These curves have slope S1 until saturation begins to set in, as predicted by the previous theory (the Boltzmann approximation) for compensated materials. However, when the compensation is negligibly small, q is very large, as in the case of curves H and H' in Fig. 2. At relatively small values of α, curve H bends from slope S1 to slope S2, the slope predicted by the previous theory for noncompensated materials, and curve H' has slope S2 down to zero field.

A more sophisticated model was derived in which the wells were assumed to be discretely or continuously distributed in energy in the band gap. The main characteristics of the one-level model are generally preserved (two linear parts with slopes S1 and S2, and saturation), showing that the Poole-Frenkel effect alone cannot be used to distinguish a unique level from a distribution of levels.

Theory of Poole effect. In a Poole-Frenkel plot, the experimental curves often display a more weakly rising initial part (**Fig. 3**). Two main interpretations of this behavior have been proposed. One interpretation postulates a three-dimensional Poole-Frenkel effect, where the electrons are assumed to be ejected in all space directions. From this behavior, various relationships are derived, according to the definition adopted for the inverse barrier height. For example, with an inverse barrier independent of E, Eq. (10) is

$$\sigma_{PF} \propto \frac{1}{2} + \alpha^{-2}\left[1 + (\alpha - 1)e^{\alpha}\right] \qquad (10)$$

derived for the conductivity.

In the other interpretation, the Poole effect results from interaction between the Coulomb potentials. This interaction, which can arise only for concentrations of donors higher than expected in Poole-Frenkel theories, is arbitrarily restricted in published research to pairs of nearest neighbors separated by a distance a. The enhancement of the probability of thermal excitation is then taken to be proportional to expression (11), where $\Delta\Phi_P$, given by Eq. (12), is the field lowering of

$$\exp\left(\frac{\Delta\Phi_P}{kT}\right) \qquad (11)$$

$$\Delta\Phi_P = \frac{1}{2}eaE \qquad (12)$$

the intersite barrier height. This concept is just Poole's law, and the model provides an interpretation of this law for the low-to-medium field range. The transition from Poole's law to the Poole-Frenkel law at a field E_t implies that pairs connected at low fields become disconnected at high fields. Also, the interaction reduces the effective zero-field activation energy to the value given by Eq. (13).

$$\Phi - \Phi' = \Phi - \frac{\beta^2}{ea} \qquad (13)$$

Synthetic Poole-Frenkel effect. An attempt was made recently, in the one-level, one-dimen-

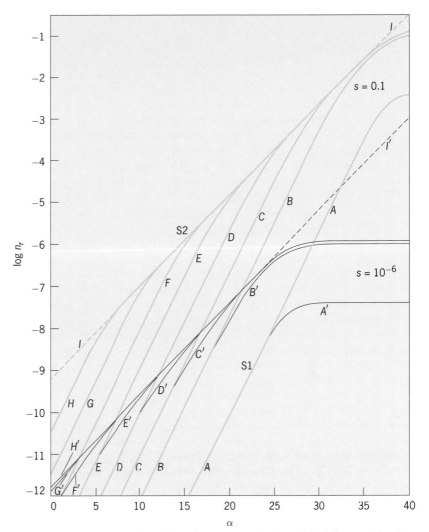

Fig. 2. Poole-Frenkel plot of the relative free electron density n_r (which is proportional to the conductivity of σ_{PF} insofar as mobility is independent of field) for a set of values of s and q, and for $\epsilon = 2.2$, temperature $T = 300\ K = 80°F$, and $\eta = 40$. Symbols are explained in the text. Curves A–I have $s = 0.1$, and curves A'–I' have $s = 10^{-6}$. (Curves A'–E' are indistinguishable from curves A–E at lower values of α.) For A, A', $q = 1.04$; B, B', $q = 10.4$; C, C', $q = 104$; D, D', $q = 10^3$; E, E', $q = 10^4$; F, F', $q = 10^5$; G, G', $q = 10^6$; and H, H', $q = 10^7$. Curves I and I' give Boltzmann approximation for large values of q. (*After R. Ongaro and A. Pillonnet, Poole-Frenkel (PF) effect for high field saturation, Rev. Phys. Appl. (France), 24:1085–1095, 1989*)

sional, Poole-Frenkel case, to generalize the interactions to a finite arbitrary number of coulombic sites, and to an infinite number of noncoulombic sites. The barrier lowering $\Delta\Phi_z$ for a set of $2Z$ Coulomb potentials can be expressed in the form of a nonconverging series of Z terms. This series can be approximated in a reduced form by Eq. (14), where f is given by Eq. (15).

$$\Delta A_z = \frac{\Delta\Phi_z}{\Phi'} = \frac{f}{2 + \sqrt{f}} + \frac{0.173\, f^{1.43}}{1 + 0.5\, f^{1.2}} \qquad (14)$$

$$f = \left(\frac{ea}{\beta}\right)^2 E = \frac{E}{E_t} \qquad (15)$$

Such a generalization sets up a smooth transition from the Poole effect to the Poole-Frenkel effect at $E = 4E_t$ (Fig. 3), and is therefore called the synthetic Poole-Frenkel effect. The nonconvergence of the sum over Z for Coulomb sites makes a pre-

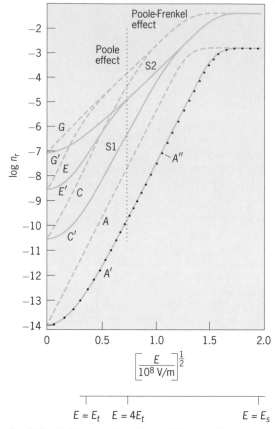

Fig. 3. Poole-Frenkel plots of the relative free electron density n_r under various theoretical assumptions. Symbols are explained in the text. Curves *A, C, E,* and *G* are derived from the theory of the Poole-Frenkel effect (as in Fig. 2); curves *A', C', E'* and *G'* are derived from the theory of the synthetic Poole-Frenkel effect; and the dots labeled *A''* are derived from the approximate Eq. (14) in the text. For *A, A', A''*, $q = 1.04$; *c, C'*, $q = 104$; *E, E'*, $q = 10^4$; and *G, G'*, $q = 10^6$. (*After A. Pillonnet and R. Ongaro, Simulated behaviour of field-assisted ionisation in the theory of synthetic Poole-Frenkel effect, J. Phys. III (France), 1:1449–1454, 1991*)

cise determination of the effective zero-field activation energy impossible, in contrast with other types of wells.

Difficulties. The above synthesis of the Poole and Poole-Frenkel effects is open to criticism, as were the previous theories. Strong objections can be raised to the semiclassical model based on the construction of potential wells in the band gap:

1. The concept of barrier lowering makes no sense when E exceeds the saturation field E_s given by Eq. (16), because the potential maximum then lies

$$\left(\frac{\Phi}{\beta}\right)^2 = E_s \qquad (16)$$

below the ground state (as in curve C in Fig. 1). This situation can arise in the medium range of electric fields usually applied in the study of the Poole-Frenkel effect, especially for shallow states.

2. The construction of a potential well in the band gap is not self-consistent, as it confuses the total and potential energies. Thus, the walls of the well are explicitly referred to the potential energy, while the infinite branches of the well and the ground state Φ

refer to the total energy. Two attempts have been made to resolve this discrepancy. First, $\Delta\Phi_{PF}$ was determined by means of an estimate of total-energy variation versus E, derived as usual at the potential-energy maximum. This determination results either in Eq. (17) for large compensation or in Eq. (18) for

$$\Delta\Phi_{PF}^{MR} = \frac{1}{2}\Delta\Phi_{PF}^{(1)} = \frac{\beta}{2}\sqrt{E} \qquad (17)$$

$$\Delta\Phi_{PF}^{MR} = \frac{1}{2}\Delta\Phi_{PF}^{(2)} = \frac{\beta}{4}\sqrt{E} \qquad (18)$$

noncompensation. Second, a Rutherford-Bohr model was substituted for the usual potential well, resulting in Eq. (19). Both these models still leave standing the first

$$\Delta\Phi_{PF}^{OP} = \frac{\Delta\Phi_{PF}^{(1,2)}}{\sqrt{2}} \qquad (19)$$

objection discussed above.

3. The treatment of interactions is carried out as an electrostatic problem, with a moving particle placed in the vicinity of two or more fixed point charges. Therefore, the neutrality of a filled center is not duly accounted for.

The available Poole-Frenkel theories assume, in effect, that applying an electric field E does not change the wave function of a particle in a potential well, for the field-dependent energy term is simply added to the zero-field energy. Now, the wave functions in the presence or absence of a field necessarily differ, since they are solutions of two different Schrödinger equations, written with or without field-related terms. However, there is no purely quantum theory presently available that can provide an unquestionable interpretation of the experimentally observed exponential variation of the conductivity with field. Consequently, experimentalists should be provisionally allowed to invoke the Poole and Poole-Frenkel effects, especially in their recently improved formulations, as a simple means of characterizing material properties.

For background information *SEE BOLTZMANN STATISTICS; ELECTRICAL INSULATION; ELECTRICAL RESISTIVITY; FERMI-DIRAC STATISTICS; NONRELATIVISTIC QUANTUM THEORY; SCHOTTKY EFFECT; SEMICONDUCTOR* in the McGraw-Hill Encyclopedia of Science & Technology.

R. Ongaro; A. Pillonnet

Bibliography. G. A. Dussel and K. W. Böer, Field-enhanced ionization, *Phys. Status Solidi*, 39:375–389, 1970; J. L. Hartke, The three-dimensional Poole-Frenkel effect, *J. Appl. Phys.*, 39:4871–4873, 1968; R. Ongaro and A. Pillonnet, Generalized Poole-Frenkel (PF) effect with donors distributed in energy, *Rev. Phys. Appl. (France)*, 24:1097–1110, 1989; R. Ongaro and A. Pillonnet, Synthetic theory of Poole and Poole-Frenkel (PF) effects, *IEE Proc. A*, 138:127–137, 1991.

Sound recording

Digitally recorded music first became widespread in the late 1970s. The early digital recordings were still converted to analog for mastering and pressing in vinyl. These recordings were followed in the early 1980s by compact disks, in which format digital sound has made its greatest impact. Subsequently, the near-instantaneous companded and multiplexed (NICAM) format for broadcast television sound has been introduced, and the most recent development is digital audio tape. The major recording studios now record most material digitally.

For the end consumer, the main advantages of digital recording over analog are the greater dynamic range (the difference in loudness between the quietest and loudest notes) and the very low background noise (that is, the absence of residual tape hiss from the analog master and of vinyl surface noise or playback tape hiss). For postrecording production and eventual manufacture, there are further advantages. Since the music is recorded as a digital sequence of ones and zeros, the recording can be copied, transmitted over long distances, edited, mixed, and remixed without loss of sound quality, while the storage or transmission medium also has no effect. Of course, in practice maintaining fidelity depends critically upon error correction from the original recording right through to playback by the consumer. These error-correction systems allow lost ones and zeros to be detected and reconstituted, or the sound may be muted for a short period in cases where the error is too large for correction. Since most error correction must be carried out in real time, there is a limit on the time allowed to correct errors and hence on the maximum size of error, but loss of sound quality due to poor error correction is normally negligible.

Analog-to-digital conversion. A digital representation of a sound is obtained by first converting the sound waves to an analog electrical signal by using a microphone, resulting in a voltage that varies with time. The analog signal, which is entirely contin-uous and able to assume any value between zero and the maximum provided by the microphone, may then be quantized or allocated one of a number of discrete voltage levels between zero and the maximum. If the analog signal at that instant is not exactly one of the quantization levels (as will usually be the case), the nearest level is allocated. Because of these differences between the real signal and the quantized signal, a digital signal always contains a degree of quantization noise with a maximum of one-half the difference between the discrete quantization levels. The accuracy of the digital representation and hence the quantization noise depends on the number of quantization levels, with more levels giving greater accuracy but also increasing complexity and cost. The number of quantization levels is always a power of 2, depending on the number of bits of resolution required. For example, a 16-bit system has 2^{16} levels evenly spaced between 0 and the maximum analog value, and each quantization level is represented by a binary number in the range $0-2^{16}$ ($0-65,536$), so that the digital representation of the sound signal is a series of 16-bit binary numbers. The domestic reproduction formats of compact disk and digital autio tape are 16-bit, but studios record with 16- or 20-bit systems.

Converting an analog signal to digital form is not an instantaneous process but takes a finite amount of time. Therefore, it is not possible for the output of an analog-to-digital converter to continuously track the variations of the input, and the analog signal must be periodically sampled. The analog input is therefore applied to a sample-and-hold circuit that takes the instantaneous value of the input and holds it constant for a short time during which the analog-to-digital converter performs the quantization and binary number allocation. The sample-and-hold circuit then allows its output to change to sample the next value.

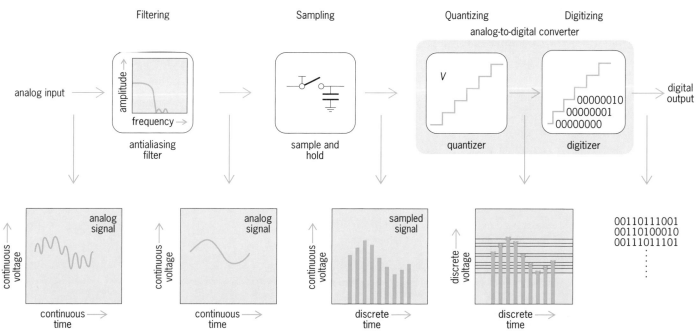

Fig. 1. Sequence for digitizing an analog signal.

An important feature of a digital recording system is the frequency at which it samples the analog input, since a fundamental criterion of sampling theory specifies that it is not possible to sample an analog signal that has a frequency greater than half the sampling frequency without generating audible distortion through an effect known as aliasing. The upper limit of audible frequencies can be assumed to be 20 kHz, so a minimum sampling frequency of 40 kHz is required. Professional recording systems have two widely accepted sampling frequencies: 44.1 kHz, used by compact disks, and 48 kHz. However, aliasing of frequencies greater than the audible limit will still produce distortion in the audible band, and a filter, known as an anti-aliasing filter, is used to remove the frequencies above 20 kHz. Some systems employ a technique known as oversampling, where the signal is sampled at a much higher frequency than required. Often a 96-kHz frequency is used; this technique is called 2× oversampling because sampling is carried out at about twice the minimum acceptable rate. The effect is to reduce the performance specification of the antialias filter, since only signals of frequency greater than 48 kHz must now be removed and the frequency discrimination need not be as sharp. **Figure 1** shows the full process for converting an analog signal to a digital representation.

Production. While digital sound may reach the consumer through a number of media, the original recording still relies on the use of magnetic tape, which is the only medium that can provide the high storage density required economically and enable the synchronized recording of 16 and 32 tracks. When the raw recording has been obtained, the production sequence uses a number of processes that can broadly be categorized as editing, mixing, and the application of audio effects such as echo.

Editing. With an analog recording, the simplest edit is the tape splice, where the tape is physically cut and then joined to another piece of tape. Since the tape is cut at an angle of 45°, there is a fade-out and fade-in over the two sections of tape and the edit is generally inaudible. However, this method does not work well with digital recordings because the digital data are interleaved. The binary numbers representing the recording are not stored on tape in the same order as required for playback, but are rearranged in a predetermined manner, and the original sequence is restored during playback. Therefore, so-called burst errors in the recording (due to tape damage, for example), which would otherwise cause the loss of many adjacent bits, can be more easily corrected because the missing bits

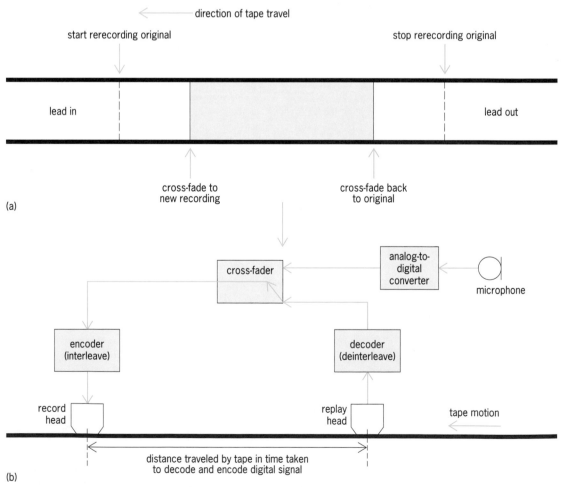

Fig. 2. Performing a digital tape edit. (*a*) Tape sequence. (*b*) Equipment.

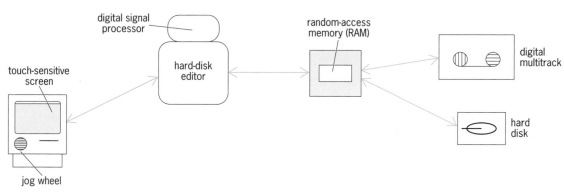

Fig. 3. Main elements of a digital audio workstation. Jog wheel is used to scroll data forward or backward until the exact edit point is found.

are distributed through the recording rather than forming a continuous string of lost data. The disadvantage is that a single point on the tape no longer represents a single point in the recording. A crude tape splice will therefore be audible. Instead (**Fig. 2**), editing must be accomplished by playing back the recording and immediately rerecording the original onto the same tape. (The distance between the record and playback heads is critical and must equal the distance traveled by the tape in the time taken to deinterleave and reinterleave the data read from the tape.) At a suitable point, the original signal is cross-faded to a new source, which could be another recording or might derive directly from musicians playing along with the recording, until the desired edit is complete. The recorded signal is then cross-faded back to the orignal recording, the recorder is switched off, and an inaudible edit results.

Mixing. Mixing is the process of blending a number of different tracks to produce a single stereo recording, where the loudness and the equalization or tonal balance of each track must be adjusted relative to the others. Digital faders for adjusting volume and digital filters for setting equalization are now used, enabling the entire postrecording production process to be carried out in the digital domain. The process can have important consequences for the ultimate quality of sound because each time an analog signal undergoes such a step it suffers a small degradation. The total degradation after a large number of edits, mixes, and copies is significant, so that producers using analog techniques become accustomed to minimizing the degree of acceptable postrecording processing. This restriction does not apply to a digital recording that can (in theory at least) be processed an unlimited number of times with no degradation.

However, degradation can occur during digital editing and mixing, operations because additional bits are generated that are not ultimately required. (Mixing two 16-bit digital recordings will generate digital data of up to 32 bits.) The desired number of bits is obtained by discarding the least significant, but the new least-significant bit must be rounded up or down depending on the value of the discarded bits, resulting in the loss of some very low level musical information. If this happens many times during postrecording production,

where more than 50 operations of this type are not unusual, the cumulative effect can be a small loss in fidelity. The good practices adopted for analog recordings can therefore be applied to digital recordings for the best fidelity.

Effects. Effects are used to increase the ambience, interest, or novelty of a recording. The most important are echo, reverberation, and flanging. Echo is the introduction of single or multiple echoes to a recording. Reverberation processing is the simulation of the great number of multiple echoes that occur in a large concert hall. Flanging is the effect generated by mixing the outputs of two identical recordings while retarding one with respect to the other. (It was initially performed by tape operators touching the flange of one tape spool to slow it down.) All of these effects can be realized in the digital domain by using combinations of signal delay, addition, and multiplication. The sophistication of reverberation processing is greatly superior to the crude analog equivalents, and it is now possible to program the reverberation characteristics of any concert venue in the world.

Digital audio workstations. A major advantage in the arrival of digital techniques is that digital systems do not have to be dedicated but can be programmed by software. In audio workstations (**Fig. 3**), where user-friendly interfaces such as menus and a mouse or light pen are used to allow simple cut, paste, and menu-selection functions to perform the complex tasks of editing, mixing, and applying effects, a hardware digital signal processor is often employed. Favorite settings for faders and equalizers can be stored and reimplemented in an instant, resulting in a console that can be used by a number of people, each with different setups, so that studios do not require capital investment in many consoles.

In the late 1970s, digital recording meant that only the master tape was digital, with all the editing, mixing, and effects applied by using traditional analog technology. Now, digital alternatives for the whole production sequence have been developed so that all processes can take place in the digital domain, and each compact disk is truly a result of 100% digital technology. The high flexibility and negligible loss of fidelity offer a standard of reproduction previously unavailable to most people and realized each time a compact disk is played.

For background information SEE ANALOG-TO-DIGITAL CONVERTER; COMPACT DISK; MAGNETIC RECORDING; SOUND RECORDING in the McGraw-Hill Encyclopedia of Science & Technology.

L. I. Haworth

Bibliography. K. B. Benson, *Audio Engineering Handbook*, 1988; W. L. Sinclair and L. I. Haworth, Digital recording in the professional industry, Part 1: Recording technology, *IEE Electr. Commun. Eng. J.*, 3(3):108–118, 1991; W. L. Sinclair and L. I. Haworth, Digital recording in the professional industry, Part 2: Studio techniques, *IEE Electr. Commun. Eng. J.*, 3(4):177-184, 1991; J. Watkinson, *The Art of Digital Audio*, 1989.

Space flight

During 1992 nearly three dozen countries cooperated in approximately 170 formal space projects and educational programs as part of the International Space Year (ISY). The designation of 1992 as such commemorated the 500th anniversary of Columbus's first voyage to the New World.

The year was especially fruitful for the United States, and particularly for the National Aeronautics and Space Administration (NASA) science missions. Astronomers came closer to understanding the mysterious black holes when the Hubble Space Telescope uncovered evidence of possible massive black holes in the cores of two galaxies. The orbiting telescope also provided the first direct view of an immense ring of dust, which may fuel a massive black hole at the heart of another galaxy. Six scientific spacecraft were launched to explore the universe, the solar system, the Earth, and the Earth-Sun environment. Among them was the *Mars Observer*, the United States' first mission to the Red Planet since *Viking* began its journey in 1975.

NASA flew eight shuttle missions in 1992. The shuttle *Endeavor*'s maiden voyage (May 7, STS 49) highlighted shuttle program activities. In a dramatic rescue effort, the crew reclaimed a malfunctioning communications satellite and, in the process, set three new records for space flight: four space walks on a single mission, the longest space walk ever conducted (8 h 29 min), and the first three-person space walk (**Fig. 1**).

Three shuttle missions (January 22, STS 42; June 25, STS 50; and September 12, STS 47) featured use of the pressurized spacelab module. Experiments conducted on those flights were precursors to activities that will be undertaken on Space Station *Freedom*.

The shuttle system demonstrated its ability to perform a wide variety of functions. It served as an orbiting observatory (March 24, STS 45) and demonstrated new technology in space with the Tethered Satellite System payload (August 1, STS 46; **Fig. 2**). The shuttle *Columbia* and its crew (October 22, STS 52) demonstrated the orbiter's ability to fly a combi-

Fig. 1. Astronauts Richard J. Hieb, Thomas D. Akers, and Pierre J. Thuot cooperating in an effort to attach a specially designed grapple bar to the *INTELSAT VI* communications satellite in order to move the satellite into the cargo bay of the space shuttle *Endeavor*.

nation mission, deploying the *LAGEOS* satellite and also conducting microgravity research with the U.S. Microgravity Payload. NASA flew the last dedicated Department of Defense mission (December 2, STS 53).

The post-Soviet space program, still the world's largest, continued major retrenchment during 1992. Formation in 1991 of a Russian Space Agency, and agreement to a Commonwealth of Independent States (C.I.S.) space accord, now provide the basis for management and coordination of the former Soviet space program. Important funding and policy issues, however, remain to be resolved.

Other republics in the C.I.S. that boast major space facilities, such as Kazakhstan, have formed their own space agencies. These agencies function, in part, by leasing the space infrastructure located on their territories to other republics, notably to Russia. Contrary

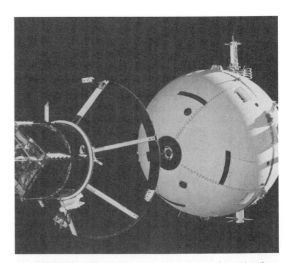

Fig. 2. Tethered Satellite System moving away from the ring structure on the boom device in the cargo bay of the space shuttle *Atlantis*.

Table 1. Some significant space launches in 1992

Payload or vehicle	Date	Country or organization	Purpose or outcome
JERS	Feb. 11	Japan	Japan's first Earth resources remote-sensing satellite.
Soyuz 14	Mar. 17	Russia/C.I.S.	Three cosmonauts to *Mir* to continue experiments.
STS 49	May 7	United States	Shuttle *Endeavor* crew rescue of *INTELSAT VI* communications satellite involving first three-person space walk. Satellite was refurbished and launched to geosynchronous orbit.
Soyuz 15	July 27	Russia/C.I.S.	Three cosmonauts to *Mir* to continue experiments.
STS 51	July 31	United States	Shuttle *Atlantis* launched *Eureca* (recoverable European Space Agency satellite) and conducted tethered satellite experiment by using Italian-built spacecraft. *Eureca* is to be recovered in the future.
Topex/Poseidon	Aug. 10	European Space Agency	Ariane launched two ocean remote-sensing spacecraft in a United States–France combined effort.
Hispasat	Sept. 10	European Space Agency	Ariane launched first Spanish communications satellite.
STS 47	Sept. 12	United States	*Endeavor* crew conducted Japanese-designed life sciences experiments. Crew included first black woman, first married couple, and first Japanese person aboard shuttle.
Mars Observer	Sept. 25	United States	First United States spacecraft to Mars since *Viking 1* and 2.
Freja	Oct. 6	China	China launched science spacecraft for Sweden.
Ekran 20	Oct. 30	Russia/C.I.S.	First television satellite of this kind by C.I.S. in 2 years.
Resurs 500	Nov. 15	Russia/C.I.S.	Launch by private business in C.I.S., carrying gifts for United States citizens. Recovered off coast of Seattle and brought to Seattle by ship *Marshall Krylov*.
Cosmos 2219	Nov. 17	Russia/C.I.S.	First successful Zenit launch after three consecutive failures. Zenit is intended for commercial launches.

to earlier indications, Ukraine, where booster rockets and other major items of space hardware are built, has refused to sign the C.I.S. space accord.

Significant space launches in 1992 are listed in **Table 1**. The total numbers of payloads launched by various countries are given in **Table 2**.

United States Space Activity

An Earth-bound NASA project may yet prove to be among the most significant scientific searches ever undertaken. On October 12, 1992 (Columbus Day), the High Resolution Microwave Survey (HRMS) was initiated from NASA's Deep Space Network in Goldstone, California, and the Arecibo Observatory in Puerto Rico. The project involves a search for signals transmitted from other civilizations. The search is being conducted in two modes: a sky survey, sweeping the celestial sphere for signals, and a targeted search that will look at 800 nearby stars with characteristics similar to those of the Sun.

Hubble Space Telescope. The Hubble Space Telescope yielded important information throughout 1992. The Space Telescope conducted long-term observations of global weather changes on Mars and measured the extent of the atmosphere of the Jovian Satellite Io. It provided the first clear view of one of the hottest known stars ($3.6 \times 10^5\,^\circ$F or 2.0×10^5K), which lies at the center of the Butterfly Nebula (NGC 2440). It discovered a cataclysmic variable star in the core of a globular cluster (47 Tucanae), the first known optical counterpart to an x-ray source in a globular cluster.

Ulysses. The *Ulysses* spacecraft received a gravity assist as it flew by Jupiter on February 8, 1992, at 280,000 mi (450,000 km) from the planet's center. *Ulysses*, which is designed to study the Sun's magnetic

field and solar wind, used Jupiter's gravity assist to gain the momentum needed to break into a solar polar orbit. During the hazardous Jupiter fly-by, scientists investigated the interaction of the giant planet's magnetic field and the solar wind.

Magellan. The *Magellan* spacecraft, which has been mapping the hidden surface of Venus with radar since August 1990, approached Venus at its closest altitude on September 14, 1992, when it began a full 243-day cycle of gravity mapping. *Magellan* has completed three cycles of mapping with its radar, covering 99% of the surface of Venus. The objective of the fourth cycle, which was scheduled to begin on May 15, 1993, is to obtain a global map of the Venus gravity field from its elliptical orbit.

Galileo. The *Galileo* spacecraft flew by the Earth on December 8, 1992, at an altitude of 189 mi (304 km), completing a 3-year gravity-assist trajectory. This latest gravity assist added about 8300 mi/h (13,300 km/h) to the spacecraft's speed and changed its direction slightly, putting it on an elliptical trajectory to the orbit of Jupiter, about 4.8×10^8 mi (7.8×10^8 km) from the Sun. The spacecraft will arrive at Jupiter on December

Table 2. Successful launchings conducted in 1992*

Country or organization	Number of launches
United States	28[†]
Russia (C.I.S.)	54
People's Republic of China	4
European Space Agency	7
Japan	1
India	1
Total	95

* Launchings achieved Earth orbit or beyond.
[†] Includes 11 NASA launches, 12 Department of Defense launches, and 5 commercial launches from Cape Canaveral, Florida.

7, 1995, and will relay data from a probe launched into the planet's atmosphere to obtain direct measurements of that environment for the first time.

Galileo's main antenna, which failed to deploy after launch, appears to have seriously and permanently degraded the scientific value of the spacecraft's mission. For 3 weeks during 1992, engineers cycled the antenna's deployment motor on and off more than 12,000 times in an unsuccessful attempt to drive three stuck ribs free of the central shaft. A last attempt to open the antenna, by increasing the spacecraft's spin from a normal 3 revolutions per minute to 10, was scheduled for March 1993. Unless this procedure works, which is unlikely, the spacecraft will return only 2000–4000 pictures instead of the planned 50,000, because it will be using a less powerful, slower-transmitting antenna.

Expendable launch missions. NASA launched five expendable launch vehicles without mishap during 1992, maintaining a 5-year perfect launch record. The first launch was on June 7, when a Delta 2 placed the *Extreme Ultraviolet Explorer*, an astrophysics satellite, into low Earth orbit. On July 3, a Scout placed *SAMPEX*, a small Explorer-class space-physics satellite, into low Earth orbit.

A Delta 2 carried the Japanese *Geotail* satellite into space on July 24. On September 25, a Titan III lifted the *Mars Observer* into Earth orbit, where the Transfer Orbit Stage, in its maiden flight, ignited, sending the spacecraft on to Mars. On November 21, a Scout placed a Strategic Defense Initiative payload into orbit.

Russian/C.I.S. Space Activity

During 1992, Russia and other parties to the C.I.S. space accord successfully launched 54 space missions, with eight taking place in December. This total reflects a further decline in launchings (down from 59 in 1991) because of the continuing political and economic crisis in the former Soviet Union.

Mir space station. The Russian space platform, *Mir*, remained fully funded and crewed during 1992. Thus, *Mir* was able to meet earlier commitments to host several foreign cosmonauts, in separate missions.

On March 17, a German guest cosmonaut was launched to *Mir* aboard the *Soyuz TM-14*. The mission was the first Russian crewed space flight since the collapse of the Soviet Union. The *TM-14* crew reached *Mir* on March 19 and docked successfully with the aft port. During a 1-week stay aboard *Mir*, German and Russian cosmonauts conducted medical and other scientific experiments using 200 lb (90 kg) of German equipment previously delivered to the space station by a crewless Progress vehicle.

On March 25, the German guest cosmonaut returned to Earth aboard the *Soyuz TM-13* spacecraft, piloted by two cosmonauts who had been working aboard *Mir* since before the dissolution of the Soviet Union. The crew members' departure from the Soviet Union and return to independent Kazakhstan is only one of many anomalies associated with the rapid pace of political change.

On July 27, a French cosmonaut was launched to *Mir* on *Soyuz TM-15* to conduct a series of biological and other scientific experiments from the vantage of the space platform. The flight was one of five joint Franco-Russian commercial missions planned through the year 2000.

The launch of two *Mir* building-block modules, including the Priroda for Earth resources and ecological monitoring, which had been planned for 1992, was delayed until 1993 or even later. The delays were said to be related to both technical and budgetary difficulties.

Mir, which was launched in February 1986, has an expected operational lifetime of at least 10 years. Russia has abandoned the concept of replacing *Mir* with a totally redesigned space platform. Instead, *Mir*'s central core will be replaced with one incorporating only incremental improvements (probably in 1997 or 1998); *Mir 1* building-block modules will be replaced on an individual basis, as conditions warrant.

Buran space shuttle. *Buran*, the Russian shuttle or winged reusable spacecraft, received only minimal funding in 1992. *Buran* has not been flown since its maiden launch in 1988. A second *Buran* mission is intended for 1993, but given the budgetary uncertainties that beset Russia's space program, this schedule may slip; indeed, some analysts believe that *Buran* may even be abandoned. Alternatively, managers of Russia's space program are said to be considering the use of *Buran* only for crewless flights.

United States–Russian cooperation. In a remarkable display of cooperation, U.S. Defense Department experiments were carried to orbit in August aboard a Russian spacecraft. The U.S. Air Force Space Test Program and the U.S. Naval Research Laboratory sponsored the experiments, which were carried aboard a Russian imaging-film return spacecraft. Also in August, Russian scientists at the World Space Congress in Washington, D.C., presented information on former Soviet tracking radar capabilities and military space surveillance techniques. Such information may assist the world community in coordinating space debris calculations for safer space operations. SEE SPACE TECHNOLOGY.

In October, NASA and the Russian Space Agency signed an agreement with a number of unprecedented provisions, including the flight of a Russian cosmonaut on the United States space shuttle, the flight of a United States astronaut on Russia's *Mir* space station, and a joint mission featuring rendezvous and docking of the space shuttle with *Mir*. Another agreement also was signed in October for the flight of two NASA scientific instruments aboard the Russian *Mars '94* mission.

European Space Activity

European activity included the Ariane and Spot programs.

Ariane. The first flight of 1992 for Ariane took place February 26, when an Ariane V49 booster successfully launched two communications satellites (*Superbird B* and *Arabsat 1C*) into geostationary transfer orbits.

On August 10, the *Topex/Poseidon* satellite was successfully launched aboard an Ariane 4 from the Guiana Space Center. Topex/Poseidon is a joint French-NASA scientific program to study ocean circulation and its role in regulating global climate.

In April, with the maiden launch of the new V50 booster, Arianespace fully instituted new operational procedures that were described as a turning point in Europe's Ariane program. The new procedures, which reduce the number of launch operations from 1200 to 700, are designed to lower mission costs, reduce the length of launch campaigns, and increase operational availability and flexibility of the new three-stage vehicle.

The outlook for Arianespace, the marketing-management organization for Europe's Ariane booster, appears bright. By midyear, the company had booked a total of more than 100 payload launch contracts since its founding in 1980. With a current backlog of several dozen satellites to be launched, Arianespace expects to sign up an average of 10 payloads annually for the next several years.

Spot. Another of Europe's commercial space programs, the *Spot* Earth resources satellite, also had a highly successful year. Imagery is currently being provided by *Spot 2* and supplemented by the 6-year-old *Spot 1*, which was reactivated in 1992 to help meet growing consumer demand.

Spot Image announced plans to provide higher-resolution imagery and upgraded stereo data-gathering capabilities for its next-generation satellites. The improved spacecraft, which will feature 16-ft (5-m) image resolution as well as stereo imaging along its flight track, is expected to be ready for launch in the late 1990s. Spot Image, which is responsible for commercial marketing and sales of Spot data, was organized by CNES (Centre National d'Etudes Spatiales or National Center for Space Studies), the French national space agency.

Asian Space Activities

Japan and China continued their ambitious efforts to develop comprehensive commercial space infrastructures, and it has become clear that India aims to achieve a larger role in the commercial satellite market.

Japan. Development launches of Japan's H-2 heavy booster were delayed at least a year, to January or February 1994. The postponement is the second major slippage in the booster's operational schedule. This delay is a result of changes required to correct for a design and manufacturing flaw that led to a fire during a June 18 test run of the booster's LE-7 first stage. The cause of the mishap appeared to be temperature-related fatigue of a weld on an elbow joint connecting the engine's liquid hydrogen turbopump to the main injector. A crack in the elbow allowed hydrogen gas to leak and ignite 5 s into the planned 20-s test. Design changes included adding an interior liner to the elbow and flattening the weld to better distribute the heat load. Improved heat treatment of the weld during the manufacturing process will also be specified.

The H-2 had already undergone minor modifications to its first-stage oxygen-hydrogen LE-7 engine to correct problems that caused four successive premature shutdowns of firing tests in April. The two-stage H-2 is 164 ft (50 m) in overall height and weighs 260 tons (236 metric tons).

Japan's *JERS* Earth resources remote-sensing satellite was launched on February 11 aboard an H-1 booster. After some initial difficulty in deploying one of the antennas and an extended check-out period, the satellite began normal operation in August. The optical observation path surveyed by *JERS 1* is 46.5 mi (75 km) in width. Data collected by the satellite will be used to survey Earth resources and provide useful information for environmental protection agencies, as well as for the agricultural, forestry, and fisheries industries. Distribution of *JERS 1* data and images is being coordinated by the Remote Sensing Technology Center in Tokyo. Approximately 300 researchers worldwide have been approved to receive information from *JERS*, and data from the satellite will also be exchanged with other national space agencies.

Beyond *JERS*, Japan has begun development of a family of next-generation satellites and sensors designed to monitor the global environment into the twenty-first century. Work is progressing on the *Advanced Earth Observing Satellite (ADEOS)*, presently scheduled for launch in February 1996, although problems with the H-2 could result in some delay. *ADEOS* will feature eight sensors, including two from the United States and one from France. The French POLDER sensor will observe solar radiation reflected by the Earth's atmosphere. The United States–NASA instruments include the NSCAT scatterometer, for measuring wind speed and direction over the Earth's oceans, and the total ozone mapping spectrometer (TOMS), for recording ozone changes, sulfur dioxide levels, and changes in ultraviolet radiation.

On December 1, Japan's *Superbird A* communications satellite, built by Space Systems/Loral, was launched from Guiana aboard an Ariane 42P with two solid rocket boosters. The successful launch ended a succession of costly disasters for the satellite's owner, Japan's Space Communications Corporation. Although the original *Superbird A* was launched successfully in 1989, *Superbird B* was destroyed when its Ariane booster exploded just after takeoff in February 1990. A few months later, the original *Superbird A* was lost in orbit as a result of ground control errors. The destroyed *Superbird B* spacecraft was replaced in February by a new *Superbird B* satellite, also launched by Ariane.

China. A continuing bid by the People's Republic of China to become a major commercial player in the space field gathered momentum in 1992. The first successful Chinese launch of the year took place August 8 from the Jiuquan space facility in north-central China. The launch was the first mission to use the new Chinese CZ-2D booster, with a standard Long March 2 first stage and a stretched second stage to accommodate more propellant and achieve greater performance. The flight also involved the first mission for the new Chinese *FSW 2* recoverable imaging spacecraft.

The year began on an inauspicious note for China with an aborted Long March 2E mission in March, intended to place a Hughes-Australian *Optus B1* (formerly designated *Aussat B1*) into geosynchronous orbit. The Long March rocket malfunctioned on the launch pad 2–3 s into the ignition sequence, resulting in a spectacular fire. A damage control and rescue effort requiring 39 h resulted in recovery of the *Optus B1* in undamaged condition. Several technicians who were engaged in controlling the blaze were subjected to poisoning from toxic gases, although no fatalities were reported.

Launch of the recovered *Optus B1* was successfully achieved on August 14 by using the same (repaired) Long March 2E vehicle with four large liquid-fuel strap-on boosters. This flight marked the first operational mission for the 2E, which had undergone only partially successful flight tests in 1990.

On October 6, Chinese and Swedish spacecraft were successfully launched by China from Jiuquan. The Swedish *Freja* spacecraft features scientific instruments supplied by the United States, Canada, and Germany for recording auroral images and magnetospheric phenomena. The Chinese *FSW 1* imaging spacecraft, which went aloft with the *Freja*, was the second imaging satellite launched by China within a 2-month period. A principal purpose of the imaging craft is to monitor progress in reforestation of portions of China's Gobi Desert and to evaluate new forest growth.

Although China has not yet gained a significant share of the commercial launch market, clearly its intention is to do so. The APT Satellite Company, which is 75% owned by firms backed by the Chinese government, will likely use a Long March booster to launch the Hughes HS 376 telecommunications satellite into orbit in 1994. Designated *APStar 1*, the satellite will compete directly with the Asian Satellite Telecommunications Company's *AsiaSat 1*, also a Hughes HS 376. A commitment to launch the *APStar 1*, facilitated by the Bush Administration's 1992 relaxation of restrictions on launches from China with United States–made components, is regarded as a major step toward achieving China's commercial space ambitions.

During 1992, facilities were expanded at the Xichang launch site in anticipation of future launch business, with the construction of an impressive new launch complex for the Long March 2E heavy booster rocket.

India. *Insat 2a*, a multimission satellite owned by the Indian Satellite Research Organization (ISRO), was launched aboard an Ariane 4 from the Guiana Space Center on July 9. The satellite is designed to perform telecommunications, television, and data-relay missions, as well as to provide meteorological imaging and satellite-aided search and rescue services. *Insat 2A* is the first in the series of *Insat 2* satellites to be developed by using primarily Indian technology and hardware. The *Insat 2* craft are successors to *Insat 1* satellites, provided by Ford Aerospace.

ISRO has indicated its hope, based on success of the *Insat 2*, of becoming a component supplier to French or other satellite manufacturers. India has already served as a small-scale supplier of electrooptical sensors and other components to Brazil. For telecommunications and television relay, *Insat 2* features two 42-dBW (decibels above 1 watt), S-band channels; twelve 32-dBW, C-band channels; and two 34-dBW extended C-band channels. The ISRO *Insat 2B* spacecraft is scheduled to be launched, also on Ariane, in 1993.

India's launch program scored an important success with the May launch of its Augmented Satellite Launch Vehicle (ASLV). The ASLV payload, a scientific spacecraft, functioned properly, although the orbit perigee is lower than intended, so premature decay of the orbit will occur. The ASLV is designed to place 330-lb (150-kg) payloads into near-circular orbit. The successful launch followed two ASLV failures.

ISRO also announced that it has concluded an agreement to acquire a Russian upper stage for use as its future Geosynchronous Satellite Launch Vehicle.

For background information *SEE APPLICATIONS SATELLITES; COMMUNICATIONS SATELLITE; EXTRATERRESTRIAL INTELLIGENCE; SATELLITE ASTRONOMY; SCIENTIFIC SATELLITES; SPACE FLIGHT; SPACE PROBE; SPACE SHUTTLE; SPACE STATION* in the McGraw-Hill Encyclopedia of Science & Technology.

Robert J. Griffin, Jr.

Bibliography. M. Giget (ed.), *World Space Industry Survey: Ten-Year Outlook, 1991–1992,* 1991; Russians to keep *Mir* manned this year, *Aviat. Week Space Technol.,* 136(2):22, January 13, 1992; J. N. Wilford, Scientists set to give up on *Galileo*'s antenna, *The New York Times,* p. A20, January 21, 1993; A. Wilson (ed.), *Space Directory 1992–93,* 8th ed., 1992.

Space technology

Space vehicles and satellites operate in an environment of high vacuum, temperature extremes, particle radiation, atomic oxygen, human-made debris, micrometeoroids, and solar radiation. Spacecraft materials are affected by this environment, even to the point of failure. The first section of this article discusses the effects of the space environment on spacecraft materials, focusing on investigations that were carried out with the *Long Duration Exposure Facility* (*LDEF*) satellite. The second section discusses one feature of the space environment, space debris, and its effects on spacecraft.

Effects of Space Environment on Spacecraft Materials

The *LDEF* satellite of the National Aeronautics and Space Administration (NASA) was built to investigate the effects of the low-Earth-orbit space environment on spacecraft materials and systems. Launched aboard the shuttle *Challenger* in April 1984 (**Fig. 1**), the

Fig. 1. *Long Duration Exposure Facility (LDEF)* satellite suspended from the space shuttle *Challenger*'s remote manipulator arm, prior to release into space on April 7, 1984. (*NASA*)

LDEF spent nearly 6 years in space before its retrieval by the shuttle *Columbia* in January 1990. Knowledge of space-induced materials degradation was greatly expanded by *LDEF* experiments. Understanding the space environment and its effect on materials is essential to the development of future space systems. One example is Space Station *Freedom*, which must operate in low Earth orbit for 15 years or more.

LDEF. The *LDEF* was developed as an inexpensive, passive, free-flying platform that could be placed into orbit by the space shuttle and retrieved on a later flight. The initial *LDEF* mission was planned to investigate the effects of long-term exposure in low Earth orbit on several types of materials, as well as to conduct a variety of other experiments, including exposing millions of tomato seeds to the same space radiation hazards. The *LDEF* carried a total of 57 experiments in 86 desktop-sized trays around the twelve sides and two ends of its aluminum frame (Fig. 1). More than 200 principal investigators, representing universities, corporations, governments, and international organizations, contributed to the 10,000-plus specimens and experiments carried aboard the *LDEF*. The 21,400-lb (9700-kg), 30-ft-long (9.1-m) satellite was carried into space by *Challenger* in April 1984. The mission was originally planned for 10 months, but the *Challenger* disaster delayed recovery of the *LDEF* until January 1990. The nearly 6-year stay in space significantly increased the value of the *LDEF* experiments.

Space environment. Several features of the space environment at low Earth orbit must be considered in the selection of materials for spacecraft use. At low Earth orbit the space environment includes the Earth's residual outer atmosphere (principally atomic oxygen), ultraviolet electromagnetic radiation from the Sun, particle radiation (high-energy electrons and protons) from the solar wind trapped in the Earth's magnetic field, micrometeoroids and cosmic dust, and debris from previous spacecraft. In the high vacuum of space, materials such as lubricants, adhesives, paints, and polymers will outgas and contaminate sensitive spacecraft components. In addition, a satellite orbiting the Earth is periodically heated and cooled as it moves into and out of the Earth's shadow. Each of these elements of the space environment affects materials in different ways.

Atomic oxygen. The Earth's atmosphere extends tenuously into outer space. Atomic oxygen, produced by dissociation of molecular oxygen by solar ultraviolet radiation, is the predominant constituent of the atmosphere from 100 to 350 nautical miles (200 to 650 km) above the Earth's surface. This range includes the nominal orbital altitude for low-Earth-orbit satellites as well as the space shuttle. As a spacecraft orbits the Earth at a speed of 5 mi/s (8 km/s), it collides with trillions of oxygen atoms every second. The energy from these collisions (4–5 electronvolts per collision) causes chemical and physical reactions on the exposed surfaces of the spacecraft. The principal result is erosion of surface material, although the amount of erosion will vary from material to material. Polymeric materials are affected the most by atomic oxygen. Thermal control blankets used on spacecraft are made of very thin polymer films (Kapton, Teflon, or Mylar) coated on the back with a reflective metal such as aluminum or silver. On the *LDEF* some of the thinner Kapton blankets were completely eroded away by atomic oxygen. Some Teflon blankets became cloudy because of the erosion of surface material. Specimens of uncoated polymer composites similarly lost the equivalent of one ply of material. Among metallic specimens on the *LDEF*, both silver and copper were heavily oxidized by atomic oxygen.

Solar ultraviolet radiation. Although the majority of the Sun's energy reaching the Earth is in the infrared and visible portions of the electromagnetic spectrum, it is ultraviolet radiation that concerns spacecraft designers. Like atomic oxygen, solar ultraviolet and shorter-wavelength radiation exhibits the most pronounced effects on polymeric materials. Ultraviolet radiation breaks the bonds of polymer molecules, reducing the polymer's tensile strength. Teflon films flown on the *LDEF* exhibited up to a 30% decrease in tensile strength because of ultraviolet degradation. Ultraviolet radiation also results in the production of molecules of small molecular weight, such as water or carbon dioxide, which then cause internal defects (crazing) or add to spacecraft contamination. Finally, ultraviolet radiation can discolor a material. *LDEF* specimens of thermal control paints containing polymeric binders exhibited significant discoloration and associated changes in emissivity and absorptivity. The discoloration reduces the effectiveness of the paint in reflecting solar radiation.

Particle radiation. Particle radiation (energetic electrons, protons, and neutrons) has always been a particular concern for astronauts and other living organisms because of the potential for tissue damage. Thus, crewed spacecraft are shielded to reduce the level of radiation exposure. Particle radiation affects most materials. Metals can become slightly radioactive from the collision of high-energy protons. Polymers are affected in much the same way as they are by solar ultraviolet radiation. Perhaps the most severely affected materials are semiconductors such as those used in solar cells and integrated circuits. High-energy protons, electrons, and cosmic rays ionize atoms in semiconductor materials, causing solar cells to lose efficiency and electronic components to experience shorts or logic errors. The *LDEF* exhibited minimal degradation from particle radiation. It had no extensive solar arrays, and the few electronic instruments for recording data were unsophisticated and well shielded. Even the tomato seeds flown on the *LDEF* easily survived the space journey, as attested by thousands of healthy plants grown by students across the United States. *LDEF*, the space shuttle, and other space vehicles in low Earth orbit are not subjected to the more intense regions of the Van Allen radiation belts. However, geostationary and polar orbiting spacecraft operate in regions of the radiation belts that are orders of magnitude more intense. Electronics in these spacecraft must be especially designed to operate in this radiation environment, and perhaps must be shielded from the higher-flux, lower-energy particles.

Micrometeoroids and debris. Collision with micrometeoroids or space debris is another hazard of the space environment for crewed space flight, as will be discussed in detail in the second part of this article. Even small particles can damage spacecraft because of high impact velocities of about 10 mi/s (16 km/s). Windows on the space shuttle have been replaced numerous times after becoming pitted by micrometeoroids, flecks of paint, or other debris. More than 34,000 impact features were examined on the *LDEF*. The largest was an impact crater 0.2 in. (5 mm) in diameter. Several composite-material specimens on the *LDEF* utilized protective layers to limit damage from atomic oxygen or ultraviolet radiation. In many cases, these protective coatings were penetrated by micrometeoroids or debris, exposing substrate material to attack by atomic oxygen and ultraviolet radiation. Solar-cell covers and optical materials were also damaged by micrometeoroids and debris, reducing their efficiency.

Vacuum-induced contamination. At low Earth orbit, the ambient pressure of the outer atmosphere is 10^{-5} to 10^{-3} pascal (10^{-7} to 10^{-5} torr), compared to standard sea-level pressure of 10^5 Pa (760 torrs). In this vacuum, volatile materials, such as lubricants, adhesives, and solvents, on spacecraft surfaces can evaporate or outgas. Some of this evaporated material will condense on the spacecraft surface. Contamination reduces the reflectance of mirrors, clouds lenses and solar cell covers, and changes the radiative properties of thermal control surfaces. On the *LDEF*, two major sources of contamination were urethane paint and silicone-containing materials. The widespread contamination on the *LDEF* was one of the major surprises of the mission. Scientific study of the sources, mechanisms, and effects of such contamination is difficult because atomic oxygen, solar ultraviolet radiation, and thermal cycling play a large role in contamination sources and mechanisms.

Thermal cycling. As the *LDEF* circled the Earth every 90 min over a nearly 6-year period, it experienced roughly 32,000 temperature cycles as it moved from orbital day to night. Space Station *Freedom*, which could orbit the Earth for 30 years, would experience 175,000 thermal cycles. The difference in temperature from hot to cold on the surface of a spacecraft as it orbits the Earth can be of the order of 100°F (60°C). This thermal cycling can produce stresses in materials, resulting in mechanical defects or failure.

Combined effects. The combined effect of all the various elements of the space environment tended to accelerate the degradation of many of the materials on the *LDEF*. Micrometeoroid damage to surface films allowed atomic oxygen and ultraviolet radiation to attack underlying material. Thermal cycling, atomic oxygen, and ultraviolet radiation worked in concert to damage polymeric materials. Small molecules produced by ultraviolet irradiation of polymer materials added to the overall contamination. In another case, the effects of atomic oxygen and ultraviolet radiation canceled each other. Thermal control paints were first discolored by ultraviolet radiation. Then atomic oxygen "sandblasted" away the brownish organic contaminants, leaving relatively clear oxides behind. Spacecraft designers, therefore, need to consider all aspects of the space environment when selecting materials.

Although many of the material specimens on the *LDEF* suffered some level of damage from the space environment, several materials survived the nearly 6-year exposure with little or no degradation. The anodized aluminum structure fared very well in the low-Earth-orbit environment. Fiber-reinforced aluminum composites, materials protected by stable ceramic oxide coatings, and several types of thermal control paints displayed excellent resistance to the effects of the space environment. As a test-bed for future spacecraft materials, the *LDEF* easily exceeded its mission objectives. The materials data generated by the *LDEF* experiments will be instrumental in designing future systems to withstand long-term exposure to the space environment.   *James N. Bower, Jr.*

Space Debris

Since 1961, more than 100 of the spacecraft and rocket bodies placed in Earth orbit have exploded, leaving satellite fragments in Earth orbit. Beginning in the late 1970s, NASA mathematical models predicted a large number of undetected orbiting fragments that could impact other spacecraft at an average speed of 6 mi/s (10 km/s), causing an increasing hazard to space missions. Since the early 1980s, NASA has worked with other national agencies and with other governments to better characterize the hazard to spacecraft from orbital debris and to minimize the possibility of future satellite breakups. These efforts have been partially successful. The number of smaller fragments in Earth orbit has now been measured, and these measurements agree with earlier model predictions. Also, a consensus has developed among the world space community that the possibility of satellite breakups must be minimized. This consensus has led to operational and spacecraft design changes. In the future, more such changes are likely to be necessary.

Cataloged objects. The U.S. Space Command, located in Colorado Springs, Colorado, has the responsibility of maintaining a catalog of all human-made objects in space. However, this goal is not possible since the radars used to maintain this catalog operate at a wavelength that is too long to detect objects smaller than about 4 in. (10 cm). Even so, the catalog includes about 5000 objects orbiting the Earth below 1000 mi (1600 km) altitude. About half of these objects are the result of explosions of spacecraft or rocket bodies in orbit. Another 2000 cataloged objects

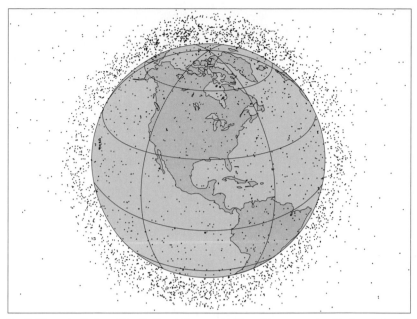

Fig. 2. "Snapshot" view of the Earth-orbiting population of human-made objects cataloged by the U.S. Space Command.

are above 1000 mi (1600 km), some in interplanetary space. Although orbital debris is an issue at some of these higher altitudes, such as around the geostationary orbital arc, it is in the region of space below 1000 mi (1600 km) that orbital debris is of immediate concern. **Figure 2** is a computer-generated "snapshot" of this region, illustrating the position of the cataloged population at a given time. **Figure 3** quantifies the hazard from the cataloged population by giving the average rate that a cataloged object will pass within 100 yards (100 m) of an orbiting satellite during 1977 and 1992.

Figure 3 illustrates three properties of the orbital debris population:

1. The effects of varying atmospheric density at altitudes below 500 mi (800 km). Atmospheric density

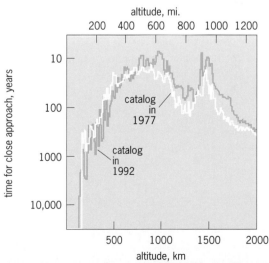

Fig. 3. Time within which a cataloged object is expected to pass within 100 yd (100 m) of an orbiting spacecraft (measured from center of object to center of spacecraft).

decreases with altitude and increases as solar activity increases. Solar activity was at a minimum in 1977, and 1½ solar cycles later, in 1992, solar activity was at a maximum. A higher atmospheric density causes more objects to slow down and fall out of orbit.

2. The increasing accumulation of orbital debris above 500 mi (800 km). At these higher altitudes, orbital lifetimes are very long because of the extremely thin atmosphere, so the accumulation of cataloged debris has about doubled since 1977.

3. The necessity for a satellite to be relatively large to have a significant chance of colliding with a cataloged piece of debris. If a satellite, or group of satellites, is sufficiently large to have an area equivalent to a spherical satellite of 100 yd (100 m) radius, that satellite can expect to collide with a cataloged object as frequently as once every 10 years, depending upon its altitude. The 5000 cataloged objects have a total area this large; consequently a collision between two cataloged objects would be expected to be a likely event. To date, none has occurred. A more detailed calculation that considers the distribution of the cataloged objects with altitude predicts a collision between two cataloged objects once every 20 years. If the accumulation of cataloged objects again doubles, both the target area and the number double, so that the rate of collisions between cataloged objects quadruples to once every 5 years. Consequently, although none has occurred, the odds are increasing that such an event will occur in the near future. Collisions between cata-

loged objects would not likely involve an operational satellite, but the collisions would generate an even larger population of small, uncataloged debris.

Uncataloged debris. Uncataloged debris represents the greater hazard to operational satellites of average size. At 6 mi/s (10 km/s), a 0.4-in.-diameter (1-cm) aluminum sphere will hit with the same energy as a 400-lb (180-kg) safe falling at 60 mi/h (100 km/h). Debris as small as 0.04 in. (1 mm) in diameter will penetrate the surfaces of most unshielded spacecraft. When rocket bodies explode on the ground, they produce a large number of fragments in this size range. An even larger number of small fragments is produced when a satellite breaks up as a result of a collision at 6 mi/s (10 km/s). By collecting fragments from ground tests where objects were fragmented by either exploding them or impacting them with high-speed projectiles, early NASA models predicted a significant and growing population of uncataloged orbital debris.

Early data. In 1983, NASA began to look for techniques to characterize the uncataloged population in low Earth orbit. Two types of data quickly became available. One type was planned, while the other resulted from circumstance.

In 1983, NASA contracted with the Lincoln Laboratory of the Massachusetts Institute of Technology to use ground telescopes to search for small orbital debris. These telescopes were capable of detecting objects larger than about 2 in. (5 cm) as the objects passed through the telescope field of view at altitudes below 1000 mi (1600 km). They detected nearly three times as many objects as were cataloged.

Also in 1983, a space shuttle window was replaced because it contained the largest impact crater ever found on a spacecraft surface. This crater, which measured 0.16 in. (4 mm) across, was analyzed and found to have resulted from an orbiting paint fleck, about 0.01 in. (0.2 mm) across, hitting the window at a speed of about 3 mi/s (5 km/s). This finding suggested an orbiting population of very small debris millions of times greater than the cataloged population. Data from other returned spacecraft surfaces confirmed this conclusion.

Recent data. The best data so far on the uncataloged population come from two sources: the *LDEF* satellite and large ground radars, operating at a shorter wavelength than is used to catalog objects. Both sets of data statistically sample the environment, rather than attempt to catalog an object.

The *LDEF* was hit by both human-made orbital debris and natural meteoroids. Natural meteoroids are in orbit about the Sun and pass through the Earth's orbital space at an average velocity of 10 mi/s (16 km/s). Consequently, for a meteoroid impact of given size, the damage to spacecraft is about the same as an orbital debris impact. Whether *LDEF* impact craters are due to meteoroids or orbital debris can be determined from chemical and orbital properties of these two groups. Orbital debris impacts dominated the small-size craters on the *LDEF*, while meteoroids dominated the large-size craters.

The best statistical samples of the environment with

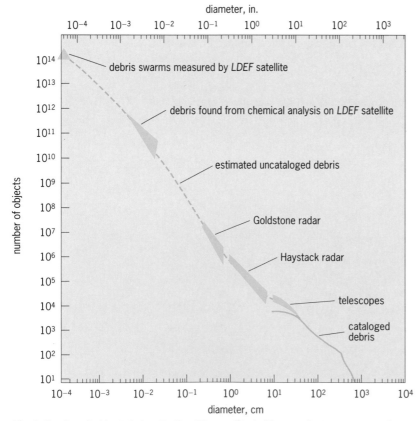

Fig. 4. Number of objects in low Earth orbit as estimated from various measurements.

ground radars have been with the Goldstone radar, located near Los Angeles, California, and the Haystack radar, located near Boston, Massachusetts. These radars are normally used to track deep-space missions, such as the *Galileo* mission to Jupiter, or to image the surface of planets, such as Venus. For purposes of detecting small orbital debris, these radars are best operated in a mode where they point vertically and simply count the number of objects passing through their field of view. When operated in this mode, they can detect orbiting objects as small as 0.08 in. (2 mm). Most of the data collection has been with the Haystack radar, which detects objects as small as 0.25 in. (6 mm) across at a rate approximately 60 times the rate that the cataloged objects pass through the field of view. **Figure 4** summarizes the measurements to date by giving the inferred number of objects in low Earth orbit. The measurements indicate that there are millions of fragments orbiting the Earth that are 0.04 in. (1 mm) or larger. An orbiting population of this magnitude means that human-made orbital debris represents a greater hazard than meteoroids over most size intervals, a hazard that is very close to that predicted by early NASA models.

Control. NASA has also been working with other space-faring nations to better understand orbital debris and to agree on the best ways to control the future environment. NASA currently has working relationships with the European Space Agency, Japan, and Russia, and is developing working relationships with other nations, including China. All countries have agreed to alter their operational practices to minimize the possibility of future accidental explosions. As a result, over the last few years, there have been fewer explosions, reducing the expected rate of increase of the small debris population. However, within the next few decades the amount of small orbital debris could again begin to increase significantly as a result of random collisions rather than chemical explosions. The only way to control this type of debris is to eliminate from orbit objects that are likely to collide, such as rocket stages and payloads left in orbit at the end of the operational life. The challenge of future debris control will be to obtain international agreement not to leave these nonoperational objects in orbit.

For background information *SEE AMOSPHERE; INTERPLANETARY MATTER; RADIATION HARDENING; SPACE BIOLOGY; SPACE STATION; SPACE TECHNOLOGY; VAN ALLEN RADIATION* in the McGraw-Hill Encyclopedia of Science & Technology.

Donald J. Kessler

Bibliography. N. L. Johnson and D. S. McKnight, *Artificial Space Debris*, 1987; D. J. Kessler, J. C. Zarnecki, and D. L. Matson (eds.), Space dust and debris, *Adv. Space Res.*, vol. 11, no. 12, 1991; A. S. Levine (ed.), *LDEF-69 Months in Space, 2d Post-Retrieval Symposium*, NASA Conf. Pub. 1394, June 1993; J. P. Loftus, Jr. (ed.), *Orbital Debris from Upper-Stage Breakup*, Progress in Astronautics and Aeronautics, vol. 121, 1989; *69 Months in Space: A History of the First LDEF (Long Duration Exposure Facility)*, NASA Pub. NP 149, 1990.

Speech

The period of speech development in humans begins with the earliest cries and vocalizations of infants and extends throughout early childhood until all the speech sounds or phonemes of a language are produced with the use of real words in a generally acceptable adult manner. Speech is only part of oral language, specifically its sound system or phonology. Other aspects of language include form or syntax, meaning or semantics, and function or pragmatics. In the 1940s, linguists believed that early infant vocalizations or babbling had no relationship to later real speech except that infant babble contained all the speech sounds that humans were capable of producing. Thus, infant vocalizations were not considered interesting by many investigators until the 1970s. Since that time, research in the development of speech, particularly in infant vocalizations, has blossomed.

Stages of vocal development. Many investigators have proposed models of vocal development that show stages of speech production that change with age (see the **table**). In the first 2 months of life, infants cry and produce sounds that can be called vegetative, such as burps, grunts, sneezes, and hiccoughs. These earliest productions are reflexive in nature (that is, related to biological functions other than speech development) and do not resemble later speech in any clear way.

After the first month of life, infants begin to laugh and to produce sounds that are like very early, primitive syllables. Adults generally imitate these as "coos and goos." The initial consonantlike portion is made by the velum, or soft palate, pushing against the back of the vocal tract. The vowellike portion is made with the mouth closed or partially closed at times, with an almost humlike quality.

By age 3 months, infants engage in vocal play and begin to expand their repertoire of sounds. They produce squeals, growls, "raspberries" (by placing both lips together and blowing air), and trills. They may appear to whisper or to yell with a loud voice. During this period, infants begin to make a different kind of primitive syllable that sounds more like speech. For some of these productions, the consonants are longer than any made by adults and may sound somewhat like a long w-sound in the word "we" or like a long y-sound in "yo-yo."

Between the ages of 5 and 10 months, infants begin reduplicated babbling in which they produce repetitive strings of sounds, such as "dadada" or "mamama." Parents may interpret these as real words at times because the consonants and vowels resemble adult models so closely in timing and quality. In the same time period, infants make other multisyllabic utterances in which they vary the consonants or vowels within a string. They may produce strings such as "daladaladala" or "numanuma." Investigators have called this variegated babbling to suggest the variety of sounds produced. At one time, stage models of infant vocal development indicated that simpler reduplicated babbling precedes variegated babbling. More recent research has shown that once multisyllables are produced by infants, they may be either reduplicated or varied. By age 10 months, infants who are developing normally should be producing sounds that are very speechlike to any adult listener.

Many aspects of early speech sounds produced by infants are determined by the size and shape of the infant's vocal folds and vocal tract and by the spatial relationships of the tongue, palate, pharynx, and epiglottis. These spatial relationships are quite different for adults. Acoustic analyses of infants' vocalizations compared to those of adults show differences in the patterns of resonances or formants as well as the fundamental frequency or voice pitch that reflect the anatomical and functional differences.

Between the ages of 9 and 18 months, infants begin to use real words, that is, utterances that are meaningful and refer to something or someone in their environment. Although these words may not sound like the adult word (for example, a child may say "baba" to refer to the bottle), both the child and the interacting adults understand the reference for the utterance. Infants during this period also use jargon, long strings of meaningless utterances with pitch changes and timing markers mimicking those in adult sentences.

It is now evident that individual infants develop preferences for sound production during the latter half of the first year of life. These preferences appear to determine to some extent the first words an individual child uses. That is, if an infant produced a high proportion of "d" sounds, the first words might be "dada" for "daddy" or "dadi" for "doggie." In the earliest stages of word development, more individual differences among children are noted. During the second year of life language experience grows, and more similarities are observed in the speech sounds children use. The first real words usually have one or two syllables rather than long syllable strings.

Environmental influences. During the period that infants begin reduplicated babbling, they also develop a number of rhythmic motor behaviors such as hand banging, hand clapping, and rocking. This co-occurrence had led some investigators to believe that babbling is an exercise of the speech motor system. However, if babbling has only motoric components, it should be relatively unaffected by the language environment of the infant, that is, infants around the world should produce the same babbling repertoire. Recent evidence from a team of international researchers shows subtle differences in babbling in infants from different language-learning environments, such as French, Thai, English, Arabic, and Cantonese households. These differences are observed more in vowel than in consonant elements. Nevertheless, there remain striking similarities in the babbling of infants from different language environments.

Hearing impairment. A more dramatic influence on infant vocal development occurs when an infant is deaf and has little or no auditory input from the environment. Deaf infants do vocalize in some of the same ways that infants with normal hearing do: they may yell, squeal, and produce some vowellike sounds. However, recent evidence has shown that infants with near or total deafness do not develop well-formed reduplicated syllables (called canonical syllables) until 11 months of age or greater, much later than infants with normal hearing do. Infants with lesser degrees of hearing loss produce a smaller variety of vowellike and consonantlike sounds than infants with normal hearing. These studies of hearing-impaired infants as well as infants from a variety of language groups suggest that babbling has more than simply motor origins. They also show that hearing-impaired infants begin the real-word stage with fewer speech sounds in their repertoire.

Recently, investigators have begun to explore the idea of manual babbling in deaf infants born to deaf parents who use sign language for communication. If these infants demonstrate strings of hand and finger movements that precede the use of true, meaningful signs (the parallel to true meaningful words in oral languages), babbling may be a universal precursor of language itself and may not be tied to oral language only. Early results show that manual babbling does occur for this subset of deaf infants exposed to native sign language by their parents and that manual babble is related to early sign languages in ways similar to oral babble's being related to first speech.

Cognition and experience. Other investigators have studied vocal development in infants with cognitive deficits such as Down syndrome. These infants began the onset of reduplicated or canonical babbling over the same age range as normally developing infants despite their other delays, including a delay in overall language development (specifically, the use

Speech development in infants		
Age, months	Speechlike behavior	Description
0–2	Reflexive crying	
	Vegetative sounds	Grunts, burps, sneezes, hiccoughs
1–4	Primitive syllables	Coos and goos
	Laughter	
3–8	Vocal play	Squeals, growls, "raspberries," trills
	Expanding sounds	Vowellike sounds with some consonantlike sounds
5–10	Reduplicated babbling	Strings of consonants and vowels such as "dada" or "mama"
	Variegated babbling	Strings of varied consonants and vowels such as "daladaladala"
9–18	Jargon	Consonants and vowels in nonsense strings of words
	Real words	Meaningful words even though they do not sound like the adult model

of meaningful utterances). Infants with tracheostomy tubes that prevent oral production of speech have been studied as well. These infants received the normal input of their environment but were unable to use their vocal tract for speech while the tube was in place. Some delays in speech production were noted and attributed to lack of motor practice. Overall, results of this area of research suggest that infant vocal development is a complex phenomenon, with strong motoric and auditory influences and weaker intellectual influences.

Later speech development. Children develop both vowel and consonant sounds. Some consonants develop early in children's speech, for example, "p," "m," "h," and "n." Approximately 90% of children have developed these sounds by age $2^1/2$ years. Other consonants develop much later, for example, "ch" as in "chair," "th" as in "this" and "thin," "z" as in "zoo," "r" as in "rabbit." Only about 50% of children produce these sounds like adults by ages 3 through 4 years. The speech produced by young children has an organization that can be described with rules. This organized system is called the child's phonology. The complexity of this system changes over time until the child's phonology approximates that of the adult.

Development of animal vocalizations. Vocalization development in animals does not completely parallel that in humans because animals do not develop meaningful speech for communication. However, evidence exists in some species of birds that deafening at an early stage of song development prohibits the later production of a mature song. In contrast, nonhuman primates tend to show the characteristics of their species-specific vocalization pattern even when raised by a foster mother of a different species with a different pattern of vocalization.

For background information SEE LINGUISTICS; PHONETICS; PSYCHOLINGUISTCS; SPEECH in the McGraw-Hill Encyclopedia of Science & Technology.

Arlene Earley Carney

Bibliography. B. Lindblom and R. Zetterstrom (eds.), *Precursors of Early Speech*, 1986; J. L. Lockc, *Phonological Acquisition and Change*, 1983; J. Miller (ed.), *Research on Child Language Disorders*, 1991; D. K. Oller and R. E. Eilers, The role of audition in infant babbling, *Child Develop.*, 59:441–449, 1988.

Speech recognition

Speech is a natural form of communication for humans. The recognition of speech by machines, with a performance equal to or better than that of humans, has challenged scientists for many years. The technology is developing rapidly and is reaching a level of sophistication that can support a wide range of applications.

Research in speech recognition since the 1970s has produced solutions for increasingly challenging tasks, from recognition of a few isolated words from one speaker to recognition of fluent speech from virtually any speaker. The availability of fast processors and high-density memories on inexpensive digital circuits offers new possibilities for creative applications of speech recognition technology. Potential applications include voice-activated telephone dialing, automated operator assistance, voice input to computers, voice-activated bank and trade transactions, and dictation machines with written output.

The progress in automatic recognition of speech continues. Recognition of a small number of words in continuous speech or a large vocabulary of isolated words has become practical. The research frontiers are shifting toward the solution of an even more difficult problem, that of an interactive dialogue with machines to access information without imposing unnecessary constraints on the user.

Current status. The performance of a speech recognition system depends upon many factors, such as the size of the vocabulary, the complexity of the task, and whether the system is usable by many speakers without additional training. The discussion here will be limited to speaker-independent systems, which can handle speech from any speaker.

A perspective on the current status of automatic speech recognition is diagrammed in **Fig. 1**, showing various tasks in a space of two dimensions: speaking mode and size of vocabulary. The speaking mode covers a wide range of spoken material, from isolated words to spontaneous speech. Current systems can properly recognize vocabulary of as many as a few thousand words. Generally, the number of confusable words increases with the size of vocabulary.

Examples of speech recognition tasks that can be handled by the current technology are shown on the left side of the diagonal line in Fig. 1. In one task, word spotting, occurrences of a few key words in fluent speech can be recognized. Commercial products are available for isolated word recognition (or for speech with pauses between words) with vocabularies up to about several thousand words and for connected digit strings.

The items on the right of the diagonal line in Fig. 1 are examples of tasks that need more research to become useful. Recognition of continuously spoken (fluent) speech is significantly more difficult than that of isolated words. In isolated words, or speech where words are separated by distinct pauses, the beginnings and ends of words are clearly marked. In fluent speech, such boundaries are blurred. Thus, machine recognition of fluent speech with a large vocabulary is not feasible unless constraints on the syntax or semantics are introduced. The recognition of spontaneous speech, such as is produced by a person talking to a friend on a well-known subject, is even harder.

Figure 2 shows the progress made between 1980 and 1992 in improving the recognition accuracy for different tasks. The figure shows a steady reduction in the error rate for all tasks. Of course, the error rate is not the only measure of performance. Other issues, such as the constraints imposed on the user by the system and robustness of the system under different speaking conditions, must be considered.

Process. Speech is language in acoustic form. Speech recognition is essentially the process of rec-

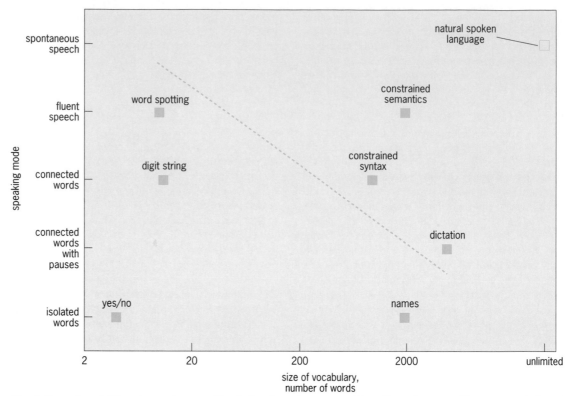

Fig. 1. Examples of different speech recognition tasks shown in a space of two dimensions, speaking mode and size of vocabulary. Examples of tasks that can be handled by current technology are shown to the left of the diagonal line. Items to the right of the line are examples of tasks that need more research to bring performance to a useful level.

ognizing acoustic patterns. There are at present three principal approaches to speech recognition. The first approach is based on statistical techniques of pattern recognition that use a training set of speech data to learn important statistical information about the speech signal. The second approach, commonly known as the acoustic-phonetic approach, uses knowledge of the relationship between the acoustic and phonetic structures of the language. The third and relatively new approach uses artificial neural networks. The greatest success

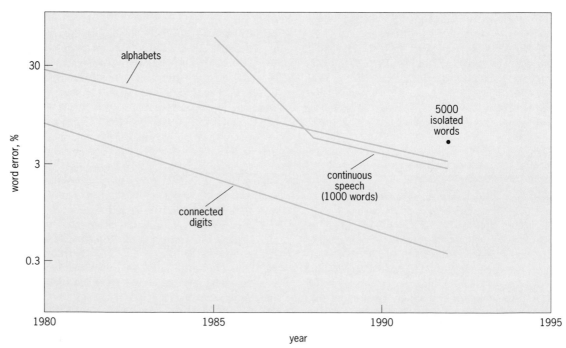

Fig. 2. Decrease in the word error rate between 1980 and 1992 for different speech recognition tasks.

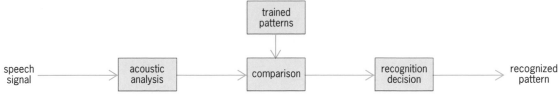

Fig. 3. Basic pattern recognition model used in automatic speech recognition.

has been achieved with statistical pattern recognition.

The basic steps of the pattern recognition approach are illustrated in **Fig. 3**. The speech signal is analyzed to provide a parametric representation at the acoustic level. These parameters (features) are then compared to a prestored set of patterns derived from a large training set of speech utterances from many speakers. This comparison provides a set of scores representing the similarity between the unknown pattern and each of the prestored patterns. The last step uses these scores together with other knowledge about the speech utterance, such as the language and semantics, to provide the best transcription.

Acoustic representation of speech. The selection of proper acoustic features is important for achieving high performance in a speech recognition system. A great deal of research has therefore been done to identify these features. In general, those representations that eliminate information not pertinent to phonetic differences are effective. The short-time spectral envelope of speech, obtained either by filtering or by linear prediction analysis, is still considered the most effective representation for speech recognition. The spectra are computed sequentially in time at intervals of 10–20 milliseconds and are usually converted into cepstral coefficients (the cepstrum is the inverse Fourier transform of the logarithm of the power spectrum) to provide a compact set of 10–20 such coefficients.

The cepstral coefficients are instantaneous (static) features. One of the most important advances in the acoustic representation of speech has been the introduction of dynamic features, such as first- and second-order derivatives of the cepstrum, which represent the temporal changes in the cepstrum. The instantaneous and dynamic features are generally combined to form a large feature set; a smaller set can be obtained by proper selection from the larger set. New representations based on human hearing have been proposed, but they have not yet been found to have a significant advantage over the cepstral representation.

Pattern-matching procedures. The objective in pattern matching is to compare the unknown test pattern with a set of prestored reference patterns (templates) established from the training data and to provide a set of scores representative of the similarity between the test and reference patterns. An important problem in matching speech patterns is caused by the variability due to changes in the speaking rate. Speaking rate varies from one utterance to another, even for the same word. Pattern-matching procedures use dynamic time warping to achieve time alignment

between the unknown and reference utterances.

Template-based pattern matching formed the basic approach of the early isolated-word recognition systems, but this approach is not suitable for the recognition of continuous speech. Therefore, most speech recognition systems now use statistical pattern matching based on hidden Markov models.

Hidden-Markov-model techniques. The development of statistical hidden-Markov-model-based speech recognition techniques represented a major step in the automatic recognition of continuous speech. Because of its statistical nature and simple algorithmic structure for handling large variability, the hidden-Markov-model approach has found widespread use for automatic speech recognition. Most successful systems are now based on this approach.

Models and states. In the hidden-Markov-model technique, each word or sound in the vocabulary is characterized by a model containing a number of states. Each state usually represents a cluster of similar observations (acoustic feature vectors) that occur sequentially in time. The probability that an acoustic vector belongs to a state is characterized by a probability density function with a number of unknown parameters that are estimated from a training set of speech data. The probability density is often approximated as a weighted mixture of gaussian densities.

For small vocabularies, the hidden-Markov-model technique can be applied to whole words, without further division into smaller subword units. Such an approach is not practical for large-vocabulary continuous speech recognition because of the large number of word models that must be trained. Phonemes or phonemelike elements form a natural choice for subword units, resulting in a much smaller number of models. There are approximately 50 phonemes in English, and thus the entire vocabulary can be represented by only 50 models. Each subword is modeled by about three hidden Markov-model states, and the hidden Markov model for each word is constructed by concatenating models of subword units.

Training of models. The unknown parameters of the probability density function for each state of a hidden Markov model are estimated from a large database covering the multiple ways a sentence or word might be uttered. An important reason for the success of hidden Markov models is the availability of an efficient training procedure for estimating the unknown state parameters. There are two important methods used for training the hidden Markov models. The first method is based on the forward-backward algorithm, which provides a mathematically elegant solution to the training problem. A second, simpler method is the segmental k-means training algorithm. Each step in this procedure modifies the parameters to increase the probability (or the likelihood) of realizing the hidden-Markov-model state sequence for the utterance. The phoneme boundaries in continuous speech are not automatically detectable, and training procedures for speech recognition often require manual segmentation and labeling of speech data. The segmental k-means training eliminates the problem of manual segmentation of speech utterances. An orthographic transcription of the utterance is usually sufficient.

The segmental k-means training procedure consists of three parts: (1) Each utterance in the training set is divided uniformly into a number of segments equal to the number of states predicted by the orthographic transcription. (2) The acoustic vectors belonging to each state are separated into several clusters, with each cluster representing a mixture component of the probability distribution. The clustered data are used to estimate the parameters of the probability density function, such as the mean vectors, covariance matrices, and the weights of the mixtures for each state of the hidden Markov models. (3) Each training utterance is segmented into states by using the state models determined in the second part. These steps are repeated until the total likelihood for the training data based on the estimated models is maximized.

Recognition decision. The trained subword unit models are used to recognize an unknown utterance. In the recognition phase, the utterance is decoded by determining the optimal sequence of hidden-Markov-model states and the corresponding subword units based on the observed sequence of acoustic feature vectors in the utterance. Search procedures based on dynamic programming methods are used to find the sequence of states with the maximum likelihood. Additional information based on the syntax and semantics of the language is included in the recognition decision in order to produce outputs that are admissible in the language.

Prospects. The advances in digital technology are rapidly changing the fabric of telecommunications and the way that information is accessed. A new mode of interacting with computers through voice is emerging. When combined with video, the voice mode offers an easy natural communication interface with computers. Speech recognition technology is a key component of such an interface. Human speech communication is a complex process, and scientific understanding of many other issues beyond acoustics and pattern recognition will be required to mimic this process in computers. Speech science is expanding its frontiers to answer the basic question of how words are put together to express ideas in the spoken language.

For background information SEE SPEECH RECOGNITION in the McGraw-Hill Encyclopedia of Science & Technology.

Bishnu S. Atal

Bibliography. D. A. Berkley and J. L. Flanagan, Humanet: An experimental human-machine communications network based on ISDN wideband audio, *AT&T Tech. J.*, 69(5):87–89, 1990; J. L. Flanagan and C. J. Del Riesgo, Speech processing: A perspective on the science and its application, *AT&T Tech. J.*, 69(5):2–13, 1990; S. Furui and M. M. Sondhi (eds.), *Advances in Speech Signal Processing*, 1992; L. R. Rabiner and B. H. Juang, An introduction to hidden Markov models, *IEEE ASSP Mag.*, 3(1):4–16, 1986.

Steel

Recently a number of alloys known as supersteels have been developed. These are ferrous alloys with highly improved combinations of mechanical properties compared to the steels in general use. These steels can have both ultrahigh strength and damage tolerance or both high strength and formability instead of having to trade one desirable property for another. The development and practical production of these materials have been made possible by the evolution of modeling and analytical techniques and by improvements in control of melting and thermomechanical processing.

Iron-based materials made possible the Industrial Age, which began in the early nineteenth century, and steel has been the metallic material of choice for critical structural applications for more than 100 years. In recent years, improvements in competing materials have led to substitution of polymers, aluminum alloys, and titanium alloys for steel in many applications. However, steel remains the material of choice in critical applications where high strength and high stiffness are required in a load-bearing component that must be small in overall size or thickness. These are the applications for which the supersteels were developed.

Nickel-cobalt steels. The most notable supersteels belong to a family of nickel-cobalt (Ni-Co) steels based on an alloy known as HY-180 (see **table**). Developed under the sponsorship of the U.S. Navy, HY-180 (high yield strength of 180,000 lb/in.^{2} or 1,240,000 kilopascals) is a high-strength, high-fracture-toughness, weldable material for submarine hull applications. Hulls of HY-180 would be about half the weight of current hulls: however, the high costs of the material and its fabrication and the doubts about the ability of such hulls to withstand underwater explosions have kept the material from being used for its intended application.

Alloy development. HY-180 is a material of interest to aircraft designers and users, who, in contrast to submarine designers, have a lesser requirement for fracture toughness in thick sections and no requirements for fabrication by welding in uncontrolled environments. Material of the original HY-180 composition was used in a few highly loaded, flight-critical applications in the U.S. Air Force F-16 fighter; but it was generally considered to be too low in strength to save weight in aircraft structures. However, the potential of the alloy system for higher strengths with minimal penalty in fracture toughness was recognized. The first modified alloy was AF1410 (Air Force alloy with 14% Co and 10% Ni) with a tensile strength of

250,000 lb/in.^{2} (1,724,000 kilopascals). Subsequently, a modified AF1410 was developed with a higher carbon content and a tensile strength of 260,000–280,000 lb/in.^{2} (1,793,000–1,931,000 kPa). Most recently a still higher strength derivative, AerMet 100, has been developed with a tensile strength of 280,000–300,000 lb/in.^{2} (1,931,000–2,068,000 kPa).

These materials have been developed to maximize fracture toughness in an ultrahigh-strength martensitic alloy system. Normally, fracture toughness decreases with increasing strength. The composition and processing of these materials are designed to mitigate this relationship. To this end, substantial amounts of nickel are added to suppress the tendency for cleavage fracture. Cleavage is a brittle fracture node in which fracture occurs along crystallographic planes. This type of fracture is experienced by carbon steels at low temperatures. Cobalt is added to inhibit dislocation recovery for an alloy with high strength and a low carbon content. Dislocations, which are defects in the metal crystals, strengthen metals by interacting and inhibiting plastic flow. Recovery is the process of combining and annihilating dislocations, thus reducing dislocation density. Cobalt atoms substitute for iron atoms on lattice sites, but since their size is slightly different from that of iron atoms, they impede dislocation motion and recovery. Chromium and molybdenum are added as alloy carbide formers.

Maximizing toughness. With cleavage suppressed, the normal fracture mode for high-strength steels is microvoid coalescence, or dimpled rupture. This mode of fracture occurs by plastic deformation and shear from multiple internal initiation sites. Toughness can be maximized by minimizing the number of initiation sites. In conventional low-alloy martensitic steels, these initiation sites are generally iron carbides: either epsilon carbides ($Fe_{2.4}C$) from decomposition of martensite upon tempering or relatively coarse theta cementite (Fe_3C). In HY-180 and its derivatives, the composition and processing are controlled to minimize the formation of iron carbides and to enhance the formation of fine metastable M_2C-type carbides. To this end, the carbon content is held to the minimum necessary for strengthening, and chromium and molybdenum are added to combine preferentially with carbon.

In order to achieve the optimum combination of ultrahigh strength and toughness, it is further necessary to produce these alloys by vacuum induction melting followed by vacuum arc remelting; this process minimizes the content of hydrogen, oxygen, and nitrogen and the formation of nonmetallic inclusions. Composi-

tion must be closely controlled to ensure that only the desirable carbides are formed. An example of the degree of composition control required is that the carbon content must be within 0.01 wt % of the targeted level. The steels must then be thermomechanically processed to produce a fine-grain lath-type martensitic structure in the end product.

Hardening. The steels are hardened by austenitizing, quenching, and aging; that is, the material is heated to a temperature of approximately 1600°F (871°C) in order to transform it to austenite, a high-temperature phase of iron. Carbon dissolves and is held in solution in the austenite. The material is then quenched in air or oil to transform the austenite to a low-temperature, metastable martensitic phase in which carbon is trapped in the crystal lattice. Quenching is followed by refrigeration at −100°F (−73°C) to transform any retained austenite, followed by aging at 900–1000°F (480–538°C) to induce precipitation of M_2C carbides. The primary hardening mechanism is transformation of austenite to low-carbon martensite. Aging induces secondary hardening via the dispersion strengthening effect of the M_2C carbide precipitates. Softening for machining can be achieved by overaging at 1300°F (704°C).

Properties. In their final heat-treated form, these materials exhibit fracture toughness values two to three times as high as conventional low-alloy steels at the same strength levels. In terms of damage tolerance, these high levels of toughness mean that structures made from these materials can tolerate cracks four to nine times as deep as those tolerated by low-alloy steels without catastrophic failure. Linked to the high fracture toughness is a corresponding increase in resistance to hydrogen embrittlement and stress corrosion cracking. Additionally, these materials can tolerate higher temperatures than low-alloy steels without softening, and they have marginally higher resistance to general corrosion, the latter being due to their lower iron content.

Applications. The first application of AF1410 steel was the arresting hook shank for the U.S. Navy F/A-18 fighter aircraft. The hook shank is a tensile member that takes the full impact load of stopping the aircraft on an arrested landing on an aircraft carrier deck. The application requires extreme damage tolerance and reliability. Failure of the member would prevent the aircraft from landing on the ship, and the aircraft would have to be diverted to a field landing; if fuel was insufficient to reach the field, the aircraft would be lost. AF1410 steel replaced a lower-strength alloy steel in this application, with an improvement in

Table 1. Compositions of nickel-cobalt steels based on HY-180, in weight percent

Material	Carbon	Nickel	Cobalt	Chromium	Molybdenum
HY-180	0.10	10	8	2	1
AF1410	0.16	10	14	2	1
Modified AF1410	0.20	10	14	2	1
AerMet 100	0.23	11.1	13.4	3.1	1.2

damage tolerance and reduction of weight.

The higher-carbon Modified AF1410 and AerMet 100 alloys are prime candidates for landing gears, particularly those on carrier-based aircraft. Landing gears on high-performance, carrier-based aircraft must meet unique, severe, and conflicting requirements. The main landing gear must withstand impact loads from sink rates (the vertical component of landing velocity) in excess of 25 knots (12.86 m/s), while the nose gear must bear the entire force of catapulting the aircraft from the ship. These requirements dictate stronger, more massive gears than those required for aircraft designed for field operation.

However, once the aircraft is in flight the landing gears represent parasitic weight. Thus, the weight and, more importantly, the volume that the gears occupy when retracted must be kept to a minimum, because stowage space in the fuselages of modern aircraft is limited. Lightweight landing gears can be made from low-density aluminum and titanium alloys, but these materials lack the high absolute strength and stiffness of steel and cannot be configured into small landing gear. The ultrahigh strength of the two steel alloys permits design of landing gears that are not only lightweight but also small-size.

AerMet 100 has been demonstrated to be suitable for armor for military vehicles, as it has excellent ballistic tolerance. AerMet 100 has been used also in a critical nonmilitary application, replacing an ultrahigh-strength low-alloy steel as the material for the rear axles of Indianapolis-type racing cars. The previous material softened during use because of impingement of high-temperature exhaust gas.

Further improvements are being made in the fracture toughness of the nickel-cobalt steels by means of alterations in the standard heat-treatment process to reduce retained austenite and by additions of rare-earth elements to the composition to control the shape of nonmetallic inclusions. Newer materials are being developed from the basic alloy system, owing to the availability of analytical tools and computer modeling. Auger spectroscopy and field-ion microscopy make it possible for researchers to analyze small constituents of the steel microstructure to determine methods that will ensure the presence of desirable constituents and the absence of undesirable ones. Computer modeling can be used to determine the effects of alloying additions on the nature of bonding in the metal crystal lattice.

Probably the next alloy of this type will be a high-performance bearing steel for the cryogenic pumps in the space shuttle. The materials used at present are only marginally adequate, and reliable lives for the bearings fabricated for them are measured in seconds. The pumps must be overhauled for every flight.

Obviously the nickel-cobalt steel alloy system originally conceived for the development of HY-180 has by no means been fully exploited.

Dual-phase steels. The newly developed dual-phase steels are supersteels with applications for commercial products. These steels combine the strength of high-strength low-alloy (HSLA) steels with

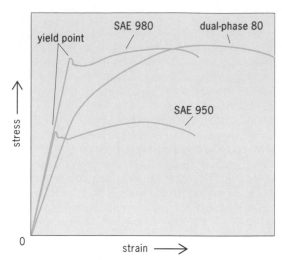

Stress versus strain behavior of two types of Society of Automotive Engineers high-strength low-alloy steels and one dual-phase steel.

the elongation and easy formability of low-strength low-carbon steels.

HSLA sheet steels that have been used in the automotive industry can have tensile strengths as high as 80,000 lb/in.^{2} (551,600 kPa), but they have poor elongation and pronounced yield points (abrupt breaks in their stress versus strain behavior at the transition between elastic and plastic strain). These two factors severely limit the extent and quality of cold forming that can be performed on these materials (see **illus.**). In contrast, dual-phase steels exhibiting the same tensile strength have elongations of about 30% and exhibit a smooth transition between elastic and plastic behavior. Dual-phase steels do not achieve this desirable combination of properties through exotic alloy additions, but through a carefully tailored composition and innovative heat treatment.

Ferrite phase. The predominant phase in dual-phase steels is ferrite, the low-temperature equilibrium phase of iron. As carbon has very low solubility in ferrite, the carbon content of the steels is very low, typically 0.05–0.11 wt %. Manganese is added in the range 0.90–1.70% in order to strengthen the ferrite, provide stability to the austenite phase, and combine with any residual sulfur in the iron. Additionally, silicon and vanadium, chromium, or molybdenum are added to refine the grain size, strengthen the ferrite, and increase hardenability (the ability to form martensite upon quenching from the austenite phase).

Intercritical annealing. The dual-phase condition is produced by an intercritical annealing heat treatment. In intercritical annealing, the material in sheet form is heated for a short time to an intermediate temperature, approximately 1450°F (788°C), for 30 s to 3 min so that it is partially transformed to austenite. This treatment produces a microstructure of 80–90% ferrite. The remaining 10–20% is carbon-enriched austenite that transforms to a mixture of martensite and retained austenite upon cooling. A volume change associated with the martensitic transformation induces residual stresses in the ferrite that suppress the abrupt yield-point behavior when the material is stretched. Finally, the material is strained 2–3% to further suppress the yield-point behavior and to increase strength by cold working of the ferrite. The strength after processing increases with formation of increasing amounts of martensite, and thus can be controlled by composition and heat treatment.

More recently it has been shown that with proper control of the composition of the steel the same dual-phase microstructure can be achieved by rapid cooling from the austenite range. In this way the desired microstructure is produced during the normal steel-mill practice of making sheets by hot rolling and air cooling, with no additional heat treatment being required.

Application. Dual-phase steels are used extensively for the structural components of automobiles and trucks in order to achieve weight reductions for improved fuel economy.

For background information *SEE EMBRITTLEMENT; HEAT TREATMENT (METALLURGY); IRON ALLOYS; STEEL; STRESS AND STRAIN* in the McGraw-Hill Encyclopedia of Science & Technology.

Charles Edwin Neu

Bibliography. American Society for Metals, *ASM Handbook*, 10th ed., vol. 1: *Properties and Selection: Irons, Steels, and High-Performance Alloys*, 1990; G. B. Olson, M. Azrin, and E. S. Wright (eds.), *Innovations in Ultrahigh-Strength Steel Technology*, Sagamore Army Materials Research Conference Proceedings, 1987.

Stratigraphy

Sequence stratigraphy is the study of rock relationships within a time-stratigraphic framework wherein the succession is cyclic and is composed of genetically related stratal units called sequences and systems tracts. Punctuating the rock succession and bounding the sequences are surfaces of erosion (unconformities) or nondeposition, or their correlative surfaces. Time-correlative surfaces that are independent of lithology form the basis for this type of stratigraphic interpretation; in contrast, correlation of like lithologies yields a rock-stratigraphic (lithostratigraphic) interpretation. Modern sequence stratigraphy has its roots in the development of seismic stratigraphy. P. R. Vail and colleagues developed stratigraphic techniques and principles based on time-stratigraphic rather than rock-stratigraphic relationships using seismic data.

It has been observed recently that the same stratigraphic principles that were applied to seismic analyses could also be applied to studies of outcrops, well logs, cores or cuttings from a well bore, and high-resolution seismic data. Clearly, many of the principles developed on the basis of large-scale seismically observed stratal geometries were applicable also to smaller-scale stratal geometries. Thus, sequence concepts were said to be scale-independent both spatially and temporally. At the same time there evolved an increased realization that stratal geometries developed in response not strictly to eustatic change but rather to relative sea-level change, a function of both eustasy and tectonics.

The two interrelated yet distinct applications of sequence stratigraphic concepts are (1) establishment of age models and (2) lithologic prediction. Each is based on the establishment of an unconformity-punctuated time-stratigraphic framework, and each has its unique shortcomings as well as strong points.

Establishment of age models. The establishment of age models is based on correlation of the coastal onlap curve for rock successions of unknown age, with a global coastal onlap curve. Subsequently, the best fit between the two curves yields an approximate age model. This procedure necessarily also involves the integration of other geochronologic (that is, geologic means of age determination) data such as isotope geochemistry, paleontology, and paleomagnetic information.

The use of sequence concepts for age prediction has been the subject of intense scrutiny and debate in recent years. At the heart of this discussion are two key questions. The first concerns whether or not the global sea-level curves are valid or have been developed by summing sea-level events from different basins that may not be coeval. It has been argued that the error bars associated with geochronologic control commonly exceed the duration of published third-order sea-level cycles ($1–3 \times 10^6$ years). Consequently, age correlations of coastal onlap from basin to basin may lack the precision that would permit valid correlations.

The second question concerns whether or not third-order sea-level events can be distinguished from fourth-, fifth-, or even sixth-order events. With the advent of high-resolution sequence stratigraphy, it has been observed that the rock record seems to be a function of relative sea-level cyclicity of significantly higher frequency than was previously thought. Some disagreement arises as to whether third-order sequences, as shown on published coastal onlap and sea-level curves, can in fact be differentiated from similar-appearing higher-order sequences. Consequently, different order sequences may be virtually indistinguishable, thus calling into question both the validity of global sea-level curves and the certainty that locally observed sequences have been correctly identified with regard to order of sea level. Some of these issues may be resolved as more precise geochronologic indicators are developed.

Ideally, the age-modeling aspect of sequence stratigraphy finds its greatest utility in areas where little or no "solid" age information is available. In petroleum exploration, this situation occurs most commonly in frontier basins where few wells exist or where geochronologic data are not reliable. Age information constitutes an essential part of petroleum system evaluation, especially with regard to petroleum maturation history, fluid migration studies, and structural timing.

Lithologic prediction. The lithologic prediction aspect of sequence stratigraphy is useful both for exploration and for field development as well as for enhanced understanding of general sedimentary basin fill evolution. The sequence stratigraphic approach

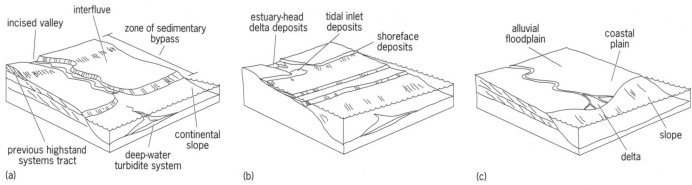

Fig. 1. Three-dimensional views of the sequence stratigraphic model on a passive or trailing continental margin. (*a*) Lowstand systems tract. (*b*) Transgressive systems tract. (*c*) Highstand systems tract.

provides geologic models that enable prediction of lithofacies distribution both vertically and horizontally.

Depositional sequences are composed of a succession of three systems tracts. Systems tracts comprise linked coeval depositional systems bounded by unconformities, transgressive surfaces of erosion, or maximum flooding surfaces. Depositional sequences that can be observed are characterized as type I and type II. Type I sequences consist of lowstand, transgressive, and highstand systems tracts; type II sequences consist of shelf-margin, transgressive, and highstand systems tracts. The precise expression of each systems tract depends on a number of factors, including local physiography, sediment flux, and the rate and amount of both eustatic change and subsidence or uplift.

Type I sequences. These sequences occur in response to cycles of change in relative sea level that begin with an interval of relative fall in sea level. Relative sea-level fall occurs if the rate of eustatic fall is greater than the rate of total subsidence. During the interval of relative sea-level fall, shorelines migrate rapidly seaward as the sea regresses and the sea floor is progressively exposed. If the gradient of this freshly exposed surface is greater than the gradient of the alluvial/coastal plain floodplain, the river systems flowing across this freshly exposed surface cut down into the underlying substrate. Such incised valleys cause sediment to bypass the former alluvial/coastal plain and be delivered directly to the shoreline and beyond. However, if the gradient of this freshly exposed surface is the same as or less than the gradient of the alluvial/coastal plain floodplain, no incised valleys accompany the development of this zone of sedimentary bypass. Landward of the shoreline, within this bypass zone, a type I unconformity develops. Sediments delivered to the shoreline and beyond make up deposits of the lowstand systems tract (**Fig. 1***a*). The lowstand systems tract can include basin-isolated shoreface, submarine fan, lowstand delta, and incised-valley-fill fluvial deposits. A zone of sedimentary bypass characterized by a subaerial unconformity commonly separates depocenters from sediment source areas. Deposition of the systems tract occurs during the interval of relative fall in sea level and subsequent slow relative rise in sea level. Sedimentation may occur at the shoreline in the form of lowstand shoreface and deltaic deposits,

or it may occur predominantly on slope and deeper-water settings to form deep-water turbidite systems if a shelf/slope physiography is present.

Lowstand deposition ends with the first significant flooding event across the shelf. This cessation occurs when a relative rise in sea level produces accommodation (space available for sediment to fill) at a rate that exceeds the sediment flux at the shoreline. The resulting transgressive surface commonly is characterized by erosion of upper shoreface and coastal/alluvial/delta plain deposits of as much as 10–15 m (30–50 ft) by wave action associated with the landward passage of shoreline. This surface, referred to as a transgressive surface of erosion, marks the upper boundary of the lowstand systems tract and the lower boundary of the transgressive systems tract (Fig. 1*b*). The transgressive systems tract is characterized by backstepping (landward-stepping) depocenters. If incised valleys formed during the preceding lowstand, estuarine-to-open marine deposits subsequently fill these valleys. During this interval of shelf flooding, a starved or condensed section is deposited on the shelf and in the deep water. These deposits are the coeval toes of the backstepping depositional units as well as of the later forestepping highstand systems tract units. Condensed sections typically can be recognized paleontologically by faunal abundance and diversity peaks that characterize this interval. Condensed section deposits commonly are rich in organic materials and may form excellent source rocks for petroleum generation.

Deposition of the transgressive systems tract ends and initiation of the highstand systems tract occurs when shelf flooding (transgression) ceases and regression resumes. The surface that characterizes the time of maximum flooding is referred to as the maximum flooding surface. The highstand systems tract is characterized by widespread fluvial and coastal-plain deposition linked to prograding shoreface deposits (Fig. 1*c*). The tract comprises a linkage of the fluvial depositional system, coastal/delta-plain depositional systems, marginal marine/shoreface depositional systems, and shelf, slope, and basinal depositional systems.

Type II sequences. These sequences occur in response to cycles of relative changes in sea level that do not include any interval of relative fall in sea level. Without relative fall in sea level, incised valleys and zones of sedmentary bypass do not develop. Rather, these cycles are characterized by decelerating and then accelerating rates of relative rise in sea level. The characteristic stratal response to these sea-level fluctuations is progradation during intervals of slow relative sea-level rise, followed by aggradation (shoreline stillstand) during intervals of accelerating relative rise in sea level, and finally shoreline backstepping during intervals of rapid relative rise in sea level.

Three systems tracts characterize type II sequences: the shelf-margin, transgressive, and highstand systems tract. The transgressive and highstand systems tract are similar to those described for the type I sequence. The shelf-margin systems tract comprises a middle-to-outer shelf wedge of sediments that are deposited during intervals of increasingly rapid relative rise in sea level. Separating it from the underlying highstand systems tract is a cryptic surface that is identified on the basis of change in stratal stacking pattern from increasingly progradational during the highstand systems tract to decreasingly progradational and aggradational during the self-margin systems tract. The shelf-margin systems tract is separated from the overlying transgressive systems by the transgressive surface of erosion (ravinement surface). Subsequently, the transgressive systems tract is separated from the overlying highstand by the maximum flooding surface.

Interpretations. **Figure 2** illustrates sequence stratigraphic principles at extremely small scale (the

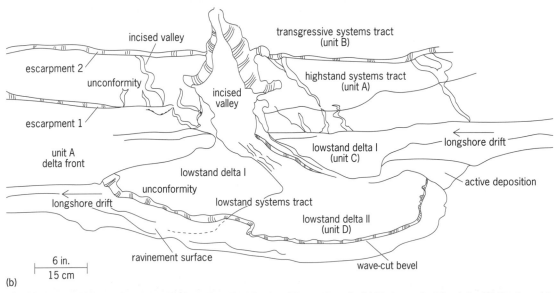

Fig. 2. Sequence stratigraphy of a small fan delta at East Coulee, Alberta, Canada. (a) Photograph of the delta. (b) Stratigraphic interpretation. (After H. W. Posamentier, G. P. Allen, and D. P. James, High resolution sequence stratigraphy: The East Coulee Delta, J. Sedimentol. Petrol., 62:310–317, 1992)

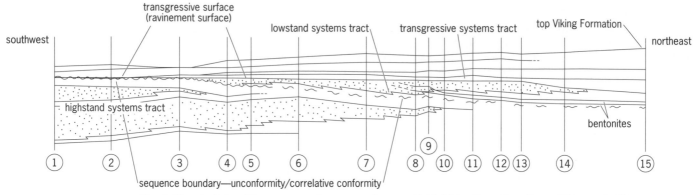

Fig. 3. Dip-oriented well-log cross section across Joarcam Field (Viking Formation: Upper Albian age), Alberta, Canada, illustrating lowstand shoreface progradation. Circled numbers indicate wells. (*After H. W. Posamentier and C. J. Chamberlain in H. W. Posamentier et al., eds., Sequence Stratigraphy and Facies Associations, Int. Ass. Sedimentol. Spec. Pub. 18, pp. 469–485, 1993*).

East Coulee Delta, Alberta, Canada). The delta measured about 2.3 m (7.6 ft) across and 1.8 m (5.9 ft) along dip, and prograded into a roadside drainage ditch about 2.8 m (9.2 ft) wide and 30 m (100 ft) long during a short interval following a rainstorm. Highstand, transgressive, and lowstand systems tracts can be recognized even at this small scale. When the water level was high a broad prograding delta formed. Subsequently, with lowered water level, an incised valley formed, serving to deliver all sediment from the hinterlands directly to the shelf edge and beyond, thus bypassing the shelf. At this time an unconformity develops at the interfluve tops as well as within the valley. The lowstand delta that develops at the mouth of the valley is deposited in response to a relative fall in sea level, producing a forced regression. This depositional unit is physically isolated from the shoreface deposits of the immediately preceding highstand systems tract. If the sea level were to have risen subsequent to deposition of this lowstand delta, the upper part of the lowstand delta would likely have been eroded off and a

thin veneer of transgressive deposits and subsequently condensed section facies would overlie the delta. The incised valley system would likely have been filled with either estuarine or open marine deposits. At larger scale, units analogous to these small-scale systems tracts would represent fluid flow units separated from each other by stratal discontinuities.

Figure 3 illustrates an example of a high-resolution sequence stratigraphic interpretation across an oil field where systems tracts have been identified. Initial deposition of highstand systems tract deposits are separated from subsequent lowstand systems tract deposits by an unconformity and correlative conformity. The transgressive systems tract follows the lowstand and is separated from it in the proximal areas by a transgressive surface of erosion and in the distal areas by a drowning surface. Landward of where the lowstand systems tract pinches out, the transgressive systems tract directly overlies the highstand systems tract and is separated from it by a transgressive surface.

Each systems tract has its own unique "plumbing" attributes and to some extent its own unique lithofacies. Recognition of these high-resolution flow units can have significance with regard to primary as well as secondary field development schemes.

Figure 4 compares the response to relative fall in sea level of a passive margin and of a ramp margin. On passive margins, where relative fall in sea level exposes the relatively steep sloping upper slope, most sediments are presumed to bypass the shelf through incised valleys. Although these sediments may temporarily be deposited at the shoreline, they commonly are remobilized because of relatively steep gradients that occur at the outer shelf/upper slope. Ultimately these sediments are deposited on the downdip slope and basin as deep-water turbidite systems. In contrast, on the ramp margin, basinally isolated lowstand shorelines rather than deep-water turbidite systems are a common response to intervals of relative fall in sea level. These lowstand shorelines may be separated from preceding highstand deposits by broad expanses of the shelf across which sediment passes with no net deposition, and are characterized by little or no associated incised valley formation.

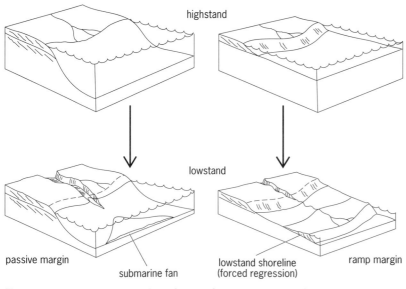

Fig. 4. Three-dimensional view of passive margin versus ramp margin response to relative fall in sea level.

As sequence stratigraphic concepts continue to be applied to a greater variety of tectonic settings, new variations on the general sequence stratigraphic model will continue to emerge. It may be said that there are no templates for applying the concepts of sequence stratigraphy, as stratal geometries can be highly variable. First principles upon which the concepts are based must be understood, and locally important variables must be accounted for before correctly developing a sequence stratigraphic interpretation. Correct interpretation of stratal pattern geometry is critical in assessing hydrocarbon stratigraphic trapping potential, determining fluid-flow dynamics, recognizing bounding stratal discontinuities, and interpreting depositional systems.

For background information SEE SEA-LEVEL FLUCTUATIONS; SEISMIC EXPLORATION FOR OIL AND GAS; SEISMIC STRATIGRAPHY; STRATIGRAPHY in the McGraw-Hill Encyclopedia of Science & Technology.

Henry W. Posamentier

Bibliography. B. U. Haq, J. Hardenbol, and P. R. Vail, Chronology of fluctuating sea levels since the Triassic, *Science*, 235:1156–1167, 1987; C. E. Payton (ed.), *Seismic Stratigraphy: Applications to Hydrocarbon Exploration*, Amer. Ass. Petrol. Geol. Mem. 2b, 1977; H. W. Posamentier et al., Forced regressions in a sequence stratigraphic framework: Concepts, examples, and exploration significance, *Amer. Ass. Petrol. Geol. Bull.*, 76:1687-1709, 1992; J. C. Van Wagoner, K. M. Campion, and V. D. Rahmanian, *Siliciclastic Sequence Stratigraphy in Well Logs, Core, and Outcrops: Concepts for High-Resolution Correlation of Time and Facies*, Amer. Ass. Petrol. Geol., Methods Explor. Ser. 7, 1990; C. K. Wilgus (ed.), *Sea Level Change: An Integrated Approach*, Soc. Econ. Paleontol. Mineral. Spec. Pub. 42, 1988.

Stream systems

A striking feature of stream systems is the simultaneous presence of order and apparent randomness. For example, obvious similarities exist in spatial and temporal patterns of channel network morphology and sediment yields; yet these patterns are quite complex and are often accurately modeled as though they were random. Recent applications of nonlinear dynamical systems theory to stream systems offer possibilities for explaining commonalities (and perhaps universalities) in structure among the irregular observed patterns and for reconciling the simultaneous existence of deterministic order and apparent randomness.

Instability and chaos. Various aspects of stream systems are unstable in that they are vulnerable to disturbances. This vulnerability is often a manifestation of asymptotic instability. A dynamic system is asymptotically stable when small perturbations do not persist and feedbacks operate to restore the system to or close to its predisturbance condition. Asymptotic instability means that systems will not be restored after disturbances: perturbations will persist or grow. The distinction between large and small perturbations is not important in the case of instability, because a system is unstable in response to perturbations of any magnitude.

Dynamic asymptotic instability in nonlinear systems is linked to deterministic chaos, which can be practically defined as sensitive dependence upon initial conditions. In a chaotic system, initially minuscule differences are magnified over time, producing irregular, apparently random behavior from a (sometimes simple) deterministic system. Asymptotic instability is a necessary condition for chaos, and it is sufficient to show that chaotic behavior is possible under certain conditions or parameter values. This link exists because an unstable system has at least one positive eigenvalue. In a classical nonlinear dynamical system, the real parts of the eigenvalues are equivalent to the Lyapunov exponents (λ) of the system. Lyapunov exponents measure the convergence or divergence of neighboring trajectories. If Δ_0 is the initial difference between two system states and Δ_t the difference at time t, the general solution for determining the fate of perturbations in the system is given by expression (1).

$$\Delta_t \sim \left(\Delta_o e^{\lambda t} \right) \tag{1}$$

A positive Lyapunov exponent leads to exponential divergence, sensitive dependence on initial conditions, and chaos.

The stability properties of a linearized system are the same as those of the nonlinear parent system. Fluvial geomorphologists have exploited this property to apply linear stability analyses to fluvial problems. Fluvial systems are inherently nonlinear because of thresholds; therefore asymptotic instability indicates a potentially chaotic system.

Instability of fluvial systems. One example of unstable and potentially chaotic behavior is known as at-a-station hydraulic geometry, that is, the problem of determining how hydraulic parameters [principally, flow velocity, width, depth, energy grade slope (energy gradient), and frictional resistance] at a channel cross section respond to changes in flow. Qualitative asymptotic stability analyses based on a set of conceptual equations linking the major hydraulic variables, and based on a system network model (bow-and-arrow diagram), have found that hydraulic geometry to be unstable.

The problem can also be examined by using the D'Arcy-Weibach equation. This standard kinematic flow equation is usually written to solve for velocity V, Eq. (2a); but it can be rearranged as in Eqs. (2b)–(2d)

$$V = \left(\frac{8gRS}{f} \right)^{0.5} \tag{2a}$$

$$S = \frac{V^2 f}{(8g)R} \tag{2b}$$

$$f = \frac{(8gRS)}{V^2} \tag{2c}$$

$$R = \frac{V^2 f}{(8g)S} \tag{2d}$$

to solve for the other critical hydraulic variables (slope S, friction factor f, and hydraulic radius R, which is a function of width and depth) [g is the acceleration due to gravity].

Regardless of the values used, the qualitative relationships among the critical hydraulic variables are the same, and the system is unstable. The instability and occasional chaotic behavior are manifested not only in the form of complex responses to imposed flows but also as opposite-from-expected behavior, as when the velocity, hydraulic radius, or slope decrease or the friction factor increases in response to an increase in imposed discharge (**Fig. 1**).

Other aspects of fluvial systems have also been found to be asymptotically unstable. These include the response of channel longitudinal profiles to drops in the base level or to tectonic uplift; the partitioning of fluvially eroded sediment among colluvial storage, alluvial storage, and sediment yield; and the formation of rill and incipient channel networks.

Sensitive dependence on initial conditions. Because of the time scales involved in fluvial geomorphology (decades to millennia), direct observation and field experiments to test for sensitive dependence on initial conditions are not feasible. However, the problem may be addressed via modeling (**Fig. 2**). The time step in the model is 1 year, with 1000 iterations per model year. The flat input line reflects an assumption of a constant rate of sediment input to the channel. The irregular output line shows that feedbacks involving sediment storage and channel adjustments result in irregular output (sediment yield; sediment transported through the channel) even though input is constant.

A model of the evolution of a river basin developed recently is built upon the same basic geomorphic concepts as models used since the early 1960s. Unlike earlier models, however, the network extension process is governed by the drainage pattern on the hillslopes and by local slopes around channel heads, incorporating the roles of preexisting topography and flow convergence. The heart of the model is a channel initiation function a, based on the classical premise that erosive energy of flow shown is a power function of discharge Q and slope gradient S, as shown in Eq. (3),

$$a = BQ^m S^n \qquad (3)$$

where B is a constant. The model produces realistic simulated basins and accurately reproduces statistical properties of real basins; in addition, it is sensitively dependent upon initial conditions. Minor variations in initial elevation grids, with all the other physical characteristics and gross physical attributes of model runs being identical, produced basins with significant and often striking differences in mean stream network size and geometric properties, drainage densities, elevation distributions, and simulated sediment yields.

Self-organized criticality. Chaotic evolution of nonlinear dynamical systems can produce patterns (such as drainage basins) with power-law distributions that are readily described as multifractal structures. The theory of self-organized criticality was developed to explain the ubiquity of such structures in nature. Generally, self-organized criticality holds that many systems evolve to a marginally stable state, so that the system organizes itself to be perpetually in a critical state (the angle of repose in slopes is one geomorphic example). The concepts of chaos and self-organized criticality are not inconsistent. The fact that both chaos and self-organized criticality result from underlying deterministic dynamics of a complex system and display self-similarity (geometric properties are identical at different scales) leads to a natural association between the two concepts, although details of the relationship are not yet well known.

There is accumulating evidence that fluvial systems are self-organizing critical systems that exhibit power-law and multifractal distributions. Although this evidence does not prove that fluvial systems are, or can be, chaotic, it is consistent with, and indeed could be produced by, chaos. Distributions of mass (discharge) and energy expenditure in drainage basins have been found to follow power-law distributions and to have multifractal structures. Multifractal structures differ from simple fractal distributions in that the fractal scaling and self-similarity relationships are not constant at all spatial scales.

A recent statistical model of drainage network evolution was based on the premise that stream heads branch and propagate at a rate dependent solely on the local resistance of the substrate. This invasion-percolation model explains and generates the observed fractal geometry and scaling properties of real net-

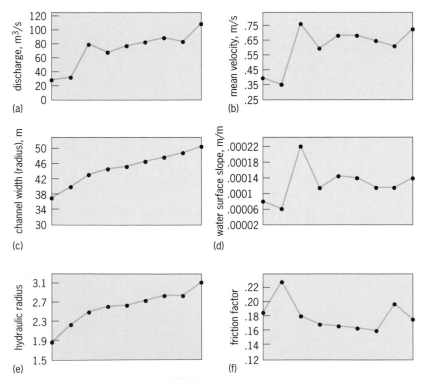

Fig. 1. Variation of discharge and hydraulic variables over nine flow events of the Bogue Phalia River, Mississippi, indicating opposite-from-expected behavior, a result of instability. (*a*) Discharge. (*b*) Mean velocity. (*c*) Channel width. (*d*) Water surface slope. (*e*) Hydraulic radius. (*f*) Friction factor. 1 m = 3.28 ft. 1 m³ = 264 gal. (*After J. D. Phillips, The instability of hydraulic geometry, Water Resour. Res., 26:739–744, 1990*)

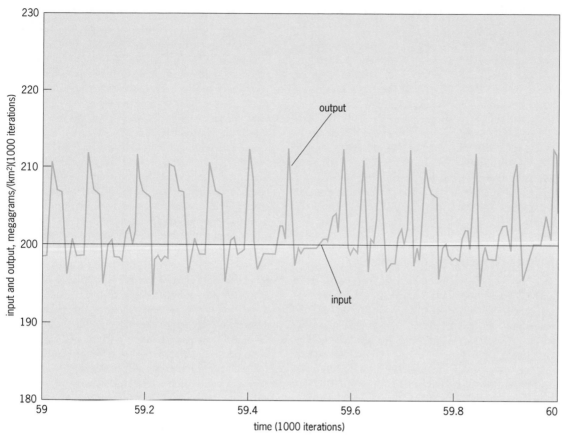

Fig. 2. Results from a stream profile model showing how feedback (in this case, positive feedback between channel size and aggradation or degradation) can lead to complex, pseudo-random behavior from a simple deterministic system. Inputs and outputs refer to sediment supplied to the stream channel and sediment yield. There are 1000 iterations per model year. Megagram = metric ton. (*After W. H. Renwick, Equilibrium, disequilibrium, and nonequilibrium landforms in the landscape, Geomorphology, 5:265–276, 1992*)

works, and it produces a self-organized critical system.

Implications. Water and land are the fundamental resources, and the actions of stream systems are critical to both. The need to predict the response of stream systems to changes in climate, land use, or other factors is obvious. The emerging evidence of chaotic behavior in stream systems has important implications for prediction.

On the negative side, long-term deterministic prediction is impossible for a chaotic system, because initial conditions can never be precisely known, or even specified with adequate precision. There are also special problems for interpreting statigraphic evidence such as alluvial sediments. If the processes of sediment storage and transport are chaotic, given that the initial conditions can never be known, there are obvious barriers to inferring past conditions. Chaos also implies that some complexities in fluvial sytems are inherent in system dynamics. This situation limits what can be achieved via reductionist approaches to the study of microprocesses in streams.

However, there are several positive implications. First, evidence suggests that many (perhaps most) fluvial systems are potentially, but not generally, chaotic. They will behave chaotically under certain circumstances (in models and equations, under certain parameter values), but will not always or even routinely

do so. Second, deterministic chaos, while preventing long-term deterministic prediction, does not preclude short-term deterministic prediction. In fact, discovery of a low-dimensional chaotic attractor may actually improve short-term deterministic modeling, because a small number of critical controlling variables can be identified. Also, chaos does not preclude prediction in a probablistic mode, because stochastic methods work equally well whether randomness is real or apparent.

Chaos also implies an underlying order and universality in the observed complexity of stream systems. Both instability and chaos in hydraulic geometry and power-law multifractal distributions in drainage basins, for example, have been linked to (and can be derived from) underlying, traditional geomorphic principles of energy dissipation.

Several major challenges remain. It is still extremely difficult in real-world data to isolate chaotic signals from the nonchaotic stochastic ones. The long time spans involved in the evolution of stream systems make field confirmation of model and theoretical results highly problematic. Finally, it is difficult to link the chaos observed in models that are low-order analogs of real fluvial systems to the much more complicated real landscape. *SEE GEOGRAPHY.*

For background information *SEE CHAOS; FLUID MECHANICS; GEOMORPHOLOGY; MODEL THEORY; TURBU-*

LENT FLOW in the McGraw-Hill Encyclopedia of Science & Technology.

Jonathan D. Phillips

Bibliography. M. J. Haigh, Evolution of an anthropogenic desert gully system, *Eros. Transp. Deposit. Proc.*, 189:65–77, 1990; J. D. Phillips, Qualitative chaos in geomorphic systems, with an example from wetland response to sea level rise, *J. Geol.*, 100:365–374, 1992; C. P. Stark, An invasion percolation model of drainage network evolution, *Nature*, 352:423–425, 1991; G. Willgoose, R. L. Bras, and I. Rodriguez-Iturbe, A coupled channel network growth and hillslope evolution model, 2. Nondimensionalization and applications, *Water Resour. Res.*, 27:1685–1696, 1991.

Superconducting devices

The discovery of superconductivity in ceramic oxide materials with transition temperatures T_c to a zero-resistance state as high as 128 K ($-145°$C or $-229°$F) has stimulated much effort to apply these new materials to small-scale electronic devices as well as large-scale electrical engineering systems. Drastically reduced refrigeration costs should allow much wider exploitation of superconducting technology. This article outlines some of the basic principles on which electronic applications of superconductivity can be based, as well as summarizing devices that may prove useful and available in the reasonably near future. All of these devices have already been developed by using conventional, low-T_c superconductors such as niobium and lead, and efforts are well under way to use the new high-temperature superconductors also.

The zero-resistance property of superconductivity is well known, and underlies expected applications in power transmission and high-field magnets, for example. Some passive electronic devices are also based on low-loss conduction, but active superconducting electronic devices rely in general on the principle of flux quantization and the Josephson effects.

Passive devices. The resistance of a superconductor to an impressed current at a high frequency ω is not zero because of the inertia of the paired charge carriers responsible for superconductivity. The pairs present sufficient "kinetic" inductance to prevent complete screening of electromagnetic fields from the superconductor. However, the resistance to alternating current is still very low; it varies as ω^2, and falls rapidly with reducing temperature, at least for the purest superconductors. This alternating-current loss is best described by a surface resistance R_s, there also being a reactive component of the impedance arising from the inertia of the pairs. The two components are not equal in magnitude as they are for normal conductors, since for superconductors the conductivity is essentially complex.

Thin films of the high-temperature superconductors $YBa_2Cu_3O_7$ (**Fig. 1**) and $Tl_2Ba_2Ca_2Cu_3O_{10}$ already show at least 100 times lower loss than does copper at the same temperature and at 10 GHz. As a result, one of the first applications of the cuprate superconductors is expected to be as passive microwave circuit components made from thin- or thick-film superconductors. This expectation has led to the development of microstrip and coplanar thin-film filters and antennas that easily outperform their normal-metal counterparts, giving much lower insertion loss and more sharply defined cutoff. High-Q planar resonators have been built with Q values up to 70,000. Cavity resonators have also been constructed from the new materials with Q values at 5 GHz and 77 K ($-196°$C or $-321°$F) of up to 700,000, far higher than can be achieved with normal metals. These resonators are important microwave components that provide narrow-band filters and frequency references to be used in low-noise inputs of microwave receivers, for example.

In addition, the surface reactance of superconductors may be utilized in a novel way: The electromagnetic skin depth of a superconductor is considerably smaller than that of a normal conductor, and also differs in being almost entirely independent of frequency. As a result, dispersionless propagation (that is, the wave velocity is frequency-independent) occurs in a medium bounded by superconductors. Such thin-film transmission lines should prove useful for propagating fast pulses between components or boards in advanced digital circuits. If the dielectric spacing between the superconducting wires of the transmission line is sufficiently thin, the wave velocity along the line is dominated by the kinetic inductance of the superconductors, and may be reduced to perhaps 0.1 of the free-space velocity c. Propagation is still dispersionless,

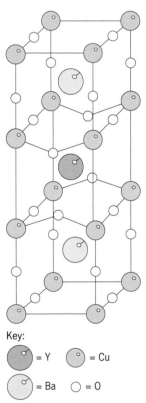

Key:

Fig. 1. Structure of $YBa_2Cu_3O_7$, the most widely used superconductor, with $T_c = 93$ K ($-180°$C or $-292°$F).

and these two properties, combined with the narrow linewidths that may be patterned lithographically in thin films of high-temperature superconductors, allow nondispersive delay lines of very compact size to be produced.

The passive components discussed above are the simplest to manufacture. They are expected to go into use shortly in space applications (telecommunication satellites, for example), where refrigeration is not a major problem and both size and weight savings made possible by replacing bulk normal-metal cavity resonators with thin-film superconducting equivalents are particularly cost-effective.

Active high-frequency devices. Prototype computers, far-infrared detectors, and analog-to-digital converters that use low-temperature superconducting materials have already been developed to a high degree of sophistication, employing thousands of carefully designed active components. In the longer term, useful active devices made from the new superconductors are expected. In general, these are not simple analogs of conventional semiconducting components but rely on the unique macroscopic quantum nature of superconductivity. The realization that essentially all of the paired carriers are combined in a single wave function or order parameter leads, through simple application of quantum mechanics to this wave function, to the realization that the magnetic flux penetrating a superconductor must be quantized in bundles of minimum size, $\Phi_0 = h/2e$, where h is Planck's constant and e is the charge of the electron.

Individual fluxons may be stored or moved around within holes within the superconductor one by one, provided a switch known as a Josephson junction can be incorporated in the circuits. **Figure 2** shows three ways of realizing such a junction. This junction carries a supercurrent for low values of current, but when some critical value i_c is exceeded a voltage appears across the switch and this resistive state allows several novel features. First, if i_c is designed so that it is approximately equal to Φ_0/L (where L is the inductance of the hole), fluxons will be allowed in or out of the loop one at a time. Such a general

circuit can be used in the analog mode as the most sensitive detector known for signals at frequencies up to the order of 200 MHz; this detector is known as a SQUID (superconducting quantum interference device). Although it is basically a detector of magnetic flux changes, by using a suitable input transducer a SQUID may be converted to measure almost any physical parameter with extraordinary sensitivity. Examples (with sensitivity values in parentheses) are a magnetic field (1 femtotesla), voltage (10^{-21} V), current (1 femtoampere), displacement (10^{-19} m), and cryogenic temperature (0.1 millikelvin). A SQUID can be used in a small-signal, low-noise amplifier configuration at frequencies up to at least 200 MHz, and has been used for some very sensitive nuclear resonance measurements.

Fluxons can also penetrate thin films of superconducting materials rather easily. Likewise, fluxons can exist in a very mobile form within the tunnel barriers of Josephson junctions, provided these barriers are wider than a characteristic length (approximately 10 micrometers). A Josephson junction biased by a constant voltage V also carries an alternating supercurrent oscillating at a high frequency, $\omega = 2\pi V/\Phi_0$. Similarly, fluxon oscillation within a long junction barrier can be tuned by a magnetic field. These two systems provide voltage- or current-tunable millimeter-wave oscillators that may prove useful as local oscillators to be used with existing superconducting tunnel-junction arrays employed in many radio telescopes in this part of the spectrum. Although the formidable materials fabrication problems involved in making tunnel barriers between two thin films of the new superconductors have not yet been solved, the high mobility of fluxons in thin films has allowed a novel active three-terminal device, the flux-flow transistor, to be produced. Here, a current applied to a control line regulates the ease with which fluxons pass through a weak superconducting link, and in turn the voltage across the device is changed. Such transistors have been used as low-noise microwave amplifiers, operating at 77 K ($-196°$C or $-321°$F).

In a digital manifestation, the presence and the absence of a single fluxon in a ring can be used as the 1 state and the 0 state of a binary logic gate or of a memory cell. As a result of the very low dissipation combined with the very fast time scale on which fluxons can move and the wave function can change, a single logic gate is very fast (2 picoseconds), and thousands of gates may be integrated into a very small space without the usual problems of extracting the heat. Conventional superconducting Josephson junctions can be made with the high reproducibility required for large-scale integrated circuits, but so far such reproducibility is not true of high-temperature junctions. However, much effort is being devoted to the solution of this materials problem, and high-speed superconducting digital circuits operating at 77 K ($-196°$C or $-321°$F) are already being produced in prototype form.

For background information SEE CAVITY RESONATOR; ELECTRICAL IMPEDANCE; JOSEPHSON EFFECT; MICROWAVE TRANSMISSION LINES; SQUID; SUPERCONDUCTING

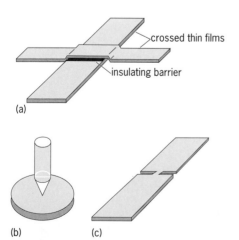

Fig. 2. Three types of Josephson junction. (a) Thin-film tunnel junction. (b) Point contact. (c) Thin-film microbridge.

DEVICES; SUPERCONDUCTIVITY in the McGraw-Hill Encyclopedia of Science & Technology.

John C. Gallop

Bibliography. J. C. Gallop, *SQUIDs, the Josephson Effects and Superconducting Electronics*, 1991; S. T. Ruggiero and D. A. Rudman, *Superconducting Devices*, 1990.

Superconductivity

The 1986 discovery of high-temperature superconductivity in the layered cuprates has spurred an intensive effort to understand these materials. The high critical temperatures T_c (up to 130 K or $-226°F$) suggest that a new mechanism for superconductivity, quite unlike the phonon-driven mechanism in low-T_c superconductors, exists in the cuprates. Investigators currently believe that a broad understanding of the electronic properties of the cuprates, especially in the so-called normal state above T_c, is needed to pin down the superconducting mechanism. Since 1989, experimental research on these issues has been greatly aided by the availability of high-quality single crystals of the key compounds $YBa_2Cu_3O_7$, $Bi_2Sr_2CaCu_2O_8$, and $La_{2-x}Sr_xCuO_4$.

Structure of cuprate superconductors. The universal structure shared by the 40 (or more) cuprates known to be superconducting is a two-dimensional square lattice made up of repeating CuO_2 units (**Fig. 1**). Aside from the CuO_2 planes, individual compounds may have ancillary layers containing barium, bismuth, or thallium, or (in $YBa_2Cu_3O_7$) one-

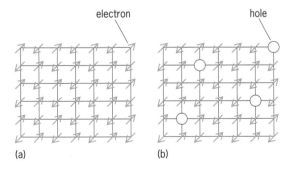

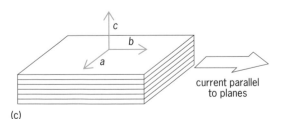

Fig. 1. Structure of cuprate superconductors. (a) Two-dimensional model of the CuO_2 plane of a parent compound (filling factor $\nu = 1$). Each site is occupied by one electron. Arrows show the electrons spins. (b) Compound in which doping has created vacancies (holes) by removing some electrons. (c) Crystal morphology. The CuO_2 layers lie in the a-b plane. The c axis is perpendicular to the planes. In the normal state, electrical conductivity is high parallel to the planes (large arrow) but low along the c axis.

dimensional CuO chains. Most of the relevant physics of the cuprates are contained in a two-dimensional model that replaces the CuO_2 unit at each site of the square lattice with a simple well (the ancillary layers are ignored). Figure 1a shows a lattice in which each well is filled with exactly one electron, that is, $\nu = 1$, where the filling factor ν is the ratio of number of electrons to sites. This arrangement corresponds to the electron population in the parent compounds La_2CuO_4 and $YBa_2Cu_3O_6$.

Conventionally, a $\nu = 1$ filling is expected to behave like a metal, because electrons can hop to adjacent sites, leaving a vacant site next to one that is doubly occupied. [Two electrons may occupy the same well if their spins are opposed, and below 240 K ($-28°F$) the spins of electrons assume an alternating up-down pattern known as an antiferromagnetic state (Fig. 1a) that allows electrons to hop to neighboring sites.] Instead of metallic behavior, however, all the cuprates are observed to be insulators at this filling. It turns out that, unlike in ordinary solids, Coulomb repulsion between electrons in the CuO_2 plane is very strong. Placing two electrons in the same site incurs a prohibitively large energy cost. Thus, at $\nu = 1$, motion of all electrons is effectively precluded by the energy cost of such double occupancies.

Correlated carrier motion. A remarkable property of the cuprates is that the electron population can be reduced readily by a process called doping. In La_2CuO_4, doping is accomplished by substituting some of the lanthanum with strontium, the dopant (**Fig. 2**). In $YBa_2Cu_3O_6$ the dopant is oxygen. When ν is reduced slightly from 1, some of the sites become vacant (Fig. 1b). In the presence of an electric field, electrons may now hop to the vacant sites, so that a charge current flows. The process of an electron hopping to the right to fill a vacant site may be viewed, equivalently, as a vacancy or hole hopping to the left. For ν slightly less than 1, the picture of holes diffusing in a sea of fixed electrons is the more natural one. As the hole population increases, the planes become more conducting. Empirically, it is found that superconductivity appears when 10–25% of the sites are vacant (Fig. 2). Throughout this range of doping, the electrons maintain strict observation of the no-double-occupancy rule. The point bears repeating that strong Coulomb repulsion is responsible for freezing the electrons in the insulating state of the CuO_2 plane at $\nu = 1$. When holes are introduced, the Coulomb force remains important in correlating their motion (keeping them as far apart as possible). This strong correlation effect is what makes the problem challenging and, to date, intractable.

Strong evidence for this filling scenario in the cuprates comes from Hall effect and reflectivity experiments. (A Hall experiment detects the deflection of charge carriers by the Lorentz force in a perpendicular magnetic field. The Hall coefficient R_H yields the sign of the carriers as well as their density.) By measuring how R_H varies with doping in the cuprates (increasing the strontium concentration in La_2CuO_4, for example), investigators have found that the mea-

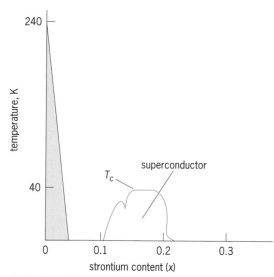

Fig. 2. Phase diagram of the cuprate $La_{2-x}Sr_xCuO_4$, **showing how the transition temperature** T_c **varies with the strontium content** x. **Compositions with** x **less than** ~ 0.04 **show very poor conductivity. The parent compound** ($x = 0$) **is an insulator at low temperatures. In the shaded region, electrons on the copper ions are in the antiferromagnetic state. Outside the shaded region, the cuprate is in the normal state, except for the region where it is a superconductor.**

sured carrier density matches the dopant concentration: one hole is introduced per strontium atom, up to a vacancy rate of 15%. (The behavior of the Hall effect at high concentrations of holes is less well understood.) In a series of strontium-doped La_2CuO_4 crystals, infrared reflectivity measurements confirm the picture. (By monitoring how the reflectivity of a sample varies with frequency, the density of carriers may be inferred from the low-frequency end of the spectrum.) As the strontium concentration increases from 0 to 10%, the so-called spectral weight, which is proportional to the carrier density, increases linearly from zero, again implying that each strontium dopant creates a hole. These results confirm the general validity of the strong-correlation picture. While many theorists seek to learn how the strong repulsion between holes can lead to superconductivity, there is also intense investigation of the charge transport behavior.

Anomalous normal-state properties. Increasingly, the evidence points to a metallic state in the cuprates that is highly unconventional.

Anisotropic conductivity. One of the anomalous properties is the anisotropy of the normal-state conductivity. The electrical conductivity perpendicular to the planes (along the c axis) is surprisingly poor compared with that parallel to the planes (Fig. 1c). The ratio of the latter to the former varies from 100 to 10,000, depending on the compound and temperature. Moreover, in almost all cuprates the resistivity observed with current parallel to the c axis increases with decreasing temperature, whereas the in-plane resistivity decreases. This unusual trend implies that the stack of putatively metallic CuO_2 planes approaches purely two-dimensional behavior at zero temperature. (The onset of superconductivity interrupts this trend.) Hall experiments reveal that the two-dimensional na-

ture above T_c is even more perplexing. The sign of the carriers is positive when the Lorentz force lies parallel to the planes but negative when it is aligned with the c axis. At present, the poor conduction along the c axis and other two-dimensional aspects of the normal state are not understood. In contrast, once superconductivity appears, the supercurrent flows freely in all directions. The striking change from insulator to superconductor shows up clearly in the infrared reflectivity. In the normal state, crystals of $La_{2-x}Sr_xCuO_4$ reflect like a good metal when the electric field E of the incident light is polarized parallel to the planes, but reflect like an insulator when E is parallel to the c axis. Below T_c, however, the flow of supercurrent along the c axis converts the insulatorlike reflectivity to that characteristic of a metal. Such a change from quasi-two-dimensional to three-dimensional behavior at T_c was anticipated by the theory of P. Anderson.

Intense scattering. Yet another unusual aspect of the normal state is the intense scattering rate of carriers. In ordinary metals such as lead and tin, the electrical conductivity at temperatures above T_c is limited by the vibrations of the lattice (phonons). An electron accelerates in an electric field until it is scattered by a phonon. Since elevating the temperature increases the phonon population and hence the scattering rate, the resistivity of a metal increases with temperature. Careful measurements in the cuprates showed that, while the in-plane resistivity increases with T, the increase is linear over an unusually large range of temperature. Moreover, the scattering rate is so intense that the average lifetime of the carriers (time between scattering events) in the cuprates is unusually short, prompting suggestions that the normal-state conductivity is limited by the strong Coulomb force between electrons rather than by scattering by phonons. Recent high-frequency measurements offer strong support for this conjecture.

In the superconducting state, almost all the electrons pair up to form Cooper pairs. However, just below T_c, a small fraction of the electrons remains in unpaired states (called quasiparticles). Unlike Cooper pairs, the quasiparticles may be scattered, for example, by phonons. It turns out that a good deal may be learned about how electrons are scattered above T_c by studying the quasiparticles. A way to proceed is to expose the superconductor to microwave radiation. The microwave fields penetrate a slight distance below the sample's surface and accelerate the quasiparticles. Repeated scattering and acceleration of the quasiparticles leads to microwave absorption, that is, heating of the superconductor. Thus, measurement of the dissipation allows the scattering rate to be determined. Experiments carried out with a crystal of $YBa_2Cu_3O_7$ show a dramatic increase in the electron lifetime when the sample enters the superconducting state. Apparently, changes in the electronic properties at T_c remove the source of the intense scattering that exists in the normal state. As a result, the lifetime of the quasiparticles becomes very long. Since scattering by phonons is incompatible with such a change, the experiment strongly supports a scattering mechanism in the normal state that derives from the Coulomb interaction. These results further undercut the case for phonons as the cause for superconductivity but strengthen the case for an electronic mechanism.

For background information SEE ANTIFERROMAGNET-ISM; ELECTRICAL CONDUCTIVITY OF METALS; ELECTRICAL RESISTIVITY; HALL EFFECT; REFLECTION OF ELECTROMAGNETIC RADIATION; SUPERCONDUCTIVITY in the McGraw-Hill Encyclopedia of Science & Technology.

N. Phuan Ong

Bibliography. K. S. Bedell et al. (eds.), *High Temperature Superconductivity: Proceedings of the Los Alamos Symposium 1989*, 1990; R. J. Cava, Superconductors beyond 1-2-3, *Sci. Amer.*, 263(2):42–49, August 1990; D. M. Ginsberg (ed.), *Physical Properties of High Temperature Superconductors*, 3 vols., 1989, 1990, 1992; Special issue on superconductivity, *Phys. Today*, vol. 44, no. 6, June 1991.

Supercritical fluids

Solid, liquid, and gas are the three phases of matter that are easily recognized. Depending on the physical conditions, such as the temperature and pressure, these phases exist alone or in equilibrium with each other, as illustrated in the pressure-temperature projection of the phase diagram for a pure substance in **Fig. 1a**. In this diagram, the sublimation, melting, and vapor pressure curves represent the pressure-temperature conditions when solid and gas, solid and liquid, and liquid and gas phases coexist in equilibrium. At the triple point, all three phases, solid, liquid, and gas, exist in equilibrium. The end point of the vapor pressure curve is known as the gas-liquid critical point, at which the coexisting liquid and gas phases become identical. At the critical point, in addition to the densities of liquid and gas becoming equal, heat of vaporization goes to zero and liquid–gas interfacial tension disappears. Above the critical point, there is no distinction between a liquid and a gas, and it is impossible to liquefy a sub-

stance by simply increasing the pressure. A substance that has been heated above and compressed beyond its critical temperature and pressure is in a homogeneous fluid state and is known as a supercritical fluid.

Tunable fluids. Supercritical fluids can be compressed from low to high density without entering two-phase regions, and they may assume property values ranging from gaslike to liquidlike in response to simple manipulations of pressure. Many physicochemical properties, such as dissolving power, viscosity, diffusivity, and dielectric constant of fluids, depend on the density of the fluid. Therefore, by changing the degree of compression, the same homogeneous fluid may be turned into a specific solvent or into a nonsolvent for a given substance or process. These fluids are thus distinguished from ordinary solvents. With their adjustable properties, they represent a new class of solvents and reaction media for the chemical process industry. Simply stated, they are tunable solvents.

The tunable nature of these fluids is better illustrated by the pressure-volume projection of the phase diagram in Fig. 1b. The diagram shows the changes in volume (or density) and the various phases that are entered when pressure of a pure fluid is changed along an isotherm (constant temperature paths shown as broken lines). At temperatures above the critical ($T > T_c$), fluid can be compressed from the low-density (large-volume) gaseous state to high-density liquidlike conditions (that is, from point A to B) without actually passing through the gas-liquid two-phase region, which would not be avoidable if the path followed were below the critical temperature (T_c). The pressure variation required to bring about a sufficient change in density depends on the temperature, and it decreases as the critical temperature is approached. The isotherms just above the critical temperature are relatively flat, and the switchover from gaslike to liquidlike densities can be realized with only small changes in pressure. Another mode by which the density of the fluid may be changed from gaslike to liquidlike values without entering the two-phase regions would be to lower the

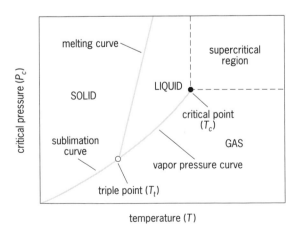

(a)

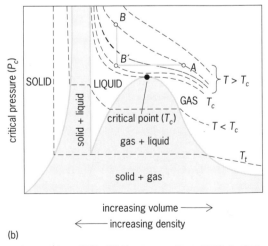

(b)

Fig. 1. Projections of phase diagrams for a pure fluid. (a) Pressure-temperature (P-T). (b) Pressure-volume (P-V); broken lines are isotherms. At temperatures above T_c, gas-liquid two-phase regions are not entered with change in pressure. The solid lines BB', AB', and the portion of the isotherm AB at T are related to different process paths discussed in Fig. 2.

temperature at a constant pressure (that is, from point A to B') above the critical pressure.

The basic philosophy in the utilization of supercritical fluids as process solvents is to adjust the pressure (or density) of the fluid so as to make it behave as a solvent for a substance at one stage of the process stream and then, at another stage, to change its density to make it behave as a nonsolvent. The dissolved substances are precipitated while the conditions supercritical for the fluid are maintained. Since two-phase regions for the fluid are not entered, solvent recovery and product separation are facilitated. The homogeneous fluid can simply be recompressed and sent back to the process cycle. **Figure 2** illustrates the two general approaches, based on operations at constant temperature (such as the path from B to A in Fig. 1) operations at constant pressure (such as the path from B' to A in Fig. 2). Actual processes may of course involve multiple stages or combinations of pressure and temperature changes.

Critical temperatures and pressures.
Critical temperatures and pressures depend on the composition of the fluids.

Pure fluids. **Table 1** shows the critical properties of selected fluids. For the majority of pure fluids, critical pressures are below 100 bars (10 megapascals). Critical temperatures, however, show a greater variation. For fluids such as carbon dioxide, the critical temperature is not much above the room temperature, but for other, more polar fluids such as ethanol or water, it is relatively high.

Among the various pure fluids, supercritical carbon dioxide and water have received the greatest attention. Carbon dioxide is a gas under normal conditions, is nontoxic and nonflammable, and has a low critical temperature; thus, it is ideal for applications such as manufacture of food and pharmaceuticals, where thermal stability of materials during processing and the residual solvent left in the products after processing are normally of great concern.

Supercritical water has the unique property of readily dissolving organic matter but not the inorganic salts; it finds special use in applications related to

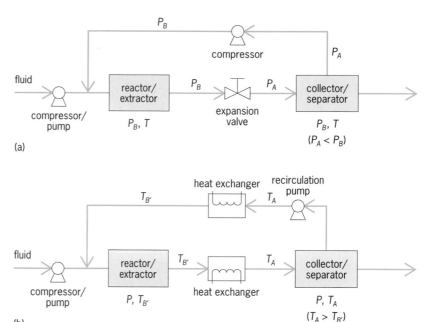

(a)

(b)

Fig. 2. Utilizing supercritical fluids as process solvents. (*a*) Diagram of an approach based on operations at constant temperature (*T*). (*b*) Diagram of an approach based on operations at constant pressure (*P*). Subscripts *A*, *B*, and *B'* refer to points on the phase diagram in Fig. 1.

treatment and destruction of hazardous waste.

Binary mixtures. Whereas pure substances are characterized by a critical point, binary fluid mixtures display a critical line. The critical line is the locus of critical points corresponding to different compositions. In the simplest form of binary mixtures, the critical line is continuous between the critical points of the pure components. Critical temperatures assume intermediate values between the critical temperatures of the pure components. The critical pressures often take values higher than the critical pressures of the pure components. These values are illustrated for the binary mixtures of carbon dioxide and n-butane in **Fig. 3**. However, with increasing dissimilarity of the constituents making up the mixture, departures from simple binary behavior set in and critical lines become

---

Table 1. Boiling points and the critical properties of pure fluids*

Fluid	Boiling point, °C	Critical temperature, °C	Critical pressure, atm
Ethylene	−103.7	9.9	50.5
Xenon	−107.1	16.6	58
Carbon dioxide	−78.5 (sublimes)	31	72.9
Propane	−42.1	96.8	42
Ammonia	−33.34	132.5	112.5
n-Butane	−0.5	152	37.5
Sulfur dioxide	−10	157.8	77.7
n-Pentane	36.1	196.6	33.3
Acetone	56.2	235.5	47
Methanol	65.15	240	78.5
Ethanol	78.5	243	63
Toluene	110.6	320.8	41.6
Water	100	374.1	218.3

*°F = (°C × 1.8) + 32; 1 atm = 1.01325 bars; 1 bar = 10^5 pascals; 10 bars = 1 MPa.
SOURCE: After *CRC Handbook of Chemistry and Physics*, 71st ed., CRC Press, 1990.

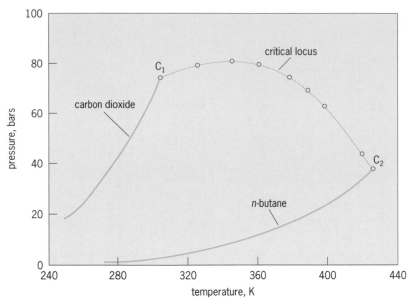

Fig. 3. Pressure-temperature plot for the binary mixture of carbon dioxide and *n*-butane. Solid lines represent the vapor pressure curves for pure carbon dioxide and pure *n*-butane. They terminate at the respective critical points C_1 and C_2. Critical points for the binary mixtures of these fluids lie on the critical locus. The open circles are the critical points corresponding to the binary mixture compositions indicated in Table 1.

water. **Table 2** shows the critical temperatures and pressures for three representative binary mixtures of carbon dioxide with polar and nonpolar fluids as a function of composition.

Solvent characteristics of a pure fluid can be altered upon addition of a second component (a co-solvent). Since critical temperature and pressure are influenced by fluid composition, fluid mixtures can be used not only to regulate specific interactions with a given solute but also to regulate the critical properties and hence operational conditions of a process. For example, addition of about 5 mol % ethanol to carbon dioxide would result in a mixture with a critical temperature of about 42°C (108°F) but a critical pressure of about 86 bars (8.6 MPa), compared to 243°C (469°F) and 63.8 bars (6.38 MPa) for ethanol. Such a mixture offers the specific feature of having polar character in the fluid at a much lower temperature than would be possible by using the pure alcohol. Despite the resulting increase in critical pressure, it is important to have the ability to adjust and lower the operational temperature for applications involving thermally labile compounds such as biomaterials. Similarly, intermediate critical temperatures with only modest increase in pressures are also achieved in binary mixtures of carbon dioxide with alkanes, which will display the features of alkanes at lower operational temperatures.

discontinuous. For example, critical lines are continuous for mixtures of carbon dioxide with butane (Fig. 3) but discontinuous for mixtures of carbon dioxide with

Table 2. Critical data for selected binary mixtures of carbon dioxide*

Mole fraction of carbon dioxide	Critical temperature, K	Critical pressure, bars
Carbon dioxide + *n*-butane[†]		
1.000 (pure carbon dioxide)	304.19	73.82
0.839	325.01	78.38
0.720	344.25	82.17
0.643	357.17	80.70
0.501	377.55	74.43
0.422	387.62	69.53
0.328	397.89	63.58
0.088	418.48	44.05
0.000 (pure butane)	425.16	37.96
Carbon dioxide + ethanol[‡]		
0.990	305.8	76.5
0.979	308.4	78.3
0.963	312.1	82.5
0.954	315.2	86.1
0.936	320.1	91.9
0.927	325.1	97.4
Carbon dioxide + water[§]		
0.997	304.6	74
0.435	541.2	3600
0.430	540.2	3140
0.415	539.2	2450
0.395	540.2	1890
0.375	542.2	1470
0.347	548.2	1080
0.295	563.2	716
0.242	583.2	539
0.190	603.2	427
0.125	623.2	333
0.025	643.2	235

*°F = (K × 1.8) − 459.67; 1 bar = 10^5 pascals.

[†] After M. E. P. de Fernandez, J. A. Zollweg, and W. B. Streett, Vapor-liquid equilibrium in the binary system carbon dioxide + *n*-butane, *J. Chem. Eng.*, 34:324,1989).

[‡] After G. S. Gurdial et al., Phase behavior of supercritical fluid-entrainer systems, in E. Kiran and J. F. Brennecke (eds.), *Supercritical Fluid Engineering Science: Fundamentals and Applications*, ACS Symp. Ser. 514, 1993.

[§] After C. P. Hicks and C. L. Young, The gas-liquid critical properties of binary mixtures, *Chem. Rev.*, 75:119, 1975.

Applications. At supercritical conditions, fluids have greater dissolving power compared to the gaseous state, and display greater diffusivity and lower viscosities compared to the liquid state. **Table 3** shows some representative values. All these properties are further tunable by changing the pressure, the temperature, or the composition of the fluid. With such adjustable characteristics, these fluids are ideal for use in a wide range of physiochemical processes, including selective extractions, fractionations, purifications, material deposition and impregnations, nucleation and particle-size regulation, and chemical reactions and synthesis. Lower viscosities and higher diffusivities at supercritical conditions become especially useful in penetration and extraction from complex networks.

Explorations for industrial use of supercritical fluids are rapidly expanding, especially in the areas of natural products, biochemicals, food, pharmaceuticals, petroleum, fuel, polymers, electronics, and specialty chemicals. Processes for removing caffeine from coffee beans and obtaining extracts of hops by using supercritical carbon dioxide are among commerical applications. Supercritical fluid chromatography using carbon dioxide and carbon dioxide plus small amounts of cosolvent (such as ethanol) for separation and analysis of complex mixtures is used in commercial applications, and it also has become a widely accepted analytical technique. In the polymer industries and in the electronics industry, studies are in progress seeking to replace chlorofluorocarbons suspected of degrading the Earth's ozone layer with environmentally benign supercritical fluids such as carbon dioxide. Supercritical carbon dioxide is being used to replace organic solvents in spray painting in order to minimize atmospheric contamination and exposure of personnel. Supercritical fluids are being considered for soil cleanup and removal of hazardous contaminants from wastes. A commercial wastewater treatment unit is already in operation in Baltimore, Maryland, where a variety of wastes and organic compounds are extracted with supercritical carbon dioxide.

Supercritical water oxidation is another approach used in environmental remediation. Complete miscibility of oxygen and organic compounds in supercritical water creates a single-fluid phase, providing favorable reaction environments for destruction of industrial wastes. Even though it is not yet used commercially on a large scale, the technology itself is at an advanced stage. Oxidation carried out in supercritical water has also been shown to be effective in the treatment of human metabolic wastes, and it is currently being considered for use in life-support systems on long-term space explorations.

Another novel application is the use of supercritical carbon dioxide in the dissolution and rapid expansion of pharmaceutical compounds and biocompatible polymers to form controlled-release polymer-drug microspheres. Other applications being explored include the formation of microcellular foams (aerogels) with small pore sizes and low densities, polymer synthesis and the

Table 3. Comparison of the properties of supercritical fluids with gases and liquids*

	Density, g/cm^3	Viscosity, g/cm-s	Diffusivity, cm^2/s
Gas	0.001	0.0001	0.1
Supercritical fluid	0.1–1	0.0001–0.001	0.001–0.01
Liquid	1	0.01	0.00001

*1 g = 0.04 oz; 1 cm = 0.4 in.; $1 cm^2$ = 0.16 $in.^2$; $1 cm^3$ = 0.06 $in.^3$

regulation of molecular weight, polymer processing by lowering the glass transition temperature or viscosity, fiber formation, fiber dyeing, enzymatic reactions, wood modifications and the removal of lignin from wood, and the liquefaction of biomass and coal.

For background information SEE INTERFACE OF PHASES; PHASE EQUILIBRIUM; SUPERCRITICAL-FLUID CHROMATOGRAPHY in the McGraw-Hill Encyclopedia of Science & Technology.

Erdogan Kiran

Bibliography. T. J. Bruno and J. F. Ely (eds.), *Supercritical Fluid Technology: Reviews in Modern Theory and Applications*, 1991; E. Kiran and J. F. Brennecke (eds.), *Supercritical Fluid Engineering Science: Fundamentals and Applications*, ACS Symp. Ser. 514, 1993; M. A. McHugh and V. Krukonis, *Supercritical Fluid Extraction: Principles and Practice*, 1986; E. Stahl, K.-W. Quirin, and D. Gerard, *Dense Gas Extraction and Refining*, 1988.

Surface and interfacial chemistry

Interest in the formation, structure, and interfacial properties of organic films has exploded in the last several years. The primary reason is that these films play central roles in the structures of numerous materials and interfacial processes, such as adhesion, biocompatibility, catalysis, corrosion control, and tribology (friction and wear). For example, schemes to enhance tissue integration while minimizing bacterial adhesion are critical to the implantation of prosthetic devices, such as artificial hearts and joint replacements. In contrast, approaches seeking to enhance the adhesive strength at the interface formed by the metallization of a fluoropolymer surface promise to advance the large-scale integration of microelectronic devices. Unraveling the underlying chemistry and physics that control the structure and interfacial properties of organic films is therefore of fundamental importance.

Monolayers. The development of fundamental descriptions of organic films has proven elusive, largely because of the poor morphological and compositional definition at the surface of these films (**Fig. 1a**). Recently, however, opportunities to pursue fundamental studies involving organic films have emerged through the use of self-assembled monolayer films. Figure 1b

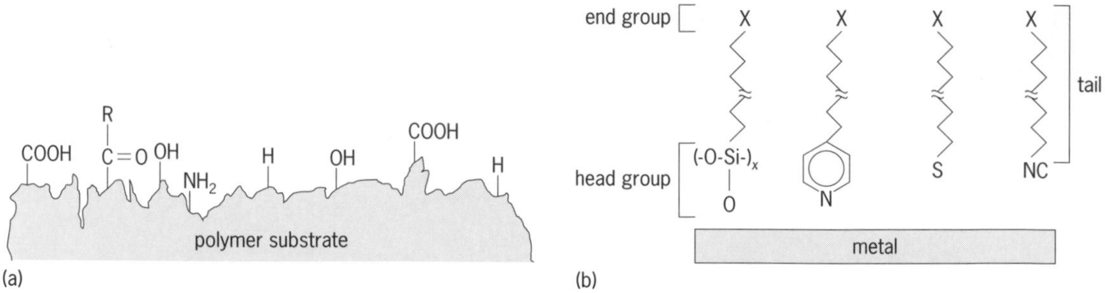

Fig. 1. Monolayers. (*a*) Organic film. R = functional group; OH = hydroxyl; NH₂ = amino group; COOH = carboxyl group; H = hydrogen. (*b*) Self-assembled monolayer. O = oxygen; Si = silicon; S = sulfur; N = nitrogen; C = carbon.

represents a few of the types of self-assembled monolayers. In contrast to the structure in Fig. 1*a*, monolayer films are structurally simple, since their thickness is limited to one molecular layer. In addition, through the synthetic alteration of the end group X the composition of the interface can be systematically altered. Monolayers can therefore serve as model systems for probing details of relationships between structure and function at organic surfaces.

Sulfur-containing compounds at gold. Perhaps the most exciting recent finding is that the strong specific interaction of sulfur in organosulfur compounds at gold (Au) leads to the formation of a monomolecular film. Since few other ligands compete favorably for binding sites at gold, strategies based on immobilization through sulfur linkages have emerged as effective routes to the construction of interfaces with a specific architecture. Of the types of sulfur-containing compounds, monolayers formed from thiols [$X(CH_2)_nSH$] are the most attractive. The primary reason is that monolayers from $X(CH_2)_nSH$ for $n \geq 10$ are more densely packed and free of pinholes than those from disulfide and sulfide precursors, yielding an interface with properties governed predominantly by the end group X and not the gold substrate or the gold–sulfur interface. Monolayers with $n < 10$ are less well defined structurally, and are susceptible to decomposition through oxidation of the sulfur head group upon exposure to the ambient atmosphere in the laboratory.

Preparation. To prepare a monolayer, a gold substrate is immersed into a dilute ($\sim 1\ mM$) solution of the adsorbate precursor for a few minutes to several days. Immersion times are dependent on the structure of the adsorbate, with the longer immersion times required to form layers with bulky end groups. The substrates are generally prepared by the evaporation of gold films (100–2000-nanometer thicknesses) onto a silicon, glass, or mica support. The surface of the gold at all three supports exhibits a pronounced (111) crystallinity. At silicon and glass, a thin ($\sim$150–30 nm) layer of chromium or titanium is deposited prior to gold to promote adhesion. Gold deposited at mica is generally subjected to a postevaporation annealing procedure, resulting in a surface that is smoother at an atomic level than those at silicon and glass.

The formation and structure of monolayers from $X(CH_2)_nSH$ have been characterized by a variety of macroscopic and microscopic techniques. Most of the findings deal with long-chain structures ($n \geq 10$) with end group X similar in size to the cross section of the underlying polymethylene spacer chain [for example, X = methyl (CH_3), carboxyl (COOH), or hydroxyl (OH)]. The results from macroscopic probes (for example, infrared, x-ray photoelectron, and Raman spectroscopies; electrochemistry; and contact angles) indicate that monolayers from $X(CH_2)_nSH$ have interfacial properties governed largely by end group X, are densely packed polymethylene chains tilted $\sim$30° from the surface normal, and form as the corresponding thiolates.

Characterizations using techniques that reveal the microscopic structure of surfaces largely support the findings of the macroscopic studies. Both diffraction and atomic-scale scanning microscopic studies have found a nearest-neighbor separation distance of 0.5 nm for monolayers of *n*-alkanethiolates, a spacing indicative of a ($\sqrt{3} \times \sqrt{3}$)R30° adlayer at a Au(111) lattice. This notation signifies an adsorbate that is separated from its nearest neighbor by the square root of three times that of the underlying gold lattice. **Figure 2** summarizes the macroscopic and microscopic results, which provide the beginnings of a model for the prediction of the average structures of such monolayers. The model relies largely on considerations of the large driving force for the formation of thiolates at gold (desorption activation energies of $\sim$ 30–40 kcal/mol) and the packing limitations imposed by end group X. Effectively, the energetics of chemisorption lead to a structure that is limited by the van der Waals diameter (the diameter of the electron cloud of the alkyl chain) of end group X, thereby establishing the average tilt of the chains. For *n*-alkanethiolates at Au(111), the model predicts an average tilt of 27°, which is in reasonable agreement with the above structural descriptions. Thiolates with other small end groups (for example, X = OH, COOH), as well as those formed with a long perfluorocarbon tail [$CF_3(CF_2)_7(CH_2)_2SH$], follow the expected predictions. Those with bulkier end groups (for example, ferrocenyl groups) also appear to follow the expected predictions, but not always as rigorously.

Applications of thiolate monolayers. The descriptions of sulfur-containing compounds at gold provide a framework for the use of thiolate monolayers as model systems for studies in the interfacial sciences.

As an example, long-range electron transfer plays a crucial role in several biological areas (such as photosynthesis) and technological areas (such as batteries). However, fundamental issues that relate the rate of electron transfer to the separation distance between a redox group and the electron surface, and to the molecular structure of the interface, remain largely unanswered. Progress in this area is beginning to appear through studies designed to take advantage of the structural definition of thiolate monolayers. One strategy involves the coassembly of a ferrocene-terminated alkanethiol and an inert methyl-terminated alkanethiol. This procedure leads to the formation of spatially isolated ferrocenyl groups with a well-defined separation distance between the redox group and the electrode—key features for interfaces designed to test theories of electron transfer. Through studies of the forward and reverse electron-transfer rates, results confirming the prediction of Marcus theory were obtained. Studies along these lines promise new and exciting insights into electron transfer processes and should be forthcoming in the near future. *SEE NOBEL PRIZES.*

Several strategies have also been devised that use monolayers as components of devices for selective chemical detection. As shown in **Fig. 3**, the coassembly of octadecanethiol with a ligating sulfide leads to the creation of an ion-selective monolayer for the cupric ion (Cu^{2+}). The tetradentate sulfide ligand binds divalent cations such as Cu^{2+} but not trivalent cations

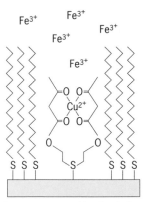

Fig. 3. A monolayer, a coassembly of octadecanethiol with a ligating sulfide, that is ion-selective for the cupric ion (Cu^{2+}).

such as the ferric ion (Fe^{3+}), which require octahedral coordination. A monolayer of the sulfide alone failed to form a layer sufficiently compact to discriminate against ions other than Cu^{2+}, an observation attributed to imperfections in the packing of the single-component monolayers. Coassembly of the long-chain alkanethiol with the sulfide, however, results in the successful formation of a layer with ion-selective sites embedded in a relatively defect-free, inert matrix.

Monolayer films of specific architecture also promise new insights into reactions that take place in unique chemical environments, such as in membrane materials. One approach is through the construction of multicomponent monolayers. For example, a two-component monolayer can be fabricated that is composed of a reactive end group X and an inert end group X. By variation of the length of the spacer group on either of the components, or of end group X on the inert component, the immediate environment that surrounds the reactive group can be systematically altered. To this end, the effect of surrounding a ferrocene-terminated alkanethiolate in an increasingly hydrocarbonlike environment by a systematic increase in the length of an *n*-alkanethiolate coadsorbate has been examined. The results indicate that the generation of a positive charge in the layer through the oxidation of the ferrocenyl group is increasingly impeded as the environment becomes more hydrocarbonlike. Similar trends have been observed for the reactivity of acid-terminated monolayers. Studies of more complex reactions may reveal the importance of steric constraints.

Self-assembled multimolecular films. Recent studies have also focused on the use of self-assembly techniques for the preparation of multilayered structures. The goal is to gain control of the interfacial structure in three dimensions, analogous to that obtained via Langmuir-Blodgett deposition techniques. Approaches to date have relied on a sequential adsorption process whereby either silane or thiol head groups are used to tether the first layer at the substrate. Subsequent formation of layers generally proceeds following a chemical activation of an end group or through electrostatic binding. **Figure 4** shows

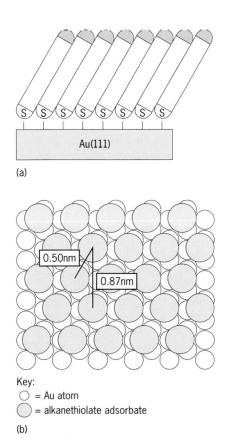

Key:
○ = Au atom
◔ = alkanethiolate adsorbate

(b)

Fig. 2. Sulfur-containing monolayer at gold. (*a*) Macroscopic structure. (*b*) Microscopic structure. The distance 0.5 nm represents the square root of three times the gold-gold spacing in the underlying lattice. Au = gold; S = sulfur.

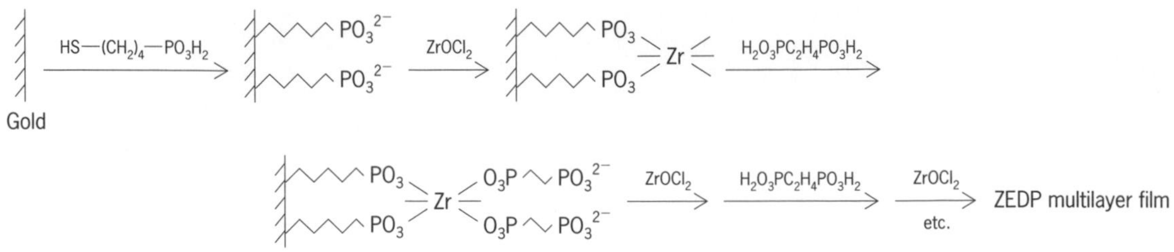

Fig. 4. Reaction scheme for the preparation of a multilayered structure based on a zirconium 1,2-ethanediylbis(phosphonate) repeating unit.

the preparation of a multilayered structure based on a zirconium 1,2-ethanediylbis(phosphonate) repeating unit. Both approaches have met with varied degrees of success, with the limiting factor being the increase in disorder as the number of layers increases. Though self-assembled multimolecular films are only beginning to be exploited as materials for electrochemical purposes, control of the interfacial architecture in three dimensions may provide routes to devices whose function is based on the directional control of charge transfer and light propagation.

For background information SEE FILM (CHEMISTRY); INTERFACE OF PHASES; MONOMOLECULAR FILM; SURFACE AND INTERFACIAL CHEMISTRY in the McGraw-Hill Encyclopedia of Science & Technology.

Marc D. Porter

Bibliography. C. D. Bain and G. M. Whitesides, Modeling organic surfaces with self-assembled monolayers, *Adv. Mater.*, 4:110–116, 1989; L. H. Dubois and R. G. Nuzzo, Synthesis, structures, and properties of model organic surfaces, *Annu. Rev. Phys. Chem.*, 43:437–463, 1992; A. Ulman, *An Introduction to Ultrathin Organic Films from Langmuir-Blodgett to Self-Assembly*, 1991.

Swim bladder

Achieving neutral buoyancy is a great advantage for any organism in water, preventing it from sinking or rising without expenditure of energy. Fish are usually about 5% denser than the surrounding water, resulting in a net downward force that requires a formidable energy expenditure to overcome. The swim bladder, a device that enables fish to become neutrally buoyant, occurs in many marine fish, mostly in inhabitants of the upper 660 ft (200 m) of free ocean water and in species living near the bottom to about 6560 ft (2000 m), the depth record being 22,960 ft (7000 m).

Reduction in density. An air-filled swim bladder reduces the density of fish without adding much weight, since the density of air is only 0.00125 g/cm^3 at 1 atmosphere (10^{12} kilopascals), and only 0.7 g/cm^3 at 700 atm, a pressure experienced in the sea at a depth of 22,960 ft (7000 m).

There are, however, problems with such a floating device. For example, a given swim bladder volume, which apparently obeys Boyle's law, will give the fish neutral buoyancy only at one particular depth. Elevation results in increased swim bladder volume

and thus in a lifting force, and vice versa. Since fish vary their depth, often in a diurnal rhythm, there must be provision for adjustments of swim bladder volume, that is, deposition or resorption of gas in the swim bladder. There must also be provision for keeping the gases inside the swim bladder.

Architecture. The swim bladder may be viewed as a gas-containing sac. Despite differences in details, it is functionally very similar in most fishes. The two main types of bladder are known as physostome, and physoclist, the former with an open connection to the digestive tract and the latter without one. The eel has served as an experimental animal in many studies on swim bladder function. Its swim bladder consists of an elongated sac connected to the esophagus through the pneumatic duct. The opening to the esophagus is too small to be of significance for gas filling or emptying of the bladder. The swim bladder and the pneumatic duct serve quite different functions, the former as the site for gas deposition, the latter for gas resorption. The opening between both can be occluded.

The swim bladder is lined by a cuboidal epithelium, whereas the pneumatic duct has a flat epithelium. Both structures are supplied with blood through a common artery; but whereas the capillaries of the pneumatic duct are drained by small veins that enter directly into large veins, a countercurrent capillary arrangement, the retia mirabilia, exists in the swim bladder. Arterial blood enters the arterial capillary supply of the rete, an arrangement of some 35,000 arterial capillaries, before it reassembles into postrete arterial vessels that feed the capillaries of the swim bladder wall. Blood draining from these vessels into veins in the swim bladder wall, in turn, feed the venous capillaries in the rete, about 20,000 capillaries that run strictly parallel and in checkerboard arrangement with the arterial capillaries. Venous blood continues to the liver (again a capillary network) before returning to the main circulation. The rete presents the basis for a hairpin countercurrent system, similar to that of the mammalian kidney. This system is of functional significance for the gas deposition.

Composition of gas. The total pressure of the swim bladder gas at any depth is in equilibrium with the hydrostatic pressure. Analysis of swim bladder gas from fish at various depths has shown that oxygen (O_2), carbon dioxide (CO_2), nitrogen (N_2), and other gases occur in varying fractions: in shallow water the fractional composition resembles that of air, whereas in deeper regions O_2 becomes the major component. There is, furthermore, a time variation in the gas fraction: CO_2 is enriched in freshly deposited gas, but is later replaced by O_2 and other gases.

At any depth, the partial pressure of any gas component exceeds that in the surrounding water because the partial pressures of gases dissolved in the surrounding water hardly increase with ocean depth.

The gases in the swim bladder can be under tremendous pressure. Three factors are important in preventing gases from leaving the swim bladder: (1) The entrance to the pneumatic duct may be closed off by smooth muscle in the wall. (2) The swim bladder wall is covered with a silvery layer impregnated with crystals of guanine and some hypoxanthine, constituting an impermeable coat. Removing this layer increases the gas permeability of the swim bladder wall some 100 times. (3) The rete countercurrent system acts as a barrier in that gas is shunted from the rete venous capillaries to the rete arterial capillaries. The rete thus plays a key role not only in gas deposition but also in gas maintenance.

When fish move to lesser depth, the gas volume of the swim bladder must be reduced. In physostome fish, gas can exit to the gut, and bubbles can indeed be observed leaving the fish's mouth. In physoclist fish, where there is no bladder-gut connection, gas is resorbed in structures such as the pneumatic duct or in parts of the swim bladder wall normally separated from the main swim bladder by muscular activity. These resorptive epithelia are well vascularized, and the resorptive process apparently employs the same mechanisms as the alveolar lung.

Mechanisms for deposition of gas. It is now generally accepted that passive diffusion, rather than active secretion, governs gas deposition. For gases to be deposited by passive diffusion from the blood in the capillaries of the swim bladder into the latter's gas phase, high partial pressures must be created in the blood. Two independent mechanisms are thought to be responsible for the formation of high gas partial pressures in the swim bladder blood: reduction of gas solubility in blood by the metabolic action of the swim bladder epithelium and countercurrent concentration of gases in the rete mirabile.

Reduction of gas solubility. The swim bladder epithelium forms lactic acid, even in the presence of high O_2 concentrations. As a result, the pH of the capillary blood is lowered, and its salt (lactate) concentration is increased. This change acts on the solubility of certain gases such as O_2 and CO_2. These gases are known to display a salting-out effect, that is, a reduction of solubility with increased salt concentration. This reduction may be very small under the physiological conditions encountered in the swim bladder, but it will be enough for an efficient concentrating action in the rete.

Reduction in pH leads to reduced O_2 affinity via the Bohr effect, and thus to increased partial pressure of O_2. In some fish there is also a reduction in O_2 capacity when pH is reduced (Root effect), leading to a particularly large increase in the partial pressure of O_2 upon acidification.

Acidification converts bicarbonate ion (HCO_3^-) into CO_2, and the partial pressure of CO_2 is thus also increased. A further substantial increase in the partial pressure of CO_2 of the swim bladder blood is achieved by anoxidative CO_2 formation from glucose in the pentose phosphate cycle.

Countercurrent enhancement in rete. When blood with increased gas partial pressure enters the rete, there will be a partial pressure head from the venous capillaries to the arterial capillaries, and gas will diffuse back. The partial pressure will thereby be reduced in the rete venous blood on its way out of the swim bladder, but will be increased in the rete arterial blood on its way toward the swim bladder. The amount of increase of the partial pressures along the rete capillaries supplying the swim bladder wall depends on a number of factors, among them the rete blood perfusion and the rate of passive diffusion between the afferent and efferent rete capillaries. However, it appears from calculations based on morphometric and physiologic parameters that a substantial rise in partial pressures can indeed be produced by the rete countercurrent system.

Additional factors. Two additional factors appear to be important for gas deposition into the swim bladder: the occurrence and significance of water shifts in the rete, and the effect of back diffusion of salt in the rete.

Flow balance in rete countercurrent. Loss of fluid from the vascular bed in the swim bladder, such as by lymph flow, would yield a lower flow rate in the venous than in the arterial rete capillaries. Such a flow imbalance in a hairpin countercurrent system would largely compromise its concentrating efficiency, since the imbalance would effectively wash out any solute concentration. The water shift in the eel from the capillaries to the swim bladder epithelium appears to be very small, at most 5% of the total flow, indicating virtually balanced rete flows.

Back diffusion of salt. Most studies on rete function have concentrated on back diffusion of gas, while back diffusion of solutes has been assumed to exert negligible or detrimental effects. There is, however, undebatable evidence for back diffusion of lactate from venous to arterial capillaries in the rete. A recent analysis shows that, contrary to the classical view, back diffusion of salt significantly enhances the concentrating efficiency for certain gases in the rete; the increased salt concentration in the arterial capillaries will enhance gas partial pressure by the salting-out effect.

For background information SEE BUOYANCY; OS-TEICHTHYES; SWIM BLADDER in the McGraw-Hill Encyclopedia of Science & Technology.

Peter Scheid; Bernd Pelster

Bibliography. E. J. Denton, The buoyancy of fish and cephalopods, *Prog. Biophys. Chem.*, 11:177–234, 1961; R. Fänge, Gas exchange in fish swimbladder, *Rev. Physiol. Biochem. Pharmacol.*, 97:111–158, 1983; B. Pelster and P. Scheid, Countercurrent concentration and gas secretion in the fish swim bladder, *Physiol. Zool.*, 65:1–16, 1992.

Systems engineering

Invention is the creation of a new device, system, or process. Innovation is the creation of change via something new. Although innovations may depend on one or more inventions, the vast majority of inventions do not contribute to innovations. Most inventions do not contribute to job creation, economic growth, and prosperity. Innovations have the potential to do so.

Virtually every country, in one way or another, is trying to facilitate the process of innovation. It has been recognized that innovation is most successful if it is managed. Managing innovation involves establishing the intention to innovate, developing both short-term and long-term plans that reflect this intention, paying careful attention to executing these plans, and updating the plans regularly.

These steps are simple in general and very difficult in particular. Various concepts, principles, methods, and tools are available for supporting the innovation process. Numerous workers in the field have developed methodologies, which in some cases are facilitated by computer-based support. However, no single approach is clearly preferable.

In fact, it appears that the greatest leverage is obtained by simply having a formalized process—almost any process—for managing innovation instead of expecting innovation to happen on its own. In addition, people must be trained to develop skill in this process, as well as to appreciate its general value. The enterprise should support this process with methods, tools, and information.

Beyond the formalized management process, innovation increasingly requires focusing on understanding the world and transforming the organization. Thus, the innovation management process must be targeted appropriately, and the organization must be redesigned to match form to function or, in other words, to facilitate pursuit of innovation.

Understanding the world. In the past, a company had to have a good understanding of its own market sector and perhaps the general economic conditions in its own country in order to be innovative. Innovation now increasingly requires understanding of the global marketplace, in part because of the tremendous opportunities provided by global markets. Numerous major United States corporations, for example, make most of their sales and profits outside the United States.

A second motivation for understanding the world is the fact that a company is increasingly likely to have to compete with foreign companies in its own local market. Computer and communications technologies have made it possible to conduct business almost anywhere in the world almost instantaneously. Consequently, foreign companies can close sales and service accounts, perhaps via local sales representatives, almost as fast as local companies can.

A third motivation is the fact that local competitors are adopting global perspectives. These competitors avail themselves of technology sources, labor, materials, and especially knowledge and expertise that are inherently broader and deeper than what can be obtained within any single locality. To remain strong players, companies are required to seek for themselves the same resources.

Methods and tools for gaining an understanding of the world are beginning to emerge. They are based in part on traditional approaches to market analysis, but they also rely on methodologies from anthropology, sociology, and related disciplines. These methods and tools include questionnaires, interviews, focus groups, and observational techniques that are applied to studying consumer preferences, organizational cultures, and environments, and the needs, beliefs, and values within particular market sectors.

Government, particularly state government, is playing a role in helping companies to understand global markets. Seminars, workshops, and advice are readily available. The motivation for states providing such support is the desire to foster economic development.

Companies are increasingly likely to send personnel overseas to gather information and foster new relationships. Corporate staff members, for example, consequently better understand the business and ethnic cultures of operating units. People in new product development thereby gain greater understanding of users' needs, customers' preferences, maintenance practices, and so on.

Of course, the need to understand the market is a time-honored principle. What is different now is the breadth with which this principle is being applied, as well as the formalization of methods and tools, typically supported by computers, to facilitate its application.

Transforming the organization. While the scope with which companies now need to understand the world is unparalleled, the organizational changes necessary to facilitate the innovation are far more daunting. Worldwide, companies are struggling to transform themselves. They must overcome the inertia of technology-driven institutionalization to become vehicles for rapid, innovative change.

Companies are trying to move away from technology-driven development and design to satisfy fixed requirements. They want to emphasize market-oriented prototyping and collaboration with customers. A change of methods and tools, as well as subtle and difficult cultural changes, is necessary. However, companies often find themselves unable to overcome inertia.

For example, methodologies and philosophies such as total quality management (TQM), concurrent engineering (CE), and quality function deployment (QFD) are being adopted by many companies. However, most companies have great difficulty implementing these approaches.

Total quality management emphasizes continuous process improvement, frequently via statistical techniques. Concurrent engineering attempts to pursue marketing, engineering, and manufacturing in parallel to decrease time to market and to produce a higher-quality product. Quality function deployment focuses on customers' preferences and trade-offs among these preferences relative to competitors' products. It provides a graphical picture of linkages among relevant quality elements.

Virtually all companies want to adopt these or equivalent approaches. However, endorsing them in general is much easier than implementing them in particular. Typically, the problems are not technical; the organization is the problem.

Successfully transforming an organization and empowering it with new methods and tools require understanding the needs, beliefs, and values that underlie the organization and the mental models whereby organization members understand the world. Inconsistent and incompatible mental models across functional areas within an enterprise can result in functions working at cross purposes, expending much energy and not making progress.

Training and appropriate methods and tools can help to uncover the needs, beliefs, values, and mental models within an organization, as well as to slowly modify these attributes. However, it is necessary to have commitment to change. Ideally, this commitment should be made on the basis of an understanding of fundamental changes in the nature of business and competition. The apparent stability of past decades—with the employee having a career with one company and slowly but surely making it up the corporate ladder—is becoming less and less a reality.

Changing a corporate culture is very difficult. Past relationships and roles, as well as hard-won positions, may no longer make sense. However, people cannot change instantaneously. Vision and leadership can compel them in the right direction, but considerable patience is also needed.

Downsizing has been dictated by economic problems and facilitated by information technology. Product life cycles have shrunk with the rapid introduction of new technologies. Consumers have come to expect and demand products and systems tailored to their needs. The overall result is a rapid rate of change. Organizations, including the people in them, find it difficult to adapt to this rate of change.

The traditional hierarchical and bureaucratic structure of many enterprises also contributes to this difficulty. Corporate review committees and long chains of approval are no longer tenable when a short time to market is important. The people who populate these committees and have approval authority may be reluctant to give up this power, but the value of position is giving way to the importance of contribution.

This trend is being accelerated by developments in computer and communications technologies. Low-cost, microcomputer-based workstations and networking are enabling provision of information throughout the enterprise and, consequently, distribution of the power associated with access to information. The resulting delayering of organizations is adding to the impetus for change. SEE INFORMATION MANAGEMENT.

Prospects. Many enterprises have come to understand quite clearly the situations they face. They now accept the need for high-quality innovations, with minimal time to market. They understand that new concepts, principles, methods, and tools such as total quality management, concurrent engineering, and quality function deployment are needed. They realize that the inertia of their organization is a primary impediment to change.

Such insights are not limited to industry. Government agencies, educational institutions, and nonprofit enterprises are reaching similar conclusions. It is common to hear government agencies discuss how they can best serve their constituencies—their customers. Educational institutions strategically plan how to attract their customers—students and organizations that hire graduates. Nonprofit enterprises, such as fund-raising groups, now talk about how to provide benefits that will entice more benefactors.

These insights are expanding the notion of managing innovation. It is no longer simply a matter of being a good steward of the technology base. It is now realized that managing innovation involves managing the processes of invention, understanding the world, and transforming the organization. This realization is an excellent first step; the next steps will be easier.

For background information SEE QUALITY CONTROL; SYSTEMS ENGINEERING; VALUE ENGINEERING in the McGraw-Hill Encyclopedia of Science & Technology.

William B. Rouse

Bibliography. H. R. Booher (ed.), *MANPRINT: An Approach to Systems Integration*, 1990; W. B. Rouse, *Design for Success: A Human-Centered Approach to Designing Successful Products and Systems*, 1991; W. B. Rouse, *Strategies for Innovation: Creating Successful Products, Systems, and Organizations*, 1992; P. Senge, *The Fifth Discipline: The Art and Practice of the Learning Organization*, 1990.

Telephone service

Advances in technology are enabling more powerful telecommunications services to be developed. For example, personal communications networks allow maintenance of mobile communications wherever the users are located; and satellite communication systems based on aeronautical phased-array antennas allow aircraft passengers and crew to communicate via telephone, facsimile, or data channels.

Personnel Communications Networks

Representing the most advanced public telecommunications service yet developed, personal communications networks (PCNs) hold the promise of enabling users to maintain high-quality, two-way mobile communications whether at home, at work, or en route anywhere in between. Unlike current cellular networks, which are basically just an extension of the existing public switched telephone network, the personal communications network is to be a self-contained intelligent network capable of providing service options beyond those immediately foreseeable with any existing communications system. Through the use of smart cards—cards that can be programmed for specific user needs and then inserted into the personal communications network handsets—each user's service can be individualized. The result is a technological innovation that could enable every user of a personal communications network to carry a customized means of exchanging computer data, facsimile data, digitized voice, and possibly even compressed video as easily as a beeper is now carried.

Cellular structures. Based on a concept similar to that underlying current cellular networks, the personal communications network service will rely on radio base stations, operating in a cellular structure and capable of contacting users handsets regardless of their location in the network. Within this network, made flexible by innovative microcell digital technology that allows base stations to be located wherever there are user populations, the user of a personal communications network handset will be free to move both indoors and outdoors without suffering degraded transmission or a dropped call. This transportability of equipment will be made possible through the installation of residential, transitional (that is, mounted on telephone poles, buildings towers, and so forth), and office-building base stations to provide local, wireless access to all in-range users and then, as the user transits through the various cell areas, hand off the signals to contiguous base stations. The result will be continuous accessibility with the capability to contact an individual, instead of only a geographic location.

Spectrum reallocation. As desirable as actualization of the personal communications network technology would appear, development was initially hampered by practical difficulties such as where on the already-crowded frequency spectrum a personal communications network could best be accommodated and whether a personal communications network could coexist on a frequency already occupied by another service. The need to resolve such issues slowed development in the United States, while personal communications services (PCSs), which include personal communications networks as well as second-generation cordless telephones and wireless local-area network systems, proliferated in other countries where the problem of spectrum shortage was far less acute. Recognizing the revolutionary potential of personal communications services and the broad public interest in the development of these technologies, however, the Federal Communications Commission (FCC) responded to the need for action by announcing an aggressive frequency reallocation proposal in February 1992. Under this proposal, 220 MHz of spectrum, located between 1850 and 2200 MHz, would be designated specifically for use by personal communications services and other emerging telecommunications technologies.

The 1850–2200-MHz band is divided into a plurality of fixed-service, microwave channels and is currently designated for private, fixed line-of-sight microwave communications links. These microwave communications links employ analog voice channels, and channels with relatively low data rates, such as 9.6 kilobits per second, which are multiplexed to form a signal with a bandwidth of 10 MHz or less. Public utilities; highway administrations; public safety orga-

nizations, such as police, fire, and ambulance; and so on use the spectrum in this band.

The FCC proposal to relocate these users outlined a three-stage transition plan. First, existing users would be given co-primary status with new services for a transition period of sufficient length to allow the useful life of existing microwave equipment to be exhausted. Second, existing users would be allowed to continue to use their frequencies indefinitely but on a secondary basis, that is, causing no interference but being subject to it. Third, governmental microwave links related to public safety would be allowed to remain but with a ban on additional expansion. In addition to this phased approach, the FCC proposed a voluntary migration plan whereby new technology licensees would be encouraged to negotiate privately with existing microwave users to reach mutually beneficial arrangements for use of the spectrum based on actual market realities. *SEE ELECTRICAL UTILITY INDUSTRY; RADIO SPECTRUM ALLOCATIONS.*

Use of spread-spectrum modulation. As an alternative to this relocation plan, or at least as a means of facilitating the introduction of personal communications networks, several companies in the vanguard of those seeking licenses to provide personal communications services submitted proposals to the FCC. They promised to obviate the need to displace existing users by the unprecedented application of spread spectrum modulation to wireless communications systems.

The availability of spread spectrum technology for civil use is a fairly recent development. FCC authorization to use spread spectrum and other wide-bandwidth emissions for nonmilitary applications was granted for the first time only in 1985. Since then, the FCC has authorized the use of spread spectrum in the police radio service; the industrial, scientific, and medical (ISM) bands; and the amateur radio service.

Advantages. The appeal of spread spectrum modulation for application to personal communications networks lies in its ability to reduce the power per unit bandwidth by spreading the transmitted power over whatever bandwidth has been assigned. The result is a low signal-to-noise ratio as compared with that obtained with conventional radio, which concentrates the transmitted power into a small frequency range. In the context of personal communications networks, this spreading of the transmission would serve both to reduce the amount of interference received by the unit since the personal communications network receiver would spread the interfering signal and therefore receive only a portion of the original signal's total power, and to reduce the interference potentially transmitted to other users under the same rationale of dilution in power per unit bandwidth. In addition to reducing interference, signal spreading contributes to spread spectrum's characteristic low-probability-of-interception (LPI) and low-probability-of-detection (LPD) features.

Low probability of interception and low probability of detection refer to the difficulty encountered in attempting to detect the presence of a spread spectrum signal. As a result of these characteristics, the spread spectrum signals of a personal communications network system should be virtually transparent to existing microwave users. Provided the power level of each spread spectrum signal is kept below a predetermined level, the total power from all of the spread spectrum users within a geographic region should not interfere with the microwave users in the fixed-service microwave system, enabling both services to coexist on the same band. Such coexistence through the overlaying of a personal communications network system in a spectrum having already existing users would increase communications capacity and yield superior efficiency in spectrum use. Spread spectrum's low-probability-of-detection feature would also promise greater communications privacy through increased impregnability of the signals to airwave eavesdropping.

CDMA technology. At the heart of spread spectrum modulation is code-division multiple-access (CDMA) technology. The code-division multiple-access concept enables multiple users to operate on the same frequency band simultaneously through the assignment of a specific code sequence to each user. This code sequence is used to delimit the desired signal from the barrage of incoming but undesired transmissions. In practical application, this identification through code assignment would work as follows: A given receiver on the emerging-technologies band would receive all of the energy sent by every user transmitting. To translate this energy into a usable signal, only the signal corresponding to the specific user's code would be despread. As a result of this selective despreading, the receiver would see all the energy transmitted by the target user, while reception of extraneous signals would be limited to that fraction of the remaining transmission occupying the targeted subsegment of the bandwidth.

Problems. Spread spectrum capacity for multiple simultaneous communication through code designation is limited, however, by the total interference power that all the users generate in the receiver. If there are code-division multiple-access transmitters in proximity to the receiver, the operational threshold can be exceeded even though the total number of users is well below the theoretical maximum possible. Known as the near-far effect, this overwhelming interference created by transmitters close to the receiver is a problem likely to be encountered repeatedly in a mobile environment and represents an obstacle that must be overcome if the concept of a personal communications network is to realize its full multiple-user-support potential.

Despite the presence of unresolved details in system performance, the FCC has demonstrated the depth of its committment to the development of personal communications networks by its decision to affirmatively allocate a band of spectrum to the emerging technologies. By giving providers of personal communications services the freedom to develop personal communications networks on a primary-user basis, it has brought the commercial introduction of these communications services much closer to becoming a reality.

David B. Newman, Jr.

Aeronautical Phased-Array Antennas for Satellite Communications

Electronically scanned antennas known as phased arrays are now being installed on commercial aircraft. These antennas continually track satellites as the aircraft maneuvers, providing a reliable communication channel for voice, data, and facsimile. The reliable communications offer improved safety and operational efficiency for airliners. New services are available to passengers that will make the time spent aboard an airliner more productive. As new satellites are launched, costs will decrease and the services available will increase. At the end of 1992, approximately 200 aircraft had been equipped with aeronautical satellite communications equipment; by the year 2000, 10,000 aircraft could be equipped.

Principles of phased-array antennas. An array antenna is composed of a number of identical radiators that are generally spaced in a uniform manner. Each radiator, or element, receives a signal. The signals of the individual elements are input to a power combiner, which adds the individual signals at a common input point.

In **Fig. 1**, phase front 1 is parallel to the surface of the antenna and is associated with signal 1. This signal is received by a beam that is normal to the surface of the antenna, known as the boresight beam. The signal of this phase front arrives at all elements at the same time, and the signals are summed constructively at the antenna input.

For observation angles that are increasingly away from the antenna boresight, the signals of the individual elements no longer add constructively. The signals travel different distances from the source to the individual elements. The individual signals tend to add destructively, and the antenna gain is

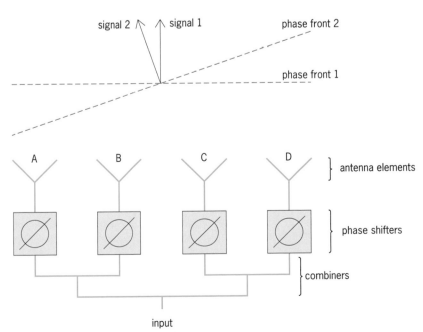

Fig. 1. Diagram of phased-array antenna.

diminished in these directions.

Signal 2 is associated with phase front 2. The signal from this direction arrives first at element A, then B, C, and D. In order to maximize the strength of a signal received from this direction, the phase shifters must delay the input signal of elements A, B, and C so that the signals to all elements are summed at the antenna input.

The required phase relationship between adjacent elements is defined by Eq. (1), where $\Delta\phi$ is the

$$\Delta\phi = 360° \times (d/\lambda) \times \sin(\theta_0) \qquad (1)$$

interelement phase shift, d is the distance between adjacent elements in meters, λ is the wavelength in meters, and θ_0 is the steering angle from boresight. For example, if the values of Eqs. (2) are assumed, then

$$d = 0.09375 \text{ m (3.691 in.)}$$
$$\lambda = 0.18750 \text{ m (7.382 in.)} \qquad (2)$$
$$\theta_0 = 30°$$

by using Eq. (1), the interelement phase shift required to steer 30° from boresight is $\Delta\phi = 90°$. Therefore, the phase shifter of element C must be set to delay the input signal by 90° with respect to element D. In a similar manner, the phase shifter of element B must be set to delay the input signal 90° with respect to element C and 180° with respect to element D. To steer the antenna beam by 30° from boresight, the phase shifters must be set to the states given by Eqs. (3).

$$\text{Element A} = -270°$$
$$\text{Element B} = -180°$$
$$\text{Element C} = -90° \qquad (3)$$
$$\text{Element D} = 0°$$

The steering angle depends upon the relative phase difference between the elements, and not upon the absolute phase of the elements. Therefore, all phases may be offset by a common amount and the antenna will still steer to the same direction. Phased-array antennas are generally steered about the physical center of the antenna rather than about an individual element. To steer about the center of the array and maintain the same 30° steering angle, the element phases would be given by Eqs. (4).

$$\text{Element A} = -135°$$
$$\text{Element B} = -45°$$
$$\text{Element C} = +45° \qquad (4)$$
$$\text{Element D} = +135°$$

The interelement phase shift is still 90°. The antenna beam is steered 30° in the opposite direction with the phases given by Eqs. (5).

$$\text{Element A} = +135°$$
$$\text{Element B} = +45°$$
$$\text{Element C} = -45° \qquad (5)$$
$$\text{Element D} = -135°$$

In this manner, the beam of the antenna is steered without physically moving the antenna. The beam is formed as the interference pattern between the individual elements of the antenna. For the sake of simplicity, the example was shown for a linear array. However, the same principle applies to a two-dimensional antenna. Also, antennas are reciprocal devices, so the phase settings are the same for a transmitting antenna.

Advantages of phased-array antennas. A phased-array antenna offers several advantages over a mechanically steered antenna. The solid-state electronics that are used to steer the antenna are far more reliable than mechanical devices. The electronic phased-array antenna can also be steered much faster than a mechanical antenna.

Most importantly, phased-array antennas are ideally suited for installation on aircraft. The radiating elements can be constructed of thin microstrip devices. This thin array can be attached to the aircraft skin with minimal mechanical modifications. This configuration is known as a conformal antenna. The electronics can be enclosed within the aircraft away from the harsh outside environment.

By contrast, installation of a mechanically steered antenna would require major structural modifications to the airframe. The mechanical antenna and modifications would add complexity and weight and would create much more aerodynamic drag than a conformal phased-array antenna. The added weight and drag would increase the fuel consumption of the aircraft. These considerations favor the use of a conformal phased array for aircraft installations.

Aircraft earth stations. The equipment installed on board the aircraft is referred to as an aircraft earth station (AES) and comprises the antenna subsystem, the avionics subsystem, and the cabin equipment (**Fig. 2**).

The antenna subsystem must acquire and track a desired satellite as the aircraft changes speed, direction, and attitude. The avionics subsystem must modulate the voice and data into a format suitable for transmission, upconvert and downconvert the signals to radio frequencies, and provide pointing commands to the antenna subsystem. The cabin equipment consists of the telephones, data ports, and facsimile machines that connect the passengers and crew to the satellite communication (SATCOM) system.

The avionics subsystem stores the locations of the system satellites in memory. The aircraft inertial navigation system provides the aircraft pitch, roll, heading, latitude, and longitude to the avionics. Based on this information, the avionics subsystem calculates the required antenna pointing angles, which are forwarded to the antenna subsystem. After receipt of pointing commands, the antenna subsystem points the antenna beam in the direction of the desired satellite. The pointing commands are sent to the antenna system at a rate of 15–25 words per second, allowing the system to continuously track the satellite.

Satellite communication systems. In order to appreciate what satellite communication systems

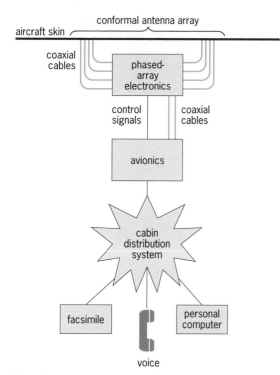

Fig. 2. Components of aircraft earth station.

can do for airliners, it is helpful to understand the existing communication facilities. Over land, aircraft communicate to fixed radio sites with very high frequency (VHF) radio. Passenger telephones are also available over the continental United States through an ultrahigh-frequency (UHF) cellular network. Over the ocean, VHF radios are out of range, and high-frequency (HF) radios must be used. The high-frequency radios are not reliable, as they are subject to atmospheric conditions. It is possible for a transoceanic aircraft to be out of contact for hours.

The overall aeronautical satellite communication system is illustrated in **Fig. 3**. Satellite communication systems utilize *INMARSAT* satellites that orbit 22,300 mi (35,900 km) above the Equator. The systems operate in the L band, 1530–1660.5 MHz. These frequencies are not subject to the previously noted problems of terrestrial radios. For this reason, toll-quality voice (the sound quality provided by ordinary telephone service) and reliable data (from aircraft equipped with aircraft earth stations) are available virtually everywhere on Earth. SEE COMMUNICATIONS SATELLITE.

The aircraft communicates to a dedicated earth station through the satellite. The earth station connects the aircraft to a leased network or the public switched telephone network, allowing the aircraft to communicate to any place on Earth from any other place on Earth.

Use in air-traffic control. The system is also used for air-traffic control purposes. At the present time, aircraft are tracked by fixed radar sites at approach, enroute, and departure areas. As the aircraft leaves the range of radar, positive tracking is not possible. Using the satellite communications system, the aircraft can periodically report its position to air-traffic

control centers, allowing better control of aircraft that are not within radar range.

Use in cockpit communications. Satellite communication systems offer vastly improved cockpit communications, particularly for transoceanic flights. Crews have access to more timely and detailed weather reports, allowing aircraft to avoid bad weather or even to take advantage of more favorable weather during the flight. Clearances to more favorable altitudes can be requested, allowing the aircraft to take advantage of prevailing winds along the route with resultant fuel savings. In-flight emergencies such as equipment failures or hijackings can be reported quickly, improving safety and also resulting in cost savings to the airlines.

Use in equipment monitoring. Equipment on board the aircraft can be monitored during flight. Any faulty equipment can be reported to the ground crews that service the aircraft. The result will be better aircraft maintenance, shorter turnaround time for aircraft, and safer aircraft for passengers. Flight attendants will be able to use the system to perform ticketing and reservations aboard the aircraft for other flights, hotels, and rental cars.

Passenger use. The system is available for passengers, creating an "office in the sky." Quality voice and facsimile transmissions are available. Data-communication interfaces will enable a passenger to connect a laptop computer.

Channels. The systems offer one to six channels, depending upon the capability of avionics. Voices are encoded at 9600 bits per seconds (bps). Data communication is available at rates from 300 to 10,500 bps. These rates are available with current technology and satellites that are already in orbit. As new, larger satellites are launched, new services will be available. More powerful satellites will allow increased capability, such as more channels. The cost of using the satellite resources will also decrease.

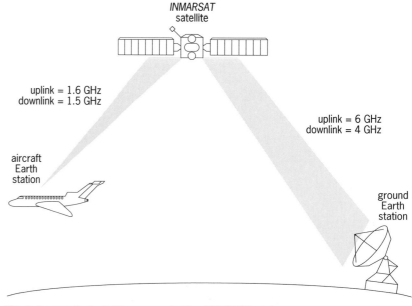

Fig. 3. Aeronautical satellite communication (SATCOM) system.

For background information SEE AIR-TRAFFIC CONTROL; ANTENNA; COMMUNICATIONS SATELLITE; MOBILE RADIO; MULTIPLEXING; RADIO SPECTRUM ALLOCATIONS; RADIO-WAVE PROPAGATION; SPREAD SPECTRUM COMMUNICATION in the McGraw-Hill Encyclopedia of Science & Technology.

Brian J. Cox

Bibliography. E. Brookner, Phased-array radars, *Sci. Amer.*, 252(2):94–102, February 1985; B. J. Cox, Worldwide aeronautical communications, *Avionics*, 16(8):38–46, August 1992.

Television

The application of digital techniques to the production, transmission, and reception of television signals is being actively discussed in business and technical circles. Digital techniques are expected to have a pervasive impact on television in the near future, similar to the revolutionary effect of digital audio compact disks on the recording industry. There are several reasons for this development. First, the Federal Communications Commission (FCC) is in the process of establishing an all-digital standard for the terrestrial transmission of high-definition television (HDTV). Second, cable and direct broadcasting satellite (DBS) companies are increasing their efforts to deliver conventional (that is, non-HDTV) television programming material to homes by digital means, with the added benefit of being able to increase the number of channels available for use. These additional channels are likely to be used for special programming such as pay-per-view. Third, with the recent ruling of the courts allowing telephone companies to provide other services within the existing telephone infrastructure, several regional telephone operating companies are studying the possibility of delivering digital video-on-demand services to homes. All these activities are made possible by significant advances in digital-image-compression techniques.

Lossy versus lossless compression. Compression methods take advantage of redundancies in the data inherent in the image structure and the nonlinearities of the human visual system. Most of the practical compression techniques used in digital television are lossy in that the reconstructed image sequences are not exactly identical to the original data. Lossless approaches exist, but these usually achieve no better than a 3:1 compression whereas the lossy methods can result in compression ratios of 100:1 without perceptible differences to the observer. Thus, for economically viable systems in commercial, industrial, and consumer applications, lossy algorithms are preferred because they save on data storage and communication bandwidth.

Luminance and chrominance signals. An anomaly of the human eye is that it is more receptive to the luminance (or brightness) signal than to the chrominance (or color) signal. Hence, most compression methods subsample the color information prior to performing the compression procedure. For example, in conventional broadcast-quality television,

the digital resolution of the sampled luminance signal is 720×480 pixels, while the color signals may be only 360×240 pixels. In addition, the encoded representation of the luminance signal is assigned more bits per pixel (that is, a larger dynamic range) than are the chrominance signals.

Spatial redundancy. Of the several methods of exploiting the spatial redundancy inherent in images, the one most commonly used is transform-based coding, where the discrete cosine transform is often employed. In this method, the sampled data of the image sequences are partitioned into blocks, typically 8×8 pixels. A two-dimensional discrete cosine transform operation is then applied to transform the data into the frequency domain.

To reduce the amount of data needed to represent an image sequence, advantage is taken of the fact that the eye is less sensitive to energy with high spatial frequencies. For example, consider an image on a television set formed by an alternating sequence of spatial signals representing white and black. The human observer will see a uniform gray instead of the alternating checkerboard pattern. In a sense, the eye does not have the discrimination necessary to resolve the alternating pattern and instead averages the high-frequency signals over spaces. This deficiency is exploited by coding the high-frequency coefficients that result from the discrete cosine transform operation with fewer bits and the low-frequency coefficients with more bits. This procedure is referred to as quantization.

Temporal redundancy. Temporal redundancy refers to redundancy in time. Within a sequence of images, there is significant replication of data between adjacent frames of images, especially since the successive frames are typically only 1/30 s apart. Of course, this replication is not true when there is a scene change, but such events occur less frequently.

Two techniques are commonly applied to achieve data compression in this regime. In predictive coding, the system codes the difference between the current value and a value predicted from past data, instead of coding the sequences directly. The rationale is that if the prediction is effective, the difference between the current and predictive value is small. Therefore, the dynamic range of bit assignments needed to represent the difference is small, resulting in compression.

A technique called motion compensation is also commonly applied. The premise is that an arbitrary block in the current image frame is the result of a translatory motion from a block in the prior image frame. Hence, if the motion vector for this block translation is computed, data reduction is achieved because it is necessary to code only the motion vector, not the block of image data. The motion vector is represented by only two numbers, one for the x displacement and the other for the y displacement, while coding of an 8×8 block requires transmission of 64 values.

Entropy coding. After the image sequence has been coded by using the techniques of removing spatial and temporal redundancies, the events in the resultant bit stream are compressed by a process called entropy coding. Entropy coding is an information-preserving transformation (that is, the inverse process at the receiver recovers the exact bit stream that was sent to the entropy coder). The resultant codes are variable-length. The principle of entropy coding is rather like Morse code for signaling, where events (letters) that occur often are represented by few bits and events that occur rarely are assigned more bits. The actual compression ratio achieved by this process is dependent on the statistical behavior of the input bit stream, but a compression ratio of 3:1 is typically achievable.

Standards. Both open and proprietary standards are being proposed for digital television. The open standard that has the largest support base is MPEG (Motion Pictures Expert Group). MPEG is a full-motion compression standard that was initiated by the International Standards Organization, and major corporations around the world are actively participating in its development. The formulation of MPEG is proceeding in several phases. The first phase (MPEG1) was targeted for video compression of digital storage media with data transfer rates up to 1.5 megabits/s. The designers of the second phase (MPEG2) are currently investigating broadcast applications with data rates around 5–10 megabits/s. Such rates are likely to be suitable for use in digital television.

Among the proprietary television standards, three all-digital proposals are under consideration by the FCC for adoption as the HDTV standard. All three proposals make use of the basic principles of spatial and temporal redundancies described above. They differ in the details of where the techniques are applied in the overall system architecture, and also in other aspects such as the frame resolution supported, the modulation schemes for transmission, the structure (that is, grammar) of the compressed bit stream, the details of the variable-length codes, and the error protection and concealment methods.

Applications. The latest developments in compression techniques have effectively removed the major technological barriers to new digital video applications. Although digital television may serve as the first springboard for rapid market growth that utilizes this technology, other applications are on the horizon. With the reduction in storage requirements, desktop image publishing will allow users to seamlessly integrate high-quality photographs into their documents. With the reduction in communication bandwidth requirements, the picture phone may become commercially viable, since the rapid increase in picture quality and decrease in transmission costs should stimulate increased consumer demand. Finally, multimedia applications may become a reality as the digital integration of video, audio, graphical, and textual information becomes possible.

For background information SEE INFORMATION THEORY; TELEVISION; VIDEO TELEPHONE; VISION in the McGraw-Hill Encyclopedia of Science & Technology.

Peng H. Ang

Bibliography. P. H. Ang, A. Reutz, and D. Auld, Video compression makes big gains, 27(10):16-19, *IEEE Spect.*, October 1991; Special issue on digital multimedia, *Commun. ACM*, vol. 34, no. 4, April 1991.

Transgenic organisms

Extraordinary advances have occurred in the use and production of transgenic plants since 1987. The list of plants that have been altered by the introduction of novel genes into their chromosomes now includes many crop plants. This increase is due to new gene-transfer techniques, better methods for the selection of plant cells that carry the introduced gene, and increases in the efficiency of regenerating whole plants from cells. The success in producing transgenic plants has been paralleled by a better understanding of the mechanisms that control gene expression and the identification and isolation of coding sequences from genes. Many genes that are developmentally regulated, expressed in specific tissues, and contain coding regions for proteins with known function have been isolated and characterized. The ability to manipulate these genes and introduce them into plants has led to the production of transgenic plants with a variety of improved traits.

Transformation methods. The ability to produce transgenic plants has been limited by the lack of success in introducing genes into a plant cell or regenerating a whole plant from that cell. The natural gene-transferring system of *Agrobacterium* has been limited by the host range of the bacterium. In general, *Agrobacterium*-based transformation systems are not useful for monocots (which include the important cereal crops of the world) because they are not hosts for the bacterium. Methods for direct deoxyribonucleic acid (DNA) delivery to a plant cell have been developed to avoid using *Agrobacterium*. Two methods that require the isolation of plant cells without walls (protoplasts) to allow DNA uptake have been developed. The protoplasts are formed from plant tissue by using a combination of enzymes to degrade the cell wall. The protoplasts must be maintained in the proper osmotic conditions to prevent water from entering the cell. Without the structural confinement of the cell wall, cells would rupture from the influx of water. Purified DNA containing the gene to be introduced is added to the protoplasts. DNA uptake is facilitated by the addition of polyethylene glycol to the mixture or by an electric pulse (electroporation) across the solution. Both approaches make the cell membrane more permeable to the large DNA molecule. The mechanisms of DNA entry into the cell and integration into a chromosome are not understood completely. However, the genes that are introduced by these methods are stably inherited. These transformation protocols are limited by the inability to regenerate whole plants from protoplasts of many plant species.

For microinjection of DNA, which is used extensively to transform animals, a pipette that is much smaller than the plant cell is used to deliver the gene to the cell's cytoplasm or nucleus. Its use in transforming plant cells has been limited because of difficulties in injecting the cells and regenerating a whole plant from a single transformed cell. Although routine procedures are not available for the microinjection of plant cells, the method may be useful in the transformation of pollen. The transformed pollen could be used to fertilize a flower, eventually leading to a whole plant.

Microparticle bombardment to introduce DNA has been adapted successfully to many organisms, including corn, green algae, and yeast. Microscopic (approximately 1 micrometer) tungsten or gold particles are coated with DNA (the gene to be introduced), and the particles are shot into the intact plant cells. The particles pierce the cell wall and enter the cytoplasm, where the gene is released and stably integrated into the cell's chromosomes in the nucleus. Cells survive microparticle bombardment because of the particles' minute size. The velocity of the particles can be varied to allow cells with different structure or size to be penetrated, or to allow particles to traverse several layers of cells so as to introduce the gene to interior cell layers. Many forms of plant tissue have been transformed by accelerated microparticles, including unorganized clumps of cells from culture and from the growing apex of a plant.

Selection of transformed cells. The transformation methods described usually result in a mixed population of cells that either do or do not carry the gene of interest. An essential step in the regeneration of a transformed plant from a mixed population is allowing the transformed cell to grow and divide while slowing the growth of cells that are not transformed. This selection of the transformed cells is made possible by physically linking the DNA of the gene to be introduced with DNA that contains a gene for resistance to an antibiotic or a herbicide. Since many of the genes for resistance have been isolated from bacteria, they must be made functional by fusing them with a promoter that is expressed in plants. During regeneration of the whole plant from the transformed cell, the mixed population of cells is placed in a regeneration medium with the antibiotic or herbicide. The growth of cells that do not carry the gene for resistance is inhibited or the cells are killed, allowing preferential growth of the transformed cells. In this manner, the selection greatly increases the efficiency of the transformation process. The identification of new genes for resistance and the promoters to express them in plants have made it possible to select transformed cells in more species.

Identification and characterization of genes. The development of methods to introduce genes and the characterization of genes for expression in plants have advanced together. As coding regions for antibiotic and herbicide resistance were identified in bacteria, it became necessary to isolate the promoter regions from other genes to express these coding regions in plants. One way to analyze a promoter region of a plant gene is to remove it from its normal coding sequence and use it to drive the expression of a reporter gene whose product is easy to follow. Visual markers have been especially helpful. The bacterial gene β-glucuronidase has been used extensively as a reporter to determine where, to what level, and when, during a plant's life cycle, the promoter region is activated. β-Glucuronidase is simple to use, sensitive, and quantitative, and no β-glucuronidase activity is found in most nontransformed plant cells. The promoter being studied is fused to the β-glucuronidase coding region, so that the promoter is responsible for the expression of β-glucuronidase when the gene is introduced into plants. To test if the promoter is active within a tissue of the plant, the tissue is sampled and added to a colorless substrate that, when acted on by β-glucuronidase, turns into a blue precipitate. The substrate enters the cell, and if β-glucuronidase is present in the cell it becomes blue. Each tissue of the plant can be quickly analyzed for the level of this enzyme. In this way, many promoters that are activated only in specific tissues, activated at specific times during development, or induced by specific treatments have been identified, allowing the tailoring of gene expression in a transgenic plant.

Uses of transgenic plants. Several strategies for the production of transgenic plants with improved characteristics, such as resistance to virus infection and insect attack, alterations in fruit development, control of male fertility, and tolerance to herbicides, have advanced to the stage of field testing. These modifications have usually involved the introduction of a gene that is the result of fusing a promoter that is expressed at the correct time and in the appropriate tissue with a coding sequence that modifies the plant. For example, plants resistant to insect attack have been produced by using a promoter that is active at high levels in most plant tissues to express a gene from the bacterium *Bacillus thuringiensis*. The gene encodes for a protein that, when ingested by some insects, is toxic. When tomato hornworms are released on nontransgenic plants, the plants are severely damaged whereas transgenic tomato plants, expressing the *B. thuringiensis* protein, are virtually untouched.

Male sterility of oilseed rape (*Brassica napus*) was engineered by producing transgenic plants that expressed a ribonuclease (RNase). The RNase is an enzyme that degrades ribonucleic acid (RNA) and when expressed results in cell death, because protein synthesis from RNA templates is blocked. Killing the entire plant is avoided by using a promoter that is active only in the stamens (the organs that produce the pollen). Thus, the pollen grains are killed because the RNase is expressed in the stamen, while the other portions of the plant remain normal because no RNase is produced in other tissues. These examples demonstrate the power of using transgenic plants to understand how the expression of a coding region can modify a plant and how a promoter region controls the expression pattern of a gene. The ability to introduce genes into plants makes it possible to test the effects of combining coding and promoter regions from different genes or from different organisms on the growth and development of a plant and to improve plants by using novel strategies.

For background information SEE BIOTECHNOLOGY; BREEDING (PLANT); GENETIC ENGINEERING in the McGraw-Hill Encyclopedia of Science & Technology.

Alan G. Smith

Bibliography. A. B. Bennett and S. D. O'Neill, *Horticultural Biotechnology*, 1990; M. W. Fowler and

G. S. Warren, *Plant Biotechnology*, 1992; C. S. Gasser and R. T. Fraley, Genetically engineering plants for crop improvement, *Science*, 244:1293–1299, 1989; R. S. Sangwan and B. S. Sangwan-Norreel (eds.), *The Impact of Biotechnology in Agriculture*, 1990.

Transplantation biology

Transplantation between genetically dissimilar individuals or animals elicits rapid organ rejection. In all species, a specific genetic region called the major histocompatibility complex (MHC) produces cell-surface proteins, which differ markedly between the recipient and the donor, and these products are rapidly recognized as foreign by the recipient.

Graft rejection theories. The types of MHC proteins are class I and class II. In about 1970 Peter Medawar proposed that MHC antigens themselves constitute the barrier to successful transplantation. According to this theory, successful transplantation is feasible with perfect MHC matching such as in identical-twin transplants, or with suppression of the transplant recipient's immune system to hinder recognition of foreign MHC. This premise is the basis of the approach to most clinical transplantation. MHC antigen matching is attempted between the donor and the recipient, and lifelong immunosuppression with drugs such as Cyclosporin A, FK506, or Azathioprine is initiated to prevent the host's lymphocytes from recognizing and attacking foreign MHC antigens.

In 1957, George Snell proposed the concept that MHC graft antigens expressed on different cell types were not equally efficient at eliciting graft rejection. Rather, so-called passenger lymphocytes, which innocently existed in donor grafts, were responsible for vigorously triggering the immune rejection. These passenger lymphocytes were from the blood and happened to be migrating through the tissue at the time of harvesting. The implication was that the graft antigens on graft cells were perhaps weak stimulators of the host's immune response.

Interestingly, these two opposing theories on the mechanisms of graft rejection have simultaneously existed in the scientific literature. The first purpose of this article is to present recent scientific evidence demonstrating that passenger lymphocytes are present in donor tissues, and that selective depletion of passenger lymphocytes is sufficient for tissue survival without immunosuppression. These data provide evidence in favor of the Snell hypothesis. The second purpose is to review recent scientific work demonstrating that MHC graft antigens can trigger xenograft (cross-species) rejection but that the direct concealment of foreign MHC itself is possible and averts rejection in experimental models.

Passenger lymphocyte experiments. Because lymphocytes have a limited life-span, in contrast to the cells of an organ, which are present throughout life, initial experiments attempting to deplete grafts of passenger lymphocytes used pretransplant culture. Using ovarian tissue as well as thyroid tissue, organ culture of the donor grafts at 99°F (37°C) prior to transplantation significantly prolonged survival when combined with minimal immunosuppression of the recipients. Room-temperature culture inactivated the ability of lymphocytes to subsequently stimulate immune responses. Other workers demonstrated that low-temperature pretransplanted culture of donor-rat islets (the insulin-secreting cells of the pancreas) in combination with brief immunosuppression of the recipients prolonged survival of the islets. Although the most successful pretransplant culture condition varied between tissues, the common belief existed that culture either depleted or incapacitated the passenger lymphocytes, thereby hindering the recipient's recognition of the target.

The ability to selectively rid donor tissue of passenger lymphocytes prior to transplantation became feasible with the discovery of discordant cell-surface molecules selectively expressed on lymphocytes but absent from graft cells. The indication was that direct depletion of passenger lymphocytes prior to transplantation could possibly be achieved by selective antibody treatment of grafts prior to transplantation. This brief treatment that aimed at killing only passenger lymphocytes directly tested the Snell hypothesis. The first tested antibody pretreatment of donor tissue was with MHC class II antibodies (Ia antibodies), owing to the demonstrated absence of the corresponding antigen on freshly isolated islets but its known expression on almost all lymphoid cells. A 30-min pretreatment of freshly isolated foreign islets with donor-specific MHC class II antibodies allowed islet survival as well as heart tissue survival without immunosuppression of the nonidentical mouse recipients.

The ability to transplant tissue successfully after pretreatment with antibody perfusion strongly supported the Snell hypothesis. Furthermore, since these grafts expressed only MHC class I, it appeared that this strong transplantation antigen expressed on parenchymal cells was not, in itself, necessarily immunogenic when presented in the absence of donor cells that were positive for MHC class II antigen.

Passenger lymphocyte depletion with antibody pretreatment of the donor tissue has recently been applied to human kidney transplantation. Human kidney transplantation represents a challenging barrier to application of antibody-mediated donor lymphocyte depletion owing to the large mass of foreign tissue and the large burden of donor lymphocytes. Thus, success of this clinical trial substantiates the clinical importance of passenger cells in eliciting the rejection cascade.

Masking donor tissue. Basic immunology has contributed significantly to the understanding of mechanisms that trigger cytotoxic T cells, the lymphocytes that are believed to directly kill foreign tissue. However, the mechanisms of T-cell adherence to target cells and T-cell-receptor activation must be better understood to prevent graft rejection.

In a most basic way, T cells are attracted to and adhere to their target through reciprocal lock-and-key adhesion molecules. One of the most important interactions involves the adhesion of the CD8 molecule on lymphocytes to the ubiquitous target MHC class I molecule. This critical interaction has been investigated to prevent graft rejection. Highly purified preparations of masking fragments of antibody to MHC class I antigens were produced to conceal the donor MHC class I molecules prior to transplantation. The masked donor tissue was human islets transplanted into nonimmunosuppressed mice. Autopsy results as well as secretion of human insulin by the mice demonstrated that the masked foreign islets survived up to 200 days after transplantation.

Although it might be argued that the mechanism of graft prolongation in this new model is secondary to MHC class I protein masking of the passenger lymphocytes within the islets, masking of all surface epitopes on cultured tumor cells that do not contain passenger lymphocytes also allowed cross-species transplant survival beyond 200 days. Since established tumor lines lack passenger lymphocytes, the primary role of graft antigens in eliciting direct activation of T cells seemed the most likely explanation. Additionally, since successfully transplanted tumor cells proliferate after masking, the induction of systemic tolerance sufficient for new donor antigen exposure without rejection must develop.

Tolerance mechanisms. Although pathways toward tolerance are certainly distinct from those responsible for graft rejection, the exact mechanisms are unknown. Multiple T-cell structures are necessary for T-cell activation and killing; in contrast, engagement of an isolated T-cell structure might induce tolerance. Support for this concept is substantiated by the frequent ability to trigger cultured T-cell unresponsiveness by this mechanism. It is hypothesized that the most likely outcome of MHC class I masking of donor grafts is inadequate T-cell-receptor engagement. Two recent experimental reports add support to this hypothesis. In a challenging heart transplant model, it was demonstrated that masking of a different donor heart antigen known as LFA-1 allowed graft survival as well as specific induction of T-cell tolerance in the recipient. In an islet transplant model, the same was accomplished through masking of the CD28 molecule of the T cell and the target B7 molecule of the B cell. Therefore, successful transplantation may be possible by depriving the recipient's T cells of some but not all of the necessary triggering events. It is suspected that selective inhibition of T-cell binding can produce specific T-cell tolerance in masked transplants, and may perhaps explain why graft survival as well as systemic tolerance can develop after transplantation in all these experimental models. SEE CELLULAR IMMUNOLOGY.

Although the masking approach to transplantation is a new way to prolong transplants, perhaps this technique has already been encountered by the immune system. A wealth of data on tumors has documented the frequent occurrence of deficient MHC class I expression on malignant cells. Furthermore, some tumor cells appear to be protected by host-produced proteins that coat the tumor cells and protect them from attack via attachment of cytotoxic T lymphocytes. These observations may explain the paradox of malignant

tumor growth without immune rejection. Interestingly, a recent report contends that early experiments using pretransplant culture to reduce graft foreignness by presumed passenger lymphocyte depletion actually may have worked by decreasing the MHC class I expression on the cells. These artificial systems to reduce graft antigens by culture or masking, and the natural avoidance of tumor cell rejection by elimination of MHC class I expression, should aid in the discovery of new ways to evade immune detection in the transplant field. SEE IMMUNOSUPPRESSION.

For background information SEE HISTOCOMPATIBILITY; HYPERSENSITIVITY; IMMUNOSUPPRESSION; TRANSPLANTATION BIOLOGY in the McGraw-Hill Encyclopedia of Science & Technology.

Denise L. Faustman

Bibliography. Y. Brewer et al., Effect of graft perfusion with two CD45 monoclonal antibodies on incidence of kidney allograft rejection, *Lancet*, 2:935–937, 1989; D. Faustman et al., Prolongation of murine islet allograft survival by pretreatment of islets with antibody directed to Ia determinants, *Proc. Nat. Acad. Sci.*, 78:5156–5159, 1981; D. Faustman and C. Coe, Prevention of xenograft rejection by masking donor HLA class I antigens, *Science*, 252:1700–1702, 1991; D. J. Lenschow et al., Long-term survival of xenogeneic pancreatic islet grafts induced by CTLA4Ig, *Science*, 257:789–792, 1992.

Tree rings

Reconstructing past climates is a puzzle that must often be solved despite missing pieces. Paleoclimate information essentially comprises evidence from the physical environment, including data from sediments, and evidence from the biosphere, in the form of floral and faunal composition and diversity. If floras include anatomically preserved specimens, the plant structures themselves may offer clues to the paleoenvironment. For example, the presence of a thick cuticle or numerous epidermal hairs on leaves can be indicative of arid environments. One of the best-studied aspects of plant anatomy as related to climate is the growth rings in the wood (secondary xylem). The periodicity of paleoclimates is recorded in the structure and size of these growth rings. Structurally preserved wood is common in many different types of depositional environments. In fact, because of the resistant nature of tracheid cell walls, wood is probably the plant part (except for pollen) most often preserved in the fossil record.

Tree-ring data have become an important interdisciplinary tool in interpreting past climates. Such data were initially studied as a record of Holocene climate information and have been especially useful in the dating of archeological sites, especially in arid regions. Since the early 1980s, techniques that were initially used to analyze Pleistocene wood have been extended to analysis of Mesozoic and Paleozoic material.

Tree-ring formation. The periodic growth of the vascular cambium in woody plants results in tree-ring formation. The vascular cambium, or lateral meristem, is responsible for the increased diameter of woody plants, in which new wood cells (secondary xylem) are produced to the inside and bark cells (secondary phloem) are produced to the outside. In temperate regions, the cambium exhibits a slowing of growth in late summer followed by a cessation of cell production when the leaves abscise in the fall (in deciduous plants) or when the temperature drops below a certain level (in evergreens). The following spring, cell production begins again. The wood cells produced in the early spring (called springwood or earlywood), when the water supply is generally plentiful, are larger and more thin-walled than those produced in late summer (summerwood or latewood), when available water is limited. The regular alternation of these cell types produces the annual rings that are apparent in temperate woods; each year's ring boundary represents the cessation of cambial growth in the fall. If the tree is known to produce a single ring each year (as in most temperate trees), it is possible to determine its age by counting the rings. Although growth rings are produced by some tropical trees, they may not be annual rings and therefore are difficult to use for dating or correlation.

Factors affecting wood growth. Cambial activation and wood production are mediated by hormone levels in the plant, but the effects of environment upon plant growth in general, and wood growth in particular, are very complex. The timing and extent of cambial activity are influenced by a complex interaction of light regime, temperature (for example, nighttime versus daytime, mean annual high versus low), length of growing season, water availability, and genetics. Phytohormones in the cambium are produced in the apical meristem, but exactly how these physiological events respond to environmental cues and what role genetics plays remain poorly understood.

Certain ecological events leave a distinct signature in the wood. For example, if a late freeze occurs in the spring after cambial activation or in the fall prior to growth cessation, a region of disorganized growth and irregularly shaped cells will be formed in the ring. This structure, termed a frost ring, is also recognizable in fossil woods. Changes in the climate that lead to an increase in the number of late freezes, and therefore the occurrence of more frost rings, are especially noticeable in trees growing at the limit of their climatic tolerance—for example, at the tree line in alpine regions or at high latitudes. Recent modeling of tree growth at high northern latitudes, however, suggests that climate warming may cause an increase in the number of frost rings in the wood. As the climate warms, the cambium breaks dormancy earlier in the year (possibly even in midwinter) and is therefore more likely to experience a hard frost event. Severe damage to the crown of the plant or defoliation by insects can also result in a zone of disrupted cells in the tree ring.

In some cases, there may appear to be more than one tree ring within a single season. These false rings, or intraannual bands, result from the production of latewood in the middle of the growing season, often due to a temporary water shortage. When water again becomes available, earlywood production is resumed. Intraannual bands in modern plants are often found in trees from arid or semiarid climates but may also result from insect defoliation or damage to the crown of the tree.

Climate periodicity and ring structure. Since ring structure is affected by climate in specific ways, it is possible to extrapolate paleoclimate information from an examination of ring growth. Some parameters that can easily be measured in the laboratory in fossil woods include changes in radial cell diameter within individual rings (that is, the proportion of earlywood to latewood), annual variation in ring width, and the prevalence of deviations in ring growth, such as the response to frost or false rings. An important indicator of climatic variability and the effect of environment on growth is the climate sensitivity of a series of growth rings. The mean sensitivity is seen as a measure of change in the environment: a complacent series of rings results from periods of stable growth conditions, and a sensitive series from periods of variable conditions.

In addition to the climatically caused variation in wood production, there is an endogenous (that is, genetic) factor in the structure and periodicity of the wood. In dealing with fossil specimens in which growth parameters may not be known, it is important to have an assemblage of wood specimens from each locality, so that environmental factors can be separated from genetic ones. Data from fossil tree rings are most useful when coupled with information from megafloras, microfloras, and the sedimentology of the site. This approach has been utilized in studies of Late Cretaceous floras from the North Slope of Alaska.

Fossil forests. Specific problems are associated with examining tree rings in fossil wood. Many of these problems are more pronounced in older fossils, since the possibility of diagenetic effects is greater in progressively older rocks. Most are inherent in any study of fossil material. The problems include incomplete preservation, such as only a portion of an axis being preserved, the remainder having been destroyed by fungi or bacteria prior to fossilization; incomplete mineralization, that is, the mineralizing fluid (containing silica, calcium carbonate, or other minerals) did not penetrate the entire specimen or penetrated unevenly; and crushing, that is, the specimen was crushed or otherwise altered in shape in the process of fossilization. The best specimens to study are permineralized (minerals have filled intercellular spaces and cell lumens, while the cell walls remain organic) rather than petrified (intercellular spaces, as well as cell walls, have been replaced by minerals).

Some of the first work that applied modern tree-ring techniques to pre-Pleistocene wood was done on material from the Antarctic peninsula region, specifically a lower Cretaceous fossil forest on Alexander Island. Parameters that were examined included annual and mean sensitivity, mean and maximum ring width, and evidence of periodicity in wood growth within the forest as a whole. By using these data, it was possible to reconstruct this high-latitude forest and to address the problems of paleogeographic reconstructions of the Antarctic peninsula region.

During the 1990–1991 austral field season, two fossil forests near the Beardmore Glacier in the central Transantarctic Mountains, Antarctica, were examined. One of these is late Permian (about 260 million years ago) and the other Middle Triassic (about 235 m.y.a.). The Permian deposit includes 15 permineralized trunks in growth position at a paleolatitude approximately 80–85°S. The Triassic site is larger, consisting of 99 permineralized trunks (paleolatitude approximately 70–75°S). Based on compressed leaves preserved in the shale in which the trunks are rooted, the older forest consists of trees belonging to the seed fern *Glossopteris*, and the Triassic trunks bear the common Triassic foliage type *Dicroidium*. Tree rings reveal that both forests were growing in an equable, strongly seasonal climate. The trees show uniform growth throughout the growing season and a rapid cessation of cambial activity in the fall with little latewood production, an arrangement that is typical of fossil and extant tree rings at high latitudes. Growth rings are relatively wide for such high-latitude forests. These fossil forests provide an important source of paleoclimate data that can be integrated with climate models based on physical parameters. In addition, the existence of forests at high latitudes where no forests occur today expands historical knowledge of plant growth and can help to further understanding of current climatic changes.

For background information SEE DENDROCHRONOLOGY; PALEOCLIMATOLOGY; TREE-RING HYDROLOGY in the McGraw-Hill Encyclopedia of Science & Technology.

Edith L. Taylor

Bibliography. G. T. Creber and W. G. Chaloner, Influence of environmental factors on the wood structure of living and fossil trees, *Bot. Rev.*, 50:357-448, 1984; H. C. Fritts, *Tree Rings and Climate*, 1976; T. H. Jefferson, Fossil forests from the Lower Cretaceous of Alexander Island, Antarctica, *Palaeontology*, 25:681–708, 1982; E. L. Taylor, T. N. Taylor, and N. R. Cuneo, The present is not the key to the past: A polar forest from the Permian of Antarctica, *Science*, 257:1675–1677, 1992.

Tuberculosis

Tuberculosis is the leading infectious cause of death in the world, killing millions of people each year. In the United States, the incidence of tuberculosis decreased greatly between 1953 and 1984, raising the hope that the disease could be eliminated by the year 2010. However, a resurgence in tuberculosis has occurred and is complicated by the appearance of cases with drug-resistant organisms, which are both difficult and costly to treat.

Development of disease. Tuberculosis is caused by the microorganism *Mycobacterium tuberculosis*, the tubercle bacillus. Although tuberculosis can affect any site in the body, it most commonly is found in the lungs. It is spread by inhalation of infectious airborne particles that are produced when a person with tuberculosis of the lungs coughs. Usually, after a person is infected with tubercle bacilli, the immune system responds, halting the multiplication of the bacilli. During this time, the infected person usually has a positive skin test for tuberculosis but shows no sign of disease and is not infectious.

Reactivation of the infection and development of active disease can occur at any time. The average person who is infected with the tubercle bacillus has a 10% chance of developing active disease in his or her lifetime; but certain conditions, such as infection with the human immunodeficiency virus (HIV), significantly increase that chance. In such conditions, the immune system may not be able to contain the bacilli, and active disease can develop. Progression to active disease can be prevented in most cases if an infected person is given medication for 6–12 months.

History. Tuberculosis has caused disease in humans since ancient times. With the onset of the industrial revolution, overcrowding and poor sanitary conditions promoted its spread. Indeed, one-quarter of all adult deaths in Europe in the eighteenth century were attributable to tuberculosis.

Before effective medications were available, many tuberculosis patients were isolated in specialized hospitals, called sanatoria, where they remained for months to years. Treatment consisted of enhanced nutrition and prolonged bed rest. In some institutions, patients slept adjacent to an open window or on a sleeping porch, where they were exposed to the open air even in very cold weather. Life expectancy for patients with minimal disease was similar to that of the general population. However, approximately one-half of those with extensive disease died within 5 years of diagnosis.

Treatment of tuberculosis changed in 1944 with the discovery of the antibiotic streptomycin. This drug was extremely effective against tuberculosis. Scientists discovered, however, that treatment with a single medication could lead to the development of drug resistance. After prolonged treatment, the tubercle bacilli became resistant to the medication. Fortunately, other effective medications were soon discovered, and by using two or three drugs in combination, drug resistance generally could be avoided. The medications were generally given for a period of 18–24 months, and more than 90% of patients who completed the course of treatment were cured. The introduction of more potent medications has allowed the duration of therapy to be shortened for many patients to 6–12 months, but the use of multiple drugs in combination continues to be an important principle of tuberculosis therapy.

Reemergence. Improvements in living conditions and effective drug therapy produced a steady decline in tuberculosis cases and deaths in the United States. Since 1953, when national records were first kept, the number of reported tuberculosis cases declined an average of 5% each year until 1984. In 1985, the number of reported cases reached a low of 22,201. However, since then the number of reported tuberculosis cases has increased each year. Between 1985 and 1991, there were 39,000 more cases of tuberculosis in the United States than expected based on previous trends.

The increased cases have occurred more frequently in certain groups. Greater than two-thirds of cases now occur among blacks, Hispanics, Asians, and Native Americans. A significant component of the recent rise in tuberculosis cases is its association with HIV infection and acquired immunodeficiency syndrome (AIDS). Persons with HIV infection have a weakened immune system and are much more likely to develop active tuberculosis if they are infected. Tuberculosis is also more difficult to diagnose in the group (partly due to simultaneous occurrence with other lung infections), and patients may be ill for long periods before tuberculosis is diagnosed and treatment is begun.

Social and economic factors have also contributed to the rise in tuberculosis cases. Low-income populations, alcoholics, and substance abusers have a greater risk of developing tuberculosis than the general population. Homelessness has been associated with high rates of tuberculosis in several areas. Recent immigrants from countries with a high prevalence of tuberculosis and residents of correctional facilities and nursing homes are at high risk. However, although the highest rates occur in low-income groups, nearly one-third of the tuberculosis cases in the United States occur in middle- and upper-income groups.

Control. The steps in controlling tuberculosis in a community include identifying and treating all active cases quickly in order to prevent transmission to others, identifying contacts of those cases to determine whether infection has occurred, and giving preventive therapy to infected contacts.

Many patients are reluctant to take medication for the prescribed prolonged periods (6–24 months). Care of patients in high-risk groups can be complicated by factors such as poor access to health care, lack of stable housing, and language barriers. Patients who do not take their medication properly can continue to be infectious for longer periods and may cause drug-resistant organisms to emerge. A proven method of ensuring completion of therapy is for a public health worker to observe the ingestion of each dose of medication, either in the clinic or at the patient's residence. This practice, called directly observed therapy, has been successful in controlling tuberculosis in the past. However, federal, state, and local resources for public tuberculosis control programs have been inadequate to implement this approach for all patients who need it.

Drug-resistant tuberculosis. Accompanying the increase of tuberculosis cases is an apparent increase in cases infected with drug-resistant organisms. Low levels of drug resistance occur naturally because of random mutations in mycobacteria. In a population of tubercle bacilli, resistance to an individual drug occurs in about 1 in 10^6–10^8 organisms. The probability of naturally occurring resistance to two drugs is extremely low. A patient with severe tuberculosis has approximately 10^9 bacilli in the lungs; one or two of these bacilli may be resistant to any one drug. If the patient is treated with a single drug, the resistant bacilli will continue to multiply. As they become more numerous, the single drug will no longer be effective and the patient's disease will progress. If multiple drugs are used, resistant bacilli will be susceptible to one of the drugs and all of the bacilli will be killed.

A person may be infected with a drug-resistant organism through contact with an active case of drug-resistant tuberculosis. Drug-resistant tuberculosis is much more difficult to cure and more complicated to treat than drug-susceptible tuberculosis. Drug-resistant tuberculosis requires longer therapy (24 instead of 6–12 months) and medications that are more toxic and costly than those used for drug-susceptible tuberculosis. Often, drug resistance is not suspected, and improper choice of therapy can lead to further drug resistance.

For background information SEE ACQUIRED IMMUNE DEFICIENCY SYNDROME; DRUG RESISTANCE; MYCOBACTERIAL DISEASES; TUBERCULOSIS in the McGraw-Hill Encyclopedia of Science & Technology.

Michael Iseman; Patricia Simone

Bibliography. P. F. Barnes et al., Tuberculosis in patients with human immunodeficiency virus infection, *N. Engl. J. Med.*, 324:1644–1650, 1991; K. Brudney and J. Dobkin, Resurgent tuberculosis in New York City: Human immunodeficiency virus, homelessness, and the decline of tuberculosis control programs, *Amer. Rev. Respir. Dis.*, 144:745–749, 1991; Centers for Disease Control, A strategic plan for the elimination of tuberculosis in the United States, *MMWR*, 38:1–25, 1989; M. D. Iseman and L. A. Madsen, Drug-resistant tuberculosis, *Clinics Chest Med.*, 10:341–353, 1989.

Tumor viruses

The first part of this article discusses recent developments suggesting that the interaction of viral infection, alteration in genes governing cellular metabolism, and the body's own immune system can produce perhaps 15% of human cancers. The second part of this article discusses biological aspects of papillomaviruses and their importance in human disease.

Tumor Viruses and Malignant Transformation

Growth and differentiation are the basic features of normal development of all cells. Uncontrolled cell proliferation, characterizing the cancerous cell, is the consequence of several genetic alterations in cellular metabolism. Several factors are thought to contribute to the development of animal cancers and tumors, including physical irritation, environmental chemicals (carcinogens), viral infection, normal genetic variation influencing the immune response, and mutational changes in the germ line. Tumor viruses may act in concert with one or more of these factors to promote the malignant transformations that result in cancer. The most likely candidates for tumor viruses in humans include the Epstein-Barr virus, human papillomavirus, hepatitis B virus, and human T-cell leukemia-lymphoma virus.

Human tumor viruses. Epstein-Barr virus (EBV), a member of the herpes virus family, is found in up to 100% of adult populations worldwide. This high prevalence is partly attributable to the infectious particles being shed in the saliva of the host, creating highly favorable conditions for viral transmission. Acquisition of the virus is most common during adolescence and may be asymptomatic or result in mononucleosis. Epstein-Barr virus has been linked to several malignant tumors, including Burkitt's lymphoma, nasopharyngeal cancer, and Hodgkin's disease. All cells sampled from Burkitt's lymphoma demonstrate characteristic genetic changes in the cancerous cells, including genetic modifications of the oncogene *c-myc* found on chromosome 8. Oncogenes, such as *c-myc*, produce uncontrolled cell proliferation when mutated. The genetic changes in Burkitt's lymphoma also include mutation in one of the immunoglobulin genes, the molecular products of which are antibodies. Nasopharyngeal cancer, especially prevalent in parts of eastern Asia, has been linked to carcinogen exposure in preserved food; in the United States it has been linked to tobacco and alcohol use.

The papillomaviruses have an affinity for epithelial cells in the vertebrate host, and are associated with benign tumors such as warts, as well as with malignant cancers in both animals and humans. The human papillomaviruses (HPV) have been linked to several anogenital cancers, the most common of which is cervical carcinoma. Some transmission of this virus is clearly due to sexual contact. Several dozen specific genotypes have been identified, some associated with particular cancers.

Both hepatitis B virus (HBV) and hepatocellular carcinoma (a form of liver cancer) have a relatively high incidence in some regions of eastern Asia and Africa. The frequent concidence of the virus and liver cancer in the same individual led to the notion of a causal link between the two. A common prelude to this form of liver cancer is chronic hepatitis lasting several decades, during which continuous destruction and regeneration of liver tissue occur. Hepatitis B virus may lead to neoplastic transformation in hepatocytes of the liver. Specific chromosomal changes and mutations in the p53 gene are characteristic of hepatocellular carcinoma.

The retrovirus human T-cell leukemia-lymphoma virus (HTLV) also has well-substantiated connections to human neoplasia. Adult T-cell leukemia-lymphoma (ATLL) is prevalent in some areas of Japan, the Caribbean, and central Africa. The virus is spread through sexual contact, breast feeding, blood transfusions, and needle sharing. There is typically a decades-long delay between viral infection and expression of the cancer. Laboratory studies have shown the ability of human T-cell leukemia-lymphoma virus to immortalize T lymphocytes (that is, induce a state of continuous growth, a signal feature of malignant transformation). The virus also induces interleukin-2, a growth promoter of the immune system, thereby enhancing the further prolif-

eration of transformed T cells. The identification of human T-cell leukemia-lymphoma virus and its involvement in adult T-cell leukemia-lymphoma laid the foundation for the identification of the human immunodeficiency virus (HIV).

Influence of immune system. Besides its role in direct defense against infectious disease, the immune system functions in locating and eliminating cells that have deviated from their normal metabolic activities. These cells, if not eliminated, are subject to a series of metabolic changes that sometimes lead to cancer. This elimination activity of the immune system is called immune surveillance. It occurs through the continuous examination of all bodily tissue to which the immune system has access (the blood-brain barrier, for example, greatly restricts immune surveillance of the central nervous system). Divergent cells, expressing markers previously unseen by the immune system, are destroyed. Some of the altered cells not detected by this immune system vigil become cancerous. The existence of this generally silent surveillance operation is demonstrated by the development of tumors in immune-suppressed individuals. The occurrence of Kaposi's sarcomas and B-cell lymphomas in the immunosuppressed environment of the individual infected with HIV is an example. This case also points out the quite indirect role by which viruses might influence cancer development.

The immune system also plays a much more direct role in virus-mediated carcinogenesis through the action of the major histocompatibility gene complex of humans (known as the HLA system). The HLA genes code for proteins expressed on the cell surface. These molecules present foreign antigen to activate the effector components of the immune system and thereby initiate the host's defense to microbial or parasitic attack. The enormous polymorphism of the histocompatibility genes, with up to 60 alleles present at some loci, is thought to be a consequence of the continuous, and often invisible, evolutionary battle between organisms causing infectious disease and the vertebrate (in this case, human) host. The tremendous genetic variability, guaranteeing differences among most individuals, promotes the effectiveness of the recognition and control of an infectious agent in the host population. In the case of abnormal cell growth and differentiation, this variability influences the recognition of the novel antigens produced during the course of malignant transformation.

An integrated model for the association of virus, cancer, and HLA variation follows. Individual A has HLA genes expressing proteins that can recognize virally altered cells. The antigen that is actually recognized might be a viral product itself or a product of cell metabolism expressed abnormally, as a consequence of the virus infection. The effector components of the immune system, including cytotoxic T cells, specifically locate and destroy the altered cell in the infected tissue. A second individual, B, has a normally functioning immune system that is unable to effectively recognize and eliminate the altered cells. Some of these cells might survive to undergo the series of transformations leading to the cancerous state.

HLA system and cancer. In women, certain HLA alleles confer a higher risk of cervical carcinoma. However, the many tightly linked poymorphic loci of the HLA region make it difficult to assign the specific gene responsible for the effect. An animal model may help in studying this complex situation. For instance, genotypic variation in the major histocompatibility system of the rabbit is associated with both protection from and susceptibility to the cellular transformation caused by the rabbit papillomavirus. The rabbit virus-tumor system may be a useful experimental model for human cervical carcinoma.

In addition to its clear link with environmental stress (carcinogens) and Epstein-Barr infection, nasopharyngeal cancer is associated with HLA variation. A second cancer associated with Epstein-Barr virus, Hodgkin's lymphoma, also has HLA associations in Caucasians of European origin. Some alleles confer greater susceptibility, some appear protective, and others are neutral with respect to the risk of Hodgkin's disease. Hepatocellular carcinoma has also revealed associations with particular HLA alleles. The HLA associations may, however, reflect an indirect response to metabolic disturbances induced by hepatitis B virus cells, rather than a direct effect of the virus itself. This suggestion is supported by the fact that other diseases of the liver also caused by hepatitis B virus do not have HLA associations.   *William Klitz*

Papillomaviruses

Papillomaviruses are the causative agent of warts in both humans and animals. Warts are generally benign lesions that occur on the skin or on mucosal surfaces such as the oral region and genital tract. For many years, human warts were considered relatively harmless and only of cosmetic concern. Although the majority of warts are still considered harmless, there is increasing evidence that some human papillomavirus infections are potentially more serious. Beginning in 1982, the viruses have been isolated from a variety of cancers of the skin and mucosa. The presence of viral deoxyribonucleic acid (DNA) in these tumors raises the possibility that the virus is contributing to the formation of the malignancy. Confirmation of the role of specific human papillomaviruses in cancers has proven difficult, but epidemiological studies have now strongly linked cervical carcinoma to infection by certain sexually transmitted human papillomaviruses. Similar associations between human papillomavirus infections and other types of human cancers have also been noted, but the evidence is less compelling than for cervical carcinoma. It is likely, however, that additional studies will confirm a role for papillomaviruses in several types of human cancers, thus solidifying the status of human papillomaviruses as true human tumor viruses.

Viral life cycle. Papillomaviruses are members of the Papovaviridae family and consist of a small (approximately 8000 base pairs), circular, double-stranded DNA genome surrounded by an icosahedral protein coat. The DNA encodes the proteins necessary for reproduction of the virus, and the coat proteins protect the DNA from the environment. Infection is initiated when the virus encounters and penetrates a susceptible host cell. The viral DNA enters the nucleus of the cell and begins to express its proteins. Although not yet completely defined, human papillomavirus DNA appears to encode 10–15 different proteins. The viral proteins produced early after infection are referred to as E proteins, while the proteins expressed late after infection are referred to as L proteins. The E proteins regulate the interaction of the virus with the host cell and also direct viral DNA replication, while L proteins are structural, making up the viral coat.

Upon entering the host cell, the infecting DNA molecule is replicated numerous times to yield approximately 200 identical copies per cell. These copies remain as independent, circular DNA molecules in the nucleus and do not become part of the host-cell chromosome. After this initial amplification period, replication of the viral DNA becomes coordinated with cellular DNA replication. At this stage, the viral DNA molecules replicate only each time the cell DNA replicates. When the cell divides, the viral DNA molecules are distributed between the daughter cells, resulting in a fairly constant viral DNA copy number per cell. Viral DNA may persist in this relatively inactive state for long periods with little or no visible lesion; this stage is referred to as a latent or subclinical infection. Ultimately, the virus may be triggered to complete its replication process. Viral DNA synthesis becomes extensive again and the structural L proteins are expressed. This stage is known as productive infection and is generally associated with development of a detectable wart or lesion. The newly synthesized coat proteins and DNA molecules associate to form new viral particles, and the finished viruses are released to infect new cells.

Clinical distribution. Comparison of numerous human papillomaviruses isolated from different clinical lesions reveals that all human papillomaviruses are not identical. Although all are similar in size, structure, and genome organization, there is considerable diversity of the nucleotide sequence in the DNA of these viruses. By convention, any new isolate that differs by more than 50% from all known isolates (as measured by solution hybridization) is designated a new type. By using this scheme, more than 60 different human papillomavirus types have been identified, and the number continues to grow. Different types can be divided into three general groups based on the anatomical site of infection: skin, oral/nasal region, and genital region. There can, however, be significant overlap between the types found in each group, particularly for the oral and genital types. Within each group, most types are considered relatively benign, but some specific types have been found to have oncogenic potential. For example, among the more than 20 human papillomaviruses that can infect the genital mucosa, types 16 and 18 are most strongly associated with progression to cervical carcinoma. Approximately six to eight types are less commonly associated with cervical cancer, and the remainder of the genital types appear to pose little or no risk.

Transmission. Human papillomaviruses are transmitted from person to person (horizontal transmission) via direct contact or from infected objects such as clothing and jewelry. For typical skin infections, the virus must penetrate the protective layers of dead skin cells to reach the replicating basal cells of the epidermis. Conditions such as moisture and abrasion facilitate entry of the virus through the protective layer. Consequently, skin infections commonly occur on areas of the body that regularly suffer minor trauma such as the hands, feet, knees, and elbows. Nonetheless, human papillomavirus lesions can be found virtually anywhere on the surface of the skin. Spread of infection on the same individual (that is, self-inoculation) is also common. Once an infection has occurred, there is an incubation period ranging from one to several months before the wart may appear. Many infections, however, remain latent and probably never give rise to visible lesions. If a wart does appear, it can persist indefinitely, but may spontaneously regress within 2 years.

The transmission of genital warts occurs by three distinct pathways. The primary and most well-documented path is by sexual contact, with recent studies indicating that genital human papillomaviruses are the most frequently transmitted venereal disease. Estimates of the frequency of these infections among the general population vary widely, but it is likely that 30–40% of young adults carry some type of genital human papillomavirus. In addition to horizontal sexual transmission, there is good evidence for vertical transmission of the viruses from infected mothers; apparently, newborns may contract human papillomavirus infection during passage through an infected birth canal, accounting for at least some cases of the infection in very young children. However, in some cases no infection of the mother can be documented. Infection of newborns in these cases must come through nonsexual, horizontal transmission, although the conditions for this transmission are unknown.

Among the oral/nasal human papillomaviruses the laryngeal viruses have the same transmission pattern as genital ones. Laryngeal papillomas are typically caused by types 6 and 11, which are also frequently associated with benign genital lesions. Childhood onset of laryngeal papillomatosis results from infection during birth when the infant's oral region is exposed to infected maternal fluids. Adult onset likely results from oral sexual contact with an infected individual. Although progression to malignancy is uncommon for both childhood and adult onset forms of laryngeal papillomatosis, both cause significant illness and mortality due to physical obstruction of the airways. Transmission of other types of oral/nasal papillomaviruses is not as well as understood, but presumably also results from either maternal infection or direct contact.

Role in malignancy. The evidence for an association of human papillomaviruses with human cancer derives from epidemiological studies of human diseases and from molecular characterizations of the biologic and biochemical properties of the virus. It has long been known that cervical cancer had a venereal component; that is, increased numbers of sexual partners increase the risk for developing this disease. Over the years, numerous sexually transmitted microorganisms (such as *Neisseria gonorrhoeae, Treponema pallidum*, and herpes simplex virus) have been examined to determine if prior infection correlated with the subsequent occurrence of cervical carcinoma. To date, of all the venereal disease agents examined, only certain human papillomavirus types have consistently been found to be associated with cervical malignancy. Women with evidence of genital type 16 or 18 infection are at a tenfold greater risk of developing cervical carcinoma than uninfected women. When cervical carcinomas are directly tested, human papillomaviruses can be detected in more than 90% of the samples. Types 16 and 18 are the most commonly detected in these tumors, with types 31, 33, 35, 39, 45, 51, 52, and 56 found to a lesser extent. Other genital types, such as 6 and 11, are rarely found in malignant lesions, suggesting that only certain types possess a malignant potential. The viral DNA in malignant tissue is often found integrated into the host DNA, unlike the situation in nonmalignant lesions where the viral DNA does not integrate. The significance of this observation is unknown.

A role for human papillomaviruses in cervical cancer also is supported by studies of the virus in cell culture. Human papillomaviruses encode at least two proteins, E6 and E7, that can transform normal cells into ones with properties of malignant cells. Each of these viral proteins interacts with host-cell proteins and presumably disrupts the normal function of the cell proteins in growth regulation. Both E6 and E7 are typically expressed in cervical carcinomas, an observation that is consistent with a presumed role for these viral products in the development or maintenance of the neoplasia.

Although human papillomavirus infection is clearly a risk factor for development of cervical carcinoma, it is not the sole component of this disease; many women with type 16 or 18 infections never develop malignant disease. In addition, the time between infection and occurrence of the carcinoma is typically 10–20 years, suggesting that multiple events are occurring over a long period of time before the cancer forms. The factors responsible for the outcome of infection are relatively unknown at this time, and may include individual genetic susceptibility as well as exposure to other risk factors. For example, smoking and use of oral contraceptives have also been epidemiologically related to cervical cancer. Therefore, human papillomavirus infection may be a necessary but not sufficient factor in the development of malignant disease. Whether or not a particular individual develops cancer subsequent to infection would reflect the specific combination of factors to which that person was later exposed.

Treatment. To date, there are no drugs that eliminate the virus from infected host cells. Instead, treatment of papillomas requires the physical destruction of the infected tissue. Current procedures for this process include surgical excision, cryotherapy, treatment with toxic chemicals, and laser ablation. However, all of these procedures produce significant physical trauma in the patient, and are subject to recurrence of the lesion. Often the skin surrounding a visible lesion may also be latently infected, with no clinically observable sign. Wound healing following destruction of the visible papilloma may trigger surrounding latent infections to become active, yielding a recurrence of the lesion. Development of specific antiviral drugs or an effective vaccine would greatly facilitate the future treatment of these viruses and reduce the incidence of associated malignancies.

For background information SEE ANIMAL VIRUS; HERPES; SEXUALLY TRANSMITTED DISEASE; TUMOR; TUMOR VIRUSES; VIRUS CLASSIFICATION in the McGraw-Hill Encyclopedia of Science & Technology.

Van G. Wilson

Bibliography. E. L. Franco, Viral etiology of cervical cancer: A critique of the evidence, *Rev. Infect. Dis.*, 13:1195–1206, 1991; H. zur Hausen, Viruses in human cancers, *Science*, 254:1167–1173, 1991; M. Kaelbling et al., Loss of heterozygosity on chromosome 17p and mutant p53 in HPV-negative cervical carcinomas, *Lancet*, 340:140–142, 1992; W. Klitz, Immunogenetics: Viruses, cancer and the MHC, *Nature*, 356:17–18, 1992; H. Pfister (ed.), *Papillomaviruses and Human Cancer*, 1990; M. H. Schiffman, Recent progress in defining the epidemiology of human papillomavirus infection and cervical neoplasia, *J. Nat. Cancer Inst.*, 84:394–398, 1992.

Turbulence

Turbulence is a state of fluids (liquids, gases, and plasmas) in which the flow velocity and other associated physical quantities randomly vary in both space and time. In contrast, laminar fluid motion is characterized by well-organized flows with smooth variations in space and time. The transport coefficients, such as mass, momentum, and energy diffusivities, in a turbulent fluid are in general much enhanced over those in a laminar flow, and better understanding of turbulence-induced transport is one of the fundamental objectives of turbulence studies.

The transition from laminar to turbulent motion in a fluid usually involves the development of instabilities. For example, a water flow in a pipe becomes unstable when the flow velocity exceeds a critical value. (In fluid dynamics, the critical condition is expressed in terms of the Reynolds number.) As the flow velocity is further increased, a large number of eddies (vortices) are excited. A model for the onset of turbulence based on the collection of many harmonic oscillators was originally put forward by L. D. Landau. However, it is now well recognized that there are several alternative routes to turbulence. For example, a system of a few variables, which are interrelated through nonlinear equations, has been discovered to exhibit stochastic or turbulent behavior

(for example, the Lorenz model of atmospheric turbulence). Turbulence in low-dimensional systems is characterized by nonlinear instabilities in the phase space of the variables, namely, exponential growth in the separation distance of two trajectories. In the phase space, stochastic and laminar regions may coexist, leading to the manifestation of intermittent turbulence. In the intermittent turbulence, the probability distribution function is not normal (nongaussian), and the degree of deviation from gaussianity (kurtosis) is an important parameter in turbulence analysis.

Diffusion processes in turbulent fluids depend on the effective dimensionality of the geometrical structure associated with turbulence. The dimension often becomes nonintegral (a fractal dimension), and the variance of the migration distance, which is a measure of diffusivity, may not be linearly proportional to time as in classical brownian diffusion processes.

Weak and strong turbulence. When the amplitude of fluctuations is small, the fluctuating quantity (for example, the density) can be described by superposition of many Fourier components, n_k, as in Eq. (1), where ω_k is the mode frequency correspond-

$$n(x, t) = \sum_k n_k(x, t) = \sum_k N_k e^{i(kx - \omega_k t)} \quad (1)$$

ing to the wave number k, $|N_k|$ is the amplitude, and the summation of k is from $k_{\min} \simeq \pi/L$ (where L is the system size) to $k_{\max}$, at which turbulence decays significantly. Each mode independently satisfies the linear harmonic oscillation equation (2). Turbulence

$$\frac{\partial^2 n_k}{\partial t^2} + \omega_k^2 n_k = 0 \quad \text{or} \quad \frac{\partial n_k}{\partial t} + i\omega_k n_k = 0 \quad (2)$$

that can be described by superposition of individual harmonic modes is referred to as weak turbulence. The quasilinear and weak turbulence theories have been developed in the past under these restrictions.

As the amplitude of each harmonic oscillator increases, nonlinear coupling among them becomes more pronounced. The frequency of each harmonic is modified by nonlinearity, as in Eq. (3), where $\Delta\omega_k(t)$ is a

$$\frac{\partial n_k}{\partial t} + i(\omega_k + \Delta\omega_k)n_k = 0 \quad (3)$$

nonlinear frequency shift, which is a complex, time-dependent quantity. Its magnitude increases with the amplitude of harmonics, and Eq. (3) no longer describes a simple harmonic oscillation. Weak turbulence corresponds to the case $|\Delta\omega_k| \ll \omega_k$, and strong turbulence is characterized by $|\Delta\omega_k| \lesssim \omega_k$. In strong turbulence, the frequency of a given spatial harmonic k becomes ill-defined because of effective broadening about the linear frequency ω_k. The wave nature that is manifested in the weak turbulence limit may be destroyed, and often coherent macroscopic structures emerge from strong turbulence. This process is called self-organization and plays important roles in the realization of highly nonequilibrium, yet macroscopically stable states. Examples are Langmuir solitons and drift-wave vortices in magnetically confined plasmas.

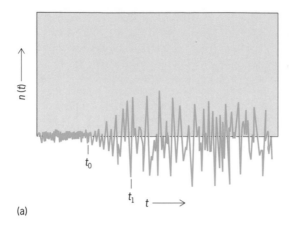

(a)

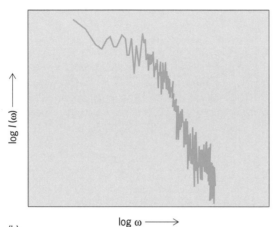

(b)

Fig. 1. Electron density perturbation at a tokamak edge. (*a*) Plot of density perturbation *n* versus time *t*. An instability sets in at *t* = *t*$_0$, grows during the period *t*$_0$ < *t* < *t*$_1$, and enters a saturated, quasisteady, turbulent phase at *t* > *t*$_1$. (*b*) Frequency spectrum of the density perturbation during *t* > *t*$_1$. The plot is on a log-log graph and shows log *I*(ω) versus log ω, where *I*(ω) is the intensity of the component at frequency ω. The spectrum in the high-frequency region follows a power law, *I*(ω) $\propto$ $\omega^{-2.6}$ in this example.

Figure 1 shows the time evolution of the electron density fluctuations observed at the edge of a tokamak discharge. (A tokamak is a toroidal magnetic fusion device. A high-temperature, high-density plasma can be produced and confined with the aid of a large toroidal current, and the confinement times in large tokamak devices are approaching that required in fusion reactors.) At time t_0, an instability is triggered. After exponential (or sometimes algebraic) growth, the instability becomes saturated and a steady turbulent state is reached. During the growth period, the frequency spectrum is relatively narrow and is dominated by the fastest-growing modes. After saturation, the spectrum is broadened considerably, and even those modes that are linearly stable can be excited through nonlinear couplings.

Turbulence in plasmas. As in ordinary gases, local plasma parameters (such as plasma density and electric current) fluctuate irregularly even when a plasma is in thermal equilibrium, which is characterized by maxwellian velocity distributions of electrons and ions with negligible spatial gradients. Of-

ten, however, plasmas in the laboratory and space are far from thermal equilibrium primarily because plasmas are produced and maintained by injecting a large amount of energy into a finite volume of gas. The amplitude of fluctuations can be enhanced, by orders of magnitude, from the intrinsic thermal level. Such plasmas are referred to as turbulent. Since density and current fluctuations cause corresponding fluctuations in the electromagnetic fields, the particle motion in a turbulent plasma is much more irregular and random than in a quiescent plasma, which can cause enhanced (anomalous) transport of particles and thermal energy in both configuration and velocity space. One principal objective of studying plasma turbulence is to elucidate how turbulence affects the transport coefficients, such as the particle and thermal diffusivities.

Plasmas can support various electromagnetic waves. In thermal equilibrium, the wave amplitude remains at a small level determined through competition between thermal excitation and dissipation. In nonequilibrium plasmas, some of the waves can become unstable, and their amplitude grows with time. The growth based on linear analysis continues until nonlinear effects set in, and the amplitude becomes saturated. Turbulence in plasmas is in general the asymptotic nonlinear stage of linearly unstable waves and is characterized by a continuous, broad-frequency spectrum.

Plasma instabilities can be classified into two major types, configuration-space instabilities and velocity-space instabilities. The configuration-space instabilities are driven by the spatial nonuniformity of plasma density, temperature, and current. Any plasma that is spatially confined must have a pressure gradient, which provides a source of free energy for the growth of instabilities. In magnetically confined plasmas, the so-called drift instability is a likely candidate for the agent responsible for the anomalously short confinement times in tokamaks, stellarators, and other devices. The drift instability (and its variations) is driven by the pressure gradient, and its frequency and growth rate are characterized by the diamagnetic drift frequency, given by Eq. (4), where $K_B = 1.38 \times 10^{-23}$ J/K is the

$$\omega_* = \frac{K_B T}{eBL_n} k_\perp = V_* k_\perp \qquad (4)$$

Boltzmann constant, T (K) is the plasma temperature, $k_\perp$ (m^{-1}) is the wave number perpendicular to the magnetic field B (T), $e = 1.6 \times 10^{-19}$ C is the electronic charge, L_n (m) is the density-gradient scale length, and $V_* = K_B T/(eBL_n)$ (m/s) is the diamagnetic drift velocity. **Figure 2** shows the propagation direction of the electron diamagnetic wave in a magnetically confined plasma. In toroidal devices (such as tokamaks), several drift-type instabilities have been predicted with frequencies and growth rates of the order of the diamagnetic frequency. The drift instability is a typical configuration-space instability and occurs even when the velocity distribution functions of electrons and ions are maxwellian.

In contrast, the velocity-space instabilities are caused by velocity distributions strongly deviated from maxwellian. For example, when an electron beam is injected into a plasma, the beam interacts with the plasma electrons and excites the Langmuir wave at the plasma frequency given by Eq. (5), where

$$\omega_{pe} = \sqrt{\frac{ne^2}{m\epsilon_0}} \qquad (5)$$

$m = 9.1 \times 10^{-31}$ kg is the electron mass and $\epsilon_0 = 8.85 \times 10^{-12}$ F/m is the vacuum permittivity. The instability has a large growth rate and rapidly enters the nonlinear regime in which ion dynamics becomes important too. Another well-investigated example is the current-driven ion instability. When electrons are forced to drift relative to ions (for example, by applying a large electric field), ion waves characterized by the propagation velocity c_s (the ion acoustic velocity), given by Eq. (6), are excited; here T_e is the electron

$$c_s = \sqrt{\frac{K_B T_e}{M}} \qquad (6)$$

temperature and M is the ion mass. The instability is known to enhance, by orders of magnitude, the plasma resistivity over the classical resistivity that prevails in a quiescent plasma, and efficient plasma heating has been realized through anomalously large plasma resistivity. This process is call turbulent heating.

Anomalous transport in plasmas. In magnetic confinement fusion research, it has long been recognized that the energy confinement times are substantially shorter than that predicted from the classical theory based on the Coulomb collisions. Experimental observations in tokamaks and stellarators clearly reveal the presence of drift-type, low-frequency density fluctuations well above the thermal level, and it is generally accepted that they are responsible for the anomalous particle and thermal diffusivities.

The diffusivity predicted from the electrostatic, long-wavelength drift-type instabilities is of the order of the Bohm-like diffusivity, given by Eq. (7). [The original form proposed by D. Bohm is Eq. (8), which was

$$D_\perp = \frac{K_B T}{eB} \frac{1}{k_\perp L_n} \qquad (7)$$

nal form proposed by D. Bohm is Eq. (8), which was

$$D_\perp = \frac{1}{16} \frac{K_B T}{eB} \qquad (8)$$

based on dimensional analysis.] The diffusivity in Eq. (7) contains the diamagnetic drift velocity V_* defined in Eq. (4), and is in the form $D =$ (turbulent velocity) $\times$ (correlation length $1/k_\perp$). For short-wavelength, electromagnetic drift modes, the diffusivity takes the form of the Rudakov-Galeev diffusivity, given by Eq. (9), where $c = 3.0 \times 10^8$ m/s is the

$$D_\perp = \left(\frac{c}{\omega_{pe}}\right)^2 \frac{c_s}{L_n} \qquad (9)$$

speed of light. The quantity c/ω_{pe}, the skin depth in a collisionless plasma, is a measure of electromagnetic field penetration, and in Eq. (9) this quantity plays the role of random walk distance for the diffusion process, and c_s/L_n gives an estimate of the growth rate that

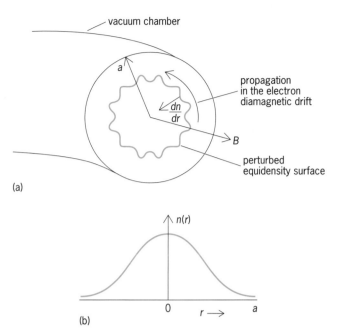

(a)

(b)

Fig. 2. Drift instability in a magnetically confined plasma. (a) Cross section of a plasma, showing a perturbed equidensity surface and its propagation in the direction of the electron diamagnetic drift. Here a = radius of chamber, B = magnetic field, dn/dr = rate of change of density n with respect to radial distance r. **(b)** Radial profile of the density of an unperturbed plasma.

may be taken as the effective decorrelation rate.

Diffusion in velocity space. As the drift instability causes particle and thermal diffusion in configuration space, velocity-space instabilities cause anomalous diffusion in the velocity space. In the one-dimensional case, the process can be described by the diffusion equation for the velocity distribution function $f(v, t)$, given by Eq. (10), where $D(v, t)$ is the diffu-

$$\frac{\partial}{\partial t} f(v, t) = \frac{\partial}{\partial v}\left(D(v, t)\frac{\partial f}{\partial v}\right) \qquad (10)$$

sivity in the velocity space. In weak turbulence, $D(v)$ is time-independent and proportional to the turbulence energy density, given by Eq. (11), where E_k is the

$$W = \sum_k \frac{1}{2}\epsilon_0 |E_k|^2 \qquad (11)$$

Fourier component of the fluctuating electric field. However, as the amplitude E_k increases, the linear relation $D \propto W$ no longer holds, and the exponent α in the dependence $D \propto W^\alpha$ becomes fractional. More important, in strong turbulence, the diffusivity becomes time-dependent and decreases with time. Recent theoretical studies on the velocity diffusion in one-dimensional turbulent electric fields have revealed the dependence on W and time t given by Eq. (12), where $\alpha = 1/2 \sim 2/3$ and $\beta = 1/3 \sim 1/2$, and a

$$D = \text{const}\frac{W^\alpha}{t^\beta} \qquad (12)$$

corresponding velocity variance proportional to $t^{1-\alpha}$, in contrast to the usual brownian diffusion process in which the variance increases linearly with time. The subdiffusive nature has been attributed to the nonmarkovian statistics of motion of charged particles even in the relatively simple case of a one-dimensional turbulent field.

For background information *SEE DIFFUSION; FOURIER SERIES AND INTEGRALS; FRACTALS; HARMONIC OSCILLATOR; PLASMA PHYSICS; TURBULENT FLOW; WAVES AND INSTABILITIES IN PLASMAS* in the McGraw-Hill Encyclopedia of Science & Technology.

A. Hirose; A. Smolyakov; O. Ishihara

Bibliography. A. Aharony, Percolation, fractals, and anomalous diffusion, *J. Statist. Phys.*, 34:931–939, 1984; H. de Kluiver, N. F. Perepelkin, and A. Hirose, Experimental results on current-driven turbulence in plasmas, *Phys. Rep.*, 199:281–381, 1990; O. Ishihara, H. Xia, and A. Hirose, Resonance broadening theory of plasma turbulence, *Phys. Fluids B*, 4:349–362, 1992; E. N. Lorenz, Deterministic nonperiodic flow, *J. Atm. Sci.*, 20:282–293, 1963.

Underground mining

Maintaining an opening in an underground mine can be difficult. Once an opening is formed, the surrounding rock will usually move and, in some cases, fall. Accident statistics show that approximately 35% of all accidents in underground metal mines are associated with rock falls.

Stabilization. The stability of the rock around an underground opening depends on many factors, such as the strength of the rock, the depth of the opening, and the number and size of fractures in the rock. To stabilize the rock, miners use artificial supports such as rock bolts, which are steel rods anchored into the rock (**Fig. 1**).

Cable bolts. Miners have also begun to use flexible steel cables, known as cable bolts, grouted into drill holes to reinforce the rock (**Fig. 2**). Generally the steel cables consist of seven wires, have a load-carrying capacity of approximately 58,000 lb (26,309 kg), and are 0.6–0.625 in. (15.24–15.875 mm) in diameter. Because of its flexibility, the steel cable can be easily bent and pushed into the drill hole, allowing a 60-ft-long (18.3-m) cable bolt to be installed in an opening with less than 8 ft (2.4 m) of headroom.

Cable bolts are an established ground-control technique in the mining industry, permitting a variety of approaches to reinforcing rock masses. They are easy to install, can be placed at any angle in the rock, and can be very long. Properly installed, cable bolts provide a safe and effective ground-control system under many mining conditions.

Various techniques exist for installing cable bolt supports. A typical procedure for installation in a vertical hole involves drilling a hole about 2.25 in. (57 mm) in diameter to a desired length and then inserting the cable into the hole along with a plastic breather tube 0.5 in. (12.70 mm) in diameter (Fig. 2). Then a grout tube is inserted into the hole, the end of the hole is plugged, and grout is pumped into the hole through the grout tube. The air being displaced by the grout is forced out through the

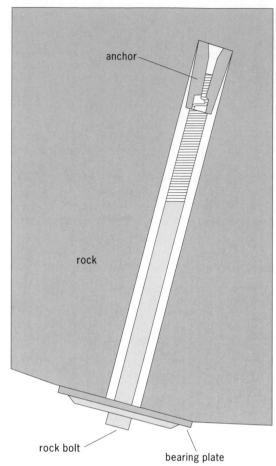

Fig. 1. A rock bolt.

breather tube. The hole is completely filled when the air stops flowing. The grout begins to harden in about 2 h.

Cut-and-fill stoping. **Figure 3** illustrates how cable bolts are applied during an underground mining operation known as cut-and-fill stoping. Long cable bolts are installed at specified intervals in a fan-shaped configuration that supports the rock in the roof as well as in the hanging wall. Next, blast holes are drilled into the rock and charges are set off, thereby breaking the rock. Miners remove the ore, fill the stope with sand (backfilling), cut the ends of the cables exposed by the blast, and install conventional rock bolts. As a result of shock from the blast, a zone of fractured and weak rock several feet thick usually extends just above the opening. However, because cable bolts have been installed, the rock is supported and the area remains safe as mining continues. The cable bolts in the hanging wall also help prevent rock that contains little or no ore from falling on the pile of high grade ore and diluting it. This process of removing the ore in layers (lifts) continues for four or five lifts, and then another set of cable bolt supports is installed.

Variables. Many variables are associated with the use of cable bolt supports, such as the number, type, and length of cables used; the type of group; and the type of rock being supported. A single cable in each hole allows large amounts of rock to deform, thereby redistributing stress in the rock to the walls on each side of the opening. The use of two cables in each hole, however, provides very high load-carrying capacities at low levels of displacement and can restrict rock movement.

Cable bolt supports have been tested to understand what impact these variables have on the strength of the bolts. For example, researchers were interested in the relationship between the zone of fractured and weak rock left after blasting in a cut-and-fill operation and the load-carrying capacity of bolts. Laboratory pull tests were conducted on cable bolts at various embedment lengths corresponding to the fracture zone of rock. The results showed that for embedment lengths between 8 and 32 in. (0.20 and 0.81 m), the load-carrying capacity can be represented by the equation below, where P is the load-carrying capacity and L

$$P = 5537 + 1244 \times L$$

is the embedment length. Knowing how much a cable bolt can carry for a given rock thickness and weight is important in determining how far apart to place cable bolts.

Additional laboratory tests showed that using two cables in a single hole can be beneficial because the maximum load-carrying capacity for double cable bolts is achieved at shorter displacement lengths than for single cables. Because double cable supports are stiffer than single cable supports, they allow less rock movement.

The grout used for cable bolts consists of portland cement and water at ratios of 0.3–0.45 part of water to 1 part of cement by weight. The amount of water used greatly affects the strength of the grout, which in turn influences the load-carrying capacity of the cable bolt. Laboratory pull tests showed that bolts installed with a grout having a water:cement ratio of 0.3:1 had approximately an 86% higher load-carrying capacity

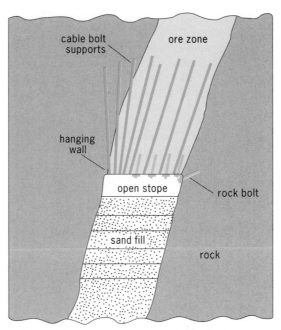

Fig. 3. Application of cable bolts during cut-and-fill stoping.

than bolts installed with a grout having a water:cement ratio of 0.45:1. Grouts with lower water:cement ratios also helped to reduce the separation of water from the cement; this process, known as bleeding, can weaken the strength of a support system. A major disadvantage of using grouts with lower water:cement ratios is that they are stiff and therefore are difficult to pump into a hole, necessitating the use of strong pumps or a chemical additive to increase pumping efficiency.

The breather tube used during the placement of cable bolts is left in the hole and becomes part of the support. Laboratory tests showed that the presence of the breather tube does not influence the strength of the bolts, provided that grout remained in the tube after the bolt had been installed. In addition, tests showed that the size of the breather tube does not influence the strength of the cable bolt.

The laboratory evaluations demonstrated that it is important for engineers designing rock support systems to recognize and understand the impact of each component of a cable bolt support on the entire support. As many beneficial properties as possible should be incorporated into a support system.

For background information SEE ROCK MECHANICS; UNDERGROUND MINING in the McGraw-Hill Encyclopedia of Science & Technology.

John M. Goris

Bibliography. J. M. Goris, Laboratory evaluation of cable bolt supports: I. Evaluation of supports using conventional cables, II. Evaluation of supports using conventional cables with steel buttons, birdcage cables, and epoxy-coated cables, U. S. Bureau of Mines, RI 9308 and RI 9342, 1990; J. M. Goris, F. Duan, and J. Pfarr, Evaluation of cable bolt supports at the Homestake Mine, *Can. Min. Metallurg. Bull.*, 84 (947):146–150, March 1991.

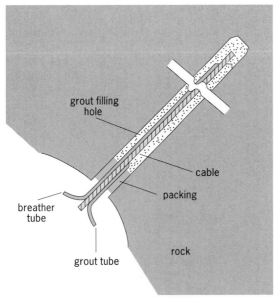

Fig. 2. Installation of a cable bolt support.

Underwater acoustics

An area of active investigation in underwater acoustics is the development of capable, efficient, and accurate computer-implemented methods for determining the propagation of a signal through realistic ocean environments. The propagation problem is central to traditional sonar applications, including resource discovery and source localization as well as newer applications such as acoustic tomography and ocean monitoring. Advances in the capabilities and use of computers in attacking the formidable propagation problem have led to progress in new directions.

Complexity of propagation. An acoustic signal is a pressure disturbance generated by a source in the ocean volume or on its boundaries. For most underwater acoustics situations, an appropriate mathematical model is that the pressure, as a function of three spatial coordinates and time, satisfies a classical wave equation throughout the ocean volume. General properties and special solutions are known for this linear, second-order, partial differential equation. However, the sound speed appearing in it can vary in a complicated way with all three spatial coordinates and time, because of changes in ocean temperature, salinity, and pressure. Additionally, waves and wave-induced bubble layers near the ocean surface can affect the pressure field. Near the ocean bottom, strong influences can arise from variability in the shape of the water–sediment interface and in the bottom itself, including coupling to other wave mechanisms occurring in the bottom. Consequently, determination of the pressure field as a function of four variables can be very complicated.

Monte Carlo method. This discussion concentrates on propagation problems for so-called deterministic models, in which the sound speed and other ocean quantities are specified through data or simplifying assumptions. These deterministic models contrast with stochastic models, for which randomness in the variabilities is somehow incorporated. One method of treating random propagation problems is the Monte Carlo (simulation) method: Random inputs are given representative values within their anticipated ranges, and the resulting problems are solved by using a computationally based propagation method. Results are suitably averaged in order to get predictions for the statistics of the stochastic model solutions. The success of this procedure relies critically on the ability to use computers efficiently to obtain solutions to each of the sample problems.

3-D propagation problem. The usual approach to finding the pressure field is to consider the propagation of one acoustic frequency at a time. This approach is useful for sources that broadcast at a single frequency, or for those with signals that have been Fourier-analyzed to permit identification of specific frequencies of interest. Since sound travels rapidly in the ocean, time variations in the sound speed can usually be neglected during transmissions from source to receiver. Provided that variations with time at or near the ocean surface are also neglected, the pressure field associated with any frequency is a function of three spatial variables and can be shown to satisfy the Helmholtz second-order, partial differential equation. This equation, along with information about the ocean bottom, a condition at the ocean surface, and required specifications concerning the source, represents the three dimensional (3-D) propagation problem to be solved for any frequency component of the acoustic pressure. One way that modern computing power can contribute is immediately apparent. If an approximate solution to the problem can be obtained by employing a high-speed computer, approximate solutions can also be found for other frequencies. The results from these computations can be Fourier-synthesized, in accord with the distribution of source frequencies, to produce the time behavior of received pressure that results from a multiple-frequency source. Just as for Monte Carlo simulations, this scheme is ideal for parallel processing, which is of extraordinary importance in advanced scientific computation.

Mechanisms of 3-D variation. Propagation problem input with three-dimensional variations can arise from several mechanisms.

The first mechanism is three-dimensional changes in the shape of the water–bottom interface. This mechanism has significant effects in many situations, including both shallow and deep ocean channels and most frequencies of interest. The importance of the mechanism in any situation depends on scales and other features of the propagation. For example, an ocean seamount may substantially change long-range, low-frequency sound transmission over it while negligibly affecting sufficiently short-range, high-frequency propagation. The opposite conclusions may be true for a patch of bottom roughness with scales of a few meters.

A second mechanism is three-dimensional changes in the structure of the ocean bottom itself. Horizontal variations in the sediments that constitute the water–bottom interface can influence acoustic propagation. Moreover, except for cases such as an extremely hard bottom, acoustic waves typically penetrate to a depth of up to several wavelengths. Horizontal changes in bottom structure that affect sound speed, density, and attenuation properties to these depths should not be ignored. Horizontal bottom variability can arise, for example, from different formation patterns of geological deposits.

A third mechanism is three-dimensional sound-speed variations throughout the ocean volume, occurring on scales ranging from ocean front and eddy systems (hundreds of kilometers) down to instability processes (scales of meters). These oceanographic effects have significant thermal signatures, and by this means they affect sound speed. Their influence on propagation has been the subject of considerable experimental and theoretical study.

A fourth mechanism is three-dimensional variability near the ocean surface. The rapid variations of ocean–air processes mean that these are usually included by means of stochastic models. Still another mechanism is from horizontally nonuniform sound-source output.

Propagation methods. Capable propagation methods (that is, methods for solving the propagation problem) and efficient computational implementations have proven essential for investigation of these mechanisms. A variety of such procedures exist, many more than the two that were mainstays through the mid-1970s. These latter two procedures, geometrical (ray) acoustics and normal modes, have traditional limitations, respectively, to high-frequency problems and to problems with variability only in depth.

Extensions of older methods. Modern computing power has permitted extensions of both these older methods. For example, new implementations of geometrical acoustics permit treatment of three-dimensional ocean variability, are efficient even for long-range calculations, and approximately account for realistic frequency effects. Other recent implementations based on normal modes account for three-dimensional variability by decomposing the problem domain into many subdomains, which are chosen so that efficient normal-mode methods can be applied on each one. Computational power is essential both for finding the solutions efficiently and for combining them.

Parabolic equation methods. Among newer propagation procedures, those receiving the widest acceptance are known as parabolic equation methods. Their first appearance, in the 1940s, was for atmospheric electromagnetic propagation, but adoption was handicapped by the absence of sufficiently powerful computers. The key idea is that for any selected horizontal line, the solution to the Helmholtz equation contains waves that propagate in both directions. A parabolic equation is based on a modification of the Helmholtz equation to account for propagation in only one of the directions along the selected line. The parabolic label is based on terminology that characterizes partial differential equations of similar type, as opposed to the Helmholtz equation, which is elliptic. (This terminology, in turn, has a connection with properties of conic sections.)

Unlike traditional but much less capable ray or normal-mode methods, the effectiveness of parabolic equations is based entirely on the availability of high-speed computers to perform pressure-field calculations. Parabolic equations also differ from earlier methods in being useful for low-frequency propagation in ocean environments with three-dimensional variations. Since their introduction to underwater acoustics, parabolic equations have proven extraordinarily successful in treating complicated situations. They have undergone many extensions of their capabilities, and implementations have been improved in accuracy and efficiency. At least three versions have demonstrated their capabilities for treating certain three-dimensional ocean variations, and continued advances are expected.

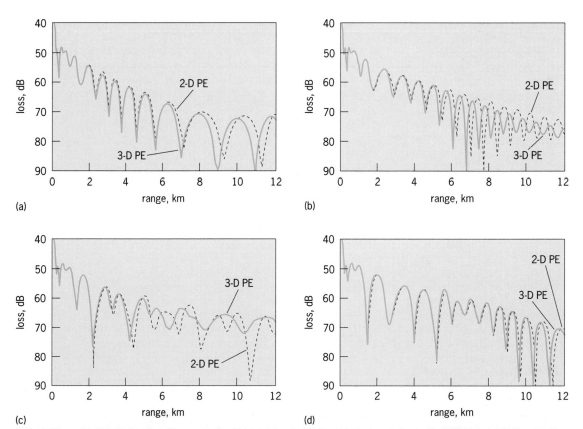

Calculations of transmission loss in an ocean with a corrugated bottom, based on a three-dimensional parabolic equation (3-D PE) and a two-dimensional parabolic equation (2-D PE). The loss is expressed in decibels with respect to the sound intensity at a distance of 1 m (3.3 ft) from the source. 1 km = 0.6 mi. (*a*) $\theta = 60°$. (*b*) $\theta = 80°$. (*c*) $\theta = 100°$. (*d*) $\theta = 120°$. (*After M. D. Collins and S. A. Chin-Bing, A three-dimensional parabolic equation model that includes the effects of rough boundaries, J. Acoust. Soc. Amer., 87:1104–1109, 1990*)

If past progress is a guide, future improvements will develop in concert with better numerical algorithms and more powerful computers.

Azimuthal coupling problem. Results from solving three-dimensional propagation problems provide new insights on the importance of the mechanisms described previously. In addition, significant scientific issues can be addressed. For example, propagation can be described by using a cylindrical coordinate system where z is depth, r is the distance in the horizontal propagation direction, and θ is the angle in the azimuthal cross-propagation direction. A three-dimensional propagation situation in which net acoustic energy transfer occurs across one or more vertical planes, θ = constant, is called azimuthally coupled. When no net energy transfer occurs across any vertical plane, the propagation is called azimuthally uncoupled (or N × 2-D). The question is, for which three-dimensional mechanisms does azimuthally coupled propagation occur, or, equivalently, when is a method with azimuthal coupling capability necessary for accurate predictions? The resolution of this question has many implications for applications such as accurate comparisons between experimental data and model simulations. Results so far for lower-frequency propagation indicate that three-dimensional variations at and below the bottom interface often require incorporation of azimuthal coupling, while three-dimensional sound speed variations generally do not. For example, calculations of transmission losses have been performed in a model ocean, having a shallow isospeed channel, a flat upper surface, a corrugated bottom, and an isospeed sediment level below. The **illustration** shows that solutions based on a three-dimensional parabolic equation differ significantly from those based on a two-dimensional parabolic equation, which neglects azimuthal coupling. (The large difference between the solutions for θ near 90° arises because energy gets trapped and channeled in the deep parts of the corrugated bottom.) However, the issue has not been resolved, particularly in terms of quantifying acoustically critical magnitudes of three-dimensional environmental variations.

Applications. New applications of three-dimensional propagation methods to underwater acoustics problems will occur as capabilities, algorithms, and computers continue to improve. Two areas that have already experienced applications are matched field processing and the development of ocean prediction models.

Matched field processing. In this source localization method, many possible sound source positions are selected and corresponding propagation calculations are performed by using a model for the actual ocean environment. Predictions from the simulations are correlated with data from the actual source, from which an optimal estimate for the source location is determined. This method is obviously computationally intensive and is a natural beneficiary of improvements in three-dimensional propagation methods.

Ocean prediction models. A serious limitation to the accuracy of three-dimensional propagation cal-

culations for any ocean region is the absence of adequate three-dimensional environmental data. Ocean prediction models seek to remedy part of this problem by using up-to-date observations along with oceanographic databases and feature models to provide sound-speed values over a wide area. The coupling of three-dimensional acoustic propagation methods to ocean prediction models is in its infancy but holds promise of substantial benefits for both physical oceanography and ocean acoustics.

For background information SEE DIFFERENTIAL EQUATION; MONTE CARLO METHOD; SUPERCOMPUTER; UNDERWATER SOUND in the McGraw-Hill Encyclopedia of Science & Technology.

William L. Siegmann

Bibliography. M. D. Collins and S. A. Chin-Bing, A three-dimensional parabolic equation model that includes the effects of rough boundaries, *J. Acoust. Soc. Amer.*, 87:1104–1109, 1990; W. A. Kuperman, M. B. Porter, and J. S. Perkins, Rapid computation of acoustic fields in three-dimensional ocean environments, *J. Acoust. Soc. Amer.*, 89:125–133, 1991; D. Lee, G. Botseas, and W. L. Siegmann, Examination of three-dimensional effects using a propagation model with azimuth-coupling capability (FOR3D), *J. Acoust. Soc. Amer.*, 91:3192–3202, 1992; D. Lee, A. Cakmak, and R. Vichnevetsky (eds.), *Computational Acoustics: Ocean-Acoustic Models and Supercomputing*, 1990.

Vapor condenser

When vapor is placed in contact with a wall that is subcooled below the local vapor saturation temperature, condensation will occur. The resulting liquid that is formed, the condensate, will collect in one of two ways. If the liquid wets the cold surface, the condensate will form in a continuous film; this mode is referred to as film condensation. If the liquid does not wet the cooled wall, the condensate will form into numerous discrete droplets; this mode is called dropwise condensation. Although all vapor condensers are now designed to operate in the filmwise mode, it is well known that condensation rates in them could be dramatically improved if dropwise conditions occurred. For example, for steam condensers, overall heat transfer rates could be doubled if dropwise conditions could exist. Dropwise designs are not yet commercially viable because nonwetting, drop-forming surfaces tend to lose their effectiveness over time.

Dropwise condensation mechanism. Dropwise condensation has been studied since about 1930. During this time, much progress has been made in understanding how it can be achieved and maintained, but so far there has been little success in putting this mode of condensation into practice. Dropwise condensation is a stochastic process involving a series of randomly occurring subprocesses as droplets grow, coalesce, and depart from a cold surface. The sequence of these subprocesses forms a dynamic life cycle. The cycle begins with the nucleation of microscopic droplets at discrete locations on the condenser surface. These

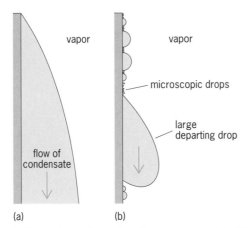

Fig. 1. Comparison of condensation processes. (*a*) Film condensation. (*b*) Dropwise condensation.

droplets grow very rapidly because of intense condensation of vapor on them, and they begin to merge with neighboring droplets, constantly shifting in position in the process. As a result, rapid local surface temperature fluctuations occur. This active growth and coalescence continues until larger drops are formed. Condensation contributes relatively little to the larger drops, but they continue to grow because of coalescence with neighboring smaller droplets. Eventually, these large "dead" drops will merge to form a very large drop whose size is sufficient that the adhesive force due to surface tension is overcome either by gravity or by vapor shear. This very large drop then departs from the surface, sweeping away all condensate droplets in its path and allowing fresh microscopic droplets to begin to grow again and start another cycle.

Figure 1 shows some relative drop sizes during the dropwise condensation life cycle as compared to a representative film thickness during filmwise condensation. The size of the microscopic active drops creates little conduction resistance in comparison to the relatively large thermal resistance across the condensate during film condensation. However, if the departure drop size is too large, the effectiveness of this mode of condensation diminishes. It is well known that the heat transfer coefficient in dropwise condensation falls off with increasing departure drop diameter, so keeping the departure drop diameter as small as possible

is important. This goal is usually accomplished with high vapor shear forces (that is, high vapor velocities). **Figure 2***a* shows the typical behavior of steam during dropwise condensation on a silver-plated horizontal tube.

The dropwise condensation process is very similar to the nucleate boiling process, where bubbles form at discrete nucleation sites in a similar, cyclic process. As a result, in recent years several investigators have demonstrated that there exists a condensation curve similar to the well-known boiling curve. **Figure 3** shows a typical condensation curve for propylene glycol vapor condensing on a vertical octadecanethiol-treated copper block. The heat flux increases linearly in the dropwise region at low surface subcoolings. The inset at point *A* shows the appearance of the surface during "perfect" dropwise conditions. It is evident that as the large drops sweep across the surface, numerous new microscopic droplets begin to grow. In the transition region, the rate of heat flux increase becomes smaller as the condensate begins to show signs of rivulet formation (point *B*). A maximum heat flux is reached beyond which further increases in subcooling reduce performance, until at point *C* a minimum heat flux is reached with the onset of film condensation, where a continuous film of condensate covers the surface. The exact shape of this condensation curve will depend on a variety of operating conditions such as operating pressure and the surface condition of the condenser wall, and a hysteresis pattern may develop depending on whether the subcooling is increasing or decreasing. Thus, under certain circumstances film conditions may exist at subcoolings below point *C* (indicated by the broken line in Fig. 3).

Use of promoter surfaces. In order for dropwise conditions to occur, the condensate must be prevented from wetting the solid surface. Therefore, the surface tension of the liquid must be greater than the critical surface tension of the solid surface, which is defined as the surface tension at which the contact angle between the liquid and the solid is zero. [The contact angle is defined as the angle between the liquid–vapor interface and the solid surface, as measured through the liquid at a point where all three phases (solid, liquid, and vapor) meet.] Water has one of the highest surface tensions of ordinary liquids,

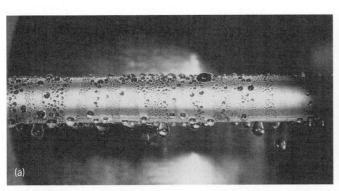
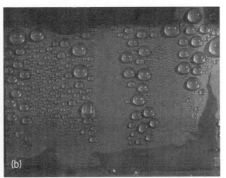

Fig. 2. Dropwise condensation of steam. (*a*) Condensation on a silver-coated, copper-nickel tube. (*b*) Condensation on a Teflon-coated, copper surface after 800 h of operation.

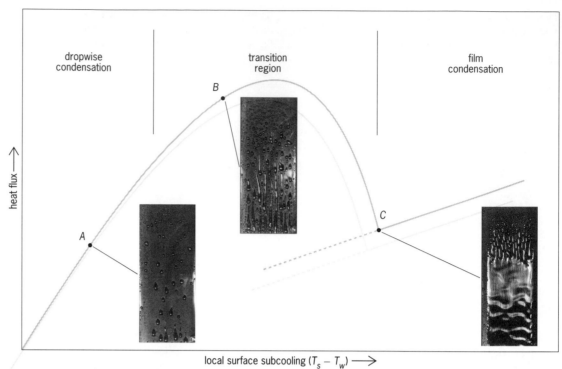

Fig. 3. Typical condensation curve. Insets show the appearance of the surface at points *A*, *B*, and *C* on the curve. T_S = vapor saturation temperature; T_w = wall temperature. (*After Y. Utaka et al., The effect of condensate flow in the dropwise condensation curves, ASME-JSME Thermal Eng. Proc., 2:451–458, 1991*)

yet water generally wets most metal surfaces. The implication is that metal surfaces have high critical surface tensions, and dropwise condensation of steam will occur only if a suitable promoter is used to create a nonwetting condition on the surface. These promoters are generally organic materials that are either injected directly into the vapor or deposited onto the condenser surface. The **table** lists critical surface tensions of some representative solid surfaces. Since the surface tension of water (68 millinewtons per meter at 50°C or 122°F) is larger than these critical surface tensions, water will not wet these surfaces but will form drops. Thus, steam condensation on these solid surfaces will occur in the dropwise mode.

In recent years, numerous investigations have been carried out to find a steam promoter that will last under industrial conditions. These so-called permanent coatings include noble metals (silver and gold), organic polymers (Teflon, Parylene, and other fluorinated organic molecules), inorganic compounds (metal sulfides), and alloys (copper-chromium and copper-iron). Recent research using these promoters has been

encouraging, bringing their use closer to reality. However, no one has demonstrated yet that these promoters can be applied in a reliable, cost-effective way to last for durations of several years or more. Organic coatings may show the most promise. However, because of their poor thermal conductivity, these coatings must remain ultrathin (of the order of 1 micrometer or less), yet be thick enough to create a strong adherent bond with the cooled wall. Figure 2*b* shows dropwise condensation of steam on a Teflon-coated, smooth copper surface after 800 h of operation. The appearance of mixed dropwise and film condensation in the lower right indicates either that the film is being removed locally or that the very thin coating of Teflon absorbs sufficient water with time to reduce its effectiveness. New methods of forming ultrathin polymer coatings are being developed continually and could lead to a dramatic breakthrough in this technology in the future.

Theoretical studies. The theory of dropwise condensation heat transfer is not as well developed as theoretical aspects of film condensation. Although understanding of the various dropwise mechanisms has improved significantly since the 1960s, many questions involving drop formation, growth, and departure remain because of the randomness of the processes.

The first comprehensive dropwise condensation heat transfer theory in 1966 used a quasiequilibrium approach. Consideration was given to the heat transfer through a single drop of a given size, and the distribution of drop sizes on a surface. The average heat transfer rate for a surface was then determined by integrating the individual drop heat transfer rates over all drop sizes. The analysis included an average con-

Critical surface tensions

Solid	σ_{cr}/ mN/m
Nylon	46
Polyethylene plastic	31
Teflon	18
Platinum with perfluorobutyric acid monolayer	10

duction resistance across each drop, as well as a thermal resistance due to interphase mass transfer on the surface of the drop. The vapor-to-surface temperature difference across a drop was also corrected for drop curvature effects. This theory was subsequently shown to accurately predict experimental data for steam and organic vapors, provided slight adjustments were made to several empirical coefficients. When comparison was made to data for dropwise condensation of mercury, somewhat different values of the coefficients had to be used because of large differences between the properties of water and mercury.

In 1973–1974, a more elaborate theoretical model was developed that considered the transient changes in local drop size distribution due to coalescence and growth. Despite several questionable assumptions, this model gives the most complete description of the dropwise condensation process.

For background information SEE CONVECTION (HEAT); SURFACE TENSION; VAPOR CONDENSER in the McGraw-Hill Encyclopedia of Science & Technology.

Paul J. Marto

Bibliography. P. J. Marto et al., Evaluation of organic coatings for the promotion of dropwise condensation of steam, *Int. J. Heat Mass Transfer*, 29:1109–1117, 1986; *Proceedings of the 8th International Heat Transfer Conference*, San Francisco, vol. 4, 1986; J. W. Rose, Some aspects of condensation heat transfer theory, *Int. Comm. Heat Mass Transfer*, 15:449–473, 1988; I. Tanasawa, Advances in condensation heat transfer, *Adv. Heat Transfer*, 21:55–139, 1991.

Virtual acoustics

Virtual acoustics, also known as three-dimensional (3-D) sound and auralization, is the simulation of the complex acoustic field experienced by a listener within an environment. Going beyond the simple left-right volume adjustment of normal stereo techniques, the goal is to process sounds so that they appear to come from particular locations in three-dimensional space. Although loudspeaker systems are being developed, much of the recent work focuses on using headphones for playback and is the outgrowth of earlier analog techniques. For example, in binaural recording, the sound of an orchestra playing classical music is recorded through small microphones in the two imitation ear canals of an anthropomorphic artificial or dummy head placed in the audience of a concert hall. When the recorded piece is played back over headphones, the listener passively experiences the illusion of hearing the violins on the left and the cellos on the right, along with all the associated echoes, resonances, and ambience of the original environment. Current techniques use digital signal processing to synthesize the acoustical properties that people use to localize a sound source in space. Thus, they provide the flexibility of a kind of digital dummy head, allowing a more active experience in which a listener can both design and move around or interact with a simulated acoustic environment in real time. Such simulations are being developed for a variety of application areas, including architectural acoustics, advanced human–computer interfaces, telepresence, virtual reality, navigation aids for the visually impaired, and as a test bed for psychoacoustical investigations of complex spatial cues.

Psychoacoustical cues. The success of virtual acoustics is critically dependent on whether the acoustical cues used by humans to locate sounds have been adequately synthesized. Much of the present understanding of sound localization is based on Lord Rayleigh's classic duplex theory of 1907, which describes the role of two primary cues (**Fig. 1**), interaural time differences and interaural intensity differences. The original conception, based on experiments with pure tones (sine waves), was that interaural intensity differences determine localization of high frequencies because wavelengths smaller than the human head create an intensity loss or head shadow at the ear farthest from the sound source. Conversely, interaural time differences were thought to be important for low frequencies since interaural phase (delay) relationships are nonambiguous only for frequencies below about 1500 Hz (wavelengths larger than the head). The duplex theory, however, cannot account for the ability to localize sounds on the vertical median plane, where interaural cues are minimal. Also, when subjects listen to sounds over headphones, the sounds usually appear to be inside the head even though interaural

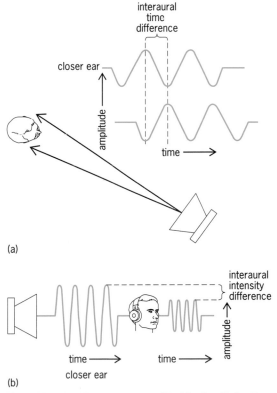

Fig. 1. The two primary cues postulated by the duplex theory of sound localization. (*a*) Interaural time differences. Sources off to one side arrive sooner at the closer ear. (*b*) Interaural intensity differences. Sources off to one side are louder at the closer ear because of head shadowing. (*After E. M. Wenzel, Localization in virtual acoustic displays, Presence: Teleoper. Virt. Environ., 1(1):80–107, 1992*)

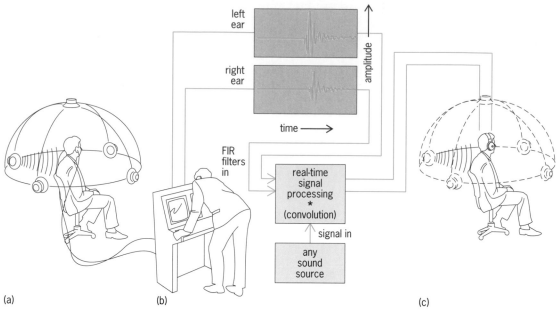

Fig. 2. Technique for synthesizing virtual acoustic sources with measurements of the head-related transfer function. (*a*) Pinnae (outer-ear) responses are measured with probe microphones. Placement of loudspeakers is illustrative only. (*b*) Pinnae transforms are digitized as finite impulse response (FIR) filters. An example of a pair of finite impulse responses measured for a source location at 90° to the left and 0° elevation (at ear level) is shown in the insets for the left and right ears. (*c*) Synthesized cues are played over headphones.

time differences and interaural intensity differences appropriate to an external source location are present. Many studies now indicate that these deficiencies of the duplex theory reflect the important contribution to localization of the direction-dependent filtering that occurs when incoming sound waves interact with the outer ears or pinnae. Experiments have shown that spectral shaping by the pinnae is highly direction-dependent, that the absence of pinna cues degrades localization accuracy, and that pinna cues are at least partially responsible for externalization or the outside-the-head sensation. Such data suggest that perceptually realistic localization over headphones may be possible if both the pinna and interaural difference cues can be adequately measured and reproduced.

There may be many cumulative effects on the sound as it makes its way to the eardrum, but it turns out that all of these effects can be expressed as a single filtering operation much like the effects of a graphic equalizer in a stereo system. The exact nature of this filter can be measured by a simple experiment in which an impulse (a single, very short sound pulse or click) is produced by a loudspeaker at a particular location. The acoustic shaping by the two ears is then measured by recording the outputs of small probe microphones placed inside the ear canals of an individual or an artificial head (**Fig. 2**). If the measurement of the two ears occurs simultaneously, the responses, when taken together as a pair of filters, include an estimate of the interaural differences as well. Thus, this technique makes it possible to measure all of the relevant spatial cues together for a given source location, for a given listener, and in a given room or environment.

Figure 3 illustrates these effects. Figure 3*a* shows equivalent representations, via a mathematical opera-

tion known as the Fourier transform, in the time and frequency domains for an acoustic signal before interaction with the outer ear (and other body) structures. Figure 3*b* and *c* show what happens to an impulse delivered from two different locations after interaction with the outer ears. Figure 3*d* and *e* show the same interaction in the frequency domain, where the differences between the left and right intensity curves are the interaural intensity differences at each frequency. Spectral phase effects (frequency-dependent phase, or time, delays) are also present in the measurements but are not shown here for clarity. The filters constructed from these ear-dependent characteristics are examples of finite impulse response filters, and the characteristics themselves are often referred to as head-related transfer functions (HRTFs). Filtering in the frequency domain is a point-by-point multiplication of the input signal with the left and right head-related transfer functions, while filtering in the time domain by the finite impulse response filters occurs via a somewhat more computationally intensive multiply-and-add operation known as convolution.

By using these filters based on head-related transfer functions, it should be possible to impose spatial characteristics on a signal so that the signal apparently emanates from the originally measured location. Of course, the localizability of the sound will also depend on other factors such as its original spectral content. Narrow-band sounds (sine waves) are generally hard to localize, while broad-band, impulsive sounds are the easiest to locate. Use of filters based on head-related transfer functions cannot increase the bandwidth of the original signal; the filtering merely transforms the frequency components that are already present. The spatial cues provided by head-related transfer functions, especially those measured in simple anechoic (free-field or echoless) environments, are not the only cues likely to be necessary to achieve accurate localization. For example, acoustic features such as the ratio of direct to reflected energy in a reverberant field can provide a cue to distance (closer sources correspond to larger ratios) as well as enhance the sensation of externalization.

Psychophysical validation. The only conclusive test of the adequacy of the simulation technique is an operational one in which the localization of real and synthesized sources is directly compared in psychophysical studies. Two kinds of errors are usually observed when subjects are asked to judge the position of a stationary sound source in the free field. One is a relatively small error in resolution on the order of about 5–20°. Another type of error observed in nearly all localization studies is the front-back reversal. These judgments, which occur about 5% of the time for real sources, indicate that a source in the front (or rear) was perceived by the listener as if it were in the rear (or front). Occasionally, reversals in elevation are also observed.

Recent studies have compared localization judgments of stationary broad-band sources in the free field with judgments of virtual (headphone-presented) sources synthesized from the subjects' own head-related transfer functions. Presumably, synthesis using personalized head-related transfer functions would be the most likely to replicate the free-field experience for a given listener. In general, localization accuracy is comparable for the free-field and headphone stimuli, although source elevation is less well defined and reversal rates increase to about 10% for the virtual sources. Other experiments suggest that it may be feasible to use nonpersonalized head-related transfer functions to synthesize spatial cues, so long as they come from a "good localizer," who accurately localizes both real and virtual sources. Again, localization is comparable for free-field and headphone stimuli, although, for some subjects, source elevation is completely disrupted and reversal rates increase substantially when virtual sources are generated from nonpersonalized head-related transfer functions. Interestingly, some people show little ability to localize source elevation for both real and virtual sounds, as if they

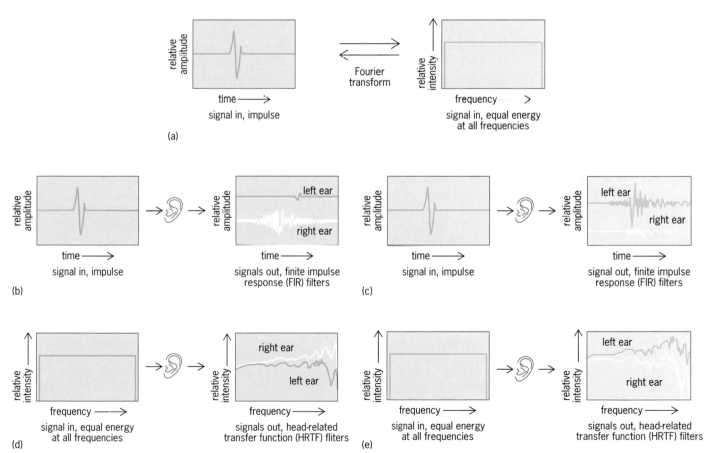

Fig. 3. Effects of spectral shaping by the pinnae (outer-ear structures). (a) Equivalent representations, via the Fourier transform, of a broad-band acoustic signal before interaction with the outer-ear structures, in the time domain (where the acoustic signal is an impulse) and the frequency domain (where only intensity is shown here). **(b)** Effect of the pinnae on an impulse delivered from a loudspeaker directly to the right (+90° azimuth, 0° elevation), as measured in the left-ear and right-ear canals of the individual. Relative amplitude scales have arbitrary origins and units. The larger signal in the right ear is reflected in the larger fluctuation of the relative amplitude. **(c)** Effect of the pinnae on an impulse delivered from a loudspeaker 60° to the left (−60° azimuth, 0° elevation). **(d)** Frequency-domain representation of the effect on the signal from 90° azimuth, 0° elevation. **(e)** Frequency-domain representation of the effect on the signal from −60° azimuth, 0° elevation. **(After E. M. Wenzel, Localization in virtual acoustic displays, Presence: Teleoper. Virt. Environ., 1(1):80–107, 1992)**

have poorly designed pinna structures that have little impact on the spectra of incoming sounds. Listening "through" these people's head-related transfer functions results in a loss of elevation perception for any listener. In general, the psychophysical data suggest that most listeners can obtain useful directional information from an auditory display without requiring the use of individually tailored head-related transfer functions, particularly for the left-right dimension of azimuth.

Real-time implementation. Techniques for creating digital filters based on head-related transfer functions have been under development since the late 1970s, but only with the advent of powerful new digital signal processing chips have a few real-time systems begun to appear. In general, these systems are intended for headphone delivery and use time-domain convolution to achieve real-time performance. Usually, one or more sources are simulated in an anechoic environment by using filters based on head-related transfer functions and chosen according to the output of a head-tracking device in order to compensate for the listener's position and orientation. Motion trajectories and static locations at finer resolutions than those provided by empirical data can be simulated by interpolation between impulse responses originally measured at intervals of $10°$ or more. Also, a simple distance cue may be provided via real-time scaling of amplitude. While such systems implement only the direct paths from each virtual source to the listener, they possess a high degree of interactivity. That is, it is possible to perform free manipulation of source position, listener position, and listener orientation in a dynamic, interactive display.

The same level of interactivity is desirable in a more complex system since perceptual research suggests that errors like front-back reversals and failures of externalization can be mitigated by providing the acoustic cues available in reverberant environments. Of particular interest for implementation is the development of the image model and related ray-tracing techniques that simulate room characteristics by using precomputed impulse responses, which in turn are based on calculated early reflections. Recent extension of these techniques implement the directional characteristics of source reflections in real time by convolution with filters based on head-related transfer functions. For example, in one implementation of the image model the walls, floor, and ceiling in an environment are simulated by placing the mirror image of a sound source behind each surface to account for the specular reflection of the source signal. The filtering effect of surfaces such as wood or drapery can also be modeled with a separate finite-impulse-response filter whose output is delayed by the time required for the sound to propagate from each image source being represented. Finally, the delayed signals are processed with the filters based on head-related transfer functions appropriate to the direction of each reflection to create the components of the binaural output. The superposition of each of these components with the original source (the direct path from source to listener) filtered via head-related transfer functions is then presented over headphones. Such dynamic modeling requires enormous computational resources for real-time implementation in a truly interactive (head-tracked) display. Thus, it is not currently practical to render more than the first one or two reflections from a very small number of reflecting surfaces. Future work will no doubt extend this approach to more realistic models of acoustic environments. In addition to increasing the number of modeled reflections, some of the issues that need to be addressed are nonuniform radiators, diffuse reflections, scattering reflectors, diffraction and partial obscuration by walls or other objects, spreading loss and high-frequency absorption, and simulation of late reverberation or the nondirectional ambience of an environment.

Applications. Applications of virtual acoustics include the direct representation of acoustic environments. For example, in architectural design the ability to simulate room acoustics interactively could be of great use in exploring and avoiding possible acoustically undesirable effects that may not be obvious from the visual design. Similarly, when coupled with some type of range-finding device, artificial acoustical environments could be used as aids for visually impaired people by audibly representing the surfaces and obstacles through which the people must navigate. Other such applications include virtual reality, telerobotic control in remote or hazardous situations, and, as the technology becomes more cost-effective, entertainment environments.

Another application of virtual acoustics develops because spatial sound is potentially useful in a wide variety of advanced information displays. The omnidirectional nature of acoustic signals is especially useful in inherently spatial tasks, particularly when visual cues are limited and workload is high, as in air-traffic control displays. An example is an air-traffic control display in which the controller hears communications from incoming traffic in positions that correspond to their actual location in the terminal area. In such a display, it should be more immediately obvious to the controller when aircraft are on a potential collision course because they would be heard in their true spatial locations and their routes could be tracked over time. Another advantage of binaural systems, often referred to as the cocktail party effect, is that the spatial separation of sounds improves the intelligibility of speech in a background of noise or other speech. Thus, virtual acoustics could be quite useful in cockpit communications systems or in advanced teleconferencing environments that require listening to multiple speakers.

Perhaps most important, virtual acoustics enables experiments in human sound localization that were previously impossible because of a lack of control over the stimuli. Real-time control systems further expand the scope of this research, allowing the study of dynamic, intersensory aspects of localization that may substantially reduce some perceptual problems encountered in trying to produce the realistic experience that is critical for many applications.

For background information *SEE CONTROL SYSTEMS; EAR; INTEGRAL TRANSFORM; SOUND* in the McGraw-Hill

Encyclopedia of Science & Technology.

Elizabeth M. Wenzel

Bibliography. J. Blauert, *Spatial Hearing: The Psychophysics of Human Sound Localization*, 1983; E. M. Wenzel, Localization in virtual acoustic displays, *Presence: Teleoper. Virt. Environ.*, 1(1):80–107, 1992; F. L. Wightman and D. J. Kistler, Headphone simulation of free-field listening, I: Stimulus synthesis, *J. Acous. Soc. Amer.*, 85(2):858–867, 1989.

Wave motion

The propagation of waves through a disordered medium is an important subject that cuts across many disciplines. Examples include the propagation of seismic waves through geological formations, the propagation of microwaves and light through clouds and fog, and the potential use of light waves for imaging objects such as tumors in biological tissues. In addition to these classical waves, physicists have realized that the conduction of electrons through disordered metals at low temperatures is yet another example of the phenomenon. The point is that at low temperatures quantum mechanics plays an important role, and the propagation of an electron is described by the Schrödinger wave. In the past few years, significant progress has been made in the case in which the disorder is strong and the waves are multiply scattered by the disordered medium. The recent advances are the result of a fruitful cross-fertilization between the studies of classical and Schrödinger waves, which traditionally have been pursued as separate disciplines.

Localization and coherent backscattering. In a pure metal, the electron wave propagates as a plane wave. In reality, impurity atoms or defects in the crystal structure create centers that scatter the plane waves. The scattered wave is in turn scattered by other defects, resulting in a complicated wave structure that is distorted compared with the plane wave. This structure is shown schematically in **Fig. 1a**. The distance over which the wavefront is shifted by approximately a wavelength relative to the plane wave is called a mean free path l. Another way of representing this phenomenon is shown in Fig. 1b, where a wavepacket that represents the electron is scattered in different directions each time it encounters an impurity atom. Thus, the electron exhibits a kind of random walk with step size given by the mean free path l as it traverses the sample. If the sample size L is much greater than l, multiple scattering takes place. A similar picture holds for the propagation of classical waves. For example, white paint is a suspension of titanium oxide (TiO_2) spheres, which scatter visible light strongly because of their large dielectric constant. The multiple scattering of the light can also be visualized as the random walking of the light beam.

A crucial point is that there are many paths that the light beam can take through the sample, and these different paths can interfere constructively or destructively with each other. Thus, even though the wavefronts appear random because of multiple scattering, the wave remains coherent in the sense that interference phenomena still occur. This coherence is destroyed only by absorption in the case of classical waves or by inelastic scattering such as collisions with other electrons or with lattice vibrations in the case of electron waves. At low temperatures, such collisions are rare and the coherent wave nature of the electron becomes manifest.

In 1957, P. W. Anderson posed the question: What would happen if the disorder gets progressively stronger so that the mean free paths get progressively shorter? His answer is that if the mean free path is so short that it becomes comparable to the wavelength of the Schrödinger wave, a new phenomenon sets in. The wave packet cannot propagate at all, and instead the envelope of the wave decays exponentially about certain centers. This phenomenon is called Anderson localization, and the length scale of exponential decay is called the localization length, which in general can be much longer than the mean free path. When the electron wave becomes localized, the conductivity of the material vanishes and the material undergoes a phase transition from a metal to an insulator. Many experimental systems, such as alloys of germanium and gold or silicon doped with phosphorus, exhibit this metal-to-insulator transition, and the concept of Anderson localization is a key ingredient in the understanding of these phenomena.

More recently, physicists have turned their attention to the case in which the disorder is still relatively weak so that the multiple scattering picture is still valid. It was discovered that a precursor of the localization phenomenon, now called weak localization, occurs. It manifests itself in unusual temperature and magnetic field dependence of the conductivity at low temperatures. For example, a logarithmic temperature dependence of the conductivity of thin metallic films was

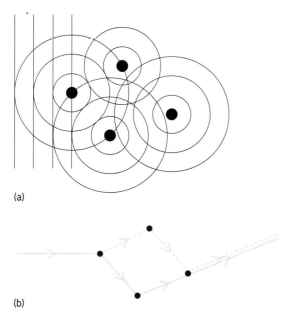

(a)

(b)

Fig. 1. Representations of multiple scattering. (a) Multiple scattering of an incident plane wave. (b) Different multiple scattering paths that can interfere with each other.

predicted and observed experimentally. The physical origin of this phenomenon was understood to be the interference between a multiple scattering path and its time-reversed path, that is, the path of the electron when it undergoes the same random walk but in the reverse direction. An analogous phenomenon in classical wave motion was investigated some time earlier and is now known as coherent backscattering. In this case, a plane wave is incident upon a disordered medium. The prediction is that within a narrow cone in the backscattering direction the scattering intensity is enhanced by a factor of 2. The origin of this factor is illustrated in **Fig. 2**, which shows that the plane wave that is multiply scattered and reemerges in exactly the opposite direction can interfere constructively with the conjugate beam, that is, a multiply scattered beam whose path is the time reversal of the first one. Slightly away from the exact backscattering direction, a random phase appears in the interference and the enhancement disappears. The discovery of weak localization stimulated experiments to test carefully the coherent backscattering idea by using laser beams scattered by disordered media such as titanium oxide spheres. Excellent agreement between theory and experiment was found.

Physicists have also been searching for an experimental realization of the localization of light, but a medium that is sufficiently strongly scattering but weakly absorbing to exhibit this phenomenon has not been found.

Conductance fluctuations and speckle.
In the 1980s physicists began studying the properties of small metallic structures (with linear dimensions between 0.1 micrometer and a few micrometers) at low temperatures (typically below 1 K). It was discovered that the conductivity exhibits reproducible but ran-

dom structures as a function of magnetic field. It was soon found theoretically that, because of interference between multiply scattered paths, the conductance of the sample (defined as the ratio of the current to the electrical potential) should exhibit such reproducible fluctuations with a root-mean-square magnitude of $e^2/h \approx (25{,}800 \text{ ohm})^{-1}$, where e is the electron charge and h is Planck's constant, regardless of the sample size or the amount of disorder, as long as the temperature is low enough. This prediction, called universal conductance fluctuation, was confirmed experimentally. The idea has been further expanded to discuss the influence of the motion of a single impurity atom on the conductance of metallic films and to interpret the $1/f$ noise of disordered films below liquid nitrogen temperatures. The study of small structures at low temperatures is now called mesoscopic physics, referring to the intermediate scale between atomic dimensions and everyday macroscopic dimensions. This is an area where quantum-mechanical coherence is manifest.

Physicists had long known that the interference between different multiply scattered paths manifests itself in the phenomenon of the speckle pattern, where a laser beam transmitted or reflected by a disordered medium exhibits random but reproducible patterns of strong and weak intensities. However, they believed that such random intensities obeyed simple gaussian statistics and that different spots were uncorrelated with each other. It was realized that this description of the speckle pattern, when applied to electron waves, was incompatible with the universal conductance fluctuation idea. Further studies have revealed that subtle but detectable correlations should exist between speckle spots that are far apart. This phenomenon is under investigation using light beams and microwaves, and some of the predicted correlations have been observed.

Diffusive wave spectroscopy. Yet another area where the understanding of multiply scattered light is important is in dynamic light-scattering experiments, in which the dynamics of a system are investigated by studying the time correlation of the scattered light. Light scattering is an established spectroscopic tool when the medium is transparent, that is, when the scattering of the light by the medium is weak. Only recently have experiments been pursued in the multiple scattering limit, where the medium is murky. This method is known as diffusive wave spectroscopy. One of the applications is based on the sensitivity of the multiply scattered light to the location of the individual scatterers. This sensitivity enables the light-scattering experiment to probe dynamics on very short time scales, when the individual scatterer in a fluid has moved only a short distance. Thus, time scales much shorter than the diffusion time of the medium can be probed.

Imaging through a murky medium. A potentially important application of multiple scattering theory is the possibility of imaging objects hidden behind a strongly scattering medium, such as submarines in murky water or cancerous tumors in breast

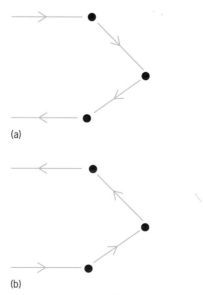

(a)

(b)

Fig. 2. Coherent backscattering. (*a*) A backscattering path. (*b*) Its time-reversed path. These paths interfere with each other constructively, leading to a factor-of-2 enhancement in the backscattering amplitude. If the backscattering amplitude deviates from 180°, a random phase appears in the interference between the two paths.

tissues. One idea is simply that the object will cast a shadow of either enhanced or reduced scattering intensity. The object can be thought of as a dipolar source of waves, which in turn propagate diffusively through the medium. The resulting shadow is larger than the object, and detailed features are smeared out in the diffusion process. Nevertheless, it is found that the intensity of the shadow decreases only as a power of the distance between the object and the surface of the medium, as opposed to the exponential decrease that occurs in the case of an absorbing medium. Thus, the detection of a hidden object may be quite feasible even though its identification may be very difficult.

Another detection scheme makes use of the idea that the propagation of waves can be decomposed into a variety of multiple scattering paths, including a path that is not scattered at all by the medium. The probability and therefore the intensity of this path is exponentially small in a multiply scattering medium, but its transit time across the medium is the shortest among the paths, and therefore it can be detected by using ultrafast optical techniques, which block out the slow paths. The advantage is that the selected path should produce a faithful geometrical image of the object. This technique has been demonstrated in the laboratory to be capable of imaging an object hidden several mean free paths behind a multiply scattering medium.

For background information *SEE COHERENCE; FREE-ELECTRON THEORY OF METALS; INTERFERENCE OF WAVES; QUANTUM MECHANICS; SCATTERING OF ELECTROMAGNET-IC RADIATION; SPECKLE; WAVE MOTION* in the McGraw-Hill Encyclopedia of Science & Technology.

Patrick A. Lee

Bibliography. R. Alfano, P.-P. Ho, and K.-M. Yoo, Photons for prompt tumour detection, 5(1):37–40, January 1992; S. Feng and P. A. Lee, Mesoscopic conductors and correlations in laser speckle patterns, *Science*, 251:633–639, 1991; P. Sheng (ed.), *Scattering and Localization of Classical Waves in Random Media*, 1990.

Waveriders

A waverider is a supersonic or hypersonic vehicle that, at the design point, has an attached shock wave along its entire leading edge (**Fig. 1***a*). Thus, the vehicle appears to be riding on top of its shock wave. In contrast, in a more conventional hypersonic vehicle the shock wave is usually detached from the leading edge (Fig. 1*b*). The future of waveriders looks promising for numerous hypersonic missions in the twenty-first century. Many aerospace engineers feel that the waverider is the configuration of choice for advanced hypersonic vehicles.

Aerodynamic advantages. To better understand the nature of waverider vehicles, it is helpful to consider the following aspects of shock waves in a gas. A shock wave is a very thin region in a flow field (much thinner than the thickness of this page)

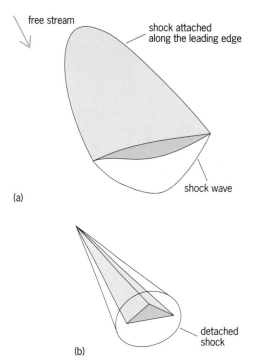

Fig. 1. Comparison of supersonic and hypersonic vehicles. (*a*) Waverider. (*b*) Generic hypersonic vehicle configuration.

across which the pressure greatly increases. A shock wave is a boundary between the lower-pressure gas in front and the higher-pressure gas behind; it is nature's own compression process. For a shock wave to occur in a steady flow, the flow in front of the wave must be moving faster than the speed of sound; that is, the upstream flow velocity must be supersonic. In other words, the Mach number, defined as the ratio of the flow speed divided by the speed of sound, must be greater than 1. All flight vehicles moving faster than Mach 1 generate one or more shock waves in the flow around the vehicle. For a given vehicle at a given angle of attack to the flow, the higher the Mach number, the stronger is the shock wave; that is, the larger is the pressure ratio across the shock wave.

The aerodynamic advantage of the waverider in Fig. 1*a* is that the high pressure behind the shock wave under the vehicle does not leak around the leading edge to the top surface; because the shock wave is attached along the entire leading edge, the flow field over the bottom surface is contained, and the high pressure exerted on the bottom surface of the vehicle is preserved. In contrast, for the vehicle in Fig. 1*b*, because the shock wave is detached from the leading edge, there is communication between the flows over the bottom and top surfaces; the pressure tends to leak around the leading edge, and the general integrated pressure level on the bottom surface of the vehicle is reduced, resulting in less lift. By comparison, waveriders are high-lift vehicles.

As a result, waveriders have an additional aerodynamic advantage that can be understood by considering the two vehicles shown in Fig. 1 in steady, level flight, both with equal weights. Since, for the level flight, lift

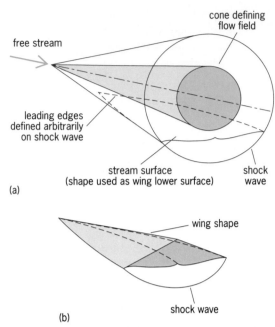

(a)

(b)

Fig. 2. Waverider design. (a) Construction of a waverider from a known flow field. (b) Resulting wing and shock wave.

must equal weight, both vehicles have the same lift. However, because of the pressure containment on the bottom surface, the waverider (Fig. 1a) will produce the required lift at a smaller angle of attack than the generic vehicle (Fig. 1b). The benefit of this smaller angle of attack is that the drag will be smaller. Hence, the ratio of lift to drag for the waverider will be higher than for the generic vehicle, both at the same lift. In airplane aerodynamics, the ratio of lift to drag is an important measure of the aerodynamic efficiency of the vehicle; the higher this ratio, the less is the cost of producing lift. Hence, waveriders have the potential for decreasing the cost of travel at supersonic and hypersonic speeds.

Design philosophy. The design philosophy for waveriders, that is, how the vehicle shape is designed to ensure that the shock wave is indeed attached along the entire leading edge, is also an example of a mathematical advantage of the waverider. Waveriders are generated from known flow fields established by other basic shapes different from that of the waverider itself. For example, in the case of supersonic or hypersonic flow over a sharp, right-circular cone at zero angle of attack (**Fig. 2a**), a conical shock wave is wrapped around this cone, with its vertex at the nose of the cone. The flow field between the shock and the cone is well known; there is an exact solution for this flow that was obtained in the early 1930s. If an arbitrary curve is drawn on the surface of this conical shock (Fig. 2a), all the streamlines that trail downstream from this curve form a continuous surface, called a stream surface. If this stream surface is replaced by a solid surface, the flow between the surface and the shock wave does not know the difference. Hence, the stream surface in Fig. 2a can represent the bottom compression surface of a vehicle. Moreover, the curve originally drawn on the conical shock will constitute

the leading edge of the vehicle. If a vehicle is designed with this surface shape, the physics of the flow will guarantee that the shock wave will be attached along the entire leading edge of the vehicle. The resulting vehicle is sketched in Fig. 2b. This vehicle is the resulting waverider, which is generated from the known flow field in Fig. 2a. Of course, there is an infinite number of different curves that can be drawn on the conical shock wave in Fig. 2a; hence, there is an infinite number of different waveriders that can be generated from this generating flow field, each one with a different ratio of lift to drag. One of these waveriders will have a maximum value of the lift-to-drag ratio, and for many applications this will be the waverider shape of choice.

The known conical flow field in Fig. 2a results in only one family of waveriders. Other known flow fields can also be used to generate other families of waveriders. An even simpler flow is that behind a straight oblique shock wave, for which an exact solution has been known since 1908. Any curve traced on the surface of a plane oblique shock wave, with the resulting stream surface that trails downstream, will lead to a waverider configuration in the same vein as discussed above. Indeed, this type of generating flow was used by T. Nonweiler in 1959 to introduce the waverider concept, so waveriders are sometimes called Nonweiler wings.

Hypersonic versus supersonic flight. Conventional wisdom in the design of supersonic aircraft is to avoid strong shock waves, which produce large wave drag. Aircraft designed for Mach 2 or 3 are slender configurations, which tend to generate fairly weak shock systems. However, as the Mach number is increased to substantially larger values, the shock waves become stronger no matter what ploys are used in the design of the aircraft. Flight at Mach 5 or higher is called hypersonic flight. Hypersonic vehicles must contend with strong shock waves. Because the waverider concept makes good use of the high pressure behind a shock wave, waveriders appear to be more advantageous for hypersonic flight than supersonic flight.

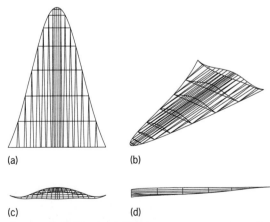

(a)

(b)

(c)

(d)

Fig. 3. Typical viscous-optimized hypersonic waverider, designed for Mach 6. The maximum lift-to-drag ratio is 8.2. (a) Top view. (b) Perspective view. (c) Front view. (d) Side view.

Indeed, most waverider research has been carried out in the hypersonic regime.

However, the XB-70 supersonic bomber, designed in the late 1950s, had variable-geometry wing tips that could be canted down. In this configuration at supersonic speeds, the XB-70 benefited from the type of pressure containment on its bottom surface associated with waveriders. For this reason, the XB-70 is sometimes pointed out as the first (and so far the only) crewed waverider vehicle to fly.

A modern hypersonic waverider configuration is shown in **Fig. 3**. This shape was obtained from a conical generating flow field such as that in Fig. 2. It represents the optimum shape for maximum lift-to-drag ratio at Mach 6.

Research history. Research on waveriders is chronologically divided into two periods. The first period extends from the inception of the waverider by Nonweiler in 1959 to the advent of a new approach in 1986. During this first period, the concept of the waverider received sporadic attention, with the majority of the research carried out in Europe, and waverider configurations were optimized by accounting for wave drag only. (Drag on a flight vehicle is due to both the pressure distribution over the surface and the frictional shear stress over the surface. At supersonic and hypersonic speeds, the pressure distribution is dominated by the presence of shock waves, and hence such drag is called wave drag. The drag due to frictional shear stress is called skin friction drag.) Since optimization of waveriders took into account wave drag only, the predicted lift-to-drag ratios were optimistic (and, of course, very exciting). However, waveriders have a large amount of wetted surface area (Fig. 3), and hence the skin friction drag can be important. Indeed, when the real-life aspect of skin friction was added after the fact to these waverider designs (either by humans in aerodynamic analyses or by nature in wind tunnel tests), the performance of such waveriders was severely degraded. For this reason, serious interest in waveriders by the aerodynamic community was weak, and sometimes even negative.

The second period of waverider research stems from 1986, when waverider configurations were first optimized by taking into account both wave drag and skin friction drag. Consequently, new families of waveriders called viscous optimized waveriders emerged. These waveriders (Fig. 3) exhibited shapes and surface areas that led to a trade-off between both wave drag and skin friction drag. The resulting lift-to-drag ratios are both realistic (as confirmed by recent wind tunnel tests) and high enough to exceed those of earlier waveriders as well as other generic hypersonic configurations. As a result, interest in waveriders as advanced hypersonic configurations has substantially increased. Increased research has led to substantial improvements in the understanding of waverider properties, including effects of aerodynamic heating, high altitude, and chemically reacting flow over the waverider surface. Applied work has shown the way toward air-breathing propulsion integrated with the waverider airframe, and has examined various flight applications such as hyper-

sonic cruise vehicles (including hypersonic transports), maneuvering reentry vehicles, and foreign planetary atmosphere vehicles for missions to the far planets in the solar system.

For background information SEE AERODYNAMIC WAVE DRAG; AIRFOIL; SHOCK WAVE; SKIN FRICTION; SUPERSONIC FLIGHT in the McGraw-Hill Encyclopedia of Science & Technology.

John D. Anderson, Jr.

Bibliography. J. D. Anderson, Jr., *Introduction to Flight*, 3d ed., 1989; J. D. Anderson, Jr., et al., Hypersonic waveriders for planetary atmospheres, *J. Spacecraft Rockets*, 28(4):401–410, 1991; D. Kuchemann, *The Aerodynamic Design of Aircraft*, 1978.

X-ray astronomy

X-ray astronomy is the study of celestial objects in the x-ray portion of the electromagnetic spectrum. Virtually all celestial objects, from nearby stars to the most distant quasars, are natural sources of x-rays, which are the natural radiation produced by either very hot or very energetic objects and are to the short-wavelength (blue) side of the ultraviolet portion of the electromagnetic spectrum. Thus, x-ray observations of cosmic sources, like ultraviolet or gamma-ray observations, must be carried out above the Earth's atmosphere, which screens these higher-energy radiations (higher than the eye can see) from reaching the Earth's surface.

X-ray astronomy, therefore, began only in 1962, with the launch of a sounding rocket carrying a simple x-ray detector that revealed the first bright cosmic sources of x-rays. X-ray astronomy developed rapidly throughout the 1970s, with the discoveries of a wide range of types of cosmic sources, and came of age in the early 1980s, with intensive studies of astronomical objects in x-rays using the first true imaging cosmic x-ray telescope, the *Einstein Observatory*. The spectacular success of *Einstein* was followed by the *EXOSAT* observatory (1983–1987), as well as an important series of three nonimaging x-ray satellites launched and operated by Japan throughout the 1980s. These satellites made important additional discoveries and provided further impetus for the next x-ray telescope on the *Roentgen Satellite* (*ROSAT*), launched in June 1990.

ROSAT carries a larger and higher-resolution x-ray telescope than did the *Einstein Observatory*. Most importantly, *ROSAT* has carried out the first all-sky imaging survey in x-rays, thereby discovering and mapping nearly 100,000 cosmic x-ray sources. This is nearly 100 times the number of sources discovered by the previously most sensitive all-sky x-ray survey carried out with the *HEAO 1* satellite in 1978–1980. Because the telescope and supporting satellite were developed in Germany, the primary control and operation remains there. However, the *ROSAT* observatory also has major contributions from both the United Kingdom and the United States. A small ultraviolet telescope, the Wide-Field Camera, was provided by the United

Kingdom, whereas one of the two x-ray detectors at the focus of the telescope was provided by the United States. Thus, *ROSAT* is a trinational collaboration and is used by astronomers all over the world.

ROSAT x-ray telescope. This grazing-incidence telescope focuses x-rays from a nested set of four hyperboloid and paraboloid mirrors to a focal plane 2.4 m (8 ft) behind the mirrors. This arrangement gives a high-energy cutoff of about 2.4 keV for the x-ray telescope (versus 4.5 keV for *Einstein*); the low-energy limit of about 0.1 keV is set by the transmission of the window materials of the detectors at the focus of the telescope. The *ROSAT* telescope and observatory are shown in **Fig. 1**.

The detectors at the focus are of two types: the position-sensitive proportional counter (PSPC), with moderate spatial resolution (about 1 arc-minute) but higher spectral resolution; and the high-resolution imager (HRI), with high spatial resolution (about 5 arc-seconds) but low spectral resolution. The PSPC was designed and built in Germany and the HRI in the United States. Both imaging detectors are significant advances over their counterparts on the *Einstein Observatory*. The PSPC has significantly better spectral resolution and lower background, and the HRI has some energy (spectral) resolution, which was totally absent on the *Einstein* HRI detector. The *ROSAT* satellite was launched in June 1990 by the National Aeronautics and Space Administration (NASA) on a Delta rocket into a high-inclination orbit that passes over the primary ground station in Germany, from which the mission is controlled and data are received.

After initial processing, the data are forwarded to centers in the United Kingdom and United States for further analysis and distribution.

Results from ROSAT. The first imaging all-sky survey in soft x-rays was completed by *ROSAT* in its first 7 months in orbit. More than 60,000 sources were detected, with more than 100,000 expected in the final analysis of the data. This first imaging view of the x-ray sky (obtained with the PSPC detector, and therefore multicolored and with about 1 arc-minute resolution) is to x-ray astronomy what the Palomar Sky Survey has been to optical astronomy: an atlas of the deep sky that can be consulted for both countless individual objects and large-scale extended sources. Entire constellations, such as Orion, can now be seen for the first time in the light of the x-rays naturally produced by their stars (**Fig. 2**). The differences from the familiar optical view are sometimes striking: the bright star Betelgeuse (the right shoulder of the hunter Orion) is faint in x-rays, whereas relatively inconspicuous faint red stars show up as bright in the x-ray image.

Stars. At least half of the sources in the sky survey are stars (in the Milky Way Galaxy), and so will greatly extend the studies of stellar x-ray emission begun with the *Einstein Observatory*. The *ROSAT* results have already greatly narrowed the range of stars that do not emit x-rays to a subset of so-called A-stars (like the familiar bright star Vega). Why these stars should seem so dark in x-rays is still a mystery but is probably related to the lack of convection and magnetic fields in their atmospheres. For other stars, with

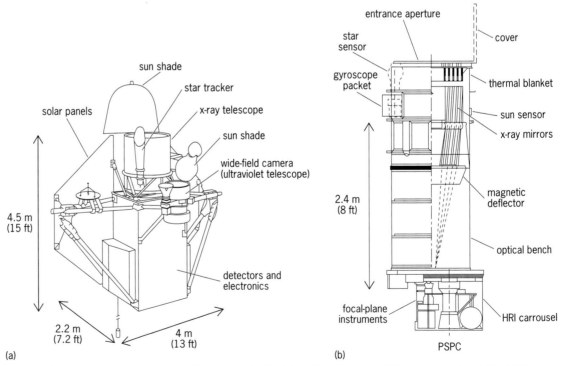

(a)

(b)

Fig. 1. *ROSAT* **satellite.** (*a*) **Spacecraft configuration, with ultraviolet telescope and satellite support system (***after J. Trümper, The ROSAT mission, Adv. Space Res., 2(4):241–249, 1983***). (***b***) Cross section of x-ray telescope, with PSPC and HRI detectors (***after B. Aschenbach, First results from the x-ray astronomy mission ROSAT, Fall Meeting, Astronomische Gesellschaft, Freiberg, Germany, September 1990***).**

Fig. 2. *ROSAT* PSPC view of the constellation Orion. (*Max Planck Institut, Garchingbei München, Germany*)

much more active chromospheres, the enhanced sensitivity of *ROSAT* operated as a pointed observatory has shown x-ray variations that are probably due to the presence of giant star spots (analogous to sunspots) on the stellar surface.

ROSAT has been operated as an observatory, with pointed observations of carefully selected fields, ever since the sky survey was completed. Although individual stars and stellar clusters have also been studied in detail, and spectra (and variability) derived, high-sensitivity observations of compact objects have provided some of the most spectacular early results.

X-ray binaries. The most luminous sources of x-rays in the Milky Way Galaxy are binary systems containing a neutron star accreting matter from a binary companion. These low-mass x-ray binaries have poorly understood formation and evolution histories, which may relate to the oldest stellar populations in the Milky Way Galaxy. Almost as conspicuous are the accreting white dwarf binaries, or so-called cataclysmic variables. *ROSAT* pointed observations have had the sensitivity to extend the studies of both types of objects to much greater distances: cataclysmic variables in (relatively) distant globular clusters of stars within the Milky Way Galaxy, and low-mass x-ray binaries in external galaxies such as the Andromeda Galaxy (M31).

The high spatial resolution of the HRI detector on *ROSAT* is being put to the extreme test with high-sensitivity observations of several globular clusters in the Milky Way Galaxy. Globular clusters are the oldest objects known in the Milky Way Galaxy and may be closely related to its formation. Some 10 globulars were known to contain individual bright low-mass x-ray binaries from studies with the *Einstein Observatory* and other x-ray satellites. The *ROSAT* sky survey has added two more, showing how extremely variable the entire population is. However, it is the population of much fainter sources in globulars, originally discovered with *Einstein*, that has spurred high-sensitivity *ROSAT* pointed observations of several relatively nearby globular clusters, such as NGC 6397 (the second closest globular cluster), shown in **Fig. 3**.

These sources are all of the luminosity expected for cataclysmic variables, and are remarkable for being so numerous (for example, at least four in the core of the cluster NGC 6397). If they are confirmed as cataclysmic variables with follow-up optical images and spectra with the high resolution required and available from the Hubble Space Telescope, they may provide an important link between cataclysmic variables and low-mass x-ray binaries in globular clusters and the Milky Way Galaxy.

Hot gas in galaxy clusters. One of the most surprising and fundamental discoveries of the first x-ray astronomy satellite mission, *Uhuru*, in 1971 was that rich clusters of galaxies are filled with a low-density gas at very high temperatures (10^8 K) that is readily detectable as a bright diffuse source of x-rays. *ROSAT* has contributed enormously to the study of x-ray emission from galaxy clusters as well, and will measure the spatial variation of the gas temperature for at least some clusters. A surprising discovery was the detection of hot gas in a relatively sparse group of galaxies, the NGC 2300 group. This group of galaxies is so sparse, or poor, that it should not be massive enough to retain such a bright source of x-ray-emitting gas. The discoverers, R. Mushotzky and D. Burstein,

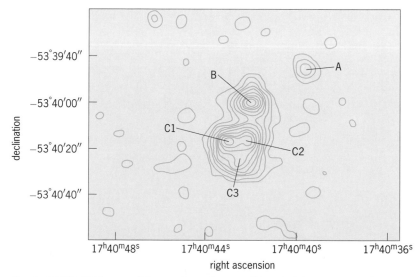

Fig. 3. *ROSAT* HRI image of the nearby and core-collapsed globular cluster NGC 6397, showing four separate sources (B, C1, C2, and C3) within about 20 arc-seconds of the cluster center (near C3), and a fifth source (A) that is also probably a cluster member.

who reported the results in January 1993, therefore concluded that the cluster must contain dark matter with 10 times the mass of the visible matter in the constituent galaxies. Confirmation by further studies would have major impact, since it would imply that the ubiquitous dark matter, long suspect by cosmologists, may be concentrated in poor groups of galaxies rather than in rich clusters. *SEE COSMOLOGY.*

Quasars and high-sensitivity surveys. *ROSAT* has already contributed greatly to the understanding of the x-ray properties of quasars and active galaxies. PSPC observations have shown low-energy absorption features in the spectra of distant quasars. Not only are these the first x-ray spectra of distant quasars, but also these objects may be unlike the more familiar lower-redshift quasars in that they contain substantial amounts of gas and dust, which absorb the low-energy x-rays.

It has long been suspected that the diffuse cosmic x-ray background may be due to the superposition of discrete extragalactic x-ray sources, primarily quasars. High-sensitivity *ROSAT* observations on a relatively clear patch of extragalactic sky (near the pole of the Milky Way Galaxy) have shown that about 40% of the diffuse background in fact does become resolved into individual sources. These observations also reveal important new findings on the spectrum of the background at the soft x-ray energies accessible to *ROSAT*: an emission line of highly ionized oxygen appears to be detected in all viewing directions, suggesting the existence of a hot gas at temperatures of $2-3 \times 10^6$ K distributed over the whole sky.

ROSAT operation. The *ROSAT* observatory is expected to be operated for as long as possible; no follow-on x-ray observatory in space is planned until the first part of the large United States permanent x-ray telescope in orbit, the *Advanced X-Ray Astrophysics Facility (AXAF)*, is launched in 1998. At present, *ROSAT* is limited by the lifetime of its detectors and control systems. The PSPC will exhaust its gas supply sometime in 1993, but the HRI is expected to operate much longer.

For background information *SEE BINARY STAR; COSMOLOGY; NEUTRON STAR; NOVA; QUASAR; STAR CLUSTERS; X-RAY ASTRONOMY; X-RAY STAR; X-RAY TELESCOPE* in the McGraw-Hill Encyclopedia of Science & Technology.

Jonathan E. Grindlay

Bibliography. A. Cool et al., Discovery of multiple low-luminosity x-ray sources in NGC 6397, *Astrophys. J. Lett.*, 1993; J. Trümper, The *ROSAT* mission, *Adv. Space Res.*, 2(4):241–249, 1983; B. Wilkes et al., A high-redshift quasar with strong x-ray absorption, *Astrophys. J. Lett.*, 393:L1–L4, 1992.

Contributors

The affiliation of each Yearbook contributor is given, followed by the title of his or her article. An article title with the notation "in part" indicates that the author independently prepared a section of an article; "coauthored" indicates that two or more authors jointly prepared an article or section.

A

Adams, Dr. Michael W. W. *Department of Biochemistry, University of Georgia, Athens.* MICROBIAL ECOLOGY.

Adrian, Prof. Ronald J. *Department of Theoretical and Applied Mechanics, University of Illinois at Urbana-Champaign.* FLOW MEASUREMENT.

Agee, Prof. James K. *Chair, Forest Resources Management Division, University of Washington, Seattle.* FOREST FIRE.

Allen, Prof. Leland C. *Department of Chemistry, Princeton University, Princeton, New Jersey.* PERIODIC TABLE.

Andersen, Dr. David R. *Department of Electrical and Computer Engineering, University of Iowa, Iowa City.* QUANTUM INTERFERENCE DEVICES.

Anderson, Prof. John D., Jr. *Department of Aerospace Engineering, University of Maryland, College Park.* WAVERIDERS.

Ang, Dr. Peng H. *Vice President and General Manager, Digital Signal Processing Division, Microprocessor & DSP Products Group, LSI Logic Corporation, Milpitas, California.* TELEVISION.

Angell, Prof. Charles A. *Department of Chemistry, Arizona State University, Tempe.* GLASS.

Armbruster, Prof. W. Scott. *Biology and Wildlife, University of Alaska, Fairbanks.* FLOWER—in part.

Arnold, Dr. Frances H. *Division of Chemistry and Chemical Engineering, California Institute of Technology, Pasadena.* CHEMICAL SYNTHESIS—coauthored.

Atal, Dr. Bishnu S. *Head, Speech Research Department, AT&T Bell Laboratories, Murray Hill, New Jersey.* SPEECH RECOGNITION.

B

Balibar, Dr. Sebastien. *Directeur de Recherches, Laboratoire de Physique Statistique, Paris, France.* HELIUM.

Ball, Prof. Donald M. *Department of Agronomy and Soils, Auburn University, Auburn, Alabama.* FESCUE—coauthored.

Ball, Prof. Robert E. *Department of Aeronautics and Astronautics, Naval Postgraduate School, Monterey, California.* AIRCRAFT.

Banerjee, Dr. Partha P. *Department of Electrical and Computer Engineering, University of Alabama, Huntsville.* ACOUSTOOPTICS.

Beaumont, Andy R. *School of Ocean Sciences, University of Wales, Bangor, Gwynedd.* BIVALVIA—in part.

Beerbower, Prof. James Richard. *Department of Geological Sciences and Environmental Studies, State University of New York, Binghamton.* REPTILIA.

Beitz, Prof. Donald C. *Departments of Animal Sciences and Biochemistry-Biophysics, Iowa State University, Ames.* NUTRITION.

Bennett, Prof. Albert F. *Department of Ecology and Evolutionary Biology, University of California, Irvine.* LOCOMOTION (VERTEBRATE)—in part.

Bennett, Dr. John. *Senior Molecular Biologist, Division of Plant Breeding, Genetics and Biochemistry, International Rice Research Institute, Manila, Philippines.* RICE—coauthored.

Berman, Dr. Robert G. *Geological Survey of Canada, Ottawa, Ontario.* CHEMICAL THERMODYNAMICS.

Bianco, Dr. Antonio C. *Department of Physiology, University of São Paulo, Brazil.* ADIPOSE TISSUE—coauthored.

Billig, Dr. Frederick S. *Applied Physics Laboratory, Johns Hopkins University, Laurel, Maryland.* COMBUSTION.

Blobel, Dr. Carl P. *Cellular Biochemistry and Biophysics Program, Memorial Sloan-Kettering Cancer Center, New York, New York.* CELL MEMBRANES.

Bloor, Prof. David. *Department of Physics, Applied Physics Group, Science Laboratories, University of Durham, England.* ELECTRONICS.

Bone, Dr. Quentin. *Marine Laboratory, Plymouth, England.* ASCIDIACEA.

Bower, James N., Jr. *Adjunct Faculty, Florida Institute of Technology, Melbourne.* SPACE TECHNOLOGY—in part.

Bowles, Dr. Thomas J. *Physics Division, Subatomic*

Research and Applications, Los Alamos National Laboratory, Los Alamos, New Mexico. SOLAR NEUTRINOS.

Bremel, Prof. Robert D. *Department of Dairy Science, University of Wisconsin, Madison.* BIOTECHNOLOGY.

Brown, Prof. Kenneth M. *Department of Zoology and Physiology, Louisiana State University, Baton Rouge.* SNAIL.

Bryant, Prof. Vaughn M., Jr. *Chair, Department of Anthropology, Texas A&M University, College Station.* PALEONUTRITION—coauthored.

Burke, Dr. Laura. *Department of Industrial Engineering, Harold S. Mohler Laboratory, Lehigh University, Bethlehem, Pennsylvania.* NEURAL NETWORKS.

Busscher, Dr. Warren J. *Agricultural Research Service, U.S. Department of Agriculture, Florence, South Carolina.* AGRICULTURAL SOIL AND CROP PRACTICES.

C

Carleton, Dr. Andrew M. *Department of Geography, Indiana University, Bloomington.* CLIMATIC CHANGE.

Carney, Dr. Arlene Earley. *Boys Town National Research Hospital, Omaha, Nebraska.* SPEECH.

Castleman, Prof. A. W., Jr. *Department of Chemistry, Pennsylvania State University, University Park.* METALLO-CARBOHEDRENE.

Chellemi, Dr. Daniel O. *Plant Pathologist, Florida Cooperative Extension Service, University of Florida, Quincy.* ANTHRACNOSE DISEASE.

Cholewiak, Dr. Roger W. *Senior Research Psychologist, Department of Psychology, Princeton University, Princeton, New Jersey.* CUTANEOUS SENSATION.

Chu, Prof. Steven. *Department of Physics, Stanford University, Stanford, California.* ATOM.

Clarricoats, Prof. Peter. *Head, Department of Electronic Engineering, Queen Mary and Westfield College, London, England.* ANTENNA (ELECTROMAGNETISM).

Cox, Brian J. *Principal Engineer, Ball Communication Systems Division, Broomfield, Colorado.* TELEPHONE SERVICE—in part.

Crabtree, Dr. Gerald R. *Associate Investigator, Howard Hughes Medical Institute Research Laboratories, Stanford University School of Medicine, Stanford, California.* IMMUNOSUPPRESSION.

Crane, Dr. Peter R. *Department of Geology, Field Museum of Natural History, Chicago, Illinois.* MAGNOLIOPHYTA.

Cross, Dr. Frederick R. *Rockefeller University, New York, New York.* CELL CYCLE.

Crystal, Dr. Ronald G. *National Heart, Lung, and Blood Institute, National Institutes of Health, Bethesda, Maryland.* GENE THERAPY—coauthored.

Currie, Dr. Philip J. *Head, Dinosaur Research, Royal Tyrrell Museum of Palaeontology, Drumheller, Al-* *berta, Canada.* DINOSAUR.

Curtis, Dr. Corrine L. *Department of Chemistry, University of California, San Diego.* SILICON—coauthored.

D

Dalla Betta, Dr. Ralph A. *Catalytica, Inc., Mountain View, California.* GAS TURBINE—coauthored.

DeFranco, Dr. Anthony L. *Department of Microbiology and Immunology, University of California, San Francisco.* CELLULAR IMMUNOLOGY.

Dengler, Prof. Nancy G. *Department of Botany, University of Toronto, Ontario, Canada.* LEAF.

Doan, Dr. Vincent V. *Department of Chemistry, University of California, San Diego.* SILICON—coauthored.

Dooley, Prof. David M. *Department of Chemistry, Amherst College, Amherst, Massachusetts.* COENZYME.

Dryden, Dr. Robert. *Department of Industrial and Systems Engineering, Virginia Polytechnic Institute and State University, Blacksburg.* HUMAN-MACHINE SYSTEMS—coauthored.

Duke, Dr. Stephen O. *Director, Southern Weed Science Laboratory, U.S. Department of Agriculture, Stoneville, Mississippi.* HERBICIDE.

E

Edwards, Dr. Linnell M. *Agriculture Canada Research Station, Charlottetown, Prince Edward Island.* SOIL MECHANICS.

Ehrlich, Prof. Robert. *Department of Geological Sciences, University of South Carolina, Columbia.* PETROGRAPHIC IMAGE ANALYSIS.

F

Faustman, Dr. Denise. *Director, Immunobiology Laboratories, Massachusetts General Hospital, Charlestown.* TRANSPLANTATION BIOLOGY.

Filisko, Dr. Frank E. *Department of Materials and Metallurgy, University of Michigan, Ann Arbor.* ELECTRORHEOLOGICAL MATERIAL.

Fishman, Dr. Gerald J. *Space Science Laboratory, NASA, Marshall Space Flight Center, Huntsville, Alabama.* GAMMA-RAY ASTRONOMY.

Foote, Prof. Christopher S. *Department of Chemistry and Biochemistry, University of California, Los Angeles.* FULLERENES—in part.

Forshay, Steven E. *Vice President Engineering, Dolby Laboratories, Inc., San Francisco, California.* AUDIO COMPRESSION.

Franklin, Dr. Michael J. *Research Microbiologist, Department of Microbiology and Immunology, University of Tennessee, Memphis.* MICROBIAL BIOFILM.

Fryrear, Dr. D. W. *Research Leader/Agricultural Engineer, Agricultural Research Service, U.S. Department of Agriculture, Big Spring, Texas.* SOIL EROSION—in part.

Fuller, Prof. Christopher R. *Director, Vibration and Acoustics Laboratories, Virginia Polytechnic Institute and State University, Blacksburg.* ADAPTIVE SOUND CONTROL.

G

Gallop, Dr. John C. *Division of Quantum Metrology, National Physical Laboratory, Teddington, Middlesex, England.* SUPERCONDUCTING DEVICES.

Geisler, Kenneth I. *Preisdent, Configured Energy Systems, Inc., Plymouth, Minnesota.* ELECTRIC DISTRIBUTION SYSTEMS.

Gertler, Prof. Janos John. *Department of Electrical and Computer Engineering, George Mason University, Fairfax, Virginia.* FAULT DETECTION.

Gilden, Dr. David L. *Department of Psychology, University of Texas, Austin.* PERCEPTION.

Glantz, Dr. Michael. *National Center for Atmospheric Research, Boulder, Colorado.* CLIMATOLOGY.

Goris, Dr. John M. *Mining Engineer, Spokane Research Center, U.S. Bureau of Mines, Spokane, Washington.* UNDERGROUND MINING.

Gorur, Dr. Ravi S. *Department of Electrical Engineering, Arizona State University, Tempe.* ELECTRICAL INSULATION.

Gorzelnik, Eugene F. *North American Electric Reliability Council, Princeton, New Jersey.* ELECTRICAL UTILITY INDUSTRY.

Griffin, Robert J., Jr. *Science Writer, Washington, D.C.* SPACE FLIGHT.

Grindlay, Dr. Jonathan E. *Department of Astronomy, Harvard University, Cambridge, Massachusetts.* X-RAY ASTRONOMY.

Grossman, Dr. Lawrence I. *Department of Molecular Biology and Genetics, Wayne State University School of Medicine, Detroit, Michigan.* ISCHEMIA—coauthored.

H

Halpern, Prof. Diane F. *Department of Psychology, California State University, San Bernadino.* COGNITION.

Haworth, Dr. L. I. *Department of Electrical Engineering, University of Edinburgh, Scotland.* SOUND RECORDING.

Heath, Michelle. *Director, Domestic Oil and Transportation Fuels, Canadian Energy Research Institute, Calgary, Alberta, Canada.* FUEL—in part.

Heller, Dr. Adam. *Department of Chemical Engineering, University of Texas, Austin.* ELECTRODE—coauthored.

Hirose, Dr. Akira. *Department of Physics, University of Saskatchewan, Saskatoon, Saskatchewan, Canada.* TURBULENCE—coauthored.

Hofmann, Prof. David J. *U.S. Department of Commerce, National Oceanic and Atmospheric Administration, Boulder, Colorado.* ATMOSPHERIC OZONE—in part.

Holton, Prof. James R. *Department of Atmospheric Sciences, University of Washington, Seattle.* ATMOSPHERIC OZONE—in part.

Hoover, Dr. Donald B. *Geological Survey, Branch of Geophysics, U.S. Department of the Interior, Denver, Colorado.* GEOCHEMICAL PROSPECTING—coauthored.

Höss, Dr. Matthias. *Zoologisches Institut der Universität München, Germany.* DEOXYRIBONUCLEIC ACID (DNA)—coauthored.

Howard, Dr. Richard M. *Department of Aeronautics and Astronautics, Naval Postgraduate School, Monterey, California.* DRONE.

Hume, Dr. Ian D. *School of Biological Sciences, University of Sydney, Australia.* DIGESTIVE SYSTEM.

Hutley, Dr. Michael. *National Physical Laboratory, Teddington, Middlesex, England.* INTEGRATED OPTICS.

Hwang, Dr. Patrick. *Rockwell International, Cedar Rapids, Iowa.* SATELLITE NAVIGATION SYSTEMS.

I

Iseman, Dr. Michael D. *Chief, Clinical Mycobacteriology Service, National Jewish Center for Immunology and Respiratory Medicine, and Professor of Medicine, University of Colorado School of Medicine, Denver.* TUBERCULOSIS—coauthored.

Ishihara, Dr. O. *Department of Physics, University of Saskatchewan, Saskatoon, Saskatchewan, Canada.* TURBULENCE—coauthored.

J

Jajodia, Prof. Sushil. *Department of Information and Software Systems Engineering, George Mason University, Fairfax, Virginia.* COMPUTER SECURITY.

Johnson, Dr. D. Barrie. *School of Biological Sciences, University of Wales, Bangor, Gwynedd.* SOIL MICROBIOLOGY.

Jones, Prof. William D. *Department of Chemistry, University of Rochester, New York.* ORGANOMETALLIC CHEMISTRY.

K

Katz, Prof. J. Lawrence. *Department of Biomedical Engineering, Case Western Reserve University, Cleveland, Ohio.* BIOMIMETIC MATERIAL.

Kenrick, Dr. Paul. *Department of Palaeobotany, Swedish Museum of Natural History, Stockholm.* LAND PLANTS, ORIGIN OF.

Keshavan, Dr. Matcheri S. *Schizophrenia Treatment and Research Center, University of Pittsburgh, Pennsylvania.* MAGNETIC RESONANCE SPECTROSCOPY.

Kessler, Dr. Donald J. *NASA Senior Scientist for Orbital Debris Research, Lyndon B. Johnson Space Center, Houston, Texas.* SPACE TECHNOLOGY—in part.

Khush, Dr. Gurdev S. *Principal Plant Breeder and Head, Division of Plant Breeding, Genetics and Biochemistry, International Rice Research Institute, Manila, Philippines.* RICE—coauthored.

Kiran, Dr. Erdogan. *Department of Chemical Engineering, University of Maine, Orono.* SUPERCRITICAL FLUIDS.

Klitz, Dr. William. *Department of Integrative Biology, University of California, Berkeley.* TUMOR VIRUS—in part.

Knoll, Prof. Andrew H. *Botanical Museum of Harvard University, Cambridge, Massachusetts.* PRECAMBRIAN.

Kofman, Dr. Lev. *Princeton University Observatory, Princeton, New Jersey.* COSMOLOGY—coauthored.

Kort, Dr. John. *Shelterbelt Biologist, Investigation Section, Prairie Farm Rehabilitation Administration, Shelterbelt Centre, Indian Head, Saskatchewan, Canada.* SOIL EROSION—in part.

Krause, Dr. Gary S. *Department of Emergency Medicine, Wayne State University School of Medicine, Detroit, Michigan.* ISCHEMIA—coauthored.

Krogfelt, Dr. Karen A. *Department of Bacteriology, Statens Seruminstitut, Copenhagen, Denmark.* CELLULAR ADHESION.

Kuehn, Prof. Thomas H. *Department of Mechanical Engineering, University of Minnesota, Minneapolis.* COMFORT HEATING.

L

LaBarbera, Dr. Michael. *Department of Organismal Biology and Anatomy, University of Chicago, Illinois.* BARNACLE.

Lacefield, Prof. Garry. *Department of Agronomy, University of Kentucky, Lexington.* FESCUE—coauthored.

Lee, Prof. Patrick A. *Department of Physics, Massachusetts Institute of Technology, Cambridge.* WAVE MOTION.

Le Grice, Dr. Stuart F. J. *Division of Infectious Diseases, Case Western Reserve University School of Medicine, Cleveland, Ohio.* ACQUIRED IMMUNE DEFICIENCY SYNDROME (AIDS).

Leland, Dr. Jonathan K. *IGEN Inc., Rockville, Maryland.* CHEMILUMINESCENCE—coauthored.

Lentz, Dr. Rodrick D. *Soil Scientist, Agricultural Research Service, U.S. Department of Agriculture, Kimberly, Idaho.* SOIL EROSION—in part.

Leventis, Dr. Nicholas. *Vice President, Research and Development, Molecular Displays, Inc., Cambridge, Massachusetts.* ELECTROCHROMIC DEVICES.

Li, Dr. Wen-Hsiung. *Health Science Center, University of Texas, Houston.* MOLECULAR EVOLUTION.

Lindstedt, Dr. Stan L. *Department of Biological Sciences, Northern Arizona University, Flagstaff.* ENDURANCE PHYSIOLOGY.

Little, Dr. Colin. *Department of Zoology, School of Biological Sciences, University of Bristol, England.* LIMPET.

Liu, Prof. Henry. *Director, Capsule Pipeline Research Center, University of Missouri, Columbia.* PIPELINE.

Lounasmaa, Prof. Olli V. *Director, Low Temperature Laboratory, Helsinki University of Technology, Otakaari, Espoo, Finland.* NEGATIVE TEMPERATURE.

M

McCormick, Prof. Barnes W. *Department of Aerospace Engineering, Pennsylvania State University, University Park.* HELICOPTER—in part.

Mallouk, Prof. Thomas E. *Department of Chemistry and Biochemistry, University of Texas, Austin.* ELECTRODE—in part.

Marchbanks, Michael L. *Department of Chemistry, University of Wisconsin, Madison.* ARCHEOLOGY—in part.

Marcus, Dr. W. Andrew. *Department of Earth Sciences, Montana State University, Bozeman.* HEAVY METALS, ENVIRONMENTAL.

Marto, Prof. Paul J. *Dean of Research and Distinguished Professor of Mechanical Engineering, Department of the Navy, Naval Postgraduate School, Monterey, California.* VAPOR CONDENSER.

Marullo, Michael A. *Editor and Publisher, TechMarc Publications Group, New Orleans, Louisiana.* PETROLEUM ENGINEERING.

Meeks-Wagner, Dr. D. Ry. *Institute of Molecular Biology, University of Oregon, Eugene.* FLOWER—in part.

Melara, Dr. Robert D. *Department of Psychological Sciences, Purdue University, West Lafayette, Indiana.* SENSATION.

Michaels, Abraham. *Consulting Engineer, Osterville, Massachusetts.* ENVIRONMENTAL ENGINEERING.

Micklin, Prof. Philip P. *Department of Geography, Western Michigan University, Kalamazoo.* ARAL SEA.

Moulton, Prof. William G. *National High Magnetic Field Laboratory, Los Alamos National Laboratory, Florida State University, Tallahassee.* MAGNET.

N

Nalin, Dr. David R. *Director, Clinical Research, Merck Sharp & Dohme Research Laboratories, West Point, Pennsylvania.* HEPATITIS.

Neu, Charles E. *Materials Engineer, Naval Air Warfare Center, Warminster, Pennsylvania.* STEEL.

Newman, Dr. David B., Jr. *David Newman and Associates, La Plata, Maryland.* TELEPHONE SERVICE—in part.

Nickle, Dr. William R. *Nematology Laboratory, U.S. Department of Agriculture, Beltsville Agricultural Research Center, Beltsville, Maryland.* SOIL ECOLOGY.

Nikols, Dr. Dennis J. *Retread Resources Ltd., Sherwood Park, Alberta, Canada.* COAL MINING.

Novacek, Dr. Michael. *Dean of Science, American Museum of Natural History, New York, New York.* MAMMALIA.

O

Ong, Prof. N. Phuan. *Department of Physics, Princeton University, Princeton, New Jersey.* SUPERCONDUCTIVITY.

Ongaro, Dr. Roger. *Institut des Sciences de l'Ingénierie et du Développement Technologique, Université Claude Bernard Lyon 1, Villeurbanne, France.* SOLID-STATE PHYSICS—coauthored.

P

Pääbo, Prof. Svante. *Zoologisches Institut der Universität München, Germany.* DEOXYRIBONUCLEIC ACID (DNA)—coauthored.

Padki, Dr. M. M. *Clean Energy Research Institute, University of Miami, Coral Gables.* SOLAR ENERGY—coauthored.

Pelster, Dr. Bernd. *Institut für Physiologie, Ruhr Universität Bochum, Germany.* SWIM BLADDER—coauthored.

Peterson, Prof. G. P. *Department of Mechanical Engineering, Texas A&M University, College Station.* HEAT TRANSFER.

Phillips, Dr. Jonathan D. *Department of Geography and Planning, East Carolina University, Greenville, North Carolina.* STREAM SYSTEMS.

Pillonnet, Dr. A. *Institut des Sciences de l'Ingénierie et du Développement Technologique, Université Claude Bernard Lyon 1, Villeurbanne, France.* SOLID-STATE PHYSICS—coauthored.

Pishko, Dr. Michael V. *Department of Chemical Engineering, University of Texas, Austin.* ELECTRODE—coauthored.

Plunkett, Dr. Sean D. *Division of Chemistry and Chemical Engineering, California Institute of Technology, Pasadena.* CHEMICAL SYNTHESIS—coauthored.

Pollack, Louis. *Satellite Systems Design, Louis Pollack Associates, Rockville, Maryland.* COMMUNICATIONS SATELLITE.

Porter, Dr. Marc D. *Ames Laboratory, U.S. Department of Energy, and Department of Chemistry, Iowa State University, Ames.* SURFACE AND INTERFACIAL CHEMISTRY.

Posamentier, Dr. Henry W. *Exploration Advisor, Geological Analysis, Arco Exploration and Production Technology, Plano, Texas.* STRATIGRAPHY.

Potter, Seth D. *Department of Applied Science, New York University, New York.* SOLAR ENERGY—in part.

Pour-El, Prof. Marian B. *School of Mathematics, Institute of Technology, University of Minnesota, Minneapolis.* COMPUTABILITY (PHYSICS).

Preskill, Dr. John P. *Department of Physics, California Institute of Technology, Pasadena.* BLACK HOLE.

Primmerman, Dr. Charles A. *Lincoln Laboratory, Massachusetts Institute of Technology, Lexington.* ADAPTIVE OPTICS.

R

Reid, Dr. Robert G. B. *Biology Department, University of Victoria, British Columbia, Canada.* BIVALVIA—in part.

Ries, Dr. Allen A. *Enteric Disease Branch, National Center for Infectious Diseases, Centers for Disease Control, Atlanta, Georgia.* CHOLERA—coauthored.

Robinson, Dr. David A. *Department of Geography, Rutgers University, New Brunswick, New Jersey.* SNOW COVER.

Rogoff, Mortimer. *Digital Directions Co., Inc., Washington, D.C.* ELECTRONIC NAVIGATION SYSTEMS.

Rome, Dr. Lawrence C. *Department of Biology, University of Pennsylvania, Philadelphia.* LOCOMOTION (VERTEBRATE)—in part.

Rosenfeld, Dr. Melissa A. *National Heart, Lung, and Blood Institute, National Institutes of Health, Bethesda, Maryland.* GENE THERAPY—coauthored.

Rost, Prof. Thomas L. *Section of Botany, University of California, Davis.* ROOT (BOTANY)—in part.

Rothblatt, Martin A. *President, Multi-Technology Analysis and Research Corporation, Washington, D.C.* RADIO SPECTRUM ALLOCATIONS.

Rothwell, Prof. Gar W. *Department of Botany, University of Alberta, Edmonton, Alberta, Canada.* REPRODUCTION (PLANT).

Rouse, Dr. William B. *Search Technology, Norcross, Georgia.* SYSTEMS ENGINEERING.

Rowe, Dr. Clinton M. *Department of Geography, University of Nebraska, Lincoln.* GEOGRAPHY.

S

Saam, Prof. William F. *Department of Physics, Ohio State University, Columbus.* PHASE TRANSITIONS.

Sailor, Prof. Michael J. *Department of Chemistry, University of California, San Diego.* SILICON—coauthored.

Salisbury, Dr. Alan B. *Executive Vice President and Chief Operating Officer, Microelectronics and Computer Technology Corporation, Washington, D.C.* INFORMATION MANAGEMENT.

Satpathy, Dr. Sashi. *Department of Physics and Astronomy, University of Missouri, Columbia.* FULLERENES—in part.

Scheid, Dr. Peter. *Institut für Physiologie, Ruhr Universität Bochum, Germany.* SWIM BLADDER—coauthored.

Schenker, Prof. Marc B. *Division of Occupational/Environmental Medicine and Epidemiology, University of California, Davis.* AGRICULTURAL HEALTH—coauthored.

Schiefelbein, Dr. John W. *Department of Biology, University of Michigan, Ann Arbor.* ROOT (BOTANY)—in part.

Schlatter, Dr. James C. *Catalytica, Inc., Mountain View, California.* GAS TURBINE—coauthored.

Schon, Dr. Eric A. *Columbia University College of Physicians and Surgeons, New York, New York.* MITOCHONDRIA.

Sereno, Dr. Paul. *Department of Organismal Biology and Anatomy, University of Chicago, Illinois.* FLIGHT.

Shavlik, Dr. Jude W. *Department of Computer Sciences, University of Wisconsin, Madison.* MACHINE LEARNING.

Shaw, Kate. *Museum of Natural History, University of Kansas, Lawrence.* PHYLOGENETIC ANALYSIS.

Shelton, Dr. Peter M. J. *Department of Zoology, School of Biological Sciences, University of Leicester, England.* SHRIMP.

Sherif, Dr. S. A. *Department of Mechanical Engineering, University of Florida, Gainesville.* SOLAR ENERGY—coauthored.

Shilling, Prof. John J. *College of Computing, Georgia Institute of Technology, Atlanta.* OBJECT-ORIENTED PROGRAMMING.

Shoji, Dr. Toru. *Tanaka Kikinzoku Kogyo K. K., Tokyo, Japan.* GAS TURBINE—coauthored.

Siegmann, Prof. William L. *Department of Mathematical Sciences, Rensselaer Polytechnic Institute, Troy, New York.* UNDERWATER ACOUSTICS.

Silva, Dr. J. Enrique. *Division of Endocrinology, Department of Medicine, Sir Mortimer B. Davis Jewish General Hospital and McGill University, Montreal, Quebec, Canada.* ADIPOSE TISSUE—coauthored.

Silver, Dr. Deborah. *Laboratory for Visiometrics and Modeling, Department of Mechanical and Aerospace Engineering & CAIP Center, Rutgers University, Piscataway, New Jersey.* COMPUTER GRAPHICS—coauthored.

Simchi-Levi, Dr. David. *Department of Industrial Engineering and Operations Research, Columbia University, New York, New York.* INDUSTRIAL ENGINEERING.

Simone, Dr. Patricia M. *Medical Officer, Program Services Branch, Division of Tuberculosis Elimination, Centers for Disease Control, Atlanta, Georgia.* TUBERCULOSIS—coauthored.

Smith, Dr. Alan G. *Department of Horticultural Science, University of Minnesota, St. Paul.* TRANSGENIC ORGANISMS.

Smith, David B. *Geological Survey, Branch of Geochemistry, U.S. Department of the Interior, Denver, Colorado.* GEOCHEMICAL PROSPECTING—coauthored.

Smith, Dr. Lloyd M. *Department of Chemistry, University of Wisconsin, Madison.* DEOXYRIBONUCLEIC ACID (DNA)—in part.

Smith, Prof. Ronald W. *Director, Department of Materials Engineering, Drexel University, Philadelphia, Pennsylvania.* METAL COATINGS.

Smolyakov, Dr. A. *Department of Physics, University of Saskatchewan, Saskatoon, Saskatchewan, Canada.* TURBULENCE—coauthored.

Sobolik, Dr. Kristin D. *Department of Anthropology, University of Maine, Orono.* PALEONUTRITION—coauthored.

Speakman, Dr. John R. *Department of Zoology, University of Aberdeen, Scotland.* ECHOLOCATION.

Speer, Prof. C. A. *Head, Veterinary Molecular Biology Laboratory, Montana State University, Bozeman.* CRYPTOSPORIDIOSIS.

Spergel, Dr. David N. *Princeton University Observatory, Princeton, New Jersey.* COSMOLOGY—coauthored.

Stanford, Prof. John L. *Department of Physics and Astronomy, Iowa State University, Ames.* ATMOSPHERE.

Sternberg, Prof. Robert J. *Department of Psychology, Yale University, New Haven, Connecticut.* HUMAN INTELLIGENCE.

Stiehm, Dr. E. Richard. *Department of Pediatrics, University of California, Los Angeles.* IMMUNOTHERAPY.

Storer, Dr. Robert H. *Department of Industrial Engineering, Harold S. Mohler Laboratory, Lehigh University, Bethlehem, Pennsylvania.* QUALITY CONTROL.

Strayer, Dr. David L. *Institute of Ecosystem Studies, New York Botanical Garden, Millbrook, New York.* GROUNDWATER ECOLOGY.

Swerdlow, Dr. David L. *Infectious Disease Unit, Massachusetts General Hospital, Boston, Massachusetts.* CHOLERA—coauthored.

T

Tamm, Dr. Sidney L. *Marine Program, Marine Biological Laboratory, Boston University, Woods Hole, Massachusetts.* CTENOPHORA.

Taylor, Dr. Edith L. *Byrd Polar Research Center, Ohio State University, Columbus.* TREE RINGS.

Taylor, Prof. Thomas. *Department of Plant Biology, Ohio State University, Columbus.* FUNGI.

Theil, Prof. Elizabeth C. *Department of Biochemistry, North Carolina State University, Raleigh.* BIOINORGANIC CHEMISTRY—coauthored.

Thiele, Dr. Dennis J. *Department of Biological Chemistry, University of Michigan, Ann Arbor.* BIOINORGANIC CHEMISTRY—coauthored.

Thill, Richard E. *Supervisory Geophysicist, U.S. Department of the Interior, Bureau of Mines, Twin Cities Research Center, Minneapolis, Minnesota.* BOREHOLE LOGGING—coauthored.

Tholen, Dr. David J. *Institute for Astronomy, University of Hawaii at Manoa, Honolulu.* PLUTO.

Thomson, Dr. Norman, *Department of Educational Theory and Practice, State University of New York, Albany.* CELL WALLS (PLANT).

Thorpe, Dr. Patrick H. *Division of Occupational/Environmental Medicine and Epidemiology, University of California, Davis.* AGRICULTURAL HEALTH—coauthored.

Tolle, Tia Benson. *Wright Laboratory, Wright Patterson Air Force Base, Dayton, Ohio.* SMART MATERIALS.

Trivedi, Dr. Mohan M. *Electrical and Computer Engineering Department, University of Tennessee, Knoxville.* INTELLIGENT MACHINES.

Tweeton, Dr. Daryl R. *Research Physicist, U.S. Department of the Interior, Bureau of Mines, Twin Cities Research Center, Minneapolis, Minnesota.* BOREHOLE LOGGING—coauthored.

V

Velzy, Charles O. *Retired; formerly, Charles R. Velzy Associates, Inc., Valhalla, New York.* FUEL—in part.

W

Warfield, Dr. Carol A. *Director, Pain Management Center, Beth Israel Hospital, Boston, Massachusetts, and Professor of Anaesthesia, Harvard Medical School, Boston, Massachusetts.* PAIN.

Wayne, Dr. Randy. *Division of Biological Sciences, Section of Plant Biology, Cornell University, Ithaca, New York.* PLANT MOVEMENTS.

Wenzel, Dr. Elizabeth M. *NASA-Ames Research Center, Moffett Field, California.* VIRTUAL ACOUSTICS.

Weymouth, Prof. John W. *Department of Physics and Astronomy, Behlen Laboratory of Physics, University of Nebraska, Lincoln.* ARCHEOLOGY—in part.

White, Dr. Blaine C. *Department of Emergency Medicine, Wayne State University School of Medicine, Detroit, Michigan.* ISCHEMIA—coauthored.

Wikswo, Prof. John P., Jr. *Department of Physics & Astronomy, Vanderbilt University, Nashville, Tennessee.* BIOMAGNETISM—coauthored.

Wilkinson, Chris F. *South Coast Computing, Eastbourne, England.* ELECTRONIC MAIL.

Williamson, Dr. Samuel J. *Department of Physics & Astronomy, Vanderbilt University, Nashville, Tennessee.* BIOMAGNETISM—coauthored.

Wilson, Dr. Van G. *Department of Medical Microbiology and Immunology, Texas A&M University Health Science Center, College Station.* TUMOR VIRUS—in part.

Woldstad, Dr. Jeffrey C. *Department of Industrial and Systems Engineering, Virginia Polytechnic Institute and State University, Blacksburg.* HUMAN-MACHINE SYSTEMS—coauthored.

Wood, Prof. E. Roberts. *Department of the Navy, Naval Postgraduate School, Monterey, California.* HELICOPTER—in part.

Wrangell, Maurice. *Landscape Architects/Site Planners, New York, New York.* LANDSCAPE ARCHITECTURE.

Y

Yang, Dr. Hongjun. *IGEN Inc., Rockville, Maryland.* CHEMILUMINESCENCE—coauthored.

Young, Wayne. *Program Officer, Marine Board, National Research Council, National Academy of Sciences, Washington, D.C.* MARINE NAVIGATION.

Z

Zabusky, Prof. Norman J. *Laboratory for Visiometrics and Modeling, Department of Mechanical and Aerospace Engineering & CAIP Center, Rutgers University, Piscataway, New Jersey.* COMPUTER GRAPHICS—coauthored.

Index

Asterisks indicate page references to article titles.

T

U

V